Monographs in Mathematics
Vol. 101

Managing Editors:
H. Amann
Universität Zürich, Switzerland
J.-P. Bourguignon
IHES, Bures-sur-Yvette, France
K. Grove
University of Maryland, College Park, USA
P.-L. Lions
Université de Paris-Dauphine, France

Associate Editors:
H. Araki, Kyoto University
F. Brezzi, Università di Pavia
K.C. Chang, Peking University
N. Hitchin, University of Warwick
H. Hofer, Courant Institute, New York
H. Knörrer, ETH Zürich
K. Masuda, University of Tokyo
D. Zagier, Max-Planck-Institut Bonn

Vitaly Volpert

Elliptic Partial Differential Equations

Volume 1:
Fredholm Theory of Elliptic Problems
in Unbounded Domains

Vitaly Volpert
Institut Camille Jordan, CNRS
Université Claude Bernard Lyon 1
43 boulevard du 11 novembre 1918
69622 Villeurbanne cedex
France
volpert@math.univ-lyon1.fr

2010 Mathematics Subject Classification: 35J, 47F, 47A53, 47J05

ISBN 978-3-0348-0322-9   ISBN 978-3-0346-0537-3 (eBook)
DOI 10.1007/978-3-0346-0537-3

© Springer Basel AG 2011
Softcover reprint of the hardcover 1st edition 2011
This work is subject to copyright. All rights are reserved, whether the whole or part of the material is concerned, specifically the rights of translation, reprinting, re-use of illustrations, recitation, broadcasting, reproduction on microfilms or in other ways, and storage in data banks. For any kind of use permission of the copyright owner must be obtained.

Cover design: deblik

Printed on acid-free paper

Springer Basel AG is part of Springer Science+Business Media

www.birkhauser-science.com

To my mathematical family

Tous les géomètres et physiciens semblent aujourd'hui d'accord que le domaine d'application des mathématiques n'a d'autres limites que les limites de nos connaissances mêmes. Il serait pourtant bien téméraire d'affirmer que nous nous trouvons déjà en possession des symboles les mieux appropriés pour interpréter simplement les phénomènes de la nature. I1 est au contraire beaucoup plus probable que bien des théories mathématiques aujourd'hui en estime ne seront admirés plus tard que comme des chefs d'oeuvre historiques, et d'autres théories aussi belles et aussi parfaites, mais d'une application plus large viendront les remplacer.

Serge Bernstein, 1904 ([63])

Contents

Preface xv

1 Introduction
   1 Elliptic problems in applications 1
       1.1 Heat and mass transfer 1
       1.2 Electrostatic and gravitational fields 4
       1.3 Hydrodynamics 5
       1.4 Biological applications 7
   2 Classical theory of linear elliptic problems 10
       2.1 Function spaces 10
       2.2 Elliptic problems 13
       2.3 A priori estimates 20
       2.4 Fredholm operators 25
   3 Elliptic problems in unbounded domains 30
       3.1 Function spaces 31
       3.2 Limiting problems 32
       3.3 Fredholm property and solvability conditions 36
       3.4 Index 39
       3.5 A priori estimates 42
       3.6 Non-Fredholm operators 42
   4 Nonlinear Fredholm operators 43
       4.1 Topological degree 43
       4.2 Existence and bifurcations of solutions 45

2 Function Spaces and Operators
   1 The space E 48
   2 Systems of functions 49
   3 The space E_p 50
   4 The space E_∞ 55
   5 Completeness of the space E_∞ 57
   6 Other definitions of the space E_∞ 59

		7	Bounded sequences in E_∞	62
		8	The space $E_p(\Gamma)$	65
		9	Spaces in unbounded domains	73
		10	Dual spaces	78
		11	Dual spaces in domains	89
		12	Spaces $W_q^{s,p}(\mathbb{R}^n)$	93
		13	Local operators	100

3 A Priori Estimates
   1 Model problems in a half-space 106
      1.1 Formulation of the main result 106
      1.2 Auxiliary results 107
      1.3 Reduction to homogeneous systems 112
      1.4 Proof of Theorem 1.2 113
      1.5 Proof of Theorem 1.1 116
   2 A priori estimates in the spaces $W_\infty^{s,p}$ 116
   3 A priori estimates for adjoint operators. Model systems 120
   4 The general problem in a half-space 123
   5 The general problem in unbounded domains 130

4 Normal Solvability
   1 Limiting domains 137
      1.1 General properties 137
      1.2 Examples 142
   2 Sobolev spaces 146
      2.1 A priori estimates in the spaces $W_\infty^{k,p}$ 146
      2.2 Limiting problems 147
      2.3 A priori estimates with condition NS 154
      2.4 Normal solvability 163
   3 Hölder spaces 170
      3.1 Operators and spaces 170
      3.2 Normal solvability 171
      3.3 Dual spaces. Invertibility of limiting operators 176
      3.4 Weighted spaces 180

5 Fredholm Property
   1 Estimates with Condition NS* 183
   2 Abstract operators 193
   3 Elliptic problems in spaces $W_\infty^{s,p}(\Omega)$ 198
   4 Exponential decay 203
   5 The space E_q 205
   6 The space E_q. Continuation 210

		7	Weighted spaces	215
		8	Hölder spaces	217
		9	Examples	220
			9.1 Bounded domains	220
			9.2 Unbounded domains	222

6 Formally Adjoint Problems
1. Definitions 227
2. Spaces $V^{2m}(\Omega)$ 231
3. Fredholm theorems 237
4. Spaces E_q 243
5. Smoothness of solutions 246
6. Examples 247

7 Elliptic Problems with a Parameter
1. Parameter-elliptic boundary value problems 252
2. Spaces 254
3. Model problem in $\mathbb{R}^n$ 256
4. Model problem in a half-space 258
5. Problem in $\mathbb{R}^n$ 272
6. Problem in $\mathbb{R}^n_+$ 275
7. Problem in Ω 276
8. Generation of analytic semigroups 284
9. Elliptic problems with a parameter at infinity 285
10. Examples 286
    10.1 Invertibility of second-order operators 286
    10.2 Operators with a parameter 288

8 Index of Elliptic Operators
1. Definitions 292
   1.1 Limiting domains 293
   1.2 Limiting operators and problems 295
   1.3 Conditions NS and estimates 296
   1.4 Results and examples 298
2. Stabilization of the index 302
3. Formally adjoint problems 313
4. Convergence of solutions 319
5. Cauchy-Riemann system 324
6. Laplace operator with oblique derivative 331
7. Cauchy-Riemann system and Laplace operator 336
   7.1 Reduction of the problems to each other 336
   7.2 Formula for index 345

	8	General first-order systems on the plane 346
		8.1 Reduction to canonical systems 346
		8.2 Index in bounded domains 350
		8.3 Index in unbounded domains 352
	9	Examples 355

9 Problems in Cylinders
 1 Ordinary differential operators on the real line 360
   1.1 Representation of solutions 360
   1.2 Calculation of the index 362
 2 Second-order equations. 365
   2.1 Reduction to first-order equations. 365
   2.2 Solvability conditions 368
 3 Second-order problems in cylinders 375
   3.1 Normal solvability 375
   3.2 Calculation of the index 380
 4 Lower estimates of elliptic operators 388
   4.1 Operators with constant coefficients 389
   4.2 Variable coefficients. 393
   4.3 Weighted spaces 397
   4.4 Fredholm property and applications 399

10 Non-Fredholm Operators
 1 Weakly non-Fredholm operators. 402
   1.1 Some properties of operators 402
   1.2 Solvability conditions 408
   1.3 Examples of weakly non-Fredholm operators 418
 2 Non-Fredholm solvability conditions 426
   2.1 Example of first-order ODE 426
   2.2 Ordinary differential systems on the real line 428
   2.3 Second-order equations. 432
 3 Space decomposition of operators 434
   3.1 Normal solvability on a subspace 436
   3.2 Scalar equation 440
 4 Strongly non-Fredholm operators 442

11 Nonlinear Fredholm Operators
 1 Nonlinear elliptic problems 446
 2 Properness 448
   2.1 Lemma on properness of operators in Banach spaces 448
   2.2 Properness of elliptic operators in Hölder spaces 449
   2.3 Operators depending on a parameter 453

		2.4	Example of non-proper operators	454
		2.5	Properness in Sobolev spaces	455
	3	Topological degree	468
		3.1	Definition and main properties of the degree	468
		3.2	Orientation of operators	470
		3.3	Degree for Fredholm and proper operators	474
		3.4	Uniqueness of the degree	480
		3.5	Degree for elliptic operators	484
		3.6	Non-existence of the degree	487
	4	Existence and bifurcations of solutions	489
		4.1	Methods of nonlinear analysis	489
		4.2	Existence of solutions	493
		4.3	Elliptic problems in unbounded domains	501
		4.4	Bifurcations	518

Supplement. Discrete operators 527

	1	One-parameter equations	527
		1.1 Limiting operators and normal solvability	527
		1.2 Solvability conditions	533
		1.3 Spectrum of difference and differential operators	534
	2	First-order systems	535
		2.1 Solvability conditions	536
		2.2 Higher-order equations	540
	3	Principal eigenvalue	541
		3.1 Auxiliary results	543
		3.2 Location of the spectrum	545
	4	Stability of finite difference schemes	549

Historical and Bibliographical Comments 553

	1	Historical Notes	554
		1.1 Beginning of the theory	554
		1.2 Existence of solutions of boundary value problems	556
		1.3 Other elliptic equations	559
		1.4 Analyticity	561
		1.5 Eigenvalues	563
		1.6 Fredholm theory	564
	2	Linear equations	568
		2.1 A priori estimates	568
		2.2 Normal solvability and Fredholm property	570
		2.3 Index	576
		2.4 Elliptic problems with a parameter	581

3	Decay and growth of solutions	583
4	Topological degree	586
5	Existence and bifurcation of solutions	591
6	Concluding remarks	593

Acknowledgement 595

Bibliography 597

Notation of Function Spaces 635

Index 637

Preface

Investigation of elliptic partial differential equations began more than two centuries ago with the mathematical theories of gravimetry, fluid dynamics, electrostatics and heat conduction due to works by Euler, Laplace, Lagrange, Poisson, Fourier, Green and Gauss. In the second half of the XIXth century, Schwarz, Neumann, Harnack, Poincaré and Picard developed efficient methods to study existence of solutions of boundary value problems for linear and nonlinear equations. The basis of the spectral theory was established by Schwarz, Poincaré, Steklov and Zaremba. Elliptic equations had important development in the XXth century. Its beginning can be related to the mathematical congress in Paris in 1900 where Hilbert had posed 23 problems, including the 18th and 19th devoted to elliptic boundary value problems. They stimulated works by Bernstein followed by Caccioppoli, Schauder, Leray and other authors who established the foundations of modern analysis. The method of Fredholm integral equations was developed in 1900–1903. It had important applications to elliptic partial differential equations.

Today's understanding of elliptic partial differential equations begins with a priori estimates of solutions of linear problems. They provide normal solvability, the Fredholm property and solvability conditions. Important achievements were related to index theories. The structure of the spectrum of linear elliptic problems provides their sectorial property. It allows introduction of analytic semi-groups and investigation of parabolic problems. The properties of linear operators play an important role for investigation of nonlinear problems. In particular, for construction of the topological degree, which is a powerful tool to study existence and bifurcations of solutions. Many of these methods and results are related to the Fredholm property of elliptic operators. Little can be done in elliptic partial differential equations without it. This determines the presentation of the material of the book around this fundamental property of elliptic problems.

Under the influence of numerous applications, the theory of elliptic partial differential equations continues to attract much attention. During the last several decades, travelling wave solutions of parabolic systems have been intensively studied in relation with combustion problems, population dynamics and many other applications. Travelling waves are solutions of elliptic problems in unbounded domains. These studies intensified the development of the theory of elliptic operators in unbounded domains.

It should be noted that the classical theory of elliptic problems has been developed in the case of bounded domains with a sufficiently smooth boundary. In this case, their Fredholm property is provided by the ellipticity condition, proper ellipticity and the Lopatinskii condition. In the case of unbounded domains, these conditions are no longer sufficient and the Fredholm property may not be satisfied. This is related to a lack of compactness. In order to satisfy the Fredholm property, we need to impose an additional condition, which can be formulated in terms of limiting problems. They characterize the behavior of the operator at infinity and determine the location of the essential spectrum.

In this book, we present a systematic investigation of general elliptic problems applicable both for bounded and unbounded domains. We pay more attention to the case of unbounded domains which is not yet sufficiently well presented in the existing literature. We introduce and use in essential ways some special function spaces well adapted for problems in unbounded domains. They generalize Sobolev spaces by specifying the behavior of functions at infinity. Another point we emphasize is related to limiting domains. This notion is necessary to define limiting operators. Moreover, there are two types of limiting domains. One of them determines the Fredholm property, another one some properties of the index. The proof of the Fredholm property is based on the properties of function spaces, on limiting domains and operators and on some special a priori estimates.

If the operator satisfies the Fredholm property, then its index is well defined. Computation of the index is well known in the case of bounded domains. It is also possible for some classes of operators in unbounded domains. We will develop a new method based on approximation of unbounded domains by a sequence of bounded domains. Under some conditions formulated in terms of limiting operators, the index in the sequence of bounded domains stabilizes to the index in the unbounded domain.

Elliptic operators with a parameter is an important class of operators essentially used in the theory of elliptic and parabolic problems. They are related to sectorial operators and to analytic semi-groups. We will extend the theory of elliptic operators with a parameter to general elliptic problems in unbounded domains.

If limiting problems have nonzero solutions, then the operator does not satisfy the Fredholm property, and the usual solvability conditions are not applicable. We introduce a class of operators, weakly non-Fredholm operators, for which it appears to be possible to formulate solvability conditions. In some cases, these conditions are similar to the usual ones and consist in orthogonality to solutions of the homogeneous adjoint problem. In some other cases, the solvability conditions are different.

Methods of nonlinear analysis, such as asymptotic methods, bifurcation theory or topological degree are based on the solvability conditions. We will define the topological degree for Fredholm and proper operators with the zero index and will discuss some of its applications including the Leray-Schauder method and bifurcations of solutions.

The book is organized as follows. In the first chapter we present a brief introduction to the classical theory of elliptic problems and to the theory of elliptic problems in unbounded domains. Chapters 2–8 are devoted to the theory of general linear elliptic problems (a priori estimates, normal solvability and the Fredholm property, operators with a parameter, index). In many cases we present it directly for unbounded domains, keeping in mind that it remains valid and even simpler for bounded domains. In the next chapter we deal with second-order operators in cylinders where some additional, more explicit methods can be used. Non-Fredholm operators are discussed in Chapter 10. The theory of linear operators will be used in Chapter 11 devoted to nonlinear Fredholm operators. Infinite-dimensional discrete operators are discussed in the supplement. Historical and bibliographical comments presented at the end of this monograph can help to acquire a general view of the theory of elliptic equations, its evolution and perspectives.

The presentation of the results is mostly self-contained. We should note that some proofs are rather technical. However, the formulations of the results and their application are quite clear. The results of the first volume will be used in the second volume to study reaction-diffusion problems.

Most of the results of this book devoted to linear and nonlinear elliptic problems in unbounded domains were obtained in our works with Aizik Volpert. We began to write this book together. It concludes many years of our collaboration.

Chapter 1

Introduction

1 Elliptic problems in applications

We briefly present here some physical, chemical and biological problems described by elliptic partial differential equation. This list of applications is necessarily incomplete and the choice of them is determined by the purposes of presentation. We return to some of these problems below in this and in the second volume. Some other important applications, such as for example elasticity problems, remain out of the scope of this short introduction. We will use some classical examples in order to show the physical importance of elliptic boundary value problems, including the questions of existence of solutions, spectral properties and bifurcations.

1.1 Heat and mass transfer

Consider a stationary heat distribution in a homogeneous isotropic medium. According to the Fourier law, the heat flux q is proportional to the temperature gradient, $q = -\kappa \nabla T$. If V is a domain in $\mathbb{R}^3$ with the surface S, then the heat flux through this surface, in the absence of heat sources or sinks, equals zero:

$$\kappa \int_S \frac{\partial T}{\partial n} ds = 0.$$

Here $T(x)$ is the temperature at the point x, κ is the coefficient of heat conduction, $\frac{\partial T}{\partial n}$ is the normal derivative at the surface S, n is the outer normal vector. From Green's formula

$$\int_V (T \Delta w - w \Delta T) dv = \int_S \left(T \frac{\partial w}{\partial n} - w \frac{\partial T}{\partial n} \right) ds,$$

where we put $w = 1$, follows

$$\int_V \Delta T \, dv = 0. \qquad (1.1)$$

Here

$$\Delta T = \sum_{i=1}^{3} \frac{\partial^2 T}{\partial x_i^2}$$

is the Laplace operator. Since the volume V in (1.1) is arbitrary, then we obtain the equality

$$\Delta T = 0 \qquad (1.2)$$

called the Laplace equation. If there are heat sources in the medium, then the stationary heat distribution can be described by the Poisson equation

$$\Delta T = f. \qquad (1.3)$$

In order to find temperature distribution in a domain Ω, we need to complete equations (1.2) or (1.3) by some boundary conditions at the boundary $\partial\Omega$. Most often considered are Dirichlet (a), Neumann (b) and third order (also called Robin or mixed) (c) boundary conditions:

$$(a)\ T(x) = \phi(x),\ (b)\ \frac{\partial T}{\partial n} = \phi(x),\ (c)\ \frac{\partial T}{\partial n} + \beta(x)T = \phi(x),\quad x \in \partial\Omega. \qquad (1.4)$$

Here $\phi(x)$ and $\beta(x)$ are some given functions.

Similar boundary value problems can be considered for mass diffusion. In the case of chemically reacting media, we should take into account heat and mass production (consumption):

$$\kappa \Delta T + \sum_{i=1}^{n} q_i W_i(A, T) = 0,$$

$$d_j \Delta A_j + \sum_{i=1}^{n} \gamma_{ij} W_i(A, T) = 0,\ j = 1, \ldots, m,$$

where $A_1, \ldots, A_m$ are the concentrations of reacting species, $A = (A_1, \ldots, A_m)$, $W_i, i = 1, \ldots, n$ are the reaction rates which represent given functions of A and T, d_j are the diffusion coefficients and γ_{ij} the stoichiometric coefficients. Problems of this type arise in combustion and in chemical kinetics.

If the mass transport coefficients are not constant, then the Laplace operator in the first equation above should be replaced by

$$-\operatorname{div}\ q = \operatorname{div}\ (\kappa\ \nabla\ T) = \sum_{i=1}^{3} \frac{\partial}{\partial x_i}\left(\kappa\ \frac{\partial T}{\partial x_i}\right),$$

and similarly for the second equation. If heat and mass transfer takes place in a moving reacting medium, then instead of the previous equations we consider the equations

$$\kappa \Delta T - u.\nabla T + \sum_{i=1}^{n} q_i W_i(A, T) = 0,$$

1. Elliptic problems in applications

$$d_j \Delta A_j - u.\nabla A_j + \sum_{i=1}^{n} \gamma_{ij} W_i(A, T) = 0, \quad j = 1, \ldots, m,$$

where $u = (u_1, u_2, u_3)$ is the velocity vector, and

$$u.\nabla T = \sum_{i=1}^{3} u_i \frac{\partial T}{\partial x_i}, \quad u.\nabla A_j = \sum_{i=1}^{3} u_i \frac{\partial A_j}{\partial x_i}$$

are the convective terms. The velocity of the medium can be a given function of space and time or should be found as a solution of equations of motion.

More precise description of heat and mass transfer should take into account complex multi-component diffusion where the matrix of transport coefficients is not diagonal. In the case of a binary mixture the heat and diffusion fluxes are, respectively:

$$q = -\kappa \nabla T - k_2 \nabla C + Hj, \quad j = -d\left(\nabla C + k_1 \nabla T\right),$$

where the coefficients k_1 and k_2 depend on T and C, H is related to the enthalpy [182]. Stationary temperature and concentration distributions are described by the system of two equations:

$$\text{div } q = 0, \quad \text{div } j = 0.$$

Heat explosion. One of the most well-known problems in combustion theory is the problem of heat explosion [591]. Under certain simplifications, it is described by the elliptic problem

$$\kappa \Delta T + K(T) = 0, \quad T|_{\partial \Omega} = T_0. \tag{1.5}$$

Here T is the temperature, $K(T)$ heat production due to an exothermic chemical reaction, κ the coefficient of thermal conductivity, Ω is a bounded domain in $\mathbb{R}^n$. Since the function $K(T)$ is positive, a solution of problem (1.5) may not exist. In this case, the solution of the corresponding parabolic initial boundary value problem becomes unbounded (blow-up solution). This unbounded temperature increase is interpreted as heat explosion. Solutions of problem (1.5) may not be unique. Their uniqueness and stability determine the behavior of solutions of the evolution problem.

Thus, conditions of heat explosion are related to the existence, stability and bifurcations of solutions of the elliptic problem. There exists an extensive physical and mathematical literature devoted to this question.

Flame propagation. Another classical problem of combustion theory concerns propagation of flames. In the simplest case, stationary flame propagation is described by the elliptic system of equations

$$\kappa \Delta T + c \frac{\partial T}{\partial x_1} + qW(T, \alpha) = 0, \tag{1.6}$$

$$d\Delta\alpha + c\frac{\partial\alpha}{\partial x_1} + W(T,\alpha) = 0 \tag{1.7}$$

considered in unbounded cylinders. Here α is the depth of conversion, W the reaction rate, q the adiabatic heat release, c the speed of the flame. This is an unknown parameter which should be found as a solution of the problem. Existence, uniqueness or non-uniqueness, speed of propagation, stability and bifurcations of combustion waves are among the major topics in combustion theory.

1.2 Electrostatic and gravitational fields

For a constant electrostatic field, Maxwell equations have the form

$$\text{div } E = 4\pi\rho, \quad \text{rot } E = 0, \tag{1.8}$$

where E is the electric field and ρ the density of the charge. The electric field can be expressed through the potential ϕ, $E = -\nabla\phi$. Substituting this expression into the first equation in (1.8), we obtain the Poisson equation:

$$\Delta\phi = -4\pi\rho. \tag{1.9}$$

In the case of a point charge e located at the origin, that is $\rho = e\delta(x)$, in $\mathbb{R}^3$, $\phi(x) = e/R$, where $R = \sqrt{x_1^2 + x_2^2 + x_3^2}$. In the case of two space dimensions, up to a constant factor $\phi(x) = \ln(1/R)$.

Gravitational potential $\psi(x)$ of a point mass m located at the space point ξ equals $-\kappa m/r$, where κ is a constant and $r = |x - \xi|$. The difference in sign with respect to the electric potential reflects the fact that the gravitational force is attracting while electric charges of the same sign are repulsing. If the mass is distributed with the density ρ in some domain V, then up to a constant factor the corresponding potential is given by the expression

$$U_N(x) = \int_V \frac{\rho(\xi)}{r} d\xi.$$

It is similar for the electric potential. It is called the Newton potential. Under certain conditions on the function ρ, for example if it is bounded and continuous together with its first derivatives and decays at infinity at least as $1/|\xi|^2$, then the Newton potential has continuous first and second derivatives. It satisfies the Poisson equation, $\Delta U_N = -4\pi\rho$.

Consider a surface S with the density charge $\rho_s(\xi)$, $\xi \in S$. The potential of this field equals up to a constant factor

$$U_S(x) = \int_S \frac{\rho_s(\xi)}{r} d\xi,$$

where $r = |x - \xi|$. This is the simple layer potential. If there is a layer of dipoles at the surface S with the axis in the direction of outer normal vectors and with

1. Elliptic problems in applications

the density $\rho_d(\xi)$, then the potential is given by the integral

$$U_D(x) = \int_S \rho_d(\xi) \frac{\partial}{\partial n}\left(\frac{1}{r}\right) d\xi.$$

This is the double layer potential.

The theory of potential is applied in numerous problems of electrostatics and gravimetry. On the other hand, it plays an important role in the theory of elliptic problems.

1.3 Hydrodynamics

Many problems in hydrodynamics are described by parabolic and, in the stationary case, by elliptic problems. Among commonly used models are Navier-Stokes equations and Darcy's law in the case of a porous medium.

Viscous incompressible fluid. Motion of a viscous incompressible fluid is described by the Navier-Stokes equations:

$$\frac{\partial v}{\partial t} + v.\nabla v = -\frac{1}{\rho}\nabla p + \nu \Delta v. \tag{1.10}$$

Here $v = (v_1, v_2, v_3)$ is the velocity vector, p is the pressure, ρ and ν are the density and kinematic viscosity supposed to be constant, $v.\nabla v$ denotes the scalar product of the two vectors. The incompressibility condition implies that

$$\nabla.v \equiv \sum_{i=1}^{3} \frac{\partial v_i}{\partial x_i} = 0. \tag{1.11}$$

In the stationary case, problem (1.10), (1.11) is elliptic. It should be completed by boundary conditions. In the case of the no-slip boundary condition the fluid velocity at the boundary equals the velocity of the boundary itself. Another often used boundary condition is the free surface boundary condition.

Navier-Stokes equations admit some explicit solutions. In the two-dimensional strip $0 \leq x_2 \leq 1$ with the no-slip boundary condition, the horizontal component of the velocity v_1 depends only on x_2, $v_1(x) = ax_2(1-x_2)$, the vertical component of the velocity v_2 equals zero. This is the Poiseuille profile. Flow in diffusor is another example of explicit solutions (see, e.g., [289]). In some other cases, as for example fluid motion over a rotating disc (Kármán problem) or some boundary layer problems, the Navier-Stokes equations can be reduced to ordinary differential equations.

Stream function. In the two-dimensional case it is often convenient to introduce the stream function ψ related to the fluid velocity $v = (v_x, v_y)$ as follows:

$$v_x = \frac{\partial \psi}{\partial y}, \quad v_y = -\frac{\partial \psi}{\partial x}.$$

The space variables are denoted here by x and y. In the stationary case the Navier-Stokes equations can be reduced to the fourth-order equation

$$\nu\Delta\Delta\psi - v.\nabla\Delta\psi = 0.$$

In the case of the free surface boundary condition for the velocity, the boundary conditions for the stream function become $\psi = 0$ and $\Delta\psi = 0$.

Natural convection. A nonuniform temperature or density distribution under the gravity field can result in fluid motion. It can be described by the Navier-Stokes equations under the Boussinesq approximation

$$\frac{\partial v}{\partial t} + v.\nabla v = -\frac{1}{\rho}\nabla p + \nu\Delta v + g\beta\gamma(T - T_0), \tag{1.12}$$

$$\frac{\partial T}{\partial t} + v.\nabla T = \kappa\Delta T, \tag{1.13}$$

$$\nabla.v = 0, \tag{1.14}$$

where the density of the fluid is considered to be constant everywhere except for the buoyancy term, the last term in the right-hand side of the first equation describing the action of gravity. Here g is the gravity acceleration, β the coefficient of thermal expansion, γ the unit vector in the vertical direction, T_0 a reference temperature.

Among many problems related to natural convection, let us mention the Rayleigh-Benard problem about convection in a horizontal layer of a liquid heated from below. In the layer $0 \leq y \leq 1$ with the Dirichlet boundary condition for the temperature and the free surface boundary condition for the velocity,

$$y = 0: T = T_0, \ \frac{\partial v_x}{\partial y} = 0, \ v_y = 0; \quad y = 1: T = T_1, \ \frac{\partial v_x}{\partial y} = 0, \ v_y = 0,$$

problem (1.12)–(1.14) has a stationary solution $T(y) = T_0 + (T_1 - T_0)y$, $v = 0$. Linearization about this solution gives an eigenvalue problem which admits an analytical solution. If the eigenvalue with the maximal real part crosses the origin, then the stationary solution loses its stability resulting in appearance of a convective motion (see, e.g., [107], [195]).

Binary fluids. Nonuniform concentration distribution in binary fluids can create additional volume forces in the equations of motion:

$$\frac{\partial v}{\partial t} + v.\nabla v = -\frac{1}{\rho}\nabla p + \nu\Delta v + \frac{K}{\rho}\nabla c\Delta c. \tag{1.15}$$

The last term in the right-hand side of this equation is called Korteweg stress [266]. Equation (1.15) should be completed by the equation for the concentration. It can be the Cahn-Hilliard

$$\frac{\partial c}{\partial t} + v.\nabla c = M\Delta\mu \tag{1.16}$$

or diffusion equation. Here μ is the chemical potential $\mu = \frac{df_0}{dc} - \lambda \Delta c$, f_0 the free energy density,

$$\frac{df_0}{dc} = -K_1 + K_2 c + n_0 k_b T (\ln c - \ln(1 - c)),$$

T the temperature, $\lambda, K_1, K_2, n_0, k_b$ some thermodynamical parameters. This model describes various capillary phenomena in miscible and immiscible liquids and phase separation.

Porous medium. Fluid motion in a porous medium can be described by Darcy's law

$$v = -\frac{K}{\mu} \nabla p,$$

where v is the velocity vector, p the pressure, K permeability and μ viscosity. If the fluid is incompressible, that is $\nabla . v = 0$, then the pressure satisfies the Laplace equation

$$\Delta p = 0.$$

In the case of the presence of sources or sinks, this equation becomes inhomogeneous. Let us consider as an example spreading of the fluid in the Hele-Shaw cell from a point source (Stokes-Leibenzon problem, see [223] and the references therein). We obtain the free boundary problem for the equation

$$\Delta p = \delta(x), \quad x \in \Omega, \quad p = 0, \quad x \in \partial \Omega,$$

where $\delta(x)$ is the Dirac δ-function, the domain Ω changes in time according to the equation $v_n = -\nabla p$. Here v_n denotes the speed of the boundary in the direction of the outer normal vector.

1.4 Biological applications

Mathematical models in biology are often described by reaction-diffusion systems and by equations of fluid and solid mechanics. On the other hand, complexity of biological phenomena requires the development of new approaches. Among them hybrid and multi-scale models.

Population dynamics. Many problems in population dynamics are described by reaction-diffusion equations and systems. The scalar reaction-diffusion equation

$$\frac{\partial u}{\partial t} = \frac{\partial^2 u}{\partial x^2} + F(u) \tag{1.17}$$

was first introduced in population dynamics by Fisher [172] and Kolmogorov, Petrovskii, Piskunov [263] as a model of propagation of a dominant gene. The variable u is the density of a population, the nonlinearity $F(u)$ gives the rate of its reproduction. In the case of the logistic equation, $F(u) = ku(1 - u)$, the reproduction is proportional to the density of the population u and to available resources $(1-u)$.

If we consider this equation on the whole axis, $x \in \mathbb{R}$, then there exist travelling wave solutions, the solutions of the form $u(x,t) = w(x - ct)$. Here c is the wave speed, the function $w(x)$ satisfies the equation

$$w'' + cw' + F(w) = 0.$$

Travelling wave solutions of parabolic systems arise in many applications including population dynamics and combustion. They have stimulated the development of the theory of elliptic problems in unbounded domains. From the mathematical point of view, elliptic problems in unbounded domains possess some specific features related to the essential spectrum. We will discuss them in Section 3. Another particularity of travelling waves is related to the fact that c is an unknown parameter.

Among many reaction-diffusion systems in population dynamics we can mention competition of species and prey-predator models. More recent development is related to nonlocal reaction-diffusion equations. The model with nonlocal consumption of resources

$$\frac{\partial u}{\partial t} = \frac{\partial^2 u}{\partial x^2} + ku\left(1 - \int_{-\infty}^{\infty} \phi(x-y)u(u,t)dy\right)$$

describes reproduction with the same phenotype, random mutations and intraspecific competition. These three conditions determine the process of speciation including the emergence of biological species [194], [560].

Biological pattern formations. One of the most interesting and intriguing questions of mathematical biology is related to biological pattern formation. Consider the following idealized biological situation. At the first stage of the development of an embryo, all its cells are identical. After some time, they differentiate and lead to appearance of various organs. What is the mechanism of the transition from identical cells to differentiated? A possible explanation was suggested by Turing [522]. Consider a reaction-diffusion system

$$\frac{\partial u}{\partial t} = d\Delta u + F(u) \qquad (1.18)$$

in a bounded domain Ω with no-flux boundary conditions

$$\frac{\partial u}{\partial n} = 0, \quad x \in \partial\Omega. \qquad (1.19)$$

Here u and F are vectors, d is a square matrix, n the outer normal vector. Suppose next that $F(u_0) = 0$ for some constant vector u_0 and that this is a stable stationary point of the ordinary differential system of equations

$$\frac{du}{dt} = F(u), \qquad (1.20)$$

that is the matrix $F'(u_0)$ has all eigenvalues in the left half-plane. At the same time, u_0 can be unstable as a stationary solution of problem (1.18), (1.19). This instability takes place if the eigenvalue problem

$$d\Delta v + F'(u_0)v = \lambda v, \quad \frac{\partial v}{\partial n}|_{\partial\Omega} = 0$$

has some eigenvalues in the right half-plane. In this case, a nonhomogeneous in space stationary solution can appear. These are so-called Turing (or dissipative) structures. It appears that the instability conditions depend on the size of the domain. When it is sufficiently small, the solution u_0 remains stable. When the domain becomes large enough, it loses its stability resulting in appearance of spatially distributed stationary solutions.

From the biological point of view, the components of the vector u are concentrations of some bio-chemical substances, morphogenes. When the embryo is small, their distribution is uniform in space and all cells are identical. When the embryo becomes sufficiently large, the distribution of morphogenes becomes nonuniform due to the instability described above. This leads to a spatially nonhomogeneous gene expression and, as a consequence, to cell differentiation.

Thus, in the framework of this approach, biological pattern formation is related to the existence, stability and bifurcations of solutions of elliptic problems. Other biological mechanisms of morphogenesis are also discussed in the literature [345], [363], [394].

Multi-scale modelling. Modern approaches to modelling of living organisms require simultaneous description of physical and bio-chemical phenomena at different levels [22, 23]. They include intra-cellular regulatory networks, cell populations (tissue), extra-cellular matrix, organs or the whole organism.

Cell populations can be considered as a continuous medium and described by partial differential equations of continuum mechanics. Another approach is to consider cells as individual objects. In the simplest case, they are viewed as mathematical points with a pairwise force between them which depends on the distance between the two points (soft sphere model). Their motion is described by Newton's second law. If we consider a system of N particles and take into account energy dissipation, then we obtain the system of second-order ordinary differential equations which represents a particular case of elliptic systems:

$$m\ddot{x}_i - F_i(x, \dot{x}, c) = 0, \quad i = 1, \ldots, N. \tag{1.21}$$

Here x_i is the coordinate of the ith particle, $\dot{x}_i$ is its speed, $\ddot{x}_i$ acceleration, m is the mass of the particles, $x = (x_1, \ldots, x_N)$, F_i is the sum of forces acting on the ith particle, $c = (c_1, \ldots, c_m)$ are the concentrations of bio-chemical substances in the extra-cellular matrix. The dependance of the force on these concentrations takes into account a possible chemotactic motion.

The concentrations in the extra-cellular matrix can be described by the diffusion equation
$$\frac{\partial c}{\partial t} = d\Delta c + W(c, n), \tag{1.22}$$
where the rate of their production or consumption W depends on the local cell concentrations $n = (n_1, \ldots, n_k)$. Here the subscript corresponds to different cell types. We do not take into account convective terms in equation (1.22) which can appear because of cell motion.

Intra-cellular regulatory networks can be described by ordinary differential systems of equations:
$$\frac{ds^{(i)}}{dt} = \Phi(s^{(i)}, c), \quad i = 1, \ldots, N, \tag{1.23}$$
where $s^{(i)}$ is the vector of intra-cellular concentrations for the ith cell. The rate Φ of their production depends on the local concentrations $c(x, t)$ in the extra-cellular matrix. Cell division, differentiation, death, adhesion to other cells depend on the intra-cellular concentrations. In particular, cell division leads to the evolution of cell population where cells interact chemically and mechanically. These mechanical forces influence cell motion.

System (1.21)–(1.23) is a multi-scale model of cell dynamics. It represents a parabolic or elliptic system coupled with ordinary differential equations. Depending on specific applications, it can be completed by various regulatory mechanisms. For example, production of red blood cells in the bone marrow is regulated by the hormone erythropoietin produced by the kidney and liver. In response to hypoxia (lack of oxygen) it decreases cell apoptosis in the bone marrow and, as a consequence, increases the quantity of erythrocytes in blood and reinforces the transport of oxygen.

2 Classical theory of linear elliptic problems

2.1 Function spaces

In this section we briefly recall the main definitions and some properties of function spaces used in this book. More detailed presentation can be found elsewhere [1], [277], [285], [318]. *Sobolev space* $W^{l,p}(\mathbb{R}^n)$, where $1 < p < \infty$ and l is an integer, consists of all functions which belong to $L^p(\mathbb{R}^n)$ together with their derivative up to the order l. The derivatives are understood in the sense of generalized functions. The norm in this space is given by the equality
$$\|u\|_{W^{l,p}(\mathbb{R}^n)} = \left(\sum_{|j| \leq l} \|u^{(j)}\|^p_{L^p(\mathbb{R}^n)} \right)^{1/p}.$$

2. Classical theory of linear elliptic problems

In order to define *Sobolev-Slobodetskii spaces* (l is not integer), consider first the case $0 < l < 1$. The space $W^{l,p}(\mathbb{R}^n)$ consists of all functions from $L^p(\mathbb{R}^n)$ for which the integral

$$I(u) = \int_{\mathbb{R}^n} \int_{\mathbb{R}^n} \frac{|u(x) - u(y)|^p}{|x - y|^{n+pl}} dx dy$$

is bounded. The norm is given by the equality

$$\|u\|_{W^{l,p}(\mathbb{R}^n)} = \left(\|u\|^p_{L^p(\mathbb{R}^n)} + I(u)\right)^{1/p}.$$

If $l > 1$, then we put $l = [l] + \lambda$, $0 \le \lambda < 1$. The space $W^{l,p}(\mathbb{R}^n)$ consists of all functions for which all derivatives of the order $[l]$ belong to the space $W^{\lambda,p}(\mathbb{R}^n)$,

$$\|u\|_{W^{l,p}(\mathbb{R}^n)} = \left(\|u\|^p_{W^{[l],p}(\mathbb{R}^n)} + \sum_{|j|=[l]} \|u^{(j)}\|^p_{W^{\lambda,p}(\mathbb{R}^n)}\right)^{1/p}.$$

Finally, if $l < 0$, then

$$W^{l,p}(\mathbb{R}^n) = \left(W^{-l,p'}(\mathbb{R}^n)\right)^*,$$

where $1/p + 1/p' = 1$ and $*$ denotes the dual space.

The *space of Bessel potentials* $H^{l,p}(\mathbb{R}^n)$ ($-\infty < l < \infty, p > 1$) consists of all generalized functions u from $S'(\mathbb{R}^n)$ for which

$$F^{-1}\left((1 + |\xi|^2)^{l/2} F(u)\right) \in L^p(\mathbb{R}^n).$$

Here F denotes the Fourier transform. The norm is given by

$$\|u\|_{H^{l,p}(\mathbb{R}^n)} = \|F^{-1}\left((1 + |\xi|^2)^{l/2} F(u)\right)\|_{L^p(\mathbb{R}^n)}.$$

For $l = 0$ the space of Bessel potential coincides with L^p, $H^{0,p}(\mathbb{R}^n) = L^p(\mathbb{R}^n)$. The space $H^{-l,p}(\mathbb{R}^n)$ can be identified with the dual space to the space $H^{l,p'}(\mathbb{R}^n)$.

Besov spaces $B^{l,p}(\mathbb{R}^n)$ coincide with $W^{l,p}(\mathbb{R}^n)$ for noninteger positive l. For integer $l \ge 1$ and $p > 1$ it consists of functions $u \in W^{l-1,p}(\mathbb{R}^n)$ for which the integrals

$$I_j(u) = \int_{\mathbb{R}^n} \int_{\mathbb{R}^n} \frac{|u^{(j)}(x) - 2u^{(j)}((x+y)/2) + u^{(j)}(y)|^p}{|x - y|^{n+p}} dx dy, \quad |j| = l$$

are finite. The norm in this space is defined as follows:

$$\|u\|_{B^{l,p}(\mathbb{R}^n)} = \left(\|u\|^p_{W^{l-1,p}(\mathbb{R}^n)} + \sum_{|j|=l} I_j(u)\right)^{1/p}.$$

For $l < 0$, by definition $B^{l,p}(\mathbb{R}^n) = \left(B^{-l,p'}(\mathbb{R}^n)\right)^*$.

The spaces $W^{l,p}(\mathbb{R}^n), H^{l,p}(\mathbb{R}^n), B^{l,p}(\mathbb{R}^n)$ are reflexive. The space $D(\mathbb{R}^n)$ of infinitely differentiable functions with bounded supports is dense in each of these spaces. For $p = 2$ all three spaces coincide for any positive l. For integer l, $W^{l,p}(\mathbb{R}^n)$ and $H^{l,p}(\mathbb{R}^n)$ coincide, for noninteger l and $p \neq 2$ they are different. For noninteger l, $W^{l,p}(\mathbb{R}^n) = B^{l,p}(\mathbb{R}^n)$, for integer l and $p \neq 2$ they are different.

We next define these spaces in domains $\Omega \subset \mathbb{R}^n$. Let its boundary Γ be an oriented infinitely differentiable manifold of the dimension $n - 1$. The domain Ω is located from one side of Γ (locally). The space $W^{l,p}(\Omega)$ consists of all functions $u(x)$ defined in Ω and such that they can be extended to some functions $\tilde{u}(x)$ defined in $\mathbb{R}^n$ and $\tilde{u}(x) \in W^{l,p}(\mathbb{R}^n)$. The norm in the space $W^{l,p}(\Omega)$ is defined as infimum with respect to all possible extensions,

$$\|u\|_{W^{l,p}(\Omega)} = \inf \|\tilde{u}\|_{W^{l,p}(\mathbb{R}^n)}.$$

The spaces $H^{l,p}(\Omega)$ and $B^{l,p}(\Omega)$ are defined similarly. The spaces $W^{l,p}(\Omega)$ and $B^{l,p}(\Omega)$ can also be defined as in the case $\mathbb{R}^n$ with the integrals taken over Ω. The assumption that the boundary of the domain is infinitely differentiable is not necessary and weaker assumptions will be used.

For any function u from $W^{l,p}(\Omega)$ ($H^{l,p}(\Omega)$), $B^{l,p}(\Omega)$), $l > 1/p$ its trace at the boundary Γ is defined and belongs to the space $B^{l-1/p,p}(\Gamma)$. On the other hand, for any $\phi \in B^{l-1/p,p}(\Gamma)$, there exists its continuation u from $W^{l,p}(\Omega)$ ($H^{l,p}(\Omega)$) such that

$$\|u\|_{W^{l,p}(\Omega)(H^{l,p}(\Omega),B^{l,p}(\Omega))} \leq C\|\phi\|_{B^{l-1/p,p}(\Gamma)},$$

where the constant C does not depend on ϕ.

A function $u(x)$ satisfies the Hölder condition with the exponent $\alpha \in (0,1)$ if the quantity

$$\langle u \rangle_\Omega^\alpha = \sup_{x,y \in \Omega, x \neq y} \frac{|u(x) - u(y)|}{|x-y|^\alpha}$$

is bounded. The *Hölder space* $C^{(\alpha)}(\bar{\Omega})$ consists of all functions continuous in Ω and such that $\langle u \rangle_\Omega^\alpha$ is bounded. The norm in this space is given by the equality

$$\|u\|_{C^{(\alpha)}(\bar{\Omega})} = \sup_\Omega |u(x)| + \langle u \rangle_\Omega^\alpha.$$

The space $C^{(k+\alpha)}(\bar{\Omega})$ consists of all functions continuous in Ω together with their derivatives up to the order k and such that the derivatives of the order k satisfy the Hölder condition with the exponent α. The norm in this space is given by the equality

$$\|u\|_{C^{(k+\alpha)}(\bar{\Omega})} = \sum_{|j|=0}^{k} \sup_\Omega |D^j u(x)| + \sum_{|j|=k} \langle D^j u \rangle_\Omega^\alpha.$$

2.2 Elliptic problems

2.2.1. Main definitions. We present here the main definitions and some properties of elliptic problems (see, e.g., [7], [11], [318], [542]).

Scalar operators. Consider the linear differential operator

$$Au = \sum_{|\alpha| \leq r} a^\alpha(x) D^\alpha u, \qquad (2.1)$$

with complex coefficients $a^\alpha(x)$, where $\alpha = (\alpha_1, \ldots, \alpha_n)$ is the multi-index, $D^\alpha = D_1^{\alpha_1} \ldots D_n^{\alpha_n}$, D_i denotes the partial derivative with respect to x_i, $x \in \Omega \subset \mathbb{R}^n$. The smoothness of the coefficients and of the boundary $\partial \Omega$ will be specified below. Let

$$A_0(x, \xi) = \sum_{|\alpha|=r} a^\alpha(x) \xi^\alpha,$$

where $\xi \in \mathbb{R}^n$. The operator A is called *elliptic* in $\bar{\Omega}$ if

$$A_0(x, \xi) \neq 0, \quad \forall x \in \bar{\Omega}, \, \xi \in \mathbb{R}^n, \, \xi \neq 0.$$

The integer r is the order of the operator. It is known that for $n > 2$ every elliptic operator has an even order. This assertion remains valid for $n = 2$ if the coefficients of the operator are real. If they are complex, then it is not true. The Cauchy-Riemann operator is a first-order elliptic operator in $\mathbb{R}^2$.

The operator A is called *properly elliptic* if it is elliptic and the equation

$$A_0(x, \xi + \tau \eta) = 0$$

with respect to the complex number τ has the same number $m = r/2$ of solutions with positive and negative imaginary parts for any $\xi, \eta \in \mathbb{R}^n$, $x \in \partial \Omega$. For $n > 2$, proper ellipticity follows from ellipticity. If $n = 2$ and the boundary is connected, then the condition of proper ellipticity is satisfied everywhere if it is satisfied at one point of the boundary.

Consider next the boundary operators

$$B_j u = \sum_{|\beta| \leq r_j} b_j^\beta(x) D^\beta u, \quad j = 1, \ldots, m. \qquad (2.2)$$

We assume that the order r of the operator A is even. The number m of the boundary operators equals $r/2$. In order to introduce the Lopatinskii (or Shapiro-Lopatinskii) condition, let us consider a point $x \in \partial \Omega$ and local coordinates (x', x_n) where x' are the coordinates in the tangent plane. Consider the boundary problem on the half-line:

$$\tilde{A}_0(x', 0, \xi', D_n) v(t) = 0, \qquad (2.3)$$

$$\tilde{B}_{j,0}(x', 0, \xi', D_n) v(t)|_{t=0} = h_j, \quad j = 1, \ldots, m, \qquad (2.4)$$

where $(x', 0)$ is the point at the boundary, D_n is the derivative with respect to the variable $x_n = t$, the functions $\tilde{A}_0$ and $\tilde{B}_{j,0}$ are obtained from (2.1), (2.2) by freezing the coefficients, omitting the lower-order terms and applying the formal Fourier transform with respect to the variables x'. For any $\xi' \neq 0$ and any complex numbers h_j, problem (2.3), (2.4) is supposed to have a unique bounded solution. There exist equivalent formulations of the Lopatinskii condition (see [7], [11], [318], [579] and the next subsection).

The boundary value problem

$$Au = f, \quad x \in \Omega, \quad B_j = g_j, \ j = 1, \ldots, m, \ x \in \partial\Omega \tag{2.5}$$

is called elliptic if the operator A is elliptic, properly elliptic and the Lopatinskii condition is satisfied.

General elliptic problems. Consider the differential operators

$$A_i u = \sum_{k=1}^{N} \sum_{|\alpha| \leq \alpha_{ik}} a_{ik}^{\alpha}(x) D^{\alpha} u_k, \quad i = 1, \ldots, N, \ x \in \Omega \tag{2.6}$$

in a domain $\Omega \subset \mathbb{R}^n$ and at its boundary $\partial\Omega$:

$$B_j u = \sum_{k=1}^{N} \sum_{|\beta| \leq \beta_{jk}} b_{jk}^{\beta}(x) D^{\beta} u_k, \quad j = 1, \ldots, m, \ x \in \partial\Omega. \tag{2.7}$$

Here $u = (u_1, \ldots, u_N)$, $\alpha = (\alpha_1, \ldots, \alpha_n)$, $D^{\alpha} = D_1^{\alpha_1} \ldots D_n^{\alpha_n}$, $D_k = \partial/\partial x_k$, $|\alpha| = |\alpha_1| + \cdots + |\alpha_n|$, D^{β} and $|\beta|$ are defined similarly. These operators can also be written in the matrix form

$$A = \begin{pmatrix} A_{11} & \ldots & A_{1N} \\ & \ldots & \\ A_{N1} & \ldots & A_{NN} \end{pmatrix}, \quad B = \begin{pmatrix} B_{11} & \ldots & B_{1N} \\ & \ldots & \\ B_{m1} & \ldots & B_{mN} \end{pmatrix},$$

where

$$A_{ik} = \sum_{|\alpha| \leq \alpha_{ik}} a_{ik}^{\alpha}(x) D^{\alpha}, \quad B_{jk} = \sum_{|\beta| \leq \beta_{jk}} b_{jk}^{\beta}(x) D^{\beta}.$$

Conditions on the coefficients of the operators and on the domain Ω will be specified below. At the moment, we can suppose for simplicity that the coefficients and the boundary of the domain are infinitely differentiable.

According to the definition of elliptic operators in the Douglis-Nirenberg sense we suppose that

$$\alpha_{ik} \leq s_i + t_k, \ i, k = 1, \ldots, N, \ \beta_{jk} \leq \sigma_j + t_k, \ j = 1, \ldots, m, \ k = 1, \ldots, N$$

for some integers s_i, t_k, σ_j such that $s_i \leq 0$, $\max s_i = 0$, $t_k \geq 0$. The number $\sum_{i=1}^{N}(s_i + t_i)$ is called the order of the operator. It is supposed to be equal to $2m$, that is twice the number of the boundary operators.

2. Classical theory of linear elliptic problems

Denote by A_{ik}^0 the principal symbol of the operator A_{ik}, that is

$$A_{ik}^0(x,\xi) = \sum_{|\alpha|=\alpha_{ik}} a_{ik}^\alpha(x)\xi^\alpha,$$

where $\xi = (\xi_1,\ldots,\xi_n) \in \mathbb{R}^n$, $\xi^\alpha = \xi_1^{\alpha_1} \ldots \xi_n^{\alpha_n}$. Let $A^0(x,\xi)$ be the matrix with the elements $A_{ik}^0(x,\xi)$ if $\alpha_{ik} = s_i + t_k$ and 0 if $\alpha_{ik} < s_i + t_k$.

The operator is called *elliptic in the Douglis-Nirenberg sense* if

$$\det A^0(x,\xi) \neq 0, \quad \forall\, x \in \bar{\Omega},\ \xi \in \mathbb{R}^n,\ |\xi| \neq 0.$$

We next define the condition of *proper ellipticity*. We introduce local coordinates (x', x_n) at the boundary of the domain where x' are the coordinates in the tangent plane, the vector $(0, x_n)$ is normal to the boundary, $(x', 0) \in \partial\Omega$. Consider the equation

$$\det A^0(x', 0, \xi', \zeta) = 0$$

with respect to ζ. By virtue of the ellipticity condition this equation does not have solutions for real ζ. It is supposed to have the same number of roots m with positive and negative imaginary parts. This formulation of the condition of proper ellipticity is equivalent to the formulation given above for the scalar operator. It is known that for $n > 2$ the condition of proper ellipticity follows from the ellipticity condition (see, e.g., [11]).

We finally introduce the *Lopatinskii condition*. Consider the principal terms ($|\alpha| = \alpha_{ik}, |\beta| = \beta_{jk}$) of problem (2.6), (2.7) with constant coefficients fixed at some boundary point $x \in \partial\Omega$. Consider, next, the local coordinates $x = (x', x_n)$ in the neighborhood of this point and apply the Fourier transform with respect to the variables x'. We suppose that for each point of the boundary and for each vector h the corresponding one-dimensional problem

$$A^0(x', 0, \xi', D_n)v(t) = 0, \quad B^0(x', 0, \xi', D_n)v(t)|_{t=0} = h$$

considered on the half-axis $t > 0$ ($t = x_n$) has a unique solution $v(t)$ such that $|v(t)| \to 0$ as $t \to \infty$. Here B^0 is the matrix with the elements

$$B_{jk}^0 = \sum_{|\beta|=\beta_{jk}} b_{jk}^\beta(x) D^\beta$$

if $\beta_{jk} = \sigma_j + t_k$ and 0 if $\beta_{jk} < \sigma_j + t_k$.

If the ellipticity condition, proper ellipticity and Lopatinskii conditions are satisfied then the problem

$$A_i u = f_i, \quad B_j u = g_j, \quad i = 1,\ldots,N,\ j = 1,\ldots,m$$

is called *elliptic in the Agmon-Douglis-Nirenberg* (ADN) *sense* or in the general sense. A particular case of elliptic problems in the ADN sense is an elliptic problem in the *sense of Petrovskii* where $s_1 = \cdots = s_N$. This means that the orders of the operators A_{ik} in the principal part of the operator A are the same for all i.

We will also use another form of the Lopatinskii condition. In order to formulate it, for any point $x \in \partial\Omega$ consider the local coordinates (ξ', ν), where $\xi' = (\xi_1, \ldots, \xi_{n-1})$ are the coordinates in the tangential hyperspace, ν is the normal coordinate. Let $\hat{A}(x, \xi', \nu)$ and $\hat{B}(x, \xi', \nu)$ be the matrices with the elements

$$\hat{A}_{ij}(x, \xi', \nu) = \sum_{|\alpha'|+\alpha_n=\alpha_{ij}} a_{ij}^{\alpha'\alpha_n} \xi'^{\alpha'} \nu^{\alpha_n}, \quad i, j = 1, \ldots, N,$$

$$\hat{B}_{ij}(x, \xi', \nu) = \sum_{|\alpha'|+\alpha_n=\beta_{ij}} b_{ij}^{\alpha'\alpha_n} \xi'^{\alpha'} \nu^{\alpha_n}, \quad i = 1, \ldots, m, \, , j = 1, \ldots, N.$$

The Lopatinskii matrix is defined by the equality

$$\Lambda(x, \xi') = \int_{\gamma_+} \hat{B}(x, \xi', \mu) \hat{A}^{-1}(x, \xi', \mu) \Phi(\mu) d\mu,$$

where

$$\Phi(\mu) = (E, \mu E, \ldots, \mu^{s-1} E),$$

E is the identity matrix of the order N, $s = \max_{i,j} \alpha_{ij}$, γ_+ is a Jordan curve in the half-plane $\operatorname{Im} \mu > 0$ enclosing all the roots of $\det \, A(x, \xi', \mu)$ with positive imaginary parts. By virtue of the condition of proper ellipticity there are m such roots, and there are no roots on the real axis. The Lopatinskii condition implies that the rank of the matrix $\Lambda(x, \xi')$ equals m for all $|\xi'| \neq 0$.

We will use the notation

$$e_d^0 = \inf_{x \in \Omega, |\xi|=1} |\det A^0(x, \xi)|, \quad e_\Gamma^0 = \inf_{x \in \Gamma, |\xi'|=1} \sum_\alpha |\mu_\alpha(x, \xi')|,$$

where $\mu_\alpha(x, \xi')$ are all m-minors of the Lopatinskii matrix in the local coordinates (ξ', ν) at the point x. Problem (2.6), (2.7) is called *uniformly elliptic* if $e_d^0 > 0$, $e_\Gamma^0 > 0$. In what follows, when we assume that the problem is uniformly elliptic, we will understand that the condition of proper ellipticity is also satisfied.

2.2.2. Examples. Laplace operator. Consider the Laplace operator

$$Lu = \Delta u \equiv \frac{\partial^2 u}{\partial x_1^2} + \frac{\partial^2 u}{\partial x_2^2}$$

in the half-plane $\mathbb{R}_+^2 = \{(x_1, x_2) : x_2 > 0\}$. Its principal symbol $A^0(\xi) = \xi_1^2 + \xi_2^2$ is different from zero for $|\xi| \neq 0$. Therefore the ellipticity condition is satisfied. In order to verify the condition of proper ellipticity, consider the equation

$$\xi_1^2 + \zeta^2 = 0$$

2. Classical theory of linear elliptic problems

with respect to ζ. For any real $\xi_1 \neq 0$ it has one solution with positive and another one with negative imaginary part. The condition of proper ellipticity is also satisfied.

We discuss next the Lopatinskii condition. We carry out the partial Fourier transform with respect to x_1 and obtain the ordinary differential equation

$$\frac{d^2\tilde{v}}{dt^2} - \xi_1^2 \tilde{v} = 0 \tag{2.8}$$

on the half-axis $t > 0$. In the case of the Dirichlet boundary condition we have

$$\tilde{v}(0) = h. \tag{2.9}$$

Here t replaces x_2, $\tilde{v}$ denotes the Fourier transform. For any real $\xi_1 \neq 0$ and any h problem (2.8), (2.9) has a unique solution decaying at infinity. It is also the case for the Neumann boundary condition:

$$\tilde{v}'(0) = h.$$

In the case of the boundary operator with oblique derivative,

$$Bu \equiv a\frac{\partial u}{\partial x_1} + b\frac{\partial u}{\partial x_2}, \quad a^2 + b^2 \neq 0;$$

after the partial Fourier transform we obtain the boundary condition for $\tilde{v}$:

$$ai\xi_1 \tilde{v}(0) + b\tilde{v}'(0) = h. \tag{2.10}$$

Problem (2.8), (2.10) has a unique decaying solution for any $\xi_1 \neq 0$. This means that the Lopatinskii condition is satisfied.

In the case of the tangent derivative ($b = 0$) the Lopatinskii condition remains valid in the 2D case but not in 3D. For the boundary operator

$$Bu \equiv a\frac{\partial u}{\partial x_1} + b\frac{\partial u}{\partial x_2} + c\frac{\partial u}{\partial x_3}, \quad a^2 + b^2 + c^2 \neq 0$$

we obtain

$$ai\xi_1 \tilde{v}(0) + bi\xi_2 \tilde{v}(0) + c\tilde{v}'(0) = h. \tag{2.11}$$

If $c = 0$, the equation

$$\frac{d^2\tilde{v}}{dt^2} - (\xi_1^2 + \xi_2^2)\tilde{v} = 0$$

with boundary condition (2.11) may not have solutions for $|\xi_1| + |\xi_2| \neq 0$. Indeed, this is the case if $a\xi_1 + b\xi_2 = 0$.

Cauchy-Riemann system. Consider the Hilbert problem for the Cauchy-Riemann system:
$$\frac{\partial u}{\partial x} - \frac{\partial v}{\partial y} = 0, \quad \frac{\partial u}{\partial y} + \frac{\partial v}{\partial x} = 0 \tag{2.12}$$
in the half-plane $\mathbb{R}_+^2 = \{(x,y) : y > 0\}$ with the boundary condition
$$au + bv = h(x), \tag{2.13}$$
where a and b are real numbers, $a^2 + b^2 = 1$. We have
$$A^0(\xi) = \begin{pmatrix} \xi_1 & -\xi_2 \\ \xi_2 & \xi_1 \end{pmatrix}, \quad \det A^0(\xi) \neq 0 \text{ for } |\xi| \neq 0.$$

Hence the ellipticity condition is satisfied. To verify the condition of proper ellipticity, we obtain the same equation as for the Laplace operator. There is one root with a positive and one with a negative imaginary part.

Applying the partial Fourier transform with respect to x to system (2.12), we obtain the system
$$\begin{cases} \tilde{u}' + i\xi_1 \tilde{v} = 0, \\ \tilde{v}' - i\xi_1 \tilde{u} = 0 \end{cases}$$
on the half-axis $y > 0$. Here the tilde denotes the partial Fourier transform and prime denotes the derivative with respect to y. There exists a bounded solution for $y > 0$:
$$\begin{pmatrix} \tilde{u} \\ \tilde{v} \end{pmatrix} = \begin{pmatrix} p_1 \\ p_2 \end{pmatrix} e^{-|\xi_1| y},$$
where
$$p_2 = -i \frac{\xi_1}{|\xi_1|} \cdot p_1.$$
From the boundary condition (2.13) we have
$$\left(a - bi \frac{\xi_1}{|\xi_1|} \right) p_1 = \tilde{h}(\xi_1),$$
where $\tilde{h}$ is the Fourier transform of the function h. Since a and b are real, then this equation has a solution for any $\xi_1 \neq 0$. The Lopatinskii condition is satisfied.

Stokes equations. Let us now consider the Stokes equations for an incompressible fluid:
$$\Delta u_i + \frac{\partial p}{\partial x_i} = 0 \ (i = 1, 2, 3), \quad \frac{\partial u_1}{\partial x_1} + \frac{\partial u_2}{\partial x_2} + \frac{\partial u_3}{\partial x_3} = 0. \tag{2.14}$$
In this case
$$A = \begin{pmatrix} \partial^2 & 0 & 0 & \partial_1 \\ 0 & \partial^2 & 0 & \partial_2 \\ 0 & 0 & \partial^2 & \partial_3 \\ \partial_1 & \partial_2 & \partial_3 & 0 \end{pmatrix},$$

2. Classical theory of linear elliptic problems

where

$$\partial_i = \frac{\partial}{\partial x_i} \quad (i = 1, 2, 3), \quad \partial^2 = \partial_1^2 + \partial_2^2 + \partial_3^2.$$

The orders α_{ik} of the operators A_{ik} differ from each other. According to the definition of elliptic operators in the Douglis-Nirenberg sense, we introduce integers s_i, t_k, $i, k = 1, 2, 3, 4$ such that $\alpha_{ik} \leq s_i + t_k$. If the inequality is strict, then in the definition of the principle symbol the corresponding operator is replaced by zero. We put $s_1 = s_2 = s_3 = 0, s_4 = -1, t_1 = t_2 = t_3 = 2, t_4 = 1$. Then

$$A^0(\xi) = \begin{pmatrix} |\xi|^2 & 0 & 0 & \xi_1 \\ 0 & |\xi|^2 & 0 & \xi_2 \\ 0 & 0 & |\xi|^2 & \xi_3 \\ \xi_1 & \xi_2 & \xi_3 & 0 \end{pmatrix}, \quad \det A^0(\xi) = -(|\xi_1|^2 + |\xi_2|^2 + |\xi_3|^2)^3.$$

Therefore the ellipticity condition is satisfied.

Consider system (2.14) in the half-space $\mathbb{R}_+^3 = \{(x_1, x_2, x_3), x_3 \geq 0\}$ with the Dirichlet boundary conditions for the velocity. Applying the partial Fourier transform with respect to x_1, x_2 we obtain the problem on the half-axis $x_3 \geq 0$:

$$\tilde{u}_i'' - \xi^2 \tilde{u}_i + i\xi_i \tilde{p} = 0, \quad i = 1, 2, \tag{2.15}$$

$$\tilde{u}_3'' - \xi^2 \tilde{u}_3 + \tilde{p}' = 0, \tag{2.16}$$

$$i\xi_1 \tilde{u}_1 + i\xi_2 \tilde{u}_2 + \tilde{u}_3' = 0, \tag{2.17}$$

$$\tilde{u}_i(0) = h_i, \quad i = 1, 2, 3, \tag{2.18}$$

where tilde denotes the Fourier transform and prime the derivative with respect to $t(= x_3)$, $\xi^2 = \xi_1^2 + \xi_2^2$. We multiply (2.15) by $i\xi_i$, differentiate (2.16) and take their sum. By virtue of (2.17) we obtain

$$\tilde{p}'' - \xi^2 \tilde{p} = 0.$$

Let, for certainty, $\xi > 0$. Then

$$\tilde{p}(t) = ce^{-\xi t},$$

where c is a constant which will be found below. From (2.15), (2.16), (2.18),

$$\tilde{u}_i(t) = h_i e^{-\xi t} + \frac{\xi_i}{2\xi} icte^{-\xi t}, \quad i = 1, 2,$$

$$\tilde{u}_3(t) = h_3 e^{-\xi t} - \frac{1}{2} cte^{-\xi t}.$$

Finally, from (2.17),

$$c = 2\xi(ih_1 + ih_2 - h_3).$$

Hence, there exists a unique solution decaying at infinity for any h_i. Therefore the Lopatinskii condition is satisfied.

2.3 A priori estimates

A priori estimates of solutions play an important role in the theory of linear elliptic problems. They imply normal solvability and Fredholm property of elliptic operators. We will sketch the derivation of the Schauder estimates for second-order problems and will give their formulation for general elliptic operators. The detailed proof of these estimates can be found elsewhere [7], [285]. The derivation of integral estimates will be presented in Chapter 3 in more detail. We will use them to investigate the Fredholm property of elliptic operators in unbounded domains.

For a bounded operator L acting from a Banach space $E(\Omega)$ into another Banach space $F(\Omega)$, where $\Omega \subset \mathbb{R}^n$ is a domain with a sufficiently smooth boundary, a priori estimates can be written in the form

$$\|u\|_{E(\Omega)} \leq k \left(\|Lu\|_{F(\Omega)} + \|u\|_{E'(\Omega)} \right), \tag{2.19}$$

where $E'(\Omega)$ is some other Banach space such that

$$E(\Omega) \subset E'(\Omega), \tag{2.20}$$

and the inclusion is understood in the algebraic and in the topological sense. Such estimates are obtained for general elliptic operators acting in Sobolev (bounded domains) and in Hölder (bounded or unbounded domains) spaces (see [7], [318], [457], [542] and the references therein). For the second-order elliptic operators and the spaces

$$E(\Omega) = C^{2+\alpha}(\bar{\Omega}), \quad F(\Omega) = C^{\alpha}(\bar{\Omega}), \quad E'(\Omega) = C(\bar{\Omega}), \tag{2.21}$$

(2.19) corresponds to the Schauder estimate.

2.3.1. Schauder estimates for second-order problems. In this section we consider the second-order operator

$$Lu = \sum_{i,j=1}^{n} a_{ij}(x) \frac{\partial^2 u}{\partial x_i \partial x_j} + \sum_{i}^{n} b_i(x) \frac{\partial u}{\partial x_i} + c(x) u$$

with real coefficients which belong to the space $C^{l-2+\alpha}(\bar{\Omega})$, $l \geq 2$. The matrix $a = (a_{ij})$ is symmetric and satisfies the condition of strong ellipticity:

$$(a\xi, \xi) \geq \nu |\xi|^2, \quad \forall \xi \in \mathbb{R}^n, \quad \nu > 0.$$

The domain $\Omega \subset \mathbb{R}^n$ is bounded with the boundary $\partial\Omega$ of the class $C^{l+\alpha}$. Then the solution of the Dirichlet problem

$$Lu = f, \quad u|_{\partial\Omega} = g \tag{2.22}$$

satisfies the estimate

$$\|u\|_{C^{l+\alpha}(\bar{\Omega})} \leq K \left(\|f\|_{C^{l-2+\alpha}(\bar{\Omega})} + \|g\|_{C^{l+\alpha}(\partial\Omega)} + \sup_{x \in \bar{\Omega}} |u| \right), \tag{2.23}$$

2. Classical theory of linear elliptic problems

where the constant K depends only on l, α, ν, on the norms of the coefficients and on the boundary of the domain.

The proof of such estimates is based on the following approach. At the first step, the Newton and double layer potentials are estimated. Then these estimates are used in order to estimate the solution of the Poisson equation. Finally, freezing the coefficients of problem (2.22), we reduce it locally to the Poisson equation.

We briefly present here this construction (see, e.g., [285]). Consider the Newton potential:

$$w(x) = \tau_n^{-1} \int_{\mathbb{R}^n} \frac{f(y)}{|x-y|^{n-2}} dy,$$

and the double layer potential in the half-space $\mathbb{R}^n_+ = \{x \in \mathbb{R}^n, x_n \geq 0\}$:

$$v(x) = 2\tau_n^{-1} \int_{y_n=0} \frac{\partial}{\partial x_n} \frac{\phi(y')}{|x-y|^{n-2}} dy'.$$

Here $\tau_n = n(n-2)\kappa_n$, κ_n is the volume of the unit ball.

If $f \in C^\alpha(\mathbb{R}^n)$ has a bounded support, then the Newton potential $w(x)$ is continuous together with its derivatives up to the second order and

$$\langle D^2 w \rangle^\alpha_{\mathbb{R}^n} \leq c \langle f \rangle^\alpha_{\mathbb{R}^n} \tag{2.24}$$

(see the notation in Section 2.1). If $\phi \in C^{2+\alpha}(\mathbb{R}^{n-1})$ has a bounded support, then the double layer potential $v(x)$ is continuous together with its second derivatives and

$$\langle D^2 v \rangle^\alpha_{\mathbb{R}^n_+} \leq c \langle D^2 \phi \rangle^\alpha_{x_n=0}. \tag{2.25}$$

The constant c in these estimates depends only on n and α.

The estimates of the potentials allow us to estimate solutions of the model problems in the whole space and in the half-space. The solution of the problem

$$\Delta u = f, \quad x \in \mathbb{R}^n, \tag{2.26}$$

where the function $f \in C^\alpha(\mathbb{R}^n)$ has a bounded support, is given by the Newton potential. Therefore, we can use estimate (2.24). The solution of the model problem in the half-space

$$\Delta u = \hat{f}, \quad x_n > 0, \quad u = \psi(x'), \quad x = x_n, \tag{2.27}$$

where the functions $\hat{f} \in C^\alpha(\mathbb{R}^n_+)$ and $\phi \in C^{2+\alpha}(\mathbb{R}^{n-1})$ have bounded supports, can be sought in the form $u = v + w$. The Newton potential w is a solution of (2.26) where the function f is defined in the whole space, coincides with $\hat{f}$ for $x_n \geq 0$ and has the same norm. Then v is a solution of the Laplace equation in the half-space with the boundary value $v(x', 0) = \psi(x') - w(x', 0)$. It is given by the double layer potential for which we can use estimate (2.25). Thus, we can estimate the solutions of the model problems.

Let us return to problem (2.22). Consider a partition of unity ζ_k, $\sum_{k=1}^N \zeta_k = 1$ in $\bar{\Omega}$ and put $u_k = u\zeta_k$. We suppose that the functions ζ_k are uniformly bounded in the norm $C^{2+\alpha}(\bar{\Omega})$. Multiplying equation $Lu = f$ by ζ_k, we obtain after some simple calculations

$$\sum_{i,j=1}^n a_{ij}(x_0)\frac{\partial^2 u_k}{\partial x_i \partial x_j} = F_k, \qquad (2.28)$$

where

$$F_k = \sum_{i,j=1}^n (a_{ij}(x_0) - a_{ij}(x))\frac{\partial^2 u_k}{\partial x_i \partial x_j} + \sum_{i,j=1}^n a_{ij}(x)\left(\frac{\partial^2 \zeta_k u}{\partial x_i \partial x_j} - \zeta_k \frac{\partial^2 u}{\partial x_i \partial x_j}\right)$$

$$- \left(\sum_i^n b_i(x)\frac{\partial u}{\partial x_i} + c(x)u\right)\zeta_k + f\zeta_k.$$

Without loss of generality we can suppose that the support of ζ_k is a ball with the center at x_0. We denote it by B_k. If it is completely inside Ω, then we will use estimate (2.24). If it crosses the boundary, we map it on the half-ball and use estimate (2.25). Here the smoothness of the boundary is used. In both cases we apply the change of variables in order to reduce the operator in the left-hand side of (2.28) to the Laplace operator. It is supposed that B_k is sufficiently small such that we can apply the estimates above and that the first term in the expression for F_k is sufficiently small in B_k. The second and the third terms in the expression for F_k contain the first-order derivatives. They can be estimated with the help of the inequality

$$\|u\|_{C^s(\bar{\Omega})} \leq \epsilon \sum_{|j|=s} \|D^j u\|_{C^\alpha(\bar{\Omega})} + c_\epsilon \max_\Omega |u|,$$

where $\epsilon > 0$ is arbitrarily small, c_ϵ depends on ϵ and on the domain Ω. Thus, we obtain the following estimate of u_k:

$$\|u_k\|_{C^{2+\alpha}(\bar{\Omega})} \leq K\left(\epsilon \|u\|_{C^{2+\alpha}(\bar{\Omega})} + \|\phi\zeta_k\|_{C^{2+\alpha}(\partial\Omega)} + \|f\zeta_k\|_{C^\alpha(\bar{\Omega})} + \max_\Omega |u|\right).$$

Taking a sum of such estimates with respect to k, we easily obtain estimate (2.23). Similar estimates can be obtained in the case of other boundary conditions.

2.3.2. General operators in Hölder spaces. Consider the operators

$$A_i u = \sum_{k=1}^p \sum_{|\beta| \leq \beta_{ik}} a_{ik}^\beta(x) D^\beta u_k \quad (i = 1, \ldots, p), \quad x \in \Omega, \qquad (2.29)$$

$$B_i u = \sum_{k=1}^p \sum_{|\beta| \leq \gamma_i k} b_{ik}^\beta(x) D^\beta u_k \quad (i = 1, \ldots, r), \quad x \in \partial\Omega, \qquad (2.30)$$

where $\beta = (\beta_1, \ldots, \beta_n)$ is a multi-index, β_i nonnegative integers, $|\beta| = \beta_1 + \cdots + \beta_n$, $D^\beta = D_1^{\beta_1} \ldots D_n^{\beta_n}$, $D_i = \partial/\partial x_i$. According to the definition of elliptic

2. Classical theory of linear elliptic problems

operators in the Douglis-Nirenberg sense we consider integers $s_1, \ldots, s_p$; $t_1, \ldots, t_p$; $\sigma_1, \ldots, \sigma_r$ such that

$$\beta_{ij} \leq s_i + t_j, \ i, j = i, \ldots, p; \ \gamma_{ij} \leq \sigma_i + t_j, \ i = 1, \ldots, r, j = 1, \ldots, p, \ s_i \leq 0.$$

We suppose that the number $m = \sum_{i=1}^{p}(s_i + t_i)$ is even and put $r = m/2$. We assume that the problem is uniformly elliptic, that is the ellipticity condition

$$\det \left(\sum_{|\beta|=\beta_{ik}} a_{ik}^{\beta}(x)\xi^{\beta} \right)_{ik=1}^{p} \neq 0, \ \beta_{ik} = s_i + t_k$$

is satisfied for any $\xi \in R^n$, $\xi \neq 0$, $x \in \bar{\Omega}$, as well as the condition of proper ellipticity and the Lopatinskii conditions. Here $\xi = (\xi, \ldots, \xi_n)$, $\xi^{\beta} = \xi_1^{\beta} \ldots \xi_n^{\beta}$. Moreover, the condition of uniform ellipticity implies that the last determinant is bounded from below by a positive constant for all $|\xi| = 1$ and $x \in \bar{\Omega}$, as well as the minors of the Lopatinskii matrix (Section 2.2.1).

Denote by E_0 the space of vector-valued functions $u(x) = (u_1(x), \ldots, u_p(x))$, where $u_j \in C^{l+t_j+\alpha}(\bar{\Omega})$, $j = 1, \ldots, p$, l and α are given numbers, $l \geq \max(0, \sigma_i)$, $0 < \alpha < 1$. Therefore

$$E_0 = C^{l+t_1+\alpha}(\bar{\Omega}) \times \cdots \times C^{l+t_p+\alpha}(\bar{\Omega}).$$

The domain Ω is supposed to be of the class $C^{l+\lambda+\alpha}$, where $\lambda = \max(-s_i, -\sigma_i, t_j)$, and the coefficients of the operator satisfy the following regularity conditions:

$$a_{ij}^{\beta} \in C^{l-s_i+\alpha}(\bar{\Omega}), \ b_{ij}^{\beta} \in C^{l-\sigma_i+\alpha}(\partial\Omega).$$

The operator A_i acts from E_0 into $C^{l-s_i+\alpha}(\Omega)$, and B_i from E_0 into $C^{l-\sigma_i+\alpha}(\partial\Omega)$. Let $A = (A_1, \ldots, A_p)$, $B = (B_1, \ldots, B_r)$. Then $A : E_0 \to E_1$, $B : E_0 \to E_2$, $L = (A, B) : E_0 \to E$, where $E = E_1 \times E_2$,

$$E_1 = C^{l-s_1+\alpha}(\bar{\Omega}) \times \cdots \times C^{l-s_p+\alpha}(\bar{\Omega}), \ E_2 = C^{l-\sigma_1+\alpha}(\partial\Omega) \times \cdots \times C^{l-\sigma_r+\alpha}(\partial\Omega).$$

Then the following estimate holds [7], [8]:

$$\|u\|_{E_0} \leq K(\|Lu\|_E + \|u\|_C). \tag{2.31}$$

Here the constant K is independent of the function $u \in E_0(\Omega)$ and $\| \ \|_C$ is the norm in $C(\bar{\Omega})$.

2.3.3. Sobolev spaces. Denote by E the space of vector-valued functions $u = (u_1, \ldots, u_N)$, where u_j belongs to the Sobolev space $W^{l+t_j, p}(\Omega)$, $j = 1, \ldots, N$, $1 < p < \infty$, l is an integer, $l \geq \max(0, \sigma_j + 1)$, $E = \Pi_{j=1}^{N} W^{l+t_j, p}(\Omega)$. The norm in this space is defined as

$$\|u\|_E = \sum_{j=1}^{N} \|u_j\|_{W^{l+t_j, p}(\Omega)}.$$

The operator A_i acts from E into $W^{l-s_i,p}(\Omega)$, the operator B_j from E into $W^{l-\sigma_j-1/p,p}(\partial\Omega)$. Set

$$L = (A_1, \ldots, A_N, B_1, \ldots, B_m), \qquad (2.32)$$

$$F = \Pi_{i=1}^{N} W^{l-s_i,p}(\Omega) \times \Pi_{j=1}^{m} W^{l-\sigma_j-1/p,p}(\partial\Omega).$$

Then $L : E \to F$.

We suppose that the coefficients of the operator are defined for $x \in \mathbb{R}^n$ and

$$a_{ij}^{\alpha\beta}(x) \in C^{l-s_i+\theta}(\mathbb{R}^n), \quad b_{kj}^{\alpha\beta}(x) \in C^{l-\sigma_k+\theta}(\mathbb{R}^n), \quad 0 < \theta < 1.$$

We also assume that the domain Ω satisfies the following condition.

Condition D. For each $x_0 \in \partial\Omega$ there exists a neighborhood $U(x_0)$ such that:

1. $U(x_0)$ contains a sphere with the radius δ and the center x_0, where δ is independent of x_0,
2. There exists a homeomorphism $\psi(x; x_0)$ of the neighborhood $U(x_0)$ on the unit sphere $B = \{y : |y| < 1\}$ in R^n such that the images of $\Omega \cap U(x_0)$ and $\partial\Omega \cap U(x_0)$ coincide with $B_+ = \{y : y_n > 0, |y| < 1\}$ and $B_0 = \{y : y_n = 0, |y| < 1\}$ respectively,
3. The function $\psi(x; x_0)$ and its inverse belong to the Hölder space $C^{m_0+\theta}$, $0 < \theta < 1$, $m_0 = \max_{i,j,k}(l + t_i, l - s_j, l - \sigma_k)$. Their $\|\cdot\|_{m_0+\theta}$-norms are bounded uniformly in x_0.

Theorem 2.1. *Let $\Omega \subset \mathbb{R}^n$ be a bounded domain satisfying Condition D and $u \in \Pi_{j=1}^{N} W^{l_1+t_j,p}(\Omega)$, where $l_1 = \max(0, \sigma_j + 1)$. If L is a uniformly elliptic operator, then for any $l \geq l_1$ the following estimate holds:*

$$\|u\|_E \leq C \left(\|Lu\|_F + \|u\|_{L^p(\Omega)} \right).$$

Here the constant C depends on the domain and on the coefficients of the operator.

This a priori estimate plays a fundamental role in the theory of elliptic problems. It is proved in [7] under slightly different conditions on the coefficients and on the domain.

One of the main properties of elliptic problems is given by the following theorem (see the definitions in the next section).

Theorem 2.2. *Let $\Omega \subset \mathbb{R}^n$ be a bounded domain satisfying Condition D. If the operator L is elliptic, that is the ellipticity condition, proper ellipticity and the Lopatinskii condition are satisfied, then it satisfies the Fredholm property.*

There are many variants of this theorem. Rather often it is formulated for the scalar case or for $p = 2$ (see the bibliographical comments). The formulation above is close to that in [542] where it is supposed for simplicity that the coefficients are infinitely differentiable. We will see below that this result may not be valid for unbounded domains. To provide the Fredholm property in this case, an additional condition, which determines the behavior at infinity, should be imposed.

2.4 Fredholm operators

2.4.1. Abstract operators. We recall some notions and results from operator theory. We say that a linear operator L acts from a Banach space E into another Banach space F if the domain of definition $D(L)$ of the operator L belong to E and its image $R(L)$ belongs to F. The manifold $\mathrm{Ker}(L)$ of solutions of the equation $Lu = 0$ is called the kernel of the operator L. The subspace $R^{\perp}(L)$ of all linear functionals ϕ from the dual space F^* such that $\phi(f) = 0$ for any $f \in R(L)$ is called the defect subspace of the operator L. The dimension of the kernel of the operator will be denoted by $\alpha(L)$, the dimension of the defect subspace by $\beta(L)$. We will also call it the codimension of the image. Their difference is, by definition, the index of the operator, $\kappa(L) = \alpha(L) - \beta(L)$.

If $\alpha(L) = 0$, then there exists an inverse operator L^{-1} defined on $R(L)$ and acting onto $D(L)$. An operator L has a bounded inverse on $D(L)$ when and only when there exists a constant c such that $\|Lu\|_F \geq c\|u\|_E$ for any $u \in D(L)$. We say that the operator L is continuously invertible if $R(L) = F$ and the inverse operator exists and is bounded.

An operator L is said to be normally solvable if the equation $Lu = f$ is solvable if and only if $\phi(f) = 0$ for all $\phi \in R^{\perp}(L)$. The operator is normally solvable if and only if its range is closed.

We recall that an operator L is called closed if from $u_n \to u$ ($u_n \in D(L)$) and $Lu_n \to f$ it follows that $u \in D(L)$ and $Lu = f$. According to the Banach theorem a closed linear operator defined on all of a Banach space is continuous.

A closed linear operator L is called a Fredholm operator if it is normally solvable, its kernel has a finite dimension and the codimension of its image is also finite. In this case we also say that the operator L satisfies the Fredholm property. The following properties of Fredholm operators are known (see, e.g., [207]).

Theorem 2.3. *Let L and M be two Fredholm operators acting in a Banach space E and let $D(L)$ be dense in E. Then the product LM is also a Fredholm operator and $\kappa(LM) = \kappa(L) + \kappa(M)$.*

An operator $\tilde{L}$ is called an extension of the operator L if $D(L) \subset D(\tilde{L})$ and $\tilde{L}u = Lu$ for $u \in D(L)$. If the dimension of the factor space $D(\tilde{L})/D(L)$ is finite and equals k, then it is a k-dimensional extension. In this case $D(\tilde{L})$ can be represented as a direct sum, $D(\tilde{L}) = D(L) \oplus \mathcal{M}$, where $\mathcal{M}$ is a k-dimensional subspace, and
$$\tilde{L}(u+v) = Lu + Kv, \quad \forall u \in D(L), v \in \mathcal{M},$$
where K is a finite-dimensional operator defined on $\mathcal{M}$. Let $\tilde{L}$ be a k-dimensional extension of a Fredholm operator L. Then $\tilde{L}$ is also a Fredholm operator and $\kappa(\tilde{L}) = \kappa(L) + k$, $\alpha(\tilde{L}) \leq \alpha(L) + k$.

Theorem 2.4. *Let L be a Fredholm operator acting from E into F. Then there exists a positive number ρ such that for any bounded linear operator $B: E \to F$ for which $\|B\| < \rho$, the operator $L + B$ is Fredholm and $\kappa(L+B) = \kappa(L)$, $\alpha(L+B) \leq \alpha(L)$.*

It is also known that if L if a Fredholm operator acting from E into F and $T : E \to F$ is a compact linear operator, then the operator $L + T$ is also Fredholm and $\kappa(L + T) = \kappa(L)$.

If an operator L acts in a Banach space E, and for some complex number λ the operator $L - \lambda I$ satisfies the Fredholm property, then λ is called a Φ-point of the operator L.

Theorem 2.5. *The set of Φ-points of a linear closed operator L is open. For all λ of a connected component G of this set, the index of the operator $L - \lambda I$ is constant, and for all $\lambda \in G$ with a possible exception of certain isolated points, $\alpha(L - \lambda I)$ has a constant value n. At these isolated points $\alpha(L - \lambda I) > n$.*

This theorem admits a generalization for holomorphic operator functions L_λ whose values are closed linear operators acting from E into F. They are represented in the form of the series

$$A_\lambda = A_{\lambda_0} + \sum_{k=1}^{\infty} (\lambda - \lambda_0)^k C_k$$

in the neighborhood of each point $\lambda_0 \in G$. Here C_k are bounded linear operators acting from E to F, the series converges in the operator norm.

We continue this short review with some properties of semi-Fredholm operators. A closed linear operator $L : E \to F$ will be called a Φ_+-operator if it is normally solvable, the dimension $\alpha(L)$ of its kernel is finite and the codimension $\beta(L)$ of its image is infinite. Similarly, for Φ_--operators, $\alpha(L)$ is infinite and $\beta(L)$ is finite. The results above remain valid for semi-Fredholm operators. Namely, if L is a Φ_+-operator and B is a bounded linear operator with a sufficiently small norm, then $L + B$ is also a Φ_+-operator, and $\alpha(L + B) \leq \alpha(L)$. Similarly, for Φ_--operators, $\alpha(L + B) = \alpha(L)$, $\beta(L + B) \leq \beta(L)$. Adding a compact operator to a $\Phi_\pm$-operator leaves it a $\Phi_\pm$-operator.

Let $L : E \to F$ be a linear operator with the dense domain of definition. An adjoint operator $L^* : F^* \to E^*$ is defined by the equality $\langle L^* \phi, u \rangle_E = \langle \phi, Lu \rangle_F$, where $u \in E$, $\phi \in F^*$, $\langle \cdot, \cdot \rangle$ denotes the duality in the corresponding spaces. It is known that if L is a bounded operator, then L^* is also bounded and $\|L^*\| = \|L\|$. Moreover, $L^* \phi = 0$ if and only if $\phi(f) = 0$ for any $f \in R(L)$. Hence, $\alpha(L^*) = \beta(L)$. On the other hand, $\beta(L^*) \geq \alpha(L)$. In the case of reflexive spaces E and F, the last relation becomes equality. We finally note that if a closed operator L with a dense domain of definition is normally solvable, then the operator L^* is also normally solvable.

2.4.2. Solvability of elliptic problems. Under some natural assumptions, a priori estimate (2.19) implies that the operator L is normally solvable with a finite-dimensional kernel (see, e.g., [7]). Suppose that the domain Ω is bounded and inclusion (2.20) is compact, that is a closed bounded set in $E(\Omega)$ is compact in $E'(\Omega)$ (cf. (2.21)). Let B be a unit ball in the kernel $\operatorname{Ker} L$ of the operator L,

2. Classical theory of linear elliptic problems

Ker $L \subset E(\Omega)$. Consider a sequence $u_n \in \text{Ker } L$. It has a subsequence, for which we keep the same notation, converging in $E'(\Omega)$ to some $u_0 \in E'(\Omega)$. From (2.19) it follows that the sequence u_n is fundamental in $E(\Omega)$. Properties of elliptic operators and of the function spaces allow us to conclude that $u_n \to u_0$ in $E(\Omega)$ and $u_0 \in \text{Ker } L$. Therefore the unit ball is compact and, consequently, the kernel of the operator L is finite dimensional.

A similar construction can be used to prove that the image of the operator is closed. Consider the equations $Lu = f_n$, where $f_n \to f_0$ in $F(\Omega)$, f_n belongs to the image Im L of the operator L. We should verify that f_0 is also in Im L. Let $E(\Omega) = \text{Ker } L \oplus E_1$ and let $u_n \in E_1$ be such that $Lu_n = f_n$. If the sequence u_n is bounded, then it has a subsequence converging to some u_0 in $E'(\Omega)$. It follows from (2.19) that it converges also to $u_0 \in E(\Omega)$, and $Lu_0 = f_0$. If the sequence is not bounded, we consider the equation for the functions $v_n = u_n/\|u_n\|_{E(\Omega)}$. As before, we obtain $v_n \to v_0$. Moreover, $Lv_0 = 0$, that is $v_0 \in \text{Ker } L$. On the other hand, $v_0 \in E_1$. Therefore the assumption that the sequence is unbounded gives a contradiction.

Normal solvability of the operator L means that the equation

$$Lu = f \qquad (2.33)$$

is solvable if and only if $\langle f, \phi \rangle = 0$ for all functionals ϕ from some subspace Φ of the dual space $F^*(\Omega)$. It is equivalent to the closeness of the image. Since $\Phi = \text{Ker } L^*$, where $L^* : F^*(\Omega) \to E^*(\Omega)$ is the adjoint operator, then we obtain the solvability conditions: equation (2.33) is solvable if and only if $\langle f, \phi \rangle = 0$ for all solutions ϕ of the homogeneous adjoint equation $L^*\phi = 0$.

The number of linearly independent solvability conditions is determined by the dimension of the kernel of the adjoint operator L^*. A priori estimates for the adjoint operator similar to (2.19) allow one to prove that its kernel has a finite dimension. In some cases, it is possible to replace the adjoint operator by a formally adjoint operator for which the solvability conditions can be written as orthogonality in L^2.

One of the simplest examples of elliptic problems is given by the Laplace operator with the Dirichlet boundary condition

$$\Delta u = f, \quad u|_{\partial\Omega} = 0. \qquad (2.34)$$

Here Ω is a bounded domain with a sufficiently smooth boundary. If $f \in L^2(\Omega)$, then we look for a solution in $H^2(\Omega)$. The formally adjoint operator is determined by the equality

$$\int_\Omega (Lu)\bar{v}\,dx = \int_\Omega u\overline{(L^*v)}\,dx.$$

The problem is formally self-adjoint, and the homogeneous formally adjoint problem

$$\Delta v = 0, \quad v|_{\partial\Omega} = 0$$

has only zero solution. Therefore, problem (2.34) is solvable for any right-hand side. If we consider the Neumann boundary condition, $\partial u/\partial n = 0$, where n is the outer normal vector, then the homogeneous formally adjoint problem has a nonzero solution $v = $ const. It provides one solvability condition $\int_\Omega f dx = 0$. In both cases, the numbers of linearly independent solutions of the homogeneous problem and of solvability conditions for the nonhomogeneous problem are equal to each other. Therefore the index equals zero.

Thus, a priori estimates for the direct and adjoint operators imply the Fredholm property and solvability conditions. The construction briefly described above uses the compact embedding of $E(\Omega)$ into $E'(\Omega)$. It may not take place for unbounded domains. This is why the ellipticity condition, proper ellipticity and the Lopatinskii condition may not be sufficient in this case for the Fredholm property. We will discuss this question in Section 3.

2.4.3. Index. If the operator satisfies the Fredholm property, then its index is well defined. It equals the difference between the dimension of the kernel and the codimension of the image (or the number of linearly independent solvability conditions). The index is stable with respect to small perturbations of the operator. If $L : E(\Omega) \to F(\Omega)$ satisfies the Fredholm property, $B : E(\Omega) \to F(\Omega)$ is bounded and has a sufficiently small norm, then $L + B$ is also Fredholm and its index equals the index of the operator L. Hence the index does not change under a continuous deformation of the operator in the class of Fredholm operators. This means that if the family of operators L_τ is continuous with respect to τ in the operator norm, then the index does not depend on τ. It remains also true for semi-Fredholm operators. In this case the index does not change under a continuous deformation in the class of normally solvable operators with a finite-dimensional kernel.

These properties can be used to study the index of elliptic operators. One of the possible approaches consists in constructing a continuous deformation to some model operator for which the index is known. All operators from the family should be normally solvable with a finite-dimensional kernel. For elliptic operators in bounded domains this is provided by ellipticity conditions (ellipticity, proper ellipticity, Lopatinskii condition). A possible approach to find the index in the case of two-dimensional elliptic problems can be schematically described as follows (see Chapter 8 and the bibliographical comments). It is known that general elliptic problems can be reduced to first-order problems, while the latter to canonical problems. In their turn, canonical problems can be reduced by a continuous deformation, which preserves the index, to the problems of the type

$$i\frac{\partial u_1}{\partial x_1} - \frac{\partial u_1}{\partial x_2} = 0, \quad -i\frac{\partial u_2}{\partial x_1} - \frac{\partial u_2}{\partial x_2} = 0,$$

$$b_1(x)u_1 + b_2(x)u_2 = g.$$

The final step of this construction is to represent the last problem in the form of a Hilbert problem, that is to find a holomorphic function in the inner and in the outer domains with a given jump at the boundary. The index of the Hilbert problem can be found by means of singular integral equations [190], [366], [532].

A Laplace operator with oblique derivative

$$\Delta u = 0, \quad a(x)\frac{\partial u}{\partial x} - b(x)\frac{\partial u}{\partial y} \bigg|_{\partial \Omega} = h(x), \quad a^2(x) + b^2(x) \neq 0 \qquad (2.35)$$

provides an example of problems which can have a nonzero index. Let N denote the number of rotations of the vector $(a(x), b(x))$ around the origin when x goes around the boundary $\partial \Omega$ counterclockwise. Positive N corresponds to a counterclockwise rotation. The index of this problem equals $2N + 2$. It equals zero in the case of the Neumann boundary conditions ($N = -1$).

The index theories for elliptic problems had an important development beginning from the 1960s due to the Atiyah-Singer theory.

2.4.4. Spectrum and invertibility. Consider an unbounded operator $L : E \to E$ acting in a Banach space E with a dense domain of definition. We introduce the operator $L_\lambda = L - \lambda I$, where λ is a complex parameter and I is the identity operator. The set $\Phi(L)$ of all λ, for which L_λ is a Fredholm operator, is open. Its complement is called the essential spectrum of the operator L. The index $\kappa(L_\lambda)$ of the operator L_λ is constant in each connected component of the set $\Phi(L)$. The dimension of the kernel $\alpha(L_\lambda)$ is constant except for some isolated points where its value is greater. These are eigenvalues of finite multiplicity. They can accumulate to the points of the essential spectrum or to infinity. This is the structure of the spectrum of abstract Fredholm operators. It is applicable for elliptic operators.

If the index of the operator L_λ is zero in some connected component of the set $\Phi(L)$, then it is invertible for all λ from this component except possibly for some isolated points. To prove the invertibility in this case it is sufficient to verify that the kernel of the operator is empty. Various specific methods can be used for some particular classes of operators. For example, in order to prove that the equation $\Delta u - \lambda^2 u = 0$ with the Dirichlet boundary condition and with a real λ has only a zero solution, it is sufficient to multiply it by u and integrate over Ω. In the general case, the invertibility can be proved for elliptic problems with a parameter

$$A(x, \lambda, D)u = f, \quad x \in \Omega, \quad B(x, \lambda, D)u = g, \quad x \in \partial\Omega$$

introduced by Agranovich and Vishik [13]. Precise definitions are given in Chapter 7. Some special a priori estimates of solutions allow one to prove the existence and the uniqueness of solutions for λ in a given sector S, and $|\lambda| \geq \lambda_0 > 0$.

3 Elliptic problems in unbounded domains

One of the main results of the theory of elliptic problems affirms that they satisfy the Fredholm property. If we consider unbounded domains, then the ellipticity condition, proper ellipticity and the Lopatinskii condition are not sufficient, generally speaking, in order for the operator to be Fredholm. Some additional conditions formulated in terms of limiting problems should be imposed. The typical result says that the operator satisfies the Fredholm property if and only if all its limiting operators are invertible. The question is about the classes of operators for which this result is applicable.

Limiting operators and their inter-relation with solvability conditions and with the Fredholm property were first studied in [160], [309], [310] (see also [491]) for differential operators on the real axis, and later for some classes of elliptic operators in R^n [48], [361], [362], in cylindrical domains [114], [567], or in some specially constructed domains [45], [46]. Some of these results are obtained for the scalar case, some others for the vector case, under the assumption that the coefficients of the operator stabilize at infinity or without this assumption. This theory is also developed for some classes of pseudo-differential operators [153], [295], [440]–[442], [475], [490] and discrete operators [13], [443].

In spite of the considerable progress in the understanding of properties of elliptic operators in unbounded domains, this question is not yet completely elucidated. In this book we will systematically develop Fredholm theory of general elliptic problems. As it is often the case for elliptic problems, these studies will be based on a priori estimates of solutions of direct and adjoint operators and on the introduction of special function spaces. The results will be formulated in terms of the unique solvability of limiting problems.

We will introduce two types of limiting domains. Limiting domains of the first type are defined through translations of a given unbounded domain Ω. Limiting domains of the second type are introduced for a sequence of bounded domains Ω_n approximating an unbounded domain Ω. We next define limiting operators. In order to do this, we consider sequences of shifted coefficients and choose locally convergent subsequences. Limiting operators are operators with limiting coefficients. According to the type of limiting domains, we obtain two types of limiting problems.

Unique solvability of limiting problems of the first type determine the Fredholm property of elliptic operators in unbounded domains. Unique solvability of limiting problems of the second type determine stabilization of the index. This means that the indices of the operators in the domains Ω_n become independent of n for n sufficiently large. They are equal to each other and to the index of the operator L.

If one of the limiting operators of the first type is not uniquely solvable, then the operator L does not satisfy the Fredholm property. The theory of such operators is not yet sufficiently well developed. In some cases it is possible to reduce

3. Elliptic problems in unbounded domains

them to Fredholm operators on some subspaces or by some special constructions. An interesting question is about solvability conditions for non-Fredholm operators. We will discuss this question for some particular classes of operators.

3.1 Function spaces

Sobolev spaces $W^{s,p}$ proved to be very convenient in the study of elliptic problems in bounded domains. In order to study elliptic problems in unbounded domains, we will introduce some generalization of the spaces $W^{s,p}$. They will coincide with $W^{s,p}$ in bounded domains and will have a prescribed behavior at infinity in unbounded domains. It turns out that such spaces can be constructed for arbitrary Banach spaces of distributions (not only Sobolev spaces) as follows. Consider first functions defined in $\mathbb{R}^n$. As usual, we denote by D the space of infinitely differentiable functions with compact support and by D' its dual. Let $E \subset D'$ be a Banach space, where the inclusion is understood both in an algebraic and a topological sense. Denote by E_{loc} the collection of all $u \in D'$ such that $fu \in E$ for all $f \in D$. Let us take a function $\omega(x) \in D$ such that $0 \leq \omega(x) \leq 1$, $\omega(x) = 1$ for $|x| \leq 1/2$, $\omega(x) = 0$ for $|x| \geq 1$.

Let $\{\phi_i\}$, $i = 1, 2, \ldots$ be a partition of unity. Then, by definition, E_p is the space of all $u \in E_{\text{loc}}$ such that $\sum_{i=1}^{\infty} \|\phi_i u\|_E^p < \infty$, where $1 \leq p < \infty$, with the norm

$$\|u\|_{E_p} = \left(\sum_{i=1}^{\infty} \|\phi_i u\|_E^p \right)^{1/p};$$

E_∞ is the space of all functions $u \in E_{\text{loc}}$ such that $\sup_i \|\phi_i u\|_E < \infty$, with the norm

$$\|u\|_{E_\infty} = \sup_i \|\phi_i u\|_E.$$

These spaces do not depend on the choice of the partition of unity. We will also use an equivalent definition: E_q ($1 \leq q \leq \infty$) is the space of all $u \in E_{\text{loc}}$ such that

$$\|u\|_{E_q} := \left(\int_{\mathbb{R}^n} \|u(.)\omega(.-y)\|_E^q dy \right)^{1/q} < \infty, \quad 1 \leq q < \infty,$$

$$\|u\|_{E_\infty} := \sup_{y \in \mathbb{R}^n} \|u(.)\omega(.-y)\|_E < \infty.$$

It will be shown that E_q is a Banach space.

If Ω is a domain in $\mathbb{R}^n$, then by definition $E_q(\Omega)$ is the space of restrictions of E_q to Ω with the usual norm of restrictions. It is easy to see that if Ω is a bounded domain, then

$$E_q(\Omega) = E(\Omega), \quad 1 \leq q \leq \infty.$$

In the particular case where $E = W^{s,p}$, we set $W_q^{s,p} = E_q$ ($1 \leq q \leq \infty$). It is proved that

$$W_p^{s,p} = W^{s,p} \quad (s \geq 0, \ 1 < p < \infty).$$

Hence the spaces $W_q^{s,p}$ generalize the Sobolev spaces ($q < \infty$) and the Stepanov spaces ($q = \infty$) (see [309], [310]). We will study elliptic operators acting in these spaces.

3.2 Limiting problems

The notion of limiting domains and of limiting problems or operators will be essentially used throughout this book. They will determine normal solvability and Fredholm property of elliptic operators in unbounded domains. Limiting operators were first considered in [160], [309], [310] for differential operators on the real axis with quasi-periodic coefficients, and then for elliptic operators in $\mathbb{R}^n$ or for domains cylindrical or conical at infinity [48], [361], [362], [491]. In the general case limiting operators and domains are introduced in [565], [564].

3.2.1. Limiting domains of the first type. We will illustrate construction of limiting domains with some examples. Consider first a cylinder Ω in $\mathbb{R}^n$ with the axis along the x_1-direction and with a bounded cross-section $G \subset \mathbb{R}^{n-1}$. Let $x^k = (x_1^k, y^k) \in \Omega$ be a sequence of points for which the first coordinate x_1^k tends to infinity. Denote by Ω_k the shifted cylinders for which the point x^k is moved to the origin. More precisely, let $\chi(x)$ be the characteristic function of the domain Ω. Then Ω_k is the domain with the characteristic function $\chi(x + x^k)$.

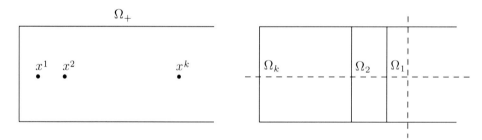

Figure 1: Construction of the limiting domain for a half-cylinder.

The sequence of points $y^k \in G$ is bounded. Let y^{k_l} be a subsequence converging to some $y^0 \in \bar{G}$. Consider the sequence of domains Ω_{k_l}. Since the shift with respect to x_1 does not change the cylinder, then the sequence of domains converges to the domain Ω_0 with the characteristic function $\chi(x + x^0)$, where $x^0 = (0, y^0)$. The cylinder Ω_0 is a limiting domain for the domain Ω. If the sequence y^k has another converging subsequence, then we obtain another limiting domain, which is also a cylinder that differs from Ω_0 by a shift. We then consider all sequences $x^k \in \Omega$ with the first coordinate going to $\pm\infty$ and the corresponding limiting domains. All of them can be obtained from the cylinder Ω by a shift. We will not distinguish limiting domains obtained from each other by a finite shift. Therefore, an infinite cylinder has a single limiting domain, the cylinder itself.

3. Elliptic problems in unbounded domains

If the domain Ω_+ is a half-cylinder, $\Omega_+ = \{x \in \Omega, x_1 \geq 0\}$, then it has a single limiting domain, the whole cylinder Ω (Figure 1). We shift the half-cylinder in such a way that the points x^k coincide with the origin. Therefore we obtain a sequence of half-cylinders Ω_k. The left boundary of Ω_k moves to $-\infty$ as k increases. Passing to the limit, we obtain the whole cylinder Ω, which is the only limiting domain in this case.

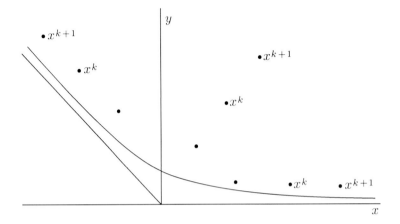

Figure 2: Construction of limiting domains for a hyperbolic domain.

In the general case, we have a sequence of domains obtained as translations of the same unbounded domain, and we choose all locally convergent subsequence of this sequence. More precisely, the construction is as follows. Let $x_k \in \Omega$ be a sequence which tends to infinity. Consider the shifted domains Ω_k corresponding to the shifted characteristic functions $\chi(x + x_k)$, where $\chi(x)$ is the characteristic function of the domain Ω. Consider a ball $B_r \subset R^n$ with the center at the origin and with the radius r. Suppose that for all k there are points of the boundaries $\partial\Omega_k$ inside B_r. If the boundaries are sufficiently smooth, then from the sequence $\Omega_k \cap B_r$ we can choose a subsequence that converges to some limiting domain $\hat{\Omega}$. After that we take a larger ball and choose a convergent subsequence of the previous subsequence. The usual diagonal process allows us to extend the limiting domain to the whole space.

We consider two more examples to explain the local convergence of the sequence of domains. Consider the half-space $\mathbb{R}^n_+ = \{x \in \mathbb{R}^n, x_n \geq 0\}$. Limiting domains depend now on the choice of the sequence. Let us first take a sequence x^k for which its last coordinate is fixed, $x^k_n = h$ for some constant h. Then each domain in the sequence Ω_k is the half-space $\{x \in \mathbb{R}^n, x_n \geq -h\}$. It depends on the value of h but, as it is indicated above, we do not distinguish limiting domains obtained from each other by translation. Therefore, the first limiting domain is a half-space. Consider next a sequence x^k for which the coordinate x^k_n tends to infinity as $k \to \infty$. Then the limiting domain is the whole space $\mathbb{R}^n$. Indeed, we fix a ball B_{R_1} with the radius R_1. There exists k_1 such that $B_{R_1} \subset \Omega_k$ for all $k \geq k_1$. We next take a ball B_{R_2} with $R_2 > R_1$ and choose k_2 such that $B_{R_2} \subset \Omega_k$ for all $k \geq k_2$. We continue this construction with a sequence of growing balls. Each of them belongs to the limiting domain. Hence the limiting domain is the whole $\mathbb{R}^n$.

The next example illustrates the convergence of the boundaries of the domains. Consider the domain above the curve $y = f(x)$ (Figure 2). It has two asymptotes, $y = -x$ and $y = 0$. Depending on the choice of the sequence there are three limiting domain: the whole $\mathbb{R}^2$ and two different half-planes. Let us describe the construction of the limiting domain for the sequence of points (x_k, y_k), where $x_k \to \infty$, $y_k = h$ with some constant h. If we translate the domain in such a way that the point (x_k, y_k) coincides with the origin, then we obtain the domain $\Omega_k = \{(x, y), \; y \geq f_k(x)\}$, where $f_k(x) = f(x + x_k) - h$. The sequence of functions $f_k(x)$ locally converges to $f_0(x) \equiv -h$. Therefore, the limiting domain is the half-plane $y \geq -h$. If we take the left sequence of points shown in Figure 2, we obtain the limiting domain $y \geq -x - h$. For the sequence in the middle, the limiting domain is the whole $\mathbb{R}^2$.

We note that limiting domains for the same sequence of points may be nonunique. In the previous example, we can choose a function $f(x)$ in such a way that the sequence $f_k(x)$ would have more than one local limit.

3.2.2. Limiting domains of the second type. In the previous section we have defined limiting domains by means of all possible sequences of domains Ω_k obtained as translations of an unbounded domain Ω. In this section we will introduce another type of limiting domains. It will be also used to study elliptic operators in unbounded domains.

Consider now a sequence of arbitrary domains Ω_k. They can be bounded or unbounded, and they are not necessarily obtained from each other by translation. As above, we will assume that their boundaries are sufficiently smooth.

In what follows we will be interested by sequences of growing bounded domains, $\Omega_1 \subset \Omega_2 \subset \cdots$ which locally converge to an unbounded domain Ω. By local convergence we understand the convergence of the boundaries of the domains in every bounded ball. Precise definitions will be given in Chapters 4 and 8.

We will illustrate the definition of limiting domains with the example shown in Figure 3. We consider a sequence of growing rectangles (with smoothed angles) whose length tends to infinity and the width to some constant (Figure 3a). We denote these domains by Ω_k and choose points $x^k \in \Omega_k$ in such a way that $|x^k| \to \infty$. Consider the sequence of points in Figure 3a going to the left. We translate the domains Ω_k n such a way that these points coincide with the origin. Hence we obtain a new sequence of domains shown in Figure 3b. As before, these are bounded rectangles but now their left boundary is fixed while the right boundary moves to infinity. The local limit of this sequence of domains is a right half-cylinder. Similarly, if we consider the sequence of point going to the right, we will obtain

3. Elliptic problems in unbounded domains

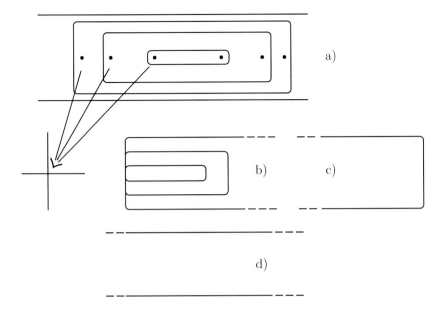

Figure 3: Construction of limiting domains of the second type. There are three limiting domains: left half-cylinder, right half-cylinder, and the whole cylinder.

another limiting domain, a left half-cylinder. Finally, we can choose a sequence of points for which the distance to both side boundaries of the rectangles grows to infinity. Then the limiting domain is the whole cylinder.

Thus, there are three limiting domains for the sequence of growing rectangles. At the same time, the unbounded domain Ω, that is the infinite cylinder, has only one limiting domain, the cylinder itself. We call the sequence of bounded domains Ω_k an approximating sequence for the unbounded domain Ω. The definition of limiting domains in the second sense is applicable for approximating sequences. It appears that there can be more limiting domains in the second sense for an approximating sequence than of limiting domains in the first sense for the unbounded domain itself.

We finish the description of limiting domains in the second sense with one more example shown in Figure 4. The domains Ω_k in this case are circles with growing diameters. Translating the circles in such a way that they have the same tangent, we obtain a limiting domain, which is the half-plane bounded by the same tangent. Thus, we have a continuum of limiting domains, which are all possible half-planes. The whole $\mathbb{R}^2$ is also a limiting domain.

3.2.3. Limiting problems. In order to define limiting operators we consider shifted coefficients $a^\alpha(x+x_k)$, $b_j^\alpha(x+x_k)$ and choose subsequences that converge to some

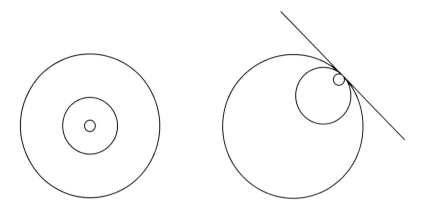

Figure 4: Construction of limiting domains of the second type. There is a continuum of limiting domains: all half-planes and the whole plane.

limiting functions $\hat{a}^\alpha(x)$, $\hat{b}_j^\alpha(x)$ uniformly in every bounded set. The limiting operator is the operator with the limiting coefficients. Limiting operators considered in limiting domains constitute limiting problems. It is clear that the same problem can have a family of limiting problems depending on the choice of the sequence x_k and on the choice of converging subsequences of domains and coefficients. Limiting problems for the same sequence of points x^k may be nonunique.

3.3 Fredholm property and solvability conditions

For elliptic problems in bounded domains, their ellipticity, proper ellipticity and the Lopatinskii condition determine their Fredholm property. In the case of unbounded domains we should introduce one more condition related to the invertibility of limiting problems.

3.3.1. One-dimensional second-order problems. We begin with the one-dimensional scalar operator
$$Lu = a(x)u'' + b(x)u' + c(x)u,$$
where $x \in \mathbb{R}$. Suppose that the coefficients are sufficiently smooth real-valued functions and that there exist limits at infinity:
$$a_\pm = \lim_{x \to \pm\infty} a(x), \quad b_\pm = \lim_{x \to \pm\infty} b(x), \quad c_\pm = \lim_{x \to \pm\infty} c(x).$$
Then we can define the limiting operators
$$L^\pm u = a_\pm u'' + b_\pm u' + c_\pm u$$

3. Elliptic problems in unbounded domains

and consider the corresponding eigenvalue problems $L^\pm u = \lambda u$. Since the limiting operators have constant coefficients, we can apply the Fourier transform to obtain

$$\lambda(\xi) = -a_\pm \xi^2 + ib_\pm \xi + c_\pm, \quad \xi \in \mathbb{R}.$$

These are two parabolas in the complex plane (Figure 5). It will be proved that the operator L satisfies the Fredholm property if and only if they do not pass by the origin. Therefore, the operator $L - \lambda I$, where I is the identity operator, satisfies the Fredholm property for the values of λ, which do not belong to the parabolas. They form the essential spectrum of the operator L, that is the set of points where the operator $L - \lambda I$ does not satisfy the Fredholm property.

We note that if the essential spectrum contains the origin, then the limiting equation has a nonzero bounded solution $u(x) = \exp(i\xi x)$ for some real ξ (cf. Condition NS below).

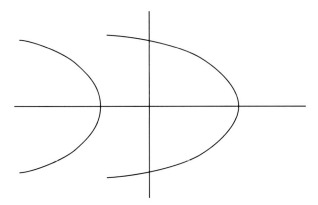

Figure 5: Essential spectrum is formed by two parabolas
$\lambda = -a_\pm \xi^2 + b_\pm i\xi + c_\pm$.

3.3.2. General problems. For general elliptic problems, limiting operators can have variable coefficients and limiting domains may not be translation invariant. Conditions of normal solvability can be formulated using the notion of limiting problems.

Condition NS. Any limiting problem

$$\hat{L}u = 0, \quad x \in \hat{\Omega}, \quad u \in E_\infty(\hat{\Omega})$$

has only the zero solution.

This condition is necessary and sufficient for general elliptic operators to be normally solvable with a finite-dimensional kernel [564], [565], [570]. Similarly to Condition NS for the direct operators, we introduce Condition NS* for the adjoint operators.

Condition NS*. Any limiting problem $\hat{L}^*v = 0$ has only the zero solution in $(F^*(\hat{\Omega}))_\infty$. Here $\hat{L}^*$ is the operator adjoint to the limiting operator $\hat{L}$.

A priori estimates for adjoint operators and Condition NS* imply that the operator $L^* : (F^*(\Omega))_\infty \to (E^*(\Omega))_\infty$ is normally solvable with a finite-dimensional kernel. However, this is not yet sufficient to affirm that the operator L satisfies the Fredholm property because the adjoint operator L^* acts here not in the dual spaces, from $(F_\infty(\Omega))^*$ into $(E_\infty(\Omega))^*$ but from $(F^*(\Omega))_\infty$ into $(E^*(\Omega))_\infty$. These spaces are different. In fact, we will show that $(E^*)_\infty = (E_1)^*$, $(F^*)_\infty = (F_1)^*$. This property will allow us to establish a relation between the operators $(L^*)_\infty : (E^*)_\infty \to (F^*)_\infty$ and $(L_\infty)^* : (E_\infty)^* \to (F_\infty)^*$. We will use it to prove the Fredholm property of the operator L in the spaces $W^{s,p}$, $1 < p < \infty$ and $W^{s,p}_q$ for some q.

We arrive here at the notion of local operators and realization of operators. An operator L is called local if for any $u \in E$ with a bounded support, we have $\operatorname{supp} Lu \subset \operatorname{supp} u$. Differential operators satisfy obviously this property. If an operator L is local, then the adjoint operator is also local. For local operators we can define their realization in different spaces, $L_q : E_q \to F_q$, $1 \leq q \leq \infty$. We will also consider the operator $L_D : E_D \to F_D$, where E_D and F_D are the spaces obtained as a closure of functions from D in the norms of the spaces E_∞ and F_∞, respectively. We will first prove that Conditions NS and NS* imply the Fredholm property of the operator L_D and then of the operators L_∞ and L_q. The exact formulations of the results are given in Chapter 5.

One of the main properties of Fredholm operators is related to solvability conditions of nonhomogeneous equations. If the operator $L : E_\infty(\Omega) \to F_\infty(\Omega)$ satisfies the Fredholm property, then the equation $Lu = f$ is solvable if and only if $\phi(f) = 0$ for all $\phi \in (F_\infty(\Omega))^*$ such that $L^*\phi = 0$. Similar solvability conditions are valid for the operators acting in other spaces.

3.3.3. Formally adjoint problems. Solvability conditions can be represented in a simpler form in terms of a formally adjoint operator. This property is well known for elliptic operators in bounded domains. It is not directly applicable in the case of unbounded domains because of the additional conditions on limiting operators.

We will illustrate it with the operator L considered in Section 3.3.1. Suppose that the coefficients $a(x), b(x)$ and $c(x)$ of the operator L are sufficiently smooth and consider the operator

$$L^*v = (a(x)v)'' - (b(x)v)' + c(x)v.$$

Both operators, L and L^* act from $H^2(\mathbb{R})$ into $L^2(\mathbb{R})$. The operator L^* is a formally adjoint operator to the operator L. It satisfies the equality

$$\int_{-\infty}^{\infty} (Lu)v\,dx = \int_{-\infty}^{\infty} u(L^*v)\,dx$$

for any $u, v \in H^2(\mathbb{R})$. We will use the same notation for formally adjoint operators and for adjoint operators but we should keep in mind that they can be different and act in different spaces.

We will introduce Condition NS* for the formally adjoint operator similarly to Condition NS for the direct operator. It appears that if both of them are satisfied, then the operator L satisfies the Fredholm property and the equation $Lu = f$ is solvable if and only if $\int_{-\infty}^{\infty} fv\,dx = 0$ for all solutions v of the homogeneous formally adjoint equation $L^*v = 0$.

3.4 Index

The index theories developed for elliptic operators in bounded domains may not be applicable for unbounded domains. In some cases, the index can be computed by rather simple and explicit methods. Among them, some problems in cylinders which can be reduced to one-dimensional equations by a spectral decomposition (Chapter 9).

In Chapter 8, we will develop the method based on the approximation of operators in unbounded domains by operators in bounded domain. Under some conditions, the value of the index for the approximating domains stabilizes and equals the index for the limiting unbounded domain. There are counterexamples which show that the stabilization may not take place. The conditions under which the stabilization occurs can be formulated in terms of limiting problems. We recall that limiting problems of the first type determine the Fredholm property. If it is satisfied, then the index is well defined. Conditions of stabilization of the index are formulated in terms of limiting problems of the second type. They are more restrictive than the conditions which provide the Fredholm property. Indeed, there are more limiting domains of the second type than of the first type for the same unbounded domain.

3.4.1. One-dimensional operators. We consider the same operator as in Section 3.3.1 as acting in Sobolev or in Hölder spaces. We suppose that the essential spectrum of this operator does not contain the origin, which implies Conditions NS and NS*. Then its index is defined and equals

$$\kappa = n_+ + n_- - 2,$$

where n_+ is the number of solutions of the equation $L_+ u = 0$ bounded at $+\infty$, n_- is the number of solutions of the equation $L_- u = 0$ bounded at $-\infty$ (Chapter 9) They can be easily expressed through the coefficients of the limiting operators.

The value of the index is related to the location of the spectrum shown in Figure 5. If both of the parabolas are in the left half-plane, then the index equals zero. If one of them is partially in the right half-plane, as it is shown in the figure, then the index is ± 1. If the vertices of both parabolas are on the positive half-axis, then the index is zero or ± 2. In fact, each time when one of the parabolas crosses the origin, the index changes by ± 1.

Let us now consider the problem

$$Lu = f, \quad u(\pm N) = 0 \tag{3.1}$$

in the interval $[-N, N]$. It can be shown that its index is zero for any finite N. On the other hand, as it is discussed before, the index of the problem on the whole axis can have any value between -2 and 2. We will use this example to discuss the conditions of stabilization of the index.

We have a sequence of bounded domains $\Omega_N = [-N, N]$ approximating the unbounded domain $\Omega = \mathbb{R}$. The limiting domain of the first type is the same Ω, the limiting operators $L^\pm$ are introduced in Section 3.3.1. Let us now construct limiting domains and operators of the second type. Consider a sequence $x_N \in \Omega_N$. Put $x_N = \sqrt{N}$. Then the shifted domains, for which x_N is moved to zero, are $\tilde{\Omega}_N = [-N-\sqrt{N}, N-\sqrt{N}]$. The limiting domain in this case is $\hat{\Omega} = \mathbb{R}$. The limiting operator is obtained as before: we consider the sequence of shifted coefficients and choose locally convergent subsequences. We obtain $\hat{L} = L^+$, that is the limiting operator is the same as the limiting operator of the first type. If we take the sequence $x_N = -\sqrt{N}$, then we will obtain the limiting operator $\hat{L} = L^-$. We recall that the limiting problems $L^\pm u = 0$ do not have nonzero bounded solutions (Condition NS) if and only if $c_\pm < 0$ or $c_\pm > 0$ and $b_\pm \neq 0$.

We now consider the sequence $x_N = N - 1$. Then the limiting domain is the half-axis $\{x \leq 1\}$ and the limiting problem:

$$a_+ u'' + b_+ u' + c_+ u = 0, \quad u(1) = 0. \tag{3.2}$$

The formally adjoint problem is

$$a_+ u'' - b_+ u' + c_+ u = 0, \quad u(1) = 0. \tag{3.3}$$

We require that both of them have only zero bounded solutions. This condition is satisfied if and only if $c_+ < 0$. Hence the corresponding parabola (essential spectrum) lies in the left half-plane.

Similarly, for the sequence $x_N = -N + 1$ we obtain the limiting problems

$$a_- u'' \pm b_- u' + c_- u = 0, \quad u(-1) = 0 \tag{3.4}$$

in the half-axis $\hat{\Omega} = \{x \geq -1\}$. This problem does not have nonzero bounded solutions if $c_- < 0$ and the second parabola of the essential spectrum is completely in the left half-plane.

Thus, there is a single limiting domain of the first type with two limiting operators. There are two more limiting domains of the second type and the operators corresponding to these domains. If the limiting problems of the first type do not have nonzero bounded solutions, then the operator satisfies the Fredholm property and its index is defined. The condition of stabilization of the index is formulated in a similar way: the limiting problems of the second type should not have nonzero bounded solutions.

3. Elliptic problems in unbounded domains

In the example considered above, this condition is satisfied if $c_\pm < 0$. In this case, the essential spectrum of the operator L considered on the whole axis lies in the left half-plane and its index is zero. It equals the indices of the problems considered on bounded intervals. Therefore, stabilization of the index takes place. We note that the index of the operator L may be equal zero but the conditions of stabilizations may not be satisfied. It can be the case for positive c_+ and c_-.

3.4.2. Stabilization of the index. To formulate conditions of stabilization of the index in the general case, consider a sequence of bounded domains Ω_n locally convergent to an unbounded domain Ω. The sequence of operators $L_n : E_\infty(\Omega_n) \to F_\infty(\Omega_n)$ converges to an operator $L : E_\infty(\Omega) \to F_\infty(\Omega)$ in the sense of local convergence of the coefficients. We suppose that the operators L_n satisfy the ellipticity condition, proper ellipticity and the Lopatinskii condition. Therefore, they satisfy the Fredholm property and their indices are well defined.

Condition NS(seq). Any limiting problem of the second type

$$\hat{L}u = 0, \quad x \in \hat{\Omega}, \quad u \in E_\infty(\hat{\Omega})$$

has only zero solution.

This condition is practically the same as Condition NS. The difference between them is that there are more limiting problems of the second type than of the first type. Therefore, Condition NS(seq) is more restrictive than Condition NS. A similar condition should be imposed on the adjoint problems.

Condition NS*(seq). Any limiting problem $\hat{L}^*v = 0$ of the second type has only zero solution in $(F^*(\hat{\Omega}))_\infty$. Here $\hat{L}^*$ is the operator adjoint to the limiting operator $\hat{L}$.

If these conditions are satisfied, then the indices of the operators L_n are equal to each other for n sufficiently large, the index of the operator L is defined and equals the indices of the operators L_n. Condition NS*(seq) for the adjoint operators can be replaced by a similar condition for formally adjoint problems if they are defined.

Stabilization of the index is related to the convergence of solutions of the nonhomogeneous problems $L_n u = f_n$ to solutions of the problem $Lu = f$, under the assumption that the right-hand sides of these equations converge. Generally speaking, we cannot expect the convergence of solutions even if the conditions of stabilization of the index are satisfied. Indeed, equality of the indices does not necessary imply equality of the dimensions of the kernels and of codimensions of the images (we recall that the index is the difference between them). Hence, if the dimension of $\operatorname{Ker} L_n$ does not stabilize for large n, then the solutions may not converge even for the homogeneous problems. If we assume that Conditions NS(seq) and NS*(seq) are satisfied and, moreover, the dimension of the kernels is the same for all n sufficiently large, then the convergence of solutions takes place (Chapter 8).

3.5 A priori estimates

In the case of bounded domains, normal solvability of elliptic operators follows from a priori estimates. If a sequence u_n is bounded in E, then we can choose a subsequence u_{n_k} convergent in a weaker norm E'. Assuming that $Lu_n \to 0$, we obtain from (2.19) convergence of u_{n_k} in the norm E. We can use it to prove that the kernel of the operator has a finite dimensional and its image is closed (Section 2.4.2).

In the case of unbounded domains, a sequence u_n bounded in E may not be compact in E'. Hence, a priori estimates presented in Section 2.3 do not ensure normal solvability with a finite-dimensional kernel. Another type of a priori estimates

$$\|u\|_{E(\Omega)} \leq k \left(\|Lu\|_{F(\Omega)} + \|u\|_{E'(\Omega_R)} \right) \tag{3.5}$$

was first suggested in [360]–[362] in the case $\Omega = \mathbb{R}^n$. Here Ω_R is the intersection of the domain Ω with a ball $B_R = \{x \in \mathbb{R}^n : |x| \leq R\}$. The difference with estimate (2.19) is in the second term in the right-hand side. This norm is now taken over a bounded subdomain of the domain Ω. Hence, local convergence in the $E'(\Omega_R)$ norm will provide strong convergence in the $E(\Omega)$ norm. We will prove such estimates for general elliptic problems.

Estimate (3.5) will allow us to show normal solvability of elliptic problems in unbounded domains. The proof of the estimate employs Condition NS. So we can use this condition to prove normal solvability directly or through a priori estimates. These two approaches are equivalent. We will obtain similar estimates for sequences of domains,

$$\|u\|_{E(\Omega_k)} \leq k \left(\|Lu\|_{F(\Omega_k)} + \|u\|_{E'(\Omega_{k_R})} \right).$$

They will be used to study stabilization of the index and convergence of solutions.

3.6 Non-Fredholm operators

If Condition NS is not satisfied, then the operator is not normally solvable with a finite-dimensional kernel. Hence, it does not satisfy the Fredholm property. One of the simplest examples of such problems is given by the equation

$$u'' = f \tag{3.6}$$

considered on the whole axis. The operator $Lu = u''$ does not satisfy Condition NS. Indeed, it coincides with its limiting operator, and the equation $Lu = 0$ has a nonzero bounded solution. It can be easily verified that the usual solvability conditions are not applicable for equation (3.6).

General theory of non-Fredholm operator does not exist. In some cases, it is possible to reduce them to Fredholm operators by a modification of function spaces or operators. For the example considered above we can introduce an exponential weight and put

$$v(x) = u(x)e^{\mu\sqrt{1+x^2}}, \quad g(x) = f(x)e^{\mu\sqrt{1+x^2}} \quad (\mu > 0).$$

Then v satisfies the equation

$$v'' - 2\mu\frac{x_1}{\sqrt{1+x_1^2}}v' + \left(\mu^2\frac{x_1^2}{1+x_1^2} - \mu\frac{1}{(1+x_1^2)^{3/2}}\right)v = g.$$

In this case the limiting problems do not have nonzero bounded solutions. The operator satisfies the Fredholm property and its index equals -2.

An interesting question is about solvability conditions different from the usual ones. Instead of the classical solvability condition $\int_{-\infty}^{\infty} f(t)v(t)dt = 0$ for the equation $Lu = f$ considered on the whole axis, the solvability condition can take the form

$$\sup_x \left|\int_0^x f(t)v(t)dt\right| < \infty.$$

In both cases, v is a solution of the homogeneous adjoint equation $L^*v = 0$. The last condition is principally different compared with the previous one. It does not have the form of a linear functional from the dual space. We will study some classes of non-Fredholm operators in Chapter 10.

4 Nonlinear Fredholm operators

Nonlinear operators are called Fredholm operators if the corresponding linearized operators satisfy the Fredholm property. Methods of analysis of nonlinear problems often use solvability conditions and other properties of linear operators. One of the most powerful methods of nonlinear analysis is related to topological degree. It applies to investigation of existence and bifurcations of solutions, their persistence, convergence of approximate methods. The topological degree theory starts with finite-dimensional mappings. It was generalized by Leray and Schauder for compact perturbations of the identity operator. They applied it to second-order elliptic operators in bounded domains. The degree generalization for Fredholm and proper operators with the zero index, which begins with the works by Caccioppoli and uses the results by Smale, makes it possible to apply it to general elliptic problems in bounded and unbounded domains.

4.1 Topological degree

The notion of the degree. Consider an operator A acting from a Banach space E into another Banach space F. Let D be a bounded domain in E. Topological degree is by definition an integer number $\gamma(A, D)$ which depends on the operator

A and on the domain D. This number should satisfy the following three properties: homotopy invariance, additivity, normalization. We will not give here the exact formulations (see Chapter 11) but rather a simple intuitive understanding of the notion of the degree, its properties and applications.

Consider first a finite-dimensional mapping and suppose that the equation $A(u) = a$ with $a = 0$ has a finite number of solutions $u_1, \ldots, u_k$ in D. Moreover, we assume that the matrices $A'(u_j)$, $j = 1, \ldots, k$ do not have zero eigenvalues. In this case we say that $a = 0$ is a regular point. If some of these matrices are degenerate, then this is a singular point. According to Sard's lemma, singular points form a set of the first category (countable union of nowhere dense sets). Therefore, regular points are dense in F.

If $a = 0$ is a regular point, then the degree can be determined as

$$\gamma(A, D) = \sum_{j=1}^{k} (-1)^{\nu_j}, \tag{4.1}$$

where ν_k is the number of real negative eigenvalues of the matrix $A'(u_j)$ together with their multiplicities. If $a = 0$ is a singular point, then we can approximate it by a regular point $\hat{a}$ sufficiently close to $a = 0$. In this case we can use the same formula (4.1) for the solutions of the equation $A(u) = \hat{a}$. This definition is correct and independent of the choice of $\hat{a}$ if $A(u) \neq 0$ at the boundary ∂D.

A similar formula remains valid for infinite-dimensional operators: compact perturbation of the identity operator (Leray-Schauder degree) and for Fredholm and proper operators with the zero index. The topological degree given by (4.1) satisfies the properties indicated above. Moreover, the degree is unique. It has the same value even if its construction is different.

Homotopy invariance of the degree means that its value does not change under a continuous deformation of the operator. More precisely, consider an operator $A_\tau : E \to F$ which depends on the parameter τ. This dependence is continuous in the operator norm. Suppose that $A_\tau(u) \neq 0$ for $u \in \partial D$. Then $\gamma(A_\tau, D)$ is independent of τ. This property of the degree is used in the Leray-Schauder method (see the next subsection).

Another important property of the degree is called the principle of nonzero rotation. It means that if $\gamma(A, D) \neq 0$, then the equation $A(u) = 0$ has at least one solution in D.

Degree for elliptic operators. Consider the semi-linear elliptic problem

$$\Delta u + g(u) = 0, \quad u|_{\partial \Omega} = 0 \tag{4.2}$$

in a bounded domain $\Omega \subset \mathbb{R}^n$ with a sufficiently smooth boundary. The function g is also supposed to be sufficiently smooth. Let us consider the auxiliary problem:

$$\Delta v = f, \quad v|_{\partial \Omega} = 0,$$

where $f \in C^\alpha(\bar{\Omega})$. It has a unique solution $v \in C^{2+\alpha}(\bar{\Omega})$. Denote by L the operator which puts in correspondence the solution of this problem to each f. It is a compact operator acting in the space $C^\alpha(\bar{\Omega})$. Multiplying equation (4.2) by L, we obtain the equation $A(u) = 0$, where $A(u) = u + Lg(u)$, $Lg(u)$ is a compact operator. This simple construction allows the application of the Leray-Schauder degree to elliptic operators in bounded domains.

The principle difference of unbounded domains is that the operator L is not compact and the Leray-Schauder degree cannot be used. We will use the degree construction for Fredholm and proper operators with the zero index. Consider the operator $A(u)$ which corresponds to the left-hand side of equation (4.2) and acts from the space $E_0 = \{u \in C^{2+\alpha}(\bar{\Omega}), u|_{\partial\Omega} = 0\}$ into the space $E_1 = C^\alpha(\bar{\Omega})$. It appears that in this case the degree may not exist (Chapter 11). If we look for solutions converging to zero at infinity and suppose that $F'(0) < 0$, then the operator is Fredholm with the zero index. However, it may not be proper. Properness is understood here in the sense that the inverse image of a compact set is compact in any bounded closed set. It is an important property which provides compactness of the set of solutions of operator equations. In order to provide the properness, we should introduce some weighted spaces. Then it holds and the degree can be defined for general nonlinear elliptic problems in bounded or unbounded domains.

4.2 Existence and bifurcations of solutions

We briefly recall some classical methods of nonlinear analysis based on the topological degree. The first one is used to prove existence of solutions, the second method to study bifurcations of solutions. We suppose that the operators and spaces are such that the topological degree can be constructed.

Leray-Schauder method. Consider an operator $A(u) : E \to F$ and the equation

$$A(u) = 0. \tag{4.3}$$

In order to prove the existence of its solutions in some domain $D \subset E$, we can use the following construction. Suppose that there exists a homotopy (or continuous deformation) $A_\tau(u)$ of the operator A to some other operator A_1 such that $A_0 = A$,

$$A_\tau(u) \neq 0, \quad u \in \partial D, \ \forall \tau \in [0,1], \tag{4.4}$$

and $\gamma(A_1, D) \neq 0$. Then, by virtue of homotopy invariance of the degree, $\gamma(A, D) \neq 0$. From the principle of nonzero rotation it follows that there exists a solution of the equation (4.3) in the domain D.

Put $D = B_R$, where B_R is a ball in the space E of the radius R. Suppose that we can obtain a priori estimates of solutions of the equation $A_\tau(u) = 0$, that is the estimate $\|u\|_E \leq M$ of all possible solutions of this equation. Here M is some positive constant. If we take $R > M$, then condition (4.4) is satisfied. As an operator A_1, we can take a simple model operator with known properties. For example, if the equation $A_1(u) = 0$ has a unique solution u_0 and the linearized operator $A'(u_0)$ is invertible, then $\gamma(A_1, B_R) \neq 0$.

The method described here was developed by Leray and Schauder in [303]. It is widely used to prove existence of solutions. When the topological degree is defined, the main difficulty of its application is related to a priori estimates of solutions.

Local bifurcations of solutions. As before, consider an operator A_τ which depends on the parameter τ. Suppose that the equation

$$A_\tau(u) = 0 \tag{4.5}$$

has a solution u_0 for all values of τ in a neighborhood $\delta(\tau_0)$ of $\tau = \tau_0$. Consider the operator $A'_\tau(u_0)$ linearized about u_0 and suppose that it has a simple real eigenvalue $\lambda_0(\tau)$ such that $\lambda_0(\tau_0) = 0$ and $\lambda_0(\tau_0) \neq 0$ for $\tau \neq \tau_0$. We also assume that there are no other eigenvalues in the vicinity of zero for $\tau \in \delta(\tau_0)$.

Under these assumptions, we can assert that $\tau = \tau_0$ is a bifurcation point. This means that, along with u_0, there are other solutions of equation (4.5) in some neighborhood $U(u_0)$ of u_0 for $\tau \in \delta(\tau_0)$. Indeed, suppose that this is not true and that u_0 is the only solution of equation (4.5) in the neighborhood $U(u_0)$ for $\tau \in \delta(\tau_0)$. Then we can express the value of the degree $\gamma(U, A_\tau)$ through the number of negative eigenvalues of the operator $A'_\tau(u_0)$:

$$\gamma(U, A_\tau) = (-1)^{\nu(\tau)}.$$

However, this value is different for $\tau < \tau_0$ and for $\tau > \tau_0$. This contradicts homotopy invariance of the degree.

Continuous branches of solutions. Topological degree implies persistence of solutions and existence of continuous branches of solutions. In order to explain these properties, we recall that the index $\text{ind}(u_\tau)$ of a solution u_τ of the operator equation $A_\tau(u) = 0$ is the value of the degree $\gamma(A_\tau, U(u_\tau))$ with respect to a small neighborhood $U(u_\tau)$ which does not contain other solutions. If $\text{ind}(u_\tau) \neq 0$, then the solution cannot disappear under a small change of the parameter. Indeed, otherwise the degree $\gamma(A_\tau, U(u_\tau))$ becomes zero. Therefore, each solution with a nonzero index forms a continuous branch in the function space. Each branch either goes to infinity or meets another branch where they can cross and continue or disappear. The points where they meet are points of local bifurcations where the linearized operator has a zero eigenvalue.

Continuous branches of solutions are related to their stability. Suppose that two branches of solutions u_τ^1 and u_τ^2 start at some point u_{τ_0} of the function space. Since they do not exist for $\tau < \tau_0$, then the sum of their indices equals zero. Hence the numbers of negative eigenvalues of the corresponding linearized operators differ from each other. Therefore, one of these two solutions is necessarily unstable, another one can be stable or unstable.

Chapter 2

Function Spaces and Operators

Sobolev spaces $W^{s,p}$ proved to be very convenient in the study of elliptic problems in bounded domains. For unbounded domains, it is also useful to introduce some generalizations of these spaces in such a way that they coincide with $W^{s,p}$ in bounded domains and have a prescribed behavior at infinity in unbounded domains. In this chapter we introduce spaces of functions in unbounded domains and study their properties.

It turns out that such spaces can be constructed for arbitrary Banach spaces of distributions, not only Sobolev spaces, as follows. Consider first functions defined on $\mathbb{R}^n$. As usual we denote by D the space of infinitely differentiable functions with compact support and by D' its dual. Let $E \subset D'$ be a Banach space, the inclusion is understood both in an algebraic and a topological sense. Denote by E_{loc} the collection of all $u \in D'$ such that $fu \in E$ for all $f \in D$. Let $\omega(x) \in D$, $0 \leq \omega(x) \leq 1$, $\omega(x) = 1$ for $|x| \leq 1/2$, $\omega(x) = 0$ for $|x| \geq 1$.

Definition. E_q ($1 \leq q \leq \infty$) is the space of all $u \in E_{\text{loc}}$ such that

$$\|u\|_{E_q} := \left(\int_{R^n} \|u(.)\omega(.-y)\|_E^q dy \right)^{1/q} < \infty, \quad 1 \leq q < \infty,$$

$$\|u\|_{E_\infty} := \sup_{y \in R^n} \|u(.)\omega(.-y)\|_E < \infty.$$

In what follows we will also use an equivalent definition based on a partition of unity. It will be proved that E_q is a Banach space. If Ω is a domain in $\mathbb{R}^n$, then by definition $E_q(\Omega)$ is the space of restrictions of E_q to Ω with the usual norm of restrictions. It is easy to see that if Ω is a bounded domain, then

$$E_q(\Omega) = E(\Omega), \quad 1 \leq q \leq \infty.$$

In particular, if $E = W^{s,p}$, then we set $W_q^{s,p} = E_q$ ($1 \leq q \leq \infty$). We will show that

$$W_p^{s,p} = W^{s,p} \quad (s \geq 0,\ 1 < p < \infty).$$

Hence the spaces $W_q^{s,p}$ generalize Sobolev spaces ($q < \infty$) and Stepanov spaces ($q = \infty$) (see [309], [310]).

1 The space E

Everywhere below we denote by $D(\mathbb{R}^n)$ the space of infinitely differentiable functions with finite supports, and by $D'(\mathbb{R}^n)$ the space of generalized functions, i.e., linear continuous functionals on $D(\mathbb{R}^n)$. In this section we consider only the whole $\mathbb{R}^n$, and we will use the notations D and D'.

Consider a Banach space E with the elements from D'. The inclusion $E \subset D'$ is understood both in the algebraic and topological sense.

Definition 1.1. The space of multipliers $M(E)$ on E is a set of infinitely differentiable functions $f(x)$, $x \in \mathbb{R}^n$ such that the operator of multiplication by f is a bounded operator in E. All functions defined in $\mathbb{R}^n$, which are infinitely differentiable and have all bounded derivatives, are multipliers in E.

We denote by $\|\cdot\|_E$ the norm in E and by $\|\cdot\|_M$ the norm in $M(E)$. By definition
$$\|fu\|_E \leq \|f\|_M \|u\|_E, \quad \forall f \in M(E), \forall u \in E.$$

Proposition 1.2. *Let E be invariant with respect to translation in $\mathbb{R}^n$ and*
$$\|\tau_h u\|_E = \|u\|_E, \quad \forall u \in E,$$
where τ_h is an operator of translation. Let further $f \in M(E)$, $\tau_h f(x) = f(x+h)$, $h \in \mathbb{R}^n$. Then
$$\|\tau_h f\|_M = \|f\|_M.$$

Proof. We first prove that
$$\tau_h(fu) = \tau_h f(\tau_h u). \tag{1.1}$$
Indeed, by definition for any $\phi \in D$ we have
$$\langle \tau_h(fu), \phi \rangle = \langle fu, \tau_{-h}\phi \rangle,$$
$$\langle \tau_h f(\tau_h u), \phi \rangle = \langle \tau_h u, (\tau_h f) \cdot \phi \rangle = \langle u, \tau_{-h}((\tau_h f) \cdot \phi) \rangle = \langle u, f \cdot \tau_{-h}\phi \rangle = \langle fu, \tau_{-h}\phi \rangle,$$
and (1.1) is proved. Further,
$$\|fu\|_E = \|\tau_h(fu)\|_E = \|\tau_h f(\tau_h u)\|_E \leq \|\tau_h f\|_M \|\tau_h u\|_E = \|\tau_h f\|_M \|u\|_E.$$
Hence
$$\|f\|_M \leq \|\tau_h f\|_M. \tag{1.2}$$
Therefore
$$\|\tau_h f\|_M \leq \|\tau_{-h}(\tau_h f)\|_M = \|f\|_M.$$
Together with (1.2) this estimate proves the proposition. □

2. Systems of functions

In what follows we suppose that for any $f \in D$,

$$\sup_h \|\tau_h f\|_M < \infty. \tag{1.3}$$

Example 1.3. If $E = H^{s,p}$ or $E = W^{s,p}$, where $-\infty < s < \infty$, $1 < p < \infty$, then any infinitely differentiable function f from $C^{[|s|]+1}(\mathbb{R}^n)$ belongs to $M(E)$ and

$$\|f\|_M \le K \|f\|_{C^{[|s|]+1}},$$

where K is a positive constant.

Definition 1.4. E_{loc} is a space of all $u \in D'$ such that $fu \in E$ for all $f \in D$.

2 Systems of functions

Definition 2.1. Partition of unity is a sequence $\{\phi_i\}$, $i = 1, 2, \ldots$ of functions $\phi_i \in D$, $\phi_i(x) \ge 0$ such that

$$\sum_{i=1}^{\infty} \phi_i(x) = 1, \quad x \in \mathbb{R}^n.$$

Condition 2.2. Let $\{\phi_i\}$, $i = 1, 2, \ldots$ be a sequence of functions $\phi_i \in D$. For some given N and any i there exists no more than N functions ϕ_j such that $\operatorname{supp} \phi_j \cap \operatorname{supp} \phi_i \ne \varnothing$.

Everywhere below we consider partitions of unity for which Condition 2.2 is satisfied.

Definition 2.3. Two systems of functions $\{\phi_i\}$, $\{\psi_j\}$, $i = 1, 2, \ldots$, $j = 1, 2, \ldots$, $\phi_i \in D$, $\psi_j \in D$ are called *equivalent* if there exists a number N such that:

- for any i there exists no more than N functions ψ_j such that $\operatorname{supp} \psi_j \cap \operatorname{supp} \phi_i \ne \varnothing$,
- for any j there exists no more than N functions ϕ_i such that $\operatorname{supp} \phi_i \cap \operatorname{supp} \psi_j \ne \varnothing$.

Proposition 2.4. *The equivalence relation introduced by Definition 2.3 is reflexive, symmetric, and transitive.*

We will also use systems of functions satisfying the following condition.

Condition 2.5. System of functions ϕ_i satisfies the following conditions:

1. $\phi_i(x) \ge 0$, $\phi_i \in D$,
2. Condition 2.2 is satisfied,
3. $\sup_i \|\phi_i\|_M < \infty$,

4. $\phi(x) = \sum_{i=1}^{\infty} \phi_i(x) \geq m > 0$ for some constant m,
5. the following estimate holds:

$$\sup_x |D^\alpha \phi(x)| \leq M_\alpha,$$

where D^α denotes the operator of differentiation, and M_α are positive constants.

3 The space E_p

Definition 3.1. Let $\{\phi_i\}$, $i = 1, 2, \ldots$ be a partition of unity. E_p is the space of all $u \in E_{\text{loc}}$ such that

$$\sum_{i=1}^{\infty} \|\phi_i u\|_E^p < \infty,$$

where $1 \leq p < \infty$, with the norm

$$\|u\|_{E_p} = \left(\sum_{i=1}^{\infty} \|\phi_i u\|_E^p \right)^{1/p}.$$

Proposition 3.2. *Let $\{\phi_i^1\}$ and $\{\phi_i^2\}$ be two partitions of unity such that*

$$\sup_i \|\phi_i^1\|_M < \infty, \quad \sup_i \|\phi_i^2\|_M < \infty.$$

Suppose that E_p^1 and E_p^2 are the spaces E_p corresponding to $\{\phi_i^1\}$ and $\{\phi_i^2\}$, respectively. If the partitions of unity are equivalent, then $E_p^1 = E_p^2$, and their norms are equivalent.

Proof. Let $u \in E_p^2$. We have

$$\phi_i^1 u = \phi_i^1 \sum_{j=1}^{\infty} \phi_j^2 u = \sum_{j'} \phi_i^1 \phi_{j'}^2 u,$$

where j' are all the numbers j such that $\operatorname{supp} \phi_i^1 \cap \operatorname{supp} \phi_j^2 \neq \emptyset$. By Definition 2.3 the number of such j' is no more than N. We have the estimate

$$\|\phi_i^1 u\|_E^p \leq \left(\sum_{j'} \|\phi_i^1 \phi_{j'}^2 u\|_E \right)^p.$$

Let $a_j \geq 0$, $j = 1, \ldots, m$. Then from convexity of the function s^p we obtain the estimate

$$\left(\sum_{j=1}^{m} a_j \right)^p = m^p \left(\sum_{j=1}^{m} \frac{1}{m} a_j \right)^p \leq m^{p-1} \sum_{j=1}^{m} a_j^p.$$

3. The space E_p

Therefore

$$\left(\sum_{j'} \|\phi_i^1 \phi_{j'}^2 u\|_E\right)^p \leq m^{p-1} \sum_{j'} \|\phi_i^1 \phi_{j'}^2 u\|_E^p,$$

where m is the number of j'. Since $m \leq N$, then

$$\|\phi_i^1 u\|_E^p \leq N^{p-1} \sum_{j'} \|\phi_i^1 \phi_{j'}^2 u\|_E^p = N^{p-1} \sum_{j=1}^{\infty} \|\phi_i^1 \phi_j^2 u\|_E^p.$$

Let k be a positive integer. We have

$$\sum_{i=1}^{k} \|\phi_i^1 u\|_E^p = N^{p-1} \sum_{i=1}^{k} \sum_{j=1}^{\infty} \|\phi_i^1 \phi_j^2 u\|_E^p = N^{p-1} \sum_{j=1}^{\infty} \sum_{i=1}^{k} \|\phi_i^1 \phi_j^2 u\|_E^p, \quad (3.1)$$

$$\sum_{i=1}^{k} \|\phi_i^1 \phi_j^2 u\|_E^p = \sum_{i'} \|\phi_{i'}^1 \phi_j^2 u\|_E^p \leq \sum_{i'} \|\phi_{i'}^1\|_M^p \|\phi_j^2 u\|_E^p,$$

where i' are those of i for which $\operatorname{supp} \phi_i^1 \cap \operatorname{supp} \phi_j^2 \neq \emptyset$. The number of such i' is less than or equal to N. Let

$$K_j = \sup_i \|\phi_i^j\|_M, \quad j = 1, 2.$$

Then

$$\sum_{i=1}^{k} \|\phi_i^1 \phi_j^2 u\|_E^p \leq NK_1 \|\phi_j^2 u\|_E^p.$$

It follows from (3.1) that

$$\sum_{i=1}^{k} \|\phi_i^1 u\|_E^p \leq N^p K_1 \sum_{j=1}^{\infty} \|\phi_j^2 u\|_E^p = N^p K_1 \|u\|_{E_p^2}^p.$$

From this we obtain

$$\sum_{i=1}^{\infty} \|\phi_i^1 u\|_E^p \leq N^p K_1 \|u\|_{E_p^2}^p.$$

Hence $u \in E_p^1$ and

$$\|u\|_{E_p^1} \leq N K_1^{1/p} \|u\|_{E_p^2}, \quad E_p^2 \subset E_p^1.$$

Similarly we get

$$\|u\|_{E_p^2} \leq N K_2^{1/p} \|u\|_{E_p^1}, \quad E_p^1 \subset E_p^2.$$

The proposition is proved. □

Proposition 3.3. *The space E_p is complete.*

Proof. Consider a fundamental sequence u_m in the space E_p. Then for any $\epsilon > 0$ there exists $N(\epsilon)$ such that

$$\sum_{i=1}^{\infty} \|(u_k - u_m)\phi_i\|_E^p \leq \epsilon \tag{3.2}$$

for any $k, m \geq N(\epsilon)$. Denote $\Phi_n = \sum_{i=1}^{n} \phi_i$. Let Ψ_n be an infinitely differentiable function with a finite support such that $\Psi_n = 1$ in the support of Φ_n. Since E is a Banach space and the sequence $\Psi_n u_m$ is fundamental with respect to m for any n fixed, then $\Psi_n u_m \to v_n$ in E as $m \to \infty$. Obviously, $\Phi_n u_m \to \Phi_n v_n$ in E as $m \to \infty$.

Consider a sequence n_j, $n_j \to \infty$ as $j \to \infty$. We construct the sequence of limiting functions v_{n_j} such that

$$\|\Phi_{n_j}(u_m - v_{n_j})\|_E \to 0 \text{ as } m \to \infty,$$

and for any $j_2 > j_1$,

$$\Phi_{n_{j_1}} v_{n_{j_1}} = \Phi_{n_{j_1}} v_{n_{j_2}}.$$

Therefore we have constructed the limiting function v defined in R^n. It coincides with v_j in the support of Φ_j. We have

$$\|\Phi_{n_j}(u_m - v)\|_E \to 0, \text{ as } m \to \infty. \tag{3.3}$$

We note that for any $\delta > 0$ there exists $N(\delta)$ and $i_0(\delta)$ such that

$$\sum_{i=i_0(\delta)}^{\infty} \|u_k \phi_i\|_E^p \leq \delta \tag{3.4}$$

for any $k \geq N(\delta)$. Indeed, we choose $N(\delta)$ such that

$$\sum_{i=1}^{\infty} \|(u_k - u_m)\phi_i\|_E^p \leq C_p \delta \tag{3.5}$$

for any $k, m \geq N(\delta)$. Here $C_p = 2^{-p}$. On the other hand, for a fixed m we can choose $i_0(\delta)$ such that

$$\sum_{i=i_0(\delta)}^{\infty} \|u_m \phi_i\|_E^p \leq C_p \delta \tag{3.6}$$

since the corresponding series converges. From (3.5) it follows that for m fixed and any $k \geq N(\delta)$,

$$\sum_{i=i_0(\delta)}^{\infty} \|(u_k - u_m)\phi_i\|_E^p \leq C_p \delta. \tag{3.7}$$

From (3.6) and (3.7) we obtain (3.4).

3. The space E_p

We prove next that
$$\sum_{i=i_0(\delta)}^{\infty} \|v\phi_i\|_E^p \leq \delta, \tag{3.8}$$
where $i_0(\delta)$ is the same as in (3.4). Suppose that this estimate is not true. Then there exists $i_1(\delta)$ such that
$$\sum_{i=i_0(\delta)}^{i_1(\delta)} \|v\phi_i\|_E^p > \delta. \tag{3.9}$$
On the other hand from (3.3) we have
$$\sum_{i=i_0(\delta)}^{i_1(\delta)} \|(u_m - v)\phi_i\|_E^p \to 0 \text{ as } m \to \infty.$$
This convergence and (3.9) contradict (3.4).

From (3.3), (3.4), and (3.8) we conclude that u_m converges to v in E_p. The proposition is proved. $\square$

Proposition 3.4. *Let $u_k = \sum_{i=1}^{k} u\phi_i$. Then $u_k \to u$ in E_q for $1 \leq q < \infty$.*

Proof. We have
$$\|u - u_k\|_{E_q}^q = \sum_{i=1}^{\infty} \|\phi_i(u - u_k)\|_E^q = \sum_{i=1}^{\infty} \|\phi_i \sum_{j=k+1}^{\infty} u\phi_j\|_E^q = \sum_{i=k'}^{\infty} \|\phi_i \sum_{j=k+1}^{\infty} u\phi_j\|_E^q \equiv S,$$
where the external sum is taken over all i such that $\operatorname{supp}\phi_i \cap \operatorname{supp}\phi_j \neq \emptyset$ for all $j \geq k+1$. The value k' depends on k, and $k' \to \infty$ as $k \to \infty$,
$$S = \sum_{i=k'}^{\infty} \|\phi_i \sum_{j'} u\phi_{j'}\|_E^q,$$
where j' denotes all j such that $\operatorname{supp}\phi_j \cap \operatorname{supp}\phi_i \neq \emptyset$ for a given i. Since the number of such j is uniformly bounded, we have the estimate
$$S \leq C_1 \sum_{i=k'}^{\infty} \|\phi_i u\phi_{j'}\|_E^q \leq C_2 \sum_{i=k'}^{\infty} \|u\phi_{j'}\|_E^q.$$
The last sum converges to zero as $k \to \infty$. The proposition is proved. $\square$

Corollary 3.5. *Infinitely differentiable functions with bounded supports are dense in E_q, $1 \leq q < \infty$.*

Proof. It is sufficient to note that D is dense in E, and $u_k \in E$. $\square$

Definition 3.6. Let $\{\phi_i\}$, $i = 1, 2, \ldots$ be a system of functions satisfying Condition 2.5. E_p is the space of all $u \in E_{\text{loc}}$ such that

$$\sum_{i=1}^{\infty} \|\phi_i u\|_E^p < \infty,$$

where $1 \leq p < \infty$, with the norm

$$\|u\|_{E_p} = \left(\sum_{i=1}^{\infty} \|\phi_i u\|_E^p \right)^{1/p}.$$

Proposition 3.7. *The spaces in Definitions 3.1 and 3.6 coincide and their norms are equivalent.*

The proof is similar to the proof of Proposition 4.4 below.

We introduce now one more definition of the norm in the space E_q. Let the norm be given by the equality

$$\|u\|_{E_q} = \left(\int_{\mathbb{R}^n} \|u(\cdot)\phi(\cdot - y)\|_E^q \, dy \right)^{1/q}. \tag{3.10}$$

We show that this norm is equivalent to the norm defined through a partition of unity. We note first of all that the function

$$s(y) = \|u(\cdot)\phi(\cdot - y)\|_E^q$$

is continuous. Indeed,

$$|s^{1/q}(y) - s^{1/q}(y_0)| \leq \|u(\cdot)(\phi(\cdot - y) - \phi(\cdot - y_0))\|_E \to 0 \text{ as } y \to y_0$$

by the properties of multipliers.

We have

$$\|u\|_{E_q}^q = \int_{\mathbb{R}^n} s(y) \, dy = \sum_{i=1}^{\infty} \int_{Q_i} s(y) \, dy,$$

where Q_i are unit cubes of the square lattice in $\mathbb{R}^n$,

$$\int_{Q_i} s(y) \, dy = s(y_i)$$

for some $y_i \in Q_i$ since $s(y)$ is continuous. Hence

$$\|u\|_{E_q}^q = \sum_{i=1}^{\infty} s(y_i). \tag{3.11}$$

This equality is obtained without specific assumptions on the function $\phi(x)$. Suppose now that it equals 1 in the ball of the radius $r = \sqrt{n}$, and 0 outside of the ball with the radius $2r$. Then for any $y_i \in Q_i$,

$$\phi(x - y_i) = 1, \quad x \in Q_i.$$

Therefore the system of functions $\phi_i(x) = \phi(x - y_i)$ satisfies the following conditions:

(1) $m \leq \sum_{i=1}^{\infty} \phi_i(x) \leq M$ for all $x \in R^n$ and some positive constants m and M,
(2) for each $x \in \mathbb{R}^n$ there exists a finite number of functions ϕ_i different from zero at this point. The estimate of this number is independent of x.

Hence the norm (3.11) is equivalent to the norm defined with any other system of functions equivalent to ϕ_i We have proved the following proposition.

Proposition 3.8. *The norm* (3.10) *is equivalent to the norm in Definition* 3.1.

4 The space E_∞

Definition 4.1. Let $\{\phi_i\}$ be a system of functions from D, $\phi_i(x) \geq 0$. E_∞ is the space of all functions $u \in E_{\mathrm{loc}}$ such that

$$\sup_i \|\phi_i u\|_E < \infty,$$

with the norm

$$\|u\|_{E_\infty} = \sup_i \|\phi_i u\|_E.$$

Proposition 4.2. *Let* $\{\phi_i^1\}$ *and* $\{\phi_i^2\}$ *be two partitions of unity satisfying Condition 2.2,*

$$\sup_i \|\phi_i^1\|_M < \infty, \quad \sup_i \|\phi_i^2\|_M < \infty.$$

Suppose that E_∞^1 and E_∞^2 are the spaces E_∞ corresponding to $\{\phi_i^1\}$ and $\{\phi_i^2\}$, respectively. If $\{\phi_i^1\}$ and $\{\phi_i^2\}$ are equivalent, then $E_\infty^1 = E_\infty^2$, and their norms are equivalent.

Proof. We have

$$\phi_i^1 u = \phi_i^1 \sum_{j=1}^{\infty} \phi_j^2 u = \sum_{j'} \phi_i^1 \phi_{j'}^2 u,$$

where j' are all the numbers j such that $\operatorname{supp} \phi_i^1 \cap \operatorname{supp} \phi_j^2 \neq \varnothing$. By Definition 2.3 the number of j' is less than or equal to N. Hence

$$\|\phi_i^1 u\|_E \leq \sum_{j'} \|\phi_i^1 \phi_{j'}^2 u\|_E \leq \|\phi_i^1\|_M \sum_{j'} \|\phi_{j'}^2 u\|_E \leq N \|\phi_i^1\|_M \|u\|_{E_\infty^2},$$

$$\sup_i \|\phi_i^1 u\|_E \leq N \sup_i \|\phi_i^1\|_M \|u\|_{E_\infty^2}.$$

Therefore $u \in E_\infty^1$ and
$$\|u\|_{E_\infty^1} \leq N \sup_i \|\phi_i^1\|_M \|u\|_{E_\infty^2}.$$

Similarly it can be proved that $E_\infty^1 \subset E_\infty^2$ with the corresponding inequality between their norms. The proposition is proved. $\square$

Example 4.3. If $E = H^{s,p}$ or $W^{s,p}$, $-\infty < s < \infty$, $1 < p < \infty$, then instead of 3 in Condition 2.5 we can require
$$\sup_i \|\phi_i\|_{C^{[|s|]+1}} < \infty.$$

Proposition 4.4. *Let $\{\phi_i^1\}$ and $\{\phi_i^2\}$ be two systems of functions satisfying Condition 2.5, E_∞^1 and E_∞^2 be two spaces E_∞ corresponding to $\{\phi_i^1\}$ and $\{\phi_i^2\}$, respectively. If $\{\phi_i^1\}$ and $\{\phi_i^2\}$ are equivalent, then $E_\infty^1 = E_\infty^2$, and their norms are equivalent.*

Proof. We introduce the system of functions $\theta_i^1(x) = \phi_i^1(x)/\phi^1(x)$, where
$$\phi^1(x) = \sum_{i=1}^\infty \phi_i^1(x).$$

Obviously θ_i^1 is a partition of unity. Denote by E_∞^3 the space which is constructed with the functions θ_i^1 according to Definition 4.1. We will prove that $E_\infty^1 = E_\infty^3$ and that their norms are equivalent. Indeed, let $u \in E_\infty^3$. Then
$$\sup_i \|\theta_i^1 u\|_E < \infty, \quad \|u\|_{E_\infty^3} = \sup_i \|\theta_i^1 u\|_E.$$

We have
$$\|\phi_i^1 u\|_E = \|\phi^1 \theta_i^1 u\|_E \leq \|\phi^1\|_M \|\theta_i^1 u\|_E \leq \|\phi^1\|_M \|u\|_{E_\infty^3}.$$
Hence $u \in E_\infty^1$ and
$$\|u\|_{E_\infty^1} \leq \|\phi^1\|_M \|u\|_{E_\infty^3}.$$
We have proved that $E_\infty^3 \subset E_\infty^1$. Conversely, let $u \in E_\infty^1$. We have
$$\|\theta_i^1 u\|_E = \|\frac{\phi_i^1}{\phi^1} u\|_E \leq \|\frac{1}{\phi^1}\|_M \|\phi_i^1 u\|_E \leq \|\frac{1}{\phi^1}\|_M \|u\|_{E_\infty^1}.$$
Hence $u \in E_\infty^3$ and
$$\|u\|_{E_\infty^3} \leq \|\frac{1}{\phi^1}\|_M \|u\|_{E_\infty^1}.$$
Therefore $E_\infty^1 \subset E_\infty^3$.

We can repeat the same construction for the second system of functions. Let $\theta_i^2(x) = \phi_i^2(x)/\phi^2(x)$, where

$$\phi^2(x) = \sum_{i=1}^{\infty} \phi_i^2(x).$$

Denote by E_∞^4 the space constructed with θ_i^2. Then we obtain $E_\infty^4 = E_\infty^2$ and the corresponding equivalence of the norms. It remains to apply the previous proposition to the spaces E_∞^3 and E_∞^4. The proposition is proved. □

5 Completeness of the space E_∞

Theorem 5.1. *The space E_∞ is complete.*

Proof. Let $u_k \in E_\infty$, $k = 1, 2, \ldots$ be a fundamental sequence. This means that for any $\epsilon > 0$ there exists $N = N(\epsilon)$ such that

$$\|u_k - u_l\|_{E_\infty} < \epsilon, \quad k, l > N. \tag{5.1}$$

Let $\{\phi_i\}$, $i = 1, 2, \ldots$ be a partition of unity in R^n such that supports of ϕ_i belong to the cubes of a lattice in R^n and $\sup_i \|\phi_i\|_M < \infty$. By Definition 4.1 and (5.1) we have

$$\sup_i \|\phi_i u_k - \phi_i u_l\|_{E_\infty} < \epsilon, \quad k, l > N. \tag{5.2}$$

It follows that for all i,

$$\|\phi_i u_k - \phi_i u_l\|_{E_\infty} < \epsilon, \quad k, l > N. \tag{5.3}$$

This implies that for any i the sequence $\phi_i u_k$, $k = 1, 2, \ldots$ is fundamental in the space E. Since E is a complete space, we conclude that there exists $u^i \in E$ such that

$$\phi_i u_k \to u^i, \quad k \to \infty \tag{5.4}$$

in E. Passing to the limit in (5.3) we get

$$\|u^i - \phi_i u_l\|_{E_\infty} \le \epsilon, \quad \forall i, \ l > N. \tag{5.5}$$

For any i we can construct a function $\psi_i \in D$ such that $\psi_i(x) = 1$, $x \in \operatorname{supp} \phi_i$. Then $\phi_i(x)\psi_i(x) = \phi_i(x)$ for $x \in R^n$.

Consider the formal sum $u = \sum_i \psi_i u^i$. We introduce the following functional: for any $\phi \in D$,

$$\langle u, \phi \rangle = \sum_{i'} \langle \psi_{i'} u^{i'}, \phi \rangle.$$

Here i' are those of i for which $\operatorname{supp} \psi_i \cap \operatorname{supp} \phi \ne \varnothing$. We note that

$$\langle u, \phi \rangle = \sum_i \langle \psi_i u^i, \phi \rangle$$

for any finite set of i which contains i' since $\langle \psi_i u^i, \phi \rangle = \langle u^i, \psi_i \phi \rangle = 0$ if $\operatorname{supp} \psi_i \cap \operatorname{supp} \phi = \emptyset$. Obviously u is a linear functional on D since for any $\phi_1, \phi_2 \in D$ we can take those i which contain i' for ϕ_1, ϕ_2, and $\phi_1 + \phi_2$.

We now prove that $u \in D'$, i.e., that the functional is continuous. Indeed, let $\phi_k \to 0$ in D. This means that $\operatorname{supp} \phi_k \subset B$ for all k and for some ball $B \subset \mathbb{R}^n$, and $\phi_k \to 0$ uniformly with all their derivatives. We take $u = \sum_{i'} \psi_{i'} u^{i'}$, where i' are those of i for which $\operatorname{supp} \psi_i \cap B \neq \emptyset$. Since $\psi_i u^i \in D'$, then $u \in D'$.

Moreover, $u \in E_{\mathrm{loc}}$. Indeed, let $f \in D$. We have for any $\phi \in D$:

$$\langle fu, \phi \rangle = \langle u, f\phi \rangle = \sum_{i'} \langle \psi_{i'} u^{i'}, f\phi \rangle = \langle f \sum_{i'} \psi_{i'} u^{i'}, \phi \rangle.$$

Hence $fu = f \sum_{i'} \psi_{i'} u^{i'}$. Here i' are those of i for which $\operatorname{supp} \psi_i \cap \operatorname{supp} f = \emptyset$. Since $u_i \in E$ and f and ψ_i are multipliers, we get $fu \in E$. Therefore, $u \in E_{\mathrm{loc}}$.

It remains to prove that $u \in E_\infty$ and $\lim_{k \to \infty} u_k = u$ in E_∞. We have

$$\phi_i u = \phi_i \sum_j \psi_j u^j, \tag{5.6}$$

where j are all of the subscripts for which

$$\operatorname{supp} \psi_j \cap \operatorname{supp} \phi_i \neq \emptyset. \tag{5.7}$$

Further, from (5.4)

$$\phi_i \sum_j \psi_j u^j = \phi_i \sum_j \psi_j \lim_{k \to \infty} \phi_j u_k. \tag{5.8}$$

Since ψ_j and ϕ_i are multipliers in E, we obtain

$$\sum_j \phi_i \psi_j \lim_{k \to \infty} \phi_j u_k = \sum_j \lim_{k \to \infty} \phi_i \psi_j \phi_j u_k = \lim_{k \to \infty} \sum_j \phi_i \phi_j u_k. \tag{5.9}$$

The subscripts j are defined by (5.7). For all other j we have $\operatorname{supp} \psi_j \cap \operatorname{supp} \phi_i = \emptyset$. Hence $\operatorname{supp} \phi_j \cap \operatorname{supp} \phi_i = \emptyset$. Therefore

$$\sum_j \phi_i \phi_j = \phi_i \sum_{j=1}^\infty \phi_j = \phi_i.$$

From (5.8), (5.9)

$$\phi_i \sum_j \psi_j u^j = \lim_{k \to \infty} \phi_i u_k = u^i.$$

From (5.6), $\phi_i u = u^i$. We get now from (5.5)

$$\|\phi_i u - \phi_i u_l\|_E = \|u^i - \phi_i u_l\|_E \leq \epsilon, \quad \forall i, \, l > N. \tag{5.10}$$

6. Other definitions of the space E_∞ 59

It follows that
$$\|\phi_i u\|_E \le \epsilon + \|\phi_i u_l\|_E \le \epsilon + \|u_l\|_{E_\infty}.$$

Hence $u \in E_\infty$.

We obtain from (5.10)
$$\sup_i \|\phi_i(u - u_l)\|_E \le \epsilon, \quad l > N.$$

Therefore
$$\|u - u_l\|_{E_\infty} \le \epsilon, \quad l > N.$$

Hence $\lim_{l \to \infty} u_l = u$ in E_∞. The theorem is proved. □

6 Other definitions of the space E_∞

Definition 6.1. Let $\eta(x) \in D$ satisfy the following conditions:

1. $0 \le \eta(x) \le 1$, $x \in R^n$,
2. $\eta(x) = 1$ in the cube $|x_i| \le a_1$, $i = 1, 2, \ldots, n$,
3. $\eta(x) = 0$ outside the cube $|x_i| \le a_2$, $i = 1, 2, \ldots, n$, where a_1 and a_2 are given numbers, $a_1 < a_2$.

Denote $\eta_y(x) = \eta(x - y)$, $y \in R^n$. The space E_∞ is the set of all $u \in E_{\text{loc}}$ such that
$$\sup_{y \in R^n} \|\eta_y u\|_E < \infty.$$

The norm in this space is given by the relation $\|u\|_{E_\infty} = \sup_{y \in R^n} \|\eta_y u\|_E$.

Proposition 6.2. *Let $\{\phi_i\}$ be a partition of unity in R^n with supports in lattice cubes, $\sup_i \|\phi_i\|_M < \infty$. Then the spaces in Definitions 4.1 and 6.1 coincide.*

Proof. Denote by E_∞^1 and E_∞^2 the spaces in Definitions 6.1 and 4.1, respectively. We will use the function $\eta(x)$ constructed in the following way. Let Q_a be the cube $|x_i| \le a$, $i = 1, \ldots, n$, $\chi(x)$ be the characteristic function of the cube Q_{2a}. Set
$$\eta(x) = \int \omega_\epsilon(x - \xi) \chi(\xi) d\xi = \int \omega_\epsilon(\tau) \chi(x + \tau) d\tau,$$

where $\omega_\epsilon(x)$ is a symmetric averaging kernel,
$$\omega_\epsilon(x) = 0 \text{ for } |x| > \epsilon, \quad \int \omega_\epsilon(x) dx = 1.$$

For $\epsilon > 0$ sufficiently small we obtain
$$\eta(x) = 1, \ x \in Q_a, \quad \eta(x) = 0, \ x \notin Q_{3a}.$$

Set $\eta_y(x) = \eta(x-y)$, $\chi_y(x) = \chi(x-y)$. Obviously,

$$\eta_y(x) = \int \omega_\epsilon(x-\xi)\chi_y(\xi)d\xi = \int \omega_\epsilon(\tau)\chi_y(x+\tau)d\tau.$$

We cover R^n with the cubes obtained by translation of the cube Q_{2a} such that they intersect each other only by their sides. Let $\chi_i(x)$ be the characteristic functions of these cubes. Then

$$\sum_i \chi_i(x) = 1 \quad \text{almost everywhere in } \mathbb{R}^n. \tag{6.1}$$

We have $\chi_i(x) = \chi(x-h_i)$ for some h_i. Hence

$$\eta_{h_i}(x) = \int \omega_\epsilon(\tau)\chi_i(x+\tau)d\tau.$$

From (6.1) it follows that $\sum_i \eta_{h_i}(x) = 1$. Therefore $\eta_{h_i}(x)$ is a partition of unity, $\operatorname{supp} \eta_{h_i}$ belong to some cubes. Moreover, $\eta_{h_i}(x) = \eta(x-h_i)$, $\sup_i \|\eta_{h_i}\|_M < \infty$.

Let $u \in E^1_\infty$. We have

$$\sup_i \|\eta_{h_i} u\|_M \leq \sup_y \|\eta_y u\|_E \leq \|u\|_{E^1_\infty}.$$

Hence $u \in E^2_\infty$ and $\|u\|_{E^2_\infty} \leq \|u\|_{E^1_\infty}$. We have proved that

$$E^1_\infty \subset E^2_\infty \quad \text{for this choice of } \eta(x). \tag{6.2}$$

Now let $u \in E^2_\infty$ and $\{\phi_i\}$ be the partition of unity in the formulation of the proposition. We have

$$\eta_y u = \sum_{i=1}^\infty \eta_y \phi_i u = \sum_{i'} \eta_y \phi_{i'} u, \tag{6.3}$$

where i' are all the numbers i for which $\operatorname{supp} \phi_i$ has a nonempty intersection with $\operatorname{supp} \eta_y$. The number of such i' is less than or equal to N, where N does not depend on y. It follows from (6.3) that

$$\|\eta_y u\|_E \leq \sum_{i'} \|\eta_y \phi_{i'} u\|_E \leq \|\eta_y\|_M \|\phi_{i'} u\|_E \leq N\|\eta_y\|_M \|u\|_{E^2_\infty}.$$

Hence

$$\sup_y \|\eta_y u\|_E \leq N \sup_y \|\eta_y\|_M \|u\|_{E^2_\infty} \leq K\|u\|_{E^2_\infty}.$$

From this estimate it follows that $u \in E^1_\infty$ and

$$\|u\|_{E^1_\infty} \leq K\|u\|_{E^2_\infty}, \quad E^2_\infty \subset E^1_\infty. \tag{6.4}$$

The proposition is proved for the special choice of η_y. Below we will prove that E^1_∞ does not depend on the choice of η_y. $\square$

6. Other definitions of the space E_∞

In what follows we use the space $E(G)$, where G is a domain in R^n. The space $E(G)$ is defined as the set of all generalized functions from D'_G which are restrictions to G of generalized functions from E. The norm in this space is

$$\|u\|_{E(G)} = \inf \|v\|_E,$$

where the infimum is taken over all those generalized functions $v \in E$ whose restriction to G coincides with u.

Definition 6.3. The space E_∞ is the set of all $u \in E_{\text{loc}}$ such that

$$\sup_{y \in R^n} \|u_y\|_{E(G_y)} < \infty, \tag{6.5}$$

where u_y is a restriction of u to G_y, $G \subset R^n$ is a bounded domain containing the origin, G_y is a shifted domain: the characteristic function of G_y is $\chi(x-y)$, where $\chi(x)$ is the characteristic function of G. The norm in E_∞ is given by

$$\|u\|_{E_\infty} = \sup_{y \in R^n} \|u_y\|_{E(G_y)}.$$

Proposition 6.4. *The spaces in Definitions 6.1 and 6.3 coincide.*

Proof. Denote by E^1_∞ and E^3_∞ the spaces given by Definitions 6.1 and 6.3, respectively. Let $u \in E^1_\infty$. We take a_1 sufficiently large such that $G \subset Q_{a_1}$, where Q_{a_1} is the cube $|x_i| < a_1$, $i = 1, \ldots, n$. Let u_y be the restriction of u to G_y. Then $\eta_y u$ is an extension of u_y to E. Hence

$$\|u_y\|_{E(G_y)} \leq \|\eta_y u\|_E \leq \|u\|_{E^1_\infty}.$$

Therefore, $u \in E^3_\infty$, and

$$\|u\|_{E^3_\infty} \leq \|u\|_{E^1_\infty}, \quad E^1_\infty \subset E^3_\infty. \tag{6.6}$$

This inequality is proved only for such a_1 that $G \subset Q_{a_1}$. From (6.4) we have $E^2_\infty \subset E^1_\infty$ for any a_1. Hence $E^2_\infty \subset E^3_\infty$ for any choice of G.

Now let $u \in E^3_\infty$. By Definition 6.3 this means that $u \in E_{\text{loc}}$ and (6.5) holds. By the definition of the norm $\|u_y\|_{E(G_y)}$, there exists a function $v \in E$ such that v is an extension of u_y and

$$\|v\|_E \leq 2\|u_y\|_{E(G_y)} < \infty. \tag{6.7}$$

We take a function $\eta(x)$ in Definition 6.1 such that $Q_{a_2} \subset G$. We have

$$\|\eta_y v\|_E \leq \|\eta_y\|_M \|v\|_E \leq K\|v\|_E \tag{6.8}$$

since $\sup_y \|\eta_y\|_M < \infty$.

Since $Q_{a_2} \subset G$, then

$$\eta_y v = \eta_y u_y \quad \text{in} \quad D'. \tag{6.9}$$

Indeed, for any $\phi \in D$ we have
$$\langle \eta_y v, \phi \rangle = \langle v, \eta_y \phi \rangle, \quad \langle \eta_y u_y, \phi \rangle = \langle u_y, \eta_y \phi \rangle.$$

From the inclusion $\operatorname{supp} \eta_y \phi \subset G_y$ follows the equality $\langle v, \eta_y \phi \rangle = \langle u_y, \eta_y \phi \rangle$ since v is an extension of u_y.

It follows from (6.9) that
$$\|\eta_y u_y\|_E = \|\eta_y v\|_E \leq K \|v\|_E \leq 2K \|u_y\|_{E(G_y)} \leq 2K \|u\|_{E_\infty^3}.$$

Therefore
$$\|\eta_y u\|_E \leq 2K \|u\|_{E_\infty^3} \tag{6.10}$$
since $\eta_y u = \eta_y u_y$ in D'. From (6.10) we conclude that $u \in E_\infty^1$ and
$$\|u\|_{E_\infty^1} \leq 2K \|u\|_{E_\infty^3}, \quad E_\infty^3 \subset E_\infty^1. \tag{6.11}$$

This result is obtained under the assumption that $Q_{a_2} \subset G$. We can take $\eta(x)$ as in the proof of (6.2). Since this result is true for any a, we obtain from (6.11) that $E_\infty^3 \subset E_\infty^2$ for any choice of G. Therefore $E_\infty^3 = E_\infty^2$ for any choice of G. We conclude that E_∞^3 does not depend on the choice of G.

Let us return to (6.6). We recall that it is obtained under the assumption that $G \subset Q_{a_1}$. But since E_∞^3 does not depend on the choice of G, a_1 can be taken arbitrarily. Similarly, a_2 can be taken arbitrary in the assumption $Q_{a_2} \subset G$. Hence (6.11) is true for any a_2. From (6.6) and (6.11) we obtain $E_\infty^1 = E_\infty^3$ for any choice of a_1 and a_2. The proposition is proved. $\square$

Remark 6.5. It follows from the proposition that E_∞ in Definition 6.1 does not depend on the choice of $\eta(x)$. The same result is true if instead of cubes in Definition 6.1 we take balls. Indeed, since we have proved that E_∞^3 does not depend on the choice of G, we can repeat the same proof.

7 Bounded sequences in E_∞

Definition 7.1. A sequence $u_k \in E_{\operatorname{loc}}$ is called locally weakly convergent to $u \in E_{\operatorname{loc}}$ if for any $\phi \in D$,
$$\phi u_k \to \phi u \quad \text{weakly in } E.$$

Lemma 7.2. *If a sequence $u_k \in E_\infty$ is bounded in E_∞ and locally weakly convergent to u, then $u \in E_\infty$.*

Proof. We use Definition 4.1 of the space E_∞. Let $\{\phi_i\}$ be a partition of unity. Then $u \in E_\infty$ if $\sup_i \|\phi_i u\|_E < \infty$. Suppose that $u \notin E_\infty$. Then there is a subsequence i_k of i such that
$$\|\phi_i u\|_E \to \infty \quad \text{as } i_k \to \infty. \tag{7.1}$$

7. Bounded sequences in E_∞

A set in a Banach space is bounded if and only if any functional from the dual space is bounded on it. Hence there exists a functional $F \in E^*$ such that $F(\phi_{i_k} u) \to \infty$ as $i_k \to \infty$. Since u_l is locally weakly convergent to u, then $F(\phi_{i_k} u_l) \to F(\phi_{i_k} u)$ as $l \to \infty$ for any i_k. Therefore we can choose l_k such that $|F(\phi_{i_k} u_{l_k}) - F(\phi_{i_k} u)| < 1$. It follows from (7.1) that

$$F(\phi_{i_k} u_{l_k}) \to \infty \text{ as } i_k \to \infty. \tag{7.2}$$

On the other hand, by assumption u_k is bounded in E_∞. Hence $\|u_k\|_{E_\infty} \leq M$, $\|\phi_{i_k} u_k\|_E \leq M$. This contradicts (7.2). The lemma is proved. $\square$

Theorem 7.3. *Let E be a reflexive Banach space. If $\{u_k\}$, $k = 1, 2, \ldots$ is a bounded sequence in E_∞, then there exists a subsequence u_{k_i} of u_k and $u \in E_\infty$ such that*

$$u_{k_i} \to u \text{ locally weakly and in } D'.$$

Proof. Denote by B_r a ball $|x| < r$ in R^n and consider the sequence B_j, $j = 1, 2, \ldots$. Suppose that f_j, $j = 1, 2, \ldots$ is a sequence of functions such that $f_j \in D$,

$$f_j(x) = 1, \; x \in B_j, \; f_j(x) = 0, \; x \notin B_{j+1}, \; j = 1, 2, \ldots$$

Let $\{\phi_i\}$, $i = 1, 2, \ldots$, $\phi_i \in D$ be a partition of unity for which E_∞ is defined. We suppose that $\operatorname{supp} \phi_i$ belong to unit cubes and $\sup_i \|\phi_i\|_M = K < \infty$. Since

$$\|\phi_i u_k\|_E \leq K \|u_k\|_{E_\infty} \leq M,$$

we get

$$\|f_j u_k\|_E \leq M_j \tag{7.3}$$

with a constant M_j independent of k. Indeed,

$$\|f_j u_k\|_E = \|\sum_{i'} f_j \phi_{i'} u_k\|_E.$$

Here i' are those of i for which $\operatorname{supp} \phi_i \cap \operatorname{supp} f_j \neq \emptyset$. The number of i' is less or equal to N_j, where N_j is a constant. Therefore

$$\|f_j u_k\|_E \leq \sum_{i'} \|f_j\|_M \|\phi_{i'} u_k\|_E \leq N_j M \|f_j\|_M,$$

and (7.3) is proved. Since E is a reflexive space, we conclude that there exists a subsequence u_i^j, $i = 1, 2, \ldots$ of u_k such that $f_j u_i^j \to v_j$ weakly in E, $v_j \in E$. This means that there exists a sequence $\tilde{u}_i$ such that

$$F(f_j \tilde{u}^i) \to F(v_j) \text{ as } i \to \infty \tag{7.4}$$

for any $F \in E^*$. Indeed, we choose u_i^j such that u_i^2 is a subsequence of u_i^1, u_i^3 is a subsequence of u_i^2 and so on. Denote by $\tilde{u}_i$ the diagonal subsequence. Then we obtain from (7.4)

$$F(f_j \tilde{u}_i) \to F(v_j) \text{ as } i \to \infty \tag{7.5}$$

for any j and any $F \in E^*$.

It follows from (7.5) that if $k > j$, then

$$f_j v_k = v_j. \tag{7.6}$$

Indeed, $F(f_j \cdot) \in E^*$. Hence $F(f_j f_k \tilde{u}_i) \to F(f_j v_k)$ as $i \to \infty$. But $f_j(x) f_k(x) = f_j(x)$. Therefore $F(f_j \tilde{u}_i) \to F(f_j v_k)$ as $i \to \infty$. From this and (7.5) we obtain (7.6).

From (7.6) it follows that

$$\langle v_j, \phi \rangle = \langle v_k, \phi \rangle \tag{7.7}$$

if $k > j$ and $\operatorname{supp} \phi \subset B_j$. Indeed,

$$\langle v_j, \phi \rangle = \langle f_j v_k, \phi \rangle = \langle v_k, f_j \phi \rangle = \langle v_k, \phi \rangle$$

since $f_j \phi = \phi$.

We introduce a generalized function $u \in D'$ such that for any $\phi \in D$,

$$\langle u, \phi \rangle = \langle v_j, \phi \rangle \tag{7.8}$$

if $\operatorname{supp} \phi \subset B_j$. The proof that u is a continuous linear functional on D is standard.

Obviously, $u \in E_{\text{loc}}$. Indeed, for any $\phi \in D$ and $f \in D$ we have

$$\langle fu, \phi \rangle = \langle u, f\phi \rangle = \langle v_j, f\phi \rangle = \langle fv_j, \phi \rangle,$$

where j is taken such that $\operatorname{supp} f \subset B_j$. Hence $fu = fv_j$ in D'. Since $v_j \in E$, we get $fu \in E$, and therefore $u \in E_{\text{loc}}$.

We prove now that

$$\tilde{u}_i \to u \text{ locally weakly as } i \to \infty.$$

We have to prove that

$$F(f\tilde{u}_i) \to F(fu) \text{ as } i \to \infty \tag{7.9}$$

for any $f \in D$ and $F \in E^*$. If $F \in E^*$, $f \in D$, then $fF \in E^*$. It follows from (7.5) that

$$fF(f_j \tilde{u}_i) \to fF(v_j) \text{ as } i \to \infty$$

or

$$F(ff_j \tilde{u}_i) \to F(fv_j) \text{ as } i \to \infty.$$

Since $f \in D$, we can take j so large that $\operatorname{supp} f \subset B_j$. Then $f(x) f_j(x) = f(x)$ for all $x \in \mathbb{R}^n$. Therefore $f f_j \tilde{u}_i = f \tilde{u}_i$ and

$$F(f\tilde{u}_i) \to F(fv_j) \text{ as } i \to \infty. \tag{7.10}$$

Further, for any $\phi \in D$ we have $\langle fv_j, \phi \rangle = \langle v_j, f\phi \rangle$. Since $\operatorname{supp} f\phi \subset B_j$, from (7.8) we obtain
$$\langle fu, \phi \rangle = \langle u, f\phi \rangle = \langle v_j, f\phi \rangle.$$

Hence
$$\langle fu, \phi \rangle = \langle fv_j, \phi \rangle, \quad \forall \phi \in D.$$

Therefore $fu = fv_j$ in D'. From (7.10) we obtain (7.9).

Since $\{\tilde{u}_i\}$ is a subsequence of $\{u_k\}$ and this latter is bounded in E_∞, we can conclude that $\{\tilde{u}_i\}$ is bounded in E_∞. From Lemma 7.2 it follows that $u \in E_\infty$.

It remains to prove that $\tilde{u}_i \to u$ in D'. Since $E \subset D'$ (inclusion with the topology), it follows that for any $\phi \in D$, $\phi(u) = \langle u, \phi \rangle \in E^*$. Let $f \in D, f(x) = 1$ in $\operatorname{supp} \phi$. Then
$$\langle \tilde{u}_i - u, \phi \rangle = \langle \tilde{u}_i - u, f\phi \rangle = \phi(f(u_i - u)) \to 0$$
as $i \to \infty$ because of the local weak convergence. The theorem is proved. $\square$

8 The space $E_p(\Gamma)$

Let Γ be an m-dimensional manifold, $\Gamma \subset \mathbb{R}^n$, $(m < n)$. We first consider the case where it is C^∞ manifold. We recall the definition of $D'(\Gamma)$ (see [243]). We are given a family J of homeomorphisms ψ, called coordinate systems, of open sets $\Gamma_\psi \subset \Gamma$ on open sets $\tilde{\Gamma}_\psi \subset \mathbb{R}^m$ such that:

(i) If ψ and ψ' belong to J, then the mapping
$$\psi'\psi^{-1} : \psi(\Gamma_\psi \cap \Gamma_{\psi'}) \to \psi'(\Gamma_\psi \cap \Gamma_{\psi'}) \tag{8.1}$$
is infinitely differentiable,

(ii) $\cup_{\psi \in J} \Gamma_\psi = \Gamma$.

We define the space $D(\Gamma)$. If to every coordinate system ψ in Γ we are given a function $\theta_\psi \in D(\tilde{\Gamma}_\psi)$ such that
$$\theta_{\psi'} = \theta_\psi \circ (\psi(\psi')^{-1}) \text{ in } \psi(\Gamma_\psi \cap \Gamma_{\psi'}),$$
we say that $\theta \in D(\Gamma)$ and set $\theta_\psi = \theta \circ \psi^{-1}$.

If to every coordinate system ψ in Γ corresponds a distribution $u_\psi \in D'(\tilde{\Gamma}_\psi)$ such that
$$u_{\psi'} = u_\psi \circ (\psi(\psi')^{-1}) \text{ in } \psi'(\Gamma_\psi \cap \Gamma_{\psi'}), \tag{8.2}$$
we call the system u_ψ a distribution u in Γ. The set of all distributions in Γ is denoted by $D'(\Gamma)$. We write also $u_\psi = u \circ \psi^{-1}$, $u = u_\psi \circ \psi$.

As in Section 1, we consider the space $E = E(\mathbb{R}^m)$, $E \subset D'(\mathbb{R}^m)$. We denote by $M(E)$ the space of multipliers of E, and we suppose that $D(\mathbb{R}^m) \subset M(E)$.

Moreover we suppose that there exist numbers $\kappa > 0$ and $\nu > 0$ such that for any $\phi \in D(\mathbb{R}^m)$,
$$\|\phi\|_{M(E)} \leq \kappa \|\phi\|_{C^\nu}. \tag{8.3}$$

Definition 8.1. A function u belongs to the space $E_{\mathrm{loc}}(\Gamma)$ if and only if $u \in D'(\Gamma)$ and for any $\theta \in D(\Gamma)$,
$$(\theta u) \circ \psi^{-1} \in E. \tag{8.4}$$

It is supposed that $(\theta u)_\psi$ is extended by zero from $\tilde{\Gamma}_\psi$ to $\mathbb{R}^m$. By definition $(\theta u)_\psi = \theta_\psi u_\psi$. We give the definition of the space $E_p(\Gamma)$. Let U_i, $i = 1, 2, \ldots$ be a covering of Γ for which a coordinate system ψ_i is introduced. We suppose that there exists a number N such that for any i there is no more than N of j such that $U_i \cap U_j \neq \emptyset$. Let $\theta_i \in D(\Gamma)$ be a partition of unity, $\mathrm{supp}\, \theta_i \subset U_i$, $\sum_i \theta_i(x) = 1$ for any $x \in \Gamma$.

Definition 8.2. A function u belongs to the space $E_p(\Gamma)$, $1 \leq p < \infty$ if and only if $u \in E_{\mathrm{loc}}(\Gamma)$ and
$$\|u\|_{E_p(\Gamma)} = \left(\sum_{j=1}^\infty \|(\theta_j u) \circ \psi_j^{-1}\|_E^p \right)^{1/p} < \infty.$$

A function u belongs to the space $E_\infty(\Gamma)$ if and only if $u \in E_{\mathrm{loc}}(\Gamma)$ and
$$\|u\|_{E_\infty(\Gamma)} = \sup_i \|(\theta_j u) \circ \psi_j^{-1}\|_E < \infty.$$

Definition 8.3. Two coverings U_i^1 and U_j^2 are called equivalent if there exists a number N such that for any i there is no more than N of j such that $U_i^1 \cap U_j^2 \neq \emptyset$; for any j there is no more than N of i such that $U_i^1 \cap U_j^2 \neq \emptyset$.

In the following theorem we prove the independence of the space $E^p(\Gamma)$ of the choice of equivalent coverings of Γ and of the choice of partition of unity.

Theorem 8.4. *Let U_i^1 and U_j^2 be two equivalent coverings of Γ, θ_i^1 and θ_j^2 be two corresponding partitions of unity. Suppose that the following conditions are satisfied:*

(α) *For any i, j such that $U_i^1 \cap U_j^2 \neq \emptyset$ the norms of the operators of change of variables*
$$\psi_j^2 (\psi_i^1)^{-1} : \psi_i^1(U_i^1 \cap U_j^2) \to \psi_j^2(U_i^1 \cap U_j^2),$$
$$\psi_i^1 (\psi_j^2)^{-1} : \psi_j^2(U_i^1 \cap U_j^2) \to \psi_i^1(U_i^1 \cap U_j^2)$$
are uniformly bounded in E,

(β) *The estimates*
$$K_1 = \sup_i \|\theta_i^1 \circ (\psi_i^1)^{-1}\|_{C^\nu} < \infty, \quad K_2 = \sup_j \|\theta_j^2 \circ (\psi_j^2)^{-1}\|_{C^\nu} < \infty,$$
hold with the same ν as in (8.3).

8. The space $E_p(\Gamma)$

Let $E_p^1(\Gamma)$ and $E_p^2(\Gamma)$ $(1 \leq p \leq \infty)$ be the spaces $E_p(\Gamma)$ that correspond to the coverings U_i^1 and U_j^2 and the partitions of unity θ_i^1 and θ_j^2, respectively. Then $E_p^1(\Gamma) = E_p^2(\Gamma)$, and their norms are equivalent.

Proof. Consider first the case $1 \leq p < \infty$. Let $u \in E_{\mathrm{loc}}(\Gamma)$ and $\|u\|_{E_p^2} < \infty$. We have

$$\theta_i^1 u = \theta_i^1 \sum_{j=1}^{\infty} \theta_j^2 u = \sum_{j'} \theta_i^1 \theta_{j'}^2 u, \tag{8.5}$$

where j' are all the numbers j for which $U_i^1 \cap U_j^2 \neq \emptyset$. By assumption, the number of j' is less than or equal to N. It follows from (8.5) that

$$\|(\theta_i^1 u) \circ (\psi_i^1)^{-1}\|_E^p \leq \left(\sum_{j'} \|\theta_i^1 \theta_{j'}^2 u) \circ (\psi_i^1)^{-1}\|_E \right)^p.$$

From convexity of the function s^p we obtain

$$\|(\theta_i^1 u) \circ (\psi_i^1)^{-1}\|_E^p \leq N^{p-1} \sum_{j'} \|(\theta_i^1 \theta_{j'}^2 u) \circ (\psi_i^1)^{-1}\|_E^p$$

$$= N^{p-1} \sum_{j=1}^{\infty} \|(\theta_i^1 \theta_j^2 u) \circ (\psi_i^1)^{-1}\|_E^p.$$

Let k be a positive integer. We have

$$\sum_{i=1}^{k} \|(\theta_i^1 u) \circ (\psi_i^1)^{-1}\|_E^p \leq N^{p-1} \sum_{j=1}^{\infty} \sum_{i=1}^{k} \|(\theta_i^1 \theta_j^2 u) \circ (\psi_i^1)^{-1}\|_E^p. \tag{8.6}$$

Further

$$\sum_{i=1}^{k} \|(\theta_i^1 \theta_{j'}^2 u) \circ (\psi_i^1)^{-1}\|_E^p \leq \sum_{i'} \|(\theta_{i'}^1 \theta_j^2 u) \circ (\psi_{i'}^1)^{-1}\|_E^p, \tag{8.7}$$

where i' are those i for which $U_i^1 \cap U_j^2 \neq \emptyset$. The number of such i' is less than or equal to N.

For any i and j such that $U_i^1 \cap U_j^2 \neq \emptyset$ we have, according to (8.2),

$$(\theta_i^1 \theta_j^2 u) \circ (\psi_i^1)^{-1} = \left((\theta_i^1 \theta_j^2 u) \circ (\psi_j^2)^{-1} \right) \circ (\psi_j^2 (\psi_i^1)^{-1}). \tag{8.8}$$

By the condition of the theorem, the norms of the operators $\psi_j^2 (\psi_i^1)^{-1}$ are uniformly bounded. Therefore we get from (8.8)

$$\|(\theta_i^1 \theta_j^2 u) \circ (\psi_i^1)^{-1}\|_E \leq M_1 \|(\theta_i^1 \theta_j^2 u) \circ (\psi_j^2)^{-1}\|_E.$$

Denote by $\tilde{\theta}_i^1(x)$ the restriction of $\theta_i^1(x)$ to $U_i^1 \cap U_j^2$. We have obviously $\theta_i^1 \theta_j^2 = \tilde{\theta}_i^1 \theta_j^2$. Hence

$$(\theta_i^1 \theta_j^2 u) \circ (\psi_j^2)^{-1} = (\tilde{\theta}_i^1 \theta_j^2 u) \circ (\psi_j^2)^{-1} = \tilde{\theta}_i^1 \circ (\psi_j^2)^{-1} \cdot \theta_j^2 u \circ (\psi_j^2)^{-1}$$
$$= \tilde{\theta}_i^2 \circ (\psi_i^1)^{-1} \cdot \theta_j^2 u \circ (\psi_j^2)^{-1}.$$

It follows that

$$\|(\theta_i^1 \theta_j^2 u) \circ (\psi_j^2)^{-1}\|_E \le \|\tilde{\theta}_i^2 \circ (\psi_i^1)^{-1}\|_{M(E)} \|\theta_j^2 u \circ (\psi_j^2)^{-1}\|_E$$
$$\le \kappa \|\tilde{\theta}_i^2 \circ (\psi_i^1)^{-1}\|_{C^\nu} \|\theta_j^2 u \circ (\psi_j^2)^{-1}\|_E$$

by virtue of (8.3). Obviously

$$\|\tilde{\theta}_i^2 \circ (\psi_i^1)^{-1}\|_{C^\nu} \le \|\theta_i^2 \circ (\psi_i^1)^{-1}\|_{C^\nu} \le K_1$$

according to condition (β). Thus we have obtained

$$\|(\theta_i^1 \theta_j^2 u) \circ (\psi_i^1)^{-1}\|_E \le \kappa K_1 M_1 \|(\theta_j^2 u) \circ (\psi_j^2)^{-1}\|_E. \qquad (8.9)$$

Let us return to (8.7). From (8.9) we get

$$\sum_{i'} \|(\theta_{i'}^1 \theta_j^2 u) \circ (\psi_{i'}^1)^{-1}\|_E^p \le N(\kappa K_1 M_1|)^p \|(\theta_j^2 u) \circ (\psi_j^2)^{-1}\|_E^p. \qquad (8.10)$$

Therefore (8.6), (8.7), and (8.10) imply

$$\sum_{i=1}^k \|(\theta_i^1 u) \circ (\psi_i^1)^{-1}\|_E^p \le c_1^p \sum_{j=1}^\infty \|(\theta_j^2 u) \circ (\psi_j^2)^{-1}\|_E^p,$$

where

$$c_1 = \kappa N K_1 M_1. \qquad (8.11)$$

Passing to the limit as $k \to \infty$, we obtain

$$\|u\|_{E_p^1(\Gamma)} \le c_1 \|u\|_{E_p^2(\Gamma)}.$$

Hence $u \in E_p^1(\Gamma)$.

Similarly we get for $u \in E_p^1(\Gamma)$ that

$$\|u\|_{E_p^2(\Gamma)} \le c_2 \|u\|_{E_p^1(\Gamma)},$$

and therefore $u \in E_p^2(\Gamma)$.

We have proved that $E_p^1(\Gamma) = E_p^2(\Gamma)$ and that the norms in these spaces are equivalent. Thus the theorem is proved for $1 \le p < \infty$.

Consider the case $p = \infty$. Let $u \in E_{\text{loc}}(\Gamma)$ and $\|u\|_{E_\infty^2(\Gamma)} < \infty$. From (8.5) we obtain

$$\|(\theta_i^1 u) \circ (\psi_i^1)^{-1}\|_E \le \sum_{j'} \|(\theta_i^1 \theta_{j'}^2 u) \circ (\psi_i^1)^{-1}\|_E.$$

From (8.9) it follows that
$$\|(\theta_i^1\theta_{j'}^2 u) \circ (\psi_i^1)^{-1}\|_E \leq \kappa K_1 M_1 \|(\theta_j^2 u) \circ (\psi_j^2)^{-1}\|_E \leq \kappa K_1 M_1 \|u\|_{E_\infty^2(\Gamma)}.$$

Hence
$$\|(\theta_i^1 u) \circ (\psi_i^1)^{-1}\|_E \leq c_1 \|u\|_{E_\infty^2(\Gamma)},$$

where c_1 is given by (8.11). Therefore
$$\|u\|_{E_\infty^1(\Gamma)} \leq c_1 \|u\|_{E_\infty^2(\Gamma)}.$$

Similarly
$$\|u\|_{E_\infty^2(\Gamma)} \leq c_2 \|u\|_{E_\infty^1(\Gamma)}.$$

The theorem is proved. □

Remark 8.5. In what follows $\Gamma = \partial\Omega$, where Ω is a domain satisfying Condition D (Section 2, Chapter 3), $E = W^{s,q}$, ($-\infty < s < \infty, 1 < q < \infty$). In this case condition (α) of the theorem is satisfied if $|s| + 1 \leq l$, where l is the number in Condition D. Indeed, from Condition D it follows that the functions $\psi_j^2(\psi_i^1)^{-1}$ and $\psi_i^1(\psi_j^2)^{-1}$ belong to C^l. It can be verified that the following proposition is true for $E = W^{s,q}$ and $|s| + 1 \leq l$.

Proposition 8.6. *Let $g : \mathbb{R}^m \to \mathbb{R}^m$ be a homeomorphism, $g \in C^l(\mathbb{R}^m)$, $g^{-1} \in C^l(\mathbb{R}^m)$. If $u \in E$, then $u \circ g \in E$, and $\|u \circ g\|_E \leq M\|u\|_E$, where M does not depend on u, and it depends continuously on $\|g\|_{C^l(\mathbb{R}^m)}$ and $\|g^{-1}\|_{C^l(\mathbb{R}^m)}$.*

Hence condition (α) of the theorem is satisfied. Moreover under the same assumptions on E and l, for any $f \in C^l(\mathbb{R}^m)$ and $u \in E$ we have $fu \in E$ and
$$\|fu\|_E \leq M_1 \|f\|_{C^l(\mathbb{R}^m)} \|u\|_E,$$

where M_1 does not depend on f and u. Therefore (8.3) is satisfied with $\nu = l$.

Consider now the case where Γ is a C^l manifold, where $l \geq 1$ is an integer. In this case the space D' cannot be used since the multiplication of elements from D' by functions from C^l is not defined. We can consider instead the spaces D_l and D'_l. Here D_l is the space of all functions $\phi \in C^l$ with compact supports. The convergence in D_l is defined as follows: $\phi_i \to 0$ in D_l if $D^\alpha \phi_i \to 0$ uniformly, $|\alpha| \leq l$, and there is a fixed compact set containing the supports of all ϕ_j. The space D'_l is defined as the space of all continuous functionals on D_l.

It is clear that the multiplication of elements of D'_l by functions $\phi \in C^l$ is defined. We can define the space $D'_l(\Gamma)$ similar to the definition of $D'(\Gamma)$ above. Using the space $D'_l(\Gamma)$ instead of $D'(\Gamma)$ we can give Definitions 8.1 and 8.2 and prove Theorem 8.3 for C^l manifolds exactly as it was done above. We consider the space $E = E(\mathbb{R}^m)$ such that $E \subset D'_l(\mathbb{R}^m)$, and we suppose that $D_l(\mathbb{R}^m) \subset M(E)$. In (8.3) we put $\nu = l$. In the definition of the space $E_p(\Gamma)$ we suppose that the partition of unity $\theta_i \in D_l(\Gamma)$.

The space $E_\infty(\Gamma)$ was defined in Definition 8.2. In what follows we need also other equivalent definitions. Suppose that $\Gamma = \partial\Omega$, where $\Omega \subset \mathbb{R}^n$ is a domain satisfying Condition D (Section 2, Chapter 3). We assume that the space E satisfies the condition of the Proposition 8.6 in the Remark 8.5, and the number ν in (8.3) is equal to l.

Let δ be the number in Condition D and $0 < \rho \leq \delta$. Consider a function $\eta(x) \geq 0$ such that $\eta \in C^\infty(\mathbb{R}^n)$, $\eta(x) = 1$ for $|x| < \rho/2$, $\eta(x) = 0$ for $|x| > \rho$. Let $\eta_z(x) = \eta(x - z)$, $z \in \mathbb{R}^n$.

Definition 8.7. $E_\infty(\Gamma)$ is the space of functions $u \in E_{\mathrm{loc}}(\Gamma)$ such that

$$\|u\|_{E_\infty(\Gamma)} = \sup_{z \in \Gamma} \|(\eta_z u) \circ \psi_z^{-1}\|_E < \infty,$$

where ψ_z is the function in Condition D which corresponds to the point $z \in \Gamma$.

In Theorem 8.4 we have proved that the space $E_p(\Gamma)$ ($1 \leq p \leq \infty$) does not depend on the choice of equivalent coverings of Γ. We now specify the equivalence class of coverings, which we shall use, by pointing out its representative: covering of Γ by cubes with the sides equal 2 and the centers at lattice points with the increment equal to 1.

Proposition 8.8. *Suppose that in Definition 8.2,*

$$\sup_i \|\theta_i \circ \psi_i^{-1}\|_{C^l(\mathbb{R}^{n-1})} < \infty. \tag{8.12}$$

Then the spaces $E_\infty(\Gamma)$ in Definitions 8.2 and 8.7 coincide.

Proof. Denote by $E_\infty^2(\Gamma)$ and $E_\infty^4(\Gamma)$ the spaces $E_\infty(\Gamma)$ in Definitions 8.2 and 8.7, respectively. Let $u \in E_\infty^4(\Gamma)$. Let us construct a covering of Γ and a partition of unity as follows. We cover Γ by balls B_j of the radii $\rho/2$ and the centers at the points $x_j \in \Gamma$ ($i = 1, 2, \dots$). Denote by $\tilde{B}_j$ the balls of the radii ρ and the centers at x_j. We suppose that the points x_j are chosen such that the covering by the balls $\tilde{B}_j$ belongs to the equivalence class under consideration. Denote by N such number that for any i there is no more than N of j such that the intersection $\tilde{B}_i \cap \tilde{B}_j$ is not empty. Let

$$\omega(x) = \sum_j \eta_{x_j}(x). \tag{8.13}$$

Obviously, $\mathrm{supp}\,\eta_{x_j} \subset \tilde{B}_j$. Hence for any $x \in \Gamma$ we have no more than N nonzero terms in sum (8.13). Moreover, $\omega(x) \geq 1$, $x \in \Gamma$.

Set

$$\theta_j(x) = \frac{\eta_{x_j}(x)}{\omega(x)}. \tag{8.14}$$

8. The space $E_p(\Gamma)$

Clearly, this is a partition of unity which corresponds to the covering by $\tilde{B}_j$. We have
$$\|(\theta_j u) \circ \psi_j^{-1}\|_E = \|((1/\omega) \circ \psi_j^{-1})((\eta_{x_j} u) \circ \psi_j^{-1})\|_E$$
$$\leq \|(1/\omega) \circ \psi_j^{-1}\|_{M(E)} \|(\eta_{x_j} u) \circ \psi_j^{-1}\|_E,$$
where $M(E)$ is the space of multipliers. By Condition D and (8.3) we have $\|(1/\omega) \circ \psi_j^{-1}\|_{M(E)} \leq K$, where the constant K does not depend on j. Hence
$$\|(\theta_i u) \circ \psi_j^{-1}\|_E \leq K \|(\eta_{x_j} u) \circ \psi_j^{-1}\|_E \leq K \|u\|_{E_\infty^4(\Gamma)}.$$

It follows that $\|u\|_{E_\infty^2(\Gamma)} \leq K \|u\|_{E_\infty^4(\Gamma)}$ and $u \in E_\infty^2(\Gamma)$. We have proved that $E_\infty^4(\Gamma) \subset E_\infty^2(\Gamma)$.

Now, let $u \in E_\infty^2(\Gamma)$ and θ_i be the partition of unity in Definition 8.2. For any $z \in \Gamma$ we have
$$(\eta_z u) \circ \psi_z^{-1} = \sum_{i'} (\eta_z \theta_{i'} u) \circ \psi_z^{-1}, \tag{8.15}$$
where i' are all of i for which $\operatorname{supp} \theta_i$ has nonempty intersection with $\operatorname{supp} \eta_z$. By the assumption above on the choice of the equivalent class of coverings, it is obvious that there exists a number N independent of z such that the number of i' is less than N. It follows from (8.15) that
$$\|(\eta_z u) \circ \psi_z^{-1}\|_E \leq \sum_{i'} \|(\eta_z \theta_{i'} u) \circ \psi_z^{-1}\|_E. \tag{8.16}$$

As in the proof of Theorem 8.4, we get
$$\|(\eta_z \theta_{i'} u) \circ \psi_z^{-1}\|_E \leq M \|(\theta_{i'} u) \circ \psi_{i'}^{-1}\|_E,$$
where the constant M does not depend on z and i'. Since $\|(\theta_{i'} u) \circ \psi_{i'}^{-1}\|_E \leq \|u\|_{E_\infty^2(\Gamma)}$, we obtain from (8.16) that
$$\|(\eta_z u) \circ \psi_z^{-1}\|_E \leq NM \|u\|_{E_\infty^2(\Gamma)}.$$

Hence $\|u\|_{E_\infty^4(\Gamma)} \leq NM \|u\|_{E_\infty^2(\Gamma)}$. Therefore $u \in E_\infty^4(\Gamma)$. It follows that $E_\infty^2(\Gamma) \subset E_\infty^4(\Gamma)$. The proposition is proved. □

It follows from the proof that this result remains true if instead of the partition of unity in Definition 8.2 we take an arbitrary system of functions $\theta_i \in D_l(\Gamma)$, $\operatorname{supp} \theta_i \subset U_i$ and such that (8.12) is satisfied. We now give another definition of the space $E_\infty(\Gamma)$ (Definition 8.9) and prove its equivalence to the previous ones. Let $G \subset R^n$ be a bounded domain, and the intersection $\Gamma \cap G$ be not empty. Denote by $D_l'(\Gamma \cap G)$ the restriction of $D_l'(\Gamma)$ to $\Gamma \cap G$.

Definition 8.9. The function $u \in E_\infty(\Gamma \cap G)$ if and only if $u \in D_l'(\Gamma \cap G)$ and there exists $v \in E_\infty(\Gamma)$ such that the restriction of v to $D_l'(\Gamma \cap G)$ coincides with u. The norm in $E_\infty(\Gamma \cap G)$ is given by
$$\|u\|_{E_\infty(\Gamma \cap G)} = \inf \|v\|_{E_\infty(\Gamma)},$$

where the infimum is taken over all v, for which u is the restriction of v to $D'_l(\Gamma \cap G)$.

Definition 8.10. The space $\tilde{E}_\infty(\Gamma)$ is the set of all $u \in E_{\text{loc}}(\Gamma)$ such that

$$\|u\|_{\tilde{E}_\infty(\Gamma)} = \sup_{y \in \Gamma} \|u_y\|_{E_\infty(\Gamma \cap G_y)} < \infty, \tag{8.17}$$

where u_y is the restriction of u to $\Gamma \cap G_y$, G_y is the shifted domain: the characteristic function of G_y is $\chi(x - y)$, where $\chi(x)$ is the characteristic function of G.

In what follows it is supposed that G contains the origin.

Proposition 8.11.

$$\tilde{E}_\infty(\Gamma) = E_\infty(\Gamma).$$

Proof. Let $u \in E_\infty(\Gamma)$, u_y be the restriction of u to $\Gamma \cap G_y$ and θ_j be the partition of unity (8.14). Further, let j' be all of j for which $\operatorname{supp} \theta_j$ has a nonempty intersection with $\Gamma \cap G_y$. According to the choice of x_j in (8.14) it is clear that the number of j' is less than a number N which does not depend on y. The function $\sum_{j'} \theta_{j'} u$ is an extension of u_y. Hence

$$\|u_y\|_{E_\infty(\Gamma \cap G_y)} \le \|\sum_{j'} \theta_{j'} u\|_{E_\infty(\Gamma)} \le K\|u\|_{E_\infty(\Gamma)},$$

where K does not depend on y. Therefore $\|u\|_{\tilde{E}_\infty(\Gamma)} \le K\|u\|_{E_\infty(\Gamma)}$. We have proved that $E_\infty(\Gamma) \subset \tilde{E}_\infty(\Gamma)$.

Now let $u \in \tilde{E}_\infty(\Gamma)$. This means that

$$u \in E_{\text{loc}}(\Gamma) \quad \text{and} \quad \sup_{y \in \Gamma} \|u_y\|_{E_\infty(\Gamma \cap G_y)} < \infty.$$

By Definition 8.9, there exists an extension $v \in E_\infty(\Gamma)$ of u_y such that

$$\|v\|_{E_\infty(\Gamma)} \le 2\|u_y\|_{E_\infty(\Gamma \cap G_y)}. \tag{8.18}$$

Since the space $E_\infty(\Gamma)$ does not depend on the choice of ρ, we can suppose that ρ is taken so small that $\operatorname{supp} \eta_y \in G_y$. We have

$$\|\eta_y v\|_{E_\infty(\Gamma)} \le K_1 \|v\|_{E_\infty(\Gamma)}, \tag{8.19}$$

where K_1 does not depend on y.

Since $\operatorname{supp} \eta_y \in G_y$, we have

$$\eta_y v = \eta_y u_y, \tag{8.20}$$

9. Spaces in unbounded domains

where $\eta_y u_y$ is extended by zero outside $\Gamma \cap G_y$. From (8.18), (8.19) and (8.20),

$$\|\eta_y u_y\|_{E_\infty(\Gamma)} = \|\eta_y v\|_{E_\infty(\Gamma)} \leq K_1 \|v\|_{E_\infty(\Gamma)}$$
$$\leq 2K_1 \|u_y\|_{E_\infty(\Gamma \cap G_y)} \leq 2K_1 \|u\|_{\tilde{E}_\infty(\Gamma)}.$$

But $\eta_y u_y = \eta_y u$. Therefore

$$\|\eta_y u\|_{E_\infty(\Gamma)} \leq 2K_1 \|u\|_{\tilde{E}_\infty(\Gamma)}. \tag{8.21}$$

By Definition 8.7, we have

$$\sup_{y \in \Gamma} \|\eta_y u\|_{E_\infty(\Gamma)} = \sup_{y \in \Gamma} \sup_{z \in \Gamma} \|(\eta_z \eta_y u) \circ \psi_z^{-1}\|_E. \tag{8.22}$$

On the other hand, since the space $E_\infty(\Gamma)$ in Definition 8.7 does not depend on the choice of η, we can take $\eta^2(x)$ instead of $\eta(x)$. Hence for some constant $\kappa > 0$ we have

$$\kappa \|u\|_{E_\infty(\Gamma)} \leq \sup_{z \in \Gamma} \|\eta_z^2 u \circ \psi_z^{-1}\|_E \leq \sup_{y \in \Gamma} \sup_{z \in \Gamma} \|(\eta_y \eta_z u) \circ \psi_z^{-1}\|_E = \sup_{y \in \Gamma} \|\eta_y u\|_{E_\infty(\Gamma)}$$

by (8.22). It follows from (8.21) that

$$\kappa \|u\|_{E_\infty(\Gamma)} \leq 2K_1 \|u\|_{\tilde{E}_\infty(\Gamma)}.$$

Therefore $u \in E_\infty(\Gamma)$. We have proved that $\tilde{E}_\infty(\Gamma) \subset E_\infty(\Gamma)$. The proposition is proved. □

Remark 8.12. The formulation of Proposition 8.11 and its proof remain the same if instead of (8.17) we suppose that

$$\|u\|_{\tilde{E}_\infty(\Gamma)} = \sup \|u_y\|_{E_\infty(\Gamma \cap G_y)} < \infty,$$

where the supremum is taken over all $y \in \Omega$ such that the $\Gamma \cap G_y$ is not empty.

9 Spaces in unbounded domains

Let Ω be a domain in R^n, E be a space in Definition 1.1. Instead of the estimate

$$\|fu\|_E \leq \|f\|_M \|u\|_E$$

for $f \in D$, $u \in E$, we have a similar estimate for the space $E(\Omega)$:

$$\|fu\|_{E(\Omega)} \leq \|f\|_M \|u\|_{E(\Omega)}, \tag{9.1}$$

where $f \in D$, $u \in E(\Omega)$. Indeed, let u^c be an extension of u to E. Then fu^c is an extension of fu to E. Hence

$$\|fu\|_{E(\Omega)} \leq \|fu^c\|_E \leq \|f\|_M \|u\|_E.$$

If we take the infimum over all extensions u^c, we obtain (9.1).

Definition 9.1. The space $E(\Omega)$ is defined as the space of the generalized functions from D'_Ω that are restrictions to Ω of generalized functions from E. The norm in $E(\Omega)$ is defined as
$$\|u\|_{E(\Omega)} = \inf \|u^c\|_E,$$
where the infimum is taken over all $u^c \in E$ whose restriction to Ω coincide with u.

Definition 9.2. The space $E^1_\infty(\Omega)$ is defined as the set of the generalized functions from D'_Ω that are restrictions to Ω of generalized functions from E_∞. The norm in $E^1_\infty(\Omega)$ is defined as
$$\|u\|_{E^1_\infty(\Omega)} = \inf \|u^c\|_{E_\infty},$$
where the infimum is taken over all $u^c \in E_\infty$ whose restriction to Ω coincide with u.

We will also give another definition of the space $E_\infty(\Omega)$. Let $\eta_y(x)$ be the same as in Definition 6.1.

Definition 9.3. The space $E^2_\infty(\Omega)$ is defined as the set of such generalized functions from D'_Ω that $\eta_y u \in E(\Omega)$ for all $y \in \mathbb{R}^n$ and
$$\sup_{y \in \mathbb{R}^n} \|\eta_y u\|_{E(\Omega)} < \infty.$$
If $\operatorname{supp} \eta_y \cap \Omega = \emptyset$, then $\eta_y u = 0$. The norm in $E^2_\infty(\Omega)$ is given by the equality
$$\|u\|_{E^2_\infty(\Omega)} = \sup_{y \in \mathbb{R}^n} \|\eta_y u\|_{E(\Omega)}.$$

Definition 9.4. Let $\{\phi_i\}$ be a system of functions satisfying Condition 2.2. The space $E^3_\infty(\Omega)$ is defined as the set of generalized functions $u \in D'_\Omega$ such that $\phi_i u \in E(\Omega)$ for all i and
$$\sup_i \|\phi_i u\|_{E(\Omega)} < \infty.$$
The norm in $E^3_\infty(\Omega)$ is given by the equality
$$\|u\|_{E^3_\infty(\Omega)} = \sup_i \|\phi_i u\|_{E(\Omega)}.$$

Definition 9.5. Denote by $B_{y,r}$ the ball $\{x : |x-y| < r\}$. The space $E^4_\infty(\Omega)$ is defined as the set of generalized functions $u \in D'_\Omega$ such that for any $y \in \Omega, \Omega \cap B_{y,r} \neq \emptyset$ the restriction of u to $\Omega \cap B_{y,r}$ satisfies the condition
$$\|u\|_{E(\Omega \cap B_{y,r})} \leq M$$
with M independent of u. The norm in $E^4_\infty(\Omega)$ is defined by the equality
$$\|u\|_{E^3_\infty(\Omega)} = \sup \|u\|_{E(\Omega \cap B_{y,r})},$$
where the supremum is taken over all $y \in \Omega$ such that $\Omega \cap B_{y,r} \neq \emptyset$.

9. Spaces in unbounded domains

Proposition 9.6. *Let $E^1_\infty(\Omega)$ and $E^3_\infty(\Omega)$ be the space defined above. Then $E^3_\infty(\Omega) \subset E^1_\infty(\Omega)$ and for any $u \in E^3_\infty(\Omega)$ we have*

$$\|u\|_{E^1_\infty(\Omega)} \leq M\|u\|_{E^3_\infty(\Omega)},$$

where M is a constant independent of u.

Proof. Let $u \in E^3_\infty(\Omega)$. Then $\phi_i u \in E(\Omega)$ for any i. By the definition of $E(\Omega)$, there exists $u_i \in E$ such that

$$\langle u_i, \phi \rangle = \langle \phi_i u, \phi \rangle \quad \text{for any} \phi : \operatorname{supp} \phi \in \Omega \tag{9.2}$$

and

$$\|u_i\|_E \leq 2\|\phi_i u\|_{E(\Omega)}. \tag{9.3}$$

We can suppose that $\{\phi_i\}$ is a partition of unity. Let $\{\psi_i\}$ be another system of functions satisfying Condition 2.2 such that $\psi_i(x) = 1$ for $x \in \operatorname{supp} \phi_i$. Then $\phi_i(x)\psi_i(x) = \phi_i(x)$ for any i and x. Set

$$u^c = \sum_{i=1}^\infty \psi_i u_i. \tag{9.4}$$

Obviously, for $\phi \in D$ the functional u^c is defined. Indeed, by definition

$$\langle u^c, \phi \rangle = \sum_{i=1}^\infty \langle \psi_i u_i, \phi \rangle = \sum_{i=1}^\infty \langle u_i, \psi_i \phi \rangle.$$

But this sum has only a finite number of nonzero terms. For definiteness we suppose that the supports of ϕ_i and ψ_i are cubes of a lattice.

We now prove that u^c is an extension of u. Indeed, if $\operatorname{supp} \phi \subset \Omega$, we have $\langle u^c, \phi \rangle = \sum_{i=1}^\infty \langle u_i, \psi_i \phi \rangle$. Since $\operatorname{supp} \psi_i \phi \subset \Omega$, we have, from (9.2),

$$\langle u_i, \psi_i \phi \rangle = \langle \phi_i u, \psi_i \phi \rangle = \langle u, \phi_i \psi_i \phi \rangle = \langle u, \phi_i \phi \rangle.$$

Hence

$$\langle u^c, \phi \rangle = \sum_{i=1}^\infty \langle u, \phi_i \phi \rangle = \langle u, \phi \rangle$$

since $\sum_{i=1}^\infty \phi_i(x) = 1$. It follows that u^c is an extension of u.

We will prove that $u^c \in E_\infty$. Indeed, we have

$$\phi_k u^c = \sum_i \phi_k \psi_i u_i = \sum_{i'} \phi_k \psi_{i'} u_{i'}, \tag{9.5}$$

where i' denotes all the subscripts for which $\operatorname{supp} \phi_k$ and $\operatorname{supp} \psi_i$ have a nonempty intersection. Let the number of such i' be no more than N, which does not depend on i and k. From (9.5) follows the estimate

$$\|\phi_k u^c\|_E \leq \sum_{i'} \|\phi_k \psi_{i'} u_{i'}\|_E. \tag{9.6}$$

By Definition 1.1 of the space E we have $\|\phi_k\psi_i u_i\|_E \leq M\|\phi_k\psi_i\|_M \|u_i\|_E$. By Condition 2.5 $\|\phi_k\psi_i\|_M \leq K$, where K is independent of k and i. Hence $\|\phi_k\psi_i u_i\|_E \leq MK\|u_i\|_E$. From (9.6) we obtain $\|\phi_k u^c\|_E \leq MK\sum_{i'}\|u_{i'}\|_E$. The inequality (9.3) implies $\|\phi_k u^c\|_E \leq 2MK\sum_{i'}\|\phi_{i'} u\|_{E(\Omega)}$. From Definition 9.4,

$$\|\phi_k u^c\|_E \leq 2MKN\|u\|_{E^3_\infty(\Omega)}.$$

Since k is arbitrary, by the definition of E_∞ we get $u^c \in E_\infty$ and

$$\|u^c\|_{E_\infty} = \sup_k \|\phi_k u^c\|_E \leq 2MKN\|u\|_{E^3_\infty(\Omega)}.$$

Since u is a restriction of u^c to Ω, we obtain that $u \in E^1_\infty(\Omega)$. Moreover,

$$\|u\|_{E^1_\infty(\Omega)} \leq \|u^c\|_{E_\infty} \leq 2MKN\|u\|_{E^3_\infty(\Omega)} = M_0\|u\|_{E^3_\infty(\Omega)}. \qquad (9.7)$$

Thus the proposition is proved for the special choice of ϕ_i. We can write (9.7) in the form

$$\|u\|_{E^1_\infty(\Omega)} \leq M_0 \sup_k \|\phi_k u\|_{E(\Omega)}. \qquad (9.8)$$

We now prove the proposition for any system of functions $\{\omega_i\}$ satisfying Definition 2.3 and equivalent to the system of functions $\{\phi_i\}$, which is considered above. We have

$$\phi_k u = \phi_k \sum_i \frac{\omega_i}{\omega} u = \sum_{i'} \phi_k \frac{\omega_{i'}}{\omega} u,$$

where i' are those subscripts i for which $\operatorname{supp}\omega_i$ has a nonempty intersection with $\operatorname{supp}\phi_k$. The number of such i' is no more than N by the definition of equivalence of systems of functions.

We now have

$$\|\phi_k u\|_{E(\Omega)} = \sum_i \|\frac{\phi_k}{\omega}\omega_{i'} u\|_{E(\Omega)} \leq K\sum_{i'}\|\omega_{i'} u\|_{E(\Omega)}.$$

Hence

$$\|\phi_k u\|_{E(\Omega)} \leq KN\|u\|_{E^3_\infty(\Omega)}, \quad \sup_k \|\phi_k u\|_{E(\Omega)} \leq KN\|u\|_{E^3_\infty(\Omega)}.$$

Therefore by the results of the first part of the proof we conclude that $u \in E^1_\infty(\Omega)$, and from (9.8) we obtain

$$\|u\|_{E^1_\infty(\Omega)} \leq M_0 KN\|u\|_{E^3_\infty(\Omega)}.$$

The proposition is proved. $\square$

Proposition 9.7. *Let $E^1_\infty(\Omega)$ and $E^2_\infty(\Omega)$ be the space in Definitions 9.2 and 9.3, respectively. Then $E^1_\infty(\Omega) \subset E^2_\infty(\Omega)$ and for any $u \in E^1_\infty(\Omega)$ we have*

$$\|u\|_{E^2_\infty(\Omega)} \leq M\|u\|_{E^1_\infty(\Omega)},$$

where M is a constant independent of u.

9. Spaces in unbounded domains

Proof. Let $u \in E^1_\infty(\Omega)$. Then there exists $u^c \in E_\infty$ such that

$$\langle u, \phi \rangle = \langle u^c, \phi \rangle \text{ for any } \phi \in D, \operatorname{supp} \phi \subset \Omega. \tag{9.9}$$

By the definition of E_∞ we have $\eta_y u^c \in E$ for any $y \in \mathbb{R}^n$, and

$$\|u^c\|_{E_\infty} = \sup_{y \in \mathbb{R}^n} \|\eta_y u^c\|_E < \infty. \tag{9.10}$$

From (9.9) it follows that $\langle \eta_y u, \phi \rangle = \langle \eta_y u^c, \phi \rangle$ for any $\phi \in D$, $\operatorname{supp} \phi \subset \Omega$. Hence $\eta_y u^c$ is an extension of $\eta_y u$ to E_∞. By the definition of $E(\Omega)$ we have $\eta_y u \in E(\Omega)$ and $\|\eta_y u\|_{E(\Omega)} \le \|\eta_y u^c\|_E \le \|u^c\|_{E_\infty}$. Therefore $u \in E^2_\infty(\Omega)$ and

$$\|u\|_{E^2_\infty(\Omega)} = \sup_{y \in \Omega} \|\eta_y u\|_{E(\Omega)} \le \|u^c\|_{E_\infty}.$$

The left-hand side here does not depend on the extension u^c. Hence

$$\|u\|_{E^2_\infty(\Omega)} \le \inf_{u^c} \|u^c\|_{E_\infty} = \|u\|_{E^1_\infty(\Omega)}.$$

The proposition is proved. $\square$

Proposition 9.8. *Let $E^2_\infty(\Omega)$ and $E^4_\infty(\Omega)$ be the space in Definitions 9.3 and 9.5, respectively. Then $E^2_\infty(\Omega) \subset E^4_\infty(\Omega)$ and for any $u \in E^2_\infty(\Omega)$ we have*

$$\|u\|_{E^4_\infty(\Omega)} \le M \|u\|_{E^2_\infty(\Omega)},$$

where M is a constant independent of u.

Proposition 9.9. *If $\{\phi^1_i\}$ and $\{\phi^2_i\}$ are two systems of functions equivalent in the sense of Definition 2.3, then the spaces $E^3_\infty(\Omega)$ corresponding to them coincide, and their norms are equivalent.*

Proposition 9.10. *Let $E^2_\infty(\Omega)$ and $E^3_\infty(\Omega)$ be the spaces in Definitions 9.3 and 9.4, respectively. Then $E^2_\infty(\Omega) = E^3_\infty(\Omega)$ and their norms are equivalent.*

Corollary 9.11. *The space $E^2_\infty(\Omega)$ does not depend on the choice of the numbers a_1 and a_2 in Definition 6.1.*

Proposition 9.12. *Let $E^2_\infty(\Omega)$ and $E^4_\infty(\Omega)$ be the spaces in Definitions 9.3 and 9.5, respectively. Then $E^4_\infty(\Omega) \subset E^2_\infty(\Omega)$ and for any $u \in E^4_\infty(\Omega)$ we have*

$$\|u\|_{E^2_\infty(\Omega)} \le M \|u\|_{E^4_\infty(\Omega)},$$

where M is a constant independent of u.

Theorem 9.13. *All definitions of the space $E_\infty(\Omega)$ are equivalent, i.e.,*

$$E^1_\infty(\Omega) = E^2_\infty(\Omega) = E^3_\infty(\Omega) = E^4_\infty(\Omega)$$

and their norms are equivalent.

10 Dual spaces

Let $E(\mathbb{R}^n)$ be a Banach space satisfying the conditions of Section 1, and $E^*(\mathbb{R}^n)$ be its dual. As above, we will denote them by E and E^*, respectively. We suppose that $D \subset E$, the inclusion being in the algebraic and topological sense, and that D is dense in E. Then $E^* \subset D'$, and this inclusion should be also understood in the algebraic and topological sense. We can define $(E^*)_q$ as it is done in Sections 3 and 4. For example the norm in the space $(E^*)_\infty$ is given by

$$\|v\|_{(E^*)_\infty} = \sup_i \|\phi_i v\|_{E^*}, \tag{10.1}$$

where ϕ_i is a partition of unity. The norm in the right-hand side of (10.1) is the norm of functionals from E^*.

Lemma 10.1. $(E_p)^* \subset E^*_{\text{loc}}$.

Proof. Let $v \in (E_p)^*$, $\phi \in D$. We should verify that $v\phi \in E^*$. For any $u \in E$, $\phi u \in E_p$. Therefore

$$|\langle \phi v, u \rangle| = |\langle v, \phi u \rangle| \leq \|v\|_{(E_p)^*} \|\phi u\|_{E_p} \leq M \|v\|_{(E_p)^*} \|u\|_E.$$

The lemma is proved. $\square$

In what follows we say that two normed spaces are equal or coincide if they are linearly isomorphic and their norms are equivalent.

Theorem 10.2. *The spaces $(E^*)_\infty$ and $(E_1)^*$ coincide.*

Proof. Let $v \in (E_1)^*$. Then for any $u \in E_1$,

$$\langle v, u \rangle \leq \|v\|_{(E_1)^*} \|u\|_{E_1}.$$

Since $v \in E^*_{\text{loc}}$ and $u \in E$, then $\langle \phi_i v, u \rangle$ is defined and

$$|\langle \phi_i v, u \rangle| = |\langle v, \phi_i u \rangle| \leq \|v\|_{(E_1)^*} \|\phi_i u\|_{E_1} \leq M \|v\|_{(E_1)^*} \|u\|_E.$$

Here $\{\phi_i\}$ is a partition of unity, $\sup_i \|\phi_i\|_M < \infty$.
Therefore

$$\|\phi_i v\|_{E^*} \leq M \|v\|_{(E_1)^*}.$$

Consequently,

$$\|v\|_{(E^*)_\infty} \leq M \|v\|_{(E_1)^*}.$$

Suppose that $v \in (E^*)_\infty$. Then $v \in E^*_{\text{loc}}$. Let $u \in E_1$, $u_k = \sum_{i=1}^k \phi_i u$. Then $u_k \in E$, and

$$|\langle v, u_k \rangle| = |\langle v, \sum_{i=1}^k \phi_i u \rangle| \leq \sum_{i=1}^k |\langle v, \phi_i u \rangle| = \sum_{i=1}^k |\langle \phi_i v, \psi_i u \rangle| \leq \sum_{i=1}^k \|\phi_i v\|_{E^*} \|\psi_i u\|_E$$

$$\leq \|v\|_{(E^*)_\infty} \sum_{i=1}^\infty \|\psi_i u\|_E \leq M \|v\|_{(E^*)_\infty} \|u\|_{E_1}.$$

Here $\psi_i \in D$, $\psi_i = 1$ in $\operatorname{supp} \phi_i$. We suppose that the system of functions ψ_i satisfies Condition 2.2. We can pass to the limit in the last estimate as $k \to \infty$. Therefore v can be considered as a functional on E_1, and

$$\|v\|_{(E_1)^*} \leq M \|v\|_{(E^*)_\infty}.$$

The theorem is proved. $\square$

We note that functionals from both spaces $(E^*)_\infty$ and $(E_1)^*$ are considered in Theorem 10.2 on functions from E_1.

Theorem 10.3. *Let $1/p + 1/q = 1$, $1 < p, q < \infty$. Then $(E^*)_p \subset (E_q)^*$.*

Proof. Let $v \in (E^*)_p$. We show that $v \in (E_q)^*$. For any $u \in D$ we have

$$|\langle v, u \rangle| = |\langle v, \sum_{i=1}^\infty \phi_i u \rangle| \leq \sum_{i=1}^\infty |\langle v, \phi_i u \rangle| = \sum_{i=1}^\infty |\langle \psi_i v, \phi_i u \rangle| \leq \sum_{i=1}^\infty \|\psi_i v\|_{E^*} \|\phi_i u\|_E$$

$$\leq \left(\sum_{i=1}^\infty \|\psi_i v\|_{E^*}^p \right)^{1/p} \left(\sum_{i=1}^\infty \|\phi_i u\|_E^q \right)^{1/q} \leq M \|v\|_{(E^*)_p} \|u\|_{E_q}.$$

Here $\psi_i \in D$, $\psi_i(x) = 1$ in $\operatorname{supp} \phi_i(x)$. Since D is dense in E_q, this estimate remains valid for all $u \in E_q$. Therefore

$$\|v\|_{(E_q)^*} \leq M \|v\|_{(E^*)_p}.$$

The theorem is proved. $\square$

Lemma 10.4. *Let $\phi \in (E_\infty)^*$, $u_n = \sum_{i=1}^n u \theta_i$, where $u \in E_\infty$, θ_i is a partition of unity. Then there exists the limit $\lim_{n \to \infty} \phi(u_n)$.*

Proof. We have

$$\|u_n\|_{E_\infty} = \sup_j \|u_n \theta_j\|_E = \sup_j \| \left(\sum_{i=1}^n u \theta_i \right) \theta_j \|_E$$

$$\leq \sup_j \left(\sum_{i: \operatorname{supp} \theta_i \cap \operatorname{supp} \theta_j \neq \emptyset} \|u \theta_i \theta_j\|_E \right) \leq MN \sup_j \|u \theta_j\| = MN \|u\|_{E_\infty}.$$

Suppose that the limit $\phi(u_n)$ does not exist. Then there exist two subsequences u_{n_k} and u_{n_m} such that

$$\phi(u_{n_k}) \to C_1, \quad \phi(u_{n_m}) \to C_2, \quad C_1 \neq C_2.$$

We will construct a bounded sequence in E_∞ such that the functional ϕ will be unbounded on it. This contradiction will prove the existence of the limit.

Without loss of generality we can assume that $C_1 > C_2$. For all k and m sufficiently large,
$$\phi(u_{nk}) \geq C_1 - \epsilon, \quad \phi(u_{nm}) \leq C_2 + \epsilon.$$
For $\epsilon \leq (C_1 - C_2)/4$,
$$\phi(u_{nk} - u_{nm}) \geq \frac{C_1 - C_2}{2} (= a > 0).$$
We choose k and m such that this estimate is satisfied and write $v_1 = u_{nk} - u_{nm}$. We note that
$$u_{nk} - u_{nm} = \sum_{i=n_m}^{n_k} u\theta_i.$$
Therefore the support of the function v_1 is inside $\bigcup_{i=n_m}^{n_k} \operatorname{supp} \theta_i$.

Similarly, we choose other values of k and m and define the function v_2, $\phi(v_2) \geq a$. Moreover, if the new values k and m are sufficiently large, then $\operatorname{supp} v_1 \cap \operatorname{supp} v_2 = \emptyset$. In the same way, we construct other functions v_l such that their supports do not intersect and $\phi(v_l) \geq a$. We finally put $w_j = \sum_{l=1}^{j} v_l$. Similar to the sequence u_n, the sequence w_j is uniformly bounded in E_∞. At the same time $\phi(w_j) \to \infty$. This contradicts the assumption that $\phi \in (E_\infty)^*$. The lemma is proved. $\square$

Consider a functional ϕ from $(E_\infty)^*$. We define a new functional $\tilde{\phi}$ as follows. For any function $u \in E_\infty$ with a bounded support we put
$$\tilde{\phi}(u) = \phi(u).$$
For any function $u \in E_\infty$, we put
$$\tilde{\phi}(u) = \lim_{n \to \infty} \phi(\sum_{i=1}^{n} u\theta_i).$$
Thus $\tilde{\phi}$ is a weak limit of $\sum_{i=1}^{n} \theta_i \phi$ in $(E_\infty)^*$. From Lemma 10.4 it follows that this limit exists. It is easy to verify that $\tilde{\phi}$ is a bounded linear functional on E_∞.

Let $\phi_0 = \phi - \tilde{\phi}$. Then $\phi_0(u) = 0$ for any function u with a bounded support. Thus we have the following result.

Lemma 10.5. *The space $(E_\infty)^*$ can be represented as a direct sum of two subspaces, $(E_\infty)^*_0$ and $(E_\infty)^*_\omega$, where $(E_\infty)^*_0$ consists of functionals equal to 0 on all functions with bounded supports, and $(E_\infty)^*_\omega$ consists of the functionals $\tilde{\phi}$ constructed above.*

Proof. It remains to prove that $(E_\infty)^*_\omega$ and $(E_\infty)^*_0$ are closed. Let $v_k \in (E_\infty)^*_\omega$, $v_k \to v$ in $(E_\infty)^*$. We have
$$\langle v_k, u \rangle = \lim_{n \to \infty} \langle v_k, u_n \rangle, \quad \forall u \in E_\infty, \qquad (10.2)$$

10. Dual spaces

where $u_n = \sum_{i=1}^{n} \theta_i u$ (see Lemma 10.8 below). We prove that we can pass to the limit with respect to k in the right-hand side of (10.2). Indeed we have

$$|\langle v_k - v, u_n \rangle| \leq \|v_k - v\|_{(E_\infty)^*} \|u_n\|_{E_\infty} \leq M\|v_k - v\|_{(E_\infty)^*}$$

since $\|u_n\|_{E_\infty}$ is bounded. Hence

$$|\lim_{n\to\infty} \langle v_k - v, u_n \rangle| \leq M\|v_k - v\|_{(E_\infty)^*} \to 0$$

as $k \to \infty$. Passing to the limit with respect to k in (10.2), we obtain

$$\langle v, u \rangle = \lim_{n\to\infty} \langle v, u_n \rangle, \quad \forall u \in E_\infty.$$

Therefore $v \in (E_\infty)^*_\omega$. The completeness of the space $(E_\infty)^*_\omega$ is proved. It can be easily verified that the second subspace is also closed. The lemma is proved. $\square$

If $\phi_0 \in (E_\infty)^*_0$ and $\theta \in D$, then $\theta\phi_0 = 0$ is an element of $(E_\infty)^*$. Therefore, if $\phi = \phi_0 + \phi_1$, then $\theta\phi = \theta\phi_1$ (see the next lemma).

Lemma 10.6. *If $\phi \in (E_\infty)^*$, then $\phi \in (E^*)_1$ and*

$$\|\phi\|_{(E^*)_1} \leq M \|\phi\|_{(E_\infty)^*}, \tag{10.3}$$

where M is a constant independent of ϕ.

Proof. We have $\theta_i \phi \in E^*$ for $\theta_i \in D$ and

$$\|\theta_i \phi\|_{E^*} = \sup_{u \in E,\ \|u\|_E = 1} |\theta_i \phi(u)|.$$

Hence there exists $u_i \in E$ such that $\|\theta_i \phi\|_{E^*} \leq 2 |\theta_i \phi(u_i)| = 2\, \theta_i \phi(\sigma_i u_i)$, where $|\sigma_i| = 1$. Therefore

$$\sum_{i=1}^{m} \|\theta_i \phi\|_{E^*} \leq 2\phi\left(\sum_{i=1}^{m} \theta_i \sigma_i u_i\right). \tag{10.4}$$

For any θ_k we have

$$\left\|\sum_{i=1}^{m} \theta_k \theta_i \sigma_i u_i\right\|_E \leq \sum_{i=1}^{m} \|\theta_k \theta_i u_i\|_E \leq \sum_{i'} \|\theta_k \theta_{i'} u_{i'}\|_E,$$

where i' are all those numbers i for which $\operatorname{supp} \theta_i \cap \operatorname{supp} \theta_k \neq \varnothing$. It follows that

$$\left\|\sum_{i=1}^{m} \theta_k \theta_i \sigma_i u_i\right\|_E \leq NK^2, \tag{10.5}$$

where N is the number from Condition 2.2 and $K = \sup_i \|\theta_i\|_{M(E)}$. Inequality (10.5) implies $\|\sum_{i=1}^{m} \theta_i \sigma_i u_i\|_{E_\infty} \leq NK^2$. From (10.4) we obtain $\sum_{i=1}^{m} \|\theta_i \phi\|_{E^*} \leq 2NK^2 \|\phi\|_{(E_\infty)^*}$ and (10.3) follows. The lemma is proved. $\square$

Lemma 10.7. *The following inclusions hold:*

$$E_1 \subset E, \quad (E^*)_1 \subset E^*.$$

Proof. Suppose that $u \in E_1$. Then

$$\|u\|_E = \left\|\sum_{i=1}^{\infty} \phi_i u\right\|_E \leq \sum_{i=1}^{\infty} \|\phi_i u\|_E = \|u\|_{E_1},$$

where ϕ_i is a partition of unity. Therefore $u \in E$. The second inclusion follows from the first one applied to the space E^*. The lemma is proved. $\square$

We will prove below that the spaces $(E_\infty)^*_\omega$ and $(E^*)_1$ coincide (Theorem 10.10). Let us introduce an operator $J : (E_\infty)^* \to (E^*)_1$ as follows. According to Lemma 10.6, to any $v \in (E_\infty)^*$ we can put in correspondence $w = Jv \in (E^*)_1$ such that

$$\langle w, u \rangle = \langle v, u \rangle, \quad \forall u \in E. \tag{10.6}$$

The right-hand side in (10.6) has sense since $E \subset E_\infty$. The left-hand side in (10.6) is well defined since $(E^*)_1 \subset E^*$. There is only one w satisfying (10.6). Indeed, let w_1 be another one and $w_0 = w_1 - w$. Then $\langle w_0, u \rangle = 0$, $\forall u \in E$. This means that w_0 is the zero element in E^* and

$$\|w_0\|_{(E^*)_1} = \sum_{i=1}^{\infty} \|w_0 \, \theta_i\|_{E^*} = 0.$$

Hence $w_1 = w$. It is clear that J is a linear operator.

Lemma 10.8. *If $v \in (E_\infty)^*_\omega$, then*

$$\langle v, u \rangle = \lim_{n \to \infty} \left\langle v, \sum_{i=1}^{n} \theta_i u \right\rangle, \quad \forall u \in E_\infty. \tag{10.7}$$

Proof. By definition of $(E_\infty)^*_\omega$, there exists $y \in (E_\infty)^*$ such that

$$\langle v, u \rangle = \lim_{n \to \infty} \left\langle y, \sum_{i=1}^{n} \theta_i u \right\rangle, \quad \forall u \in E_\infty.$$

In order to prove (10.7), it is sufficient to verify the equality

$$\langle y, \theta_i u \rangle = \langle v, \theta_i u \rangle. \tag{10.8}$$

We have

$$\langle v, \theta_i u \rangle = \lim_{n \to \infty} \left\langle y, \sum_{j=1}^{n} \theta_j \theta_i u \right\rangle.$$

10. Dual spaces

Since $\sum_{j=1}^{\infty} \theta_j \theta_i$ has only a finite number of terms, we can pass to the limit and obtain

$$\langle v, \theta_i u \rangle = \left\langle y, \sum_{j=1}^{\infty} \theta_j \theta_i u \right\rangle,$$

and (10.8) follows from this equality. The lemma is proved. □

Lemma 10.9. *The space* $(E_\infty)^*$ *can be represented as a direct sum of linear subspaces* $(E_\infty)^*_\omega$ *and* $\mathrm{Ker}\, J$:

$$(E_\infty)^* = (E_\infty)^*_\omega \oplus \mathrm{Ker}\, J. \tag{10.9}$$

Proof. Let $v \in (E_\infty)^*$ and $\tilde{v}$ be given by

$$\langle \tilde{v}, u \rangle = \lim_{n \to \infty} \left\langle v, \sum_{i=1}^{n} \theta_i u \right\rangle, \quad \forall u \in E_\infty.$$

Then $\tilde{v} \in (E_\infty)^*_\omega$. Set $v_0 = v - \tilde{v}$. For any $\phi \in D$ we have

$$\langle \tilde{v}, \phi \rangle = \left\langle v, \sum_{i=1}^{\infty} \phi \theta_i \right\rangle = \langle v, \phi \rangle$$

since the sum contains only a finite number of terms. It follows that $\langle v_0, \phi \rangle = 0$, $\forall \phi \in D$. Since D is dense in E and $E \subset E_\infty$ (inclusion with topology), it follows that $\langle v_0, u \rangle = 0$, $\forall u \in E$. From (10.6) it follows that $\langle Jv_0, u \rangle = 0$, $\forall u \in E$. Therefore $Jv_0 = 0$ in E^* and, hence, this equality also holds in $(E^*)_1$. This means that $v_0 \in \mathrm{Ker}\, J$. Thus we have proved that $v = \tilde{v} + v_0$, where $\tilde{v} \in (E_\infty)^*_\omega$ and $v_0 \in \mathrm{Ker}\, J$.

It remains to prove that (10.9) is a direct sum. Suppose that $v \in (E_\infty)^*_\omega$ and $v \in \mathrm{Ker}\, J$. We have to verify that $v = 0$. We have

$$\langle v, u \rangle = \lim_{n \to \infty} \left\langle v, \sum_{i=1}^{n} \theta_i u \right\rangle, \quad \forall u \in E_\infty \tag{10.10}$$

(see Lemma 10.8) because $v \in (E_\infty)^*_\omega$. Since $v \in \mathrm{Ker}\, J$ we conclude that $\langle v, \theta_i u \rangle = 0$, $\forall u \in F_\infty$. Indeed, if $u \in F_\infty$, then $\theta_i u \in F$. From (10.10) we obtain $\langle v, u \rangle = 0$, $\forall u \in E_\infty$. The lemma is proved. □

Theorem 10.10. $(E_\infty)^*_\omega = (E^*)_1$.

Proof. The inclusion $(E_\infty)^*_\omega \subset (E^*)_1$ follows from Lemma 10.6. Suppose now that $\phi \in (E^*)_1$. Consider the functionals $\phi_k = \sum_{i=1}^{k} \theta_i \phi$, where θ_i is a partition of unity. By the definition of the space $(E^*)_1$, the series $\sum_{i=1}^{\infty} \|\theta_i \phi\|_{E^*}$ converges. We show that ϕ_k converges to ϕ in $(E^*)_1$. Indeed,

$$\|\phi - \phi_k\|_{(E^*)_1} = \left\|\phi - \sum_{i=1}^{k} \theta_i \phi\right\|_{(E^*)_1} = \sum_{j=1}^{\infty} \left\|\theta_j \left(\phi - \sum_{i=1}^{k} \theta_i \phi\right)\right\|_{E^*}$$

$$= \sum_{j=1}^{\infty} \left\|\theta_j \phi - \sum_{i=1}^{k} \theta_i (\theta_j \phi)\right\|_{E^*} \equiv S.$$

All terms of this sum, for which $\sum_{i=1}^{k} \theta_i$ equals 1 in the support of θ_j, disappear. The remaining terms begin with some k', where k' depends on k and tends to infinity together with it.

$$S = \sum_{j=k'}^{\infty} \left\|\theta_j \phi - \sum_{i=1}^{k} \theta_i (\theta_j \phi)\right\|_{E^*} \leq \sum_{j=k'}^{\infty} \|\theta_j \phi\|_{E^*} + \sum_{j=k'}^{\infty} \sum_{i=1}^{k} \|\theta_i \theta_j \phi\|_{E^*}$$

$$= \sum_{j=k'}^{\infty} \|\theta_j \phi\|_{E^*} + \sum_{j=k'}^{\infty} \sum_{i'} \|\theta_{i'} \theta_j \phi\|_{E^*}$$

$$\leq \sum_{j=k'}^{\infty} \|\theta_j \phi\|_{E^*} + NM \sum_{j=k'}^{\infty} \|\theta_j \phi\|_{E^*} \to 0 \text{ as } k \to \infty.$$

Here i' denotes all those i for which the support of θ_i intersects the support of θ_j for each j fixed. As usual, we use the fact that their number is limited by N.

Thus, the functional ϕ can be represented in the form $\phi = \sum_{i=1}^{\infty} \theta_i \phi$. Then it is also a continuous functional on E_∞. Indeed, for any $u \in E_\infty$,

$$|\langle \phi, u \rangle| \leq \sum_{i=1}^{\infty} |\langle \theta_i \phi, \psi_i u \rangle| \leq \sum_{i=1}^{\infty} \|\theta_i \phi\|_{E^*} \|\psi_i u\|_E$$

$$\leq C \|u\|_{E_\infty} \sum_{i=1}^{\infty} \|\theta_i \tilde{\phi}\|_{E^*} \leq C \|\phi\|_{(E^*)_1} \|u\|_{E_\infty}.$$

Here $\psi_i = 1$ in the support of θ_i. Therefore $\phi \in (E_\infty)^*$, and

$$\|\phi\|_{(E_\infty)^*} \leq C \|\phi\|_{(E^*)_1}.$$

Let $u \in E_\infty$. Put $u_k = \sum_{i=1}^{k} \theta_i u$. Then $\phi(u_k) = \phi_k(u)$. Hence

$$\phi(u) = \lim_{k \to \infty} \phi_k(u) = \lim_{k \to \infty} \phi(u_k).$$

This means that $\phi \in (E_\infty)^*_\omega$.

10. Dual spaces

We prove that the spaces $(E_\infty)^*_\omega$ and $(E^*)_1$ are linearly isomorphic. Let $w \in (E^*)_1$. From the proof above it follows that there exists $v \in (E_\infty)^*_\omega$ such that (10.6) holds. This means that
$$Jv = w. \tag{10.11}$$
Denote by J_1 the restriction of the operator J to $(E_\infty)^*_\omega$. Then from (10.11) we get $J_1 v = w$. This means that the range of the operator J_1 coincides with $(E^*)_1$. Since according to Lemma 10.9 the operator J_1 is invertible, we conclude that the spaces under consideration are linearly isomorphic.

Since the operator J_1^{-1} is bounded, we can use the Banach theorem to conclude that J_1 is also bounded. The theorem is proved. □

Consider now the closure E_D of D in the norm E_∞. The proof of the following lemma is the same as the proof of Lemma 10.6.

Lemma 10.11. $(E_D)^* \subset (E^*)_1$.

Theorem 10.12. $(E_D)^* = (E^*)_1$.

Proof. Let $\phi \in (E^*)_1$, $u \in D$. Then $\phi(u)$ is well defined, and
$$|\phi(u)| \leq \sum_{i=1}^\infty |\phi(\theta_i u)| = \sum_{i=1}^\infty |\phi(\theta_i \psi_i u)|$$
$$\leq \sum_{i=1}^\infty \|\theta_i \phi\|_{E^*} \|\psi_i u\|_E \leq M \|\phi\|_{(E^*)_1} \|u\|_{E_\infty},$$
where $\psi_i = 1$ in the support of θ_i.

This estimate remains valid for $u \in E_D$. Therefore, $\phi \in (E_D)^*$. The opposite inclusion follows from the previous lemma.

We now prove the isomorphism of the spaces. Let $v \in (E_D)^*$, then $v \in D'$. As in the proof of Lemma 10.6 we obtain that $v \in (E^*)_1$. We introduce the embedding operator
$$T : (E_D)^* \to (E^*)_1.$$
This means that to any $v \in (E_D)^*$ we put in correspondence $Tv \in (E^*)_1$ such that
$$\langle Tv, \phi \rangle = \langle v, \phi \rangle, \quad \forall \phi \in D. \tag{10.12}$$
It is clear that T is a linear operator. It is easy to see that the range of the operator T coincides with $(E^*)_1$. Indeed let $w \in (E^*)_1$. Then, as in the proof above, we obtain
$$|\langle w, \phi \rangle| \leq M \|w\|_{(E^*)_1} \|\phi\|_{E_\infty}. \tag{10.13}$$
Consider $v \in D'$ such that $\langle v, \phi \rangle = \langle w, \phi \rangle$, $\forall \phi \in D$. Then by (10.13) we have $v \in (E_D)^*$. Hence by (10.12) we get
$$\langle Tv, \phi \rangle = \langle w, \phi \rangle, \quad \forall \phi \in D.$$

Since D is dense in E, we conclude that Tv and w coincide as elements of E^* and therefore as elements of $(E^*)_1$. We have proved that the equation $Tv = w$ has a solution $v \in (E_D)^*$ for any $w \in (E^*)_1$. Hence the range of the operator T coincides with $(E^*)_1$.

It remains to prove that the operator T is invertible. By definition of the operator T, the equality (10.12) holds. Hence if $Tv = 0$ in $(E^*)_1$, then $\langle v, \phi \rangle = 0$, $\forall \phi \in D$ and therefore $v = 0$ in $(E_D)^*$.

From (10.13) we have $|\langle v, \phi \rangle| \leq M\|w\|_{(E^*)_1} \|\phi\|_{E_\infty}$, and $\|T^{-1}w\|_{(E_D)^*} \leq M\|w\|_{(E^*)_1}$. Hence the operator T^{-1} is bounded. By the Banach theorem the operator T is also bounded. Therefore the norms in the spaces $(E_D)^*$ and $(E^*)_1$ are equivalent. The theorem is proved. □

The next theorem follows from Theorems 10.10 and 10.12. Nevertheless we give a direct proof of this theorem in order to obtain an explicit relation between the elements of these spaces.

Theorem 10.13. $(E_\infty)^*_\omega = (E_D)^*$.

Proof. We note that $E \subset E_D$. Indeed, let $u \in E$, then $u \in E_\infty$. Since D is dense in E, there exists a sequence $\{\phi_n\}$, $\phi_n \in D$ such that $\|\phi_n - u\|_E \to 0$. Hence

$$\|\phi_n - u\|_{E_\infty} \leq M\|\phi_n - u\|_E \to 0.$$

This means that $u \in E_D$.

Let $v \in (E_D)^*$. We introduce a functional w as follows:

$$\langle w, u \rangle = \lim_{n \to \infty} \left\langle v, \sum_{i=1}^{n} \theta_i u \right\rangle, \quad \forall u \in E_\infty. \tag{10.14}$$

We prove that the limit in (10.14) exists. Indeed, by the Hahn-Banach theorem we can extend v to a functional $\hat{v} \in (E_\infty)^*$. We have $\langle \hat{v}, u \rangle = \langle v, u \rangle$, $\forall u \in E_D$. Since $\sum_{i=1}^{n} \theta_i u \in E \subset E_D$, $\forall u \in E_\infty$, we get

$$\left\langle \hat{v}, \sum_{i=1}^{n} \theta_i u \right\rangle = \left\langle v, \sum_{i=1}^{n} \theta_i u \right\rangle, \quad \forall u \in E_\infty.$$

From Lemma 10.4 it follows that the limit in (10.14) exists and $w \in (E_\infty)^*_\omega$.

Let us introduce an operator $S : (E_D)^* \to (E_\infty)^*_\omega$ by the formula $w = Sv$, where w is given by (10.14). It is clear that S is a linear operator. We will prove that S is invertible. Indeed, let $Sv = 0$. Then from (10.14) we obtain

$$\lim_{n \to \infty} \left\langle v, \sum_{i=1}^{n} \theta_i u \right\rangle = 0, \quad \forall u \in E_\infty.$$

In particular, if $u \in D$, we get $\langle v, u \rangle = 0$, $\forall u \in D$. Since D is dense in E_D, it follows that $v = 0$.

10. Dual spaces

To prove that $(E_\infty)^*_w$ is linearly isomorphic to $(E_D)^*$, it is sufficient to verify that the range of S coincides with $(E_\infty)^*_w$. Let $w \in (E_\infty)^*_w$. Then $w \in (E_\infty)^*$. From Lemma 10.8 it follows that

$$\langle w, u \rangle = \lim_{n \to \infty} \left\langle w, \sum_{i=1}^{n} \theta_i u \right\rangle, \quad \forall u \in E_\infty. \tag{10.15}$$

Denote by v the restriction of w to E_D: $\langle v, u \rangle = \langle w, u \rangle$, $\forall u \in E_D$. From this equality it follows that

$$\left\langle w, \sum_{i=1}^{n} \theta_i u \right\rangle = \left\langle v, \sum_{i=1}^{n} \theta_i u \right\rangle, \quad \forall u \in E_\infty$$

since $\theta_i u \in E_D$. From this and (10.15) we conclude that (10.14) is true. Therefore $w = Sv$. We have proved that the range of S coincides with $(E_\infty)^*_w$ and hence $(E_\infty)^*_w$ and $(E_D)^*$ are linearly isomorphic.

We now prove that the operator S is bounded. From (10.14) we get

$$|\langle w, u \rangle| = \lim_{n \to \infty} \left| \left\langle v, \sum_{i=1}^{n} \theta_i u \right\rangle \right|, \quad \forall u \in E_\infty.$$

Further,

$$\left| \left\langle v, \sum_{i=1}^{n} \theta_i u \right\rangle \right| \leq \sum_{i=1}^{n} |\langle \theta_i v, \psi_i u \rangle| \leq \sum_{i=1}^{n} \|\theta_i v\|_{E^*} \|\psi_i u\|_E \leq M\|v\|_{(E^*)_1} \|u\|_{E_\infty}.$$

As in Lemma 10.6, we prove that $\|v\|_{(E^*)_1} \leq M_1 \|v\|_{(E_D)^*}$. Therefore

$$|\langle w, u \rangle| \leq M_2 \|v\|_{(E_D)^*} \|u\|_{E_\infty}.$$

It follows that $\|w\|_{(E_\infty)^*} \leq M_2 \|v\|_{(E_D)^*}$. Hence the operator S is bounded. The theorem is proved. □

Remark 10.14. The space E_D is a subspace of E_∞. Therefore we can expect that $(E_\infty)^* \subset (E_D)^*$. Nevertheless we obtain that $(E_D)^*$ coincides with a subspace $(E_\infty)^*_w$ of $(E_\infty)^*$. To explain this situation we note that E_D is not dense in E_∞. Therefore there exist different from zero functionals in $(E_\infty)^*$, equal zero at E_D. We call them "bad" functionals or functional with support at infinity. Each functional from $(E_\infty)^*$ can be formally considered as a functional from $(E_D)^*$. However, if we do not take into account zero functionals, then the inclusion $(E_\infty)^* \subset (E_D)^*$ does not hold. "Bad" functionals do not belong to D' and cannot be considered as generalized functions.

In the definition of the space E_p in Section 3 it is supposed that $E \subset D'$. Hence, in order to use this definition for the space $(E^*)_p$, we should assume that

$E^* \subset D'$. We will define the space $(E^*)_p$ without this assumption. We will give an intrinsic definition of the spaces E_1 and $(E^*)_p$ which coincides with the previous ones. We do not suppose now that $E \subset D'$ but assume, as before, that all $f \in D$ are multipliers in E according to Definition 1.1. Obviously, it follows that $D \subset M(E^*)$.

Definition 10.15. E_1 is the space of all $u \in E$ such that

$$\sum_{i=1}^{\infty} \|\phi_i u\|_E < \infty, \tag{10.16}$$

where $\{\phi_i\}$ is a partition of unity.

Proposition 10.16. *The spaces E_1 in Definition 10.15 and in Definition 3.1 coincide.*

Proof. The proof follows from the fact that any $u \in E_{\text{loc}}$ satisfying (10.16) belongs to E. $\square$

Thus, the space E_1 is defined. Hence the space $(E_1)^*$ is also defined. We can now define the space $(E^*)_\infty$.

Definition 10.17. $u \in (E^*)_\infty$ if and only if $u \in (E_1)^*$ and

$$\|u\|_{(E^*)_\infty} = \sup_i \|\phi_i u\|_{E^*} < \infty.$$

Proposition 10.18. *The spaces $(E^*)_\infty$ and $(E_1)^*$ coincide and their norms are equivalent.*

The proof of this theorem is the same as in Theorem 10.2. We can now give the intrinsic definition of the space $(E^*)_p$, $1 \leq p < \infty$.

Definition 10.19. $u \in (E^*)_p$ if and only if $u \in (E_1)^*$ and

$$\|u\|_{(E^*)_p} = \left(\sum_{i=1}^{\infty} \|\phi_i u\|_{E^*}^p \right)^{1/p} < \infty.$$

If the space E is reflexive, we can also give an intrinsic definition of the spaces E_p, $1 \leq p \leq \infty$. Indeed, as before we define $(E^*)_1$. Then $E_\infty = (E^{**})_\infty = ((E^*)^*)_\infty = ((E^*)_1)^*$ according to Proposition 10.18 applied to E^*. Therefore $(E^*)_1 \subset ((E^*)_1)^{**} = (E_\infty)^*$ (cf. Theorem 10.10).

Definition 10.20. $u \in E_p$, $1 \leq p < \infty$ if and only if $u \in E_\infty$ and

$$\|u\|_{E_p} = \left(\sum_{i=1}^{\infty} \|\phi_i u\|_E^p \right)^{1/p} < \infty.$$

11 Dual spaces in domains

Let $E \in D'$ be a reflexive Banach space. First of all, we will explain in what sense $\varphi \in D$ are understood as elements of the space E^*. We consider $\tilde{\varphi}(u)$ ($\varphi \in D$, $u \in E$) as the functional

$$\tilde{\varphi}(u) = u(\varphi). \tag{11.1}$$

The right-hand side in (11.1) is defined since $E \subset D'$. The inclusion $E \subset D'$ is supposed to be with the topology. Hence if $u_n \to 0$ in E, then $u_n \to 0$ in D'. Therefore $u_n(\varphi) \to 0$ for any $\varphi \in D$. From (11.1) it follows that $\tilde{\varphi}(u_n) \to 0$ and $\tilde{\varphi} \in E^*$.

Moreover, the functions $\varphi \in C_0^\infty(\Omega)$ can be understood as elements of $[E(\Omega)]^*$. We consider $\hat{\varphi}(u)$ ($\varphi \in D$, $u \in E(\Omega)$) as the functional

$$\hat{\varphi}(u) = u(\varphi). \tag{11.2}$$

We can do this since if the convergence

$$u_n \to 0 \tag{11.3}$$

holds in $E(\Omega)$, then

$$\hat{\varphi}(u_n) = u_n(\varphi) \to 0. \tag{11.4}$$

Indeed, let u_n^c be an extension of u_n such that

$$u_n^c(\varphi) = u_n(\varphi) \tag{11.5}$$

for $\varphi \in C_0^\infty(\Omega)$ and $\|u_n^c\|_E \leq 2\|u_n\|_{E(\Omega)}$. Then from (11.3) it follows that $u_n^c \to 0$ in E. Since $E \subset D'$, then we get $u_n^c(\varphi) \to 0$ for $\varphi \in C_0^\infty(\Omega)$, and (11.4) follows from (11.5).

Denote by $\hat{C}_0^\infty(\Omega)$ the set of all $\varphi \in C_0^\infty(\Omega)$ considered as functionals on $E(\Omega)$ in the sense described by (11.2).

Theorem 11.1. *Let $E \in D'$ be a reflexive Banach space. Then*

$$[E(\Omega)]^* = \hat{E}_0^*(\Omega), \tag{11.6}$$

where $\hat{E}_0^*(\Omega)$ is the closure of $\hat{C}_0^\infty(\Omega)$ in $[E(\Omega)]^*$.

Proof. From the definition of $\hat{C}_0^\infty(\Omega)$ it follows that

$$\hat{C}_0^\infty(\Omega) \subset [E(\Omega)]^*. \tag{11.7}$$

Denote by $E_{\overline{\Omega}}$ the subspace of E consisting of such generalized functions that their supports are contained in $\overline{\Omega}$. Then it is known that

$$E(\Omega) = E/E_{C\Omega} \tag{11.8}$$

(for the proof see, for example, in [543], p. 22). It follows that $E(\Omega)$ is a reflexive Banach space.

From (11.7) we conclude that $\hat{E}_0^*(\Omega) \subset [E(\Omega)]^*$. Suppose now that (11.6) is not true. Then there exists a functional $f \in [E(\Omega)]^{**}$ such that

$$f \neq 0 \tag{11.9}$$

and

$$f(v) = 0 \tag{11.10}$$

for all $v \in \hat{E}_0^*(\Omega)$. Since $E(\Omega)$ is a reflexive Banach space, then $E(\Omega)$ coincides with $[E(\Omega)]^{**}$ by the natural embedding. This means that there exists $y \in E(\Omega)$ such that

$$f(v) = v(y) \tag{11.11}$$

for any $v \in [E(\Omega)]^*$. It follows from (11.10) that $f(\hat{\varphi}) = 0$ for all $\varphi \in C_0^\infty(\Omega)$. Equality (11.11) implies $\hat{\varphi}(y) = 0$ for all $\varphi \in C_0^\infty(\Omega)$. We conclude from (11.2) that $y(\varphi) = 0$ for all $\varphi \in C_0^\infty(\Omega)$. By definition of $E(\Omega)$, this means that $y = 0$ as an element of $E(\Omega)$, and from (11.11) $f(v) = 0$ for all $v \in [E(\Omega)]^*$. This contradicts (11.9). The theorem is proved. □

It follows from Theorem 11.1 that for any $v \in [E(\Omega)]^*$ there exists $v_k \in C_0^\infty(\Omega)$ such that

$$\|\hat{v}_k - v\|_{[E(\Omega)]^*} \to 0 \tag{11.12}$$

as $k \to \infty$.

Theorem 11.2. Let $E \subset D'$ be a reflexive Banach space and $v \in [E(\Omega)]^*$. Then there exists a unique $\tilde{v} \in E^*$ such that for any $u \in E(\Omega)$ and any extension $\tilde{u}$ of u to E we have

$$\tilde{v}(\tilde{u}) = v(u). \tag{11.13}$$

Moreover,

$$\|\tilde{v}\|_{E^*} = \|v\|_{[E(\Omega)]^*}. \tag{11.14}$$

Proof. Let $\varphi \in C_0^\infty(\Omega)$, $\tilde{\varphi}$ and $\hat{\varphi}$ be as in (11.1) and (11.2), respectively. Then for any $u \in E(\Omega)$ and any extension $\tilde{u}$ of u to E we have

$$\hat{\varphi}(u) = u(\varphi) = \tilde{u}(\varphi) = \tilde{\varphi}(\tilde{u}). \tag{11.15}$$

By definition of the norm in $E(\Omega)$ we have

$$\|u\|_{E(\Omega)} \leq \|\tilde{u}\|_E. \tag{11.16}$$

For any $\tilde{u} \in E$ and its restriction u to $E(\Omega)$, (11.15) and (11.16) hold. Therefore

$$|\tilde{\varphi}(\tilde{u})| = |\hat{\varphi}(u)| \leq \|\hat{\varphi}\|_{[E(\Omega)]^*} \cdot \|u\|_{E(\Omega)} \leq \|\hat{\varphi}\|_{[E(\Omega)]^*} \cdot \|\tilde{u}\|_E.$$

Hence

$$\|\tilde{\varphi}\|_{E^*} \leq \|\hat{\varphi}\|_{[E(\Omega)]^*}. \tag{11.17}$$

11. Dual spaces in domains

Now, let $v_k \in C_0^\infty, k = 1, 2 \ldots$, be a sequence which satisfies (11.12). Then by (11.17) we obtain

$$\|\tilde{v}_k\|_{E^*} \leq \|\hat{v}_k\|_{[E(\Omega)]^*} \tag{11.18}$$

and

$$\|\tilde{v}_k - \tilde{v}_l\|_{E^*} \leq \|\hat{v}_k - \hat{v}_l\|_{[E(\Omega)]^*}. \tag{11.19}$$

It follows from (11.12) and (11.19) that the sequence $\{\tilde{v}_k\}$ is convergent in E^*. Denote by $\tilde{v}$ its limit. Then the inequality $\|\tilde{v}\|_{E^*} \leq \|v\|_{[E(\Omega)]^*}$ follows from (11.18). From (11.15) we get $\tilde{v}_k(\tilde{u}) = \hat{v}_k(u)$. Passing here to the limit we obtain (11.13).

We now show that the function $\tilde{v}$ with property (11.13) is unique. If we have two of them, $\tilde{v}_1$ and $\tilde{v}_2$, then for any $\tilde{u} \in E$ we take its restrictions u to $E(\Omega)$ and obtain $\tilde{v}_1(\tilde{u}) = v(u)$, $\tilde{v}_2(\tilde{u}) = v(u)$. Hence $\tilde{v}_1 = \tilde{v}_2$.

It remains to prove that

$$\|v\|_{[E(\Omega)]^*} \leq \|\tilde{v}\|_{E^*}. \tag{11.20}$$

It follows from (11.13) that $|v(u)| = |\tilde{v}(\tilde{u})|$ for any $u \in E(\Omega)$ and any extension $\tilde{u}$ to E. Hence

$$|v(u)| \leq \|\tilde{v}\|_{E^*} \|\tilde{u}\|_E. \tag{11.21}$$

Let $\epsilon > 0$ be an arbitrary number. Since $\|u\|_{E(\Omega)} = \inf \|\tilde{u}\|_E$, where the infimum is taken over all extensions $\tilde{u}$ of u, we can take $\tilde{u}$ such that $\|\tilde{u}\|_E \leq (1+\epsilon)\|u\|_{E(\Omega)}$. Therefore from (11.21), $\|v\|_{[E(\Omega)]^*} \leq (1+\epsilon)\|\tilde{v}\|_{E^*}$. Since $\epsilon > 0$ is arbitrary, we get (11.20). The theorem is proved. □

Denote by $\tilde{C}_0^\infty(\Omega)$ the set of functionals (11.1) and by $E_0^*(\Omega)$ its closure in E^*.

Theorem 11.3. *Let $E \subset D'$ be a reflexive Banach space. Then $[E(\Omega)]^*$ is isometrically isomorphic to the subspace $E_0^*(\Omega)$ of E^*. The correspondence is described by Theorem 11.2. More exactly. Any functional $v \in [E(\Omega)]^*$ can be represented in the form*

$$v(u) = \tilde{v}(\tilde{u}), \quad u \in E(\Omega), \tag{11.22}$$

where $\tilde{v}$ is the corresponding functional, which belongs to $E_0^*(\Omega)$, and $\tilde{u}$ is an arbitrary extension of u to E. Moreover,

$$\|v\|_{[E(\Omega)]^*} = \|\tilde{v}\|_{E^*}.$$

Proof. The representation (11.22) follows from Theorem 11.2. It remains only to prove that $\tilde{v} \in E_0^*(\Omega)$. Let $v_k \in C_0^\infty(\Omega)$ be a sequence such that (11.12) holds. Then we have

$$\tilde{v}_k(\tilde{u}) = \hat{v}_k(u). \tag{11.23}$$

Moreover, (11.19) holds. Hence $\{\tilde{v}_k\}$ is convergent in E^*. Denote by $\tilde{v}$ its limit. Obviously $\tilde{v} \in E_0^*(\Omega)$, and from (11.23) we obtain (11.22). The theorem is proved. □

Let $\phi_i \in D$ be a partition of unity in $\mathbb{R}^n$, and $\Omega \subset \mathbb{R}^n$ be an unbounded domain. We consider the space $E(\Omega)$ and its dual $(E(\Omega))^*$. For each $u \in E(\Omega)$ the product $\phi_i u$ is defined, and $\phi_i u \in E(\Omega)$. Therefore we can define the product $\phi_i v$ for $v \in (E(\Omega))^*$:

$$\langle \phi_i v, u \rangle = \langle v, \phi_i u \rangle, \quad u \in E(\Omega). \tag{11.24}$$

It is a bounded functional on $E(\Omega)$:

$$|\langle \phi_i v, u \rangle| = |\langle v, \phi_i u \rangle| \leq \|v\|_{E^*(\Omega)} \|\phi_i u\|_{E(\Omega)} \leq M \|v\|_{E^*(\Omega)} \|u\|_{E(\Omega)}. \tag{11.25}$$

Thus $\phi_i v \in (E(\Omega))^*$, and $\|\phi_i v\|_{E^*(\Omega)} \leq M \|v\|_{(E(\Omega))^*}$.

Let v be a functional on $E(\Omega)$. We do not assume a priori that it is bounded. We say that $v \in ((E(\Omega))^*)_{\text{loc}}$ if $\phi_i v \in (E(\Omega))^*$ for any i.

Definition 11.4. The space $((E(\Omega))^*)_\infty$ is the set of all functionals $v \in ((E(\Omega))^*)_{\text{loc}}$ such that

$$\|v\|_{((E(\Omega))^*)_\infty} = \sup_i \|\phi_i v\|_{(E(\Omega))^*} < \infty.$$

Theorem 11.5. *The spaces $((E(\Omega))^*)_\infty$ and $(E(\Omega))_1)^*$ coincide.*

The proof is the same as for Theorem 10.2

We have proved in Theorem 11.3 that a functional $v \in (E(\Omega))^*$ can be extended to $\tilde{v} \in E^* = (E(\mathbb{R}^n))^*$. We will use this result in order to show that a functional $v \in ((E(\Omega))^*)_\infty$ can be extended to $(E^*)_\infty$.

Theorem 11.6. *For any $v \in ((E(\Omega))^*)_\infty$ there exists an extension*

$$\tilde{v} \in (E^*)_\infty = ((E(\mathbb{R}^n))^*)_\infty$$

such that

$$\langle v, u \rangle = \langle \tilde{v}, \tilde{u} \rangle, \quad \forall u \in (E(\Omega))_1, \tag{11.26}$$

where $\tilde{u} \in E_1 = (E(\mathbb{R}^n))_1$ is an extension of u.

Proof. We can represent the functional $v \in (E(\Omega))^*$ in the form $v = \sum_{i=1}^\infty \phi_i v$ with the equality understood in the sense of equality of generalized functions. Let $\psi_i \in D$ equal 1 in the support of ϕ_i. Then $v = \sum_{i=1}^\infty \phi_i \psi_i v$. Denote by v_i the extension of $\psi_i v$ such that $\langle v_i, \tilde{u} \rangle = \langle \psi_i v, u \rangle$, where $u \in E(\Omega)$ and $\tilde{u}$ is its extension to E. We put $\tilde{v} = \sum_{i=1}^\infty \phi_i v_i$. We show that $\tilde{v} \in (E^*)_\infty$. Indeed,

$$|\langle \phi_i \phi_j v_i, \tilde{u} \rangle| = |\langle v_i, \phi_i \phi_j \tilde{u} \rangle| = |\langle \psi_i v, \phi_i \phi_j u \rangle| = |\langle \phi_i \phi_j v, u \rangle|$$
$$\leq \|\phi_i \phi_j v\|_{(E(\Omega))^*} \|u\|_{E(\Omega)} \leq \|\phi_i \phi_j v\|_{(E(\Omega))^*} \|\tilde{u}\|_E.$$

Therefore

$$\|\phi_i \phi_j v_i\|_{E^*} \leq \|\phi_i \phi_j v\|_{(E(\Omega))^*}.$$

We have
$$\|\phi_j \tilde{v}\|_{E^*} \leq \sum_{i=1}^{\infty} \|\phi_i \phi_j v_i\|_{E^*} \leq \sum_{i=1}^{\infty} \|\phi_i \phi_j v\|_{(E(\Omega))^*}$$
$$\leq KM\|\phi_j v\|_{(E(\Omega))^*} \leq KM\|v\|_{((E(\Omega))^*)_\infty}.$$

Thus
$$\|\tilde{v}\|_{(E^*)_\infty} \leq KM\|v\|_{((E(\Omega))^*)_\infty}.$$

To finish the proof of the theorem we verify equality (11.26). It is sufficient to check it for functions u with a bounded support since they are dense in $(E(\Omega))_1$. We have
$$\langle \phi_i v_i, \tilde{u} \rangle = \langle \phi_i \psi_i v, u \rangle = \langle \phi_i v, u \rangle,$$
where an extension $\tilde{u}$ can also be chosen with a bounded support. Taking a sum with respect to those i for which the support of ϕ_i has a nonempty intersection with the supports of u and $\tilde{u}$, we obtain (11.26). The theorem is proved. □

12 Spaces $W_q^{s,p}(\mathbb{R}^n)$

In this section we consider the spaces E_q in the case where E is a Sobolev-Slobodetskii space $W^{s,p}(\mathbb{R}^n)$ with a real $s \geq 0$ and $1 \leq p < \infty$. We denote them by $W_q^{s,p}(\mathbb{R}^n)$. If $s = 0$ we will use the notation $L_q^p(\mathbb{R}^n)$ and the conventional notation $L^p(\mathbb{R}^n)$. In what follows we do not specify the domain if it is the whole $\mathbb{R}^n$. Applying results of Section 10, we obtain the relations
$$(W_1^{s,p})^* = W_\infty^{-s,p'}, \quad (W_D^{s,p})^* = W_1^{-s,p'},$$
(Theorems 10.2 and 10.10). We begin with a result that shows the relation of the spaces $W_q^{s,p}(\mathbb{R}^n)$ and usual Sobolev or Sobolev-Slobodetskii spaces.

Lemma 12.1. $L_p^p = L^p$.

Proof. We have
$$\|u\|_{L_p^p}^p = \sum_{i=1}^{\infty} \|\phi_i u\|_{L^p}^p = \sum_{i=1}^{\infty} \int_{\mathbb{R}^n} |\phi_i u|^p dx,$$
where ϕ_i is a partition of unity. Denote by B_i the support of ϕ_i and recall that for any $x \in \mathbb{R}^n$ there exist no more than $N+1$ functions ϕ_i different from zero at this point. Therefore
$$\|u\|_{L_p^p}^p \leq \sum_{i=1}^{\infty} \int_{B_i} |u|^p dx \leq (N+1) \int_{\mathbb{R}^n} |u|^p dx.$$

On the other hand,
$$\int_{\mathbb{R}^n} |u|^p dx = \sum_{i=1}^{\infty} \int_{S_i} |u|^p dx,$$

where S_i are unit cubes forming a lattice in R^n,

$$\int_{S_i} |u|^p dx = \int_{S_i} |\sum_{i'} \phi_{i'} u|^p dx$$

$$\leq \int_{R^n} |\sum_{i'} \phi_{i'} u|^p dx \leq C_1 \sum_{i'} \int_{R^n} |\phi_{i'} u|^p dx$$

where $\sum_{i'} \phi_{i'}(x) = 1$ for $x \in S_i$, since the number of i' is bounded independently of i. Therefore

$$\int_{R^n} |u|^p dx \leq C_2 \sum_{i=1}^{\infty} \int_{R^n} |\phi_i u|^p dx.$$

Thus the norms in L_p and L_p^p are equivalent. The lemma is proved. □

Lemma 12.2. *Let s be a positive integer. Then $W_p^{s,p} = W^{s,p}$.*

Proof. By the definition of the norm in $W_p^{s,p}$,

$$\|u\|_{W_p^{s,p}} = \sum_{i=1}^{\infty} \|\phi_i u\|_{W^{s,p}}^p = \sum_{i=1}^{\infty} \sum_{|\alpha| \leq s} \int_{R^n} |D^\alpha(\phi_i u)|^p dx.$$

Taking into account that the derivatives of ϕ_i are uniformly bounded, we obtain the estimate

$$\sum_{i=1}^{\infty} \int_{R^n} |D^\beta \phi_i D^\gamma u|^p dx \leq C_1 \int_{R^n} |D^\gamma u|^p dx$$

in the same way as in the previous lemma. The opposite estimate

$$\int_{R^n} |D^\gamma u|^p dx \leq C_2 \sum_{i=1}^{\infty} \int_{R^n} |\phi_i D^\gamma u|^p dx$$

can be also obtained as above. The lemma is proved. □

Theorem 12.3. *Let s be real and positive. Then $W_p^{s,p} = W^{s,p}$.*

Proof. Consider first the case where $0 < s < 1$. Then

$$\|u\|_{W_p^{s,p}}^p = \sum_{i=1}^{\infty} \|\phi_i u\|_{W^{s,p}}^p$$

$$= \sum_{i=1}^{\infty} \|\phi_i u\|_{L^p}^p + \sum_{i=1}^{\infty} \int_{R^n} \int_{R^n} \frac{|\phi_i(x) u(x) - \phi_i(y) u(y)|^p}{|x-y|^{n+ps}} dx dy.$$

12. Spaces $W_q^{s,p}(\mathbb{R}^n)$

Denote by J_i the last integral in the right-hand side. Then

$$|J_i| \leq C_1 \int_{\mathbb{R}^n}\int_{\mathbb{R}^n} |\phi_i(x)|^p \frac{|u(x)-u(y)|^p}{|x-y|^{n+ps}}\, dxdy$$
$$+ C_1 \int_{\mathbb{R}^n}\int_{\mathbb{R}^n} |u(y)|^p \frac{|\phi_i(x)-\phi_i(y)|^p}{|x-y|^{n+ps}}\, dxdy$$
$$\leq C_1 \sup_x |\phi_i(x)|^p \int_{\mathbb{R}^n}\int_{B_i} \frac{|u(x)-u(y)|^p}{|x-y|^{n+ps}}\, dxdy$$
$$+ C_1 \int_{\mathbb{R}^n} |u(y)|^p \left(\int_{R^n} \frac{|\phi_i(x)-\phi_i(y)|^p}{|x-y|^{n+ps}}\, dx\right) dy,$$

where B_i is the support of ϕ_i. To estimate $\sum_{i=1}^{\infty} |J_i|$ we first use the inequality

$$\sum_{i=1}^{\infty} \int_{\mathbb{R}^n}\int_{B_i} \frac{|u(x)-u(y)|^p}{|x-y|^{n+ps}}\, dxdy \leq C_2 \int_{\mathbb{R}^n}\int_{\mathbb{R}^n} \frac{|u(x)-u(y)|^p}{|x-y|^{n+ps}}\, dxdy$$

which follows from the fact that at each $x \in \mathbb{R}^n$ the number of intersecting supports B_i is no more than $N+1$. We next estimate the function

$$\Phi(y) \equiv \sum_{i=1}^{\infty} \int_{\mathbb{R}^n} \frac{|\phi_i(x)-\phi_i(y)|^p}{|x-y|^{n+ps}}\, dx = \Phi_1(y) + \Phi_2(y),$$

where $\Phi_1(y)$ contains all i such that $y \in B_i$ or $y \notin B_i$ but the distance from y to B_i is less than 1. The function $\Phi_2(y)$ contains the remaining terms. There is a finite number of terms in $\Phi_1(y)$. For each such i we have

$$\int_{\mathbb{R}^n} \frac{|\phi_i(x)-\phi_i(y)|^p}{|x-y|^{n+ps}}\, dx = \int_{K_i} \frac{|\phi_i(x)-\phi_i(y)|^p}{|x-y|^{n+ps}}\, dx + \int_{\mathbb{R}^n/K_i} \frac{|\phi_i(x)-\phi_i(y)|^p}{|x-y|^{n+ps}}\, dx$$
$$\leq C_3 \int_{K_i} \frac{|x-y|^p}{|x-y|^{n+ps}}\, dx + \int_{\mathbb{R}^n/K_i} \frac{|\phi_i(y)|^p}{|x-y|^{n+ps}}\, dx,$$

where K_i is the ball with the same center as B_i and the radius two times greater. Both integrals in the right-hand side are bounded. Therefore $\Phi_1(y)$ is also bounded.

Consider next $\Phi_2(y)$:

$$\Phi_2(y) = \sum_{i=1}^{\infty} \int_{B_i} \frac{|\phi_i(x)|^p}{|x-y|^{n+ps}}\, dx \leq C_4 \int_{\mathbb{R}^n/S(y)} \frac{dx}{|x-y|^{n+ps}},$$

where $S(y)$ is the unit ball with the center at y. Hence $\Phi_2(y)$ is bounded. Thus, $\Phi(y)$ is bounded independently of y, and

$$\|u\|_{W_p^{s,p}}^p \leq C_5 \|u\|_{L_p^p}^p + C_5 \int_{\mathbb{R}^n}\int_{\mathbb{R}^n} \frac{|u(x)-u(y)|^p}{|x-y|^{n+ps}}\, dxdy = C_5 \|u\|_{W^{s,p}}^p.$$

We now prove the opposite inequality. The $2n$-dimensional space $\mathbb{R}^n \times \mathbb{R}^n$ is represented as a sum of two sets,

$$\Pi = \{x \in \mathbb{R}^n, y \in \mathbb{R}^n, |x-y| \leq \epsilon\},$$

and

$$\Lambda = \{x \in \mathbb{R}^n, y \in \mathbb{R}^n, |x-y| > \epsilon\}.$$

Consider a square lattice in $\mathbb{R}^n$ with distance d between its points, and the balls K_i with centers at the centers of the lattice and the radii $2d$. For $\epsilon > 0$ sufficiently small, $\Pi \subset \cup_{i=1}^{\infty} K_i \times K_i$. Let θ_i be a system of functions such that $\theta_i = 1$ in K_i. Then

$$\int_{\mathbb{R}^n}\int_{\mathbb{R}^n} \frac{|u(x)-u(y)|^p}{|x-y|^{n+ps}} \, dxdy$$
$$\leq \sum_{i=1}^{\infty} \int_{K_i}\int_{K_i} \frac{|u(x)-u(y)|^p}{|x-y|^{n+ps}} \, dxdy + \int_{\Lambda} \frac{|u(x)-u(y)|^p}{|x-y|^{n+ps}} \, dxdy,$$

$$\sum_{i=1}^{\infty} \int_{K_i}\int_{K_i} \frac{|u(x)-u(y)|^p}{|x-y|^{n+ps}} \, dxdy = \sum_{i=1}^{\infty} \int_{K_i}\int_{K_i} \frac{|\theta_i(x)u(x)-\theta_i(y)u(y)|^p}{|x-y|^{n+ps}} \, dxdy$$
$$\leq \sum_{i=1}^{\infty} \int_{\mathbb{R}^n}\int_{\mathbb{R}^n} \frac{|\theta_i(x)u(x)-\theta_i(y)u(y)|^p}{|x-y|^{n+ps}} \, dxdy,$$

$$\int_{\Lambda} \frac{|u(x)-u(y)|^p}{|x-y|^{n+ps}} \, dxdy \leq C_6 \int_{\Lambda} \frac{|u(x)|^p}{|x-y|^{n+ps}} \, dxdy + C_6 \int_{\Lambda} \frac{|u(y)|^p}{|x-y|^{n+ps}} \, dxdy,$$

$$\int_{\Lambda} \frac{|u(x)|^p}{|x-y|^{n+ps}} \, dxdy = \int_{\mathbb{R}^n} |u(x)|^p \left(\int_{|y-x|>\epsilon} \frac{dy}{|x-y|^{n+ps}} \right) dx.$$

Since the internal integral in the right-hand side of the last equality can be estimated by a constant independent of x, then

$$\int_{\Lambda} \frac{|u(x)-u(y)|^p}{|x-y|^{n+ps}} \, dxdy \leq C_7 \|u\|_{L^p}^p \leq C_8 \|u\|_{L_p^p}^p.$$

Thus,

$$\|u\|_{W^{s,p}} \leq C_9 \|u\|_{W_p^{s,p}}.$$

This estimate is obtained for the system of functions θ_i. We recall that the norms in $W_p^{s,p}$ with equivalent systems of functions are equivalent.

We have proved that the norms in $W^{s,p}$ and $W_p^{s,p}$ are equivalent in the case of positive $s < 1$. For an integer $s \geq 1$, the assertion of the theorem follows from

12. Spaces $W_q^{s,p}(\mathbb{R}^n)$

Lemma 12.2. Consider now noninteger $s > 1$. We have

$$\|u\|_{W_p^{s,p}}^p = \sum_{i=1}^{\infty} \|\phi_i u\|_{W^{s,p}}^p$$

$$= \sum_{|\alpha| \leq [s]} \sum_{i=1}^{\infty} \|D^\alpha(\phi_i u)\|_{L^p}^p$$

$$+ \sum_{i=1}^{\infty} \int_{\mathbb{R}^n} \int_{\mathbb{R}^n} \frac{|D^{[s]}(\phi_i(x)u(x)) - D^{[s]}(\phi_i(y)u(y))|^p}{|x-y|^{n+p\sigma}} \, dx dy,$$

where $\sigma = s - [s]$. The estimate of the integral

$$\int_{\mathbb{R}^n} \int_{\mathbb{R}^n} \frac{|D^\beta \phi_i(x) D^\gamma u(x) - D^\beta \phi_i(y) D^\gamma u(y)|^p}{|x-y|^{n+p\sigma}} \, dx dy$$

can be done in the same way as above in the case $s < 1$. This allows us to obtain the estimate

$$\|u\|_{W_p^{s,p}} \leq C_{10} \|u\|_{W^{s,p}}.$$

To prove the opposite inequality we use the estimate

$$\int_{\mathbb{R}^n} \int_{\mathbb{R}^n} \frac{|D^\alpha u(x) - D^\alpha u(y)|^p}{|x-y|^{n+p\sigma}} \, dx dy$$

$$\leq \sum_{i=1}^{\infty} \int_{\mathbb{R}^n} \int_{\mathbb{R}^n} \frac{|\theta_i(x) D^\alpha u(x) - \theta_i(y) D^\alpha u(y)|^p}{|x-y|^{n+p\sigma}} \, dx dy + \|u\|_{W^{[s],p}}^p \quad (12.1)$$

similar to the estimate obtained for $s < 1$. Since

$$\|u\|_{W^{[s],p}} \leq C_{11} \|u\|_{W_p^{[s],p}},$$

it remains to estimate the integral in the right-hand side of (12.1):

$$\sum_{i=1}^{\infty} \int_{\mathbb{R}^n} \int_{\mathbb{R}^n} \frac{|\theta_i(x) D^\alpha u(x) - \theta_i(y) D^\alpha u(y)|^p}{|x-y|^{n+p\sigma}} \, dx dy$$

$$\leq \sum_{i=1}^{\infty} \int_{\mathbb{R}^n} \int_{\mathbb{R}^n} \frac{|D^\alpha(\theta_i(x) u(x)) - D^\alpha(\theta_i(y) u(y))|^p}{|x-y|^{n+p\sigma}} \, dx dy$$

$$+ \sum_{i=1}^{\infty} C_{12} \sum_{|\beta|+|\gamma|=|\alpha|, |\gamma|<|\alpha|} \int_{\mathbb{R}^n} \int_{\mathbb{R}^n} \frac{|D^\beta \theta_i(x) D^\gamma u(x) - D^\beta \theta_i(y) D^\gamma u(y)|^p}{|x-y|^{n+p\sigma}} \, dx dy$$

$$\leq \|u\|_{W_p^{s,p}}^p + C_{13} \|u\|_{W^{s-1,p}}^p. \quad (12.2)$$

The second term in (12.2) is estimated similar to the estimate of J_i above. Thus

$$\|u\|_{W^{s,p}} \leq C_{14}(\|u\|_{W_p^{s,p}} + \|u\|_{W^{s-1,p}}) \leq C_{15}(\|u\|_{W_p^{s,p}} + \|u\|_{W^{[s],p}}) \leq C_{16} \|u\|_{W_p^{s,p}}.$$

The theorem is proved. $\square$

To study further properties of the spaces $W_q^{s,p}$ we use the integral definition of the norm in $W_q^{s,p}$:

$$\|u\|_{W_q^{s,p}} = \left(\int_{\mathbb{R}^n} \|u(\cdot)\phi(\cdot - y)\|_{W^{s,p}}^q dy \right)^{1/q}$$

(see Section 3). If $s = 0$ it becomes

$$\|u\|_{L_q^p} = \left(\int_{\mathbb{R}^n} \left(\int_{\mathbb{R}^n} |u(x)\phi(x - y)|^p dx \right)^{q/p} dy \right)^{1/q}.$$

We can replace $\phi(x)$ by the characteristic function of the unit ball. Then

$$\|u\|_{L_q^p} = \left(\int_{\mathbb{R}^n} \left(\int_{|x-y|\leq 1} |u(x)|^p dx \right)^{q/p} dy \right)^{1/q}.$$

We will determine conditions on $u(x)$ to belong to the space L_q^p. We have

$$\|u\|_{L_q^p}^q = \int_{\mathbb{R}^n} \left(\int_{|x-y|\leq 1} |u(x)|^p dx \right)^{q/p} dy = I_1 + I_2,$$

where

$$I_1 = \int_{|y|\leq 2} \left(\int_{|x-y|\leq 1} |u(x)|^p dx \right)^{q/p} dy, \quad I_2 = \int_{|y|>2} \left(\int_{|x-y|\leq 1} |u(x)|^p dx \right)^{q/p} dy.$$

Let $u(x) = |x|^{-\alpha}$, $\alpha > 0$. Then

$$I_1 = \int_{|y|\leq 2} \left(\int_{|x-y|\leq 1} \frac{dx}{|x|^{\alpha p}} \right)^{q/p} dy \leq \int_{|y|\leq 2} \left(\int_{|x|\leq 3} \frac{dx}{|x|^{\alpha p}} \right)^{q/p} dy.$$

If $\alpha p < n$, then

$$I_1 \leq 2^n \omega_n \left(\frac{3^{n-\alpha p} \kappa_n}{n - \alpha p} \right)^{q/p},$$

where ω_n and κ_n are the volume and the surface of the unit sphere, respectively.

Consider I_2. Since $|y| > 2$, then $|x| \geq |y| - 1 \geq \frac{1}{2}|y|$. If $n < \alpha q$, then

$$I_2 \leq \int_{|y|>2} \left(\int_{|x-y|\leq 1} \frac{2^{\alpha p} dx}{|y|^{\alpha p}} \right)^{q/p} dy = \frac{2^n \omega_n \kappa_n}{\alpha q - n}.$$

12. Spaces $W_q^{s,p}(\mathbb{R}^n)$

We have proved the following lemma.

Lemma 12.4. *If $\alpha p < n < \alpha q$, then $u(x) = 1/|x|^\alpha \in L_q^p$.*

From this lemma we easily obtain the following proposition.

Proposition 12.5. *If for some $R > 0$,*

$$\int_{|x| \leq R} |u(x)|^p dx < \infty$$

and $|u(x)| \leq K|x|^{-\alpha}$ for $|x| > R$, where K is a positive constant and $\alpha q > n$, then $u \in L_q^p$.

In the remaining part of this section we construct an example of "bad" functionals (Remark 10.12) in the space $L_\infty^2(\mathbb{R})$. Consider the subspace E_{lim} of this space that consists of functions $u(x)$ for which there exists the limit

$$\phi(u) = \lim_{x \to +\infty} \int_x^{x+1} u(s) ds.$$

We verify that E_{lim} is closed in the norm

$$\|u\|_{L_\infty^2} = \sup_x \left(\int_x^{x+1} u^2(s) ds \right)^{1/2}.$$

Let $u_n \in E_{lim}$, $u_n \to u_0$ in L_∞^2. Put

$$z_n(x) = \int_x^{x+1} u_n(s) ds, \quad a_n = \phi(u_n).$$

Then

$$|z_n(x)| \leq \left(\int_x^{r+1} u_n^2(s) ds \right)^{1/2} \leq \|u_n\|_{L_\infty^2} \leq M$$

for some positive constant M. The last inequality follows from the assumption that the sequence is convergent. The sequence $\phi(u_n)$ is fundamental. Denote by a_0 its limit. We will show that $z_0(x) \to a_0$. Indeed,

$$|z_0(x) - a_0| \leq |z_n(x) - a_n| + |a_n - a_0| + |z_0(x) - z_n(x)|.$$

For any $\epsilon > 0$ we can choose N such that for any $n \geq N$, $|a_n - a_0| < \epsilon/3$, and $|z_0(x) - z_n(x)| < \epsilon/3$ for all $x \in \mathbb{R}^1$. For a fixed $n \geq N$, we can choose x_0 such that $|z_n(x) - a_n| < \epsilon/3$ for $x \geq x_0$. Therefore $|z_0(x) - a| < \epsilon$ for $x \geq x_0$. This proves the convergence. Thus, $u_0 \in E_{lim}$.

By the Hahn-Banach theorem we can extend the functional $\phi(u)$ to the whole space L_∞^2. For any $u \in D$, $\phi(u) = 0$.

13 Local operators

1. Operators in $\mathbb{R}^n$. Let E and F be local spaces, that is the spaces of distributions introduced in Section 1. We suppose that $D \subset E$, $D \subset F$, and D is dense in F.

Definition 13.1. An operator $A : E \to F$ is called local if for any $u \in E$ with a compact support the inclusion $\operatorname{supp} Au \subset \operatorname{supp} u$ holds.

Theorem 13.2. *If $A : E \to F$ is a bounded local operator, then $A^* : E^* \to F^*$ is also a bounded local operator.*

Proof. Let $v \in F^*$ be a function with compact support. We have to prove that $\operatorname{supp} A^* v \subset \operatorname{supp} v$. Suppose that it is not the case. Then there exists a point $x_0 \in R^n$ such that $x_0 \in \operatorname{supp} A^* v$, $x_0 \notin \operatorname{supp} v$. Let B be a closed ball with the center at x_0 such that $B \cap \operatorname{supp} v = \varnothing$ and $f \in D$ be such that $\operatorname{supp} f \subset D$,

$$\langle A^* v, f \rangle \neq 0. \tag{13.1}$$

On the other hand,
$$\langle A^* v, f \rangle = \langle v, Af \rangle \tag{13.2}$$

and $\langle v, Af \rangle = 0$ since the support of Af belongs to B and it does not intersect the support of v. Here we use the density of D in F in order to approximate Af by functions from D with supports in B. This contradiction proves the theorem. $\square$

Theorem 13.3. *Let $A : E \to F$ be a local operator. Then*

$$A_{\operatorname{loc}} u = \sum_{i,j=1}^{\infty} \theta_j A(\theta_i u), \quad \forall u \in E_{\operatorname{loc}} \tag{13.3}$$

is a linear operator acting from E_{loc} to F_{loc}. Convergence of the series in (13.3) is understood in the sense of distributions, and it does not depend on the choice of the partition of unity θ_i.

Proof. Let
$$A_{m,n} u = \sum_{i=1}^{m} \sum_{j=1}^{n} \theta_j A(\theta_i u).$$

Since $u \in E_{\operatorname{loc}}$, then $\theta_i u \in E$ and $A(\theta_i u) \in F$. Moreover $\operatorname{supp} A(\theta_i u) \subset \operatorname{supp} \theta_i u \subset \theta_i$. Let $\phi \in D$. We have

$$\langle A_{m,n} u, \phi \rangle = \sum_{i=1}^{m} \sum_{j=1}^{n} \langle A(\theta_i u), \theta_j \phi \rangle. \tag{13.4}$$

Denote by N the number of functions θ_i for which $\operatorname{supp} \theta_i \cap \operatorname{supp} \phi \neq \varnothing$. Then the right-hand side of (13.4) contains no more than N^2 terms. Therefore we can pass

13. Local operators

to the limit in (13.4) as $m, n \to \infty$, and

$$\langle A_{\mathrm{loc}} u, \phi \rangle = \sum_{i=1}^{m} \sum_{j=1}^{n} \langle A(\theta_i u), \theta_j \phi \rangle \tag{13.5}$$

for m and n sufficiently large (depending on $\mathrm{supp}\,\phi$).

We show next that $A_{\mathrm{loc}} u \in F_{\mathrm{loc}}$, that is $\psi A_{\mathrm{loc}} u \in F$ for any $\psi \in D$. We have

$$\langle \psi A_{\mathrm{loc}} u, \phi \rangle = \sum_{i=1}^{m} \sum_{j=1}^{n} \langle A(\theta_i u), \theta_j \psi \phi \rangle,$$

where m and n depend on $\mathrm{supp}\,\psi$ but do not depend on $\mathrm{supp}\,\phi$. Hence for n sufficiently large,

$$\psi A_{\mathrm{loc}} u = \psi \sum_{i=1}^{m} \sum_{j=1}^{n} \theta_j A(\theta_i u) = \sum_{i=1}^{m} \left(\psi \sum_{j=1}^{\infty} \theta_j \right) A(\theta_i u) = \psi \sum_{i=1}^{m} A(\theta_i u),$$

where m depends on $\mathrm{supp}\,\psi$. Since $A(\theta_i u) \in F$, then $A_{\mathrm{loc}} u \in F_{\mathrm{loc}}$.

The fact that A_{loc} is a linear operator follows directly from (13.5). It remains to show that (13.3) does not depend on the partition of unity. From (13.5)

$$\langle A_{\mathrm{loc}} u, \phi \rangle = \sum_{i=1}^{m} \langle A(\theta_i u), \phi \rangle \tag{13.6}$$

for all m sufficiently large. Let $\tilde{\theta}_i$ be another partition of unity. Then from (13.6) we obtain

$$\langle A_{\mathrm{loc}} u, \phi \rangle = \sum_{i=1}^{m} \left\langle A\left(\sum_{j=1}^{\infty} \tilde{\theta}_j \theta_i u \right), \phi \right\rangle = \sum_{i=1}^{m} \left\langle A\left(\sum_{j=1}^{n} \tilde{\theta}_j \theta_i u \right), \phi \right\rangle, \tag{13.7}$$

where n depends on $\mathrm{supp}\,\phi$ and does not depend on i since $\mathrm{supp}\,A(\tilde{\theta}_j \theta_i u) \subset \mathrm{supp}\,\tilde{\theta}_j \theta_i$ and $\langle A(\tilde{\theta}_j \theta_i u), \phi \rangle = 0$ if $\mathrm{supp}\,\tilde{\theta}_j \cap \mathrm{supp}\,\phi = \emptyset$. From (13.7)

$$\langle A_{\mathrm{loc}} u, \phi \rangle = \sum_{j=1}^{n} \left\langle A\left(\tilde{\theta}_j \sum_{i=1}^{m} \theta_i u \right), \phi \right\rangle = \sum_{j=1}^{n} \left\langle A\left(\tilde{\theta}_j \sum_{i=1}^{\infty} \theta_i u \right), \phi \right\rangle = \sum_{j=1}^{n} \langle A(\tilde{\theta}_j u), \phi \rangle$$

for n sufficiently large. This equality together with (13.6) show that A_{loc} is independent of the partition of unity. The theorem is proved. $\square$

Definition 13.4. Operator $A_{\mathrm{loc}} : E_{\mathrm{loc}} \to F_{\mathrm{loc}}$ is called the extension of $A : E \to F$ to E_{loc}. Operator $A_q (1 \leq q \leq \infty)$ is a restriction of A_{loc} to E_q.

Theorem 13.5. *Let $A : E \to F$ be a bounded local operator. Then A_q is a bounded operator from E_q to F_q.*

Proof. We begin with the case $q = \infty$. Let θ_i be a partition of unity, $u \in E_\infty$. We have
$$\theta_i A_{\mathrm{loc}} u = \theta_i \sum_{j=1}^{m} A(\theta_j u)$$
for all m sufficiently large. Since $\mathrm{supp}\, A(\theta_j u) \subset \mathrm{supp}\, \theta_j u \subset \mathrm{supp}\, \theta_j$, then
$$\theta_i A_\infty u = \theta_i A_{\mathrm{loc}} u = \theta_i \sum_{j'} A(\theta_{j'} u),$$
where j' are all those j for which $\mathrm{supp}\, \theta_i \cap \mathrm{supp}\, \theta_j \neq \emptyset$. Therefore
$$\|\theta_i A_\infty u\|_F \leq \sum_{j'} \|\theta_i A(\theta_{j'} u)\|_F \leq \sum_{j'} \|\theta_i\|_{M(F)} \|A\| \|\theta_{j'} u\|_E$$
$$\leq N \|A\| \|\theta_i\|_{M(F)} \|u\|_{E_\infty}.$$
Let $\kappa = \sup_i \|\theta_i\|_{M(F)}$. Then
$$\|A_\infty u\|_{F_\infty} \leq \kappa N \|A\| \|u\|_{E_\infty}.$$

Consider next $1 \leq q < \infty$. We have
$$\theta_i A_q u = \theta_i A_{\mathrm{loc}} u = \theta_i \sum_{j'} A(\theta_{j'} u),$$
and for any integer m,
$$\sum_{i=1}^{m} \|\theta_i A_q u\|_F^q = \sum_{i=1}^{m} \|\theta_i \sum_{j'} A(\theta_{j'} u)\|_F^q \leq \sum_{i=1}^{m} N^{q-1} \sum_{j'} \|\theta_i A(\theta_{j'} u)\|_F^q$$
$$= N^{q-1} \sum_{j=1}^{\infty} \sum_{i=1}^{m} \|\theta_i A(\theta_j u)\|_F^q = N^{q-1} \sum_{j=1}^{\infty} \sum_{i'}^{m} \|\theta_{i'} A(\theta_j u)\|_F^q$$
$$\leq N^{q-1} \sum_{j=1}^{\infty} \sum_{i'}^{m} \|\theta_{i'}\|_{M(F)^q} \|A(\theta_j u)\|_F^q \leq N^q \kappa^q \sum_{j=1}^{\infty} \|A(\theta_j u)\|_F^q$$
$$\leq N^q \kappa^q \|A\|^q \sum_{j=1}^{\infty} \|\theta_j u\|_E^q = N^q \kappa^q \|A\|^q \|u\|_{E_q}^q.$$

Here i' are all those i for which $\mathrm{supp}\, \theta_i \cap \mathrm{supp}\, \theta_j \neq \emptyset$. The number of such i is not greater than N. Passing to the limit as $m \to \infty$, we get
$$\|A_q u\|_{F_q}^q \leq N^q \kappa^q \|A\|^q \|u\|_{E_q}^q.$$
Therefore
$$\|A_q u\|_{F_q} \leq N \kappa \|A\| \|u\|_{E_q}.$$
The theorem is proved. □

13. Local operators

2. Operators in Ω. Let Ω be a domain in $\mathbb{R}^n$, E be the space in Definition 13.1. The space $E(\Omega)$ is defined in Definition 9.1.

Definition 13.6. The space $E_q(\Omega)$ is defined as the set of those generalized functions from D'_Ω that are restrictions to Ω of generalized functions from E_q, $(1 \le q \le \infty)$. The norm in $E_q(\Omega)$ is given by the equality

$$\|u\|_{E_q(\Omega)} = \inf \|u^c\|_{E_q},$$

where the infimum is taken over all those $u^c \in E_q$, whose restriction to Ω coincide with u (cf. Definition 9.2).

Definition 13.7. Let $A : E \to F$ be a local bounded operator. Operator $A_q(\Omega)$, $(1 \le q \le \infty)$ is the restriction of A_q to $E_q(\Omega)$.

We discuss the last definition in more detail. Let $u \in E_q(\Omega)$. Then there exists $u^c \in E_q$ such that

$$\langle u^c, \phi \rangle = \langle u, \phi \rangle, \quad \forall \phi \in D_\Omega.$$

Then $A_q(\Omega)u$ is the restriction of $A_q u^c$ to Ω.

We show that $A_q(\Omega)u$ does not depend on the extension u^c. Indeed, let u_1 and u_2 be two extensions of u. Then

$$\langle u_1, \phi \rangle = \langle u_2, \phi \rangle = \langle u, \phi \rangle, \quad \forall \phi \in D_\Omega.$$

Let $z = u_1 - u_2$. Then $\langle z, \phi \rangle = 0, \forall \phi \in D_\Omega$. This means that the support $\operatorname{supp} z$ of z belongs to the complement $C\Omega$ of the domain Ω. By the definition of local operators, $\operatorname{supp} Az \subset C\Omega$. Therefore $\langle Az, \phi \rangle = 0, \forall \phi \in D_\Omega$, that is

$$\langle Au_1, \phi \rangle = \langle Au_2, \phi \rangle = 0, \quad \forall \phi \in D_\Omega.$$

Hence Au_1 and Au_2 coincide as elements of $F(\Omega)$. Thus $A_q(\Omega)$ acts from $E_q(\Omega)$ to $F(\Omega)$.

Theorem 13.8. *Operator $A_q(\Omega)$ is bounded as acting from $E_q(\Omega)$ to $F_q(\Omega)$.*

Proof. Let $u \in E_q(\Omega)$. By Definition 13.6 there exists $u^c \in E_q$ such that

$$\|u^c\|_{E_q} \le 2\|u\|_{E_q(\Omega)}.$$

Let $v^c = A_q u^c$. Then $v^c \in F_q$. We have

$$\|v^c\|_{F_q} = \|A_q u^c\|_{F_q} \le \|A_q\| \, \|u^c\|_{E_q} \le 2\|A_q\| \, \|u\|_{E_q(\Omega)}.$$

By definition, $A_q(\Omega)u$ is the restriction of v^c to Ω. Hence

$$\|A_q(\Omega)u\|_{F_q(\Omega)} \le \|v^c\|_{F_q} \le 2\|A_q\| \, \|u\|_{E_q(\Omega)}.$$

Therefore $A_q(\Omega) : E_q(\Omega) \to F_q(\Omega)$ is a bounded operator and $\|A_q(\Omega)\| \le 2\|A_q\|$. The theorem is proved. $\square$

3. Boundary operators. Let $\partial\Omega$ be a C^l manifold, where $l \geq 1$ is an integer. As in Section 8 we denote by D_l the space of all functions $\phi \in C^l(\mathbb{R}^n)$ with compact support. Let E be a local space and D_l be dense in E. We denote by $D_l(\Omega)$ the restriction of C^l to Ω. Since D_l is dense in E, then $D_l(\Omega)$ is dense in $E(\Omega)$.

For any $u \in E(\Omega)$ we can define its trace $\hat{u}$ on $\partial\Omega$. We first define the norm of traces of functions from $D_l(\Omega)$. Let $\phi \in D_l(\Omega)$. Then ϕ is defined on $\partial\Omega$. We put
$$\|\phi\|_{E(\partial\Omega)} = \inf \|\phi^c\|_{E(\Omega)}, \tag{13.8}$$
where the infimum is taken over all $\phi^c \in E(\Omega)$ such that $\phi^c(x) = \phi(x)$ for $x \in \partial\Omega$.

Definition 13.9. The space $E(\partial\Omega)$ is the closure of $D_l(\partial\Omega)$ in the norm (13.8), where $D_l(\partial\Omega)$ is the space of traces of $D_l(\Omega)$ on $\partial\Omega$.

Let $u \in E(\Omega)$. Then there exists a sequence $\phi_n \to u$ in $E(\Omega)$, $\phi_n \in D(\Omega)$. Then $\hat{u} = \lim_n \hat{\phi}_n$ in the norm (13.8), where $\hat{\phi}_n$ is the trace of ϕ_n. Obviously, $\hat{u}$ does not depend on the choice of ϕ_n.

Example 13.10. Let $E(\Omega) = W^{s,p}(\Omega)$. If $s > 1/p$, then $E(\partial\Omega) = W^{s-1/p,p}(\partial\Omega)$. Let $E(\Omega) = L^2(\Omega)$. Applying formally the definition we obtain $\|\phi\|_{E(\partial\Omega)} = 0$ for any ϕ. Therefore the norm of any function defined on $\partial\Omega$ equals zero. This means that we can formally define the space of traces if we consider equivalence classes of functions, but this definition has no sense because the space contains only the zero element. Nevertheless this definition can be useful since it allows us to consider the general case.

Definition 13.11. Linear operator $B : E \to F(\partial\Omega)$ is called local if for any $u \in E$ we have $\operatorname{supp} Bu \subset \operatorname{supp} u$, where $\operatorname{supp} Bu$ is taken in $\partial\Omega$.

It follows from the definition that if $\operatorname{supp} u \cap \partial\Omega = \varnothing$, then $\operatorname{supp} Bu = \varnothing$. Hence $Bu = 0$ as an element of $F(\partial\Omega)$. Consider the operator
$$B^* : (F(\partial\Omega))^* \to E^*.$$
We suppose that D is dense in E, and $D_l(\partial\Omega)$ is dense in $F(\partial\Omega)$.

Theorem 13.12. *Let $B : E \to F(\partial\Omega)$ be a bounded local operator. Then $B^* : (F(\partial\Omega))^* \to E^*$ is also a bounded local operator.*

The proof of this theorem is similar to the proof of Theorem 13.2. It follows from this theorem that for any $v \in (F(\partial\Omega))^*$, $\operatorname{supp} B^*v \in \partial\Omega$.

Definition 13.13. Linear operator $B : E(\Omega) \to F(\partial\Omega)$ is called local if for any $u \in E(\Omega)$ we have $\operatorname{supp} Bu \subset \operatorname{supp} u$.

Theorem 13.14. *Let $B : E(\Omega) \to F(\partial\Omega)$ be a bounded local operator. Then $B^* : (F(\partial\Omega))^* \to (E(\Omega))^*$ is also a bounded local operator.*

The proof of this theorem is similar to the proof of Theorem 13.2.

Chapter 3

A Priori Estimates

In this chapter we obtain a priori estimates for elliptic operators in bounded or unbounded domains. We will use the spaces of functions introduced in Chapter 2. Consider the operators

$$A_i u = \sum_{k=1}^{N} \sum_{|\alpha| \leq \alpha_{ik}} a_{ik}^{\alpha}(x) D^{\alpha} u_k, \quad i = 1, \ldots, N, \quad x \in \Omega, \tag{0.1}$$

$$B_j u = \sum_{k=1}^{N} \sum_{|\beta| \leq \beta_{jk}} b_{jk}^{\beta}(x) D^{\beta} u_k, \quad j = 1, \ldots, m, \quad x \in \partial\Omega, \tag{0.2}$$

where $u = (u_1, \ldots, u_N)$, $\Omega \subset \mathbb{R}^n$ is an unbounded domain that satisfies certain conditions given below. According to the definition of elliptic operators in the Douglis-Nirenberg sense [134] we suppose that

$$\alpha_{ik} \leq s_i + t_k, \; i, k = 1, \ldots, N, \quad \beta_{jk} \leq \sigma_j + t_k, \; j = 1, \ldots, m, \; k = 1, \ldots, N$$

for some integers s_i, t_k, σ_j such that $s_i \leq 0$, $\max s_i = 0$, $t_k \geq 0$. We assume that the operator is uniformly elliptic.

Denote by E the space of vector-valued functions $u = (u_1, \ldots, u_N)$, where u_j belongs to the Sobolev space $W^{l+t_j,p}(\Omega)$, $j = 1, \ldots, N$, $1 < p < \infty$, l is an integer, $l \geq \max(0, \sigma_j + 1)$, $E = \Pi_{j=1}^{N} W^{l+t_j,p}(\Omega)$. The norm in this space is defined as

$$\|u\|_E = \sum_{j=1}^{N} \|u_j\|_{W^{l+t_j,p}(\Omega)}.$$

The operator A_i acts from E to $W^{l-s_i,p}(\Omega)$, the operator B_j acts from E to $W^{l-\sigma_j-1/p,p}(\partial\Omega)$. Let

$$\begin{aligned} L &= (A_1, \ldots, A_N, B_1, \ldots, B_m), \\ F &= \Pi_{i=1}^{N} W^{l-s_i,p}(\Omega) \times \Pi_{j=1}^{m} W^{l-\sigma_j-1/p,p}(\partial\Omega). \end{aligned} \tag{0.3}$$

Then $L : E \to F$. The coefficients of the operators are in general complex-valued. We assume that $a_{ik}^\alpha \in C^m(\mathbb{R}^n)$, $m \geq 1 + l - s_i$ and $b_{jk}^\beta(x) \in C^m(\mathbb{R}^{n-1})$, $m \geq 1 + l - \sigma_j - 1/p$. The notation C_0 will be used for functions with bounded support.

The proof of a priori estimates is based on the invertibility of model problems for pseudo-differential operators introduced as a modification of elliptic differential operators. Their invertibility allows us to obtain a priori estimates both for the direct and for the adjoint operators. The invertibility of these operators was proved earlier in [457] in a more general case. Here we use another method suggested in [566] which allows a simplification of the proof. This method is an adaptation of the approach developed in [542] in order to obtain a priori estimates of solutions.

1 Model problems in a half-space

1.1 Formulation of the main result

We use the following notation:
$$D_j = i\partial/\partial x_j, \quad j = 1, \ldots, n,$$
$$\hat{D}_j = (F')^{-1} \frac{\xi_j}{|\xi'|} (1 + |\xi'|) F', \quad j = 1, \ldots, n-1, \quad \hat{D}_n = D_n,$$

where F' is the partial Fourier transform with respect to the variables $x_1, \ldots, x_{n-1}$, $\xi' = (\xi_1, \ldots, \xi_{n-1})$, $|\xi'| = (\xi_1^2 + \cdots + \xi_{n-1}^2)^{1/2}$.

Denote by $A(D)$ the square $N \times N$ matrix of linear differential operators $A_{ij}(D)$,
$$A_{ij}(D) = \sum_{|\alpha|=\alpha_{ij}} a_{ij}^\alpha D^\alpha$$

with constant coefficients. We suppose that the operator $A(D)$ is elliptic in the Douglis-Nirenberg sense and contains only the principal terms. Then (see [542])
$$A(c\xi) = S(c) A(\xi) T(c) \tag{1.1}$$

for any $\xi = (\xi_1, \ldots, \xi_n)$ and any real c. Here S and T are diagonal matrices,
$$S(c) = (\delta_{ij} c^{s_i}), \quad T(c) = (\delta_{ij} c^{t_j}), \tag{1.2}$$

where δ_{ij} is the Kronecker symbol, $s_1, \ldots, s_N, t_1, \ldots, t_N$ are given integers, $\alpha_{ij} = s_i + t_j$, $i, j = 1, \ldots, N$, $s_i \leq 0$.

We consider the system of equations
$$A(\hat{D}) u = f \tag{1.3}$$

in the half-space
$$\mathbb{R}_+^n = \{x \in \mathbb{R}^n, x = (x_1, \ldots, x_n), x_n > 0\},$$
$$u(x) = (u_1(x), \ldots, u_N(x)), \quad f(x) = (f_1(x), \ldots, f_N(x)).$$

1. Model problems in a half-space

We set the boundary conditions

$$B(\hat{D})u = g(x') \tag{1.4}$$

at the boundary Γ of R^n_+, where $g(x') = (g_1(x'), \ldots, g_m(x'))$, $B(D)$ is a rectangular $m \times N$ matrix with the elements

$$B_{kj}(D) = \sum_{|\alpha|=\sigma_{kj}} b^\alpha_{kj} D^\alpha,$$

and b^α_{kj} are some constants. The matrix $B(\xi)$ is homogeneous,

$$B(c\xi) = M(c)B(\xi)T(c),$$

where $M(c)$ is a diagonal matrix of order m,

$$M(c) = (\delta_{ij} c^{\sigma_i}), \tag{1.5}$$

$\sigma_i = \max_{1 \leq j \leq N}(\sigma_{ij} - t_j)$, $i = 1, \ldots, m$ (see [542]). We introduce the following spaces

$$E(\Omega) = \Pi_{j=1}^N W^{l+t_j,p}(\Omega),$$
$$F^d(\Omega) = \Pi_{j=1}^N W^{l-s_i,p}(\Omega),$$
$$F^b(\partial\Omega) = \Pi_{j=1}^m W^{l-\sigma_j-1/p,p}(\partial\Omega),$$

where Ω is a domain in $\mathbb{R}^n$, $\partial\Omega$ is its boundary, l is an integer, $l \geq \max_j(\sigma_j + 1)$, $1 < p < \infty$. The main result of this section is given by the following theorem.

Theorem 1.1. *For any $f \in F^d(\mathbb{R}^n_+)$ and $g \in F^b(\mathbb{R}^{n-1})$ there exists a unique solution $u \in E(R^n_+)$ of problem (1.3), (1.4).*

The proof of this theorem is based on the following result.

Theorem 1.2. *For any $u \in E(R^n_+)$ the following estimate holds:*

$$\|u\|_{E(R^n_+)} \leq \left(\|A(\hat{D})u\|_{F^d(R^n_+)} + \|B(\hat{D})u\|_{F^b(\Gamma)}\right),$$

where c is a constant independent of u.

The proof of this theorem is given in Section 1.4 and of the previous one in Section 1.5. Theorem 1.1 in more general spaces is proved in [457]. We use the approach developed in [542] to give a simpler proof for the case under consideration.

1.2 Auxiliary results

Set $\xi = (\xi_1, \ldots, \xi_n)$, $\hat{\xi} = (\hat{\xi}_1, \ldots, \hat{\xi}_n)$, where

$$\hat{\xi}_i = \frac{\xi_i}{|\xi_i|} + \xi_i, \quad i = 1, \ldots, n-1, \quad \hat{\xi}_n = \xi_n. \tag{1.6}$$

Proposition 1.3. Let $f(\xi)$ be a homogeneous function of degree $s \geq 0$ and
$$\mu = \inf_{1 \leq |\xi| \leq 2} |f(\xi)|. \tag{1.7}$$

Then for any ξ,
$$|f(\hat{\xi}(\xi))| \geq \max(1, |\xi|^s)\mu. \tag{1.8}$$

Proof. We have
$$|\hat{\xi}|^2 = 1 + 2|\xi'| + |\xi|^2. \tag{1.9}$$

Hence
$$|\hat{\xi}| \geq 1. \tag{1.10}$$

If $|\xi| \leq 1$, then $|\hat{\xi}| \leq 2$ and (1.8) follows from (1.7).

Let $|\xi| \geq 1$ and $\eta = \hat{\xi} \cdot |\xi|^{-1}$. Then $1 \leq |\eta| \leq 2$. Therefore $|f(\hat{\xi})| = |\xi|^s |f(\eta)| \geq \mu |\xi|^s$. The proposition is proved. $\square$

Proposition 1.4. Let $f(\xi)$ be a homogeneous function of degree $s \leq 0$ and $M = \sup_{1 \leq |\xi| \leq 2} |f(\xi)|$. Then for any ξ, $|f(\hat{\xi}(\xi))| \leq \min(1, |\xi|^s)M$.

The proof is similar to the proof of the previous proposition.

Denote by $\mathcal{R}$ the ring with the basis
$$1, \quad \phi_i(\xi') = \frac{\xi_i}{|\xi'|}, \quad i = 1, \ldots, n-1. \tag{1.11}$$

The elements of $\mathcal{R}$ are finite products of (1.11) and their algebraic sums. It is easy to see that if $a(\xi') \in \mathcal{R}$, then
$$\xi_k \frac{\partial a(\xi')}{\partial \xi_k} \in \mathcal{R}, \quad k = 1, \ldots, n-1.$$

This follows from the fact that
$$\xi_k \frac{\partial \phi_i(\xi')}{\partial \xi_k} = \delta_{ik}\phi_k - \phi_i \phi_k^2 \in \mathcal{R}.$$

Denote by Π_m the set of all polynomials of the variables ξ_i, $i = 1, \ldots, n$ of degree m with the coefficients from $\mathcal{R}$:
$$P(\xi) = \sum_{|\alpha| \leq m} c_\alpha(\xi')\xi^\alpha, \quad c_\alpha(\xi') \in \mathcal{R},$$

where α is a multi-index. It is easy to verify that
$$\xi_k \frac{\partial P(\xi')}{\partial \xi_k} \in \Pi_m, \quad k = 1, \ldots, n, \quad \forall P(\xi) \in \Pi_m. \tag{1.12}$$

1. Model problems in a half-space

Proposition 1.5. *Let a function $f(\xi)$ have continuous derivatives of order less than or equal to m. Then for $|\alpha| \le m$,*

$$\xi^\alpha D^\alpha f(\hat{\xi}(\xi)) = \sum_{|\beta| \le |\alpha|} a_\beta(\xi) D_\xi^\beta f(\hat{\xi}(\xi)), \tag{1.13}$$

where $a_\beta(\xi) \in \Pi_{|\beta|}$.

The proof is straightforward and can be done by induction in $|\alpha|$.

We will find conditions for $f(\hat{\xi}(\xi))$ to be a Fourier multiplier in L^p. We recall that a function $\Phi(\xi)$ is called a Fourier multiplier in L^p if for all $v \in C_0^\infty$,

$$\|F^{-1}\Phi(\xi) F v\|_{L^p} \le c\|v\|_{L^p},$$

where F is the Fourier transform. The norm $\|\Phi\|_M$ of this multiplier is the infimum of the constants c. We will use the following theorem.

Michlin's Theorem (see [322], [349]). *Suppose that the function $\Phi(\xi)$ is continuous with its mixed derivatives for $|\xi_j| > 0$, $j = 1, \ldots, n$ and*

$$|\xi^\alpha D^\alpha \Phi| < K, \quad |\alpha| \le n,$$

where $\alpha = (\alpha_1, \ldots, \alpha_n)$, and the numbers $\alpha_1, \ldots, \alpha_n$ take the values 0 or 1. Then $\Phi(\xi)$ is a Fourier multiplier in L^p and

$$\|\Phi\|_M \le c_p K,$$

where the constant c_p does not depend on the function $\Phi(\xi)$.

Proposition 1.6. *Let $f(\xi)$ be a homogeneous function of degree zero, $D^\alpha f(\xi)$, $|\alpha| \le n$ be continuous for $1 \le |\xi| \le 2$. Then $f(\hat{\xi}(\xi))$ is a Fourier multiplier in L^p.*

Proof. From (1.13) we have

$$|\xi^\alpha D^\alpha f(\hat{\xi}(\xi))| \le \sum_{|\beta| \le |\alpha|} |a_\beta(\xi)| |D_\xi^\beta f(\hat{\xi}(\xi))|. \tag{1.14}$$

The function $D_\xi^\beta f(\xi)$ is a homogeneous function of degree $-|\beta|$. Hence by Proposition 1.4

$$|D_\xi^\beta f(\hat{\xi}(\xi))| \le \min(1, |\xi|^{-|\beta|}) M_\beta,$$

where

$$M_\beta = \sup_{1 \le |\xi| \le 2} |D_\xi^\beta f(\xi)|. \tag{1.15}$$

Further, $a_\beta(\xi) \in \Pi_{|\beta|}$. Therefore

$$|a_\beta(\xi)| \le N_\beta |\xi|^{|\beta|}, \quad |\xi_j| > 0, \; j = 1, \ldots, n, \tag{1.16}$$

where N_β is a constant, since the elements of the ring $\mathcal{R}$ are bounded for $|\xi_j| > 0$. From (1.14)–(1.16) we obtain

$$|\xi^\alpha D^\alpha f(\hat{\xi}(\xi))| \le \sum_{|\beta| \le |\alpha|} M_\beta N_\beta,$$

and the conditions of Michlin's Theorem are satisfied. The proposition is proved. $\square$

Proposition 1.7. *For* $-\infty < s < \infty$, $1 < p < \infty$,

$$\|u\|_{H^{s,p}} = \|F^{-1}|\hat{\xi}(\xi)|^s Fu\|_{L^p}$$

is an equivalent norm in $H^{s,p}$.

The proof follows from the fact that the function $|\hat{\xi}(\xi)|^s(1+|\xi|^2)^{-s/2}$ satisfies the conditions of Michlin's Theorem.

Proposition 1.8. *Consider the function*

$$f(\xi', t) = a(\xi')\xi'^\alpha(1+|\xi'|^2)^{s/2}D_t^\kappa \psi(\hat{\xi}', t), \tag{1.17}$$

where

$$\psi(\hat{\xi}', t) = \int_{\gamma_+(\xi')} e^{i\lambda t} g(\xi', \lambda) d\lambda, \tag{1.18}$$

$\gamma_+(\xi')$ *is a contour lying in the half-plane* $\operatorname{Im} \lambda > 0$ *and enclosing the zeros* λ *of the polynomial* $\det A(\xi', \lambda)$ *lying in this half-plane;* $g(\xi', \lambda)$ *is a homogeneous (with respect to* (ξ', λ)*) function of degree* γ*, analytic in* λ *and infinitely differential in* ξ' *for* $\xi' \ne 0$*;* $a(\xi')$ *is an element of the ring* $\mathcal{R}$*;* α *is a multi-index;* $\kappa \ge 0$ *is an integer.*

If

$$|\alpha| + s + \kappa + \gamma \le 0, \tag{1.19}$$

then $f(\xi', t)$ *is a Fourier multiplier in* $L^p(\mathbb{R}^{n-1})$ *with respect to* ξ'*,* $t > 0$ *is a parameter. The norm of the multiplier satisfies the estimate*

$$\|f(\cdot, t)\|_M \le \frac{c}{t}, \tag{1.20}$$

where c *is a constant.*

Proof. We put $\hat{\xi}'$ instead of ξ in (1.18) and substitute $\lambda = \mu|\hat{\xi}'|$. Denoting $\hat{\eta}' = \hat{\xi}'|\hat{\xi}'|^{-1}$, we obtain

$$\psi(\hat{\xi}', t) = |\hat{\xi}'|^{\gamma+1} \int_{\tilde{\gamma}_+} e^{i\mu|\hat{\xi}'|t} g(\hat{\eta}', \mu) d\mu, \tag{1.21}$$

1. Model problems in a half-space

where $\tilde{\gamma}_+$ is a contour, which encloses the zeros μ of det $A(\hat{\eta}', \mu)$ (since det $A(\xi', \lambda)$ is a homogeneous function). Since $|\hat{\eta}'| = 1$, $\tilde{\gamma}_+$ can be taken to be independent of $\hat{\eta}'$ and situated in the half-plane Im $\mu > \delta > 0$.

From the boundedness of $|a(\xi')|$, (1.17) and (1.21) we get

$$|f(\xi',t)| \leq c_1 |\xi'|^{|\alpha|}(1+|\xi'|^2)^{s/2}|\hat{\xi}'|^{\kappa+\gamma+1}e^{-\delta|\xi'|t} \leq c_2|\hat{\xi}'|^{|\alpha|+s+\kappa+\gamma+1}e^{-\delta|\xi'|t}.$$

Hence in view of (1.19),

$$|f(\xi',t)| \leq \frac{c_3}{t}. \qquad (1.22)$$

It is easy to verify that $\xi_k \frac{\partial f(\xi',t)}{\partial \xi_k}$, $k = 1, \ldots, n-1$ is a sum of functions satisfying the conditions of the proposition. Therefore the same is true for $\xi'^\alpha D^\alpha f(\xi',t)$, where $\xi'^\alpha D^\alpha$ satisfies the conditions of Michlin's Theorem. Thus $\xi'^\alpha D^\alpha$ satisfies the estimate of type (1.22). The proposition follows from Michlin's Theorem. $\square$

Proposition 1.9. (cf. [542], Proposition 5.2). *Consider the operator*

$$T\phi = \int_0^\infty F'^{-1}(H(\xi',t+\tau)F'\phi(\xi',\tau))d\tau,$$

where

$$M_p(H(\cdot,t)) \leq \frac{c}{t},$$

M_p *being the norm of the Fourier multiplier in* $L^p(\mathbb{R}^{n-1})$, $\phi(x',t) \in L^p(\mathbb{R}^n_+)$. *Then*

$$\|T\phi\|_{L^p(\mathbb{R}^n_+)} \leq c_1 \|\phi\|_{L^p(\mathbb{R}^n_+)}.$$

Proof. We have

$$\|T\phi\|_{L^p(\mathbb{R}^n_+)} = \left(\int_0^\infty \left\|\int_0^\infty F'^{-1}(H(\xi',t+\tau)F'\phi(\xi',\tau))\,d\tau\right\|_{L^p(\mathbb{R}^{n-1})}^p dt\right)^{1/p}. \qquad (1.23)$$

Further

$$\left\|\int_0^\infty F'^{-1}(H(\xi',t+\tau)F'\phi(\xi',\tau))\,d\tau\right\|_{L^p(\mathbb{R}^{n-1})}^p$$
$$\leq \int_0^\infty \|F'^{-1}(H(\xi',t+\tau)F'\phi(\xi',\tau))\|_{L^p(\mathbb{R}^{n-1})}^p d\tau \qquad (1.24)$$
$$\leq c \int_0^\infty (t+\tau)^{-1}\|\phi(\cdot,\tau)\|_{L^p(\mathbb{R}^{n-1})} d\tau.$$

Set

$$\Phi(\tau) = \|\phi(\cdot,\tau)\|_{L^p(\mathbb{R}^{n-1})}, \ \tau \geq 0, \ \Phi(\tau) = 0, \ \tau < 0.$$

Then in the right-hand side of (1.24) we have the Hilbert transform of the function $\Phi(-\tau)$. Applying the theorem of M. Riesz on the boundedness in L^p of the Hilbert transform (see, e.g., [504]), we find from (1.23)

$$\|T\phi\|_{L^p(\mathbb{R}^n_+)} \leq c \left(\int_0^\infty \left(\int_{-\infty}^\infty (t+\tau)^{-1}\Phi(\tau)d\tau \right)^p dt \right)^{1/p}$$
$$\leq c \left(\int_{-\infty}^\infty \left| \int_{-\infty}^\infty (t+\tau)^{-1}\Phi(\tau)d\tau \right|^p dt \right)^{1/p}$$
$$\leq c_1 \|\Phi\|_{L^p(\mathbb{R}^1)} = c_1 \|\phi\|_{L^p(\mathbb{R}^n_+)}.$$

The proposition is proved. □

1.3 Reduction to homogeneous systems

Consider problem (1.3), (1.4). Let $f_*(x)$ be an extension of $f(x)$ to $\mathbb{R}^n$, $f_* \in F^d(\mathbb{R}^n)$. Consider the equation

$$A(\hat{D})u_* = f_* \text{ in } \mathbb{R}^n.$$

After the Fourier transform we obtain

$$A(\hat{\xi})\tilde{u}_*(\xi) = \tilde{f}_*(\xi), \tag{1.25}$$

where $\tilde{u}_* = Fu_*$, $\tilde{f}_* = Ff$, $\hat{\xi}$ is given by (1.6).

It follows from (1.1) that $\det A(\xi)$ is a homogeneous function of degree $\sum_1^N (s_i + t_i) \geq 0$. By the ellipticity condition we have

$$\mu := \inf_{1 \leq |\xi| \leq 2} |\det A(\xi)| > 0.$$

Therefore Proposition 1.3 implies that for all ξ, $|\det A(\hat{\xi}(\xi))| \geq \mu$. Thus

$$\tilde{u}_*(\xi) = A^{-1}(\hat{\xi}(\xi))\tilde{f}_*(\xi). \tag{1.26}$$

Let S and T be matrices (1.2). It follows from (1.26) that

$$F^{-1}|\hat{\xi}|^l T(|\hat{\xi}|) F u_* = F^{-1} T(|\hat{\xi}|) A^{-1}(\hat{\xi}) S(|\hat{\xi}|) F F^{-1} |\hat{\xi}|^l S^{-1}(|\hat{\xi}|) F f_*. \tag{1.27}$$

Obviously the elements of the matrix $T(|\xi|)A^{-1}(\xi)S(|\xi|)$ are homogeneous functions of degree zero (see (1.1)). By Proposition 1.6 they are Fourier multipliers in L^p. Therefore taking the L^p norm of (1.27), we get

$$\|u_*(\xi)\|_{E(R^n)} \leq c \|f_*\|_{F^d(R^n)}. \tag{1.28}$$

Set

$$v(x) = u(x) - u_*(x), \quad x \in \mathbb{R}^n_+. \tag{1.29}$$

1. Model problems in a half-space

Then
$$A(\hat{D})v = 0, \qquad x \in \mathbb{R}_+^n, \tag{1.30}$$
$$B(\hat{D})v = h(x'), \quad x' \in \Gamma, \tag{1.31}$$

where
$$h = g - B(\hat{D})u_* \in F^b(\Gamma). \tag{1.32}$$

We have reduced problem (1.3), (1.4) to problem (1.30), (1.31).

1.4 Proof of Theorem 1.2

Since $D(\overline{\mathbb{R}_+^n})$ is dense in $W_N^{s,p}(\mathbb{R}_+^n)$, it is sufficient to prove this theorem for $u \in \Pi_{j=1}^N D(\overline{\mathbb{R}_+^n})$. For such u let
$$A(\hat{D})u = f. \tag{1.33}$$

Obviously, $f \in F^d(\mathbb{R}_+^n)$. We repeat the construction of the previous subsection and consider problem (1.30), (1.31). We will prove the estimate
$$\|v\|_{E(\mathbb{R}_+^n)} \le c \|B(\hat{D})v\|_{F^b(\Gamma)}. \tag{1.34}$$

Theorem 1.2 will follow from it. Indeed, the extension f_* of f can be done such that
$$\|f_*\|_{F^d(\mathbb{R}^n)} \le \kappa \|f\|_{F^d(\mathbb{R}_+^n)}, \tag{1.35}$$

where κ is a constant independent of f. Since $u_* \in E(\mathbb{R}_+^n)$, then we have
$$\|B(\hat{D})u_*\|_{F^b(\Gamma)} \le c_1 \|u_*\|_{E(\mathbb{R}_+^n)}. \tag{1.36}$$

From (1.29), (1.28), (1.34)–(1.36) we obtain the estimate of Theorem 1.2.

Thus, it remains to prove (1.34). We write problem (1.30), (1.31) in the form
$$A(\hat{D}', D_t)v = 0, \quad x \in R_+^n, \tag{1.37}$$
$$B(\hat{D}', D_t)v|_{t=0} = h(x'), \tag{1.38}$$

where $x_n = t$, $\hat{D} = (\hat{D}', D_t)$. We do here the partial Fourier transform F' with respect to x'. We note that since $u \in \Pi_1^N D(\overline{\mathbb{R}_+^n})$, then $f \in F^d(\mathbb{R}_+^n)$ also with $p = 2$, and we can suppose that $u_* \in E(\mathbb{R}^n)$, $v \in E(\mathbb{R}^n)$ not only for the p under consideration but also for $p = 2$. Hence we can use the L^2 theory of the Fourier transform.

After the partial Fourier transform in (1.37), (1.38) we obtain
$$A(\hat{\xi}', D_t)\tilde{v}(\xi', t) = 0, \quad t > 0, \tag{1.39}$$
$$B(\hat{\xi}', D_t)\tilde{v}(\xi', 0) = \tilde{h}(\xi'), \tag{1.40}$$

where $\tilde{v} = F'v$, $\tilde{h} = F'h$. Denote by $\omega(\xi', t)$ the stable, that is decaying at infinity, solution of the problem
$$A(\xi', D_t)\omega(\xi', t) = 0, \quad t > 0, \tag{1.41}$$
$$B(\xi', D_t)\omega(\xi', 0) = I, \tag{1.42}$$
where I is the identity matrix of order r. The solution of problem (1.39), (1.40) is
$$\tilde{v}(\xi', t) = \omega(\hat{\xi}', t)\tilde{h}(\xi') \tag{1.43}$$
(cf. [542]). The operator B acts from $E(\mathbb{R}^n_+)$ into $F^b(\Gamma)$. Hence $h \in F^b(\Gamma)$. Let
$$\phi \in \Pi_{j=1}^m W^{l-\sigma_i, p}(\mathbb{R}^n_+)$$
be any extension of $h(x')$, $h(x') = \phi(x', 0)$. From (1.43),
$$v(x) = F'^{-1}(\omega(\hat{\xi}', t)\tilde{\phi}(\xi', 0)). \tag{1.44}$$

Since the elements of the matrix $\omega(\hat{\xi}', t)$ exponentially decrease as $t \to \infty$, then we have
$$\omega(\hat{\xi}', t)\tilde{\phi}(\xi', 0) = -\int_0^\infty D_\tau \omega(\hat{\xi}', t+\tau)\tilde{\phi}(\xi', \tau)d\tau - \int_0^\infty \omega(\hat{\xi}', t+\tau)D_\tau\tilde{\phi}(\xi', \tau)d\tau. \tag{1.45}$$
Let M be matrix (1.5), $Q_\kappa(\xi') = \|\delta_{ij}^{\gamma_j}\|$, where γ_j is a multi-index, $0 \le |\gamma_j| \le l + t_j - \kappa$, $\kappa \ge 0$ is an integer. From (1.44), (1.45) we have
$$D_t^\kappa Q_\kappa(D')v(x) = -(T_1 + T_2), \tag{1.46}$$
where
$$T_1 = D_t^\kappa Q_\kappa(D')F'^{-1} \int_0^\infty D_\tau \omega(\hat{\xi}', t+\tau)\tilde{\phi}(\xi', \tau)d\tau,$$
$$T_2 = D_t^\kappa Q_\kappa(D')F'^{-1} \int_0^\infty \omega(\hat{\xi}', t+\tau)D_\tau\tilde{\phi}(\xi', \tau)d\tau.$$
We write these expressions in the form
$$T_1 = F'^{-1} \int_0^\infty H(\xi', t+\tau)\tilde{a}(\xi', \tau)d\tau,$$
$$T_2 = F'^{-1} \int_0^\infty G(\xi', t+\tau)\tilde{b}(\xi', \tau)d\tau, \tag{1.47}$$
where
$$H(\xi', t) = (1+|\xi'|^2)^{-1/2} Q_\kappa(\xi') \left(D_t^{\kappa+1}\omega(\hat{\xi}', t)\right) M(\sqrt{1+|\xi'|^2}),$$
$$G(\xi', t) = (1+|\xi'|^2)^{-(l-1)/2} Q_\kappa(\xi') \left(D_t^\kappa \omega(\hat{\xi}', t)\right) M(\sqrt{1+|\xi'|^2}), \tag{1.48}$$
$$a(x) = F'^{-1}(1+|\xi'|^2)^{1/2} M^{-1}(\sqrt{1+|\xi'|^2}) F'\phi,$$
$$b(x) = F'^{-1}(1+|\xi'|^2)^{(l-1)/2} M^{-1}(\sqrt{1+|\xi'|^2}) F' D_\tau \phi.$$

1. Model problems in a half-space

Using the Laplace transform, we find from (1.41), (1.42) that

$$\omega(\hat{\xi}', t) = \int_{\gamma_+} e^{i\lambda t} \Phi(\hat{\xi}', \lambda) d\lambda,$$

where

$$\Phi(\xi', \lambda) = A^{-1}(\xi', \lambda) K(\xi', \lambda),$$

$K(\xi', \lambda)$ is an $N \times m$ matrix which satisfies the following homogeneity condition:

$$K(\rho\xi', \rho\lambda) = \rho^{-1} S(\rho) K(\xi', \lambda) M^{-1}(\rho),$$

S and M being matrices (1.2) and (1.5) (see [542]).

Direct calculations show that the elements H_{jk} and G_{jk} of the matrices H and G satisfy conditions of Proposition 1.8. According to this proposition we conclude that $H_{jk}(\xi', t)$ and $G_{jk}(\xi', t)$ are Fourier multipliers with respect to ξ' in $L^p(\mathbb{R}^{n-1})$ and the norms of these multipliers admit the estimates

$$\|H_{jk}(\cdot, t)\|_M \leq \frac{c}{t}, \quad \|G_{jk}(\cdot, t)\|_M \leq \frac{c}{t}, \quad t > 0. \tag{1.49}$$

We can now estimate T_1 and T_2. Let $a = (a_1, \ldots, a_m)$, $\phi = (\phi_1, \ldots, \phi_m)$, $T_1 = (T_{11}, \ldots, T_{1m})$,

$$A_{jk} = \int_0^\infty F'^{-1} H_{jk}(\xi', t+\tau) F' a_k(\xi', \tau) d\tau.$$

Then

$$T_{1j} = \sum_{k=1}^m A_{jk}. \tag{1.50}$$

From Proposition 1.9 we obtain

$$\|A_{jk}\|_{L^p(\mathbb{R}_+^n)} \leq c_{jk} \|a_{jk}\|_{L^p(\mathbb{R}_+^n)}. \tag{1.51}$$

Here and in what follows c with subscripts denotes some constants. It follows from (1.48) that

$$a_k(x) = F'^{-1}(1 + |\xi'|^2)^{(l-\sigma_k)/2} F' \phi_k. \tag{1.52}$$

We can suppose that ϕ_k is extended to $W^{l-\sigma_k, p}(\mathbb{R}^n)$ in such a way that

$$\|\phi_k\|_{W^{l-\sigma_k, p}(\mathbb{R}^n)} \leq 2 \|\phi_k\|_{W^{l-\sigma_k, p}(\mathbb{R}_+^n)}. \tag{1.53}$$

In view of (1.52), (1.53) we have

$$\|a_k\|_{L^p(\mathbb{R}_+^n)} \leq \|a_k\|_{L^p(\mathbb{R}^n)} \leq c \|\phi_k\|_{W^{l-\sigma_k, p}(\mathbb{R}^n)} \leq 2c \|\phi_k\|_{W^{l-\sigma_k, p}(\mathbb{R}_+^n)}. \tag{1.54}$$

We take such extension ϕ_k of h_k that

$$\|\phi_k\|_{W^{l-\sigma_k, p}(\mathbb{R}_+^n)} \leq 2 \|h_k\|_{W^{l-\sigma_k-1/p, p}(\Gamma)}. \tag{1.55}$$

It follows from (1.50), (1.51), (1.54), and (1.55) that

$$\|T_1\|_{L^p(\mathbb{R}^n_+)} \le c_1 \|h\|_{F^b(\Gamma)}.$$

Similarly we obtain the estimate

$$\|T_2\|_{L^p(\mathbb{R}^n_+)} \le c_2 \|h\|_{F^b(\Gamma)}.$$

Hence from (1.46) we conclude that

$$\|D_t^\kappa Q_\kappa v\|_{L^p(\mathbb{R}^n_+)} \le c \|h\|_{F^b(\Gamma)}.$$

Estimate (1.34) follows from the last estimate. Theorem 1.2 is proved. □

1.5 Proof of Theorem 1.1

Uniqueness of solutions follows from Theorem 1.2. From the results of Subsection 1.3 it follows that it is sufficient to prove existence of solutions of problem (1.3), (1.4) for the case $f(x) = 0$. By virtue of Theorem 1.2 it is sufficient to consider $g \in \Pi_1^m D(\Gamma)$ since this space is dense in $F^b(\Gamma)$. For such g we do the partial Fourier transform and prove existence of a solution in $E(\mathbb{R}^n_+)$ as it is done in the previous subsection. The theorem is proved. □

2 A priori estimates in the spaces $W_\infty^{s,p}$

In this section we define the spaces $W_\infty^{s,p}$ and obtain a priori estimates of solutions, which are similar to those in the usual Sobolev spaces. As before, denote by $W_\infty^{k,p}(\Omega)$ the space of functions defined as the closure of smooth functions in the norm

$$\|u\|_{W_\infty^{k,p}(\Omega)} = \sup_{y \in \Omega} \|u\|_{W^{k,p}(\Omega \cap Q_y)}.$$

Here Ω is a domain in $\mathbb{R}^n$, Q_y is a unit ball with the center at y, $\|\cdot\|_{W^{k,p}}$ is the Sobolev norm. We note that in bounded domains Ω the norms of the spaces $W^{k,p}(\Omega)$ and $W_\infty^{k,p}(\Omega)$ are equivalent.

We suppose that the boundary $\partial\Omega$ belongs to the Hölder space $C^{k+\theta}$, $0 < \theta < 1$, and that the Hölder norms of the corresponding functions in local coordinates are bounded independently of the point of the boundary. Then we can define the space $W_\infty^{k-1/p,p}(\partial\Omega)$ of traces on the boundary $\partial\Omega$ of the domain Ω,

$$\|\phi\|_{W_\infty^{k-1/p,p}(\partial\Omega)} = \inf \|v\|_{W_\infty^{k,p}(\Omega)},$$

where the infimum is taken with respect to all functions $v \in W_\infty^{k,p}(\Omega)$ equal ϕ at the boundary, and $k > 1/p$.

2. A priori estimates in the spaces $W_\infty^{s,p}$

The space $W_\infty^{k,p}(\Omega)$ with $k = 0$ will be denoted by $L_\infty^p(\Omega)$. We will use also the notation

$$E_\infty = \Pi_{j=1}^{N} W_\infty^{l+t_j,p}(\Omega),$$
$$F_\infty = \Pi_{i=1}^{N} W_\infty^{l-s_i,p}(\Omega) \times \Pi_{j=1}^{m} W_\infty^{l-\sigma_j-1/p,p}(\partial\Omega).$$

We consider the operator L defined by (0.3) and denote $l_1 = \max(0, \sigma_j + 1)$. We suppose that the integer l in the definition of the spaces is such that $l \geq l_1$, and the boundary $\partial\Omega$ belongs to the class $C^{r+\theta}$ with r specified in the following condition.

Condition D. For each $x_0 \in \partial\Omega$ there exists a neighborhood $U(x_0)$ such that:

1. $U(x_0)$ contains a sphere with the radius δ and the center x_0, where δ is independent of x_0,
2. There exists a homeomorphism $\psi(x; x_0)$ of the neighborhood $U(x_0)$ on the unit sphere $B = \{y : |y| < 1\}$ in $\mathbb{R}^n$ such that the images of $\Omega \cap U(x_0)$ and $\partial\Omega \cap U(x_0)$ coincide with $B_+ = \{y : y_n > 0, |y| < 1\}$ and $B_0 = \{y : y_n = 0, |y| < 1\}$ respectively,
3. The function $\psi(x; x_0)$ and its inverse belong to the Hölder space $C^{r+\theta}$, $0 < \theta < 1$. Their $\|\cdot\|_{r+\theta}$-norms are bounded uniformly in x_0.

Here $r \geq max(l + t_i, l - s_i, l - \sigma_j + 1)$, where the first expression under the maximum is required for a priori estimates of solutions (similar to [8]), the second and the third ones for the proof of convergence in Lemma 2.14 in Chapter 4. For definiteness we suppose that $\delta < 1$.

Theorem 2.1. *Let the support of the function $u \in \Pi_{j=1}^{N} W^{l_1+t_j,p}(\Omega)$ be sufficiently small. Then for any $l \geq l_1$,*

$$\|u\|_E \leq c \left(\|Lu\|_F + \|u\|_{L^p(\Omega)} \right), \tag{2.1}$$

where the constant c does not depend on u.

Proof. Consider first the operators

$$A_i^0 u = \sum_{k=1}^{N} \sum_{|\alpha|=s_i+t_k} a_{ik}^\alpha D^\alpha u_k, \quad i = 1, \ldots, N, \quad x \in \Omega,$$

$$B_j^0 u = \sum_{k=1}^{N} \sum_{|\beta|=\sigma_j+t_k} b_{jk}^\beta D^\beta u_k, \quad i = 1, \ldots, m, \quad x \in \partial\Omega,$$

with constant coefficients a_{ik}^α, b_{jk}^β and suppose that the domain Ω is the half-space $\mathbb{R}_n^+ = \{x_n \geq 0\}$. We will denote by $\hat{A}_i^0$ and $\hat{B}_j^0$ the operators obtained from A_i^0 and B_j^0, respectively, if we replace the derivatives D_i, $i = 1, \ldots, n-1$ by the operators $\hat{D}_i$. The operator

$$\hat{L}^0 = (\hat{A}_1^0, \ldots, \hat{A}_N^0, \hat{B}_1^0, \ldots, \hat{B}_m^0)$$

acts from E into $F = F^d \times F^b$. We consider the operator

$$L^0 : E \to F, \quad L^0 = (A_1^0, \ldots, A_N^0, B_1^0, \ldots, B_m^0).$$

From Theorem 1.1 it follows that the operator $\hat{L}^0$ has a bounded inverse. Therefore

$$\|u\|_E \leq C \left(\|L^0 u\|_F + \|(\hat{L}^0 - L^0) u\|_F \right). \tag{2.2}$$

Here and below we denote by C positive constants independent of u. To prove the theorem it is sufficient to estimate the second term in the right-hand side of this inequality. Since

$$(\hat{A}_i^0 - A_i^0) u \in W^{l-s_i+1,p}(\Omega) \subset W^{l-s_i,p}(\Omega), \quad (\hat{B}_j^0 - B_j^0) u \in W^{l-\sigma_j-1/p+1,p}(\partial\Omega)$$

(see [457]), then

$$\|(\hat{L}^0 - L^0) u\|_F \leq \epsilon \|u\|_E + C \|u\|_{L^p(\Omega)}. \tag{2.3}$$

Together with (2.2) this inequality proves (2.1).

In the general case, the coefficients of the operator can be variable, it can contain lower-order terms, and the domain is not necessarily a half-space. The lower-order terms can be estimated similar to (2.3). The difference between the operator with variable coefficients and the operator with constant coefficients can be estimated by $\epsilon \|u\|_E$ since the support of the function u is sufficiently small (the diameter of the support depends on ϵ). Finally, the case of arbitrary domains can be reduced to the case of a half-space or of a whole space. For this we map the support of the function u inside Ω on a ball or on a half-ball. We use here Condition D and the fact that the support is sufficiently small (see the proof of Theorem 5.1 below for more details). The theorem is proved. □

Theorem 2.2. *Let* $u \in \Pi_{j=1}^N W_\infty^{l_1+t_j,p}(\Omega)$. *Then for any* $l \geq l_1$ *we have* $u \in E_\infty$ *and*

$$\|u\|_{E_\infty} \leq c \left(\|Lu\|_{F_\infty} + \|u\|_{L_\infty^p(\Omega)} \right), \tag{2.4}$$

where the constant c *does not depend on* u.

Proof. Let $\omega(x)$ be an infinitely differentiable nonnegative function such that

$$\omega(x) = 1, \ |x| \leq \frac{1}{2}, \quad \omega(x) = 0, \ |x| \geq 1.$$

Set $\omega_y(x) = \omega(x - y)$. Suppose that $u(x)$ is a function satisfying the conditions of the theorem. Then $\omega_y u \in \Pi_{j=1}^N W_\infty^{l_1+t_j,p}(\Omega)$. Since the support of this function is bounded, we can use now a priori estimates of solutions (Theorem 2.1):

$$\|\omega_y u\|_E \leq c \left(\|L(\omega_y u)\|_F + \|\omega_y u\|_{L^p(\Omega)} \right), \quad \forall y \in \mathbb{R}^n, \tag{2.5}$$

2. A priori estimates in the spaces $W_\infty^{s,p}$

where the constant c does not depend on y. We now estimate the right-hand side of the last inequality. We have

$$A_i(\omega_y u) = \omega_y A_i u + T_i,$$

where

$$T_i = \sum_{k=1}^{N} \sum_{|\alpha| \leq \alpha_{ik}} a_{ik}^{\alpha} \sum_{\beta+\gamma \leq \alpha, |\beta| > 0} c_{\beta\gamma} D^{\beta} \omega_y D^{\gamma} u_k,$$

and $c_{\beta\gamma}$ are some constants. If $|\tau| \leq l - s_i$, then

$$\|D^\tau(\omega_y A_i u)\|_{L^p(\Omega)} \leq M \|A_i u\|_{W_\infty^{l-s_i,p}(\Omega)}.$$

For any $\epsilon > 0$ we have the estimate

$$\|T_i\|_{W^{l-s_i,p}(\Omega)} \leq \epsilon \sum_{k=1}^{N} \|u_k\|_{W^{l+t_k,p}(\Omega \cap Q_y)} + C_\epsilon \sum_{k=1}^{N} \|u_k\|_{L^p(\Omega \cap Q_y)}$$

$$\leq \epsilon \|u\|_{E_\infty} + C_\epsilon \|u\|_{L^p_\infty(\Omega)},$$

where Q_y is a unit ball with the center at y. Thus

$$\|A_i(\omega_y u)\|_{W^{l-s_i,p}(\Omega)} \leq M \|A_i u\|_{W_\infty^{l-s_i,p}(\Omega)} + \epsilon \|u\|_{E_\infty} + C_\epsilon \|u\|_{L^p_\infty(\Omega)}. \quad (2.6)$$

Consider next the boundary operators in the right-hand side of (2.5). We have

$$B_j(\omega_y u) = \omega_y \Phi_j + S_j,$$

where $\Phi_j = B_j u$,

$$S_j = \sum_{k=1}^{N} \sum_{|\beta| \leq \beta_{jk}} b_{jk}^{\beta} \sum_{\alpha+\gamma \leq \beta, |\alpha| > 0} \lambda_{\alpha\gamma} D^{\alpha} \omega_y D^{\gamma} u_k,$$

and $\lambda_{\alpha\gamma}$ are some constants.

There exists a function $v \in W_\infty^{l-\sigma_j,p}(\Omega)$ such that $v = \Phi_i$ on $\partial\Omega$ and

$$\|v\|_{W_\infty^{l-\sigma_j,p}(\Omega)} \leq 2\|\Phi_j\|_{W_\infty^{l-\sigma_j-1/p,p}(\partial\Omega)}.$$

Since $v \in W_\infty^{l-\sigma_j,p}(\Omega)$, then $\omega_y v \in W^{l-\sigma_j,p}(\Omega)$ and

$$\|\omega_y v\|_{W^{l-\sigma_j,p}(\Omega)} \leq M \|v\|_{W_\infty^{l-\sigma_j,p}(\Omega)}$$

with a constant M independent of v. Since $\omega_y v = \omega_y \Phi_j$ on $\partial\Omega$, then

$$\|\omega_j \Phi_j\|_{W^{l-\sigma_j-1/p,p}(\partial\Omega)} \leq M_1 \|\Phi_j\|_{W_\infty^{l-\sigma_j-1/p,p}(\partial\Omega)}.$$

Further,

$$\|S_j\|_{W^{l-\sigma_j-1/p,p}(\partial\Omega)} \leq \|S_j\|_{W^{l-\sigma_j,p}(\Omega)}$$
$$\leq \epsilon \sum_{k=1}^{N} \|u_k\|_{W^{l+t_k,p}(\Omega\cap Q_y)} + C_\epsilon \sum_{k=1}^{N} \|u_k\|_{L^p(\Omega\cap Q_y)}$$
$$\leq \epsilon \|u\|_{E_\infty} + C_\epsilon \|u\|_{L^p_\infty(\Omega)}.$$

Thus

$$\|B_j(\omega_y u)\|_{W^{l-\sigma_j-1/p,p}(\partial\Omega)} \leq M \|\Phi_j\|_{W_\infty^{l-\sigma_j-1/p,p}(\partial\Omega)} + \epsilon \|u\|_{E_\infty} + C_\epsilon \|u\|_{L^p_\infty(\Omega)}. \quad (2.7)$$

From (2.5)–(2.7) we obtain the estimate

$$\|\omega_y u\|_E \leq c\left(M_2 \|Lu\|_{F_\infty} + \kappa\epsilon\|u\|_{E_\infty} + C_\epsilon\|u\|_{L^p_\infty(\Omega)}\right)$$

with some constants M_2 and κ. Taking $\epsilon > 0$ sufficiently small, we obtain (2.4). The theorem is proved. $\square$

3 A priori estimates for adjoint operators. Model systems

In this section we consider the operators

$$A_i^0 u = \sum_{k=1}^{N} \sum_{|\alpha|=s_i+t_k} a_{ik}^\alpha D^\alpha u_k, \quad i = 1,\ldots,N, \quad x \in \Omega,$$

$$B_j^0 u = \sum_{k=1}^{N} \sum_{|\beta|=\sigma_i+t_k} b_{jk}^\beta D^\beta u_k, \quad i = 1,\ldots,m, \quad x \in \partial\Omega,$$

with constant coefficients a_{ik}^α, b_{jk}^β. We suppose here that the domain Ω is the half-space $R_n^+ = \{x_n \geq 0\}$. We will denote by $\hat{A}_i^0$ and $\hat{B}_j^0$ the operators obtained from A_i^0 and B_j^0, respectively, if we replace the derivatives D_i, $i = 1,\ldots,n-1$ by the operators $\hat{D}_i$. The operator

$$\hat{L}^0 = (\hat{A}_1^0,\ldots,\hat{A}_N^0,\hat{B}_1^0,\ldots,\hat{B}_m^0)$$

acts from E to $F = F^d \times F^b$. We consider the operator

$$(L^0)^* : F^* \to E^*$$

adjoint to $L^0 = (A_1^0,\ldots,A_N^0,B_1^0,\ldots,B_m^0)$. We have

$$\begin{aligned} E^* &= \Pi_{j=1}^{N} \dot{W}^{-l-t_j,p'}(\Omega), \\ F^* &= \Pi_{i=1}^{N} \dot{W}^{-l+s_i,p'}(\Omega) \times \Pi_{j=1}^{m} W^{-l+\sigma_j+1/p,p'}(\partial\Omega), \end{aligned} \quad (3.1)$$

3. A priori estimates for adjoint operators. Model systems 121

where $\Omega = R_+^n$, and $\dot{W}^{-s,p'}(\Omega)$ is the closure in $W^{-s,p'}(R^n)$ of infinitely differentiable functions with supports in Ω, $\dot{W}^{-s,p'}(\Omega) = (W^{s,p}(\Omega))^*$, $\frac{1}{p} + \frac{1}{p'} = 1$. Set

$$F_{-1}^* = \Pi_{i=1}^N \dot{W}^{-l+s_i-1,p'}(\Omega) \times \Pi_{j=1}^m W^{-l+\sigma_j+1/p-1,p'}(\partial\Omega). \tag{3.2}$$

Theorem 3.1. *The following estimate holds:*

$$\|v\|_{F^*} \leq C \left(\|(L^0)^* v\|_{E^*} + \|v\|_{F_{-1}^*} \right), \quad \forall v \in F^*, \tag{3.3}$$

where C is a constant independent of v.

Proof. From Theorem 1.1 it follows that the operator $\hat{L}^0$ has a bounded inverse,

$$(\hat{L}^0)^{-1} : F \to E.$$

Therefore the operator

$$((\hat{L}^0)^*)^{-1} : E^* \to F^*$$

is also bounded. Hence we have the estimate

$$\|v\|_{F^*} \leq C\|(\hat{L}^0)^* v\|_{E^*}, \quad \forall v \in F^*.$$

Therefore
$$\|v\|_{F^*} \leq C \left(\|(L^0)^* v\|_{E^*} + \|((\hat{L}^0)^* - (L^0)^*)v\|_{E^*} \right). \tag{3.4}$$

We estimate the second term in the right-hand side of this inequality. Let $\langle,\rangle_E$ be the duality between E and E^*. For $u \in E$ we have

$$|\langle u, ((\hat{L}^0)^* - (L^0)^*)v \rangle_E| = |\langle (\hat{L}^0 - L^0)u, v \rangle_F|. \tag{3.5}$$

Let $v = (v_1, \ldots, v_N, w_1, \ldots, w_m)$, where

$$v_i \in (W^{l-s_i,p}(\Omega))^*, \quad w_j \in (W^{l-\sigma_j-1/p,p}(\partial\Omega))^*.$$

Then we have

$$\langle (\hat{L}^0 - L^0)u, v \rangle_F = \sum_{i=1}^N \langle (\hat{A}_i^0 - A_i^0)u, v_i \rangle + \sum_{j=1}^m \langle (\hat{B}_j^0 - B_j^0)u, w_j \rangle. \tag{3.6}$$

Let
$$T_i : W^{l-s_i+1,p}(R^n) \to W^{l-s_i,p}(R^n)$$

be an isomorphism between the two spaces.

Denote by $\tilde{u}_j$ an extension of u_j to $W^{l+t_j,p}(R^n)$ such that

$$\|\tilde{u}_j\|_{W^{l+t_j,p}(R^n)} \leq 2\|u_j\|_{W^{l+t_j,p}(\Omega)}, \tag{3.7}$$

and $\tilde{u} = (\tilde{u}_1, \ldots, \tilde{u}_N)$. Then $(\hat{A}_i^0 - A_i^0)\tilde{u}$ is an extension of $(\hat{A}_i^0 - A_i^0)u$ from $W^{l-s_i+1,p}(\Omega)$ to $W^{l-s_i+1,p}(\mathbb{R}^n)$. We have

$$(\hat{A}_i^0 - A_i^0)u \in W^{l-s_i+1,p}(\Omega) \subset W^{l-s_i,p}(\Omega).$$

Hence v_i can be considered as an element of $(W^{l-s_i+1,p}(\Omega))^*$. It can be extended to an element $\tilde{v}_i \in (W^{l-s_i+1,p}(\mathbb{R}^n))^*$ such that

$$\langle (\hat{A}_i^0 - A_i^0)u, v_i \rangle = \langle (\hat{A}_i^0 - A_i^0)\tilde{u}, \tilde{v}_i \rangle$$

and

$$\|\tilde{v}_i\|_{(W^{l-s_i+1,p}(\mathbb{R}^n))^*} = \|v_i\|_{(W^{l-s_i+1,p}(\Omega))^*}. \tag{3.8}$$

Then

$$|\langle (\hat{A}_i^0 - A_i^0)u, v_i \rangle| = |\langle T_i^{-1}T_i(\hat{A}_i^0 - A_i^0)\tilde{u}, \tilde{v}_i \rangle| = |\langle T_i(\hat{A}_i^0 - A_i^0)\tilde{u}, (T_i^{-1})^*\tilde{v}_i \rangle|.$$

Since

$$(T_i^{-1})^* : W^{-l+s_i-1,p'}(\mathbb{R}^n) \to W^{-l+s_i,p'}(\mathbb{R}^n),$$

then

$$|\langle (\hat{A}_i^0 - A_i^0)u, v_i \rangle| \leq \|T_i(\hat{A}_i^0 - A_i^0)\tilde{u}\|_{W^{l-s_i,p}(\mathbb{R}^n)} \|(T_i^{-1})^*\tilde{v}_i\|_{(W^{l-s_i,p}(\mathbb{R}^n))^*}$$
$$\leq C\|(\hat{A}_i^0 - A_i^0)\tilde{u}\|_{W^{l-s_i+1,p}(\mathbb{R}^n)} \|\tilde{v}_i\|_{(W^{l-s_i+1,p}(\mathbb{R}^n))^*}$$
$$\leq C_1\|\tilde{u}\|_{E(\mathbb{R}^n)} \|\tilde{v}_i\|_{(W^{l-s_i+1,p}(\mathbb{R}^n))^*}.$$

From (3.7), (3.8)

$$|\langle (\hat{A}_i^0 - A_i^0)u, v_i \rangle| \leq C_2 \|u\|_{E(\Omega)} \|v_i\|_{(W^{l-s_i+1,p}(\Omega))^*}$$
$$= C_2 \|u\|_{E(\Omega)} \|v_i\|_{\dot{W}^{-l+s_i-1,p'}(\Omega)}. \tag{3.9}$$

Consider now the boundary operators in (3.6). We have

$$(\hat{B}_j^0 - B_j^0)u \in W^{l-\sigma_j-1/p+1,p}(\partial\Omega).$$

Let

$$S_j : W^{l-\sigma_j-1/p+1,p}(\partial\Omega) \to W^{l-\sigma_j-1/p,p}(\partial\Omega)$$

be an isomorphism between the two spaces. Then

$$|\langle (\hat{B}_j^0 - B_j^0)u, w_j \rangle| = |\langle S_j^{-1}S_j(\hat{B}_j^0 - B_j^0)u, w_j \rangle| = |\langle S_j(\hat{B}_j^0 - B_j^0)u, (S_j^{-1})^*w_j \rangle|$$
$$\leq \|S_j(\hat{B}_j^0 - B_j^0)u\|_{W^{l-\sigma_j-1/p,p}(\partial\Omega)} \|(S_j^{-1})^*w_j\|_{(W^{l-\sigma_j-1/p,p}(\partial\Omega))^*}$$
$$\leq C_3 \|(\hat{B}_j^0 - B_j^0)u\|_{W^{l-\sigma_j-1/p+1,p}(\partial\Omega)} \|w_j\|_{(W^{l-\sigma_j-1/p+1,p}(\partial\Omega))^*}$$
$$\leq C_4 \|u\|_{E(\Omega)} \|w_j\|_{W^{-l+\sigma_j+1/p-1,p'}(\partial\Omega)}. \tag{3.10}$$

From (3.5), (3.6), (3.9), and (3.10) we obtain

$$|\langle u, ((\hat{L}^0)^* - (L^0)^*)v\rangle_E|$$
$$\leq C_5\|u\|_{E(\Omega)}\left(\sum_{i=1}^{N}\|v_i\|_{\dot{W}^{-l+s_i-1,p'}(\Omega)} + \sum_{j=1}^{m}\|w_j\|_{W^{-l+\sigma_j+1/p-1,p'}(\partial\Omega)}\right)$$
$$= C_5\|u\|_{E(\Omega)}\|v\|_{F_{-1}^*}.$$

Therefore
$$\|((\hat{L}^0)^* - (L^0)^*)v\|_{E^*} \leq C_5\|v\|_{F_{-1}^*}.$$

Estimate (3.3) follows from this estimate and (3.4). The theorem is proved. □

4 The general problem in a half-space

We consider operators (0.1), (0.2) with $\Omega = \mathbb{R}^n_+$.

Theorem 4.1. Let $v \in F^*(\mathbb{R}^n_+)$ vanish outside the ball $\sigma(\rho) = \{x : |x| < \rho\}$. Then there exists $\rho_0 > 0$ such that for $\rho < \rho_0$ the following estimate holds:

$$\|v\|_{F^*(\mathbb{R}^n_+)} \leq C\left(\|L^*v\|_{E^*(\mathbb{R}^n_+)} + \|v\|_{F_{-1}^*(\mathbb{R}^n_+)}\right). \tag{4.1}$$

Proof. Let
$$A_i^0 u = \sum_{k=1}^{N}\sum_{|\alpha|=s_i+t_k} a_{ik}^\alpha(0)D^\alpha u_k, \quad i=1,\ldots,N, \quad x \in \mathbb{R}^n_+,$$
$$B_j^0 u = \sum_{k=1}^{N}\sum_{|\beta|=\sigma_j+t_k} b_{jk}^\beta(0)D^\beta u_k, \quad i=1,\ldots,m, \quad x \in \mathbb{R}^{n-1},$$

$L^0 = (A_1^0,\ldots,A_N^0, B_1^0,\ldots,B_m^0) : E \to F$, $(L^0)^*$ is the adjoint operator, $(L^0)^* : F^* \to E^*$. From Theorem 3.1 we have

$$\|v\|_{F^*(\mathbb{R}^n_+)} \leq C\left(\|(L^0)^*v\|_{E^*(\mathbb{R}^n_+)} + \|v\|_{F_{-1}^*(\mathbb{R}^n_+)}\right). \tag{4.2}$$

On the other hand,
$$\|(L^0)^*v\|_{E^*} \leq \|L^*v\|_{E^*} + \|(L^0)^*v - L^*v\|_{E^*}. \tag{4.3}$$

We estimate the second term in the right-hand side. For any $u \in E$ and $v = (v_1,\ldots,v_N,w_1,\ldots,w_m)$ we have

$$\langle u, ((L^0)^* - L^*)v\rangle_E = \langle ((L^0 - L)u, v\rangle_F = \sum_{i=1}^{N}\langle (A_i^0 - A_i)u, v_i\rangle + \sum_{j=1}^{m}\langle (B_j^0 - B_j)u, w_j\rangle. \tag{4.4}$$

Further,
$$|\langle (A_i^0 - A_i)u, v_i \rangle| \le |\langle A_i^1 u, v_i \rangle| + |\langle A_i^2 u, v_i \rangle|, \quad (4.5)$$
where
$$A_i^1 u = \sum_{k=1}^{N} \sum_{|\alpha|=\alpha_{ik}} (a_{ik}^\alpha(0) - a_{ik}^\alpha(x)) D^\alpha u_k, \quad A_i^2 u = \sum_{k=1}^{N} \sum_{|\alpha|<\alpha_{ik}} a_{ik}^\alpha(x) D^\alpha u_k. \quad (4.6)$$

We estimate first the operator A_i^1. We have

$$\langle A_i^1 u, v_i \rangle = \sum_{k=1}^{N} \sum_{|\alpha|=\alpha_{ik}} \langle (a_{ik}^\alpha(0) - a_{ik}^\alpha(x)) D^\alpha u_k, v_i \rangle,$$

$$|\langle (a_{ik}^\alpha(0) - a_{ik}^\alpha(x)) D^\alpha u_k, v_i \rangle| = |\langle D^\alpha u_k, \overline{(a_{ik}^\alpha(0) - a_{ik}^\alpha(x))} v_i \rangle|$$
$$\le \|D^\alpha u_k\|_{W^{l-s_i,p}(\mathbb{R}^n_+)} \|\overline{(a_{ik}^\alpha(0) - a_{ik}^\alpha(x))} v_i\|_{(W^{l-s_i,p}(\mathbb{R}^n_+))^*}$$
$$\le \|u_k\|_{W^{l+t_k,p}(\mathbb{R}^n_+)} \|\overline{(a_{ik}^\alpha(0) - a_{ik}^\alpha(x))} v_i\|_{(W^{l-s_i,p}(\mathbb{R}^n_+))^*},$$

$$|\langle A_i^1 u, v_i \rangle| \le \sum_{k=1}^{N} \|u_k\|_{W^{l+t_k,p}(\mathbb{R}^n_+)} \sum_{|\alpha|=\alpha_{ik}} \|\overline{(a_{ik}^\alpha(0) - a_{ik}^\alpha(x))} v_i\|_{(W^{l-s_i,p}(\mathbb{R}^n_+))^*}. \quad (4.7)$$

We estimate the second sum in the right-hand side. Let $\psi \in D$, $\psi(x) = 1$ in $\sigma(\rho)$, $\psi(x) = 0$ outside $\sigma(2\rho)$. Then we have

$$\|\overline{(a_{ik}^\alpha(0) - a_{ik}^\alpha(x))} v_i\|_{(W^{l-s_i,p}(\mathbb{R}^n_+))^*} = \|\overline{(a_{ik}^\alpha(0) - a_{ik}^\alpha(x))} \psi v_i\|_{(W^{l-s_i,p}(\mathbb{R}^n_+))^*}$$
$$= \|\overline{(a_{ik}^\alpha(0) - a_{ik}^\alpha(x))} \psi v_i\|_{\dot{W}^{-l+s_i,p'}(\mathbb{R}^n_+)} \equiv T.$$

From Lemma 4.2 (see below) we obtain

$$T \le C_1 \max_{x \in \mathbb{R}^n} |\overline{(a_{ik}^\alpha(0) - a_{ik}^\alpha(x))} \psi| \, \|v_i\|_{\dot{W}^{-l+s_i,p'}(\mathbb{R}^n_+)} + K_{\alpha\rho} \|v_i\|_{\dot{W}^{-l+s_i-1,p'}(\mathbb{R}^n_+)}.$$

For any $\epsilon > 0$ we can find $\rho_0 > 0$ such that for $0 < \rho \le \rho_0$ we have

$$T \le \epsilon \|v_i\|_{\dot{W}^{-l+s_i,p'}(\mathbb{R}^n_+)} + K_{\alpha\rho} \|v_i\|_{\dot{W}^{-l+s_i-1,p'}(\mathbb{R}^n_+)}.$$

From (4.7)

$$|\langle A_i^1 u, v_i \rangle| \le \|u\|_{E(\mathbb{R}^n_+)} \left(\kappa \epsilon \|v_i\|_{\dot{W}^{-l+s_i,p'}(\mathbb{R}^n_+)} + M \|v_i\|_{\dot{W}^{-l+s_i-1,p'}(\mathbb{R}^n_+)} \right), \quad (4.8)$$

where κ and M are some constants.

Consider now the operator A_i^2 in (4.6). We have $A_i^2 : E(\mathbb{R}^+_n) \to W^{l-s_i,p}(\mathbb{R}^n_+)$. We can extend its coefficients in such a way that the extended operator

$$\tilde{A}_i^2 : E(\mathbb{R}^n) \to W^{l-s_i,p}(\mathbb{R}^n)$$

4. The general problem in a half-space

is bounded. Let further T_i be a bounded linear operator with a bounded inverse acting from $W^{l-s_i+1,p}(\mathbb{R}^n)$ into $W^{l-s_i,p}(\mathbb{R}^n)$. Then

$$(T_i^{-1})^* : (W^{l-s_i+1,p}(\mathbb{R}^n))^* \to (W^{l-s_i,p}(\mathbb{R}^n))^*$$

is also bounded.

Let $u \in E(\mathbb{R}_n^+)$, $\tilde{u}$ be its extension to $E(\mathbb{R}^n)$ such that $\|\tilde{u}\|_{E(\mathbb{R}^n)} \leq 2\|u\|_{E(\mathbb{R}_+^n)}$. Suppose that

$$v_i \in (W^{l-s_i,p}(\mathbb{R}_+^n))^* = \dot{W}^{-l+s_i,p'}(\mathbb{R}_+^n).$$

We consider the extension $\tilde{v}_i \in W^{-l+s_i-1,p'}(\mathbb{R}^n)$. Then we have

$$|\langle A_i^2 u, v_i \rangle| = |\langle \tilde{A}_i^2 \tilde{u}, \tilde{v}_i \rangle| = |\langle T_i^{-1} T_i \tilde{A}_i^2 \tilde{u}, \tilde{v}_i \rangle| = |\langle T_i \tilde{A}_i^2 \tilde{u}, (T_i^{-1})^* \tilde{v}_i \rangle|$$
$$\leq \|T_i \tilde{A}_i^2 \tilde{u}\|_{W^{l-s_i,p}(\mathbb{R}^n)} \|(T_i^{-1})^* \tilde{v}_i\|_{W^{-l+s_i,p'}(\mathbb{R}^n)}$$
$$\leq C_1 \|\tilde{A}_i^2 \tilde{u}\|_{W^{l-s_i+1,p}(\mathbb{R}^n)} \|\tilde{v}_i\|_{W^{-l+s_i-1,p'}(\mathbb{R}^n)}$$
$$\leq C_2 \|u\|_{E(\mathbb{R}_n^+)} \|v_i\|_{\dot{W}^{-l+s_i-1,p'}(\mathbb{R}_+^n)}. \tag{4.9}$$

From this estimate, (4.5) and (4.8) it follows that

$$|\langle (A_i^0 - A_i) u, v_i \rangle| \leq \|u\|_{E(\mathbb{R}_n^+)} \left(\kappa \epsilon \|v_i\|_{\dot{W}^{-l+s_i,p'}(\mathbb{R}_+^n)} + M_1 \|v_i\|_{\dot{W}^{-l+s_i-1,p'}(\mathbb{R}_+^n)} \right). \tag{4.10}$$

Consider now the second term in the right-hand side of (4.4). We have

$$(B_j^0 - B_j) u = B_j^1 u + B_j^2 u, \tag{4.11}$$

where

$$B_j^1 u = \sum_{k=1}^N \sum_{|\beta|=\beta_{jk}} (b_{jk}^\beta(0) - b_{jk}^\beta(x)) D^\beta u_k, \quad B_j^2 u = -\sum_{k=1}^N \sum_{|\beta|<\beta_{jk}} b_{jk}^\beta(x) D^\beta u_k. \tag{4.12}$$

Consider first the operator B_j^1:

$$\langle B_j^1 u, w_j \rangle = \sum_{k=1}^N \sum_{|\beta|=\beta_{jk}} \langle (b_{jk}^\beta(0) - b_{jk}^\beta(x)) D^\beta u_k, w_j \rangle,$$

$$|\langle (b_{jk}^\beta(0) - b_{jk}^\beta(x)) D^\beta u_k, w_j \rangle| = |\langle D^\beta u_k, \overline{(b_{jk}^\beta(0) - b_{jk}^\beta(x))} w_j \rangle|$$
$$\leq \|D^\beta u_k\|_{W^{l-\sigma_j-1/p,p}(\mathbb{R}^{n-1})} \|\overline{(b_{jk}^\beta(0) - b_{jk}^\beta(x))} w_j\|_{W^{-l+\sigma_j+1/p,p'}(\mathbb{R}^{n-1})}$$
$$\leq \|u_k\|_{W^{l+t_k,p}(\mathbb{R}_+^n)} \|\overline{(b_{jk}^\beta(0) - b_{jk}^\beta(x))} w_j\|_{W^{-l+\sigma_j+1/p,p'}(\mathbb{R}^{n-1})}.$$

Hence

$$|\langle B_j^1 u, w_j \rangle| \leq \sum_{k=1}^N \|u_k\|_{W^{l+t_k,p}(\mathbb{R}_+^n)} \sum_{|\beta|=\beta_{jk}} \|\overline{(b_{jk}^\beta(0) - b_{jk}^\beta(x))} w_j\|_{W^{-l+\sigma_j+1/p,p'}(\mathbb{R}^{n-1})}. \tag{4.13}$$

We estimate the second sum in the right-hand side. We have

$$\|(\overline{b_{jk}^\beta(0)} - \overline{b_{jk}^\beta(x)})w_j\|_{W^{-l+\sigma_j+1/p,p'}(\mathbb{R}^{n-1})}$$
$$= \|(\overline{b_{jk}^\beta(0)} - \overline{b_{jk}^\beta(x)})\psi w_j\|_{W^{-l+\sigma_j+1/p,p'}(\mathbb{R}^{n-1})} \equiv T_1.$$

Then by Lemma 4.2 we have

$$T_1 \leq C_1 \max_{x \in \mathbb{R}^{n-1}} |(\overline{b_{jk}^\beta(0)} - \overline{b_{jk}^\beta(x)})\psi| \, \|w_j\|_{W^{-l+\sigma_j+1/p,p'}(\mathbb{R}^{n-1})}$$
$$+ K_{\beta\rho}\|w_j\|_{W^{-l+\sigma_j+1/p-1,p'}(\mathbb{R}^{n-1})}.$$

For any $\epsilon > 0$ we can find $\rho_0 > 0$ such that for $0 < \rho \leq \rho_0$ the following inequality holds:

$$T_1 \leq \epsilon \|w_j\|_{W^{-l+\sigma_j+1/p,p'}(\mathbb{R}^{n-1})} + K_{\beta\rho}\|w_j\|_{W^{-l+\sigma_j+1/p-1,p'}(\mathbb{R}^{n-1})}.$$

From (4.13)

$$|\langle B_j^1 u, w_j \rangle| \tag{4.14}$$
$$\leq \|u\|_{E(R_+^n)} \left(\kappa\epsilon \|w_j\|_{W^{-l+\sigma_j+1/p,p'}(\mathbb{R}^{n-1})} + M\|w_j\|_{W^{-l+\sigma_j+1/p-1,p'}(\mathbb{R}^{n-1})} \right).$$

Consider now the operator B_j^2 in (4.11),

$$B_j^2 : E(\mathbb{R}_+^n) \to W^{l-\sigma_j-1/p+1,p}(\mathbb{R}^{n-1}).$$

Let

$$S_j : W^{l-\sigma_j-1/p+1,p}(\mathbb{R}^{n-1}) \to W^{l-\sigma_j-1/p,p}(\mathbb{R}^{n-1})$$

be an isomorphism. We have

$$\langle B_j^2 u, w_j \rangle = \langle S_j^{-1} S_j B_j^2 u, w_j \rangle = \langle S_j B_j^2 u, (S_j^{-1})^* w_j \rangle,$$
$$|\langle B_j^2 u, w_j \rangle| \leq \|S_j B_j^2 u\|_{W^{l-\sigma_j-1/p,p}(\mathbb{R}^{n-1})} \|(S_j^{-1})^* w_j\|_{W^{-l+\sigma_j+1/p,p'}(\mathbb{R}^{n-1})}$$
$$\leq C\|B_j^2 u\|_{W^{l-\sigma_j-1/p+1,p}(\mathbb{R}^{n-1})} \|w_j\|_{W^{-l+\sigma_j+1/p-1,p'}(\mathbb{R}^{n-1})}$$
$$\leq C_1 \|u\|_{E(R_+^n)} \|w_j\|_{W^{-l+\sigma_j+1/p-1,p'}(\mathbb{R}^{n-1})}. \tag{4.15}$$

From this estimate, (4.11), and (4.14) we obtain

$$|\langle (B_j^0 - B_j)u, w_j \rangle| \tag{4.16}$$
$$\leq \|u\|_{E(R_+^n)} \left(\kappa\epsilon \|w_j\|_{W^{-l+\sigma_j+1/p,p'}(\mathbb{R}^{n-1})} + M_1\|w_j\|_{W^{-l+\sigma_j+1/p-1,p'}(\mathbb{R}^{n-1})} \right).$$

4. The general problem in a half-space

From (4.4), (4.10), and (4.16)

$$|\langle u, ((L^0)^* - L^*)v \rangle_E|$$

$$\leq \|u\|_{E(\mathbb{R}_n^+)} \left(\kappa \epsilon \sum_{i=1}^{N} \|v_i\|_{\dot{W}^{-l+s_i,p'}(\mathbb{R}_+^n)} + \sum_{i=1}^{N} M_1 \|v_i\|_{\dot{W}^{-l+s_i-1,p'}(\mathbb{R}_+^n)} \right)$$

$$+ \|u\|_{E(\mathbb{R}_n^+)} \left(\kappa \epsilon \sum_{j=1}^{m} \|w_j\|_{W^{-l+\sigma_j+1/p,p'}(\mathbb{R}^{n-1})} + M_1 \sum_{j=1}^{m} \|w_j\|_{W^{-l+\sigma_j+1/p-1,p'}(\mathbb{R}^{n-1})} \right).$$

Using notations (3.1) and (3.2) we can write this estimate as

$$|\langle u, ((L^0)^* - L^*)v \rangle_E| \leq \|u\|_{E(\mathbb{R}_n^+)} \left(\kappa \epsilon \|v\|_{F^*(\mathbb{R}_+^n)} + M_1 \|v\|_{F^*_{-1}(\mathbb{R}_+^n)} \right).$$

Hence

$$\|((L^0)^* - L^*)v\|_{E^*(\mathbb{R}_n^+)} \leq \kappa \epsilon \|v\|_{F^*(\mathbb{R}_+^n)} + M_1 \|v\|_{F^*_{-1}(\mathbb{R}_+^n)}.$$

Estimate (4.1) follows from the last estimate, (4.2) and (4.3). The theorem is proved. □

In the proof of the theorem we used the following lemma.

Lemma 4.2. *Let v have a bounded support, $a \in C_0^m(\mathbb{R}^n)$, $v \in H^{s,p}(\mathbb{R}^n)$, $1 - m \leq s \leq 0$. Then*

$$\|av\|_{H^{s,p}(\mathbb{R}^n)} \leq c_1 \max_{x \in \mathbb{R}^n} |a(x)| \|v\|_{H^{s,p}(\mathbb{R}^n)} + c_2(a) \|v\|_{H^{s-1,p}(\mathbb{R}^n)}, \quad (4.17)$$

where the constant c_1 does not depend on v and a, and $c_2 = 0$ if $s = 0$. A similar estimate holds for a function $v \in B^{s,p}(\mathbb{R}^{n-1})$:

$$\|av\|_{B^{s,p}(\mathbb{R}^{n-1})} \leq c_1 \max_{x \in \mathbb{R}^{n-1}} |a(x)| \|v\|_{B^{s,p}(\mathbb{R}^{n-1})} + c_2(a) \|v\|_{B^{s-1,p}(\mathbb{R}^{n-1})}. \quad (4.18)$$

Proof. The proof in [457] (Section 1.12) is given for the spaces $H^{s,p}(\mathbb{R}^n)$. Let us prove estimate (4.18) for the case of noninteger s which we use below. All necessary elements of the proof are given in [457]. We recall that the Besov spaces coincide in this case with the Sobolev-Slobodetskii spaces.

We note first of all that for positive noninteger s the estimate

$$\|av\|_{W^{s,p}(\mathbb{R}^n)} \leq c_1 \max_{x \in \mathbb{R}^n} |a(x)| \|v\|_{W^{s,p}(\mathbb{R}^n)} + c_2(a) \|v\|_{W^{s-\sigma,p}(\mathbb{R}^n)}, \quad (4.19)$$

where $\sigma = s - k$, $k = [s]$ can be verified directly from the definition of the space $W^{s,p}(\mathbb{R}^n)$. Indeed,

$$\|av\|^p_{W^{s,p}(\mathbb{R}^n)} = \|av\|^p_{W^{k,p}(\mathbb{R}^n)} + \sum_{|\alpha|=k} \int_{\mathbb{R}^n} \int_{\mathbb{R}^n} \frac{|D^\alpha(a(x)v(x)) - D^\alpha(a(y)v(y))|^p}{|x-y|^{n+p\sigma}} dx dy,$$

$$\|av\|_{W^{k,p}(\mathbb{R}^n)} \leq C \sum_{|\alpha|+|\beta|\leq k} \|D^\alpha a D^\beta v\|_{L^p(\mathbb{R}^n)}$$

$$= C \sum_{|\beta|\leq k} \|aD^\beta v\|_{L^p(\mathbb{R}^n)} + C \sum_{|\alpha|+|\beta|\leq k, |\alpha|>0} \|D^\alpha a D^\beta v\|_{L^p(\mathbb{R}^n)}$$

$$\leq c_1 \sup_x |a(x)| \|v\|_{W^{k,p}(\mathbb{R}^n)} + c_2(a) \|v\|_{W^{k-1,p}(\mathbb{R}^n)},$$

$$\int_{\mathbb{R}^n}\int_{\mathbb{R}^n} \frac{|D^\alpha a(x) D^\beta v(x) - D^\alpha a(y) D^\beta v(y)|^p}{|x-y|^{n+p\sigma}} dx dy \leq M_1(I_1 + I_2),$$

where

$$I_1 = \int_{\mathbb{R}^n}\int_{\mathbb{R}^n} \frac{|D^\alpha a(x)|^p |D^\beta v(x) - D^\beta v(y)|^p}{|x-y|^{n+p\sigma}} dx dy,$$

$$I_2 = \int_{\mathbb{R}^n}\int_{\mathbb{R}^n} \frac{|D^\beta v(y)|^p |D^\alpha a(x) - D^\alpha a(y)|^p}{|x-y|^{n+p\sigma}} dx dy.$$

If $|\alpha| = 0$, then

$$I_1 \leq \|a\|_{C^0(\mathbb{R}^n)}^p \|v\|_{W^{s,p}(\mathbb{R}^n)}^p.$$

If $|\alpha| > 0$, then

$$I_1 \leq \|a\|_{C^k(\mathbb{R}^n)}^p \|v\|_{W^{s-1,p}(\mathbb{R}^n)}^p.$$

We estimate now I_2:

$$I_2 = \int_{\mathbb{R}^n} |D^\beta v(y)|^p \left(\int_{\mathbb{R}^n} \frac{|D^\alpha a(x) - D^\alpha a(y)|^p}{|x-y|^{n+p\sigma}} dx\right) dy.$$

Let us prove that

$$J = \int_{\mathbb{R}^n} \frac{|D^\alpha a(x) - D^\alpha a(y)|^p}{|x-y|^{n+p\sigma}} dx \leq M_2,$$

where M_2 is a constant. We have

$$J = \int_{\mathbb{R}^n} \frac{|D^\alpha a(y+z) - D^\alpha a(y)|^p}{|z|^{n+p\sigma}} dz = J_1 + J_2,$$

where

$$J_1 = \int_{|z|\leq 1} \frac{|D^\alpha a(y+z) - D^\alpha a(y)|^p}{|z|^{n+p\sigma}} dz, \quad J_2 = \int_{|z|>1} \frac{|D^\alpha a(y+z) - D^\alpha a(y)|^p}{|z|^{n+p\sigma}} dz.$$

The integral J_1 is bounded since

$$|D^\alpha a(y+z) - D^\alpha a(y)| \leq K|z|, \quad |z| \leq 1, \ y \in \mathbb{R}^n,$$

and J_2 is bounded since $|D^\alpha a(x)| \leq M$. Here K and M are some positive constants.

4. The general problem in a half-space

Thus, $I_2 \leq M_2 \|v\|_{W^{k,p}(\mathbb{R}^n)}^p$. This completes the proof of estimate (4.19) for positive s. We recall that it is proved for $\sigma = s - [s]$. It can now be obtained for any positive σ with the help of the estimate

$$\|v\|_{W^{s,p}(\mathbb{R}^n)} \leq \epsilon \|v\|_{W^{s_1,p}(\mathbb{R}^n)} + C_\epsilon \|v\|_{W^{s_2,p}(\mathbb{R}^n)}$$

that holds for any $s_2 < s < s_1$ and any $\epsilon > 0$.

We now prove a similar estimate for the dual spaces. It is shown in [457] that there exists an operator χ_N which satisfies the following properties:

(i) it is a continuous operator from $B^{s,p}(\mathbb{R}^{n-1})$ into $B^{s+t,p}(\mathbb{R}^{n-1})$ for any real s and $t > 0$,

(ii) the estimate

$$\|(I - \chi_N)u\|_{B^{s,p}(\mathbb{R}^{n-1})} \leq M \|u\|_{B^{s,p}(\mathbb{R}^{n-1})}$$

holds with a constant M independent of u and N,

(iii) for any $\epsilon > 0$ and $\sigma_0 > 0$ there exists $N(\epsilon, \sigma_0) > 0$ such that

$$\|(I - \chi_N)u\|_{B^{s-\sigma,p}(\mathbb{R}^{n-1})} \leq \epsilon \|u\|_{B^{s,p}(\mathbb{R}^{n-1})}$$

for any $N \geq N(\epsilon, \sigma_0)$ and $\sigma \geq \sigma_0$.

Substituting in (4.19) p' instead of p and $(I - \chi_N)u$ instead of v and using the properties of the operator χ_N, we obtain the estimate

$$\|a(I - \chi_N)u\|_{B^{s,p'}(\mathbb{R}^{n-1})} \leq M \max_{x \in \mathbb{R}^{n-1}} |a(x)| \|u\|_{B^{s,p'}(\mathbb{R}^{n-1})} \quad (s > 0). \tag{4.20}$$

Let $u \in B^{-s,p}(\mathbb{R}^{n-1})$, $w \in B^{s,p}(\mathbb{R}^{n-1})$. Then we have from (4.20) for a positive s:

$$|\langle (I - \chi_N)au, w \rangle| = |\langle u, \bar{a}(I - \chi_N)w \rangle|$$
$$\leq \|u\|_{B^{-s,p}(\mathbb{R}^{n-1})} \|\bar{a}(I - \chi_N)w\|_{B^{s,p'}(\mathbb{R}^{n-1})}$$
$$\leq M \max_{x \in \mathbb{R}^{n-1}} |a(x)| \|u\|_{B^{-s,p}(\mathbb{R}^{n-1})} \|w\|_{B^{s,p'}(\mathbb{R}^{n-1})}.$$

Therefore

$$\|(I - \chi_N)au\|_{B^{-s,p}(\mathbb{R}^{n-1})} \leq M \max_{x \in \mathbb{R}^{n-1}} |a(x)| \|u\|_{B^{-s,p}(\mathbb{R}^{n-1})} \quad (s > 0). \tag{4.21}$$

We have finally for $s > 0$, $\sigma > 0$:

$$\|au\|_{B^{-s,p}(\mathbb{R}^{n-1})} \leq \|(I - \chi_N)au\|_{B^{-s,p}(\mathbb{R}^{n-1})} + \|\chi_N au\|_{B^{-s,p}(\mathbb{R}^{n-1})}$$
$$\leq M_1 \left(\max_{x \in \mathbb{R}^{n-1}} |a(x)| \|u\|_{B^{-s,p}(\mathbb{R}^{n-1})} + \|au\|_{B^{-s-\sigma,p}(\mathbb{R}^{n-1})} \right)$$
$$\leq M_1 \left(\max_{x \in \mathbb{R}^{n-1}} |a(x)| \|u\|_{B^{-s,p}(\mathbb{R}^{n-1})} + c_2(a) \|u\|_{B^{-s-\sigma,p}(\mathbb{R}^{n-1})} \right).$$

Here we use (4.21), the fact that the operator χ_N is continuous from $B^{-s-\sigma,p}(\mathbb{R}^{n-1})$ into $B^{-s,p}(\mathbb{R}^{n-1})$, and the estimate

$$\|au\|_{B^{-s-\sigma,p}(\mathbb{R}^{n-1})} \leq c_2(a)\|u\|_{B^{-s-\sigma,p}(\mathbb{R}^{n-1})} \quad (\sigma < 1).$$

To obtain (4.18) we use the inequality

$$\|u\|_{B^{-s-\sigma,p}(\mathbb{R}^{n-1})} \leq \epsilon \|u\|_{B^{-s,p}(\mathbb{R}^{n-1})} + c(\epsilon)\|u\|_{B^{-s-1,p}(\mathbb{R}^{n-1})},$$

where $\epsilon = (\max|a(x)|)/c_2(a) > 0$ (in the case $a(x) = 0$, (4.18) is obvious). The lemma is proved. □

5 The general problem in unbounded domains

Consider the operators A_i, B_j, and L defined by (0.1)–(0.3). We will use the spaces E and F introduced in Sections 1 and 3, and the corresponding ∞-spaces:

$$E_\infty(\Omega) = \Pi_{j=1}^N W_\infty^{l+t_j,p}(\Omega),$$
$$F_\infty(\Omega) = \Pi_{i=1}^N W_\infty^{l-s_i,p}(\Omega) \times \Pi_{j=1}^m W_\infty^{l-\sigma_j-1/p,p}(\partial\Omega),$$
$$(E^*(\Omega))_\infty = \Pi_{j=1}^N (\dot{W}^{-l-t_j,p'}(\Omega))_\infty,$$
$$(F^*(\Omega))_\infty = \Pi_{i=1}^N (\dot{W}^{-l+s_i,p'}(\Omega))_\infty \times \Pi_{j=1}^m (W^{-l+\sigma_j+1/p,p'}(\partial\Omega))_\infty,$$
$$(F_{-1}^*(\Omega))_\infty = \Pi_{i=1}^N (\dot{W}^{-l+s_i-1,p'}(\Omega))_\infty \times \Pi_{j=1}^m (W^{-l+\sigma_j+1/p-1,p'}(\partial\Omega))_\infty.$$

We suppose that the domain Ω satisfies Condition D (Section 3).

Theorem 5.1. *For any $v \in (F^*(\Omega))_\infty$ the following estimate holds:*

$$\|v\|_{(F^*(\Omega))_\infty} \leq M \left(\|L^*v\|_{(E^*(\Omega))_\infty} + \|v\|_{(F_{-1}^*(\Omega))_\infty} \right) \tag{5.1}$$

with a constant M independent of v.

The proof of the theorem will be given after some preliminary considerations. Let δ and ψ be the same as in Condition D, $B_\delta(x_0) = \{x : |x - x_0| < \delta\}$, $G_{x_0} = \psi(B_\delta(x_0))$. We introduce the operator of change of variables,

$$T : W^{s,p}(\Omega \cap B_\delta(x_0)) \to W^{s,p}(G_{x_0} \cap \{y_n > 0\}), \quad s \geq 0.$$

We will use the same notation also for the operator of change of variables in the space $W^{s,p}(\Gamma)(s \geq 0, \Gamma = \partial\Omega)$ defined on functions with support in $B_\delta(x_0)$,

$$T : W^{s,p}(\Gamma) \to W^{s,p}(\mathbb{R}_{y'}^{n-1}).$$

We have for functions with supports in $B_\delta(x_0)$:

$$T : F(\Omega) \to F(\mathbb{R}_+^n), \quad T^{-1} : F(\mathbb{R}_+^n) \to F(\Omega),$$
$$T : E(\Omega) \to E(\mathbb{R}_+^n), \quad T^{-1} : E(\mathbb{R}_+^n) \to E(\Omega),$$
$$L : E(\Omega) \to F(\Omega), \quad \tilde{L} = TLT^{-1} : E(\mathbb{R}_+^n) \to F(\mathbb{R}_+^n).$$

5. The general problem in unbounded domains

Consider the adjoint operators. We have

$$(\tilde{L})^* = (T^{-1})^* L^* T^* : F^*(\mathbb{R}^n_+) \to E^*(\mathbb{R}^n_+).$$

Here

$$T^* : F^*(\mathbb{R}^n_+) \to F^*(\Omega), \quad (T^*)^{-1} : F^*(\Omega) \to F^*(\mathbb{R}^n_+),$$
$$T^* : E^*(\mathbb{R}^n_+) \to E^*(\Omega), \quad (T^*)^{-1} : E^*(\Omega) \to E^*(\mathbb{R}^n_+).$$

Let $\tilde{v} \in F^*(\mathbb{R}^n_+)$ satisfy the conditions of Theorem 4.1, and $v = T^*\tilde{v} \in F^*(\Omega)$. From Theorem 4.1 we have

$$\|\tilde{v}\|_{F^*(\mathbb{R}^n_+)} \leq C \left(\|(\tilde{L})^*\tilde{v}\|_{E^*(\mathbb{R}^n_+)} + \|\tilde{v}\|_{F^*_{-1}(\mathbb{R}^n_+)} \right). \tag{5.2}$$

Since

$$(\tilde{L})^*\tilde{v} = (T^{-1})^* L^* v,$$

then from (5.2)

$$\|v\|_{F^*(\Omega)} \leq \|T^*\| \, \|\tilde{v}\|_{F^*(\mathbb{R}^n_+)} \leq C\|T^*\| \left(\|(T^{-1})^* L^* v\|_{E^*(\mathbb{R}^n_+)} + \|(T^*)^{-1} v\|_{F^*_{-1}(\mathbb{R}^n_+)} \right).$$

Therefore

$$\|v\|_{F^*(\Omega)} \leq C_1 \left(\|L^* v\|_{E^*(\Omega)} + \|v\|_{F^*_{-1}(\Omega)} \right). \tag{5.3}$$

Let $\phi(x) \in C^\infty(\Gamma_\delta)$, $\mathrm{supp}\, \phi \subset B_\epsilon(x_0)$, Γ_δ be the δ-neighborhood of Γ, $\epsilon > 0$ is taken such that $\epsilon \leq \delta/2$ and $\psi(B_\epsilon(x_0)) \subset \sigma_\rho$ with the same ρ as in Theorem 4.1. Then the previous estimate gives

$$\|\phi v\|_{F^*(\Omega)} \leq C_1 \left(\|L^*(\phi v)\|_{E^*(\Omega)} + \|\phi v\|_{F^*_{-1}(\Omega)} \right). \tag{5.4}$$

Let us estimate the difference $L^*(\phi v) - \phi L^* v$. We have for any $u \in E(\Omega)$.

$$\langle u, L^*(\phi v) - \phi L^* v \rangle = \langle \phi L u - L(\phi u), v \rangle$$
$$= \sum_{i=1}^N \langle \phi A_i u - A_i(\phi u), v_i \rangle + \sum_{j=1}^m \langle \phi B_j u - B_j(\phi u), w_j \rangle, \tag{5.5}$$

where $v = (v_1, \ldots, v_N, w_1, \ldots, w_m)$. We begin with the first term in the right-hand side of (5.5). The operator A_i acts from $E(\Omega)$ into $W^{l-s_i+1,p}(\Omega)$. Let

$$T_i : W^{l-s_i+1,p}(\mathbb{R}^n) \to W^{l-s_i,p}(\mathbb{R}^n)$$

be a linear isomorphism between the two spaces. Then

$$(T_i^{-1})^* : (W^{l-s_i+1,p}(\mathbb{R}^n))^* \to (W^{l-s_i,p}(\mathbb{R}^n))^*.$$

Consider a function $\psi \in D$ such that $\operatorname{supp} \psi \in B_\delta(x_0)$, $\psi(x) \geq 0$, $\psi(x) = 1$ for $x \in B_{\delta/2}(x_0)$. Denote by $\tilde{u}$ an extension of u to $E(\mathbb{R}^n)$ such that

$$\|\tilde{u}\|_{E(\mathbb{R}^n)} \leq 2\|u\|_{E(\Omega)}. \tag{5.6}$$

We have $\psi v_i \in (W^{l-s_i,p}(\Omega))^* \subset (W^{l-s_i+1,p}(\Omega))^*$. Hence there exists an extension $\widehat{\psi v_i} \in (W^{l-s_i+1,p}(\mathbb{R}^n))^*$ such that

$$\langle \phi A_i u - A_i(\phi u), \psi v_i \rangle = \langle \phi A_i \tilde{u} - A_i(\phi \tilde{u}), \widehat{\psi v_i} \rangle.$$

Here we suppose that the coefficients of the operator A_i are extended to $\mathbb{R}^n$. Hence

$$\begin{aligned}
|\langle \phi A_i u - A_i(\phi u), v_i \rangle| &= |\langle \phi A_i u - A_i(\phi u), \psi v_i \rangle| \\
&= |\langle T_i^{-1} T_i(\phi A_i \tilde{u} - A_i(\phi \tilde{u})), \widehat{\psi v_i} \rangle| = |\langle T_i(\phi A_i \tilde{u} - A_i(\phi \tilde{u})), (T_i^{-1})^* \widehat{\psi v_i} \rangle| \\
&\leq \|T_i(\phi A_i \tilde{u} - A_i(\phi \tilde{u}))\|_{W^{l-s_i,p}(\mathbb{R}^n)} \|(T_i^{-1})^* \widehat{\psi v_i}\|_{(W^{l-s_i,p}(\mathbb{R}^n))^*} \\
&\leq C_1 \|\tilde{u}\|_{E(\mathbb{R}^n)} \|\widehat{\psi v_i}\|_{(W^{l-s_i+1,p}(\mathbb{R}^n))^*} \leq C_2 \|u\|_{E(\Omega)} \|\psi v_i\|_{(W^{l-s_i+1,p}(\Omega))^*}
\end{aligned} \tag{5.7}$$

according to (5.6).

We obtain similar estimates for the operators B_j in (5.5).

$$\begin{aligned}
|\langle \phi B_j u - B_j(\phi u), w_j \rangle| &= |\langle S_j^{-1} S_j(\phi B_j u - B_j(\phi u)), \psi w_j \rangle| \\
&= |\langle S_j(\phi B_j u - B_j(\phi u)), (S_j^{-1})^*(\psi w_j) \rangle| \\
&\leq \|S_j(\phi B_j u - B_j(\phi u))\|_{W^{l-\sigma_j-1/p,p}(\Gamma)} \|(S_j^{-1})^*(\psi w_j)\|_{(W^{l-\sigma_j-1/p,p}(\Gamma))^*} \\
&\leq C_3 \|u\|_{E(\Omega)} \|\psi w_j\|_{W^{-l+\sigma_j+1/p-1,p'}(\Gamma)}.
\end{aligned} \tag{5.8}$$

From (5.5), (5.7), (5.8) we obtain

$$\begin{aligned}
|\langle u, L^*(\phi v) - \phi L^* v \rangle| &\leq C_4 \|u\|_{E(\Omega)} \left(\sum_{i=1}^{N} \|\psi v_i\|_{(W^{l-s_i+1,p}(\Omega))^*} + \sum_{j=1}^{m} \|\psi w_j\|_{W^{-l+\sigma_j+1/p-1,p'}(\Gamma)} \right) \\
&= C_4 \|u\|_{E(\Omega)} \|\psi v\|_{F_{-1}^*(\Omega)}.
\end{aligned}$$

Therefore
$$\|L^*(\phi v) - \phi L^* v\|_{E^*(\Omega)} \leq C_4 \|\psi v\|_{F_{-1}^*(\Omega)}. \tag{5.9}$$

From this estimate and (5.4) it follows that

$$\|\phi v\|_{F^*(\Omega)} \leq C_5 \left(\|\phi L^* v\|_{E^*(\Omega)} + \|\phi v\|_{F_{-1}^*(\Omega)} + \|\psi v\|_{F_{-1}^*(\Omega)} \right). \tag{5.10}$$

Proof of Theorem 5.1. Let δ be the same as in Condition D. We cover the boundary Γ of the domain Ω by a countable number of balls B_ϵ of the radius ϵ, where

5. The general problem in unbounded domains

$\epsilon \leq \delta/2$ is the number which appears in the proof of estimate (5.10), and extend this covering to a covering of Ω. Let $V_j, j = 1, 2, \ldots$ be all the balls of the covering, and $\tilde{V}_j$ be the balls with the same centers as V_j but with radius δ. We suppose that there exists a number N such that each of the balls $\tilde{V}_j$ has a nonempty intersection with at most N other balls.

Let further $\phi_j(x)$ and $\tilde{\phi}_j(x)$ be systems of nonnegative functions such that

$$\phi_j(x) \in C^\infty(\mathbb{R}^n), \ \operatorname{supp} \phi_j \subset V_j, \ \tilde{\phi}_j(x) \in C^\infty(\mathbb{R}^n), \ \operatorname{supp} \tilde{\phi}_j \subset \tilde{V}_j,$$

and $\tilde{\phi}_j(x) = 1$ for $x \in V_j$. For the balls V_j with the centers at Γ by virtue of (5.10) we have the estimate

$$\|\phi_j v\|_{F^*(\Omega)} \leq M_0 \left(\|\phi_j L^* v\|_{E^*(\Omega)} + \|\phi_j v\|_{F^*_{-1}(\Omega)} + \|\tilde{\phi}_j v\|_{F^*_{-1}(\Omega)} \right) \quad (5.11)$$

with a constant M_0 independent of j and v.

The covering of the domain Ω can be constructed in such a way that all other balls, with centers outside of Γ, do not contain points of the boundary. We can obtain a similar estimate for them. It is even simpler because we do not have to take into account the boundary operators.

By the definition of the spaces $(E^*(\Omega))_\infty$ and $(F^*_{-1}(\Omega))_\infty$ we have for any j:

$$\|\phi_j L^* v\|_{E^*(\Omega)} \leq M_1 \|L^* v\|_{(E^*(\Omega))_\infty}, \ \|\phi_j v\|_{F^*_{-1}(\Omega)} \leq M_2 \|v\|_{(F^*_{-1}(\Omega))_\infty},$$

$$\|\tilde{\phi}_j v\|_{F^*_{-1}(\Omega)} \leq M_3 \|v\|_{(F^*_{-1}(\Omega))_\infty},$$

where the constants M_1, M_2, M_3 do not depend on v and j. Therefore (5.11) gives

$$\|\phi_j v\|_{F^*(\Omega)} \leq M \left(\|L^* v\|_{(E^*(\Omega))_\infty} + \|v\|_{(F^*_{-1}(\Omega))_\infty} \right).$$

Estimate (5.1) follows from this. The theorem is proved. □

Chapter 4

Normal Solvability

As in the previous chapter, we consider the operators

$$A_i u = \sum_{k=1}^{N} \sum_{|\alpha| \leq \alpha_{ik}} a_{ik}^{\alpha}(x) D^{\alpha} u_k, \quad i = 1, \ldots, N, \quad x \in \Omega, \tag{0.1}$$

$$B_j u = \sum_{k=1}^{N} \sum_{|\beta| \leq \beta_{jk}} b_{jk}^{\beta}(x) D^{\beta} u_k, \quad j = 1, \ldots, m, \quad x \in \partial\Omega, \tag{0.2}$$

where $u = (u_1, \ldots, u_N)$, $\Omega \subset \mathbb{R}^n$ is an unbounded domain. Conditions on the operators and on the domains will be specified below. We will study normal solvability of the operator

$$L = (A_1, \ldots, A_N, B_1, \ldots, B_m) \tag{0.3}$$

considered in Sobolev or in Hölder spaces.

We recall that the operator is normally solvable with a finite-dimensional kernel if and only if it is proper on closed bounded sets, that is if the intersection of the inverse image of a compact set with any closed bounded set is compact. We understand properness only in this sense and will not necessarily explain it each time. In this chapter we obtain necessary and sufficient conditions for a general elliptic operator to satisfy this property. Consider for example the operator

$$Lu = a(x)u'' + b(x)u' + c(x)u \tag{0.4}$$

acting from $H^2(R)$ into $L^2(R)$. If we assume that there exist limits of the coefficients of the operator at infinity, then we can define the operators

$$L_{\pm} u = a_{\pm} u'' + b_{\pm} u' + c_{\pm} u,$$

where the subscripts $+$ and $-$ denote the limiting values at $+\infty$ and $-\infty$, respectively. Applying the Fourier transform to the equation

$$a_{\pm} u'' + b_{\pm} u' + c_{\pm} u = \lambda u,$$

we obtain the essential spectrum

$$\lambda(\xi) = -a_\pm \xi^2 + b_\pm i\xi + c_\pm, \quad \xi \in \mathbb{R}$$

of the operator L given by the two parabolas in the complex plane. It can be proved that if the essential spectrum does not pass through the origin, then L is a Fredholm operator. In particular, it is normally solvable with a finite-dimensional kernel.

This simple approach is not applicable for general elliptic problems where limits of the coefficients may not exist and the domain may not be translation invariant. In the next section we will begin with the construction of limiting problems. It can be briefly described as follows. Let $x_k \in \Omega$ be a sequence which tends to infinity. Consider the shifted domains Ω_k corresponding to the shifted characteristic functions $\chi(x + x_k)$, where $\chi(x)$ is the characteristic function of the domain Ω. Consider a ball $B_r \subset \mathbb{R}^n$ with center at the origin and with radius r. Suppose that for all k there are points of the boundaries $\partial \Omega_k$ inside B_r. If the boundaries are sufficiently smooth, we can expect that from the sequence $\Omega_k \cap B_r$ we can choose a subsequence that converges to some limiting domain Ω_*. After that we take a larger ball and choose a convergent subsequence of the previous subsequence. The usual diagonal process allows us to extend the limiting domain to the whole space.

In order to define limiting operators, we consider the shifted coefficients $a^\alpha(x + x_k)$, $b_j^\alpha(x + x_k)$ and choose subsequences that converge to some limiting functions $\hat{a}^\alpha(x)$, $\hat{b}_j^\alpha(x)$ uniformly in every bounded set. The limiting operator is the operator with limiting coefficients. Limiting operators and limiting domains constitute limiting problems. It is clear that the same problem can have a family of limiting problems depending on the choice of the sequence x_k and on the choice of converging subsequences of domains and coefficients.

We note that in the case where $\Omega = \mathbb{R}^n$ the limiting domain is also $\mathbb{R}^n$. In this case the limiting operators were introduced in [361], [362], [490], [491].

The following condition determines normal solvability of elliptic problems.

Condition NS. Any limiting problem

$$\hat{L}u = 0, \quad x \in \Omega_*, \quad u \in E_\infty(\Omega_*)$$

has only a zero solution.

It is a necessary and sufficient condition for general elliptic operators to be normally solvable with a finite-dimensional kernel. More precisely, it will be proved that the elliptic operator L is normally solvable and has a finite-dimensional kernel in the space $W_\infty^{l,p}$ ($1 < p < \infty$) if and only if Condition NS is satisfied. Similar results will also be obtained for Hölder spaces.

It is easy to see how this condition is related to the condition formulated in terms of the Fourier transform. The nonzero solution of the limiting problem

1. Limiting domains   137

$L_\pm u = 0$ corresponding to operator (0.4) has the form $u_0(x) = e^{i\xi x}$, where ξ is the value for which the essential spectrum passes through the origin. We note that the function $u_0(x)$ belongs to the Hölder spaces and also to the space $W^{2,p}_\infty(\mathbb{R})$. However it does not belong to the usual Sobolev space $W^{2,p}(\mathbb{R})$. Therefore it is more convenient to use here the spaces $W^{s,p}_q$.

1 Limiting domains

1.1 General properties

In this section we define limiting domains for unbounded domains in $\mathbb{R}^n$, we show their existence and study some of their properties. We consider an unbounded domain $\Omega \subset \mathbb{R}^n$, which satisfies the following condition:

Condition D. For each $x_0 \in \partial\Omega$ there exists a neighborhood $U(x_0)$ such that:

1. $U(x_0)$ contains a sphere with radius δ and center x_0, where δ is independent of x_0,
2. There exists a homeomorphism $\psi(x; x_0)$ of the neighborhood $U(x_0)$ on the unit sphere $B = \{y : |y| < 1\}$ in $\mathbb{R}^n$ such that the images of $\Omega \cap U(x_0)$ and $\partial\Omega \cap U(x_0)$ coincide with $B_+ = \{y : y_n > 0, |y| < 1\}$ and $B_0 = \{y : y_n = 0, |y| < 1\}$ respectively,
3. The function $\psi(x; x_0)$ and its inverse belong to the Hölder space $C^{r+\theta}$. Their $\|\cdot\|_{r+\theta}$-norms are bounded uniformly in x_0.

For definiteness we suppose that $\delta < 1$. We assume that

$$r \geq \max(l + t_i, l - s_i, l - \sigma_j + 1), \quad i = 1, \ldots, N, \; j = 1, \ldots, m.$$

The first expression in the maximum is used for a priori estimates of solutions, the second and the third allow us to extend the coefficients of the operator (Section 2.2.2).

In what follows we suppose that ψ is extended such that $\psi \in C^{r+\theta}(\mathbb{R}^n)$ and $\|\psi\|_{C^{r+\theta}(\mathbb{R}^n)} \leq M$ with M independent of x_0. It is easy to see that δ and ψ in Condition D can be chosen in such a way that this requirement can be satisfied. Indeed, denote by V_δ the sphere with the center at x_0 and the radius δ and let $W_\delta = \psi(V_\delta)$. Obviously, there exists a sphere Q_ε with center at $y_0 = \psi(x_0; x_0)$ and radius ε such that $Q_\varepsilon \subset W_\delta$ and ε does not depend on x_0. Indeed, let $\varphi = \psi^{-1}$ and let y_1 be an arbitrary point on the boundary of W_δ. We have $\delta = |\varphi(y_1) - \varphi(y_0)| \leq K|y_1 - y_0|$, where K is the Lipschitz constant which does not depend on x_0. Let $\varepsilon < \delta/K$. We have $|y_1 - y_0| > \varepsilon$ which proves existence of the desired sphere Q_ε. Let $\widetilde{U}(x_0) = \varphi(Q_\varepsilon)$. There exists a sphere S with center at x_0 and radius $\widetilde{\delta}$ such that $S \subset \widetilde{U}(x_0)$ and $\widetilde{\delta}$ does not depend on x_0. Indeed, let x_1 be an arbitrary point of the boundary of $\widetilde{U}(x_0)$. Then we have $\varepsilon = |\psi(x_0; x_0) - \psi(x_1; x_0)| \leq K_1 |x_0 - x_1|$, where

K_1 is the Lipschitz constant of ψ, which does not depend on x_0. So for $\tilde{\delta} < \varepsilon/K_1$ we have $|x_0 - x_1| > \tilde{\delta}$, which proves existence of the mentioned sphere S. We can take $\tilde{U}(x_0)$ as a new neighborhood of x_0 and $\tilde{\psi}(x;x_0) = \frac{1}{\varepsilon}(\psi(x;x_0) - \psi(x_0;x_0))$ as a new function ψ. Since $\tilde{\psi}(x;x_0)$ is defined in the sphere V_δ it can be extended on $\mathbb{R}^n$.

To define convergence of domains we use the following Hausdorff metric space. Let M and N denote two nonempty closed sets in $\mathbb{R}^n$. Set

$$\varsigma(M,N) = \sup_{a \in M} \rho(a,N), \quad \varsigma(N,M) = \sup_{b \in N} \rho(b,M),$$

where $\rho(a,N)$ denotes the distance from a point a to a set N, and let

$$\varrho(M,N) = \max(\varsigma(M,N), \varsigma(N,M)). \tag{1.1}$$

We denote by Ξ a metric space of bounded closed nonempty sets in $\mathbb{R}^n$ with the distance given by (1.1). We say that a sequence of domains Ω_m converges to a domain Ω in Ξ_{loc} if

$$\varrho(\bar{\Omega}_m \cap \bar{B}_R, \bar{\Omega} \cap \bar{B}_R) \to 0, \quad m \to \infty$$

for any $R > 0$ and $B_R = \{x : |x| < R\}$. Here bar denotes the closure of the domains.

Definition 1.1. Let $\Omega \subset \mathbb{R}^n$ be an unbounded domain, $x_m \in \Omega$, $|x_m| \to \infty$ as $m \to \infty$; $\chi(x)$ be a characteristic function of Ω, and Ω_m be a shifted domain defined by the characteristic function $\chi_m(x) = \chi(x + x_m)$. We say that Ω_* is a *limiting domain* of the domain Ω if $\Omega_m \to \Omega_*$ in Ξ_{loc} as $m \to \infty$.

We denote by $\Lambda(\Omega)$ the set of all limiting domains of the domain Ω (for all sequences x_m). It will be shown below that if Condition D is satisfied, then limiting domains exist and also satisfy this condition.

Theorem 1.2. *If a domain Ω satisfies Condition D, then there exists a function $f(x)$ defined in $\mathbb{R}^n$ such that:*
1. $f(x) \in C^{k+\alpha}(\mathbb{R}^n)$,
2. $f(x) > 0$ if and only if $x \in \Omega$,
3. $|\nabla f(x)| \geq 1$ for $x \in \partial\Omega$,
4. $\min(d(x), 1) \leq |f(x)|$, where $d(x)$ is the distance from x to $\partial\Omega$.

Proof. There exists a number N such that from the covering $U(x_0)$ of ∂D we can choose a countable subcovering U_i such that the following conditions are satisfied:

1) $\bigcup_i U_i$ covers the $\delta/2$-neighborhood of $\partial\Omega$,
2) Any N distinct sets U_i have an empty intersection.

Indeed, denote by V the $\delta/2$-neighborhood of $\partial\Omega$. Obviously, for any point $x_0 \in V$ there exists a point $x_0' \in \partial\Omega$ such that $B_{\delta/2}(x_0) \subset B_\delta(x_0') \subset U(x_0')$. Here and in

1. Limiting domains

what follows $B_r(x)$ denotes a ball in $\mathbb{R}_n$ with center at x and with radius r. So we have a covering $U'(x_0) = U(x_0')$ of V such that the centers of the balls are at the boundary of the domain. Write $\Gamma = \bigcup U'$.

Consider an ε-mesh in $\mathbb{R}^n$. We denote by K the union of all n-dimensional ε-intervals of this mesh which have a nonempty intersection with V. For any $Q_i \in K$ we take a point $x_i \in Q_i \cap V$ ($i = 1, 2, \ldots$) and consider the neighborhood $U_i \in \Gamma$, which contains the point x_i. We suppose that ε is taken such that the diameter of Q_i is less then $\delta/2$. Then $Q_i \subset U_i$ and

$$V \subset \Gamma_0 = \bigcup_i U_i.$$

Therefore the covering Γ_0 satisfies condition 1).

To each $Q_i \in K$ corresponds no more than one neighborhood $U_i \in \Gamma_0$. From Condition D it follows that the diameter of U_i is less than a constant independent of i. Hence 2) is also satisfied.

Let $\omega_i \in C^{k+\alpha}(\mathbb{R}^n)$ be a partition of unity subordinate to the covering Γ_0, i.e., $\mathrm{supp}\,\omega_i \subset U_i$. Denote by ψ_i the vector-valued function $\psi(x, x_0)$ which corresponds to U_i in Condition D and

$$f_0(x) = c \sum_i \psi_{in}(x)\, \omega_i(x), \tag{1.2}$$

where $\psi_{in}(x)$ is the last component of $\psi_i(x)$ and the constant c will be chosen later. We note that this sum contains no more than N terms.

For any points $x \in U_i$ and $x^1 \in \partial\Omega \cap U_i$ we have $|x - x^1| \leq M|y - y^1|$, where $y = \psi_i(x)$, $y^1 = \psi_i(x^1)$ and the constant M does not depend on i. So

$$d(x) \leq M \inf_{y_n^1 = 0} |y - y^1| = M|\psi_{in}(x)|.$$

It follows that for all $x \in V$ we have $d(x) \leq M|f_0(x)|/c$. We have used here the fact that the functions $\psi_{in}(x)$ have the same sign for all i. We take $c \geq M$. Then

$$\min(d(x), 1) \leq |f_0(x)|.$$

Therefore assertion 4 of the theorem is proved for $f_0(x)$.

We now prove statement 3 of the formulation. Set $\varphi_i = \psi_i^{-1}$ and denote by φ_i' and ψ_i' the Jacobian matrices of φ_i and ψ_i, respectively. Then for any $x \in U_i$ we have $\psi_i'(x) \cdot \varphi_i'(\psi(x)) = I$ (identity matrix). Let a_i be the kth row of ψ_i' and b_i be the kth column of φ_i', then $|a_i|\,|b_i| \geq 1$. From Condition D, $|b_i| \leq M_1$, where M_1 is a constant independent of i. So $|a_i| \geq 1/M_1$. In particular

$$|\nabla \psi_{in}(x)| \geq 1/M_1. \tag{1.3}$$

From (1.2) for $x \in \partial D$ we have

$$\nabla f_0(x) = c \sum_i \nabla \psi_{in}(x)\, \omega_i(x). \tag{1.4}$$

Let ν be the unit inward normal to ∂D. Then

$$(\nabla f_0(x), \nu(x)) = (\nabla f_0(x), \nabla f_0(x)/|\nabla f_0(x)|) = |\nabla f_0(x)|.$$

For $x \in \partial\Omega \bigcap U_i$ we have similarly $(\nabla \psi_{in}(x), \nu(x)) = |\nabla \psi_{in}(x)|$. Multiplying (1.4) by $\nu(x)$ we get

$$|\nabla f_0(x)| = c \sum_i |\nabla \psi_{in}(x)| \, \omega_i(x).$$

From (1.3), taking $c \geq M_1$ we obtain $|\nabla f_0(x)| \geq 1$, $x \in \partial\Omega$. Hence, point 3 is proved for $f_0(x)$.

We have defined the function $f_0(x)$ in a neighborhood of the boundary $\partial\Omega$. We can easily extend it on the whole $\mathbb{R}^n$ in such a way that its regularity is preserved, it is greater than a positive constant inside the domain Ω and less than a negative constant outside the domain. Multiplying it by a large positive number, we will have the last two assertions of the theorem also satisfied. The theorem is proved. $\square$

Let Ω be an unbounded domain satisfying Condition D and $f(x)$ be the function satisfying conditions of Theorem 1.2. Consider a sequence $x_m \in \Omega$, $|x_m| \to \infty$. Let $f_m(x) = f(x + x_m)$.

Theorem 1.3. *Let $f_m(x) \to f_*(x)$ in $C_{\text{loc}}^k(\mathbb{R}^n)$ (local convergence), where k is not greater than that in Theorem 1.2. Write $\Omega_* = \{x : x \in \mathbb{R}^n, \, f_*(x) > 0\}$. Then*

1) $f_*(x) \in C^{k+\alpha}(\mathbb{R}^n)$,
2) Ω_* *is a nonempty open set.*

If $\Omega_* \neq \mathbb{R}^n$, then

3) $|\nabla f_*(x)|_{\partial \Omega_*} \geq 1$,
4) $\min(d_*(x), 1) \leq |f_*(x)|$, *where $d_*(x)$ is the distance from x to $\partial \Omega_*$.*

Proof. The first assertion of the theorem is obvious. To prove the second assertion, we note that the origin O belongs to all domains Ω_m. Denote by d_m the distance from O to the boundary $\partial \Omega_m$. If $d_m \to 0$, then from the properties of the functions $f_m(x)$ it follows that

$$f_*(O) = 0, \quad |\nabla f_*(O)| \geq 1.$$

Hence there are points in a neighborhood of the origin where the function $f_*(x)$ is positive. Consequently Ω_* is nonempty.

If d_m does not converge to zero, then $d_{m_i} \geq a > 0$ for some positive a. From Theorem 1.2 we conclude that $f_{m_i}(O) \geq \min(a, 1)$. Therefore $f_*(O) \geq \min(a, 1) > 0$, and we obtain again that the set Ω_* is not empty. The fact that it is open is obvious.

We now verify the third assertion of the theorem. Let $f_*(x_0) = 0$ for some x_0. Then $f_m(x_0) \to 0$. From assertion 4 of Theorem 1.2 it follows that $d_m(x_0) \to 0$, where $d_m(x_0)$ is the distance of x_0 to $\partial \Omega_m$. So there exists $z_m \in \partial D_m$ such that

1. Limiting domains

$|z_m - x_0| \to 0$. Since $|\nabla f_m(z_m)| \geq 1$ by 3 of Theorem 1.2, then passing to the limit we get $|\nabla f_*(x_0)| \geq 1$.

We finally prove the last assertion of the theorem. For any $x_0 \in \mathbb{R}^n$ we have

$$\min(d_m(x_0), 1) \leq |f_m(x_0)|.$$

So we should verify that $d_m(x_0)$ converges to $d_*(x_0)$ as $m \to \infty$. Suppose that x_0 belongs to a ball B_R. Set

$$\Gamma_m = \{x : f_m(x) = 0, \; x \in B_R\}, \quad \Gamma_* = \{x : f_*(x) = 0, \; x \in B_R\}.$$

It is sufficient to prove that

$$\varrho(\Gamma_m, \Gamma_*) \to 0, \quad m \to \infty. \tag{1.5}$$

Let Γ_m^ϵ, Γ_*^ϵ be ϵ-neighborhoods of these sets, respectively. From the convergence $f_m(x) \to f_*(x)$ in $C^k(B_R)$ it follows that $\Gamma_m \subset \Gamma_*^\epsilon$ for m sufficiently large. We show that $\Gamma_* \subset \Gamma_m^\epsilon$ for m large. Indeed,

$$|f_m(x) - f_*(x)| < \epsilon, \quad x \in B_R$$

for $m \geq m_\epsilon$ and some m_ϵ. If $x \in \Gamma_*$, then $f_*(x) = 0$ and $|f_m(x)| < \epsilon$. From the last assertion of Theorem 1.2 it follows that $d_m(x) < \epsilon$, and $x \in \Gamma_m^\epsilon$. Convergence (1.5) follows from this. The theorem is proved. $\square$

We note that the limiting set Ω_* is not necessarily connected even if the domain Ω is connected.

Theorem 1.4. *If $f_m(x) \to f_*(x)$ in C_{loc}^k as $m \to \infty$, then $\partial \Omega_m \to \partial \Omega_*$ in Ξ_{loc}.*

The proof of the theorem follows from convergence (1.5).

Theorem 1.5. *If $f_m(x) \to f_*(x)$ in C_{loc}^k as $m \to \infty$, then the limiting domain Ω_* satisfies Condition D, or $\Omega_* = \mathbb{R}^n$.*

Proof. Suppose that $\Omega_* \neq \mathbb{R}^n$ and $x_0 \in \partial \Omega_*$. Then there exists a sequence $\hat{x}_m$ such that

$$\hat{x}_m \to x_0, \quad \hat{x}_m \in \partial \Omega_m,$$

where Ω_m are the domains where the functions $f_m(x)$ are positive. For each point $\hat{x}_m$ and domain Ω_m there exists a neighborhood $U(\hat{x}_m)$ and the function $\psi(x; \hat{x}_m)$ defined in Condition D.

Since the domain Ω satisfies Condition D, the functions $\psi(x; \hat{x}_m)$ are uniformly bounded in the $C^{k+\alpha}$-norm with $k \geq 1$. The domain of definition of each of these functions is an inverse image of the unit sphere in $\mathbb{R}^n$. Choosing a converging subsequence of the inverse images and of the functions $\psi(x; \hat{x}_m)$, we obtain a limiting neighborhood $U(x_0)$ and a limiting function $\psi(x; x_0)$ which satisfy Condition D. The theorem is proved. $\square$

From the previous theorems follows the main result of this section.

Theorem 1.6. *Let Ω be an unbounded domain satisfying Condition D, $x_m \in \Omega$, $|x_m| \to \infty$, and $f(x)$ be the function constructed in Theorem 1.2. Then there exists a subsequence x_{m_i} and a function $f_*(x)$ such that $f_{m_i}(x) \equiv f(x + x_{m_i}) \to f_*(x)$ in $C^k_{\mathrm{loc}}(\mathbb{R}^n)$, and the domain $\Omega_* = \{x : f_*(x) > 0\}$ satisfies Condition D, or $\Omega_* = \mathbb{R}^n$. Moreover $\bar\Omega_{m_i} \to \bar\Omega_*$ in Ξ_{loc}, where $\Omega_{m_i} = \{x : f_{m_i}(x) > 0\}$.*

1.2 Examples

Half-space, space, cylinders. We begin with some examples in which limiting domains can be constructed explicitly.

1. If $\Omega = \mathbb{R}^n$, then the only limiting domain is the whole $\mathbb{R}^n$. If Ω is an exterior domain for some bounded domain, then, as before, the only limiting domain is $\mathbb{R}^n$.
2. In the case of the half-space, $\Omega = \mathbb{R}^n_+$, $n > 1$, there are two limiting domains, the same half-space and the whole space. For the half-line, the limiting domain is $\mathbb{R}^1$.
3. If Ω is an unbounded cylinder with a bounded cross-section, then the only limiting domain is the same cylinder (up to a shift). It is also the limiting domain for a half-cylinder.

Plane domains bounded by a curve. Consider the following domain in $\mathbb{R}^2$:

$$\Omega = \{(x, y), y > f(x)\},$$

where $f(x)$, $x \in \mathbb{R}^1$ is a given function continuous with its first derivative. Suppose that $f(x)$ and $f'(x)$ have limits (finite or infinite) as $x \to \pm\infty$. Then the tangent to the boundary $\partial\Omega$ has limits. The half-planes limited by the limiting tangents are limiting domains. These half-planes and the whole plane form all limiting domains.

If, for example, $f(x) = x^2$, then there are three types of limiting domains: the whole plane, the right half-plane, the left half-plane. For the exponential, $f(x) = e^x$, the limiting domains are the whole plane, the left half-plane, the upper half-plane.

For a periodic function $f(x)$, for which limits for the function and for its derivative at infinity do not exist, the limiting domains are either the domain Ω itself (up to a sift) or the whole plane.

Domains slowly varying at infinity. We will introduce a special class of domains for which limiting domains can be either the whole spaces or a half-space. Limiting problems are easier to study in this case. We denote by $\nu(x)$ the internal normal unit vector to the boundary at $x \in \partial\Omega$.

Condition R. For any sequence $x_k \in \partial\Omega$, $k = 1, 2, \ldots$, $|x_k| \to \infty$ and for any given number $r > 0$ there exists a subsequence x_{k_i} such that the limit $\lim_{k_i \to \infty} \nu(x_{k_i} +$

1. Limiting domains

$h_{k_i})$ exists for all h_{k_i}: $h_{k_i} \in \mathbb{R}^n$, $|h_{k_i}| < r$, $x_{k_i} + h_{k_i} \in \partial\Omega$, and it does not depend on h_{k_i}.

We will study the structure of limiting domains Ω_* for domains Ω, which satisfy Condition R. Let Ω_* be a limiting domain and $x_k \in \Omega$, $k = 1, 2, \ldots$, $|x_k| \to \infty$ is the sequence that determines this limiting domain. Denote by d_k the distance from x_k to the boundary $\partial\Omega$. Consider the following two cases:

1. If the sequence d_k is unbounded, then there exists a subsequence $d_{k_i} \to \infty$. The sequence x_{k_i} determines the same limiting domain Ω_*. Then $\Omega_* = \mathbb{R}^n$,
2. If the sequence d_k is bounded, then we can assume, choosing a subsequence if necessary, that $d_k \to d < \infty$.

It is convenient to reformulate Condition R in the following form.

Condition R. If the boundary $\partial\Omega$ is unbounded, then for any sequence $x_k \in \partial\Omega$, $k = 1, 2, \ldots$, $|x_k| \to \infty$ there exists a subsequence x_{k_i} such that for any given number $r > 0$ the limit $\lim_{k_i \to \infty} \nu(x_{k_i} + h_{k_i})$ exists for all h_{k_i}: $h_{k_i} \in \mathbb{R}^n$, $|h_{k_i}| < r$, $x_{k_i} + h_{k_i} \in \partial\Omega$, and it does not depend on h_{k_i}.

In the other words, the subsequence can be chosen independently of r. To prove that the second definition follows from the first one, it is sufficient to take a sequence $r_j \to \infty$. For each value of j we can take a subsequence according to the first definition in such a way that it is a subsequence of the previous one. Then we use a diagonal process.

Denote by y_k the point of the boundary $\partial\Omega$ such that the distance from y_k to x_k equals d_k. Obviously, $|y_k| \to \infty$. Let y_{k_i} be a subsequence chosen according to Condition R. Instead of the sequence x_k we consider the subsequence x_{k_i}. The limiting domain Ω_* remains the same. We can use Theorem 1.2. Let $f(x)$ be a function that satisfies the conditions of the theorem. Then, taking a subsequence if necessary, according to Theorem 1.6 we can find a function $f_*(x)$ such that $f_{k_i}(x) \equiv f(x + x_{k_i}) \to f_*(x)$ in $C^1_{\text{loc}}(\mathbb{R}^n)$, and the domain $\Omega_* = \{x : f_*(x) > 0\}$ is the limiting domain under consideration.

For convenience we write k instead of k_i:

$$f_k(x) \equiv f(x + x_k) \to f_*(x) \tag{1.6}$$

in $C^1_{\text{loc}}(\mathbb{R}^n)$. The limit

$$\lim_{k \to \infty} \nu(x_k + h_k) = \mu \tag{1.7}$$

exists for all h_k: $h_k \in \mathbb{R}^n$, $|h_k| < r$, $x_k + h_k \in \partial\Omega$, where $r > 0$ is a given number, μ is some constant. This limit is independent of h_k.

It will be shown below that for any $z_0 \in \partial\Omega_*$ the internal unit normal vector equals μ. The following lemma allows us to conclude that Ω_* is a half-space.

Lemma 1.7. *If the domain $\Omega \subset \mathbb{R}^n$ satisfies Condition D and all internal normal unit vectors to the boundary $\partial\Omega$ coincide up to a shift, then Ω is a half-space.*

Proof. Let $\Gamma = \partial\Omega$. Let us show that any point $z \in \Gamma$ has a neighborhood U such that $U \cap \Gamma$ coincides with $U \cap T(z)$, where $T(z)$ is a tangent plane to Γ at the point z. Consider a local coordinate $y = (y_1, \ldots, y_n)$ in the vicinity of the point z such that the axis y_n is along the internal normal vector to Γ at the point z, and all other coordinate axes are in the tangent plane. We can assume that this system of coordinates is obtained from the original one by a translation of the origin and by rotation. In this case, if the neighborhood U is sufficiently small, then the surface Γ in U can be given by the equation

$$y_n = f(y'), \quad y' = (y_1, \ldots, y_{n-1}). \tag{1.8}$$

Since all normal vectors to Γ_* are parallel to each other, then all normal vectors to the surface (1.8) are parallel to the y_n-axis. This means that $\partial f/\partial y_i \equiv 0$, $i = 1, \ldots, n-1$, that is $f(y')$ is a constant. Since $f(0) = 0$, then $f(y') \equiv 0$. Thus, it is proved that $U \cap \Gamma = U \cap T(z)$.

Take an arbitrary point $z_0 \in \Gamma$. Denote by $\Gamma(z_0)$ the part of the manifold Γ such that its points can be connected by a continuous curve on Γ. Let $z \in \Gamma(z_0)$ and $\gamma \subset \Gamma$ be a continuous curve connecting z_0 and z. For each point $\zeta \in \gamma$ choose a neighborhood as it is indicated above. We can also choose a finite covering of the curve γ with such neighborhoods. If we consider consecutive neighborhoods from this covering and take into account that the vectors $\nu(\zeta)$ are equal, we obtain $T(z_0) = T(z)$. Therefore $\gamma \subset T(z_0)$. Since z is an arbitrary point in $\Gamma(z_0)$, then

$$\Gamma(z_0) \subset T(z_0). \tag{1.9}$$

Let us show that

$$\Gamma(z_0) = T(z_0). \tag{1.10}$$

Suppose that this equality is not true. Let $z_1 \in T(z_0), z_1 \notin \Gamma(z_0)$. Let us connect the points z_0 and z_1 by an interval l. Then there is such a point $z_* \in l$ that in each of its neighborhoods there are points from $\Gamma(z_0)$ and points that do not belong to $\Gamma(z_0)$. Since $[z_0, z_*) \in \Gamma$ and Γ is a closed set, then $z_* \in \Gamma$. As it is shown above, z_* has a neighborhood U where

$$U \cap \Gamma = U \cap T(z_*). \tag{1.11}$$

Therefore $[z_0, z_*] \subset \Gamma$. This means that $z_* \in \Gamma(z_0)$. From (1.9) it follows that $z_* \in T(z_0)$. As above, we obtain $T(z_*) = T(z_0)$. By virtue of (1.11), $U \cap \Gamma = U \cap T(z_0)$. Hence, we can conclude that all points of the interval l, sufficiently close to z_* belong to Γ and consequently to $\Gamma(z_0)$. This contradiction proves (1.10).

Denote by $\Pi_+(z_0)$ ($\Pi_-(z_0)$) the half-space bounded by the plane $T(z_0)$ and located in the direction of the internal (external) normal vector to Γ at the point z_0. Let us show that

$$\Pi_+(z_0) \subset \Omega. \tag{1.12}$$

If this is not the case, then there exists a point $z \in \Pi_+(z_0), z \notin \Omega$. Let $\tilde{z}$ be the projection of the point z on the plane $T(z_0)$. We have $\tilde{z} \in \Gamma$. Since Condition

1. Limiting domains

D is satisfied, then some interval $\nu(\tilde{z})$ belongs to Ω. Therefore, there is a point ξ in the interval $[\tilde{z}, z]$ such that $(\tilde{z}, \xi) \in \Omega$ and $\xi \notin \Omega$. Hence, $\xi \in \Gamma$. Consider the internal normal vector $\nu(\xi)$ at the point ξ. According to the condition of the lemma, $\nu(\xi)$ and $\nu(\tilde{z})$ coincide up to a shift. Moreover, by virtue of Condition D, none of the points of the external normal vector sufficiently close to ξ belong to Ω. This contradiction proves (1.12).

We show that
$$\Pi_+(z_0) = \Omega. \tag{1.13}$$

Indeed, otherwise there exists a point $z \in \Omega, z \notin \Pi_+(z_0)$. This means that $z \in \Pi_-(z_0)$. Let $\tilde{z}$ be the projection of the point z on the plane $T(z_0)$. There is a point ξ in the interval $[z, \tilde{z}]$ such that $[z, \xi)$ belongs to Ω, $\xi \notin \Omega$. Therefore, $\xi \in \Gamma$ and the internal normal vector $\nu(\xi)$ has a direction opposite to the interval $[z, \xi]$. Since it is not possible, we obtain a contradiction which proves (1.13). The lemma is proved. □

We will now show that for any point $z_0 \in \partial \Omega_*$ the internal normal unit vector equals μ. Denote $B(x_0, r) = \{x \in \mathbb{R}^n, |x - x_0| < r\}$. Let Γ_k be the intersection of $\partial \Omega$ with $B(x_k, r)$, where $r > d$. Then Γ_k is not empty. As above, we take the point $y_k \in \partial \Omega$, such that its distance to x_k equals d_k. All points of the set Γ_k have the form $y_k + h$, where $|h| < 2r$. Therefore, we have (1.7).

Let us shift the point x_k to the origin and denote the shifted domain by Ω_k. Further let $\tilde{\Gamma}_k = \partial \Omega_k \cap B(0, r)$. For any point $z \in \partial \Omega_* \cap B(0, r)$ we can indicate a sequence of points $z_k \to z$, $z_k \in \tilde{\Gamma}_k$. Moreover, $\nu_k(z_k) \to \nu_*(z)$, where $\nu_k(z_k)$ ($\nu(z_*)$) is the internal normal unit vector to $\tilde{\Gamma}_k$ ($\partial \Omega_*$) at the point z_k (z). From (1.7) it follows that $\nu_*(z) = \mu$. The assertion is proved. We have proved the following theorem.

Theorem 1.8. *If Conditions D and R are satisfied for a domain $\Omega \subset \mathbb{R}^n$, then each of its limiting domains is either the whole space $\mathbb{R}^n$ or some half-space.*

The inverse theorem also holds.

Theorem 1.9. *Suppose that domain $\Omega \subset \mathbb{R}^n$ satisfies Condition D and each of its limiting domains is either the space $\mathbb{R}^n$ or some half-space. Then Condition R is satisfied.*

Consider domains Ω in the space $\mathbb{R}^{n+1}$ with the coordinates (x, y), where $x = (x_1, \ldots, x_n)$, determined by the inequality $y > f(x)$, where $f(x)$ is a function defined for all $x \in \mathbb{R}^n$ and continuous with its first derivatives. As usual, put $\nabla f(x) = (\partial f(x)/\partial x_1, \ldots, \partial f(x)/\partial x_n)$ and consider the spherical coordinates $x = r\theta$ where $r = |x|$. We present the following theorem without proof.

Theorem 1.10. *For any $\theta_0, |\theta_0| = 1$, let one of the following conditions be satisfied:*

1. *There exists $\lim_{r \to \infty, \theta \to \theta_0} \nabla f(r\theta)$ and $\lim_{r \to \infty, \theta \to \theta_0} |\nabla f(r\theta)| < \infty$;*

2. $\lim_{r\to\infty,\theta\to\theta_0}|\nabla f(r\theta)| = \infty$ and there exists

$$\lim_{r\to\infty,\theta\to\theta_0} \frac{\nabla f(r\theta)}{|\nabla f(r\theta)|}.$$

Then the domain Ω satisfies Condition R.

Consider the following examples (see also [442]):

1. Let $f(x) = g(|x|)$, where $g(t)$ is a continuously differentiable function defined for $t \geq 0$. Suppose that $\lim_{t\to\infty} g'(t) = g'(\infty)$ exists. If $|g'(\infty)| < \infty$, then we have the first case of Theorem 1.10; if $|g'(\infty)| = \infty$, then it is case 2,
2. Let $f(x) = f_0(x) + f_1(x)$ (for $|x| \geq \sigma > 0$), where $f_0(x)$ is a homogeneous function of a positive order $\alpha \geq 1$,

$$f_0(\rho x) = \rho^\alpha f_0(x), \; \rho > 0,$$

and $f_1(x)$ satisfies the condition

$$\frac{|\nabla f_1(x)|}{|x|^{\alpha-1}} \to 0, \; |x| \to \infty$$

(for example, $f(x)$ can be a polynomial). These functions are supposed to be continuously differentiable and $|\nabla f(\theta)| \neq 0$ for all $|\theta| = 1$. Then it can be verified that the conditions of the theorem are satisfied.

2 Sobolev spaces

2.1 A priori estimates in the spaces $W_\infty^{k,p}$

In this section we obtain a priori estimates of solutions in the spaces $W_\infty^{k,p}$ similar to those in the usual Sobolev spaces. We recall that $W_\infty^{k,p}(\Omega)$, $(0 \leq k, 1 < p < \infty)$ is the space of functions defined as the closure of smooth functions in the norm

$$\|u\|_{W_\infty^{k,p}(\Omega)} = \sup_{y\in\Omega} \|u\|_{W^{k,p}(\Omega\cap Q_y)}.$$

Here Ω is a domain in $\mathbb{R}^n$, Q_y is a unit ball with its center at y, $\|\cdot\|_{W^{k,p}}$ is the Sobolev norm. We note that in bounded domains Ω the norms of the spaces $W^{k,p}(\Omega)$ and $W_\infty^{k,p}(\Omega)$ are equivalent.

We suppose that the boundary $\partial\Omega$ belongs to the Hölder space $C^{k+\theta}$, $0 < \theta < 1$, and that the Hölder norms of the corresponding functions in local coordinates are bounded independently of the point of the boundary. Then we can define the space $W_\infty^{k-1/p,p}(\partial\Omega)$ of traces at the boundary $\partial\Omega$ of the domain Ω,

$$\|\phi\|_{W_\infty^{k-1/p,p}(\partial\Omega)} = \inf \|v\|_{W_\infty^{k,p}(\Omega)},$$

2. Sobolev spaces   147

where the infimum is taken with respect to all functions $v \in W_\infty^{k,p}(\Omega)$ equal to ϕ at the boundary, and $k > 1/p$.

The space $W_\infty^{k,p}(\Omega)$ with $k = 0$ will be denoted by $L_\infty^p(\Omega)$. We will use also the following notation:
$$E_\infty = \Pi_{j=1}^N W_\infty^{l+t_j,p}(\Omega),$$
$$F_\infty = \Pi_{i=1}^N W_\infty^{l-s_i,p}(\Omega) \times \Pi_{j=1}^m W_\infty^{l-\sigma_j-1/p,p}(\partial\Omega).$$

We consider the operator L defined by (0.3) assuming that it is uniformly elliptic and put $l_1 = \max_j(0, \sigma_j + 1)$. We suppose that the integer l in the definition of the spaces is such that $l \geq l_1$, and the boundary $\partial\Omega$ belongs to the class $C^{r+\theta}$ with r specified in Condition D (Section 1).

Theorem 2.1. *Let $u \in \Pi_{j=1}^N W_\infty^{l_1+t_j,p}(\Omega)$. Then for any $l \geq l_1$ we have $u \in E_\infty$ and*
$$\|u\|_{E_\infty} \leq c \left(\|Lu\|_{F_\infty} + \|u\|_{L_\infty^p(\Omega)} \right), \tag{2.1}$$
where the constant c does not depend on u.

The proof is given in Section 2 of the previous chapter.

2.2 Limiting problems

2.2.1. Convergence. In the previous section we introduced limiting domains. We now define the corresponding limiting problems. Let Ω be a domain satisfying Condition D and $\chi(x)$ be its characteristic function. Consider a sequence $x_m \in \Omega$, $|x_m| \to \infty$ and the shifted domains Ω_m defined by the shifted characteristic functions $\chi_m(x) = \chi(x + x_m)$. We suppose that the sequence of domains Ω_m converge in Ξ_{loc} to some limiting domain Ω_*. In this section we suppose that $0 \leq k \leq l$ (see Section 2.1).

Definition 2.2. Let $u_m \in W_\infty^{k,p}(\Omega_m)$, $m = 1, 2, \ldots$. We say that u_m converges to a limiting function $u_* \in W_\infty^{k,p}(\Omega_*)$ in $W_{loc}^{k,p}(\Omega_m \to \Omega_*)$ if there exists an extension $v_m(x) \in W_\infty^{k,p}(\mathbb{R}^n)$ of $u_m(x)$, $m = 1, 2, \ldots$ and an extension $v_*(x) \in W_\infty^{k,p}(\mathbb{R}^n)$ of $u_*(x)$ such that $v_m \to v_*$ in $W_{loc}^{k,p}(\mathbb{R}^n)$.

Here and in what follows, convergence in $W_{loc}^{k,p}(\mathbb{R}^n)$ signifies local convergence in $W_\infty^{k,p}(\mathbb{R}^n)$.

Definition 2.3. Let $u_m \in W_\infty^{k-1/p,p}(\partial\Omega_m)$, $k > 1/p$, $m = 1, 2, \ldots$. We say that u_m converges to a limiting function $u_* \in W_\infty^{k-1/p,p}(\partial\Omega_*)$ in $W_{loc}^{k-1/p,p}(\partial\Omega_m \to \partial\Omega_*)$ if there exists an extension $v_m(x) \in W_\infty^{k,p}(\mathbb{R}^n)$ of $u_m(x)$, $m = 1, 2, \ldots$ and an extension $v_*(x) \in W_\infty^{k,p}(\mathbb{R}^n)$ of $u_*(x)$ such that $v_m \to v_*$ in $W_{loc}^{k,p}(\mathbb{R}^n)$.

Definition 2.4. Let $u_m(x) \in C^k(\Omega_m)$, $m = 1, 2, \ldots$. We say that $u_m(x)$ converges to a limiting function $u_*(x) \in C^k(\Omega_*)$ in $C_{loc}^k(\Omega_m \to \Omega_*)$ if there exists an extension

$v_m(x) \in C^k(\mathbb{R}^n)$ of $u_m(x)$, $m = 1, 2, \ldots$ and an extension $v_*(x) \in C^k(\mathbb{R}^n)$ of $u_*(x)$ such that
$$v_m \to v_* \text{ in } C^k_{\text{loc}}(\mathbb{R}^n).$$

Definition 2.5. Let $u_m(x) \in C^k(\partial\Omega_m)$, $m = 1, 2, \ldots$. We say that $u_m(x)$ converges to a limiting function $u_*(x) \in C^k(\partial\Omega_*)$ in $C^k_{\text{loc}}(\partial\Omega_m \to \partial\Omega_*)$ if there exists an extension $v_m(x) \in C^k(\mathbb{R}^n)$ of $u_m(x)$, $m = 1, 2, \ldots$ and an extension $v_*(x) \in C^k(\mathbb{R}^n)$ of $u_*(x)$ such that
$$v_m \to v_* \text{ in } C^k_{\text{loc}}(\mathbb{R}^n).$$

Theorem 2.6. *The limiting function $u_*(x)$ in Definitions 2.2–2.5 does not depend on the choice of extensions $v_m(x)$ and $v_*(x)$.*

Proof. Consider Definition 2.2. Any point $x \in \Omega_*$ has a neighborhood U such that $U \subset \Omega_m$ for all m sufficiently large. Since $v_m \to v_*$ in $\hat{W}^{k,p}_{\text{loc}}(\mathbb{R}^n)$, then
$$\int_U |v_m - v_*|^p dx \to 0, \quad m \to \infty.$$
Hence
$$\int_U |u_m - u_*|^p dx \to 0, \quad m \to \infty.$$
Therefore u_* does not depend on the choice of v_m.

Consider now Definition 2.3. Let x_0 be an arbitrary point of $\partial\Omega_*$. Then there exists a sequence x_m such that $x_m \to x_0$, $x_m \in \partial\Omega_m$. For each point x_m and domain Ω_m there exists a neighborhood $U(x_m)$ and a function $\psi_m(x) = \psi(x; x_m)$ from Condition D. The functions $\psi_m(x)$ are uniformly bounded in $C^{k+\theta}$-norm. The domain of definition of each of these functions is an inverse image of the unit sphere B in $\mathbb{R}^n$. Choosing a convergent subsequence of the inverse images and of the functions $\psi_m(x)$ we obtain a limiting neighborhood $U(x_0)$ and a limiting function $\psi_0(x)$ which satisfy Condition D. For this subsequence we retain the same notation as for the whole sequence. Thus we have
$$\psi_m : U(x_m) \to B, \quad \psi_0 : U(x_0) \to B.$$
Set $\phi_m = \psi_m^{-1}$, $\phi_0 = \psi_0^{-1}$. We have $\phi_m \to \phi_0$ in $C^k(B)$. Let
$$u_m \in W^{k-1/p,p}_\infty(\partial\Omega_m), \quad u_* \in W^{k-1/p,p}_\infty(\partial\Omega_*)$$
and
$$v_m \in W^{k,p}_\infty(\mathbb{R}^n), \quad v^* \in W^{k,p}_\infty(\mathbb{R}^n)$$
be extensions of these functions defined in Definition 2.3. Then
$$v_m \to v_* \text{ in } W^{k,p}_{\text{loc}}(\mathbb{R}^n). \tag{2.2}$$
It is sufficient to consider the case $1/p < k < 1$.

2. Sobolev spaces

We first prove that
$$v_m(\phi_m(y)) \to v_*(\phi_0(y)) \text{ in } W^{k,p}(B) \tag{2.3}$$
as $m \to \infty$. To do this we will prove that
$$\|v_m(\phi_m(y)) - v_*(\phi_m(y))\|_{W^{k,p}(B)} \to 0 \tag{2.4}$$
and
$$\|v_*(\phi_m(y)) - v_*(\phi_0(y))\|_{W^{k,p}(B)} \to 0 \tag{2.5}$$
as $m \to \infty$. We begin with (2.4). We will prove that
$$T_m \equiv \int_B \int_B \frac{|[v_m(\phi_m(y)) - v_*(\phi_m(y))] - [v_m(\phi_m(z)) - v_*(\phi_m(z))]|^p}{|y-z|^{n+pk}} dy dz \to 0 \tag{2.6}$$
as $m \to \infty$. We do the change of variables $y = \psi_m(\eta)$, $z = \psi_m(\zeta)$. Taking into account that for the Jacobians we have the uniform estimates:
$$|d\psi_m/d\eta| \leq M, \; |d\psi_m/d\zeta| \leq M,$$
and
$$|\eta - \zeta| = |\phi_m(y) - \phi_m(z)| \leq K|y-z|$$
with K independent of m, we get
$$T_m \leq M_1 \int_{U(x_m)} \int_{U(x_m)} \frac{|[v_m(\eta) - v_*(\eta)] - [v_m(\zeta) - v_*(\zeta)]|^p}{|\eta - \zeta|^{n+pk}} d\eta d\zeta.$$

Since $U(x_m)$ belongs to a ball V with the center at x_0 and with the radius independent of m, we get $T_m \leq M_1 \|v_m - v_*\|^p_{W^{k,p}(V)} \to 0$ as $m \to \infty$. We use the fact that $v_m \to v_*$ in $\hat{W}^{k,p}_{loc}(\mathbb{R}^n)$. To prove (2.4) it remains to verify it for $k = 0$. It can be done with the same change of variables as above. Thus (2.4) is proved.

Let us prove (2.5). Let
$$\omega_l(x) \to v_*(x) \text{ in } W^{1,p}_{loc}(\mathbb{R}^n) \text{ as } l \to \infty, \tag{2.7}$$
where $\omega_l(x)$ are smooth functions, and $\varepsilon > 0$ be a given number. We take l so large that
$$\|v_*(\phi_m(y)) - \omega_l(\phi_m(y))\|_{W^{k,p}(B)} \leq \frac{\varepsilon}{3} \text{ for all } m \tag{2.8}$$
and
$$\|v_*(\phi_0(y)) - \omega_l(\phi_0(y))\|_{W^{k,p}(B)} \leq \frac{\varepsilon}{3}. \tag{2.9}$$
It is possible by virtue of (2.7). The fact that the left-hand side in (2.8) tends to 0 as $l \to \infty$ uniformly with respect to m is proved similarly to (2.6) by the change of variables $x = \phi_m(y)$.

For a fixed l we can find m_0 such that

$$\|\omega_l(\phi_m(y)) - \omega_l(\phi_0(y))\|_{W^{k,p}(B)} \leq \frac{\varepsilon}{3} \tag{2.10}$$

for $m > m_0$. From (2.8)–(2.10) it follows that

$$\|v_*(\phi_m(y)) - v_*(\phi_0(y))\|_{W^{k,p}(B)} \leq \varepsilon$$

as $m > m_0$ and therefore (2.5) is proved.

Convergence (2.3) follows from (2.4) and (2.5). Set $f_m(y) = v_m(\varphi_m(y))$, $f_*(y) = v_*(\varphi_0(y))$. It follows from (2.3) that

$$\|f_m^+ - f_*^+\|_{L^p(B_0)} \to 0 \quad \text{as } m \to \infty. \tag{2.11}$$

Here f_m^+ and f_*^+ are the traces of $f_m(y)$ and $f_*(y)$ respectively from B_+. Suppose now that there is another limiting function $\tilde{u}_*(x) \in W_\infty^{k-1/p,p}(\partial\Omega_*)$ for the sequence u_m and other extensions $\tilde{v}_m(x) \in W_\infty^{k,p}(\mathbb{R}^n)$ of u_m, $m = 1, 2, \ldots$ and $\tilde{v}_*(x) \in W_\infty^{k,p}(\mathbb{R}^n)$ of $\tilde{u}_*(x)$ such that $\tilde{v}_m \to \tilde{v}_*$ in $W_{\text{loc}}^{k,p}(\mathbb{R}^n)$. We will prove that $\tilde{u}_* = u_*$.

Let $\tilde{f}_m(y) = \tilde{v}_m(\varphi_m(y))$, $\tilde{f}_*(y) = \tilde{v}_*(\varphi_0(y))$. Similar to (2.11) we obtain

$$\|\tilde{f}_m^+ - \tilde{f}_*^+\|_{L^p(B_0)} \to 0 \quad \text{as } m \to \infty. \tag{2.12}$$

On the other hand $\tilde{v}_m^+(x) = u_m^+(x) = v_m^+(x)$ H_{n-1}-almost everywhere in $\partial\Omega_m \bigcap U(x_m)$. Here H_{n-1} is an $(n-1)$-dimensional Hausdorff measure. Taking into account that $\psi_m : U(x_m) \to B$ is a diffeomorphism such that $\psi_m(\partial\Omega_m \bigcap U(x_m)) = B_0$, it is easy to prove that $\tilde{f}_m^+(y) = f_m^+(y)$ H_{n-1}-almost everywhere on B_0. To prove this it is sufficient to approximate $v_m(x)$ and $\tilde{v}_m(x)$ by smooth functions. It follows from (2.12) that

$$\|f_m^+ - \tilde{f}_*^+\|_{L^p(B_0)} \to 0 \quad \text{as } m \to \infty.$$

Comparing this convergence with (2.11) we obtain that

$$\tilde{f}_*^+(y) = f_*^+(y) \tag{2.13}$$

H_{n-1}-almost everywhere in B_0. Since $v_*^+(x) = u_*(x)$,

$$\tilde{v}_*^+(x) = \tilde{u}_*(x) \text{ in } \partial\Omega_* \bigcap U(x_0)$$

and $\psi_0 : U(x_0) \to B$ is a diffeomorphism such that $\psi_0(\partial\Omega_* \bigcap U(x_0)) = B_0$, we conclude from (2.13) that $\tilde{u}_*(x) = u_*(x)$ H_{n-1}-almost everywhere in $\partial\Omega_* \bigcap U(x_0)$.

Since x_0 is an arbitrary point in $\partial\Omega_*$, it follows that $\tilde{u}_*$ coincides with u_* as elements of the space $W_\infty^{k-1/p,p}(\partial\Omega_*)$.

2. Sobolev spaces

Consider Definition 2.3. For any point $x \in \overline{\Omega}_*$ there exists a sequence $\hat{x}_m \in \overline{\Omega}_m$ such that $\hat{x}_m \to x$. Therefore

$$u_*(x) = v_*(x) = \lim_{m \to \infty} v_m(\hat{x}_m) = \lim_{m \to \infty} u_m(\hat{x}_m).$$

Definition 2.5 is treated similarly. The theorem is proved. $\square$

Theorem 2.7. *Suppose that $0 < k \leq l - 1$. Let*

$$u_m \in W_\infty^{k+1,p}(\Omega_m), \quad \|u_m\|_{W_\infty^{k+1,p}(\Omega_m)} \leq M,$$

where the constant M does not depend on m. Then there exists a function $u_* \in W_\infty^{k+1,p}(\Omega_*)$ and a subsequence u_{m_i} such that $u_{m_i} \to u_*$ in $W_{\text{loc}}^{k,p}(\Omega_m \to \Omega_*)$.

Proof. Let

$$\|u_m\|_{W_\infty^{k+1,p}(\Omega_m)} \leq M.$$

It follows from Condition D that there exists an extension $v_m(x)$ of $u_m(x)$ on the whole $\mathbb{R}^n$ such that

$$\|v_m\|_{W_\infty^{k+1,p}(\mathbb{R}^n)} \leq M_1. \tag{2.14}$$

Denote by B_R a ball in $\mathbb{R}^n$, $|x| \leq R$. Then

$$\|v_m\|_{W^{k+1,p}(B_R)} \leq M_2.$$

Since the space $W^{k+1,p}(B_R)$ is reflexive, then there exists a subsequence v_{m_i} converging weakly in $W^{k+1,p}(B_R)$ to a function $v_* \in W^{k+1,p}(B_R)$. On the other hand, choosing if necessary a subsequence of v_{m_i} and keeping for it the same notation, we have the convergence

$$\|v_{m_i} - v_0\|_{W^{k,p}(B_R)} \to 0,$$

where $v_0(x)$ is a function in $W^{k,p}(B_R)$.

It is easy to see that $v_0 = v_*$. Indeed, from the strong convergence in $W^{k,p}(B_R)$ we obtain the weak convergence to v_0 in the same space. Since the weak limit is unique, and it equals v_*, then $v_* = v_0$.

By a diagonal process we can extend the function v_* to the whole $\mathbb{R}^n$. Therefore

$$\|v_{m_i} - v_*\|_{W_{\text{loc}}^{k,p}(\mathbb{R}^n)} \to 0 \tag{2.15}$$

by some subsequence v_{m_i} of v_m. From (2.14) and (2.15) it follows that $v_* \in W_\infty^{k,p}(\mathbb{R}^n)$. It remains to use Definition 2.2 for u_{m_i} and for the restriction u_* of v_* on Ω_*.

We finally prove that $u_* \in W_\infty^{k+1,p}(\Omega_*)$. It is sufficient to verify that $v_* \in W_\infty^{k+1,p}(\mathbb{R}^n)$. It is shown that $v_* \in W_\infty^{k+1,p}(B_R)$ for any R. We can put $R = 1$. Similarly we can prove that $v_* \in W_\infty^{k+1,p}(Q_y)$ for a unit ball Q_y with the center at y. Suppose that $v_* \notin W_\infty^{k+1,p}(\mathbb{R}^n)$. Then there exists a sequence B_i of the unit balls with the centers at y_i and such that

$$\|v_i\|_{W^{k+1,p}(B_i)} \to \infty, \quad i \to \infty,$$

where v_i is the restriction of v_* to B_i. On the other hand, v_i is a weak limit of a sequence v_i^m of functions defined in B_i. Indeed, this follows from (2.15) and from the definition

$$v_i^m(x) = v_m(x), \quad x \in B_i.$$

Set

$$w_i^m(x) = v_i^m(x - y_i), \quad w_i(x) = v_i(x - y_i).$$

The functions $w_i^m(x)$ are defined in the unit ball B with the center at 0. By the assumption of the theorem,

$$\|w_i^m(x)\|_{W^{k+1,p}(B)} \leq M_1$$

and w_i^m converges weakly in $W^{k+1,p}(B)$ to w_i as $m \to \infty$.

On the other hand,

$$\|w_i\|_{W^{k+1,p}(B)} \to \infty, \quad i \to \infty.$$

A set in a normed space is bounded if and only if any functional from the dual space is bounded on it. Therefore there exists a functional $\phi \in (W^{k+1,p}(B))^*$ such that $\phi(w_i) \to \infty$ as $i \to \infty$. From the weak convergence of w_i^m to w_i it follows that for each i we can choose $m(i)$ in such a way that

$$\phi(w_i^{m(i)}) \to \infty, \quad i \to \infty.$$

Hence the set w_i^m is not bounded. This contradiction proves the boundedness of v_* in the norm $W_\infty^{k+1,p}(\mathbb{R}^n)$. The theorem is proved. □

Theorem 2.8. *Suppose that $0 < k \leq l - 1$. Let $u_m \in W_\infty^{k+1-1/p,p}(\partial\Omega_m)$,*

$$\|u_m\|_{W_\infty^{k+1-1/p,p}(\partial\Omega_m)} \leq M,$$

where the constant M does not depend on m. Then there exists a function

$$u_* \in W_\infty^{k+1-1/p,p}(\partial\Omega_*)$$

and a subsequence u_{m_i} such that

$$u_{m_i} \to u_* \text{ in } W_{\text{loc}}^{k+1-\epsilon-1/p,p}(\partial\Omega_m \to \partial\Omega_*),$$

where $0 < \epsilon < k + 1 - 1/p$.

2. Sobolev spaces

Proof. By the definition of the space $W_\infty^{k+1-1/p,p}(\partial\Omega_m)$ there exists an extension w_m of the function u_m to $W^{k+1,p}(\Omega_m)$ such that

$$\|w_m\|_{W_\infty^{k+1,p}(\Omega_m)} \le 2\|u_m\|_{W_\infty^{k+1-1/p,p}(\partial\Omega_m)} \le 2M,$$

$w_m(x) = u_m(x)$, $x \in \partial\Omega_m$. It follows from Condition D that there exists an extension $v_m(x)$ of $w_m(x)$ to the whole $\mathbb{R}^n$ such that

$$\|v_m\|_{W_\infty^{k+1,p}(\mathbb{R}^n)} \le M_1,$$

$v_m(x) = w_m(x)$, $x \in \Omega_m$. As in the proof of the previous theorem, there exists a function $v_* \in W_\infty^{k+1,p}(\mathbb{R}^n)$ such that

$$v_{m_i} \to v_* \text{ in } W_{\mathrm{loc}}^{k+1-\epsilon,p}(\mathbb{R}^n).$$

The assertion of the theorem follows from Definition 2.3. The theorem is proved. $\square$

Theorem 2.9. *Let $u_m \in C^{k+\theta}(\Omega_m)$, $\|u_m\|_{C^{k+\theta}} \le M$, where the constant M is independent of m. Then there exists a function $u_* \in C^{k+\theta}(\Omega_*)$ and a subsequence u_{m_k} such that $u_{m_k} \to u_*$ in $C_{\mathrm{loc}}^k(\Omega_{m_k} \to \Omega_*)$.*

Let $u_m \in C^{k+\theta}(\partial\Omega_m)$, $\|u_m\|_{C^{k+\theta}} \le M$. Then there exists a function $u_* \in C^{k+\theta}(\partial\Omega_*)$ and a subsequence u_{m_k} such that $u_{m_k} \to u_*$ in $C_{\mathrm{loc}}^k(\partial\Omega_{m_k} \to \partial\Omega_*)$.

Proof. Let $u_m \in C^{k+\theta}(\Omega_m)$, $\|u_m\|_{C^{k+\theta}} \le M$. It follows from Condition D that there exists an extension $v_m(x)$ of $u_m(x)$ on the whole space $\mathbb{R}^n$ such that

$$v_m \in C^{k+\theta}(\mathbb{R}^n), \quad \|v_m\|_{C^{k+\theta}(\mathbb{R}^n)} \le M_0, \quad v_m(x) = u_m(x), \quad x \in \Omega_m,$$

where M_0 is independent of m. Passing to a subsequence and retaining the same notation we can suppose that there exists a function $v_*(x) \in C^{k+\theta}(\mathbb{R}^n)$ such that $\|v_*\|_{C^{k+\theta}(\mathbb{R}^n)} \le M_0$ and

$$v_m \to v_* \text{ in } C_{\mathrm{loc}}^k(\mathbb{R}^n). \tag{2.16}$$

Therefore

$$u_m \to u_* \text{ in } C_{\mathrm{loc}}^k(\Omega_m \to \Omega_*) \tag{2.17}$$

in the sense of Definition 2.2. Here $u_*(x)$ is the restriction of $v_*(x)$ on Ω_*.

The second part of the theorem for $u_m \in C^{k+\theta}(\partial\Omega_m)$ is proved similarly. The theorem is proved. $\square$

2.2.2. Limiting Operators

Suppose that we are given a sequence $\{x_\nu\}$, $\nu = 1, 2, \ldots$, $x_\nu \in \Omega$, $|x_\nu| \to \infty$. Consider the shifted domains Ω_ν with the characteristic functions $\chi_\nu(x) = \chi(x + x_\nu)$ where $\chi(x)$ is the characteristic function of Ω, and the shifted coefficients of the operators A_i and B_j:

$$a_{ik,\nu}^\alpha(x) = a_{ik}^\alpha(x + x_\nu), \quad b_{jk,\nu}^\beta(x) = b_{jk}^\beta(x + x_\nu).$$

We suppose that

$$a_{ik}^{\alpha}(x) \in C^{l-s_i+\theta}(\bar{\Omega}), \quad b_{jk}^{\beta}(x) \in C^{l-\sigma_j+\theta}(\partial\Omega), \tag{2.18}$$

where $0 < \delta < 1$, and that these coefficients can be extended to $\mathbb{R}^n$:

$$a_{ik}^{\alpha}(x) \in C^{l-s_i+\theta}(\mathbb{R}^n), \quad b_{jk}^{\beta}(x) \in C^{l-\sigma_j+\theta}(\mathbb{R}^n). \tag{2.19}$$

Therefore

$$\|a_{ik,\nu}^{\alpha}(x)\|_{C^{l-s_i+\theta}(\mathbb{R}^n)} \leq M, \quad \|b_{jk,\nu}^{\beta}(x)\|_{C^{l-\sigma_j+\theta}(\mathbb{R}^n)} \leq M \tag{2.20}$$

with some constant M independent of ν. It follows from Theorem 1.6 that there exists a subsequence of the sequence Ω_ν, for which we keep the same notation, such that it converges to a limiting domain Ω_*. From (2.20) it follows that this subsequence can be chosen such that

$$a_{ik,\nu}^{\alpha} \to \hat{a}_{ik}^{\alpha} \text{ in } C^{l-s_i}(\mathbb{R}^n) \text{ locally}, \quad b_{jk,\nu}^{\beta} \to \hat{b}_{jk}^{\beta} \text{ in } C^{l-\sigma_j}(\mathbb{R}^n) \text{ locally}, \tag{2.21}$$

where $\hat{a}_{ik}^{\alpha}$ and $\hat{b}_{jk}^{\beta}$ are limiting coefficients,

$$\hat{a}_{ik}^{\alpha} \in C^{l-s_i+\theta}(\mathbb{R}^n), \quad \hat{b}_{jk}^{\beta} \in C^{l-\sigma_j+\theta}(\mathbb{R}^n).$$

We have constructed the limiting operators

$$\hat{A}_i u = \sum_{k=1}^{N} \sum_{|\alpha| \leq \alpha_{ik}} \hat{a}_{ik}^{\alpha}(x) D^{\alpha} u_k, \quad i = 1, \ldots, N, \quad x \in \Omega_*, \tag{2.22}$$

$$\hat{B}_j u = \sum_{k=1}^{N} \sum_{|\beta| \leq \beta_{jk}} \hat{b}_{jk}^{\beta}(x) D^{\beta} u_k, \quad i = 1, \ldots, m, \quad x \in \partial\Omega_*, \tag{2.23}$$

$$\hat{L} = (\hat{A}_1, \ldots, \hat{A}_N, \hat{B}_1, \ldots, \hat{B}_m). \tag{2.24}$$

We consider them as acting from $E_\infty(\Omega_*)$ into $F_\infty(\Omega_*)$.

2.3 A priori estimates with condition NS

We have already discussed above that the ellipticity condition, proper ellipticity and the Lopatinskii condition may not be sufficient in order for elliptic operators in unbounded domains to be normally solvable. We introduce an additional condition.

Condition NS. Any limiting problem

$$\hat{L}u = 0, \quad x \in \Omega_*, \quad u \in E_\infty(\Omega_*)$$

has only the zero solution.

2. Sobolev spaces

In the next section we will prove that Condition NS is necessary and sufficient in order for the operator L to be normally solvable with a finite-dimensional kernel. In this section we will use it to obtain a priori estimates of solutions stronger than those given by Theorem 2.1. Estimates of this type were first obtained in [361], [362] for elliptic operators in the whole R^n.

Theorem 2.10. *Let Condition NS be satisfied. Then there exist numbers M_0 and R_0 such that the following estimate holds:*

$$\|u\|_{E_\infty} \leq M_0 \left(\|Lu\|_{F_\infty} + \|u\|_{L^p(\Omega_{R_0})} \right), \quad \forall u \in E_\infty. \tag{2.25}$$

Here $\Omega_{R_0} = \Omega \cap \{|x| \leq R_0\}$.

Proof. Suppose that the assertion of the theorem is not correct. Let $M_k \to \infty$ and $R_k \to \infty$ be given sequences. Then there exists $u_k \in E_\infty$ such that

$$\|u_k\|_{E_\infty} > M_k \left(\|Lu_k\|_{F_\infty} + \|u_k\|_{L^p(\Omega_{R_k})} \right).$$

We can suppose that

$$\|u_k\|_{E_\infty} = 1. \tag{2.26}$$

Then

$$\|Lu_k\|_{F_\infty} + \|u_k\|_{L^p(\Omega_{R_k})} < \frac{1}{M_k} \to 0 \text{ as } k \to \infty. \tag{2.27}$$

From Theorem 2.1 we obtain

$$\|Lu_k\|_{F_\infty} + \|u_k\|_{L^p_\infty(\Omega)} \geq \frac{1}{c}.$$

It follows from (2.27) that $\|Lu_k\|_{F_\infty} \to 0$. Hence

$$\|u_k\|_{L^p_\infty(\Omega)} > \frac{1}{2c} \text{ for } k \geq k_0 \tag{2.28}$$

with some k_0. Since

$$\|u_k\|_{L^p_\infty(\Omega)} = \sup_{y \in \Omega} \|u_k\|_{L^p(Q_y \cap \Omega)},$$

then it follows from (2.28) that there exists $y_k \in \Omega$ such that

$$\|u_k\|_{L^p(Q_{y_k} \cap \Omega)} > \frac{1}{2c}. \tag{2.29}$$

From (2.27)

$$\|u_k\|_{L^p(\Omega_{R_k})} \to 0.$$

This convergence and (2.29) imply that $|y_k| \to \infty$.

Let

$$Lu_k = f_k. \tag{2.30}$$

From (2.27) we get
$$\|f_k\|_{F_\infty} \to 0 \text{ as } k \to \infty. \tag{2.31}$$

Next, let $x = y + y_k$,
$$w_k(y) = u_k(y + y_k). \tag{2.32}$$

We rewrite (2.30) in the detailed form
$$\sum_{h=1}^{N} \sum_{|\alpha| \le \alpha_{ih}} a_{ih}^\alpha(x) D^\alpha u_{hk} = f_{ik}, \quad i = 1, \ldots, N, \quad x \in \Omega, \tag{2.33}$$

$$\sum_{h=1}^{N} \sum_{|\beta| \le \beta_{jh}} b_{jh}^\beta(x) D^\beta u_{hk} = f_{jk}^b, \quad i = 1, \ldots, m, \quad x \in \partial\Omega, \tag{2.34}$$

where
$$f_k = (f_{1k}, \ldots, f_{Nk}, f_{1k}^b, \ldots, f_{mk}^b), \quad u_k = (u_{1k}, \ldots, u_{Nk}).$$

Writing
$$a_{ih\,k}^\alpha(y) = a_{ih}^\alpha(y + y_k), \quad b_{jh\,k}^\beta(y) = b_{jh}^\beta(y + y_k),$$

we obtain from (2.33), (2.34),

$$\sum_{h=1}^{N} \sum_{|\alpha| \le \alpha_{ih}} a_{ih\,k}^\alpha(y) D^\alpha w_{hk}(y) = f_{ik}(y + y_k), \quad i = 1, \ldots, N, \quad x \in \Omega_k, \tag{2.35}$$

$$\sum_{h=1}^{N} \sum_{|\beta| \le \beta_{jh}} b_{jh\,k}^\beta(y) D^\beta w_{hk}(y) = f_{jk}^b(y + y_k), \quad i = 1, \ldots, m, \quad x \in \partial\Omega_k, \tag{2.36}$$

where Ω_k is the shifted domain. From (2.26) we have
$$\|w_k\|_{E_\infty(\Omega_k)} = 1. \tag{2.37}$$

We have $w_k = (w_{1k}, \ldots, w_{Nk})$, and (2.37) can be written in the form
$$\sum_{i=1}^{N} \|w_{ik}\|_{W_\infty^{l+t_i,p}(\Omega_k)} = 1.$$

We suppose that the functions w_{ik} are extended to $\mathbb{R}^n$ such that their $W_\infty^{l+t_i,p}(\mathbb{R}^n)$-norms are uniformly bounded. Passing to a subsequence and retaining the same notation, we can suppose that
$$w_{ik} \to w_{i0} \text{ in } W^{l+t_i-\epsilon,p}(\mathbb{R}^n) \text{ locally}, \quad (\epsilon > 0), \tag{2.38}$$
$$w_{ik} \to w_{i0} \text{ in } W^{l+t_i,p}(\mathbb{R}^n) \text{ locally weakly} \tag{2.39}$$

for some w_{i0} as $k \to \infty$, and
$$w_{i0} \in W_\infty^{l+t_i,p}(\mathbb{R}^n), \quad i = 1, \ldots, N. \tag{2.40}$$

2. Sobolev spaces

Let $w_0 = (w_{10}, \ldots, w_{N0})$. We prove that

$$\hat{L}w_0 = 0 \qquad (2.41)$$

for a limiting operator $\hat{L}$. To do this we pass to the limit in (2.35), (2.36) by a subsequence of k. We choose this subsequence such that Ω_k converges to a limiting domain, $\Omega_k \to \Omega_*$, and keep for it the same notation.

We begin with equation (2.35). For any $x_0 \in \Omega_*$ we take a neighborhood U in such a way that $U \subset \Omega_k$ for k sufficiently large. For any $\phi \in D$ with its support in U we get from (2.35):

$$\int_U \sum_{h=1}^N \sum_{|\alpha| \leq \alpha_{ih}} a_{ihk}^\alpha(y) D^\alpha w_{hk}(y) \phi(y) dy = \int_U f_{ik}(y + y_k) \phi(y) dy. \qquad (2.42)$$

We can suppose, passing to a subsequence, that

$$a_{ih,k}^\alpha(y) \to \hat{a}_{ih}^\alpha(y) \text{ in } C^{l-s_i}(\mathbb{R}^n) \text{ locally}$$

(see (2.20)), where $\hat{a}_{ih}^\alpha(y)$ are the coefficients of the limiting operator. It follows from (2.39) that $D^\alpha w_{hk}$ ($|\alpha| \leq \alpha_{ih}$) converges locally weakly in $W^{l-s_i,p}$ to $D^\alpha w_{h0}$ as $k \to \infty$. Hence we can pass to the limit in (2.42).

From (2.31) it follows that

$$\|f_{ik}(\cdot + y_k)\|_{W^{l-s_i,p}_\infty(\Omega_k)} \to 0 \text{ as } k \to \infty.$$

Hence the right-hand side in (2.42) tends to zero. Passing to the limit in this equation, we obtain

$$\sum_{h=1}^N \sum_{|\alpha| \leq \alpha_{ih}} \hat{a}_{ih}^\alpha(y) D^\alpha w_{h0}(y) = 0, \quad y \in \Omega_*. \qquad (2.43)$$

Consider now (2.36). From (2.38) it follows that $D^\beta w_{hk}$ ($|\beta| \leq \beta_{ih}$) tends to $D^\beta w_{h0}$ in $W^{l-\sigma_j-\epsilon,p}(\mathbb{R}^n)$ locally. Hence (2.20) implies that

$$\sum_{h=1}^N \sum_{|\beta| \leq \beta_{jh}} b_{jhk}^\beta(y) D^\beta w_{hk}(y) \to \sum_{h=1}^N \sum_{|\beta| \leq \beta_{jh}} \hat{b}_{jh}^\beta(y) D^\beta w_{h0}(y) \qquad (2.44)$$

in $W^{l-\sigma_j-\epsilon,p}_{\text{loc}}(\mathbb{R}^n)$. Therefore this convergence takes place also in $W^{l-\sigma_j-\epsilon,p}_{\text{loc}}(\Omega_*)$ and, consequently, in $W^{l-\sigma_j-\epsilon-1/p,p}_{\text{loc}}(\partial\Omega_*)$. In other words, we have proved that the convergence (2.44) is in $W^{l-\sigma_j-\epsilon-1/p,p}_{\text{loc}}(\partial\Omega_k \to \partial\Omega_*)$ (see Definition 2.3).

Consider next the right-hand side in (2.36). According to (2.31) we have

$$\|f_{jk}^b(\cdot + y_k)\|_{W^{l-\sigma_j-1/p,p}_\infty(\partial\Omega_k)} \to 0 \text{ as } k \to \infty.$$

We can extend $f_{jk}^b(y+y_k)$ to the whole $\mathbb{R}^n$ in such a way that

$$f_{jk}^b(\cdot + y_k) \to 0 \text{ in } W_\infty^{l-\sigma_j,p}(\mathbb{R}^n).$$

Therefore
$$f_{jk}^b(\cdot + y_k) \to 0 \text{ in } W_{\text{loc}}^{l-\sigma_j-1/p,p}(\partial\Omega_k \to \partial\Omega_*).$$

From this and convergence (2.44) follows

$$\sum_{h=1}^N \sum_{|\beta|\leq \beta_{jh}} \hat{b}_{jh}^\beta(y) D^\beta w_{h0}(y) = 0, \quad y \in \partial\Omega_*. \tag{2.45}$$

From (2.40) it follows that the left-hand side of this equality belongs to $W_\infty^{l-\sigma_j-1/p,p}(\partial\Omega_*)$. Hence it can be regarded as an equality in $W_\infty^{l-\sigma_j-1/p,p}(\partial\Omega_*)$.

From (2.43) and (2.45) we conclude that w_0 is a solution of the limiting problem (2.41). We prove now that $w_0 \neq 0$. From (2.29) and (2.32) we have

$$\|w_k\|_{L^p(\Omega_k \cap Q_0)} > \frac{1}{2c}, \tag{2.46}$$

where Q_0 is the unit ball with its center at the origin. We prove that

$$\|w_0\|_{L^p(\Omega_* \cap Q_0)} \geq \frac{1}{2c}. \tag{2.47}$$

Indeed, from (2.38),
$$w_k \to w_0 \text{ in } L^p_{\text{loc}}(\mathbb{R}^n).$$

Set $S_k = \Omega_k \cap Q_0$, $S_* = \Omega_* \cap Q_0$. Then

$$\big| \|w_k\|_{L^p(S_k)} - \|w_0\|_{L^p(S_*)} \big|$$
$$\leq \big| \|w_k\|_{L^p(S_k)} - \|w_0\|_{L^p(S_k)} \big| + \big| \|w_0\|_{L^p(S_k)} - \|w_0\|_{L^p(S_*)} \big| \equiv A_k + B_k.$$

Further

$$A_k \leq \|w_k - w_0\|_{L^p(S_k)} = \sum_{i=1}^N \|w_{ik} - w_{i0}\|_{L^p(S_k)}$$
$$= \sum_{i=1}^N \left(\int_{S_k} |w_{ik} - w_{i0}|^p dx \right)^{1/p} \to 0 \text{ as } k \to \infty,$$

$$B_k \leq \sum_{i=1}^N \big| \|w_{i0}\|_{L^p(S_k)} - \|w_{i0}\|_{L^p(S_*)} \big|$$
$$\leq M \sum_{i=1}^N \left(\int_{S_k \triangle S_*} |w_{i0}|^p dx \right)^{1/p} \to 0 \text{ as } k \to \infty$$

since the measure of the symmetric difference $S_k \triangle S_*$ converges to 0.

2. Sobolev spaces

We have proved that

$$\|w_k\|_{L^p(\Omega_k \cap Q_0)} \to \|w_0\|_{L^p(\Omega_* \cap Q_0)},$$

and (2.47) follows from (2.46). Thus there exists a limiting problem with a nonzero solution. This contradicts Condition NS. The theorem is proved. □

Write

$$\omega_\mu = e^{\mu\sqrt{1+|x|^2}},$$

where μ is a real number.

Theorem 2.11. *Let Condition NS be satisfied. Then there exist numbers $M_0 > 0$, $R_0 > 0$ and $\mu_0 > 0$ such that for all μ, $0 < \mu < \mu_0$ the following estimate holds:*

$$\|\omega_\mu u\|_{E_\infty} \leq M_0 \left(\|\omega_\mu L u\|_{F_\infty} + \|\omega_\mu u\|_{L^p(\Omega_{R_0})} \right) \quad \text{if } \omega_\mu u \in E_\infty. \tag{2.48}$$

Proof. According to (4.2) we have

$$\|\omega_\mu u\|_{E_\infty} \leq M \left(\|L(\omega_\mu u)\|_{F_\infty} + \|\omega_\mu u\|_{L^p(\Omega_{R_0})} \right). \tag{2.49}$$

The operator L has the form: $L = (A_1, \ldots, A_N, B_1, \ldots, B_m)$. Consider first the operator

$$A_i(\omega_\mu u) = \sum_{k=1}^{N} \sum_{|\alpha| \leq \alpha_{ik}} a_{ik}^\alpha(x) D^\alpha(\omega_\mu u_k), \quad i = 1, \ldots, N.$$

We have

$$A_i(\omega_\mu u) = \omega_\mu A_i(u) + \Phi_i, \tag{2.50}$$

where

$$\Phi_i = \sum_{k=1}^{N} \sum_{|\alpha| \leq \alpha_{ik}} \sum_{\beta+\gamma=\alpha, |\beta|>0} a_{ik}^\alpha(x) c_{\beta\gamma} D^\beta \omega_\mu D^\gamma u_k,$$

and $c_{\beta\gamma}$ are some constants. Direct calculations give the following estimate:

$$\|\Phi_i\|_{W_\infty^{l-s_i,p}} \leq M_1 \mu \|\omega_\mu u\|_{E_\infty(\Omega)}. \tag{2.51}$$

For the boundary operators we have

$$B_j(\omega_\mu u) = \sum_{k=1}^{N} \sum_{|\beta| \leq \beta_{jk}} b_{jk}^\beta(x) D^\beta(\omega_\mu u_k).$$

As above we get

$$B_j(\omega_\mu u) = \omega_\mu B_j(u) + \Psi_j, \tag{2.52}$$

$$\|\Psi_j\|_{W_\infty^{l-\sigma_j-1/p,p}} \leq M_2 \mu \|\omega_\mu u\|_{E_\infty(\Omega)}. \tag{2.53}$$

From (2.50)–(2.53) we obtain

$$\|L(\omega_\mu u)\|_{F_\infty} \leq \|\omega_\mu Lu\|_{F_\infty} + M\mu\|\omega_\mu u\|_{E_\infty}.$$

The assertion of the theorem follows from this estimate and (2.49). The theorem is proved. □

Theorem 2.12. *If $0 < \mu < \mu_0$ for some μ_0, $u \in E_\infty$, and $\omega_\mu Lu \in F_\infty$, then $\omega_\mu u \in E_\infty$. In particular, if $u \in E_\infty$ and $Lu = 0$, then $\omega_\mu u \in E_\infty$.*

Proof. Let $\{B_j\}(j = 1, 2, \ldots)$ be a covering of $\mathbb{R}^n$ by unit balls with centers at the points x_j. Let further θ_j be the corresponding partition of unity, $\operatorname{supp} \theta_j \subset B_j$. We introduce the norms in E_∞ and F_∞ in accordance with this partition of unity. Suppose that functions $\phi_j \in D$ are such that

$$\phi_j(x) = 1 \text{ for } |x - x_j| \leq 2, \ \operatorname{supp} \phi_j \subset \{|x - x_j| < 3\},$$

and functions $\psi_j \in D$ are such that

$$\psi_j(x) = 1 \text{ for } |x - x_j| \leq 3, \ \operatorname{supp} \phi_j \subset \{|x - x_j| < 4\}.$$

We introduce next a small parameter $\epsilon > 0$ and denote $\phi_j^\epsilon(x) = \phi_j(\epsilon x)$, $\psi_j^\epsilon(x) = \psi_j(\epsilon x)$. It follows from Theorem 2.11 that

$$\|\omega u \phi_j^\epsilon\|_{E_\infty} \leq M_0 \left(\|\omega L(u\phi_j^\epsilon)\|_{F_\infty} + \|\omega u \phi_j^\epsilon\|_{L^p(\Omega_R)} \right) \tag{2.54}$$
$$\leq M_0 \left(\|\omega \phi_j^\epsilon Lu\|_{F_\infty} + \|\omega u \phi_j^\epsilon\|_{L^p(\Omega_R)} \right) + M_0 \|\omega(\phi_j^\epsilon Lu - L(u\phi_j^\epsilon))\|_{F_\infty}.$$

Here and in what follows we write ω instead of ω_μ. We have

$$\|\omega(\phi_j^\epsilon Lu - L(u\phi_j^\epsilon))\|_{F_\infty} = \|\omega \psi_j^\epsilon(\phi_j^\epsilon Lu - L(u\phi_j^\epsilon))\|_{F_\infty}$$
$$\leq M_1 \rho_\epsilon \sup_\alpha \|\omega \psi_j^\epsilon D^\alpha u\|_{F_\infty}, \tag{2.55}$$

where

$$\rho_\epsilon = \sup_{x, 0 < |\alpha| \leq l + t_k, k = 1, \ldots, N} |D^\alpha \phi_j^\epsilon(x)|.$$

We estimate the right-hand side in (2.55):

$$\|\omega \psi_j^\epsilon D^\alpha u\|_{F_\infty} \leq K \sum_{i'} \|\omega \phi_{i'}^\epsilon u\|_{E_\infty}, \tag{2.56}$$

where K is a constant independent of ϵ, i' denotes all of i for which

$$\operatorname{supp} \phi_i^\epsilon \cap \operatorname{supp} \psi_j^\epsilon \neq \emptyset.$$

Denote the number of such i by N. It is easy to see that it does not depend on ϵ.

2. Sobolev spaces

From (2.54)–(2.56) we obtain

$$\|\omega u \phi_j^\epsilon\|_{E_\infty} \leq M_0 \left(\|\omega Lu\|_{F_\infty} + \|\omega u\|_{L^P(\Omega_R)}\right) + M_2 \rho_\epsilon \sum_{i'} \|\omega u \phi_{i'}^\epsilon\|_{E_\infty}.$$

In the last term in the right-hand side we take the maximum among the summands:

$$\|\omega u \phi_j^\epsilon\|_{E_\infty} \leq M_0 \left(\|\omega Lu\|_{F_\infty} + \|\omega u\|_{L^P(\Omega_R)}\right) + M_2 \rho_\epsilon N \|\omega u \phi_{i(j)}^\epsilon\|_{E_\infty}. \quad (2.57)$$

We rewrite this inequality in the form

$$\|\omega u \phi_{j_1}^\epsilon\|_{E_\infty} \leq M_0 \left(\|\omega Lu\|_{F_\infty} + \|\omega u\|_{L^P(\Omega_R)}\right) + \sigma \|\omega u \phi_{j_2}^\epsilon\|_{E_\infty}, \quad (2.58)$$

where σ is a small constant, and the support of the function ϕ_{j_2} is neighboring to the support of the function ϕ_{j_1}. Since the last estimate is true for any j, then we can write

$$\|\omega u \phi_{j_2}^\epsilon\|_{E_\infty} \leq M_0 \left(\|\omega Lu\|_{F_\infty} + \|\omega u\|_{L^P(\Omega_R)}\right) + \sigma \|\omega u \phi_{j_3}^\epsilon\|_{E_\infty}, \quad (2.59)$$

where the support of the function ϕ_{j_3} is neighboring to the support of the function ϕ_{j_2}. If we continue in the same way, we obtain the inequality

$$\|\omega u \phi_{j_k}^\epsilon\|_{E_\infty} \leq M_0 \left(\|\omega Lu\|_{F_\infty} + \|\omega u\|_{L^P(\Omega_R)}\right) + \sigma \|\omega u \phi_{j_{k+1}}^\epsilon\|_{E_\infty}, \quad (2.60)$$

where the support of the function $\phi_{j_{k+1}}$ is neighboring to the support of the function ϕ_{j_k}. In order to estimate the last summand in the right-hand side of inequality (2.58) we use the inequality (2.59):

$$\|\omega u \phi_{j_1}^\epsilon\|_{E_\infty} \leq M_0(1+\sigma) \left(\|\omega Lu\|_{F_\infty} + \|\omega u\|_{L^P(\Omega_R)}\right) + \sigma^2 \|\omega u \phi_{j_3}^\epsilon\|_{E_\infty}. \quad (2.61)$$

We estimate next the last summand in the right-hand side of inequality (2.61) and so on. We obtain the estimate:

$$\|\omega u \phi_{j_1}^\epsilon\|_{E_\infty} \leq M_0(1+\sigma+\cdots+\sigma^k) \left(\|\omega Lu\|_{F_\infty} + \|\omega u\|_{L^P(\Omega_R)}\right) + \sigma^{k+1} \|\omega u \phi_{j_{k+2}}^\epsilon\|_{E_\infty}. \quad (2.62)$$

Let us specify the choice of the functions ϕ_j. Let $\phi(x) \in D$, $0 \leq \phi(x) \leq 1$, $\phi(x) = 1$ for $|x| \leq 2$, $\operatorname{supp} \phi \subset \{|x| < 3\}$. Put $\phi_j(x) = \phi(x - x_j)$. The points x_j are chosen at the nodes of some orthogonal grid. Therefore, the function $\phi_{j_2}^\epsilon$ in (2.58) is shifted with respect to $\phi_{j_1}^\epsilon$ with the value of the shift that does not exceed λ/ϵ, where the constant λ does not depend on x and j. Hence the function $\phi_{j_{k+2}}^\epsilon$ in (2.62) is shifted with respect to $\phi_{j_1}^\epsilon$ with a value of the shift that does not exceed $(k+1)\lambda/\epsilon$. Thus, $\phi_{j_{k+2}}^\epsilon(x) = \phi_{j_1}^\epsilon(x - h_k)$, where

$$|h_k| \leq \frac{(k+1)\lambda}{\epsilon}. \quad (2.63)$$

We have, further,

$$\|\omega u \phi_{j_{k+2}}^\epsilon\|_{E_\infty(\Omega)} = \sup_l \|\omega u \theta_l \phi_{j_{k+2}}^\epsilon\|_{E(\Omega)}. \quad (2.64)$$

The following estimate holds:
$$\begin{aligned}S_{k,l} :&= \|\omega u \theta_l \phi^\epsilon_{j_{k+2}}\|_{E(\Omega)} = \|\omega u \theta_l \phi^\epsilon_{j_1}(x-h_k)\|_{E(\Omega)}\\ &= \|\omega(x+h_k)u(x+h_k)\theta_l(x+h_k)\phi^\epsilon_{j_1}(x)\|_{E(\Omega_{h_k})}\\ &\leq \left\|\frac{\omega(x+h_k)}{\omega(x)}\right\|_{M(E)} \|\omega(x)u(x+h_k)\theta_l(x+h_k)\phi^\epsilon_{j_1}(x)\|_{E(\Omega_{h_k})}.\end{aligned}$$

Here Ω_{h_k} is a shifted domain, $\|\cdot\|_{M(E)}$ is the norm of multiplier in the space E. It is known that this norm can be estimated by the C-norm of the corresponding derivatives. Therefore
$$\left\|\frac{\omega(x+h_k)}{\omega(x)}\right\|_{M(E)} \leq c e^{\mu|h_k|},$$
where the constant c is independent of μ and k. Let us return to the estimate of $S_{k,l}$. Since
$$\|\theta_l(x+h_k)\|_{M(E)} \leq c_1,$$
then we have
$$S_{k,l} \leq c_2 e^{\mu|h_k|}\|\omega(x)u(x+h_k)\phi^\epsilon_{j_1}(x)\|_{E(\Omega_{h_k})}. \tag{2.65}$$
Further,
$$\operatorname{supp}\phi^\epsilon_{j_1}(x) \subset \left\{|x-\frac{x_{j_1}}{\epsilon}|<\frac{3}{\epsilon}\right\},$$
such that at the support of the function $\phi^\epsilon_{j_1}$,
$$|x| \leq \frac{\rho_{j_1}}{\epsilon} \equiv \frac{3}{\epsilon}+\frac{|x_{j_1}|}{\epsilon}.$$
Let us introduce the function
$$f_\epsilon(x) = \begin{cases} 1, & |x|<\rho_{j_1}/\epsilon,\\ 0, & |x|>1+\rho_{j_1}/\epsilon.\end{cases}$$
Then
$$\begin{aligned}\|\omega(x)u(x+h_k)\phi^\epsilon_{j_1}(x)\|_{E(\Omega_{h_k})} &= \|\omega(x)u(x+h_k)\phi^\epsilon_{j_1}(x)f_\epsilon(x)\|_{E(\Omega_{h_k})}\\ &\leq \|\omega f_\epsilon\|_{M(E)}\|\phi^\epsilon_{j_1}\|_{M(E)}\|u(x+h_k)\|_{E(\Omega_{h_k})}\\ &\leq c_3\|\omega f_\epsilon\|_{M(E)}\|u\|_{E(\Omega)} \leq c_4 e^{\mu\rho_{j_1}/\epsilon},\end{aligned}$$
where the constant c_4 does not depend on ϵ for $\epsilon<1$. From the last inequality, (2.64) and (2.65) we have
$$\|\omega u \phi^\epsilon_{j_{k+2}}\|_{E_\infty(\Omega)} \leq c_5 e^{\mu(k+1)\lambda/\epsilon} e^{\mu\rho_{j_1}/\epsilon}. \tag{2.66}$$

Consider inequality (2.62). Taking into account (2.66), we have
$$\begin{aligned}\|\omega u \phi^\epsilon_{j_1}\|_{E_\infty} &\leq M_0(1+\sigma+\cdots+\sigma^k)\left(\|\omega Lu\|_{F_\infty}+\|\omega u\|_{L^P(\Omega_R)}\right)\\ &\quad + c_5 \sigma^{k+1} e^{\mu k\lambda/\epsilon} e^{\mu(\rho_{j_1}+\lambda)/\epsilon}.\end{aligned} \tag{2.67}$$

2. Sobolev spaces

Let ϵ be chosen in such a way that $\sigma \leq \frac{1}{2}$. Put $\mu_0 < \frac{\epsilon \ln 2}{\lambda}$. Then for $0 < \mu \leq \mu_0$ from (2.67) we obtain

$$\|\omega u \phi_{j_1}^\epsilon\|_{E_\infty} \leq 2M_0 \left(\|\omega L u\|_{F_\infty} + \|\omega u\|_{L^P(\Omega_R)}\right) + \frac{1}{2} c_5 \left(\frac{1}{2} e^{\mu_0 \lambda/\epsilon}\right)^k e^{\ln 2(\rho_{j_1}+\lambda)/\lambda}.$$

Passing to the limit as $k \to \infty$, we have

$$\|\omega u \phi_{j_1}^\epsilon\|_{E_\infty} \leq 2M_0 \left(\|\omega L u\|_{F_\infty} + \|\omega u\|_{L^P(\Omega_R)}\right).$$

The theorem is proved. □

For some classes of elliptic problems satisfying the Fredholm property the exponential decay of solutions is known (see [433] and the references therein). Here it is proved that solutions of general elliptic problems behave exponentially at infinity if the corresponding operator is normally solvable with a finite-dimensional kernel. It is not assumed that it satisfies the Fredholm property.

2.4 Normal solvability

We recall that an operator L acting in Banach spaces is normally solvable if its range is closed. It is sometimes called n-normally solvable if it is normally solvable and has a finite-dimensional kernel (see, for example [277]).

Theorem 2.13. *Let Condition NS be satisfied. Then the elliptic operator*

$$L : E_\infty(\Omega) \to F_\infty(\Omega)$$

is normally solvable and has a finite-dimensional kernel.

Proof. It is known that a linear bounded operator $L : E \to F$ has a finite-dimensional kernel and a closed range if its restriction to any bounded closed set is proper. Indeed, its kernel has a finite dimension because the inverse image of the set $\{0\} \subset F$ in the unit ball $B \subset E$ is compact, that is the unit ball in the kernel is compact. Let us check that the image of the operator is closed. Consider a sequence $f_n \in F$ and suppose that $f_n \to f_0$ as $n \to \infty$. Then there exists a sequence $u_n \in E$ such that $L u_n = f_n$. If this sequence is bounded in E, then by virtue of the properness of the operator, it has a convergent subsequence, $u_{n_k} \to u_0$. Therefore $L u_0 = f_0$, and f_0 belongs to the image of the operator. If the sequence u_n is not bounded, we represent it in the form $u_n = v_n + w_n$, where v_n belongs to the kernel of the operator and w_n to its supplement. We put $\tilde{w}_n = w_n / \|w_n\|_E$, $\tilde{f}_n = f_n / \|w_n\|_E$. Then $L \tilde{w}_n = \tilde{f}_n$ and $\tilde{f}_n \to 0$. From the properness of the operator it follows that the sequence $\tilde{w}_n$ is compact and has a convergent subsequence. Denote by w_0 its limit. Then $L w_0 = 0$. Hence w_0 belongs to the kernel of the operator while $\tilde{w}_n$ to its supplement. This contradiction proves that the image of the operator is closed. Hence it is normally solvable.

Thus, we should prove the properness of the operator L. Let $Lu_n = f_n$, $u_n \in E_\infty(\Omega)$, $f_n \in F_\infty(\Omega)$. Suppose that $\|u_n\|_{E_\infty} \leq M$ and f_n is convergent. It is sufficient to prove that the sequence u_n is compact. This follows from Theorem 2.10. The theorem is proved. □

In the next theorem we prove that Condition NS is necessary for the operator L to be normally solvable with a finite-dimensional kernel. To simplify the construction we impose a stronger regularity condition on the boundary of the domain, $\partial\Omega \in C^{r+1+\theta}$. We will use the following lemma.

Lemma 2.14. *Let Ω_k and Ω_* be a shifted and a limiting domain, respectively. Then for any N there exists k_0 such that for $k > k_0$ there exists a diffeomorphism*

$$h_k(x) : \bar{\Omega}_k \bigcap B_N \to \bar{\Omega}_* \bigcap B_N$$

satisfying the condition

$$\|h_k(x) - x\|_{C^{r+1+\theta_0}(\bar{\Omega}_k \cap B_N)} \to 0$$

as $k \to \infty$. Here $0 < \theta_0 < \theta$.

Proof. Consider a domain G such that $\bar{G} \subset \Omega_k \cap \Omega_*$ for all m sufficiently large. Let $x_0 \in \partial\Omega_*$. Denote by $n(x_0)$ the normal to $\partial\Omega_*$ at $x = x_0$. If m is sufficiently large, then in a neighborhood of x_0, $n(x_0)$ intersects $\partial\Omega_k$ only at one point. The domain G can be chosen such that it satisfies the same property. We put $h_k(x) = 1$ for $x \in G$. We define then $h_k(x)$ along each normal $n(x_0)$ by mapping the interval, which belongs to Ω_k, on the interval in Ω_*. It can be done in such a way that we have the required regularity. The lemma is proved. □

Theorem 2.15. *Suppose that a limiting problem for the operator L has a nonzero solution. Then the operator L is not n-normally solvable.*

Explanation. In order to prove the theorem we construct a sequence u_n such that it is not compact in $E_\infty(\Omega)$ but Lu_n converges to zero in $F_\infty(\Omega)$. The idea of the construction is rather simple but its technical realization is rather long. This is why we preface the proof by a short description of the construction.

Let us consider a ball $B_R(x_k)$ of a fixed radius R with its center at x_k. From the definition of limiting problems it follows that we can choose the sequence x_k in such a way that inside $B_R(x_k)$ the domain Ω is close to the limiting domain, and the coefficients of the operator are close to the coefficients of the limiting operator. Moreover, the domain and the coefficients converge to their limits as $k \to \infty$. Thus we move the ball $B_R(x_k)$ to infinity and superpose it on the domain Ω in the places where the operator and the domain are close to their limits and converge to them.

If u_0 is a nonzero solution of the limiting problem, then we shift it to the ball $B_R(x_k)$. Denote the shifted function by u_k. Then inside $B_R(x_k)$, Lu_k tends to zero as $k \to \infty$. The sequence u_k is not compact.

2. Sobolev spaces

If u_0 had a bounded support, the construction would be finished. Since it is not necessarily the case, we multiply u_0 by an infinitely differentiable function ϕ with a bounded support. Of course, this product is not an exact solution of the limiting problem any more. However, all terms of the difference $\hat{L}(\phi u_0) - \phi \hat{L} u_0$ contain derivatives of ϕ. If the support of ϕ is sufficiently large, then the derivatives of ϕ can be done sufficiently small. Hence when we move the ball $B_R(x_k)$ to infinity, we should also increase its radius and also increase supports of the corresponding functions ϕ_k.

Proof. Suppose that there exists a limiting operator $\hat{L}$ such that the corresponding limiting problem has a nonzero solution:

$$\hat{L} u_0 = 0, \quad u_0 \in E_\infty(\Omega_*), \quad u_0 \neq 0.$$

Consider an infinitely differentiable function $\varphi(x)$, $x \in \mathbb{R}^n$ such that $0 \leq \varphi(x) \leq 1$, $\varphi(x) = 1$ for $|x| < 1$, $\varphi(x) = 0$ for $|x| > 2$. If $\{x_k\}$ is the sequence for which the limiting operator $\hat{L}$ is defined, write $\varphi_k(x) = \varphi(x/r_k)$, where $r_k \to \infty$ and $r_k \leq |x_k|/3$. Some other conditions on the sequence r_k will be formulated below.

Let $V_j = \{y : y \in \mathbb{R}^n, |y| < j\}$, $j = 1, 2, \ldots$. Denote by n_j a number such that for $k \geq n_j$ the diffeomorphism h_k defined in Lemma 2.14 can be constructed in $\Omega_k \cap V_{j+1}$ and

$$\|h_k(y) - y\|_{C^{r+1+\theta_0}(\Omega_k \cap V_{j+1})} < \delta, \tag{2.68}$$

where $\delta > 0$ is taken so small that $|h'_k - I| < 1/2$, h'_k is the Jacobian matrix and I is the identity matrix. For arbitrary $k_j \geq n_j$ we take $r_{k_j} = \min(j/2, |x_{k_j}|/3)$. Let

$$v_{k_j}(y) = \varphi_{k_j}(y) \, u_0(h_{k_j}(y)) \quad \text{for} \quad y \in \Omega_{k_j} \cap V_{j+1},$$
$$v_{k_j}(y) = 0 \quad for \quad y \in \Omega_{k_j}, \ |y| \geq j+1.$$

Write

$$u_{k_j}(x) = v_{k_j}(x - x_{k_j}), \quad x \in \Omega. \tag{2.69}$$

It is easy to see that $u_{k_j} \in E_\infty(\Omega)$ and

$$\|u_{k_j}\|_{E_\infty(\Omega)} \leq M, \tag{2.70}$$

where M does not depend on k_j. Indeed, obviously

$$\varphi_{k_j}(y) = 0 \tag{2.71}$$

for y outside V_j. Therefore to prove (2.70) it is sufficient to show that

$$\|v_{k_j}\|_{E_\infty(\Omega_{k_j} \cap V_{j+1})} \leq M_1,$$

or

$$\|u_0(h_{k_j}(y))\|_{E_\infty(\Omega_{k_j} \cap V_{j+1})} \leq M_2,$$

where M_1 and M_2 do not depend on k_j. This follows from (2.68) and the fact that $u_0 \in E_\infty(\Omega_*)$.

We now prove that the choice of k_j in (2.69) can be specified in such a way that

(i) $Lu_{k_j} \to 0$ in $F(\Omega)$ as $k_j \to \infty$,

(ii) the sequence $\{u_{k_j}\}$ is not compact in $E_\infty(\Omega)$.

The assertion of the theorem will follow from this.

(i) We consider first the operators A_i, $i = 1, \ldots, N$, and then the operator B_j, $j = 1, \ldots, m$. For any $k = k_j \geq n_j$ we have

$$A_i u_k = A_i^1 u_k + A_i^2 u_k,$$

where

$$A_i^1 u_k(x) = \varphi_k(x - x_k) \sum_{r=1}^{N} \sum_{|\alpha| \leq \alpha_{ir}} a_{ir}^\alpha(x) D^\alpha u_{0r}(h_k(x - x_k)), \quad x \in \Omega, \quad (2.72)$$

and A_i^2 contains derivatives of φ_k. Obviously

$$\|A_i^2 u_k\|_{W_\infty^{l-s_i,p}(\Omega)} \to 0$$

as $k \to \infty$. Let $y = x - x_k$. From (2.72) we obtain

$$A_i^1 u_k(y + x_k) = \varphi_k(y) T_{ik}(y), \quad y \in \Omega_k, \quad (2.73)$$

where

$$T_{ik}(y) = \sum_{r=1}^{N} \sum_{|\alpha| \leq \alpha_{ir}} a_{ir\,k}^\alpha(y) D^\alpha u_{0r}(h_k(y)), \quad y \in \Omega_k,$$

$a_{ir\,k}^\alpha(y) = a_{ir}^\alpha(y + x_k)$. We prove that for any fixed j,

$$\|T_{ik}\|_{W_\infty^{l-s_i,p}(\Omega_k \cap V_{j+1})} \to 0 \quad (2.74)$$

as $k \to \infty$. Indeed, by the definition of u_0 the following equality holds:

$$\sum_{r=1}^{N} \sum_{|\alpha| \leq \alpha_{ir}} \hat{a}_{ir}^\alpha(x) D^\alpha u_{0r}(x) = 0, \quad x \in \Omega_*.$$

Here $\hat{a}_{ir}^\alpha(x)$ are the limiting coefficients. Hence

$$T_{ik}(y) = \sum_{r=1}^{N} \sum_{|\alpha| \leq \alpha_{ir}} [S_{ir\,k}^\alpha(y) + P_{ir\,k}^\alpha(y)],$$

2. Sobolev spaces

where
$$S_{irk}^{\alpha}(y) = a_{irk}^{\alpha}(y)[D_y^{\alpha}u_{0r}(h_k(y)) - D_x^{\alpha}u_{0r}(h_k(y))], \quad (2.75)$$
$$P_{irk}^{\alpha}(y) = [a_{irk}^{\alpha}(y) - \hat{a}_{ir}^{\alpha}(h_k(y))]D_x^{\alpha}u_{0r}(h_k(y)). \quad (2.76)$$

The first factor in the right-hand side of (2.75) is bounded in the norm $C^{l-s_i}(\Omega_k)$ since
$$\|a_{irk}^{\alpha}\|_{C^{l-s_i}(\Omega_k)} = \|a_{ir}^{\alpha}\|_{C^{l-s_i}(\Omega)}.$$
From Lemma 2.14 it follows that the second factor tends to 0 in the norm
$$W_\infty^{l-s_i,p}\left(\Omega_k \cap V_{j+1}\right)$$
as $k \to \infty$. Consequently,
$$\|S_{irk}^{\alpha}\|_{W_\infty^{l-s_i,p}(\Omega_k \cap V_{j+1})} \to 0 \text{ as } k \to \infty.$$

Consider (2.76). Using (2.68) we easily prove that
$$\|D_x^{\alpha}u_0(h_k(y))\|_{W_\infty^{l-s_i,p}(\Omega_k \cap V_{j+1})} \leq M_3$$
with M_3 independent of k.

To prove (2.74) it remains to show that
$$\|a_{irk}^{\alpha}(\cdot) - \hat{a}_{ir}^{\alpha}(h_k(\cdot))\|_{C^{l-s_i}(\Omega_k \cap V_{j+1})} \to 0 \text{ as } k \to \infty.$$

We recall that it is supposed that $a_{irk}^{\alpha}(y)$ and $\hat{a}_{ir}^{\alpha}(y)$ are defined for $y \in \mathbb{R}^n$, and
$$\|a_{irk}^{\alpha}\|_{C^{l-s_i+\theta}(R^n)} \leq M$$
with M independent of k, $\hat{a}_{ir}^{\alpha}(y) \in C^{l-s_i+\theta}(R^n)$ and
$$a_{irk}^{\alpha}(y) \to \hat{a}_{ir}^{\alpha}(y) \quad (2.77)$$
in $C_{\text{loc}}^{l-s_i}(R^n)$ as $k \to \infty$. We have
$$\|a_{irk}^{\alpha}(y) - \hat{a}_{ir}^{\alpha}(h_k(y))\|_{C^{l-s_i}(\Omega_k \cap V_{j+1})}$$
$$\leq \|a_{irk}^{\alpha}(y) - \hat{a}_{ir}^{\alpha}(y)\|_{C^{l-s_i}(\Omega_k \cap V_{j+1})} + \|\hat{a}_{ir}^{\alpha}(y) - \hat{a}_{ir}^{\alpha}(h_k(y))\|_{C^{l-s_i}(\Omega_k \cap V_{j+1})}.$$

The first term on the right tends to zero as $k \to \infty$ according to (2.77). The second term tends to zero by the properties of the function $\hat{a}_{ir}^{\alpha}$ mentioned above, by Lemma 2.14 and by inequality (2.68). Thus (2.74) is proved.

Now, we specify the choice of k_j in (2.69). According to (2.74) for any j we can take p_j in such a way that
$$\|T_{ik}\|_{W_\infty^{l-s_i,p}(\Omega_k \cap V_{j+1})} < 1/j$$
for $k \geq p_j$.

We put $k_j = \max(n_j, p_j)$. Then obviously

$$\|\varphi_{k_j} T_{ik_j}\|_{W_\infty^{l-s_i,p}(\Omega_{k_j})} \to 0 \text{ as } k \to \infty. \tag{2.78}$$

Consider now the boundary operators B_i. According to our assumptions, the coefficients $b_{ih}^\beta(x)$ of the operators B_i ($i = 1, \ldots, m$) are defined in the domain $\overline{\Omega}$ and belong to the space $C^{l-\sigma_i+\theta}(\overline{\Omega})$. By the same arguments, which we used for the operator A_i, we prove that

$$\|B_i u_{k_j}\|_{W_\infty^{l-\sigma_i-1/p,p}(\partial\Omega)} \to 0 \text{ as } k_j \to \infty.$$

We repeat the same construction as above and obtain the following operator:

$$T_{ik}(y) = \sum_{h=1}^N \sum_{|\beta| \leq \beta_{ih}} b_{ih\,k}^\beta(y) D^\beta u_{0h}(h_k(y)), \quad y \in \Omega_k,$$

where $b_{ih\,k}^\beta(y) = b_{ih}^\beta(y + x_k)$. We prove that

$$\|T_{ik}\|_{W_\infty^{l-\sigma_i-1/p,p}(\partial\Omega_k \cap V_{j+1})} \to 0. \tag{2.79}$$

Indeed, let

$$g_i(x) = \sum_{h=1}^N \sum_{|\beta| \leq \beta_{ih}} \hat{b}_{ih\beta}(y) D^\beta u_{0h}(x), \quad x \in \Omega_*.$$

This expression equals 0 only at the boundary $\partial\Omega_*$. Therefore instead of what is written above for the operator A_i, we have now

$$T_{ik}(y) = Q_{ik}(y) + g_i(h_k(y)), \tag{2.80}$$

where

$$Q_{ik}(y) = \sum_{h=1}^N \sum_{|\beta| \leq \beta_{ih}} [S_{ih\,k}^\beta(y) + P_{ih\,k}^\beta(y)].$$

Here S and P are the same as for the operator A but the coefficients a are replaced by b. Exactly as we have done it for the operator A, we prove that

$$\|Q_k\|_{W_\infty^{l-\sigma_i,p}(\Omega_k \cap V_{j+1})} \to 0 \text{ as } k \to \infty.$$

It follows that

$$\|Q_k\|_{W_\infty^{l-\sigma_i-1/p,p}(\partial\Omega_k \cap V_{j+1})} \to 0 \text{ as } k \to \infty. \tag{2.81}$$

Since for $y \in \partial\Omega_k$ we have $h_k(y) \in \partial\Omega_*$, we have $g_i(h_k(y)) = 0$ for $y \in \partial\Omega_k$. From this, (2.80) and (2.81) we get (2.79). Thus the assertion (i) is proved.

2. Sobolev spaces

(ii) We now prove that sequence (2.69) does not have a convergent subsequence. Obviously $u_{k_j}(x) = 0$ for $|x| < r_{k_j}$ and, consequently,

$$\int_\Omega u_{k_j}(x)\omega(x)dx \to 0 \tag{2.82}$$

as $k_j \to \infty$ for any continuous $\omega(x)$ with a compact support. For any subsequence s_i of k_j there exists N such that

$$\int_\Omega |u_{s_i}(x)|^p dx \geq \rho \tag{2.83}$$

for $s_i > N$ and some $\rho > 0$. Indeed, let $y = x - x_{s_i}$. Then

$$T_i \equiv \int_\Omega |u_{s_i}(x)|^p dx = \int_{\Omega_{s_i}} |v_{s_i}(y)|^p dy = \int_{\Omega_{s_i} \cap V_{j+1}} |\varphi_{s_i}(y) u_0(h_{s_i}(y))|^p dy$$

$$\geq \int_{\Omega_{s_i} \cap V_{r_{s_i}}} |u_0(h_{s_i}(y))|^p dy.$$

We do the change of variables $y = h_{s_i}^{-1}(x)$ in the last integral. Then

$$T_i \geq \int_{\Omega_* \cap W_{s_i}} |u_0(x)|^p \left|\frac{dh_{s_i}^{-1}(x)}{dx}\right| dx,$$

where $W_{s_i} = h_{s_i}(V_{r_{s_i}})$. Since $\|u_0\|_{L^p(\Omega_*)} \neq 0$, there exists a ball $B_l = \{x : |x| < l\}$ and a number $\rho_0 > 0$ such that

$$\int_{\Omega_* \cap B_l} |u_0(x)|^p dx \geq \rho_0. \tag{2.84}$$

Increasing N, if necessary, we can suppose that $B_l \subset W_{s_i}$ and $\left|\frac{dh_{s_i}^{-1}(x)}{dx}\right| \geq \varepsilon$ for $x \in B_l$ and some $\varepsilon > 0$. The last inequality follows from the fact that according to (2.68) the derivatives of $h_{s_i}(y)$ are uniformly bounded. By (2.84) we get $T_i \geq \varepsilon \rho_0$ and (2.83) is proved.

If (2.69) has a convergent subsequence: $u_{s_i} \to u_*$ in $E(\Omega)$, then this convergence holds also in $L^p(\Omega)$. From (2.82) it follows that $u_* = 0$ which contradicts (2.83). Thus the sequence (2.69) is not compact in $E(\Omega)$. The theorem is proved. □

3 Hölder spaces

3.1 Operators and spaces

Let $\beta = (\beta_1, \ldots, \beta_n)$ be a multi-index, β_i nonnegative integers, $|\beta| = \beta_1 + \cdots + \beta_n$, $D^\beta = D_1^{\beta_1} \ldots D_n^{\beta_n}$, $D_i = \partial/\partial x_i$. We consider the operators

$$A_i u = \sum_{k=1}^{p} \sum_{|\beta| \leq \beta_{ik}} a_{ik}^\beta(x) D^\beta u_k \quad (i = 1, \ldots, p), \quad x \in \Omega, \tag{3.1}$$

$$B_i u = \sum_{k=1}^{p} \sum_{|\beta| \leq \gamma_i k} b_{ik}^\beta(x) D^\beta u_k \quad (i = 1, \ldots, r), \quad x \in \partial\Omega. \tag{3.2}$$

According to the definition of elliptic operators in the Douglis-Nirenberg sense we consider integers $s_1, \ldots, s_p; t_1, \ldots, t_p; \sigma_1, \ldots, \sigma_r$ such that

$$\beta_{ij} \leq s_i + t_j, \ i, j = i, \ldots, p;$$
$$\gamma_{ij} \leq \sigma_i + t_j, \ i = 1, \ldots, r, j = 1, \ldots, p, \ s_i \leq 0.$$

We suppose that the number $m = \sum_{i=1}^{p}(s_i + t_i)$ is even and put $r = m/2$. We assume that the problem is elliptic, that is the ellipticity condition

$$\det(\sum_{|\beta|=\beta_{ik}} a_{ik}^\beta(x)\xi^\beta)_{ik=1}^p \neq 0, \quad \beta_{ik} = s_i + t_k$$

is satisfied for any $\xi \in R^n$, $\xi \neq 0$, $x \in \bar\Omega$, as well as the condition of proper ellipticity and the Lopatinskii conditions. Here $\xi = (\xi, \ldots, \xi_n)$, $\xi^\beta = \xi_1^\beta \ldots \xi_n^\beta$. The condition of uniform ellipticity implies that the last determinant is bounded from below by a positive constant for all $|\xi| = 1$ and $x \in \bar\Omega$ (see Chapter 1 for more details). We recall that $C^{k+\alpha}(\Omega)$ denotes the Hölder space of functions bounded in Ω together with their derivatives up to order k, and the latter satisfies the Hölder condition uniformly in x.

Denote by E_0 a space of vector-valued functions $u(x) = (u_1(x), \ldots, u_p(x))$, $u_j \in C^{l+t_j+\alpha}(\bar\Omega)$, $j = 1, \ldots, p$, where l and α are given numbers, $l \geq \max(0, \sigma_i)$, $0 < \alpha < 1$. Therefore

$$E_0 = C^{l+t_1+\alpha}(\bar\Omega) \times \cdots \times C^{l+t_p+\alpha}(\bar\Omega).$$

The domain Ω is supposed to be of the class $C^{l+\lambda+\alpha}$, where

$$\lambda = \max(-s_i, -\sigma_i, t_j),$$

and the coefficients of the operator satisfy the following regularity conditions:

$$a_{ij}^\beta \in C^{l-s_i+\alpha}(\bar\Omega), \quad b_{ij}^\beta \in C^{l-\sigma_i+\alpha}(\partial\Omega).$$

3. Hölder spaces

The operator A_i acts from E_0 into $C^{l-s_i+\alpha}(\Omega)$, and B_i from E_0 into $C^{l-\sigma_i+\alpha}(\partial\Omega)$. Let $A = (A_1, \ldots, A_p)$, $B = (B_1, \ldots, B_r)$. Then

$$A : E_0 \to E_1, \quad B : E_0 \to E_2, \quad L = (A, B) : E_0 \to E,$$

where $E = E_1 \times E_2$,

$$E_1 = C^{l-s_1+\alpha}(\bar{\Omega}) \times \cdots \times C^{l-s_p+\alpha}(\bar{\Omega}),$$
$$E_2 = C^{l-\sigma_1+\alpha}(\partial\Omega) \times \cdots \times C^{l-\sigma_r+\alpha}(\partial\Omega).$$

In this section we will use a priori estimates for elliptic operators in Hölder spaces [7], [8]:

$$\|u\|_{E_0} \leq K(\|Lu\|_E + \|u\|_C). \tag{3.3}$$

Here the constant K is independent of the function $u \in E_0(\Omega)$ and $\|\ \|_C$ is the norm in $C(\bar{\Omega})$.

We will also consider weighted Hölder spaces $E_{0,\mu}$ and E_μ with the norms

$$\|u\|_{E_{0,\mu}} = \|u\mu\|_{E_0}, \quad \|u\|_{E_\mu} = \|u\mu\|_E.$$

The weighted Hölder space with the norm $\|u\|_{C_\mu^{k+\alpha}} = \|u\mu\|_{C^{k+\alpha}}$ will be denoted by $C_\mu^{k+\alpha}$. We suppose that the weight function μ is a positive infinitely differentiable function defined for all $x \in \mathbb{R}^n$, $\mu(x) \to \infty$ as $|x| \to \infty$, $x \in \Omega$, and

$$\left|\frac{1}{\mu(x)} D^\beta \mu(x)\right| \to 0, \quad |x| \to \infty, \quad x \in \Omega \tag{3.4}$$

for any multi-index β, $|\beta| > 0$. In fact, we will use its derivative only up to a certain order. The operator $L = (A, B)$ considered in weighted Hölder spaces acts from $E_{0,\mu}$ into E_μ.

3.2 Normal solvability

We consider the operator $L : E_0(\Omega) \to E(\Omega)$ and introduce limiting domains and limiting operators defined above. In what follows we will use also the spaces E_0' and E', which are obtained from E_0 and E, respectively, if we put $\alpha = 0$.

From Theorem 2.9 it follows that, for any sequences $u_m \in E_0(\Omega_m)$, $f_m \in E(\Omega_m)$ with uniformly bounded norms, there exist subsequences u_{m_k} and f_{m_k} converging to some limiting functions $u_* \in E_0(\Omega_*)$ and $f_* \in E(\Omega_*)$ in $E_{0,\mathrm{loc}}'(\Omega_{m_k} \to \Omega_*)$ and $E_{\mathrm{loc}}'(\Omega_{m_k} \to \Omega_*)$, respectively. If L_m is a sequence of operators with shifted coefficients and $L_m u_m = f_m$, then there exists a limiting operator $\hat{L}$ such that $\hat{L} u_* = f_*$.

We introduce a condition similar to that in Section 2.3 but in the spaces considered here.

Condition NS. For any limiting domain Ω_* and any limiting operator $\hat{L}$ the problem
$$\hat{L}u = 0, \quad u \in E_0(\Omega_*) \tag{3.5}$$
has only the zero solution.

Theorem 3.1. *Let Condition NS be satisfied. Then the operator L is normally solvable and its kernel is finite dimensional.*

Proof. Let the limiting problems have only the zero solution. It is sufficient to prove that the operator L is proper (cf. Theorem 2.13). Consider the equation $Lu_n = f_n$, where $f_n \in E(\Omega)$ and $f_n \to f_0$. Suppose that $\|u_n\|_{E_0(\Omega)} \leq M$. We will prove that there exists a function $u_0 \in E_0(\Omega)$ and a subsequence u_{n_k} such that
$$\|u_{n_k} - u_0\|_{E_0(\Omega)} \to 0. \tag{3.6}$$
There exists a function $u_0 \in E_0(\Omega)$ such that $u_{n_k} \to u_0$ in $E'_{0,\text{loc}}(\Omega)$ and $Lu_0 = f_0$. Without loss of generality we can assume, here as well as below, that it is the same sequence. We prove first that
$$\|u_n - u_0\|_{C(\bar{\Omega})} \to 0. \tag{3.7}$$
Suppose that this convergence does not take place. Since $u_n \to u_0$ in $C_{\text{loc}}(\Omega)$, we conclude that there exists a sequence x_m, $|x_m| \to \infty$ and a subsequence u_{nm} of u_n such that
$$\|u_{nm}(x_m) - u_0(x_m)\| \geq \epsilon > 0.$$
Consider the shifted domains Ω_n with the characteristic functions $\chi(x + x_m)$, the operators with shifted coefficients and the functions $v_{nm}(x) = u_{nm}(x + x_m) - u_0(x + x_m)$. Passing to a subsequence we conclude that there exists a limiting domain Ω_*, a limiting operator $\hat{L}$, and a nonzero limiting function $v_0 \in E_0(\Omega_*)$ such that
$$\hat{L}v_0 = 0.$$
This contradiction proves (3.7). From this convergence, from the convergence $f_n \to f_0$ in $E(\Omega)$, and estimate (3.3) it follows that $u_n \to u_0$ in $E_0(\Omega)$. The theorem is proved. $\square$

The next theorem will provide a necessary condition of normal solvability. In fact, it is the same Condition NS. However we need now more restrictive conditions on the coefficients of the operator and on the domain Ω. We suppose here that
$$a_{ik}^\beta \in C^{l-s_i+\delta}(\bar{\Omega}), \quad b_{ik}^\beta \in C^{l-\sigma_i+\delta}(\partial\Omega) \text{ and the domain } \Omega \text{ is of the class } C^{r+1+\delta} \tag{3.8}$$
with $\alpha < \delta < 1$. Similar to Lemma 2.14 we suppose that Ω_m and Ω_* are shifted and limiting domains respectively. Then for any N there exists m_0 such that for $m > m_0$ there exists a diffeomorphism
$$h_m(x) : \bar{\Omega}_m \bigcap B_N \to \bar{\Omega}_* \bigcap B_N$$

3. Hölder spaces

satisfying the condition

$$\|h_m(x) - x\|_{C^{r+1+\alpha}(\bar{\Omega}_m \cap B_N)} \to 0$$

as $m \to \infty$.

Theorem 3.2. *Suppose that the problem* (3.5) *has a nonzero solution u_0 for some limiting operator $\hat{L}$ and limiting domain Ω_*. Then the operator L is not proper.*

Proof. Let $\varphi(x)$ be an infinitely differentiable function defined in $\mathbb{R}^n$ such that $0 \leq \varphi(x) \leq 1$, $\varphi(x) = 1$ for $|x| < 1$, $\varphi(x) = 0$ for $|x| > 2$. If $\{x_m\}$ is the sequence for which the limiting operator $\hat{L}$ is defined, let $\varphi_m(x) = \varphi(x/r_m)$, where $r_m \to \infty$ and $r_m \leq |x_m|/3$. Some other conditions on the sequence r_m will be formulated below.

Let $V_j = \{y : y \in \mathbb{R}^n, |y| < j\}$, $j = 1, 2, \ldots$. Denote by n_j a number such that for $m \geq n_j$ the diffeomorphism h_m defined in Lemma 2.14 can be constructed in $\Omega_m \cap V_{j+1}$ and

$$\|h_m(y) - y\|_{C^{r+1+\alpha}(\bar{\Omega}_m \cap V_{j+1})} < 1. \tag{3.9}$$

For arbitrary $m_j \geq n_j$ we take $r_{m_j} = \min(j/2, |x_{m_j}|/3)$. Let, further,

$$v_{m_j}(y) = \varphi_{m_j}(y)\, u_0(h_{m_j}(y)) \quad \text{for } y \in \Omega_{m_j} \cap V_{j+1},$$

$$v_{m_j}(y) = 0 \quad \text{for } y \in \Omega_{m_j},\ |y| \geq j+1.$$

Let

$$u_{m_j}(x) = v_{m_j}(x - x_{m_j}), \quad x \in \Omega. \tag{3.10}$$

It is easy to see that $u_{m_j} \in E_0(\Omega)$ and

$$\|u_{m_j}\|_{E_0(\Omega)} \leq M, \tag{3.11}$$

where M does not depend on m_j. Indeed, obviously

$$\varphi_{m_j}(y) = 0 \tag{3.12}$$

for y outside V_j. So to prove (3.11) it is sufficient to show that

$$\|v_{m_j}\|_{E_0(\Omega_{m_j} \cap V_{j+1})} \leq M_1,$$

or

$$\|u_0(h_{m_j}(y))\|_{E_0(\Omega_{m_j} \cap V_{j+1})} \leq M_2,$$

where M_1 and M_2 do not depend on m_j. This follows from (3.9) and the fact that $u_0 \in E_0(\Omega_*)$.

We will prove that choice of m_j in (3.10) can be specified so that

(i) $Lu_{m_j} \to 0$ in $E(\Omega)$ as $m_j \to \infty$,
(ii) the sequence $\{u_{m_j}\}$ is not compact in $E_0(\Omega)$.

The assertion of the theorem will follow from this.

(i) We consider operator A_i. The operator B_i is treated similarly. For any j and $m \geq n_j$ we have
$$A_i u_m = A_i^1 u_m + A_i^2 u_m,$$
where
$$A_i^1 u_m(x) = \varphi_m(x - x_m) \sum_{k=1}^{p} \sum_{|\beta| \leq \beta_{ik}} a_{ik}^{\beta}(x) D^{\beta} u_{0k}(h_m(x - x_m)), \quad x \in \Omega, \quad (3.13)$$

$u_0 = (u_{01}, \ldots, u_{0p})$ and A_i^2 contains derivatives of φ_m. Obviously
$$\|A_i^2 u_m\|_{C^{l-s_i+\alpha}(\bar{\Omega})} \to 0$$
as $m \to \infty$. Set $y = x - x_m$. From (3.13) we obtain
$$A_i^1 u_m(y + x_m) = \varphi_m(y) T_{im}(y), \quad y \in \Omega_m, \quad (3.14)$$
where
$$T_{im}(y) = \sum_{k=1}^{p} \sum_{|\beta| \leq \beta_{ik}} a_{ik,m}^{\beta}(y) D^{\beta} u_{0k}(h_m(y)), \quad y \in \Omega_m,$$
$a_{ik,m}^{\beta}(y) = a_{ik}^{\beta}(y + x_m)$. We will prove that for any j fixed,
$$\|T_{im}\|_{C^{l-s_i+\alpha}(\bar{\Omega}_m \cap V_{j+1})} \to 0 \quad (3.15)$$
as $m \to \infty$. By definition of u_0, the following equality holds:
$$\sum_{k=1}^{p} \sum_{|\beta| \leq \beta_{ik}} \hat{a}_{ik}^{\beta}(x) D_x^{\beta} u_{0k}(x) = 0, \quad x \in \Omega_*.$$
Here $\hat{a}_{ik}^{\beta}(x)$ are the limiting coefficients. So
$$T_{im}(y) = \sum_{k=1}^{p} \sum_{|\beta| \leq \beta_{ik}} [S_{ik,m}^{\beta}(y) + P_{ik,m}^{\beta}(y)],$$
where
$$S_{ik,m}^{\beta}(y) = a_{ik,m}^{\beta}(y)[D_y^{\beta} u_{0k}(h_m(y)) - D_x^{\beta} u_{0k}(h_m(y))], \quad (3.16)$$
$$P_{ik,m}^{\beta}(y) = [a_{ik,m}^{\beta}(y) - \hat{a}_{ik}^{\beta}(h_m(y))] D_x^{\beta} u_{0k}(h_m(y)). \quad (3.17)$$
The first factor on the right in (3.16) is bounded since
$$\|a_{ik,m}^{\beta}\|_{C^{l-s_i+\alpha}(\bar{\Omega}_m)} = \|a_{ik}^{\beta}\|_{C^{l-s_i+\alpha}(\bar{\Omega})}.$$

3. Hölder spaces

From Lemma 2.14 it follows that the second factor tends to 0 in the norm

$$C^{l-s_i+\alpha}\left(\bar{\Omega}_m \cap V_{j+1}\right)$$

as $m \to \infty$. So

$$\|S^{\beta}_{ik,m}\|_{C^{l-s_i+\alpha}(\bar{\Omega}_m \cap V_{j+1})} \to 0$$

as $m \to \infty$. Consider (3.17). Using Lemma 2.14 we easily prove that

$$\|D^{\beta}_x u_{0k}(h_m(y))\|_{C^{l-s_i+\alpha}(\bar{\Omega}_m \cap V_{j+1})} \leq M_3$$

with M_3 independent of m. To prove (3.15) it remains to show that, for any subsequence of m, T_{im} has a convergent to zero subsequence. If m_ν is a subsequence of m, then assumption (3.8) and Lemma 2.14 imply that

$$\|a^{\beta}_{ik,m}(.) - \widehat{a}^{\beta}_{ik}(h_m(.))\|_{C^{l-s_i+\alpha}(\bar{\Omega}_m \cap V_{j+1})} \to 0$$

as $m \to \infty$ by some subsequence of m_ν. So (3.15) is proved.

Now, we specify the choice of m_j in (3.10). According to (3.15), for any j we can take p_j such that

$$\|T_{im}\|_{C^{l-s_i+\alpha}(\bar{\Omega}_m \cap V_{j+1})} < 1/j$$

for $m \geq p_j$. We take $m_j = \max(n_j, p_j)$. Then obviously

$$\|\varphi_{m_j} T_{im_j}\|_{C^{l-s_i+\alpha}(\bar{\Omega}_{m_j})} \to 0 \tag{3.18}$$

as $m_j \to \infty$. It is easy to see that m_j can be chosen in the same manner in such a way that (3.18) is true for all $i = 1, \ldots, p$, and also for operators B_i. Thus the assertion (i) is proved.

(ii) We will prove that sequence (3.10) does not have a convergent subsequence. Obviously $u_{m_j}(x) = 0$ for $|x| < r_{m_j}$ and so

$$u_{m_j}(x) \to 0 \tag{3.19}$$

as $m_j \to \infty$ for any $x \in \Omega$ fixed. For any subsequence s_i of m_j there exists N such that

$$sup_{x \in \Omega}|u_{s_i}(x)| > 0 \tag{3.20}$$

for $s_i > N$. Indeed, let $y = x - x_{s_i}$. Then

$$sup_{x \in \Omega}|u_{s_i}(x)| \geq sup_{y \in \Omega_{s_i} \cap V_{j+1}}|\varphi_{s_i}(y) u_0(h_{s_i}(y))|. \tag{3.21}$$

Let $x_0 \in \Omega_*$ be a point such that $|u_0(x_0)| > 0$. Set $y_{s_i} = h^{-1}_{s_i}(x_0)$, $y_{s_i} \in \Omega_{s_i}$. From Lemma 2.14 it follows that $|y_{s_i}|$ is bounded. So there exists N such that $|y_{s_i}| < r_{s_i}$, $|y_{s_i}| < j+1$ for $s_i > N$. From (3.21) it follows that $sup_{x \in \Omega}|u_{s_i}(x)| \geq |u_0(x_0)|$, and (3.20) holds. Thus, $\|u_{s_i}\|_{C(\Omega)} > 0$. This and (3.19) imply that u_{m_j} is not compact in $E_0(\Omega)$. The theorem is proved. $\square$

3.3 Dual spaces. Invertibility of limiting operators

We now consider the space $E = E(\Omega)$ defined in Section 3.1 and the space E^0, which consists of functions $u \in E$ converging to 0 at infinity in the norm E, i.e.,

$$\|u\|_{E(\Omega \cap \{|x| \geq N\})} \to 0$$

as $N \to \infty$. We say that $u_n \to u_0$ in $E_{\text{loc}}(\Omega)$ if this convergence holds in $\Omega \cap \{|x| \leq N\}$ for any N.

Lemma 3.3. *Let ϕ be a functional in the dual space $(E^0)^*$, $u \in E$ and $u \notin E^0$, $u_n \in E^0$, $\|u_n\|_E \leq 1$, and $u_n \to u$ in E_{loc}. Then there exists a limit*

$$\hat{\phi} = \lim_{n \to \infty} \phi(u_n). \tag{3.22}$$

Proof. Since ϕ is a bounded functional, then $|\phi(u_n)| \leq K\|u_n\|_E \leq K$ with some positive constant K. Suppose that the limit (3.22) does not exist. We will construct a sequence $z_n \in E^0$, uniformly bounded in the norm E such that $\phi(z_n) \to \infty$. We can choose two subsequences $\hat{u}_n$ and $\bar{u}_n$ such that

$$\phi(\hat{u}_n) \to K_1, \quad \phi(\bar{u}_n) \to K_2, \quad K_1 \neq K_2.$$

Without loss of generality we can assume that $K_1 > K_2$ and that, for all $n \geq 1$,

$$\phi(\hat{u}_n) \geq K_1 - \delta > K_2 + \delta \geq \phi(\bar{u}_n)$$

for some positive δ. We put $v_1 = \hat{u}_{n_1} - \bar{u}_{n_1}$. Then $\phi(v_1) \geq K_1 - K_2 - 2\delta > 0$. For any given ball and any $\epsilon > 0$ we can choose n_1 sufficiently large such that the E-norm of v_1 in this ball is less than $\epsilon/2$. On the other hand, v_1 converges to 0 at infinity in the sense of the definition of the space E^0. Therefore there exists a function $\omega_1 \in E^0$, $\|\omega_1\|_E \leq \epsilon$ such that $w_1 = v_1 + \omega_1$ has a finite support.

We choose ϵ such that

$$|\phi(\omega_1)| \leq K\|\omega_1\|_E \leq K\epsilon < K_1 - K_2 - 2\delta.$$

Then $\phi(w_1) > K_1 - K_2 - 2\delta - K\epsilon > 0$. We choose the functions $\hat{u}_{n_2}, \bar{u}_{n_2}$ such that $v_2 = \hat{u}_{n_2} - \bar{u}_{n_2}$ is sufficiently small in the support of w_1. Then there exists ω_2 such that $\|\omega_2\|_E \leq \epsilon$, $\text{supp}\, w_1 \cap \text{supp}\, w_2 = \emptyset$ and $\phi(w_2) > K_1 - K_2 - 2\delta - K\epsilon > 0$. In the same manner we construct other functions of the sequence w_n. We put $z_n = \sum_{i=1}^n w_i$. Then the functions z_n are uniformly bounded in the E norm and $\phi(z_n) \to \infty$. The lemma is proved. □

Lemma 3.4. *The limit (3.22) does not depend on the sequence u_n.*

Proof. Suppose that there are two sequences $\hat{u}_n$ and $\bar{u}_n$ such that $\hat{u}_n \to u$, $\bar{u}_n \to u$ in E_{loc} and

$$\lim_{n \to \infty} \phi(\hat{u}_n) \neq \lim_{n \to \infty} \phi(\bar{u}_n).$$

Then we proceed as in the proof of the previous lemma. The lemma is proved. □

3. Hölder spaces

Corollary 3.5. *If $u_n \to 0$ in E_{loc}, then $\phi(u_n) \to 0$.*

We can now extend the functional ϕ to the space $E(\Omega)$. For any $u \in E(\Omega)$ we put $\hat{\phi}(u) = \phi(u)$ if $u \in E^0(\Omega)$ and $\hat{\phi}(u) = \lim_{n\to\infty} \phi(u_n)$, where $u_n \in E^0(\Omega)$ is an arbitrary sequence converging to u in E_{loc}. This is a linear bounded functional on $E(\Omega)$. Denote all such functionals by $\hat{E}$. It is a linear subspace in E^*. Suppose that $\hat{E} \ne E^*$. We take a functional $\psi \in E^*$, which does not belong to $\hat{E}$. Let ψ_0 be a restriction of ψ to E^0. Then $\psi_0 \in (E^0)^*$. As above we can define the functional $\hat{\psi}_0 \in (E)^*$. By assumption $\psi \ne \hat{\psi}_0$. Set $\tilde{\psi} = \psi - \hat{\psi}_0$. Then

$$\tilde{\psi} = 0, \quad \forall u \in E^0. \tag{3.23}$$

Thus we have proved the following theorem.

Theorem 3.6. *The dual space E^* is a direct sum of the extension $\hat{E}$ of $(E^0)^*$ on E and of the subspace $\tilde{E}$ consisting of all functionals satisfying (3.23).*

Remark 3.7. For any function $v \in L^1(\Omega)$ we can define the functional $\phi \in \hat{E}$ as

$$\phi(u) = \int_\Omega v(x) u(x) dx.$$

We do not know whether $\hat{E} = (C^\alpha(\Omega))^*$. However, if instead of the space $C^\alpha(\Omega)$ we take for example, the space of functions from $C^\alpha(\Omega)$ having limits at infinity, then all constructions above remain applicable and $\hat{E} \ne (C^\alpha(\Omega))^*$. Indeed, the functional $\psi(u) = \lim_{|x|\to\infty} u(x)$ does not belong to $\hat{E}$. However the following lemma shows that normal solvability is determined completely by functionals from $\hat{E}$.

Lemma 3.8. *Suppose that the operator $L : E_0 \to E$ is normally solvable with a finite-dimensional kernel, and the problem*

$$Lu = f, \quad f \in E \tag{3.24}$$

is solvable if and only if $\psi_i(f) = 0$, $i = 1, \ldots, N$, where ψ_i are linearly independent functionals in E^*. Then $\psi_i \in \hat{E}$.

Proof. Suppose that the assertion of the lemma does not hold and $\psi_1 \notin \hat{E}$, $\psi_2, \ldots, \psi_N \in \hat{E}$. We suppose first that $\psi_1 \in \tilde{E}$. We consider the functionals $\psi_i, i = 2, \ldots, N$ as functionals on E^0. They are linearly independent. There exist functions $f_j \in E^0, j = 2, \ldots, N$ such that $\psi_i(f_j) = \delta_{ij}$, $i, j = 2, \ldots, N$, where δ_{ij} is the Kronecker symbol.

Let $f^{(n)} \in E^0$, the norms $\|f^{(n)}\|_E$ be uniformly bounded and $f^{(n)} \to f$ in E_{loc}. Then the equation $Lu = g^{(n)}$, where

$$g^{(n)} = f^{(n)} - \sum_{i=2}^N \psi_i(f^{(n)}) f_i$$

is solvable in E_0 since
$$\psi_1(f^{(n)}) = 0, \ \psi_1(f_i) = 0, \ \psi_i(g^{(n)}) = 0, \ i = 2, \ldots, N.$$
Denote by $u^{(n)}$ its solution and put $u^{(n)} = v^{(n)} + w^{(n)}$, where
$$v^{(n)} \in Ker L, \ w^{(n)} \in (Ker L)^\perp,$$
and $(Ker L)^\perp$ denotes the supplement to the kernel of the operator L in the space E_0. Then
$$Lw^{(n)} = g^{(n)} \tag{3.25}$$
and the E_0 norms of the functions $w^{(n)}$ are uniformly bounded.

Indeed, if $\|w^{(n)}\|_{E_0} \to \infty$, then for the functions
$$\tilde{w}^{(n)} = \frac{w^{(n)}}{\|w^{(n)}\|_{E_0}}, \quad \tilde{g}^{(n)} = \frac{g^{(n)}}{\|w^{(n)}\|_{E_0}}$$
we have
$$L\tilde{w}^{(n)} = \tilde{g}^{(n)}, \quad \|\tilde{g}^{(n)}\|_E \to 0.$$
Since the operator L is proper, then there exists a function w_0 such that $\tilde{w}^{(n_k)} \to w_0$. Hence $w_0 \in (Ker L)^\perp$. On the other hand, $Lw_0 = 0$. This contradiction proves the boundedness of the sequence $w^{(n)}$.

Therefore there exists a subsequence $w^{(n_k)}$ converging in $E'_{0,\text{loc}}$ (see Section 2.4) to a limiting function $\hat{w} \in E_0$. Passing to the limit in (3.25), we have
$$L\hat{w} = f - \sum_{i=2}^{N} \psi_i(f) f_i. \tag{3.26}$$
Since this problem in solvable for any f, then
$$0 = \psi_1(f - \sum_{i=2}^{N} \psi_i(f) f_i) = \psi_1(f).$$
This means that for any function $f \in E$, the value of the functional ψ_1 equals zero. This contradiction proves the lemma under the assumption that $\psi_1 \in \tilde{E}$.

If $\psi_1 \notin \hat{E}$ and $\psi_1 \notin \tilde{E}$, then by virtue of Theorem 3.6, $\psi_1 = \tilde{\psi}_1 + \hat{\psi}_1$, where $\tilde{\psi}_1 \in \tilde{E}$, $\hat{\psi}_1 \in \hat{E}$. If the functionals $\hat{\psi}_1, \psi_2, \ldots, \psi_N$ are linearly dependent, we can take their linear combination and reduce this case to the case considered above. If they are linearly independent, we repeat the same construction with all N functionals, i.e., the sum in the expression for $g^{(n)}$ contains the term $\hat{\psi}_1(f^{(n)}) f_1$. The solvability condition
$$0 = \psi_1(f - \sum_{i=2}^{N} \psi_i(f) f_i - \hat{\psi}_1(f) f_1) = \tilde{\psi}_1(f)$$
gives $\tilde{\psi}_1 \in \hat{E}$. The proof remains the same if we suppose that more than one functional does not belong to $\hat{E}$. The lemma is proved. $\square$

3. Hölder spaces

Theorem 3.9. *If the operator L is Fredholm, then each of its limiting operators is invertible.*

Proof. It is sufficient to prove that the problem

$$\hat{L}u = f^* \tag{3.27}$$

is solvable for any $f^* \in E(\Omega^*)$ where $\hat{L}$ is a limiting operator and Ω^* is the corresponding limiting domain. The problem

$$Lu = f, \quad u \in E_0(\Omega), \ f \in E(\Omega)$$

is solvable if and only if $\psi_i(f) = 0$, $i = 1, \ldots, N$, where ψ_i are linearly independent functionals in $\hat{E}$ (see Lemma 3.8). Denote by $f_j \in E^0(\Omega)$, $j = 1, \ldots, N$ the functions, which form the biorthogonal system with these functionals. For any $f \in E(\Omega)$ the problem

$$Lu = f - \sum_{i=1}^{N} \psi_i(f) f_i \tag{3.28}$$

has a solution $u \in E_0(\Omega)$.

Let $\{x_m\}$ be the sequence for which the limiting operator $\hat{L}$ is defined. Denote $T_m f(x) = f(x + x_m)$ and consider the shifted problem. Then from (3.28)

$$L_m T_m u = T_m f - \sum_{i=1}^{N} \psi_i(f) T_m f_i, \tag{3.29}$$

where L_m is the operator with shifted coefficients. So for any $f \in E(\Omega_m)$ the equation

$$L_m u = f - \sum_{i=1}^{N} \psi_i(T_m^{-1} f) T_m f_i \tag{3.30}$$

has a solution $u \in E_0(\Omega_m)$.

To prove existence of solutions of (3.27) we use the construction given in the proof of Theorem 3.2. Let φ_m, V_j, n_j, m_j be the same as in Theorem 3.2. Suppose that (3.8) is satisfied. Set $g_{m_j}(y) = \varphi_{m_j}(y) f^*(h_{m_j}(y))$ for $y \in \overline{\Omega}_{m_j} \cap V_{j+1}$ and suppose that $g_{m_j}(y) = 0$ for $y \in \overline{\Omega}_{m_j}$, $|y| > j+1$. Consider the equation

$$L_{m_j} u_{m_j} = g_{m_j} - \sum_{i=1}^{N} \psi_i(T_{m_j}^{-1} g_{m_j}) T_{m_j} f_i. \tag{3.31}$$

It is of the same type as (3.30), and so it has a solution $u_{m_j} \in E_0(\Omega_{m_j})$.

Since $\|g_{m_j}\|_{E(\Omega_{m_j})}$ is bounded, we obtain from (3.31) that $\|u_{m_j}\|_{E_0(\Omega_{m_j})}$ is also bounded. By Theorem 2.9 there exists a function $u \in E_0(\Omega^*)$ and a subsequence $u_{m_{j_k}} \to u$ in $E'_{0,\text{loc}}(\Omega_{m_{j_k}} \to \Omega^*)$. Moreover the subsequence can be taken in such a way that $g_{m_{j_k}}$ is convergent in $E'_{\text{loc}}(\Omega_{m_{j_k}} \to \Omega^*, \partial\Omega_{m_{j_k}} \to \partial\Omega^*)$. Obviously the limit of $g_{m_{j_k}}$ is f^*. Passing to the limit in (3.31) by this subsequence and taking into account that $T_{m_j} f_i \to 0$, we obtain solvability of problem (3.27). The theorem is proved. □

Corollary 3.10. *If an operator L coincides with its limiting operator and satisfies the Fredholm property, then it is invertible.*

This corollary is applicable to operators with constant, periodic or quasi-periodic coefficients in domains with translation invariance.

3.4 Weighted spaces

In this section we discuss the Fredholm property of elliptic operators in weighted spaces. Consider the equation $Lu = f$, where $u \in E_{0,\mu}$, $f \in E_\mu$ (see the notation in Section 3.1). Set $v = u\mu$, $g = f\mu$. Then $Lv + Ku = g$, where $Ku = \mu Lu - L(\mu u)$.

Lemma 3.11. *Suppose that the operator $L : E_0 \to E$ is normally solvable and has a finite-dimensional kernel, the operator $Ku \equiv \mu Lu - L(\mu u) : E_{0,\mu} \to E$ is compact. Then the operator $L : E_{0,\mu} \to E_\mu$ is normally solvable and has a finite-dimensional kernel.*

Proof. Let f_k be a convergent sequence in E_μ, $Lu_k = f_k$, and $\|u_k\|_{E_{0,\mu}} \leq 1$. We will show that the sequence u_k is compact and, by this, that the operator $L : E_{0,\mu} \to E_\mu$ is proper. We have $Lv_k + Ku_k = g_k$, where $v_k = \mu u_k$, $g_k = \mu f_k$. Let $w_k = Ku_k$ and w_{k_l} be a subsequence converging in E. Then $Lv_{k_l} = g_{k_l} - w_{k_l}$, and the sequence v_{k_l} is compact in E_0 since the operator $L : E_0 \to E$ is proper. Therefore the sequence u_{k_l} is compact in $E_{0,\mu}$. The lemma is proved. □

Theorem 3.12. *Suppose that Condition NS is satisfied. Then the operator $L : E_{0,\mu} \to E_\mu$ is normally solvable and has a finite-dimensional kernel.*

Proof. We consider the operators A_i defined in Section 3.1. The boundary operators B_i are treated similarly. Let $K_i u = \mu A_i u - A_i(\mu u)$. According to Lemma 3.11 it is sufficient to prove that the operator $K_i : E_{0,\mu} \to C^{l-s_i+\alpha}(\bar{\Omega})$ is compact. Obviously

$$K_i u = \sum_{k=1}^{p} \sum_{0 < |\sigma| \leq \beta_{ik}, |\tau| < \beta_{ik}} c_{\sigma\tau}(x) \, D^\sigma \mu \, D^\tau u_k, \tag{3.32}$$

where $c_{\sigma\tau}$ is a linear combination of the coefficients $a_{ik}^\beta(x)$ of the operator A_i. So

$$c_{\sigma\tau}(x) \in C^{l-s_i+\alpha}(\bar{\Omega}).$$

Suppose that we have a sequence $\{u^\nu\}$, $\nu = 1, 2, \ldots$, such that

$$\|u^\nu\|_{E_{0,\mu}} = \|u^\nu \mu\|_{E_0} \leq M$$

3. Hölder spaces

with M independent of ν. We will prove that we can choose a subsequence of the sequence $K_i u^\nu$ convergent in $C^{l-s_i+\alpha}(\bar{\Omega})$. Indeed, write $v^\nu = \mu u^\nu$. Then $\|v^\nu\|_{E_0} \leq M$. So we can find a subsequence $w_j = v^{\nu_j}$ convergent in $\widehat{E} \equiv C^{l+t_1}(\bar{\Omega}) \times \cdots \times C^{l+t_p}(\bar{\Omega})$ locally to some limiting function $w_0 \in E_0$. Set $u_0 = \frac{w_0}{\mu}$. Then we have

$$\|K_i u^{\nu_j} - K_i u_0\|_{C^{l-s_i+\alpha}(\bar{\Omega})} = \|K_i \frac{z_j}{\mu}\|_{C^{l-s_i+\alpha}(\bar{\Omega})},$$

where
$$z_j = w_j - w_0, \quad \|z_j\|_{E_0} \leq M_1, \quad z_j \to 0 \tag{3.33}$$

in $\widehat{E}$ locally and M_1 does not depend on j. Set $y_j = K_i(\frac{z_j}{\mu})$. We have to prove that
$$\|y_j\|_{C^{l-s_i+\alpha}(\bar{\Omega})} \to 0 \tag{3.34}$$

as $j \to \infty$. It follows from (3.32) that

$$y_j = \sum_{k=1}^{p} \sum_{|\gamma| < \beta_{ik}} T_{k\gamma}(x) \, D^\gamma z_{jk}, \tag{3.35}$$

where $z_j = (z_{j1}, \ldots, z_{jp})$,

$$T_{k\gamma}(x) = \sum_{0 < |\sigma| \leq \beta_{ik}, |\tau| < \beta_{ik}, |\lambda| < \beta_{ik}} c_{\sigma\tau}(x) \, b_{\lambda\gamma} \, D^\sigma \mu \, D^\lambda \frac{1}{\mu},$$

$b_{\lambda\gamma}$ are constants. From (3.35) we get

$$\|y_j\|_{C^{l-s_i+\alpha}(G)} \leq M_2 \sum_{k=1}^{p} \sum_{|\gamma| < \beta_{ik}} \|T_{k\gamma}\|_{C^{l-s_i+\alpha}(\bar{G})} \|D^\gamma z_{jk}\|_{C^{l-s_i+\alpha}(\bar{G})}, \tag{3.36}$$

where $G = \Omega_{N+1}$ or $G = \widehat{\Omega}_N$, $\Omega_{N+1} = \Omega \cap \{|x| < N+1\}$, $\widehat{\Omega}_N = \Omega \cap \{|x| > N\}$. For any $\varepsilon > 0$ we can find N_0 such that for $N > N_0$ we have

$$\|y_j\|_{C^{l-s_i+\alpha}(\widehat{\Omega}_N)} < \varepsilon \tag{3.37}$$

for all j. This follows from the fact that

$$D^\beta \left(D^\sigma \mu(x) \, D^\lambda \frac{1}{\mu(x)} \right) \to 0$$

as $|x| \to \infty$, $x \in \Omega$ for any $|\sigma| > 0$, λ and β. Boundedness of the last norm in the right-hand side of (3.36) follows from (3.33). From (3.37), (3.33) and (3.36) with $G = \Omega_{N+1}$ we get (3.34). The theorem is proved. $\square$

Chapter 5

Fredholm Property

In the previous chapter we showed that under certain conditions elliptic operators are normally solvable with a finite-dimensional kernel. In order to prove their Fredholm property it remains to verify that the codimension of their image is also finite. For this we need to study some properties of the adjoint operators. Similar to Condition NS for the direct operators we will introduce Condition NS* for the adjoint operators.

We recall that Condition NS is a necessary and sufficient condition for the operator L to be normally solvable with a finite-dimensional kernel. A priori estimates for the adjoint operators and Condition NS* will allow us to prove that the operator $L^* : (F^*(\Omega))_\infty \to (E^*(\Omega))_\infty$ is normally solvable with a finite-dimensional kernel. We note that these spaces are different from the dual spaces $(F_\infty(\Omega))^*$ and $(E_\infty(\Omega))^*$ where the adjoint operator acts. Detailed analysis of these spaces and of the properties of the operators will allow us to prove that the kernel of the adjoint operator has a finite dimension when considered in the dual spaces. From this we will conclude the existence of the Fredholm property of the operator L in the spaces $W^{s,p}$ $(1 < p < \infty)$ and $W_q^{s,p}$.

1 Estimates with Condition NS*

Condition NS*. Any limiting homogeneous problem $\hat{L}^* v = 0$ has only the zero solution in the space $(F^*(\hat{\Omega}))_\infty$, where $\hat{L}^*$ is the operator adjoint to the limiting operator $\hat{L}$, and $\hat{\Omega}$ is a limiting domain.

The main result of this section is given by the following theorem.

Theorem 1.1. *Let L be an elliptic operator, and Condition NS* be satisfied. Then there exist positive numbers M and ρ such that for any $v \in (F^*(\Omega))_\infty$ the following* *estimate holds:*

$$\|v\|_{(F^*(\Omega))_\infty} \leq M \left(\|L^*v\|_{(E^*(\Omega))_\infty} + \|v\|_{F^*_{-1}(\Omega_\rho)} \right). \tag{1.1}$$

Here
$$F^*_{-1}(\Omega_\rho) = \Pi_{i=1}^N \dot{W}^{-l+s_i-1,p'}(\Omega_\rho) \times \Pi_{j=1}^m \dot{W}^{-l+\sigma_j+1/p-1,p'}(\Gamma_\rho),$$

Ω_ρ *and* Γ_ρ *are the intersections of* Ω *and* Γ *with the ball* $|x| < \rho$.

Proof. Suppose that the assertion of the theorem does not hold. Let $M_k \to \infty$, $\rho_k \to \infty$ be some given sequences. Then there exist a sequence $v_k \in (F^*(\Omega))_\infty$ such that

$$\|v_k\|_{((F(\Omega))^*)_\infty} \geq M_k \left(\|L^*v_k\|_{((E(\Omega))^*)_\infty} + \|v_k\|_{F^*_{-1}(\Omega_{\rho_k})} \right). \tag{1.2}$$

We can suppose that
$$\|v_k\|_{((F(\Omega))^*)_\infty} = 1. \tag{1.3}$$

Then from (1.2)

$$\|L^*v_k\|_{((E(\Omega))^*)_\infty} + \|v_k\|_{F^*_{-1}(\Omega_{\rho_k})} < \frac{1}{M_k} \to 0 \text{ as } k \to \infty. \tag{1.4}$$

From Theorem 5.1 (Chapter 3) we have

$$\|L^*v_k\|_{((E(\Omega))^*)_\infty} + \|v_k\|_{(F^*_{-1}(\Omega))_\infty} \geq \frac{1}{M}. \tag{1.5}$$

Estimate (1.4) implies

$$\|L^*v_k\|_{((E(\Omega))^*)_\infty} \to 0, \quad \|v_k\|_{F^*_{-1}(\Omega_{\rho_k})} \to 0 \text{ as } k \to \infty. \tag{1.6}$$

Hence
$$\|v_k\|_{(F^*_{-1}(\Omega))_\infty} > \frac{1}{2M} \tag{1.7}$$

for k sufficiently large. Since

$$\|v_k\|_{(F^*_{-1}(\Omega))_\infty} = \sup_{y \in \Omega} \|v_k\|_{F^*_{-1}(\Omega \cap B_y)},$$

then it follows from (1.7) that there exists $y_k \in \Omega$ such that

$$\|v_k\|_{F^*_{-1}(\Omega \cap B_{y_k})} > \frac{1}{2M}. \tag{1.8}$$

From this and (1.6) we conclude that $|y_k| \to \infty$. Set

$$L^*v_k = z_k. \tag{1.9}$$

1. Estimates with Condition NS*

From (1.6) it follows that

$$\|z_k\|_{(E^*(\Omega))_\infty} \to 0 \text{ as } k \to \infty.$$

Let T_h be the operator of translation in $(E^*(\Omega))_\infty$, $h \in R^n$. We apply T_{y_k} to (1.9):

$$T_{y_k} L^* v_k = T_{y_k} z_k. \tag{1.10}$$

The shifted functions are defined in shifted domains Ω_k. We will pass to the limit in this equality as $k \to \infty$. We have

$$T_{y_k} L^* T_{-y_k} T_{y_k} v_k = T_{y_k} z_k, \tag{1.11}$$

where the operators T_{y_k} act in the corresponding spaces. Set

$$w_k = T_{y_k} v_k, \quad L_k^* = T_{y_k} L^* T_{-y_k}, \quad T_{y_k} z_k = \zeta_k. \tag{1.12}$$

From (1.11) we obtain

$$L_k^* w_k = \zeta_k. \tag{1.13}$$

Let $v_k = (v_{1k}, \ldots, v_{Nk}, v_{1k}^b, \ldots, v_{mk}^b)$. From (1.3)

$$\|v_{ik}\|_{(\dot{W}^{-l+s_i,p'}(\Omega))_\infty} \leq 1,$$
$$i = 1, \ldots, N, \|v_{jk}^b\|_{(W^{-l+\sigma_j+1/p,p'}(\Gamma))_\infty} \leq 1, \quad j = 1, \ldots, m. \tag{1.14}$$

Writing $w_k = (w_{1k}, \ldots, w_{Nk}, w_{1k}^b, \ldots, w_{mk}^b)$, we have

$$w_{ik} = T_{y_k} v_{ik}, \quad i = 1, \ldots, N, \quad w_{jk}^b = T_{y_k} v_{jk}^b, \quad j = 1, \ldots, m.$$

Then from (1.14)

$$\|w_{ik}\|_{(\dot{W}^{-l+s_i,p'}(\Omega_k))_\infty} \leq 1, \quad i = 1, 2, \ldots, N,$$
$$\|w_{jk}^b\|_{(W^{-l+\sigma_j+1/p,p'}(\partial\Omega_k))_\infty} \leq 1, \quad j = 1, 2, \ldots, m, \tag{1.15}$$

where Ω_k is the shifted domain. Consider first the functions w_{ik}. Since

$$(\dot{W}^{-l+s_i,p'}(\Omega_k))_\infty \subset W_\infty^{-l+s_i,p'}(\mathbb{R}^n),$$

then $w_{ik} \in W_\infty^{-l+s_i,p'}(\mathbb{R}^n)$, and

$$\|w_{ik}\|_{W_\infty^{-l+s_i,p'}(\mathbb{R}^n)} \leq 1. \tag{1.16}$$

To continue the proof of the theorem we will use the following lemma.

Lemma 1.2. *Let E be a reflexive Banach space. If $\{u_k\}$, $k = 1, 2, \ldots$ is a bounded sequence in E_∞, then there exists a subsequence u_{k_i} of u_k and $u \in E_\infty$ such that*

$$u_{k_i} \to u \text{ locally weakly and in } \mathcal{D}'.$$

Proof. Let $\phi_j \in D$ have a support in the ball B_j of the radius $r_j = j+1$, and $\phi_j = 1$ inside the ball B_{j-1}. Consider the sequence $\phi_j u_k$, $k = 1, 2, \ldots$ We can choose a weakly convergent subsequence,

$$\phi_j u_k \to u_0 \text{ weakly and in } D'.$$

For any $\psi \in D$ with its support in B_{j-1},

$$\psi u_k \to \psi u_0 \text{ weakly and in } D'.$$

We begin this construction with the ball B_2. Then we consider the ball B_3, and consider a convergent subsequence of the previous subsequence, and so on. We choose the diagonal subsequence. Thus we obtain a function u_0 defined in any ball B_j and a sequence u_{k_j} such that

$$\psi u_{k_j} \to \psi u_0 \text{ weakly and in } D' \tag{1.17}$$

for any function $\psi \in D$.

We should show that $u_0 \in E_\infty$. Suppose that this is not true. Then there exists a sequence $\psi_j \in D$ such that

$$\|\psi_j u_0\|_E \to \infty, \ j \to \infty.$$

Therefore there exists a functional $F \in E^*$ such that $F(\psi_j u_0) \to \infty$. From convergence (1.17) for each ψ fixed, we can conclude that there exists a subsequence u_{k_j} such that $F(\psi_j u_{k_j}) \to \infty$. This contradicts the uniform boundedness of the functions u_k in the norm E_∞. The lemma is proved. □

It follows from Lemma 1.2 that there exists a subsequence w_{ik_j} and a function $\bar{w}_i \in W_\infty^{-l+s_i,p'}(\mathbb{R}^n)$ such that

$$\phi w_{ik_j} \to \phi \bar{w}_i \text{ weakly in } W^{-l+s_i,p'}(\mathbb{R}^n) \text{ as } k_j \to \infty \tag{1.18}$$

for any $\phi \in D$. Moreover the sequence w_{ik_j} can be chosen such that for an $\epsilon > 0$

$$\phi w_{ik_j} \to \phi \bar{w}_i \text{ strongly in } W^{-l+s_i-\epsilon,p'}(\mathbb{R}^n) \text{ as } k_j \to \infty \tag{1.19}$$

for any $\phi \in D$.

In what follows we need a *special covering* of the boundary $\partial \Omega_*$ of the limiting domain Ω_*. Let $x_0 \in \partial \Omega_*$. Then there exists a sequence $\hat{x}_k$ such that $\hat{x}_k \to x_0$, $\hat{x}_k \in \partial \Omega_k$. For each point $\hat{x}_k$ and domain Ω_k there exists a neighborhood $U(\hat{x}_k)$ and a function $\psi_k(x) = \psi(x; \hat{x}_k)$ defined in Condition D. It maps $U(\hat{x}_k)$ on the unit ball $B \subset \mathbb{R}^n$ with the center at 0. This mapping is a bijection. Write $\phi_k = \psi_k^{-1}$. By Condition D, the functions ϕ_k are uniformly bounded in $C^{r+\theta}(B)$. Hence this sequence has a convergent in $C^r(B)$ subsequence: $\phi_{k_i} \to \phi_0$ in $C^r(B)$. Write $U(x_0) = \phi_0(B)$. The mapping $\phi_0 : B \to U(x_0)$ is also a bijection. Set $\psi_0 = \phi_0^{-1}$:

1. Estimates with Condition NS*   187

$U(x_0) \to B$. Then $\psi_{k_i} \to \psi_0 \in C^{r+\theta}(U(x_0))$, $U(x_0)$ is an open set, it contains a sphere $S(x_0)$ of a radius δ.

Consider now a sequence $x_j \in \partial\Omega_*$, and denote by $S(x_j)$ the spheres of the radius $\delta/2$ and the centers at x_j. We can take the points x_j such that the union W of the spheres $S(x_j)$ covers the $\delta/4$-neighborhood of the boundary $\partial\Omega_*$. We repeat the construction above for the point x_1 of this sequence. We choose a subsequence of the previous sequence (denoted also k_i) such that

$$\psi_{k_i} \to \psi^1, \quad \phi_{k_i} \to \phi^1, \quad U(x_1) = \phi^1(B), \quad S(x_1) \subset U(x_1).$$

We then repeat the same construction for the point x_2 and so on, and take the diagonal subsequence. Therefore we construct neighborhoods $U(x_j)$ of all points x_j. Moreover this construction can be done in such a way that for some number N, any N different sets $U(x_j)$ have an empty intersection, and for any compact $K \subset \mathbb{R}^n$ the number of the sets for which $K \cap U(x_j) \neq \emptyset$ is finite.

The covering $V = \cup_{j=1}^{\infty} U(x_j)$ is called a special covering of $\partial\Omega_*$. Hence we have a sequence Ω_k, points $x_k^j \in \partial\Omega_k$, neighborhoods $U(x_k^j)$ of the points x_k^j, mappings $\psi_k^j : U(x_k^j) \to B$, $\phi_k^j = (\psi_k^j)^{-1}$, such that

$$\psi_k^j \to \psi^j, \quad \phi_k^j \to \phi^j, \quad U(x_j) = \phi^j(B).$$

Consider a sequence g_k defined on $\partial\Omega_k$ such that

$$\|g_k\|_{(E(\partial\Omega_k))_\infty} \leq K, \tag{1.20}$$

where K is a constant independent of k, $E = W^{-s,p}$, $s > 0$. The norm in (1.20) is defined as follows. Let $\eta(x) \in C^\infty(\mathbb{R}^n)$ be such that

$$\eta(x) \geq 0, \ \eta(x) = 1, \ |x| < \frac{\delta}{2}, \quad \eta(x) = 0, \ |x| > \delta.$$

Set $\eta_z = \eta(x - z)$, $z \in \mathbb{R}^n$. Then

$$\|g_k\|_{(E(\partial\Omega_k))_\infty} = \sup_{z \in \partial\Omega_k} \|(\eta_z g_k) \circ \psi_z^{-1}\|_{E(\mathbb{R}^{n-1}_{y'})}, \tag{1.21}$$

where ψ_z maps the neighborhood $U(z)$ to the ball B.

Consider the neighborhood $U(x_k^1)$ of the point $x_k^1 \in \partial\Omega_k$. The function ψ_k^1 maps $U(x_k^1) \cap \partial\Omega_k$ onto $B_0 = B \cap \{y^n = 0\}$. We can define a generalized function $\tilde{g}_k^1$ on D_{B_0} by the equality

$$\tilde{g}_k^1 = (\eta_{x_k^1} g_k) \circ (\psi_k^1)^{-1}.$$

We extend it to $D_{\mathbb{R}^{n-1}}$ by zero outside B_0. It follows from (1.20), (1.21) that

$$\|\tilde{g}_k^1\|_{E(\mathbb{R}^{n-1}_{y'})} \leq K, \tag{1.22}$$

where $E(\mathbb{R}_{y'}^{n-1}) = W^{-s,p}(\mathbb{R}_{y'}^{n-1})$. Since this space is reflexive, we can find a subsequence $\tilde{g}_{k_j}^1$ and a function $\tilde{h}^1 \in E(\mathbb{R}_{y'}^{n-1})$ such that $\tilde{g}_{k_j}^1 \to \tilde{h}^1$ weakly in $E(\mathbb{R}_{y'}^{n-1})$ and $\tilde{g}_{k_j}^1 \to \tilde{h}^1$ strongly in $E_{-1}(\mathbb{R}_{y'}^{n-1}) (= W^{-s-1,p}(\mathbb{R}_{y'}^{n-1}))$. We use here the compact embedding of E into E_{-1} in bounded domains. The generalized function $\tilde{h}^1$ is defined on D_{B_0}. Denote by h^1 the corresponding generalized function defined on $U(x_1) \cap \partial\Omega_*$: $h^1 = \tilde{h}^1 \circ \psi^1$. We extend h^1 by zero outside $U(x_1) \cap \partial\Omega_*$ on $\partial\Omega_*$.

We construct next a generalized function h^2 on $D_{U(x_2) \cap \partial\Omega_*}$. The construction is the same but we consider a subsequence of the previous subsequence. We continue this construction for all x_j and take a diagonal subsequence. Denote this subsequence by k_l.

Thus we have the following result. There exists a subsequence k_l of k such that for any j there exists a generalized function h^j on $D_{U(x_j) \cap \partial\Omega_*}$ defined by the equality $h^j = \tilde{h}^j \circ \psi^j$,

$$\|\tilde{g}_{k_l}^j - \tilde{h}^j\|_{E_{-1}(R_{y'}^{n-1})} \to 0 \text{ as } k_l \to \infty,$$

$$\tilde{g}_{k_l}^j \to \tilde{h}^j \text{ weakly in } E(R_{y'}^{n-1}) \text{ as } k_l \to \infty.$$

Moreover,

$$\|\tilde{g}_{k_l}^j\|_{E(R_{y'}^{n-1})} \leq K. \tag{1.23}$$

The points $x_j \in \partial\Omega_*$ and the functions $\eta_{x_k^j}$ can be chosen such that $g_k = \sum_j \eta_{x_k^j} g_k$, $x \in \partial\Omega_k$. Set $h = \sum_{j=1}^{\infty} h^j$. This is the limiting function for the sequence g_k. We note that for any $\phi \in D_{\partial\Omega_*}$ we have

$$\langle h, \phi \rangle = \sum_{j'} \langle h^{j'}, \phi \rangle,$$

where j' are those of j for which $\operatorname{supp} \phi \cap U(x_j) \neq \emptyset$. By construction of $U(x_j)$ the number of such j' is finite.

Lemma 1.3. *The limiting generalized function h belongs to $(W^{-s,p}(\partial\Omega_*))_\infty$, that is*

$$\|h\|_{(E(\partial\Omega_*))_\infty} = \sup_{z \in \partial\Omega_*} \|(\eta_z h) \circ \psi_z^{-1}\|_{E(\mathbb{R}_{y'}^{n-1})} < \infty. \tag{1.24}$$

Proof. Suppose that it is not true. Then there is a sequence $z_i \in \partial\Omega_*$ such that

$$\|(\eta_{z_i} h) \circ \psi_{z_i}^{-1}\|_{E(\mathbb{R}_{y'}^{n-1})} \to \infty \text{ as } i \to \infty. \tag{1.25}$$

Since $\eta_{z_i} h^j = 0$ for all j except for a finite number of them less than or equal to N, then there is a sequence h^{j_i} such that

$$\|(\eta_{z_i} h^{j_i}) \circ \psi_{z_i}^{-1}\|_{E(\mathbb{R}_{y'}^{n-1})} \to \infty \text{ as } i \to \infty.$$

1. Estimates with Condition NS*

Therefore
$$\|\eta_{z_i}(\phi_{z_i}(y))\tilde{h}^{j_i}\|_{E(\mathbb{R}^{n-1}_{y'})} \to \infty \text{ as } i \to \infty. \tag{1.26}$$

From this we can conclude that there exists a functional $F \in E^*(\mathbb{R}^{n-1}_{y'})$ such that
$$F(\eta_{z_i}(\phi_{z_i}(y))\tilde{h}^{j_i}) \to \infty \text{ as } i \to \infty.$$

From the weak convergence
$$\tilde{g}_k^{j_i} \to \tilde{h}^{j_i} \text{ as } k \to \infty$$

it follows that for some sequence k_i,
$$F(\eta_{z_i}(\phi_{z_i}(y))\tilde{g}_{k_i}^{j_i}) \to \infty \text{ as } i \to \infty.$$

This contradicts estimate (1.23). The lemma is proved. □

Thus from (1.15) and Lemma 1.3 we can conclude that there exist limiting functions
$$\bar{w}_j^b \in (W^{-l+\sigma_j+1/p,p'}(\partial\Omega_*))_\infty, \quad j = 1,\ldots,m.$$
Existence of the limits $\bar{w}_i \in W_\infty^{-l+s_i,p'}(\mathbb{R}^n)$, $j = 1,\ldots,N$ was proved above. Write $\bar{w} = (\bar{w}_1,\ldots,\bar{w}_N,\bar{w}_1^b,\ldots,\bar{w}_m^b)$.

Lemma 1.4. *The limiting function $\bar{w}$ is a solution of the problem adjoint to a limiting problem.*

Proof. Consider equation (1.13). It is supposed that we have done a special covering of $\partial\Omega_*$. Let ψ_k^i and ϕ_k^i be the functions from the special covering,
$$\psi_k^j \to \psi^j, \quad \phi_k^j \to \phi^j, \quad U(x_k^j) = \phi_k^j(B), \quad U(x_j) = \phi^j(B).$$

Let $\theta(x) = (\theta^1(x),\ldots,\theta^N(x))$, where $\theta^i(x) \in D(\mathbb{R}^n)$, $\operatorname{supp}\theta^i \subset U(x_j)$, $\theta_k(x)$ be the corresponding function with support in $U(x_k^j)$: $\theta_k = \theta(\phi^j(\psi_k^j))$. From (1.13) we have
$$\langle L_k^* w_k, \theta_k \rangle = \langle \zeta_k, \theta_k \rangle$$

or
$$\langle w_k, T_{y_k} L T_{-y_k} \theta_k \rangle = \langle \zeta_k, \theta_k \rangle.$$

We can rewrite this equality in the form
$$\langle w_k, L_k \theta_k \rangle = \langle \zeta_k, \theta_k \rangle, \tag{1.27}$$

where
$$L_k = T_{y_k} L T_{-y_k}, \quad L_k = (A_{1k},\ldots,A_{Nk},B_{1k},\ldots,B_{mk}),$$
$$A_{ik} = T_{y_k} A_i T_{-y_k}, \quad B_{jk} = T_{y_k} B_j T_{-y_k},$$

$$T_{y_k} A_i u = \sum_{l=1}^{N} \sum_{|\alpha| \leq \alpha_{il}} T_{y_k}(a_{il}^\alpha(x)) T_{y_k} D^\alpha u_l = \sum_{l=1}^{N} \sum_{|\alpha| \leq \alpha_{il}} a_{il}^\alpha(x+y_k) T_{y_k} D^\alpha u_l.$$

Hence

$$A_{ik} u = \sum_{l=1}^{N} \sum_{|\alpha| \leq \alpha_{il}} a_{ilk}^\alpha(x) D^\alpha u_l, \quad B_{jk} u = \sum_{l=1}^{N} \sum_{|\beta| \leq \beta_{jl}} b_{jlk}^\beta(x) D^\beta u_l,$$

where

$$a_{ilk}^\alpha(x) = a_{il}^\alpha(x+y_k), \quad b_{jlk}^\beta(x) = b_{jl}^\beta(x+y_k).$$

We can now write (1.27) as

$$\sum_{i=1}^{N} \langle w_{ik}, A_{ik}\theta_k \rangle + \sum_{j=1}^{m} \langle w_{jk}^b, B_{jk}\theta_k \rangle = \langle \zeta_k, \theta_k \rangle. \tag{1.28}$$

We will pass to the limit in this equality. We begin with the first term in the left-hand side. From (1.18) we have the weak convergence

$$\phi w_{ik} \to \phi \bar{w}_i \text{ in } W^{-l+s_i, p'}(\mathbb{R}^n) \text{ as } k \to \infty \tag{1.29}$$

for any $\phi \in D$ (we write k instead of k_j). By the definition of the limiting problem,

$$a_{ipk}^\alpha(x) \to \hat{a}_{ip}^\alpha(x) \text{ in } C_{\text{loc}}^{l-s_i}(\mathbb{R}^n), \quad p = 1, \ldots, N.$$

Here $\hat{a}_{ip}^\alpha(x)$ are the coefficients of the limiting operator. From the definition of $\theta_k(x)$ we have

$$\lim_{k \to \infty} \theta_k(x) = \theta(\phi^j(\psi^j(x))) = \theta(x),$$

where this limit is supposed to be in C^r. We suppose that $\psi_k^j(x)$ and $\psi^j(x)$ are extended on a ball which contains $U(x_j)$ and $U(x_k^j)$ with k sufficiently large. Then

$$A_{ik}\theta_k \to \sum_{p=1}^{N} \sum_{|\alpha| \leq \alpha_{ip}} \hat{a}_{ip}^\alpha(x) D^\alpha \theta^p = \hat{A}_i \theta, \quad k \to \infty$$

in $C^{l-s_i}(\mathbb{R}^n)$. Here $\hat{A}_i$ is the limiting operator. From (1.29) it follows that

$$\sum_{i=1}^{N} \langle w_{ik}, A_{ik}\theta_k \rangle \to \sum_{i=1}^{N} \langle \bar{w}_i, \hat{A}_i \theta \rangle, \quad k \to \infty. \tag{1.30}$$

Consider now $\langle w_{ik}^b, B_{ik}\theta_k \rangle$. Let η_z be the function which is used in the definition of the limiting function h above. Instead of the functions g_k considered

1. Estimates with Condition NS*

above we take functions w_{ik}^b and instead of k_l we write k. We obtain a sequence of functions

$$\tilde{w}_{ik}^{bj} = (\eta_{x_k^1} w_{ik}^b) \circ \phi_k^j,$$

where $\phi_k^j = (\psi_k^1)^{-1}$. As above, $\tilde{w}_{ik}^{bj} \to \tilde{w}_i^{bj}$ weakly in $W^{-l+\sigma_i+1/p,\, p'}(\mathbb{R}_{y'}^{n-1})$ as $k \to \infty$. Set $w_i^{bj} = \tilde{w}_i^{bj} \circ \psi^j$ and

$$\bar{w}_i^b = \sum_{j=1}^{\infty} w_i^{bj}.$$

This is the limiting function for the sequence w_{ik}^b.

Suppose that $x \in \partial\Omega_k \cap U(x_k^j)$. Write $f_{ik} = B_{ik}\theta_k$. Since $\operatorname{supp}\theta_k \subset U(x_k^j)$, then $\operatorname{supp} f_{ik} \subset U(x_k^j)$. We have

$$\langle \eta_{x_k^j} w_{ik}^b, f_{ik} \rangle = \langle \tilde{w}_{ik}^{bj}, (f_{ik}\rho_k) \circ \phi_k^j \rangle,$$

where ρ_k is the density for the manifold Ω_k. The density is the $(n-1)$-dimensional Hausdorff measure of $\partial\Omega_k$ written in local coordinates. Further

$$f_{ik}(\phi_k^j(y)) = \sum_{p=1}^{N} \sum_{|\beta| \leq \beta_{ip}} b_{ipk}^{\beta}(\phi_k^j(y)) D^{\beta} \theta_k^p(\phi_k^j(y)).$$

By definition of limiting problems, the functions $b_{ip}^{\beta}(x)$ are extended to $\mathbb{R}^n$ and

$$b_{ipk}^{\beta}(x) \to \hat{b}_{ip}^{\beta}(x) \text{ in } C_{\mathrm{loc}}^{l-\sigma_i}(\mathbb{R}^n) \text{ as } k \to \infty,$$

where $\hat{b}_{ip}^{\beta}(x)$ are the limiting coefficients. Since

$$\lim_{k \to \infty} \theta_k(\phi_k^j(y)) = \lim_{k \to \infty} \theta(\phi^j(\psi_k^j(\phi_k^j(y)))) = \theta(\phi^j(\psi^j(\phi^j(y)))) = \theta(\phi^j(y)),$$

then

$$f_{ik}(\phi_k^j(y)) \to \sum_{p=1}^{N} \sum_{|\beta| \leq \beta_{ip}} \hat{b}_{ip}^{\beta}(\phi^j(y)) D^{\beta} \theta^p(\phi^j(y)) \equiv f_i(\phi^j(y)).$$

This convergence is in $C^{l-\sigma_i}$.

Therefore

$$\langle \tilde{w}_{ik}^{bj}, (f_{ik}\rho_k) \circ \phi_k^j \rangle \to \langle \tilde{w}_i^{bj}, (f_i\rho_*) \circ \phi^j \rangle \text{ as } k \to \infty,$$

where ρ_* is the density for the manifold Ω_*. It follows that

$$\langle \eta_{x_k^j} w_{ik}^b, f_{ik} \rangle \to \langle \tilde{w}_i^{bj}, (f_i\rho_*) \circ \phi^j \rangle = \langle w_i^{bj}, f_i \rangle,$$

which is the duality on $\partial\Omega_*$.

Taking the sum with respect to j, we get

$$\langle w_{ik}^b, f_{ik} \rangle \to \left\langle \bar{w}_i^b, \sum_{p=1}^{N} \sum_{|\beta| \leq \beta_{ip}} \hat{b}_{ip}^\beta D^\beta \theta^p \right\rangle.$$

From this and (1.30) it follows that

$$\sum_{i=1}^{N} \langle w_{ik}, A_{ik}\theta_k \rangle + \sum_{i=1}^{m} \langle w_{ik}^b, B_{ik}\theta_k \rangle \to \sum_{i=1}^{N} \langle \bar{w}_i, \hat{A}_i\theta \rangle + \sum_{i=1}^{m} \langle \bar{w}_i^b, \hat{B}_i\theta \rangle,$$

where

$$\hat{B}_i \theta = \sum_{p=1}^{N} \sum_{|\beta| \leq \beta_{ip}} \hat{b}_{ip}^\beta D^\beta \theta^p.$$

We will prove that

$$\sum_{i=1}^{N} \langle \bar{w}_i, \hat{A}_i\theta \rangle + \sum_{i=1}^{m} \langle \bar{w}_i^b, \hat{B}_i\theta \rangle = 0. \tag{1.31}$$

It is sufficient to show that the following convergence holds:

$$\langle \zeta_k, \theta_k \rangle \to 0, \quad k \to \infty, \tag{1.32}$$

(see (1.28)). We recall that

$$\|\zeta_k\|_{(E^*(\Omega_k))_\infty} \to 0, \quad k \to \infty.$$

Since the diameters of $\operatorname{supp} \theta_k$ are uniformly bounded, convergence (1.32) follows from the last convergence and from the boundedness of the norm $\|\theta_k\|_{E(R^n)}$ independently of k. The lemma is proved. □

To finish the proof of the theorem it remains to prove the following lemma.

Lemma 1.5. *The solution $\bar{w}$ of the limiting problem* (1.31) *is different from* 0.

Proof. If $(\bar{w}_1, \ldots, \bar{w}_N) \neq 0$ then the lemma is proved. Consider the case where $(\bar{w}_1, \ldots, \bar{w}_N) = 0$. From (1.8), (1.12) we get

$$\|w_k\|_{F^*_{-1}(\Omega_k \cap B_0)} > \frac{1}{2M}. \tag{1.33}$$

Therefore

$$\sum_{j=1}^{m} \|w_{jk}^b\|_{W^{-l+\sigma_j+1/p-1,p'}(\partial\Omega_k \cap B_0)} > \frac{1}{2M}$$

for $k > k_0$, if k_0 is sufficiently large. For any $k > k_0$ there exists $j = j_k$ such that

$$\|w_{jk}^b\|_{W^{-l+\sigma_j+1/p-1,p'}(\partial\Omega_k \cap B_0)} > \frac{1}{2Mm}.$$

Passing to a subsequence, if necessary, we can suppose that j is the same for all k. Hence
$$\|\bar{w}_j^b\|_{W^{-l+\sigma_j+1/p-1,p'}(\partial\Omega_*\cap B_0)} \geq \frac{1}{2Mm}, \tag{1.34}$$
and $\bar{w}_j$ is different from 0 as an element of the space $W^{-l+\sigma_j+1/p-1,p'}(\partial\Omega_*\cap B_0)$. Consequently, it is also different from 0 as an element of $W^{-l+\sigma_j+1/p,p'}(\partial\Omega_*\cap B_0)$. Indeed, if it is not so, then
$$\langle \bar{w}_j^b, \phi \rangle = 0 \quad \forall \phi \in W^{l-\sigma_j-1/p,p}(\partial\Omega_* \cap B_0).$$
Then the same equality is true for all $\phi \in W^{l-\sigma_j-1/p+1,p}(\partial\Omega_* \cap B_0)$. But this contradicts (1.34). The lemma is proved. □

Thus, assuming that (1.1) does not hold, we have obtained a nonzero solution of a limiting problem, which contradicts Condition NS*. The theorem is proved. □

Corollary 1.6. *If Condition NS* is satisfied, then the operator $L^* : (E^*(\Omega))_\infty \to (F^*(\Omega))_\infty$ is normally solvable with a finite-dimensional kernel.*

2 Abstract operators

Let $E = E(\Omega)$ and $F^d(\Omega)$ be Banach spaces of functions defined in a domain Ω, $F^b(\partial\Omega)$ be a space of functions defined at the boundary $\partial\Omega$, $F = F^d(\Omega) \times F^b(\partial\Omega)$. Let further an operator $L : E \to F$ be local in the sense of Chapter 2. Then we can define its realization in various spaces:
$$L_\infty : E_\infty \to F_\infty, \quad L_D : E_D \to F_D, \quad L_q : E_q \to F_q, \quad 1 \leq q < \infty.$$
Here E_D and F_D is the closure of D in E_∞ and F_∞, respectively. We will also consider the adjoint operators
$$(L_\infty)^* : (F_\infty)^* \to (E_\infty)^*, \quad (L_D)^* : (F_D)^* \to (E_D)^*,$$
$$(L_q)^* : (F_q)^* \to (E_q)^*, \quad 1 \leq q < \infty$$
and the operator
$$(L^*)_\infty : (F^*)_\infty \to (E^*)_\infty.$$
We recall that
$$(E^*)_\infty = (E_1)^*, \quad (F^*)_\infty = (F_1)^*.$$
Therefore by the definition of local operators, $(L^*)_\infty = (L_1)^*$. This equality is understood as
$$\langle (L^*)_\infty w, \theta \rangle = \langle w, L_1 \theta \rangle$$
for any $w \in (F_1)^*$, and any $\theta \in E_1$. It is sufficient to consider it for $\forall \theta \in D$.

We suppose that there exist Banach spaces of distributions $\mathcal{E}$ and $\mathcal{F}$ such that the spaces E and F are embedded in them locally compactly. This means that for any ball B_ρ with radius ρ, the restriction $E(\Omega_\rho)$ of the spaces $E(\Omega)$ to $\Omega_\rho = \Omega \cap B_\rho$ is compactly embedded into the space $\mathcal{E}(\Omega_\rho)$. A similar property holds for spaces F and $\mathcal{F}$. We note that $E_q(\Omega_\rho) = E(\Omega_\rho)$. Therefore $E_q(\Omega_\rho)$ is also compactly embedded in $\mathcal{E}(\Omega_\rho)$.

Lemma 2.1. *Suppose that the following estimate*

$$\|u\|_{E_q} \leq M \left(\|L_q u\|_{F_q} + \|u\|_{\mathcal{E}(\Omega_\rho)} \right) \tag{2.1}$$

holds for some positive constants M and ρ, and any $u \in E_q$. Then the operator L_q is proper, that is the inverse image of a compact set is compact in any bounded closed ball. Here $1 \leq q \leq \infty$.

Proof. Let $L_q u_k = f_k$, $f_k \to f_0$ in F_q, and $\|u_k\|_{E_q} \leq C$ for some constant C and all k. Let us take ρ for which (2.1) is satisfied. Then there exists a subsequence u_{k_n} fundamental in $\mathcal{E}(\Omega_\rho)$:

$$\|u_{k_n} - u_{k_m}\|_{\mathcal{E}(\Omega_\rho)} \to 0 \text{ as } m, n \to \infty.$$

From (2.1) it follows that the same subsequence is fundamental in E_q. The lemma is proved. □

Corollary 2.2. *The operator L_q is normally solvable with a finite-dimensional kernel.*

(Cf. Theorem 2.13 of Chapter 4.)

We repeat the same construction for the adjoint operators. We suppose that there exist spaces $\mathcal{E}^*$ and $\mathcal{F}^*$ such that the spaces E^* and F^* are embedded in them locally compactly.

Lemma 2.1'. *Suppose that the estimate*

$$\|u\|_{(E^*)_q} \leq M \left(\|(L^*)_q u\|_{(F^*)_q} + \|u\|_{\mathcal{E}^*(\Omega_\rho)} \right) \tag{2.2}$$

holds for some positive constants M and ρ, and any $u \in (E^*)_q$. Then the operator $(L^*)_q$ is proper. Here $1 \leq q \leq \infty$ or $q = D$.

Corollary 2.2'. *The operator $(L^*)_q$ is normally solvable with a finite-dimensional kernel.*

Lemma 2.3. *Let the operator $L_\infty : E_\infty \to F_\infty$ be proper. Then the operator $L_D : E_D \to F_D$ is also proper.*

Proof. Let $L_\infty u_k = f_k$, $f_k \to f_0$ in F_∞, $u_k \in E_D$, $f_k, f_0 \in F_D$, $\|u_k\|_{E_\infty} \leq C$ for some constant C and all k. Since L_∞ is proper, then there exists a subsequence u_{k_n} and $u_0 \in E_\infty$ such that $u_{k_n} \to u_0$ in E_∞. Since $u_{k_n} \in E_D$, then u_0 also belongs to E_D. The lemma is proved. □

2. Abstract operators

Theorem 2.4. *Suppose that estimates (2.1) and (2.2) are satisfied for the operators L_∞ and $(L^*)_\infty$, respectively, in the corresponding spaces. Then L_D is a Fredholm operator.*

Proof. It follows from Lemma 2.3 that L_D is normally solvable with a finite-dimensional kernel. It remains to show that the adjoint operator $(L_D)^*$ has also a finite-dimensional kernel. We note that $\operatorname{Ker}(L^*)_\infty$ is finite dimensional by virtue of Corollary 2.2 for the operator $(L^*)_\infty$. We will show that $\operatorname{Ker}(L_D)^* \subset \operatorname{Ker}(L^*)_\infty$. Indeed, let $(L_D)^* v = 0$ for some $v \in (F_D)^*$. This means that

$$\langle v, L_D u \rangle = 0, \quad \forall u \in E_D.$$

Then for any $u \in E_1 \subset E_D$,

$$\langle v, L_1 u \rangle = \langle v, L_D u \rangle = 0.$$

The functional in the left-hand side is well defined because $v \in (F_D)^* \subset (F_1)^*$. Thus $(L_1)^* v = 0$. By definition,

$$\langle (L^*)_\infty v, u \rangle = \langle (L_1)^* v, u \rangle, \quad \forall u \in E_1. \tag{2.3}$$

Since $(L_1)^* v = 0$, then $(L^*)_\infty v$ also equals zero as an element of $(E^*)_\infty$. Indeed, if it is different from zero, then there exists $\phi \in D$ such that $\phi(L^*)_\infty v \neq 0$. On the other hand, $\phi(L^*)_\infty v \in E^*$. Hence for some $w \in E$,

$$\langle (L^*)_\infty v, \phi w \rangle = \langle \phi(L^*)_\infty v, w \rangle \neq 0.$$

This contradicts (2.3) since $\phi w \in E_1$. The theorem is proved. $\square$

Corollary 2.5. *The equation*

$$L_D u = f, \quad f \in F_D(\Omega) \tag{2.4}$$

is solvable in $E_D(\Omega)$ if and only if $\phi_i(f) = 0$ for a finite number of linearly independent functionals $\phi_i \in (F_D(\Omega))^*$ that are solutions of the homogeneous adjoint problem $(L_D)^* v = 0$.

In the remaining part of this section we study the operator L_∞. If it satisfies estimate (2.1), then it is normally solvable with a finite-dimensional kernel. We will use normal solvability of this operator and the Fredholm property of the operator L_D to show that the codimension of its image is also finite. From normal solvability of the operator L_∞ we conclude that the equation

$$L_\infty u = f, \quad f \in F_\infty \tag{2.5}$$

is solvable in E_∞ if and only if $\phi(f) = 0$ for all $\phi \in \Phi$, where Φ is a set in $(F_\infty)^*$.

Consider the functionals ϕ_i, $i = 1, \ldots, N$ that provide the solvability conditions for equation (2.4). By the Hahn-Banach theorem they can be extended from

$F_D(\Omega)$ to $(F(\Omega))_\infty$. Denote these new functionals by $\hat{\phi}_i$. Since $\hat{\phi}_i \in ((F(\Omega))_\infty)^*$, then by virtue of Lemma 10.4 (Chapter 2) we can define functionals $\tilde{\phi}_i \in ((F(\Omega))_\infty)^*$ as follows: $\tilde{\phi}_i(f) = \hat{\phi}_i(f)$ for functions $f \in (F(\Omega))_\infty$ with a bounded support,

$$\tilde{\phi}_i(f) = \lim_{k\to\infty} \hat{\phi}_i(\sum_{j=1}^{k} \theta_j f), \quad \forall f \in (F(\Omega))_\infty. \tag{2.6}$$

Here θ_j is a partition of unity. We note that the functionals $\hat{\phi}_i$ are not uniquely defined. However the functionals $\tilde{\phi}_i$ are uniquely defined. Indeed, if there are two different functionals $\hat{\phi}_i^1$ and $\hat{\phi}_i^2$ that correspond to the same $\hat{\phi}_i$, then the difference $\hat{\phi}_i^1 - \hat{\phi}_i^2$ vanishes on all functions with a bounded support. Therefore the limit in (2.6) is also zero.

By the definition of $\tilde{\phi}_i$,

$$\tilde{\phi}_i(f) = \phi_i(f), \quad \forall f \in F_D(\Omega). \tag{2.7}$$

Lemma 2.6. *The restriction ϕ_D of a functional $\phi \in \Phi$ from the solvability condition for equation (2.5) to $F_D(\Omega)$ is a linear combination of functionals ϕ_i from the solvability condition for equation (2.4).*

Proof. For any $f \in F_D(\Omega)$, the equation

$$Lu = f - \sum_{i=1}^{N} \langle \phi_i, f \rangle e_i, \tag{2.8}$$

where $e_i, i = 1, \ldots, N$ are such that $\langle \phi_i, e_j \rangle = \delta_{ij}$, $e_j \in F_D(\Omega)$, is solvable in $E_D(\Omega)$. Therefore it is also solvable in $E_\infty(\Omega)$. Hence for any $\phi \in \Phi$,

$$\phi\left(f - \sum_{i=1}^{N} \langle \phi_i, f \rangle e_i\right) = 0, \quad \forall f \in F_D(\Omega).$$

Denote $c_i = \phi(e_i)$. Then from the previous equality

$$\phi(f) = \sum_{i=1}^{N} c_i \phi_i(f), \quad \forall f \in F_D(\Omega). \tag{2.9}$$

Here $\phi_i(f) = \langle \phi_i, f \rangle$. The lemma is proved. □

Corollary 2.7. *For any $\phi \in \Phi$,*

$$\phi = \sum_{i=1}^{N} c_i \tilde{\phi}_i + \psi, \quad c_i = \phi(e_i), \tag{2.10}$$

where $\psi \in ((F(\Omega))_\infty)^*$, $\psi(f) = 0$ for any $f \in F_D(\Omega)$.

2. Abstract operators

Proof. We construct the functionals $\tilde{\phi}_i \in ((F(\Omega))_\infty)^*$ on the basis of the functionals $\phi_i \in (F_D(\Omega))^*$. Set $\psi = \phi - \sum_{i=1}^{N} c_i \tilde{\phi}_i$. From (2.7) and (2.9) we conclude that $\psi(f) = 0$ for any $f \in F_D(\Omega)$. The corollary is proved. □

Condition C. Let $Lu_n = f_n$, $(f_n - f_0)\theta \to 0$ in $(F(\Omega))_\infty$ for any infinitely differentiable function θ with a bounded support, $f_n, f_0 \in (F(\Omega))_\infty$, and $\|u_n\|_{(E(\Omega))_\infty} \leq M$. Then there exists $u_0 \in (E(\Omega))_\infty$ such that $Lu_0 = f_0$.

Lemma 2.8. *Let Condition C be satisfied. Then the functional ψ in (2.10) equals zero.*

Proof. Let $f \in (F(\Omega))_\infty$, $f_k = \sum_{i=1}^{k} \theta_i f$. The equation

$$Lu = f_k - \sum_{i=1}^{N} \langle \phi_i, f_k \rangle e_i \qquad (2.11)$$

is solvable in $E_D(\Omega)$. The operator $L_D : E_D(\Omega) \to F_D(\Omega)$ has a bounded inverse defined on the image $R(L_D) \subset F_D(\Omega)$ and acting on the subspace of $E_D(\Omega)$ supplementary to the kernel. Therefore

$$\|u_k\|_{E_D(\Omega)} \leq \|(L_D)^{-1}\| \, \|f_k - \sum_{i=1}^{N} \langle \phi_i, f_k \rangle e_i\|_{F_D(\Omega)},$$

where u_k is a solution of (2.11) in the subspace supplementary to the kernel.

We note that the norm in $F_D(\Omega)$ is the same as in $(F(\Omega))_\infty$. Hence

$$\|f_k - \sum_{i=1}^{N} \langle \phi_i, f_k \rangle e_i\|_{F_D(\Omega)} \leq C_1 \|f_k\|_{F_D(\Omega)} \leq C_2 \|f\|_{(F(\Omega))_\infty}.$$

Thus $\|u_k\|_{(E(\Omega))_\infty} \leq M$ with some constant M.

We can now use Condition C. Passing to the limit in (2.11), we obtain that the equation

$$Lu = f - \sum_{i=1}^{N} \langle \tilde{\phi}_i, f \rangle e_i \qquad (2.12)$$

is solvable in $(E(\Omega))_\infty$ for any $f \in (F(\Omega))_\infty$. Then for any $\phi \in \Phi$,

$$\phi\left(f - \sum_{i=1}^{N} \langle \tilde{\phi}_i, f \rangle e_i\right) = 0.$$

Hence

$$\phi(f) = \sum_{i=1}^{N} c_i \tilde{\phi}_i(f), \quad \forall f \in (F(\Omega))_\infty.$$

From (2.10) we conclude that $\psi = 0$. The lemma is proved. □

Thus we have proved the following theorem.

Theorem 2.9. *Suppose that the operators L_∞ and $(L^*)_\infty$ satisfy estimates (2.1) and (2.2), respectively, in the corresponding spaces, and Condition C is satisfied. Then the operator L_∞ is Fredholm. Equation (2.5) is solvable in $(E(\Omega))_\infty$ if and only if $\phi(f) = 0$ for a finite number of functionals $\phi \in ((F(\Omega))_\infty)^*$. They satisfy the homogeneous adjoint equation $(L_\infty)^*\phi = 0$. The restriction ϕ_D of these functionals to $F_D(\Omega)$ coincides with the functionals ϕ_i in the solvability conditions for equation (2.4).*

Remark 2.10. The space $((F(\Omega))_\infty)^*$ contains "bad" functionals that vanish at all functions from $F_D(\Omega)$ and do not belong to D' (Remark 10.14 of Chapter 2). Theorem 2.9 shows that these functionals do not enter the solvability conditions.

3 Elliptic problems in spaces $W_\infty^{s,p}(\Omega)$

Consider the operators

$$A_i u = \sum_{k=1}^{N} \sum_{|\alpha| \leq \alpha_{ik}} a_{ik}^\alpha(x) D^\alpha u_k, \quad i = 1, \ldots, N, \quad x \in \Omega, \tag{3.1}$$

$$B_j u = \sum_{k=1}^{N} \sum_{|\beta| \leq \beta_{jk}} b_{jk}^\beta(x) D^\beta u_k, \quad j = 1, \ldots, m, \quad x \in \partial\Omega, \tag{3.2}$$

where $u = (u_1, \ldots, u_N)$, $\Omega \subset R^n$ is an unbounded domain that satisfies Condition D. According to the definition of elliptic operators in the Douglis-Nirenberg sense we suppose that

$$\alpha_{ik} \leq s_i + t_k, \ i, k = 1, \ldots, N, \quad \beta_{jk} \leq \sigma_j + t_k, \ j = 1, \ldots, m, \ k = 1, \ldots, N$$

for some integers s_i, t_k, σ_j such that $s_i \leq 0$, $\max s_i = 0$, $t_k \geq 0$.

Denote by E the space of vector-valued functions $u = (u_1, \ldots, u_N)$, where u_j belongs to the Sobolev space $W^{l+t_j,p}(\Omega)$, $j = 1, \ldots, N$, $1 < p < \infty$, l is an integer, $l \geq \max(0, \sigma_j + 1)$, $E = \Pi_{j=1}^{N} W^{l+t_j,p}(\Omega)$. The norm in this space is defined as

$$\|u\|_E = \sum_{j=1}^{N} \|u_j\|_{W^{l+t_j,p}(\Omega)}.$$

The operator A_i acts from E into $W^{l-s_i,p}(\Omega)$, the operator B_j from E into $W^{l-\sigma_j-1/p,p}(\partial\Omega)$. Let

$$\begin{aligned} L &= (A_1, \ldots, A_N, B_1, \ldots, B_m), \\ F &= \Pi_{i=1}^{N} W^{l-s_i,p}(\Omega) \times \Pi_{j=1}^{m} W^{l-\sigma_j-1/p,p}(\partial\Omega). \end{aligned} \tag{3.3}$$

Then $L : E \to F$.

3. Elliptic problems in spaces $W^{s,p}_\infty(\Omega)$

Lemma 3.1. *The operator L_q, $1 \leq q \leq \infty$ is a bounded operator from E_q to F_q.*

The proof is standard.

We will apply the results of the previous section. The properness of the operator L_∞ is proved in Chapter 4. The estimates of the operator $(L^*)_\infty$ are obtained in Section 1 of this chapter. It remains to check Condition C (Section 2). Let $Lu_\nu = f_\nu$ ($\nu = 1, 2, \ldots$), $(f_\nu - f_0)\theta \to 0$ in $(F(\Omega))_\infty$ for any infinitely differentiable function with a bounded support as $\nu \to \infty$, $f_\nu, f_0 \in (F(\Omega))_\infty$, and

$$\|u_\nu\|_{(E(\Omega))_\infty} \leq M_0, \quad \forall \nu. \tag{3.4}$$

Let $u_\nu = (u_{1\nu}, \ldots, u_{N\nu})$. It follows from Theorem 7.3 (Chapter 2), which is also true for domains in R^n, and from (3.4) that there exists a subsequence of $u_{i\nu}$ and $u_{i0} \in W^{l+t_i,p}_\infty(\Omega)$ such that for $\epsilon > 0$,

$$u_{i\nu} \to u_{i0} \text{ in } W^{l+t_i-\epsilon,p}(\Omega) \text{ locally,} \tag{3.5}$$

$$u_{i\nu} \to u_{i0} \text{ in } W^{l+t_i,p}(\Omega) \text{ locally weakly} \tag{3.6}$$

as $\nu \to \infty$, $i = 1, \ldots, N$. We retain the same notation for the subsequence. Set $u_0 = (u_{10}, \ldots, u_{N0})$. We prove that

$$Lu_0 = f_0. \tag{3.7}$$

Indeed, we have

$$A_i u_\nu = f^d_{i\nu}, \quad i = 1, \ldots, N, \tag{3.8}$$

$$B_j u_\nu = f^b_{i\nu}, \quad i = 1, \ldots, m, \tag{3.9}$$

where $f_\nu = (f^d_{1\nu}, \ldots, f^d_{N\nu}, f^b_{1\nu}, \ldots, f^b_{m\nu})$. Write $f_0 = (f^d_{10}, \ldots, f^d_{N0}, f^b_{10}, \ldots, f^b_{m0})$. By (3.6) for any $\theta \in C^\infty_0(\Omega)$ we have

$$\theta A_i u_\nu \to \theta A_i u_0 \text{ as } \nu \to \infty \text{ weakly in } W^{l-s_i,p}(\Omega).$$

Hence

$$\theta A_i u_0 = \theta f^d_{i0} \quad (i = 1, \ldots, N).$$

Therefore

$$A_i u_0 = f^d_{i0} \quad (i = 1, \ldots, N) \text{ in } W^{l-s_i,p}(\Omega). \tag{3.10}$$

Now, consider (3.9). We can suppose that the coefficients b^β_{jk} of the operator B_j are extended to Ω in such a way that $b^\beta_{jk} \in C^{l-\sigma_j+\delta}(\Omega)$. From (3.5) it follows that for $\theta \in D$ we have

$$\theta B_j u_\nu \to \theta B_j u_0 \text{ in } W^{l-\sigma_j-\epsilon,p}(\Omega) \text{ as } \nu \to \infty.$$

Hence

$$\theta B_j u_\nu \to \theta B_j u_0 \text{ in } W^{l-\sigma_j-\epsilon-1/p,p}(\partial\Omega) \text{ as } \nu \to \infty.$$

By assumption of Condition C, $\theta f_{j\nu}^b \to \theta f_{j0}^b$ in $W^{l-\sigma_j-1/p,p}(\partial\Omega)$. Therefore

$$\theta B_j u_0 = \theta f_{j0}^b \text{ in } W^{l-\sigma_j-\epsilon-1/p,p}(\partial\Omega). \tag{3.11}$$

Since $u_0 \in (E(\Omega))_\infty$ and $f_0 \in (F(\Omega))_\infty$, then from (3.11)

$$B_j u_0 = f_{j0}^b \text{ in } W^{l-\sigma_j-1/p,p}(\partial\Omega) \quad j=1,\ldots,m.$$

This and (3.10) imply (3.7). Thus we have proved that the operator L defined by (3.3) satisfies Condition C. Hence Theorem 2.9 is applicable. We obtain the following result.

Theorem 3.2. *Let Conditions NS and NS* be satisfied. Then the realizations L_D and L_∞ of the operator L are Fredholm operators. The equation $L_D u = f$, $f \in F_D(\Omega)$ is solvable in $E_D(\Omega)$ if and only if $\phi(f) = 0$ for any solution $\phi \in (F_D(\Omega))^*$ of the problem $(L_\infty)^* \phi = 0$. The equation $L_\infty u = f$, $f \in (F(\Omega))_\infty$ is solvable in $(E(\Omega))_\infty$ if and only if $\phi(f) = 0$ for any solution $\phi \in ((F(\Omega))^*)_1$ of the problem $(L_\infty)^* \phi = 0$.*

Let v be a vector-valued function, $v \in F$, $v = (v_1^d, \ldots, v_N^d, v_1^b, \ldots, v_m^b)$. We use Definitions 2.2 and 2.3 (Chapter 4) for v_i^d in $W^{l-s_i,p}(\Omega)$ and for v_j^b in $W^{l-\sigma_j-1/p,p}(\partial\Omega)$. Denote by T_y the translation operator $T_y u(x) = u(x+y)$. Then we can define the operator with shifted coefficients, $L_y v = T_y L T_y^{-1} v$. It acts on functions defined in the shifted domain Ω_y. We will use the following condition.

Condition CL. *Let $L_{y_k} u_k = f_k$, $u_k \in (E(\Omega_{y_k}))_\infty$, $f_k \in (F(\Omega_{y_k}))_\infty$, $(f_k - f_0)\theta \to 0$ in $F(\Omega_{y_k} \to \hat{\Omega})$ for any infinitely differentiable function θ with a bounded support, $\|u_k\|_{(E(\Omega_{y_k}))_\infty} \leq M$, $L_{y_k} \to \hat{L}$. Then there exists a function $u_0 \in (E(\hat{\Omega}))_\infty$ such that $\hat{L} u_0 = f_0$.*

This condition is satisfied for the elliptic operators (cf. the proof of Theorem 2.10, Chapter 4).

Theorem 3.3. *If the operator L_∞ satisfies the Fredholm property, Condition CL and Condition NS, then any limiting operator $\hat{L}_\infty$ is invertible.*

Proof. For any function $f_0 \in (F(\hat{\Omega}))_\infty$ there exists a sequence of functions $f_k \in (F(\Omega))_\infty$, $\|f_k\|_{(F(\Omega))_\infty} \leq M$ and of points $y_k \in \Omega$ $|y_k| \to \infty$ such that

$$(f_k(x+y_k) - f_0(x))\theta \to 0 \text{ in } F(\Omega_{y_k} \to \hat{\Omega})$$

for any infinitely differentiable function θ with a finite support. Indeed, let $f_0 \in (F(\hat{\Omega}))_\infty$. Then $f_0 = (f_{10}^d, \ldots, f_{N0}^d, f_{10}^b, \ldots, f_{m0}^b)$, where

$$f_{i0}^d \in W_\infty^{l-s_i,p}(\hat{\Omega}), \ i=1,\ldots,N, \quad f_{j0}^b \in W_\infty^{l-\sigma_j-1/p,p}(\partial\hat{\Omega}), \ i=1,\ldots,m.$$

3. Elliptic problems in spaces $W^{s,p}_\infty(\Omega)$

We can extend these functions to R^n in such a way that for the extended functions $\tilde{f}^d_{i0}$ and $\tilde{f}^b_{j0}$ we have

$$\tilde{f}^d_{i0} \in W^{l-s_i,p}_\infty(R^n), \quad \tilde{f}^b_{j0} \in W^{l-\sigma_j,p}_\infty(R^n).$$

Let $y_k, k = 1, 2, \ldots$ be a sequence such that $y_k \in \Omega$, $|y_k| \to \infty$, $\Omega_{y_k} \to \hat{\Omega}$, where Ω_{y_k} are the shifted domains. Set

$$\tilde{f}^d_{ik}(x) = \tilde{f}^d_{i0}(x - y_k), \; i = 1, \ldots, N, \quad \tilde{f}^b_{jk} = \tilde{f}^b_{j0}(x - y_k), \; j = 1, \ldots, m.$$

Let $f^d_{ik}(x)$ be the restriction of $\tilde{f}^d_{ik}(x)$ to Ω, f^b_{jk} be the trace of $\tilde{f}^b_{jk}$ on $\partial\Omega$. Then it is easy to verify that the sequence

$$f_k(x) = (f^d_{1k}(x), \ldots, f^d_{Nk}(x), f^b_{1k}(x), \ldots, f^b_{mk}(x))$$

satisfies the conditions above.

Since the operator $L_\infty : E_\infty(\Omega) \to F_\infty(\Omega)$ satisfies the Fredholm property, then the equation

$$L_\infty u = f_m - \sum_{i=1}^{N} \langle v_i, f_m \rangle e_i \qquad (3.12)$$

is solvable in $(E(\Omega))_\infty$. Here v_i, $i = 1, \ldots, N$ are all linearly independent solutions of the homogeneous adjoint equation, $(L_\infty)^* v_i = 0$, and $e_i \in F_\infty$, $i = 1, \ldots, N$ are functions biorthogonal to the functionals v_j, $j = 1, \ldots, N$. We can suppose that e_i, $i = 1, \ldots, N$ have bounded supports (see Lemma 5.1 below).

Denote by u_m the solution of the equation (3.12). The numbers $a_{im} = \langle v_i, f_m \rangle$ are uniformly bounded because the sequence f_m is bounded in $(F(\Omega))_\infty$. The equation

$$L_{y_m} v - f_m(x + y_m) - \sum_{i=1}^{N} a_{im} e_i(x + y_m) \qquad (3.13)$$

has a solution $v_m(x) = u_m(x + y_m) \in (E(\Omega_{y_m}))_\infty$. Since $e_i(x + y_m) \to 0$ in $F(\Omega_{y_m} \to \hat{\Omega})$, then by virtue of Condition CL there exists a solution $v_0 \in (E(\hat{\Omega}))_\infty$ of the equation $\hat{L}_\infty v_0 = f_0$. It remains to note that the homogeneous equation has only the zero solution since Condition NS is necessary for normal solvability. The theorem is proved. □

Remark 3.4. In the proof of the theorem we use the existence of functions e_i biorthogonal to functionals v_j and such that they have bounded supports. We will prove this assertion in Lemma 5.1 below using Condition NS. Therefore we have to assume in the formulation of Theorem 3.3 that it is satisfied. Otherwise we can assume that Conditions NS and NS* are satisfied and not to assume that the operator is Fredholm (see Theorem 3.2).

If instead of the operator L_∞ we consider the operator L_D, then the functions e_i belong, by assumption, to F_D. Though they do not necessarily have bounded supports, the convergence

$$e_i(x + y_m) \to 0 \text{ in } F(\Omega_{y_m} \to \hat{\Omega})$$

still holds. This allows us to prove the following theorem.

Theorem 3.5. *If the operator L_D satisfies the Fredholm property and Condition CL, then any limiting operator $\hat{L}_D$ is invertible.*

The proof is the same as the proof of the previous theorem.

Theorem 3.6. *If all limiting operators $\hat{L}_D$ are invertible, then Conditions NS and NS* for the operator L_∞ are satisfied, and consequently the operators L_D and L_∞ are Fredholm.*

Proof. We prove first that for all limiting operators $\hat{L}$ the equation

$$\hat{L}_1 u = f, \quad u \in E_1(\hat{\Omega}), \quad f \in F_1(\hat{\Omega}) \tag{3.14}$$

is solvable. Indeed, consider the equation

$$\hat{L} u = \theta_j f, \tag{3.15}$$

where θ_j, $j = 1, 2, \ldots$ is a partition of unity, $f \in F_1(\hat{\Omega})$. Since $F_1(\hat{\Omega}) \subset F_D(\hat{\Omega})$, then there exists a solution $u = u_j \in E_D(\hat{\Omega})$ of equation (3.15).

Let $\omega_\delta(x) = e^{\delta \sqrt{1+|x|^2}}$. Then according to Lemma 5.4 below, for $\delta > 0$ sufficiently small the following estimate holds:

$$\|u_j(\cdot)\omega_\delta(\cdot - y_j)\|_{E_1(\hat{\Omega})} \leq C \|f\theta_j\|_{F(\hat{\Omega})},$$

where B_j is a unit ball with its center at y_j, $\operatorname{supp} \theta_j \subset B_j$ and the constant C is independent of j. Since

$$\|u_j\|_{E_1(\hat{\Omega})} \leq C_1 \|u_j(\cdot)\omega_\delta(\cdot - y_j)\|_{E_1(\hat{\Omega})},$$

we get

$$\|u_j\|_{E_1(\hat{\Omega})} \leq C_2 \|f\theta_j\|_{F(\hat{\Omega})}.$$

It follows that the series $u = \sum_{j=1}^\infty u_j$ is convergent in $E_1(\hat{\Omega})$, and

$$\|u\|_{E_1(\hat{\Omega})} \leq \sum_{j=1}^\infty \|u_j\|_{E_1(\hat{\Omega})} \leq C_2 \sum_{j=1}^\infty \|f\theta_j\|_{F(\hat{\Omega})} = C_2 \|f\|_{F_1(\hat{\Omega})}.$$

From (3.15) we conclude that

$$\hat{L}_1 u = \sum_{j=1}^\infty \hat{L}_1 u_j = \sum_{j=1}^\infty \theta_j f = f.$$

Therefore we have proved that equation (3.14) has a solution for any $f \in F_1(\hat{\Omega})$. Hence the equation
$$(\hat{L}_1)^* v = 0, \quad v \in (F_1(\hat{\Omega}))^*$$
has only the zero solution. Since $(\hat{L}_1)^* = (\hat{L}^*)_\infty$, then the equation
$$(\hat{L}^*)_\infty v = 0, \quad v \in (F^*(\hat{\Omega}))_\infty$$
also has only the zero solution. Thus we have proved that Condition NS* is satisfied.

We now prove that Condition NS is satisfied. Let u be a solution of the equation
$$\hat{L}_\infty u = 0, \quad u \in E_\infty(\hat{\Omega})$$
for a limiting operator $\hat{L}$. Then $\tilde{u} = S_{-\delta} u$ is a solution of the equation
$$\hat{L}_\delta \tilde{u} = 0, \tag{3.16}$$
where $\hat{L}_\delta = S_{-\delta} \hat{L} S_\delta$, and S_δ is an operator of multiplication by $\omega_\delta(x)$. Equation (3.16) can be written in the form $(\hat{L} + \delta K)\tilde{u} = 0$, where $K : E_D(\hat{\Omega}) \to F_D(\hat{\Omega})$ is a bounded operator. Since the operator $\hat{L}$ is invertible by the assumption of the theorem, then for δ sufficiently small, $\hat{L} + \delta K$ is also invertible. Hence $\tilde{u} = 0$, and consequently, $u = 0$. The theorem is proved. $\square$

4 Exponential decay

Denote by S the operator of multiplication by $\omega_\mu(x) = \exp(\mu \sqrt{1 + |x|^2})$, where μ is a complex number. Let $L_\mu = SLS^{-1}$. If we consider the differential operator L as acting from E_∞ into F_∞, then L_μ acts in the same spaces and $L_\mu = L + \mu K(\mu)$, where $K(\mu) : E_\infty \to F_\infty$ is a bounded operator, which depends on μ polynomially. Thus, L_μ is a holomorphic operator function with respect to the complex variable μ.

If L is a Fredholm operator, then L_μ is also Fredholm in some domain G of the μ-plane, and $0 \in G$. Its index $\kappa(L_\mu)$ is constant in G, $\alpha(L_\mu)$ and $\beta(L_\mu)$ are also constant with a possible exception of some isolated points where they have greater values (see Chapter 1, Section 2.4).

Lemma 4.1. *Equation*
$$L_\mu u = 0 \tag{4.1}$$
has the same number of linearly independent solutions for all $\mu \in G$.

Proof. Suppose that for some $\mu_0 \in G$ the number of linearly independent solutions of equation (4.1) in E_∞ is greater than any other μ in a small neighborhood of μ_0. Denote these solutions by $u_1, \ldots, u_m$. Then
$$\tilde{u}_i \equiv u_i e^{\delta \sqrt{1+|x|^2}} \in E_\infty, \quad i = 1, \ldots, m$$

for real negative δ. On the other hand, $\tilde{u}_i$ are linearly independent solutions of the equation $L_{\mu_0+\delta} u = 0$. Indeed,

$$L_{\mu_0+\delta}\tilde{u}_i = e^{\delta\sqrt{1+|x|^2}} L_{\mu_0} e^{-\delta\sqrt{1+|x|^2}} (u_i e^{\delta\sqrt{1+|x|^2}}) = e^{\delta\sqrt{1+|x|^2}} L_{\mu_0} u_i = 0.$$

We obtain a contradiction with the assertion that the number of solutions is the same for all μ except for isolated values. This contradiction proves the lemma. $\square$

Corollary 4.2. *There exists μ, Re $\mu > 0$ such that all solutions of the equation $Lu = 0$ can be represented in the form $u = v \exp(-\mu\sqrt{1+|x|^2})$, where $v \in E_\infty$.*

Proof. Let $u_1, \ldots, u_m$ be solutions of the equation $Lu = 0$. Consider the equation $L_\mu u = 0$ with Re $\mu > 0$. Denote its solutions by $v_1, \ldots, v_m$. Then $w_i = v_i \exp(-\mu\sqrt{1+|x|^2})$ are solutions of the equation $Lu = 0$. Since the number of linearly independent solutions of these two equations is the same, then u_i is a linear combination of w_j. The corollary is proved. $\square$

The assertion of the corollary signifies exponential decay of solutions at infinity. Consider now the adjoint operator $L^* : (F_\infty)^* \to (E_\infty)^*$. Write $L_\mu^* = S^{-1}L^*S$. This operator is adjoint to L_μ. Since the index $\kappa(L_\mu)$ is independent of μ for all $\mu \in G$, and also the dimension of the kernel $\alpha(L_\mu)$, then the codimension of the image $\beta(L_\mu)$ is also independent of μ. On the other hand, the kernel of the adjoint operator $\alpha(L_\mu^*)$ equals $\beta(L_\mu)$. Therefore we have proved the following lemma.

Lemma 4.3. *The dimension $\alpha(L_\mu^*)$ of the kernel of the operator L_μ^* is independent of μ for all $\mu \in G$.*

Corollary 4.4. *There exists μ, Re $\mu < 0$ such that all solutions of the equation $L^*\phi = 0$ can be represented in the form $\phi = \psi \exp(\mu\sqrt{1+|x|^2})$, where $\psi \in (F_\infty)^*$.*

The proof is the same as the proof of Corollary 4.2. We note that the spaces E and F are not supposed to be reflexive.

Theorem 4.5. *Let the operator $L : E_\infty \to F_\infty$ be Fredholm. Then there exists a domain G of the complex plane such that $0 \in G$, and for every $\mu \in G$ the dimensions of the kernel of the operators $L_\mu = S^{-1}LS$ and L_μ^* are independent of μ. Every solution of the equation $Lu = 0$ can be represented in the form $u = v \exp(-\mu\sqrt{1+|x|^2})$, where $v \in E_\infty$, $\mu \in G$. Every solution of the equation $L^*\phi = 0$ can be represented in the form $\phi = \psi \exp(\mu\sqrt{1+|x|^2})$, where $\psi \in (F_\infty)^*$.*

The domain G is the set of all μ which contains 0 and where the operator L_μ is Fredholm, i.e., Conditions NS and NS* are satisfied. For some classes of elliptic problems satisfying the Fredholm property, exponential decay of solutions is known (see [433] and the references therein). In Section 2.3 of Chapter 4 (see also [564]) we proved exponential decay if the operator is normally solvable with a finite-dimensional kernel but not necessarily Fredholm.

5 The space E_q

Suppose that the operator $L_\infty : E_\infty \to F_\infty$ satisfies the Fredholm property. Then the equation

$$L_\infty u = f \tag{5.1}$$

is solvable if and only if

$$\langle f, v_i \rangle = 0, \quad i = 1, \ldots, N, \tag{5.2}$$

and the homogeneous equation ($f = 0$) has a finite number of linearly independent solutions. Here v_i, $i = 1, \ldots, N$ are all linearly independent solutions of the homogeneous adjoint equation, $(L_\infty)^* v_i = 0$, and $(L_\infty)^* : (F_\infty)^* \to (E_\infty)^*$ is the adjoint operator.

We study in this section the operator L acting from E_q into F_q. To show its dependence on the spaces we denote it by L_q. We begin with some auxiliary results. Let $e_i \in F_\infty$, $i = 1, \ldots, N$ be functions biorthogonal to the functionals $v_j, j = 1, \ldots, N$,

$$\langle e_i, v_j \rangle = \delta_{ij}, \tag{5.3}$$

where δ_{ij} is the Kronecker symbol.

Lemma 5.1. *There exist functions $e_i, i = 1, \ldots, N$ with bounded supports satisfying* (5.3).

Proof. Let $e_i \in F_\infty$, $i = 1, \ldots, N$ satisfy (5.3). We will construct new functions $\tilde{e}_i \in E_\infty$, $i = 1, \ldots, N$ with bounded supports such that

$$\langle \tilde{e}_i, v_j \rangle = \delta_{ij}. \tag{5.4}$$

Denote

$$\hat{e}_i = \sum_{k=1}^{m} e_i \theta_k, \quad i = 1, \ldots, N,$$

where θ_i is a partition of unity. We put $\tilde{e}_i = c_{i1} \hat{e}_1 + \cdots + c_{iN} \hat{e}_N$. Then (5.4) is a system of equations with respect to $c_{i1}, \ldots, c_{iN}$. Its matrix has the elements

$$\langle \hat{e}_i, v_j \rangle = \left\langle \sum_{k=1}^{m} e_i \theta_k, v_j \right\rangle, \quad j = 1, \ldots, N.$$

Since $v_j \in (F_\infty)^*_\omega$, then

$$\left\langle \sum_{k=1}^{m} e_i \theta_k, v_j \right\rangle \to \delta_{ij}$$

as $m \to \infty$. Therefore for m sufficiently large the determinant of this matrix is different from 0, and the system has a solution. The lemma is proved. $\square$

Lemma 5.2. *If $w \in E_\infty$, then $u = S^{-1}w \in E_1$ for any $\mu > 0$, and $\|u\|_{E_1} \leq C(\mu)\|w\|_{E_\infty}$.*

The proof of the lemma is based on the definition of the spaces and on the properties of multiplicators.

Lemma 5.3. *Let an operator L, acting from a Banach space E into another space F, have a bounded inverse defined on its image $R(L) \subset F$. Suppose that the equation $L_\mu u = f$ has a solution, where $L_\mu = L + \mu K$, $K : E \to F$ is a bounded operator, $\|K\| \leq M$. Then for μ sufficiently small $\|u\|_E \leq C\|f\|_F$, where the constant C depends on μ and M but does not depend on the operator K.*

Proof. Since the equation $Lu + \mu K u = f$ has a solution, then $f - \mu K u \in R(L)$. Therefore $u = L^{-1}(f - \mu K u)$. The assertion of the lemma follows from the estimate

$$\|u\|_E \leq \|L^{-1}\| \|f - \mu K u\|_F \leq \|L^{-1}\|(\|f\|_F + |\mu|\|K\|\|u\|_E).$$

The lemma is proved. □

We generalize here the approach developed in [362] for the operators acting in $H^s(\mathbb{R}^n)$. As above, we use the function $\omega_\delta(x) = \exp(\delta\sqrt{1 + |x|^2})$.

Lemma 5.4. *Let θ_j be a partition of unity, v_i and e_i be the same as in Lemma 5.1, e_i have a bounded support. Then for any $f \in F_\infty$ there exists a solution u_j of the equation*

$$Lu = \theta_j f - \sum_{i=1}^{N} \langle \theta_j f, v_i \rangle e_i, \tag{5.5}$$

and for δ sufficiently small the following estimate holds:

$$\|u_j(\cdot)\omega_\delta(\cdot - y_j)\|_{E_q} \leq C\|f\theta_j\|_F, \tag{5.6}$$

where B_j is a unit ball with the center at y_j, $\operatorname{supp} \theta_j \subset B_j$, and the constant C is independent of j.

Proof. Since the operator $L : E_\infty \to F_\infty$ satisfies the Fredholm property, then the equation

$$Lu = g - \sum_{i=1}^{N} \langle g, v_i \rangle e_i \tag{5.7}$$

is solvable for any $g \in F_\infty$. Let $\operatorname{supp} g \in B_j$. Consider the function

$$\tilde{g}(x) = \left(g(x) - \sum_{i=1}^{N} \langle g, v_i \rangle e_i(x) \right) \omega_\delta(x - y_j).$$

We show that its norm in F_∞ is independent of j. We note first of all that $\omega_\delta(x - y_j)$ is bounded in B_j together with all derivatives independently of j. Therefore

$$\|g(x)\omega_\delta(x - y_j)\|_{F_\infty} \leq C\|g(x)\|_{F_\infty}.$$

5. The space E_q

with a positive constant C independent of j. We use here the fact that the norm of a multiplier in Ω and $\partial\Omega$ can be estimated by the norm in C^k. We have next

$$|\langle g, v_i \rangle| = |\langle g, \psi_j v_i \rangle| \equiv S$$

where $\psi_j = \psi(x - y_j)$ is a function with a finite support equal 1 in B_j,

$$S = |\langle g, \omega_{-\mu}(x)\psi_j w_i \rangle|$$

where $w_i \in (F_\infty)^*$ (see Corollary 4.4),

$$S \leq \|g\|_{F_\infty} \|\omega_{-\mu}(x)\psi_j w_i\|_{(F_\infty)^*} \leq \|g\|_{F_\infty} \|\omega_{-\mu}(x)\psi_j\|_M \|w_i\|_{(F_\infty)^*},$$

where $\|\cdot\|_M$ is the norm in the space of multipliers. By virtue of the properties of this norm

$$\|\omega_{-\mu}(x)\psi_j\|_M \leq K\|\omega_{-\mu}(x+y_j)\psi(x)\|_M \leq C\omega_{-\mu}(y_j)$$

with some constants K and C independent of j, $\mu > 0$. For $\delta \leq \mu$ the product $\omega_{-\mu}(y_j)\omega_\delta(x - y_j)$ is bounded independently of $y_j \in \mathbb{R}^n$ and of $x \in \operatorname{supp} e_i(x)$. Hence $\|\tilde{g}\|_{F_\infty} \leq C\|g\|_{F_\infty}$, where the constant C depends on the diameter of the supports of e_i but is independent of j.

Since u is a solution of equation (5.7), then $\tilde{u} = S_\delta u$ is a solution of the equation

$$L_\delta \tilde{u} = \tilde{g}, \tag{5.8}$$

where $L_\delta = S_\delta L S_{-\delta}$, and S_δ is the operator of multiplication by $\omega_\delta(x - y_j)$. On the other hand, $L_\delta = L + \delta K$, where K is a bounded operator, $\|K\| \leq C$, where C does not depend on j and on δ for δ sufficiently small. By virtue of Lemma 5.3 the solution of (5.8), which belongs to the subspace supplementary to the kernel of the operator L, admits the estimate

$$\|\tilde{u}\|_{E_\infty} \leq C_1 \|\tilde{g}\|_{F_\infty} \leq C_2 \|g\|_{F_\infty}$$

independent of j. Let $\delta = \delta_1 + \delta_2$, where δ_1 and δ_2 are positive. Then

$$\|u(x)\omega_{\delta_1}(x - y_j)\|_{E_q} = \|\tilde{u}(x)\omega_{-\delta_2}(x - y_j)\|_{E_q} \leq C_3 \|\tilde{u}(x)\|_{E_\infty} \leq C_4 \|g\|_{F_\infty}.$$

Applying this estimate to equation (5.5), we obtain

$$\|u(x)\omega_{\delta_1}(x - y_j)\|_{E_q} \leq C_5 \|f\theta_j\|_{F_\infty} \leq C_6 \|f\theta_j\|_F.$$

The lemma is proved. $\square$

Assumption 5.5. Let $u_j \in E_q$, $j = 1, 2, \ldots$, and

$$\sum_{j=1}^{\infty} \|u_j \, \omega_\delta(x - y_j)\|_{E_q}^q < \infty.$$

Then the series $u = \sum_{j=1}^{\infty} u_j$ is convergent, and the following estimate holds:

$$\|u\|_{E_q}^q \leq C \sum_{j=1}^{\infty} \|u_j\, \omega_\delta(x-y_j)\|_{E_q}^q. \tag{5.9}$$

If this assumption is satisfied, then from the estimate in Lemma 5.4 we obtain: $\|u\|_{E_q} \leq C\|f\|_{F_q}$. Therefore for any $f \in F_q (\subset F_\infty)$ there exists a solution $u \in E_q$ of the equation

$$Lu = f - \sum_{i=1}^{N} \langle f, v_i \rangle e_i. \tag{5.10}$$

Hence the operator L_q is normally solvable and the codimension of its image is finite. Its kernel is also finite dimensional since it is true for the operator L_∞. Hence L_q is a Fredholm operator.

We note that estimate (5.9) characterizes the function spaces and it is not related to the operators under consideration. In the remaining part of this section we show that it is satisfied for Sobolev spaces.

Lemma 5.6. (Elementary inequality). Let $u_i \geq 0$. Then

$$(u_1^s + u_2^s + \cdots)^{1/s} \leq u_1 + u_2 + \cdots \quad (s \geq 1),$$
$$(u_1^s + u_2^s + \cdots)^{1/s} \geq u_1 + u_2 + \cdots \quad (s \leq 1).$$

(See for example [226].)

Lemma 5.7. Let $u = \sum_{i=1}^{\infty} u_i$, y_i be the centers of an orthogonal lattice in $\mathbb{R}^n$, $1 \leq p < \infty$. Then the following estimate holds:

$$\|u\|_{L^p(\mathbb{R}^n)}^p \leq C \sum_{i=1}^{\infty} \|u_i\, \omega_\delta(x-y_i)\|_{L^p(\mathbb{R}^n)}^p. \tag{5.11}$$

Proof. Let $k = [p] + 1$, $p = ks$. Here k is an integer, $s < 1$. If p is an integer, we do not need to introduce k. All estimates below can be done directly for p. We have

$$\|u\|_{L^p(\mathbb{R}^n)}^p = \int_{\mathbb{R}^n} |u|^{ks} dx = \int_{\mathbb{R}^n} \left|\sum_{i=1}^{\infty} u_i\right|^{ks} dx \leq \int_{\mathbb{R}^n} \left(\sum_{i=1}^{\infty} |u_i|^s\right)^k dx$$
$$= \int_{\mathbb{R}^n} \sum_{i_1, i_2, \ldots, i_k = 1}^{\infty} |u_{i_1}|^s |u_{i_2}|^s \ldots |u_{i_k}|^s dx. \tag{5.12}$$

By virtue of the inequality between the geometrical and arithmetical mean values,

$$|u_{i_1}|^s |u_{i_2}|^s \ldots |u_{i_k}|^s \leq \frac{1}{k} \left(|u_{i_1}|^{ks} + \cdots + |u_{i_k}|^{ks}\right).$$

5. The space E_q

The same inequality with any positive a_{i_j} gives

$$|u_{i_1}|^s |u_{i_2}|^s \ldots |u_{i_k}|^s \leq \frac{1}{k}\left(|u_{i_1}|^{ks} a_{i_1}^{k-1} a_{i_2}^{-1} \ldots a_{i_k}^{-1} + \cdots + |u_{i_k}|^{ks} a_{i_1}^{-1} a_{i_2}^{-1} \ldots a_{i_k}^{k-1}\right). \tag{5.13}$$

Put $a_{i_k}(x) = \omega_{\delta_1}(x - y_{i_k})$. Then

$$\sum_{i_k=1}^{\infty} a_{i_k}^{-1}(x) \leq C \quad \forall x,$$

and substituting (5.13) into the right-hand side of (5.12) and taking into account that there are k similar summands, we obtain

$$\int_{\mathbb{R}^n} \sum_{i_1, i_2, \ldots, i_k} |u_{i_1}|^s |u_{i_2}|^s \ldots |u_{i_k}|^s dx \leq \sum_{i_1, i_2, \ldots, i_k} \int_{\mathbb{R}^n} |u_{i_1}|^{ks} a_{i_1}^{k-1} a_{i_2}^{-1} \ldots a_{i_k}^{-1} dx$$

$$\leq C^{k-1} \int_{\mathbb{R}^n} \sum_{i_1} |u_{i_1}|^p \omega_{\delta_1}(x - y_{i_1})^{k-1} dx.$$

Replacing C^{k-1} by C, we obtain (5.11) for $\delta = \delta_1(k-1)/p$. The lemma is proved. □

Similarly we prove the lemma for the spaces $W^{l,p}(\mathbb{R}^n)$ and $W^{l,p}(\Omega)$ with an integer $l \geq 0$.

Lemma 5.8. *The estimate*

$$\|u\|_{W^{k,p}(\Omega)}^p \leq C \sum_{i=1}^{\infty} \|u_i \, \omega_\delta(x - y_i)\|_{W^{k,p}(\Omega)}^p \tag{5.14}$$

holds with $1 \leq p < \infty$ and an integer $k \geq 0$.

This lemma proves that Assumption 5.5 holds for Sobolev spaces. Thus we can formulate the following theorem.

Theorem 5.9. *Suppose that Conditions NS and NS* are satisfied. Then for $q = p$ the equation $L_q u = 0$ has a finite number of linearly independent solutions in $(E(\Omega))_q$, and the equation*

$$Lu = f, \quad f \in (F(\Omega))_q$$

has a solution $u \in (E(\Omega))_q$ if and only if

$$\langle f, v_i \rangle = 0, \quad i = 1, \ldots, N,$$

where $v_i \in (F_q)^*$ are linearly independent solutions of the equation

$$(L_q)^* v = 0. \tag{5.15}$$

Proof. Equation (5.15) should be considered in $(F_\infty)^*$. However, if $v_i \in (F_\infty)^*$, then $v_i \in (F_q)^*$. Moreover, all solutions of this equation from $(F_q)^*$ belong also to $(F_\infty)^*$. Indeed, suppose that there exists $w \in (F_q)^*$ such that $L^*w = 0$ and w is not a linear combination of v_i, $i = 1, \ldots, N$. Then we can find $g \in F_q$ such that

$$v_i(g) = 0, \ i = 1, \ldots, N, \ w(g) \neq 0.$$

By virtue of the solvability conditions, the equation $L_q u = g$ has a solution in E_q. Applying the functional w to both its sides, we obtain a contradiction.

We finally recall that if $E = W^{k,q}$, then $E_q = E = W^{k,q}$, $(1 < q < \infty)$. Thus we obtain the Fredholm property for elliptic operators in Sobolev spaces. The theorem is proved. $\square$

Remark 5.10. Solvability conditions in the spaces $W^{k,q}$ do not depend on q.

6 The space E_q. Continuation

In this section we prove the main theorem about the Fredholm property of elliptic operators in the spaces E_q. We begin with the following lemma.

Lemma 6.1. *Let E be a Banach space such that D is dense in E, $\phi_i \in E^*$ be linearly independent functionals, $i = 1, \ldots, N$, and $\phi_i(f) = 0$ for some $f \in E$. Then for any $\epsilon > 0$ there exists $f_0 \in D$ such that $\|f - f_0\|_E \leq \epsilon$ and $\phi_i(f_0) = 0$, $i = 1, \ldots, N$.*

Proof. We first show that there exists a system of functions θ_j, $j = 1, \ldots, N$ biorthogonal to ϕ_i and such that $\theta_j \in D$. To do this we note that there exist functions $\theta_j \in D$ such that the matrix $\Phi_N = (\phi_i(\theta_j))$ is invertible. We prove it by induction on the number of functionals. For a single functional it is obvious. Suppose that for the functionals $\phi_1, \ldots, \phi_{N-1}$ there exist functions θ_j, $j = 1, \ldots, N-1$ such that the corresponding matrix Φ_{N-1} is invertible. We show that for a functional ϕ_N linearly independent with the functionals $\phi_1, \ldots, \phi_{N-1}$ we can choose θ_N such that the matrix Φ_N is invertible. Indeed, otherwise, from the equality of its determinant to zero we obtain

$$c_N \phi_N(\theta_N) = c_1 \phi_1(\theta_N) + \cdots + c_{N-1} \phi_{N-1}(\theta_N), \ \forall \theta_N \in D,$$

where the coefficients c_j are determined by $\phi_i(\theta_j)$ with $j = 1, \ldots, N-1$. We note that $c_N \neq 0$ since $\det \Phi_{N-1} \neq 0$. Hence ϕ_N is linearly dependent of $\phi_1, \ldots, \phi_{N-1}$ since D is dense in E. This contradiction proves the existence of functions $\theta_j \in D$ such that the matrix Φ_N is invertible.

The construction of the biorthogonal system of functions is now obvious. We put

$$\tilde{\theta}_j = k_1 \theta_1 + \cdots + k_N \theta_N$$

and choose k_i such that $\phi_i(\tilde{\theta}_j) = \delta_{ij}$. We omit the tilde in what follows.

6. The space E_q. Continuation

Let $f \in E$ be such that $\phi_i(f) = 0$, $i = 1, \ldots, N$. Consider a sequence $f_n \in D$ converging to f. Put

$$\tilde{f}_n = f_n - \sum_{j=1}^{N} \phi_j(f_n)\theta_j.$$

Then $\tilde{f}_n \in D$ and $\phi_i(\tilde{f}_n) = 0$. Moreover, $\tilde{f}_n$ converges to f. As a function f_0 from the formulation of the lemma we take $\tilde{f}_n$ for n sufficiently large. The lemma is proved. □

Since D is dense in E_q, then the lemma is applicable, and we can choose a system of functions $e_j \in D$, $j = 1, \ldots, N$ biorthogonal to functionals $v_i \in (F_q)^*$, $i = 1, \ldots, N$. We can use Lemma 5.4 for these spaces. If we assume that the operator $L_q : E_q \to F_q$ satisfies the Fredholm property, then the equation

$$Lu = \theta_j f_j - \sum_{i=1}^{N} \langle \theta_j f_j, v_i \rangle e_i \qquad (6.1)$$

is solvable in E_q for any $f_j \in F_q$, and its solution u_j satisfies the estimate

$$\|u_j(\cdot)\omega_\delta(\cdot - y_j)\|_{E_q} \leq C\|f_j \theta_j\|_F, \qquad (6.2)$$

where C depends on the diameters of the supports of e_i but is independent of j, and the support of θ_j belongs to a unit ball B_j with the center at y_j. Since $E_q \subset E_\infty$, we have also the estimate

$$\|u_j(\cdot)\omega_\delta(\cdot - y_j)\|_{E_\infty} \leq C\|f_j \theta_j\|_F. \qquad (6.3)$$

Let $\theta_0(x) \in C_0^\infty(R^n)$, supp $\theta_0 \subset B_0$, where B_0 is the unit ball with its center at the origin, $f_0 \in (F(\hat{\Omega}))_q$. We use the construction similar to that in the proof of Theorem 3.3. We extend the function f_0 to $F_q(R^n)$. Let it be $\tilde{f}_0$. Write

$$\tilde{f}_j(x) = \tilde{f}_0(x - y_j), \quad \theta_j(x) = \theta_0(x - y_j).$$

Then supp $\theta_j \subset B_j$. As functions $f_j(x)$, we take the restrictions of $\tilde{f}_j(x)$ to Ω. It can be proved that

$$(\theta_j(x + y_j)f_j(x + y_j) - \theta_0(x)f_0(x))\theta \to 0 \text{ in } F(\Omega_{y_j} \to \hat{\Omega})$$

for any $\theta \subset C_0^\infty(R^n)$, where $\hat{\Omega}$ is a limiting domain. This convergence is defined in Section 2.2 of Chapter 4.

Consider now sequences θ_j and f_j such that

$$\|\theta_j f_j\|_F \leq M$$

with some constant M, and where $y_j \to \infty$. This means that the support of θ_j moves to infinity. Instead of this, we can shift the domain Ω in such a way that

B_j does not change. Let it be the unit ball with the center at y_0. As in the proof of Theorem 3.3 we can pass to the limit in equation (6.1):

$$\hat{L}u = \theta_0 \hat{f}. \tag{6.4}$$

The second term in the right-hand side disappears since the functions e_i have bounded supports. Since the sequence u_j is uniformly bounded in E_∞, and the operator L_∞ satisfies Condition CL, then equation (6.4) has a solution $u_0 \in E_\infty(\hat{\Omega})$. The sequence u_j converges to u_0 locally weakly in E_∞.

On the other hand, the sequence $v_j(x) = u_j(x)\omega_\delta(x - y_0)$ is uniformly bounded in E_∞. Therefore, there exists its subsequence that converges locally weakly to some $v_0 \in E_\infty(\hat{\Omega})$. Hence $u_0(x)\omega_\delta(x - y_0) \in E_\infty(\hat{\Omega})$. As in the proof of Lemma 5.4 we conclude that

$$\|u_0(x)\omega_{\delta_1}(x - y_0)\|_{E_q} \leq C_1 \|\theta_0 \hat{f}\|_F$$

for some positive constant C_1 and $0 < \delta_1 \leq \delta$.

Let θ^j be a partition of unity. As above, the equation

$$\hat{L}u = \theta^j \hat{f} \tag{6.5}$$

has a solution. Denote it by u^j. Then $u = \sum_{j=1}^\infty u^j$ is a solution of the equation $Lu = f$. Lemmas 5.7 and 5.8 allow us to conclude that $u \in E_q$ for $q = p$. Thus we have proved the following lemma.

Lemma 6.2. *Let the operator L_q be Fredholm, $q = p$. Then any limiting problem $\hat{L}_q u = f$, $x \in \hat{\Omega}$ is solvable in E_q for any $f \in F_q$.*

We recall that for $q = p$ the spaces E_q and F_q coincide, respectively, with the spaces E and F.

Theorem 6.3. *Let q be a given number, $1 < q < \infty$, $q = p$, L be an elliptic operator. Then the following assertions are equivalent:*

(i) *The operator L_q is Fredholm,*
(ii) *All limiting operators $\hat{L}_q$ are invertible,*
(iii) *Conditions NS and NS* are satisfied.*

Proof. 1. (i) $\to$ (ii). Consider the equation

$$\hat{L}_q u = f_0, \quad u \in E_q(\hat{\Omega}), \quad f_0 \in F_q(\hat{\Omega}).$$

The solvability of this equation for any $f_0 \in F_q(\hat{\Omega})$ follows from Lemma 6.2. It remains to prove that the equation

$$\hat{L}_q u = 0, \quad u \in E_q(\hat{\Omega}) \tag{6.6}$$

has only the zero solution. Suppose that it is not true. To obtain a contradiction it is sufficient to prove that the operator $L_q : E_q(\Omega) \to F_q(\Omega)$ is not proper. Consider a nonzero solution $u = u_0$ of equation (6.6). We can suppose that u_0 is extended to $E_q(\mathbb{R}^n)$. Let $v_n(x) = \phi_n(x)u_0(x + x_n)$, where $\phi_n(x)$ are functions with compact supports, $x_n \in \Omega$, $|x_n| \to \infty$ is the sequence for which the shifted domains converge to the limiting domain $\hat{\Omega}$. Moreover we suppose that $\operatorname{supp} \phi_n$ are balls with radius $r_n \to \infty$, and all derivatives of $\phi_n(x)$ tend to zero as $n \to \infty$. We have

$$L_q v_n = \phi_n L_q u_0(\cdot + x_n) + \cdots . \quad (6.7)$$

The terms in the right-hand side of (6.7) that are not written tend to zero because of the assumption on ϕ_n that their derivatives tend to zero. The supports of the functions ϕ_n can be chosen in such a way that the first term in the right-hand side of (6.7) tends to zero as $n \to \infty$. Hence $L_q v_n \to 0$. It can be easily proved that the sequence v_n is not compact in $E_q(\Omega)$. Therefore the operator L_q is not proper. This contradiction shows that equation (6.6) has only the zero solution. Thus the invertibility of the operator $\hat{L}_q$ is proved.

2. (ii) $\to$ (iii). The proof is the same as the proof of Theorem 3.6.

3. (iii) $\to$ (i). This follows from Theorem 5.9.

The theorem is proved. $\square$

We will now show that if the Fredholm property is satisfied for some value of p, then it is also satisfied for other p assuming that the domain and the coefficients are sufficiently smooth. Suppose that the operator L_p is Fredholm for some $p = p^0$. Then from (i) of Theorem 6.3 we have (ii) and (iii) for the same p_0.

We can prove that Conditions NS and NS* are satisfied in other spaces. Let us begin with Condition NS. Suppose that it is not satisfied for some l^1, p^1, that is there exists a nonzero solution of the equation

$$\hat{L}u = 0, \quad u \in \Pi_{j=1}^N W_\infty^{l^1+t_j,p^1}(\hat{\Omega}).$$

Then, obviously, $u \in \Pi_{j=1}^N W_\infty^{l+t_j,p^1}(\hat{\Omega})$ for $\max(0, \sigma_j + 1) \leq l < l^1$. But also for $l > l^1$. This follows from a priori estimates of solutions in ∞-spaces. From the embedding theorems it follows that u belongs to $\Pi_{j=1}^N W_\infty^{l+t_j,p}(\hat{\Omega})$ with other p also. Hence if Condition NS is not satisfied in some space, then neither is it satisfied in other spaces.

Consider now Conditions NS*. We note first of all that from Theorem 6.3 it follows that limiting operators are invertible. Therefore the equation $\hat{L}u = f$ is solvable for any $f \in D$. Its solution belongs to $\Pi_{j=1}^N (W^{l+t_j,p}(\hat{\Omega}))_1$ for any l, p. Locally it follows from a priori estimates, behavior at infinity – since f has a bounded support. Suppose that Condition NS* is not satisfied for some l^1, p^1, that is there exists $v \neq 0$ such that

$$\hat{L}^* v = 0, \quad v \in (F^*)_\infty$$

or, consequently,
$$\hat{L}^*v = 0, \quad v \in (F_1)^*.$$

Since v belongs to the space dual to F_1, then there exists $f \in D$ such that $\langle v, f \rangle \neq 0$. Hence the equation $\hat{L}u = f$ is not solvable in E_1. Indeed, otherwise we apply the functional v to both sides of this equality and obtain a contradiction. However, it was shown above that this equation is solvable. This contradiction shows that Condition NS* is satisfied for all l, p.

This result shows in particular that if the Fredholm property is verified for elliptic problems in L^p, then it can be also used in L^2, which is sometimes more convenient. On the other hand, if the Fredholm property is verified in L^2, then it can be also done in L^p.

Let us now show that the Fredholm property holds not only for $q = p$ but also for $q \leq p$. We will verify Assumption 5.5. Let
$$E = L^p, \quad E_q = (L^p)_q.$$

Then
$$\|u\|_{E_q}^q = \sum_j \|\phi_j u\|_{L^p}^q.$$

If $u = \sum_i u_i$, then from Lemma 5.7,
$$\|\phi_j u\|_{L^p}^p \leq C \sum_i \|\phi_j u_i \omega_\delta(x - y_i)\|_{L^p}^p.$$

From this estimate and the previous equality we have
$$\|u\|_{E_q}^q \leq C^{q/p} \sum_j \left(\sum_i \|\phi_j u_i \omega_\delta(x - y_i)\|_{L^p}^p \right)^{q/p}.$$

On the other hand,
$$\sum_i \|u_i \omega_\delta(x - y_i)\|_{E_q}^q = \sum_i \sum_j \|\phi_j u_i \omega_\delta(x - y_i)\|_{L^p}^q.$$

Therefore, to verify Assumption 5.5 it is sufficient to satisfy the estimate
$$\left(\sum_i \|\phi_j u_i \omega_\delta(x - y_i)\|_{L^p}^p \right)^{q/p} \leq \sum_i \|\phi_j u_i \omega_\delta(x - y_i)\|_{L^p}^q.$$

It is satisfied if $q \leq p$ (see Lemma 5.6). We can now apply Theorem 5.9 for any $q \leq p$.

7 Weighted spaces

Let $\mu(x)$ be a positive infinitely differentiable function defined for all $x \in \mathbb{R}^n$ and satisfying the condition

$$|\frac{1}{\mu(x)} D^\beta \mu(x)| \to 0 \text{ as } |x| \to \infty$$

for any multi-index $\beta, |\beta| > 0$. We can take for example $\mu(x) = (1 + |x|^2)^s$, where $s \in \mathbb{R}$. For any normed space E we introduce the space E_μ with the norm

$$\|u\|_{E_\mu} = \|\mu u\|_E. \tag{7.1}$$

This means that $u \in E_\mu$ if and only if $\mu u \in E$. Consider weighted Sobolev spaces. Let

$$E = \Pi_{j=1}^N W^{l+t_j, p}(\Omega), \tag{7.2}$$

$$F = \Pi_{j=1}^N W^{l-s_i, p}(\Omega) \times \Pi_{j=1}^m W^{l-\sigma_j - 1/p, p}(\partial \Omega). \tag{7.3}$$

Then spaces E_μ and F_μ are defined.

Denote by S the operator of multiplication by μ. We have

$$S: E_\mu \to E, \quad S^{-1}: E \to E_\mu,$$
$$S: F_\mu \to F, \quad S^{-1}: F \to F_\mu.$$

If $v \in E_\mu$, then $\|Sv\|_E = \|\mu v\|_E = \|v\|_{E_\mu}$. Consider elliptic operators (3.1)–(3.3), $L: E \to F$, where E and F are spaces (7.2), (7.3).

Proposition 7.1. *The operator L is a bounded operator from E_μ to F_μ.*

Proof. Let $u \in W_\mu^{r,p}(\Omega)$, where r is a positive integer. Then $v = u\mu \in W^{r,p}(\Omega)$, and

$$\mu \frac{\partial u}{\partial x_i} = \frac{\partial v}{\partial x_i} - \mu^{-1} \frac{\partial \mu}{\partial x_i} \in W^{r-1,p}(\Omega).$$

Therefore $u \in W_\mu^{r-1,p}(\Omega)$, that is the operator of differentiation is bounded from $W_\mu^{r,p}(\Omega)$ into $W_\mu^{r-1,p}(\Omega)$. Hence D^α is a bounded operator from $W_\mu^{r,p}(\Omega)$ into $W_\mu^{r-|\alpha|,p}(\Omega)$, and A_i is a bounded operator from E_μ into $W_\mu^{l-s_i,p}(\Omega)$. Similarly, B_j is a bounded operator from E_μ into $W_\mu^{l-\sigma_j,p}(\Omega)$ and hence into $W_\mu^{l-\sigma_j - 1/p, p}(\partial \Omega)$. The proposition is proved. □

Theorem 7.2. *If operator $L: E \to F$ is Fredholm, then the operator $L: E_\mu \to F_\mu$ is Fredholm.*

Proof. Consider the operator L as acting from E_μ into F_μ. Then the operator $M = SLS^{-1}$ acts from E into F. We have for $u \in E$, $\omega = 1/\mu$:

$$SA_i S^{-1} u = SA_i(\omega u) = A_i u + \sum_{k=1}^{n} \sum_{|\alpha| \leq \alpha_{ik}} a_{ik}^\alpha(x) \sum_{\beta+\gamma \leq \alpha, \beta \neq 0} c_{\beta\gamma} \mu D^\beta \omega D^\gamma u_k.$$

Since for $|\beta| > 0$

$$\mu(x) D^\beta \omega(x) \to 0 \text{ as } |x| \to \infty,$$

then we conclude that limiting operators for $SA_i S^{-1}$ coincide with the limiting operators for A_i. The same is true for the boundary operators. Hence the operators $M : E \to F$ and $L : E \to F$ have the same limiting operators.

If the operator $L : E \to F$ is Fredholm, then Conditions NS and NS* are satisfied for it. Hence they are also satisfied for the operator $M : E \to F$. Therefore the operator M is Fredholm.

It remains to prove that if $M : E \to F$ is Fredholm, then the operator $L : E_\mu \to F_\mu$ also satisfies the Fredholm property. Indeed, let $u_i \in E, i = 1, \ldots, k$ be all linearly independent solutions of the equation $Mu = 0$. Then $v_i = S^{-1} u_i \in E_\mu, i = 1, \ldots, k$ are solutions of the equation

$$Lv = 0. \tag{7.4}$$

Conversely, if $v \in E_\mu$ is a solution of the equation (7.4), then $u = Sv$ is a solution of the equation $Mu = 0$. Hence $u = \sum_{i=1}^{k} c_i u_i$, and it follows that

$$v = S^{-1} u = \sum_{i=1}^{k} c_i S^{-1} u_i = \sum_{i=1}^{k} c_i v_i.$$

Therefore $v_i, i = 1, \ldots, k$ are all linearly independent solutions of (7.4).

Consider now the adjoint operators

$$L^* : F_\mu^* \to E_\mu^*, \quad M^* : F^* \to E^*.$$

We have

$$S^* : F^* \to F_\mu^*, \quad (S^{-1})^* : F_\mu^* \to F^*.$$

Let $\phi_j \in F^*, j = 1, \ldots, l$ be linearly independent solutions of the equation

$$M^* \phi = 0. \tag{7.5}$$

Then $\psi_j = S^* \phi_j \in F_\mu^*, j = 1, \ldots, l$ are solutions of the equation $L^* \psi_j = 0$ since $M^* = (S^{-1})^* L^* S^* : F^* \to E^*$. If $\psi \in F_\mu^*$ is an arbitrary solution of the equation

$$L^* \psi = 0, \tag{7.6}$$

then $\phi = (S^{-1})^*\psi \in F^*$ is a solution of the equation $M^*\phi = 0$. Hence $\phi = \sum_{j=1}^{l} c_j \phi_j$. Therefore

$$\psi = S^*\phi = \sum_{j=1}^{l} c_j S^* \phi_j = \sum_{j=1}^{l} c_j \psi_j.$$

We have proved that $\psi_j, j = 1, \ldots, l$ is a complete system of linearly independent solutions of equation (7.6).

Consider the equation

$$Lv = g, \ v \in E_\mu, \ g \in F_\mu. \tag{7.7}$$

Suppose that

$$\langle g, \psi_j \rangle = 0, \ j = 1, \ldots, l, \tag{7.8}$$

where $\psi_j \in F_\mu^*$ are all linearly independent solutions of the equation (7.6). Then

$$\phi_j = (S^*)^{-1} \psi_j \in F^*, \ j = 1, \ldots, l$$

are all linearly independent solutions of the equation $M^*\phi = 0$. It follows from (7.8) that

$$\langle g, S^*\phi_j \rangle = 0, \ j = 1, \ldots, l.$$

Consequently,

$$\langle Sg, \phi_j \rangle = 0. \tag{7.9}$$

Denote $f = Sg \in F$. Since the operator M is Fredholm, then from (7.9) it follows that the equation $Mu = f$ has a solution $u \in E$. We have $SLS^{-1}u = f$. Therefore $LS^{-1}u = S^{-1}f = g$. Hence $v = S^{-1}u \in E_\mu$ is a solution of equation (7.7). We have proved that from (7.8) it follows that equation (7.7) has a solution. Therefore the operator $L : E_\mu \to F_\mu$ is Fredholm. The theorem is proved. □

8 Hölder spaces

In this section we will prove that the Fredholm property in Hölder spaces follows from the Fredholm property in Sobolev spaces if the coefficients of the operator and the boundary of the domain are sufficiently smooth. We consider the operator L and the spaces E and F defined in Section 3. We consider also the same operator acting in Hölder spaces. Let

$$E_C = \Pi_{j=1}^{N} C^{l+t_j+\theta}(\bar{\Omega}), \ F_C = \Pi_{j=1}^{N} C^{l-s_i+\theta}(\bar{\Omega}) \times \Pi_{j=1}^{m} C^{l-\sigma_j+\theta}(\partial\Omega), \ 0 < \theta < 1.$$

Denote by L_C the realization of the operator L in these spaces, $L_C : E_C \to F_C$.

Along with the usual Hölder spaces we will consider the spaces E_G and F_G obtained as a closure of infinitely differentiable functions (not necessarily with a bounded support) in the norm of the spaces E_C and F_C, respectively. We denote by L_G the corresponding operator.

We recall that the operator $L_\infty : E_\infty \to F_\infty$ is normally solvable with a finite-dimensional kernel if and only if Condition NS is satisfied (Chapter 4). In Chapter 4 we have also proved the same result for the operator $L_C : E_C \to F_C$. In this case the space $E_\infty(\hat\Omega)$ in the formulation of Condition NS is replaced by the space $E_C(\hat\Omega)$. Obviously, the operator $L_G : E_G \to F_G$ is normally solvable with a finite-dimensional kernel if it is the case for the operator L_C.

Theorem 8.1. *Suppose that Condition NS is satisfied for the operator $L : E \to F$. Then it is also satisfied for the operator L_C.*

Proof. Suppose that u is a solution of the equation $\hat L u = 0$, $u \in E_C(\hat\Omega)$, where $\hat\Omega$ is a limiting domain, and $\hat L$ is a limiting operator. It is easy to see that

$$u \in E_\infty(\hat\Omega). \tag{8.1}$$

Indeed, let $u = (u_1, \ldots, u_N)$. Then $u_j \in C^{l+t_j+\theta}(\hat\Omega)$. Denoting by Q a unit cube with its center at y we have

$$\|D^\alpha u_j\|_{L^p(\hat\Omega \cap Q_y)} \leq M \|u_j\|_{C^{l+t_j+\theta}(\hat\Omega)},$$

where $|\alpha| \leq l+t_j$, and M is a constant. Then (8.1) follows from this estimate. Since Condition NS is satisfied for $L : E \to F$, then $u = 0$. The theorem is proved. □

In the following theorem we impose additional conditions on the smoothness of the coefficients of the operator and of the boundary of the domain. We suppose that the coefficients of the operator satisfy the following regularity conditions:

$$a_{ij}^\beta \in C^{\lambda - s_i + \alpha}(\bar\Omega), \quad b_{ij}^\beta \in C^{\lambda - \sigma_i + \alpha}(\partial\Omega),$$

and in the inequality for the number r in Condition D (Chapter 4, Section 1) we replace l by λ, where λ is an integer, $\lambda > l + 1 + n/p$.

Theorem 8.2. *If the operator $L : E \to F$ satisfies the Fredholm property, then the operator $L_G : E_G \to F_G$ also satisfies it.*

Proof. Suppose that the operator $L : E \to F$ is Fredholm. From Theorem 6.3 it follows that Condition NS is satisfied in the space E_∞. By Theorem 8.1 it is also satisfied in E_C. Hence the operator L_C is normally solvable with a finite-dimensional kernel. Therefore this is also true for the operator L_G.

Since the operator $L : E \to F$ is Fredholm, then Theorems 6.3 and 3.2 imply that the operator $L_\infty : E_\infty \to F_\infty$ is Fredholm. This means that there exist functionals $v_j \in (F_\infty)^*$, $j = 1, \ldots, N$ such that the equation

$$Lu = f - \sum_{j=1}^N \langle f, v_j \rangle e_j \tag{8.2}$$

8. Hölder spaces

has a solution $u \in E_\infty$ for any $f \in F_\infty$. Here $e_j \in F_\infty$ are elements biorthogonal to v_j.

Let $g \in F_G$. Then $g \in F_\infty$ and the expressions $\langle g, v_j \rangle$ are defined. We will prove that if

$$\langle g, v_j \rangle = 0, \quad j = 1, \ldots, N, \tag{8.3}$$

then the equation

$$Lu = g, \quad u \in E_G$$

has a solution. This will prove the theorem.

Since the space F_G is a closure of the set of infinitely differentiable functions, then F_G^λ is dense in F_G. We denote by F^λ and E^λ the spaces F and E where l is replaced by λ. Let $g_k \in F_G^\lambda$, $k = 1, 2, \ldots$ be a sequence which converges to g in F_G. Since $g_k \in F_\infty^\lambda \subset F_\infty$, then the equation

$$Lu_k = g_k - \sum_{j=1}^{N} \langle g_k, v_j \rangle e_j \tag{8.4}$$

has a solution $u_k \in E_\infty$. Without loss of generality we can assume that $e_j \in F_\infty^\lambda$. From a priori estimates (Chapter 4) we conclude that $u_k \in E_\infty^\lambda$. Therefore from the embedding theorems, $u_k \in E_G$.

Let $P: E_G \to \operatorname{Ker} L_G$ be a projector. Then

$$\|w_k\|_{E_G} \leq \operatorname{const} \|Lu_k\|_{F_G}, \tag{8.5}$$

where $w_k = u_k - Pu_k$. From (8.3) and (8.4) follows the convergence

$$Lw_k = Lu_k \to g \text{ in } F_G. \tag{8.6}$$

From (8.5) we obtain $w_k \to u$, for some $u \in E_G$. Therefore $Lu = g$. The theorem is proved. □

Theorem 8.3. *If the operator $L: E \to F$ satisfies the Fredholm property, then the operator $L_C: E_C \to F_C$ also satisfies it.*

Proof. Similar to the proof of the previous theorem we consider the spaces $F_\infty^\lambda \subset F_C \subset F_\infty$ and equation (8.2). It is solvable in E_∞ for any $f \in F_\infty$, and in E_∞^λ for any $f \in F_\infty^\lambda$. Here $v_j \in (F_\infty)^* \subset (F_\infty^\lambda)^*$.

Let $f \in F_C$, $f_k \in F_C^\lambda$,

$$\|f_k\|_{F_C} \leq M, \quad k = 1, 2, \ldots, \tag{8.7}$$

and $f_k \to f$ in the weaker norm of the space

$$F_C^0 = \Pi_{i=1}^{N} C^{l-s_i}(\bar{\Omega}) \times \Pi_{j=1}^{m} C^{l-\sigma_j}(\partial\Omega).$$

Such a sequence can be constructed as follows. We extend the function f to the space
$$\Pi_{i=1}^{N} C^{l-s_i+\theta}(\mathbb{R}^n) \times \Pi_{j=1}^{m} C^{l-\sigma_j+\theta}(\mathbb{R}^n).$$
Denote this function by $\hat{f}$. With an infinitely differentiable averaging kernel $\omega(x)$ we construct the function
$$\hat{f}_k(x) = \epsilon_k^{-n} \int \hat{f}(x-y)\omega(\frac{y}{\epsilon_k})dy.$$
Here $\epsilon_k \to 0$ as $k \to \infty$. It can be easily verified that the restriction f_k of the function $\hat{f}_k$ to the space F_C^λ has all the desired properties.

Since $f_k \in F_\infty^\lambda$, then equation (8.2) is solvable in E_∞^λ. By virtue of the embedding theorems its solution u_k belongs to E_C.

A priori estimates of solutions in Hölder spaces (Section 2.3, Chapter 1) and (8.7) imply that $\|u_k\|_{E_C} \leq M_1$, $k = 1, 2, \ldots$ Therefore there exists a subsequence of the sequence u_k converging locally in the space $E_C^0 = \Pi_{j=1}^{N} C^{l+t_j}(\bar{\Omega})$ to some function $u \in E_C$ satisfying the equation. Thus equation (8.2) is solvable in E_C for any $f \in F_C$. The theorem is proved. □

9 Examples

9.1 Bounded domains

Second-order equations in a bounded interval. Consider the equation
$$Lu \equiv a(x)u'' + b(x)u' + c(x)u = f \tag{9.1}$$
in the interval $I = [0, 1]$ with the Dirichlet boundary conditions
$$u(0) = u(1) = 0. \tag{9.2}$$
Here and in what follows we will suppose that the coefficients are real-valued sufficiently smooth functions, $a(x) > 0$ for all $x \in I$. We consider the operator L as acting from $H^2(I)$ with the Dirichlet boundary conditions into $L^2(I)$. The homogeneous formally adjoint problem is
$$L^*v \equiv (a(x)v)'' - (b(x)v)' + c(x)v = 0, \tag{9.3}$$
$$v(0) = v(1) = 0. \tag{9.4}$$

Equation (9.1) is solvable if and only if the solvability condition $\int_0^1 fv dx = 0$ is satisfied for any solution v of problem (9.3), (9.4). The number of linearly independent solutions of this problem equals the number of linearly independent solutions of the homogeneous ($f = 0$) problem (9.1), (9.2), that is the index equals zero. Indeed, the operator $L - \lambda$ is normally solvable with a finite-dimensional kernel for any λ, it is invertible for sufficiently large positive λ. Therefore its index equals zero for any λ. The same properties remain valid for the Neumann boundary conditions and for Hölder spaces.

9. Examples

Laplace operator. Consider the Laplace operator

$$Lu = \sum_{i=1}^{n} \frac{\partial^2 u}{\partial x_i^2}$$

in a bounded domain $\Omega \subset \mathbb{R}^n$ with a sufficiently smooth boundary, acting on functions from $H^2(\Omega)$. In the case of the Dirichlet boundary condition, the problem is formally self-adjoint. The nonhomogeneous problem

$$Lu = f, \quad u|_{\partial\Omega} = 0 \qquad (9.5)$$

is solvable if and only if

$$\int_\Omega fv\,dx = 0 \qquad (9.6)$$

for any solution v of the problem

$$Lv = 0, \quad v|_{\partial\Omega} = 0. \qquad (9.7)$$

This problem has the only zero solution. Indeed, multiplying the equation $Lv = 0$ by v and integrating over Ω, we obtain $\|\nabla v\|_{L^2(\Omega)} = 0$. Therefore, problem (9.5) is solvable for any $f \in L^2(\Omega)$.

In the case of the Neumann boundary condition, the homogeneous formally adjoint problem

$$Lv = 0, \quad \frac{\partial v}{\partial n}\Big|_{\partial\Omega} = 0$$

has a unique up to a constant factor nonzero solution, $v = 1$. Instead of (9.6), the solvability condition can be written as $\int_\Omega f\,dx = 0$.

Fourth-order equation. Consider the operator $Lu = \Delta^2 u$, where Δ denotes, as usual, the Laplace operator. We suppose that it acts from the space

$$E = \{u \in H^4(\Omega), \ u|_{\partial\Omega} = 0, \ \frac{\partial u}{\partial n}\Big|_{\partial\Omega} = 0\}$$

into $L^2(\Omega)$. Similar to the previous examples, the boundary conditions are included here in the definition of the function space. It is also possible to consider them as boundary operators. The nonhomogeneous problem

$$Lu = f, \quad u \in E \qquad (9.8)$$

is solvable if and only if condition (9.6) is satisfied for any solution v of the homogeneous formally adjoint problem $Lv = 0, \ v \in E$. Multiplying this equation by v and integrating, we obtain $\|\Delta v\|_{L^2(\Omega)} = 0$. Therefore, $v = 0$ and problem (9.7) is solvable for any $f \in L^2(\Omega)$.

9.2 Unbounded domains

Constant limits at infinity. We consider the same operator (9.1) acting from $H^2(\mathbb{R})$ into $L^2(\mathbb{R})$, $a(x) \geq a_0 > 0$ for all $x \in \mathbb{R}$. Suppose that there exist limits of the coefficients at infinity:

$$a_\pm = \lim_{x \to \infty} a(x), \quad b_\pm = \lim_{x \to \infty} b(x), \quad c_\pm = \lim_{x \to \infty} c(x).$$

There are two limiting operators that correspond to $+\infty$ and $-\infty$:

$$L_\pm u = a_\pm u'' + b_\pm u' + c_\pm u.$$

The essential spectrum of the operator L consists of all complex λ such that at least one of the two equations,

$$L_\pm u = \lambda u,$$

has a nonzero bounded solution. Applying the Fourier transform, we obtain

$$\lambda(\xi) = -a_\pm \xi^2 + b_\pm i\xi + c_\pm, \quad \xi \in \mathbb{R}.$$

The operator L satisfies the Fredholm property if and only if the two curves $\lambda_\pm(\xi)$ do not pass through the origin. Its index equals $n_+ + n_- - 2$, where n_+ is the number of bounded solutions of the equation $L_+ u = 0$ at $+\infty$, n_- is the number of bounded solutions of the equation $L_- u = 0$ at $-\infty$. In particular, the index equals zero if $c_\pm < 0$. It can also be zero if $c_\pm > 0$. Hence there exist different homotopy classes of such operators with the same index.

The nonhomogeneous equation $Lu = f$ is solvable if and only if

$$\int_{-\infty}^{\infty} f(x) v(x) dx = 0$$

for all solutions v of the equation $L^* v = 0$, where L^* is given by (9.3).

Slowly variable coefficients at infinity. We consider the same operator as above and assume that the coefficients have limits at infinity, except for the function $c(x)$ which does not have a limit as $x \to +\infty$. We suppose that

$$\overline{\lim}_{x \to +\infty} = c^*, \quad \underline{\lim}_{x \to +\infty} = c_*,$$

where $c_* < c^*$, and $c'(x) \to 0$ as $x \to +\infty$.

Let h_k be a sequence of numbers, $h_k \to \infty$ as $k \to \infty$. Consider the sequence of functions $\tilde{c}_k(x) = c(x + h_k)$. It has a subsequence locally convergent to a constant $\hat{c} \in [c_*, c^*]$. For any $\hat{c}$ from this interval, there exists a sequence h_k such that the sequence $\tilde{c}_k(x)$ converges to $\hat{c}$. Therefore the limiting operator has the form

$$L_- u = a_- u'' + b_- u' + c_- u \quad \text{and} \quad \hat{L} u = a_+ u'' + b_+ u' + \hat{c} u$$

for any $\hat{c} \in [c_*, c^*]$. Hence, the essential spectrum of the operator L consists of two parts: of the curve
$$\lambda = -a_-\xi^2 + b_-i\xi + c_-, \quad \xi \in \mathbb{R}$$
and of the domain filled by the curves
$$\lambda = -a_+\xi^2 + b_+i\xi + \hat{c}, \quad \xi \in \mathbb{R}$$
for all $\hat{c} \in [c_*, c^*]$. If the essential spectrum does not contain the origin, then the operator satisfies the Fredholm property. The solvability conditions are the same as in the previous example. In order to compute the index of the operator, we can use a continuous deformation which reduces it to an operator with constant coefficients at infinity.

Periodic and almost periodic coefficients. Consider the operator L given by (9.1), where $a(x), b(x)$, and $c(x)$ are periodic functions with the same period τ. In this case, all limiting operators have the form
$$\hat{L}u = a(x+h)u'' + b(x+h)u' + c(x+h)u,$$
where h is an arbitrary real number. Since the shift of the coefficients does not change the solvability of the equation $\hat{L}u = 0$, then Condition NS can be formulated in terms of the operator L: the operator L is normally solvable with a finite-dimensional kernel if and only if the equation $Lu = 0$ does not have nonzero bounded solutions. If a similar condition is satisfied for the formally adjoint operator, then the operator L satisfies the Fredholm property.

If the coefficients of the operator are almost periodic, then limiting operators also have almost periodic coefficients. Properties of almost periodic functions allow one to establish some specific properties of solutions. In particular, it is known for first-order systems with almost periodic coefficients that if the right-hand side is almost periodic, then any bounded solution is also almost periodic [160], [116]. There is an extensive literature devoted to spectral properties of periodic and quasi-periodic elliptic operators (see, for example, [17], [280], [449] and the references therein).

Problems on a half-axis. Consider the problem $Lu = f$, $u(0) = 0$ on the half-axis $x \geq 0$. The operator L is given by (9.1). The difference with respect to the previous examples is that limiting operators should be defined only at $+\infty$. In particular, if the coefficients have limits as $x \to +\infty$, then there exists a unique limiting equation
$$L_+u \equiv a_+u'' + b_+u' + c_+u = 0, \quad x \in \mathbb{R}.$$
The limiting domain here is the whole axis. The essential spectrum is the curve on the complex plane:
$$\lambda(\xi) = -a_+\xi^2 + b_+i\xi + c_+, \quad \xi \in \mathbb{R}.$$
If it is in the left half-plane, then the index equals zero.

The solvability conditions are given by the equality $\int_0^\infty f(x)v(x)dx = 0$ for all solutions v of the homogeneous formally adjoint problem $L^*v = 0$, $v(0) = 0$, where the operator L^* is defined in (9.3).

Systems of equations. Let $a(x), b(x)$, and $c(x)$ be sufficiently smooth square matrices of the order n,

$$(a(x)\zeta, \zeta) \geq a_0 > 0, \quad \forall x \in \mathbb{R}, \ \zeta \in \mathbb{R}^n, \ |\zeta| = 1.$$

If they have limits $a_\pm, b_\pm$, and $c_\pm$ as $x \to \pm\infty$, then the limiting operators are

$$L_\pm u = a_\pm u'' + b_\pm u' + c_\pm u.$$

The essential spectrum consists of all λ for which at least one of the equalities

$$\det(-a_\pm \xi^2 + b_\pm i\xi + c_\pm - \lambda E) = 0, \quad \xi \in \mathbb{R}$$

holds. Here E is the identity matrix. The index and solvability conditions for such problems are discussed in Chapter 9.

Second-order operators in $\mathbb{R}^n$. Consider the operator

$$Lu = a(x)\Delta u + \sum_{i=1}^n b_i(x)\frac{\partial u}{\partial x_i} + c(x)u$$

acting from the Hölder space $C^{2+\delta}(\mathbb{R}^n)$ into the space $C^\delta(\mathbb{R}^n)$. The square matrices $a(x), b_i(x), c(x)$ of the order m belong to $C^\delta(\mathbb{R}^n)$. The ellipticity condition

$$(a(x)\xi, \xi) \geq a_0 > 0, \quad \forall x \in \mathbb{R}^n, \ \xi \in \mathbb{R}^m, \ |\xi| = 1$$

is supposed to be satisfied. Let us consider the unit sphere S in $\mathbb{R}^n$ and represent $x \in \mathbb{R}^n$ as $x = (r, \theta)$, where $\theta \in S$ and $r \geq 0$ is a real number. Suppose that the matrix $a(x) = a(r, \theta)$ has a limit

$$\hat{a}(\theta) = \lim_{r \to \infty} a(r, \theta)$$

for any θ fixed. Similarly, we define the matrix-functions $\hat{b}_i(\theta)$ and $\hat{c}(\theta)$ and suppose that all of them are continuous with respect to θ.

If these conditions are satisfied, then there is a family of limiting operators

$$\hat{L}_\theta u = \hat{a}(\theta)\Delta u + \sum_{i=1}^n \hat{b}_i(\theta)\frac{\partial u}{\partial x_i} + \hat{c}(\theta)u, \quad \theta \in S$$

with constant coefficients. Denote by Λ the set of all complex numbers λ for which

$$\det(-\hat{a}(\theta)\xi^2 + i\xi \sum_{i=1}^n \hat{b}_i(\theta) + \hat{c}(\theta) - \lambda E) = 0$$

9. Examples

for some $\theta \in S$ and $\xi \in \mathbb{R}^n$. Here E is the identity matrix. If the set Λ does not contain the origin, then Condition NS is satisfied and the operator L is normally solvable with a finite-dimensional kernel. It can be shown for the multi-dimensional scalar case ($m = 1, n > 1$) that the Fredholm operator of this form has necessarily zero index. It can be different from zero for $n = 1$.

Other problems. Second-order operators in cylinders are studied in Chapter 9. The Cauchy-Riemann problem and the Laplace operator with oblique derivative in bounded and unbounded domains are discussed in Chapter 8.

Chapter 6

Formally Adjoint Problems

In this chapter we study the Fredholm property of regular scalar elliptic problems and formulate their solvability conditions in terms of formally adjoint problems. In bounded domains such results were obtained in [318] (L^2 theory) and in [457] (L^p theory). As we already know, in unbounded domains some conditions at infinity (Conditions NS and NS*) should be imposed. We will introduce a similar condition for formally adjoint problems and will use it to prove solvability conditions. We restrict ourselves to scalar problems though some of the results can be easily generalized for systems.

1 Definitions

We consider the boundary value problem

$$Au = f_0 \text{ in } \Omega, \quad B_j u = f_j \text{ in } \Gamma, \quad j = 1, \ldots, m. \tag{1.1}$$

Here

$$Au = \sum_{|\alpha| \leq 2m} a^\alpha(x) D^\alpha u, \quad x \in \Omega,$$

$$B_j u = \sum_{|\alpha| \leq s_j} b_j^\alpha(x) D^\alpha u, \quad x \in \Gamma, \quad j = 1, \ldots, m, \quad s_j < 2m,$$

where $\alpha = (\alpha_1, \ldots, \alpha_n)$ is a multi-index, $|\alpha| = \alpha_1 + \cdots + \alpha_n$, $D^\alpha = D_1^{\alpha_1} \ldots D_n^{\alpha_n}$, $D_i^{\alpha_i} = \partial^{\alpha_i}/\partial x_i^{\alpha_i}$, $\Omega \subset R^n$ is an unbounded domain and Γ is its boundary. The coefficients $a^\alpha(x)$ and $b_j^\alpha(x)$ in (1.1) are complex-valued functions given in $\bar{\Omega}$ which are assumed, for simplicity, to be infinitely differentiable with all bounded derivatives.

We suppose that problem (1.1) is *uniformly elliptic*. This means that (i) A is a uniformly elliptic operator, (ii) A is properly elliptic, (iii) the boundary operators satisfy uniformly Lopatinskii conditions (see Chapter 1). It is also supposed that B_j form a *normal system*, that is $0 \leq s_1 < \cdots < s_m$ and for each normal vector ν_x on Γ, where $x \in \Gamma$, it holds that

$$\sum_{|\alpha|=s_i} b_j^\alpha(x)\nu_x^\alpha \neq 0, \quad j = 1, \ldots, m$$

(see [32], [318]). Such problems are referred to as *regular* elliptic problems. Following [318], we call a system of boundary operators F_i, $i = 1, \ldots, \nu$ the Dirichlet system if it is normal and if the orders m_i of the operators take on the values $0, 1, \ldots, \nu - 1$ when i changes from 0 to $\nu - 1$.

The formally adjoint operator A^* to the operator A is given by the formula

$$A^* v = \sum_{|\alpha| \leq 2m} (-1)^{|\alpha|} D^\alpha (\overline{a^\alpha(x)} v).$$

We have

$$\int_\Omega Au\,\bar{v}\,dx - \int_\Omega u\,\overline{A^*v}\,dx = 0, \quad \forall u, v \in D(\bar{\Omega}).$$

The operator A^* satisfies the ellipticity condition if and only if the operator A satisfies it.

Consider problem (1.1). If the boundary operators form a normal system, then it can be completed to a Dirichlet system of order $2m$ by some operators S_j, $j = 1, \ldots, m$. The choice of these operators is not unique. Then there exists another Dirichlet system of order $2m$ formed by some operators C_j, T_j, $j = 1, \ldots, m$ such that the following Green's formula holds [318]:

$$\int_\Omega Au\,\bar{v}\,dx - \int_\Omega u\,\overline{A^*v}\,dx = \sum_{j=1}^m \int_\Gamma S_j u\,\overline{C_j v}\,ds - \sum_{j=1}^m \int_\Gamma B_j u\,\overline{T_j v}\,ds. \quad (1.2)$$

The boundary value problem

$$A^*v = f_0 \text{ in } \Omega, \quad C_j v = f_j \text{ in } \Gamma, \quad j = 1, \ldots, m. \quad (1.3)$$

is called *the formally adjoint problem* of (1.1).

As before, we suppose that the boundary of the domain Ω satisfies Condition D (cf. Chapter 3, Section 2):

Condition D. For each $x_0 \in \Gamma$ there exists a neighborhood $U(x_0)$ such that:

1. $U(x_0)$ contains a sphere with radius δ and center x_0, where δ is independent of x_0.
2. There exists a homeomorphism $\psi(x; x_0)$ of the neighborhood $U(x_0)$ on the unit sphere $B = \{y : |y| < 1\}$ in $\mathbb{R}^n$ such that the images of $\Omega \bigcap U(x_0)$ and

1. Definitions

$\Gamma \cap U(x_0)$ coincide with $B_+ = \{y : y_n > 0, |y| < 1\}$ and $B_0 = \{y : y_n = 0, |y| < 1\}$ respectively.
3. The function $\psi(x; x_0)$ and its inverse are infinitely differentiable and each derivative is uniformly bounded on Γ.

We recall the definitions of the function spaces introduced in Chapter 2. Consider first functions defined on $\mathbb{R}^n$. As usual we denote by D the space of infinitely differentiable functions with compact support and by D' its dual. Let $E \subset D'$ be a Banach space, the inclusion being understood in both the algebraic and topological sense. Denote by E_{loc} the collection of all $u \in D'$ such that $fu \in E$ for all $f \in D$. Let $\omega(x) \in D$, $0 \leq \omega(x) \leq 1$, $\omega(x) = 1$ for $|x| \leq 1/2$, $\omega(x) = 0$ for $|x| \geq 1$.

Definition 1.1. E_q $(1 \leq q \leq \infty)$ is the space of all $u \in E_{\text{loc}}$ such that

$$\|u\|_{E_q} := \left(\int_{\mathbb{R}^n} \|u(.)\omega(. - y)\|_E^q dy \right)^{1/q} < \infty, \quad 1 \leq q < \infty,$$

$$\|u\|_{E_\infty} := \sup_{y \in \mathbb{R}^n} \|u(.)\omega(. - y)\|_E < \infty.$$

If Ω is a domain in $\mathbb{R}^n$, then by definition $E_q(\Omega)$ is the space of restrictions of E_q to Ω with the usual norm of restrictions. It is easy to see that if Ω is a bounded domain, then

$$E_q(\Omega) = E(\Omega), \quad 1 \leq q \leq \infty.$$

In particular, if $E = W^{s,p}$, then we write $W_q^{s,p} = E_q$ $(1 \leq q \leq \infty)$. It is proved that

$$W_p^{s,p} = W^{s,p} \quad (s \geq 0, \ 1 < p < \infty).$$

If $E = H^s$, then we can introduce also spaces H_q^s, $1 \leq q \leq \infty$. We suppose that in problem (1.1),

$$u \in W_q^{l,p}(\Omega), \quad f_0 \in W_q^{l-2m,p}(\Omega), \quad f_j \in W_q^{l-s_j-1/p,p}(\Gamma), \quad j = 1, \ldots, m,$$

where $l \geq 2m$ is an integer, $1 < p < \infty$, $1 \leq q \leq \infty$.

As in previous chapters, we will use the notion of limiting domains. Their construction can be briefly described as follows. Let $x_k \in \Omega$ be a sequence, which tends to infinity. Consider the shifted domains Ω_k corresponding to the shifted characteristic functions $\chi(x + x_k)$, where $\chi(x)$ is the characteristic function of the domain Ω. Consider a ball $B_r \subset \mathbb{R}^n$ with its center at the origin and with radius r. Suppose that for all k there are points of the boundaries $\partial \Omega_k$ inside B_r. If the boundaries are sufficiently smooth, then we can expect that from the sequence $\Omega_k \cap B_r$ we can choose a subsequence that converges to some limiting domain Ω_*. After that we take a larger ball and choose a convergent subsequence of the previous subsequence. The usual diagonal process allows us to extend the limiting domain to the whole space.

To define limiting operators we consider shifted coefficients $a^\alpha(x+x_k)$, $b_j^\alpha(x+x_k)$ and choose subsequences that converge to some limiting functions $\hat{a}^\alpha(x)$, $\hat{b}_j^\alpha(x)$ uniformly in every bounded set. The limiting operator is the operator with the limiting coefficients. Limiting operators considered in limiting domains constitute limiting problems. It is clear that the same problem can have a family of limiting problems depending on the choice of the sequence x_k and on the choice of both converging subsequences of domains and coefficients.

Set $L = (A, B_1, \ldots, B_m)$ and let $\hat{L}$ be a corresponding limiting operator. We have already discussed in previous chapters that the following condition determines normal solvability of elliptic problems.

Condition NS. Any limiting problem
$$\hat{L}u = 0, \quad x \in \Omega_*, \quad u \in W_\infty^{l,p}(\Omega_*)$$
has only the zero solution.

A similar condition is formulated for formally adjoint problems. We recall that formally adjoint operators act in the same spaces as the direct operators, while the adjoint operator acts in dual spaces. We use here the same notation of this condition for formally adjoint operators as before for adjoint operators. In this chapter we understand it only in the sense of formally adjoint operators.

Condition NS*. Any limiting problem
$$\hat{L}^*v = 0, \quad x \in \Omega_*, \quad v \in W_\infty^{l,p}(\Omega_*)$$
has only the zero solution. Here $\hat{L}^*$ is the operator formally adjoint to $\hat{L}$.

We will use below the space of exponentially decreasing functions.

Definition 1.2. A function f belongs to the space $\mathcal{E}$ if $f(x)$ is defined on $\mathbb{R}^n$, infinitely differentiable, and for any multi-index α there exists $\epsilon > 0$ such that
$$|D^\alpha f(x)\, e^{\epsilon|x|}| \leq C_\alpha, \forall x \in \mathbb{R}^n.$$
$\mathcal{E}(\bar{\Omega})$ is the restriction of $\mathcal{E}$ to $\bar{\Omega}$, the definition of $\mathcal{E}(\Gamma)$ is standard.

The main result of this chapter is given by the following theorem. It will be proved in Section 4.

Theorem 1.3. *Let Conditions NS and NS* be satisfied. Suppose that*
$$f_0 \in W_q^{l-2m,p}(\Omega), \quad f_j \in W_q^{l-s_j-1/p,p}(\Gamma),$$
$$j = 1, \ldots, m, \quad 1 < p < \infty, \quad 1 < q \leq p, \quad l \geq 2m$$
is an integer. Then problem (1.1) *has a solution* $u \in W_q^{l,p}$ *if and only if*
$$\int_\Omega f_0\, \overline{v_0}\, dx + \sum_{j=1}^m \int_\Gamma f_j\, \overline{T_j v_0}\, ds = 0 \qquad (1.4)$$

for any v_0 from the finite-dimensional space

$$N^* = \{v_0 : v_0 \in \mathcal{E}(\bar{\Omega}),\ A^* v_0 = 0,\ C_1 v_0 = 0, \ldots, C_m v_0 = 0\}. \tag{1.5}$$

It follows from the theorem that if Conditions NS and NS* are satisfied, then the operator is Fredholm. Moreover its cokernel, that is the functions which provide solvability conditions, belongs to the space $\mathcal{E}$ and therefore does not depend on l, p, q. It is shown in Section 5 that the same is true for the kernel. Thus the index of the operator does not depend on l, p, q.

Set

$$E = W^{l,p}(\Omega),\quad F = W^{l-2m,p}(\Omega) \times W^{l-s_1-1/p,p}(\Gamma) \times \cdots \times W^{l-s_m-1/p,p}(\Gamma).$$

The corresponding spaces E_∞ and F_∞ are defined in Chapter 2. We will use a priori estimates of solutions obtained in Chapter 4:

Theorem 1.4. *Let Condition NS be satisfied. Then there exist numbers M_0 and R_0 such that the following estimate holds:*

$$\|u\|_{E_\infty} \leq M_0 \left(\|Lu\|_{F_\infty} + \|u\|_{L^p(\Omega_{R_0})} \right),\quad \forall u \in E_\infty. \tag{1.6}$$

Here $\Omega_{R_0} = \Omega \cap \{|x| \leq R_0\}$.

Theorem 1.5. *Let Condition NS be satisfied. Then there exist numbers $M_0 > 0$, $R_0 > 0$ and $\mu_0 > 0$ such that for all μ, $0 < \mu < \mu_0$ the following estimate holds:*

$$\|\omega_\mu u\|_{E_\infty} \leq M_0 \left(\|\omega_\mu Lu\|_{F_\infty} + \|\omega_\mu u\|_{L^p(\Omega_{R_0})} \right),\quad \forall u \in E_\infty. \tag{1.7}$$

Here $\omega_\mu = \exp(\mu\sqrt{1+|x|^2})$, μ is a real number.

Corollary 1.6. *If $0 < \mu < \mu_0$, $u \in E_\infty$, and $\omega_\mu Lu \in F_\infty$, then $\omega_\mu u \in E_\infty$. In particular, if $u \in E_\infty$ and $Lu = 0$, then $\omega_\mu u \in E_\infty$.*

We recall that the spaces $W^{s,2}$ and H^s coincide. We will use below the spaces of Bessel potentials.

2 Spaces $V^{2m}(\Omega)$

Denote by $V^{2m}(\Omega)$ all functions $u \in H^{2m}_\infty(\Omega)$ such that

$$\int_\Omega |Au|^2 dx + \sum_{j=0}^{m-1} \langle B_j u, B_j u \rangle < \infty. \tag{2.1}$$

Here $\langle , \rangle$ is duality in H^{2m-s_j}. We suppose that Condition NS is satisfied. Let

$$N = \{u \mid u \in H^{2m}_\infty(\Omega),\ Au = 0,\ B_j = 0, j = 0, \ldots, m-1\}.$$

Then N is finite dimensional and $u(x) \to 0$ exponentially as $|x| \to \infty$, $x \in \Omega$.

For $u, v \in V^{2m}(\Omega)$ we introduce the inner product

$$[u, v] = \int_\Omega Au \cdot \overline{Av}\, dx + \sum_{j=0}^{m-1} \langle B_j u, B_j v \rangle + \int_\Omega Pu \cdot \overline{Pv}\, dx. \tag{2.2}$$

Here P is the projection operator in $L^2(\Omega)$ on the subspace N. More precisely, let $u_1, \ldots, u_k$ be a basis in N, and $v_1, \ldots, v_k \in D(\Omega)$ such that

$$\int_\Omega u_i(x)\overline{v_j(x)}\, dx = \delta_{ij}, \quad i, j = 1, \ldots, k.$$

Then

$$Pu = \sum_{i=1}^{k}(u, v_i)u_i. \tag{2.3}$$

Here and in what follows

$$(u, v) = \int_\Omega u(x)\overline{v(x)}\, dx. \tag{2.4}$$

We introduce also the notation

$$[u, v]_0 = \int_\Omega Au \cdot \overline{Av}\, dx + \sum_{j=0}^{m-1} \langle B_j u, B_j v \rangle. \tag{2.5}$$

Therefore

$$[u, v] = [u, v]_0 + (Pu, Pv). \tag{2.6}$$

Proposition 2.1. *If $u \in V^{2m}(\Omega)$, then*

$$\|u\|_{H^{2m}_\infty(\Omega)} \leq C\|u\|_{V^{2m}(\Omega)}, \tag{2.7}$$

where

$$\|u\|_{V^{2m}(\Omega)} = \sqrt{[u, u]}.$$

Proof. Since Condition NS is satisfied, then

$$\|u\|_{H^{2m}_\infty(\Omega)} \leq C\left(\|Au\|_{L^2_\infty(\Omega)} + \sum_{j=0}^{m-1} \|B_j u\|_{H^{2m-s_j-1/2}_\infty(\Gamma)} + \left(\int_{\Omega_\rho} |u(x)|^2 dx\right)^{1/2}\right)$$

for some $C > 0$, $\rho > 0$, and for any $u \in H^{2m}_\infty(\Omega)$ (Theorem 1.4). Therefore

$$\|u\|_{H^{2m}_\infty(\Omega)} \leq C\left(\|Au\|_{L^2(\Omega)} + \sum_{j=0}^{m-1} \|B_j u\|_{H^{2m-s_j-1/2}(\Gamma)} + \left(\int_{\Omega_\rho} |u(x)|^2 dx\right)^{1/2}\right).$$

2. Spaces $V^{2m}(\Omega)$

It follows that

$$\|u\|_{H^{2m}_\infty(\Omega)} \leq C_1 \left(\|u\|_{V^{2m}(\Omega)} + \left(\int_{\Omega_\rho} |u(x)|^2 dx \right)^{1/2} \right). \tag{2.8}$$

It is sufficient to prove (2.7) for $\forall \|u\|_{V^{2m}(\Omega)} = 1$. In this case this estimate takes the form

$$\|u\|_{H^{2m}_\infty(\Omega)} \leq C, \quad \forall u \in V^{2m}(\Omega), \ \|u\|_{V^{2m}(\Omega)} = 1. \tag{2.9}$$

Suppose that this estimate does not hold. Then there exists a sequence $\{u_k\}$, $u_k \in V^{2m}(\Omega)$ such that

$$\|u_k\|_{H^{2m}_\infty(\Omega)} \to \infty, \quad \|u_k\|_{V^{2m}(\Omega)} = 1.$$

Set

$$w_k = \frac{u_k}{\|u_k\|_{H^{2m}_\infty(\Omega)}}.$$

Then we have

$$\|w_k\|_{H^{2m}_\infty(\Omega)} = 1, \tag{2.10}$$

$$\|w_k\|_{V^{2m}(\Omega)} \to 0 \text{ as } k \to \infty. \tag{2.11}$$

From (2.10) it follows that there exists a subsequence of w_k, still denoted by w_k such that the following convergence occurs:

$$\int_{\Omega_\rho} |w_k(x) - w_l(x)|^2 dx \to 0 \text{ as } k, l \to \infty.$$

Estimate (2.8) implies

$$\|w_k - w_l\|_{H^{2m}_\infty(\Omega)} \to 0 \text{ as } k, l \to \infty.$$

Hence there exists $w \in H^{2m}_\infty(\Omega)$ such that

$$\|w_k - w\|_{H^{2m}_\infty(\Omega)} \to 0 \text{ as } k \to \infty. \tag{2.12}$$

From (2.10)

$$\|w\|_{H^{2m}_\infty(\Omega)} = 1. \tag{2.13}$$

On the other hand, (2.11) implies

$$\int_\Omega |Aw_k|^2 dx \to 0, \tag{2.14}$$

$$\langle B_j w_k, B_j w_k \rangle \to 0, \quad j = 0, \ldots, m-1 \tag{2.15}$$

$$\int_\Omega |Pw_k|^2 dx \to 0 \text{ as } k \to \infty. \tag{2.16}$$

From (2.14) it follows that
$$Aw = 0. \tag{2.17}$$
Indeed, (2.14) implies the convergence
$$\int_\Omega Aw_k \overline{\phi(x)} dx \to 0 \text{ as } k \to \infty \tag{2.18}$$
for any $\phi \in C_0^\infty(\Omega)$. From (2.12) we can conclude that
$$\|w_k - w\|_{H^{2m}(B\cap\Omega)} \to 0 \text{ as } k \to \infty$$
for any ball $B \subset R^n$. Therefore we can pass to the limit in (2.18):
$$\int_\Omega Aw \overline{\phi(x)} dx = 0.$$
Equality (2.17) follows from the last one.

Similarly, from (2.15) we conclude that $B_j w = 0$, $j = 0, \ldots, m-1$. From this and (2.17) we get $w \in N$. Hence
$$w = Pw. \tag{2.19}$$
Now, (2.3) and (2.12) imply
$$\|Pw_k - Pw\|_{L^2(\Omega)} \leq \sum_{i=1}^k |(w_k - w, v_i)| \|u_i\|_{L^2(\Omega)} \to 0.$$
From this convergence and (2.16), $Pw = 0$, and from (2.19), $w = 0$. This contradicts (2.13). The proposition is proved. □

Proposition 2.2. $V^{2m}(\Omega)$ *is a Hilbert space.*

Proof. We have to prove the completeness of $V^{2m}(\Omega)$. Let
$$\|u_k - u_l\|_{V^{2m}(\Omega)} \to 0 \text{ as } k, l \to \infty. \tag{2.20}$$
Then from (2.7) we have
$$\|u_k - u_l\|_{H^{2m}_\infty(\Omega)} \to 0 \text{ as } k, l \to \infty.$$
Hence there exists $u \in H^{2m}_\infty(\Omega)$ such that
$$\|u_k - u\|_{H^{2m}_\infty(\Omega)} \to 0 \text{ as } k \to \infty. \tag{2.21}$$
Set $v_k = Au_k$. Then from (2.20)
$$\|v_k - v_l\|_{L^2(\Omega)} \to 0 \text{ as } k, l \to \infty.$$

2. Spaces $V^{2m}(\Omega)$

Therefore there exists $v \in L^2(\Omega)$ for which
$$\|v_k - v\|_{L^2(\Omega)} \to 0 \text{ as } k \to \infty. \tag{2.22}$$

Hence
$$\|v_k - v\|_{L^2(\Omega \cap B)} \to 0 \text{ as } k \to \infty$$
for any ball $B \subset \mathbb{R}^n$, or
$$\|Au_k - v\|_{L^2(\Omega \cap B)} \to 0 \text{ as } k \to \infty.$$

On the other hand, from (2.21) it follows that
$$\|Au_k - Au\|_{L^2(\Omega \cap B)} \to 0 \text{ as } k \to \infty.$$

Hence $v = Au$ in $\Omega \cap B$. Since B is arbitrary, then
$$v = Au \text{ in } \Omega \text{ and } Au \in L^2(\Omega). \tag{2.23}$$

From this and (2.22) we obtain
$$\|Au_k - Au\|_{L^2(\Omega)} \to 0 \text{ as } k \to \infty. \tag{2.24}$$

Consider now the boundary operators. Write $B_j u_k = w_k$. Then from (2.20) it follows that
$$\|w_k - w_l\|_{H^{2m-s_j-1/2}(\Gamma)} \to 0 \text{ as } k, l \to \infty.$$

Hence there exists $w \in H^{2m-s_j-1/2}(\Gamma)$ for which
$$\|w_k - w\|_{H^{2m-s_j-1/2}(\Gamma)} \to 0 \text{ as } k \to \infty. \tag{2.25}$$

Therefore
$$\|w_k - w\|_{H^{2m-ms_j-1/2}(\Gamma \cap B)} \to 0 \text{ as } k \to \infty \tag{2.26}$$
for any ball $B \subset \mathbb{R}^n$. From (2.21) we conclude that
$$\|B_j u_k - B_j u\|_{H^{2m-s_j-1/2}(\Gamma \cap B)} \to 0 \text{ as } k \to \infty.$$

This and (2.26) imply that $B_j u = w$ as elements of $H^{2m-s_j-1/2}(\Gamma \cap B)$. Since B is an arbitrary ball, we conclude that $B_j u = w$ as generalized functions on Γ. It follows that
$$B_j u \in H^{2m-s_j-1/2}(\Gamma) \tag{2.27}$$
and (2.25) implies that
$$\|B_j u_k - B_j u\|_{H^{2m-s_j-1/2}(\Gamma)} \to 0 \text{ as } k \to \infty. \tag{2.28}$$

From (2.23) and (2.27) we conclude that $u \in V^{2m}(\Omega)$.

It remains to consider the operator P. Obviously,
$$\|P(u_k - u_l)\|_{L^2(\Omega)} \to 0 \text{ as } k, l \to \infty.$$
Taking into account (2.3), we get
$$\|Pu_k - Pu\|_{L^2(\Omega)} \to 0 \text{ as } k \to \infty. \tag{2.29}$$
This convergence, (2.28), and (2.24) imply
$$\|u_k - u\|_{V^{2m}(\Omega)} \to 0 \text{ as } k \to \infty.$$
The proposition is proved. □

Proposition 2.3. *If $u \in H^{2m}(\Omega)$, then $u \in V^{2m}(\Omega)$, and*
$$\|u\|_{V^{2m}(\Omega)} \leq C\|u\|_{H^{2m}(\Omega)}. \tag{2.30}$$

Proof. Let $u \in H^{2m}(\Omega)$. Then $u \in H^{2m}_\infty(\Omega)$ and (2.1) is satisfied. Hence $u \in V^{2m}(\Omega)$ and
$$[u, u]_0 \leq C_1 \|u\|^2_{H^{2m}(\Omega)}. \tag{2.31}$$
From (2.3) it follows that
$$\|Pu\|^2_{L^2(\Omega)} \leq C_2 \|u\|^2_{H^{2m}(\Omega)}. \tag{2.32}$$
This estimate, (2.31) and (2.6) imply (2.30). The proposition is proved. □

It follows from Propositions 2.1 and 2.3 that
$$H^{2m}(\Omega) \subset V^{2m}(\Omega) \subset H^{2m}_\infty(\Omega),$$
the inclusions being with the topology. These spaces coincide locally. The difference between them consists in the behavior of the functions at infinity.

Proposition 2.4. *If $f \in L^2_1(\Omega)$, $v \in V^{2m}(\Omega)$, then*
$$(v, f) = \int_\Omega v(x)\overline{f(x)}\, dx$$
is a continuous functional on $V^{2m}(\Omega)$.

Proof. We have
$$|(v, f)| = \left|\int_\Omega \sum_{i=1}^\infty \phi_i(x) v(x) \overline{f(x)}\, dx\right| \equiv S,$$
where ϕ_i is a partition of unity, $\psi_i(x) \in D$, $\psi_i(x) = 1$ for $x \in \operatorname{supp} \phi_i$,
$$S \leq \sum_{i=1}^\infty \left|\int_\Omega \phi_i v \psi_i \overline{f}\, dx\right| \leq \sum_{i=1}^\infty \|\phi_i v\|_{L^2} \|\psi_i \overline{f}\|_{L^2} \leq C\|v\|_{L^2_\infty(\Omega)} \sum_{i=1}^\infty \|\psi_i f\|_{L^2(\Omega)}$$
$$\leq C_1 \|v\|_{L^2_\infty(\Omega)} \|f\|_{L^2_1(\Omega)} \leq C_1 \|v\|_{H^{2m}_\infty(\Omega)} \|f\|_{L^2_1(\Omega)} \leq C_2 \|v\|_{V^{2m}(\Omega)} \|f\|_{L^2_1(\Omega)}.$$
The proposition is proved. □

The following proposition is a generalization of Proposition 5.1 in [318].

Proposition 2.5. *Let $u \in V^{2m}(\Omega)$, $f \in L_1^2(\Omega) \cap H_{loc}^r(\Omega)$, where r is a positive integer, and*

$$[u,v]_0 = \int_\Omega f\overline{v}\,dx, \quad \forall v \in V^{2m}(\Omega), \tag{2.33}$$

then $u \in H_{loc}^{4m+r}(\Omega)$.

Proof. Choosing $v \in D(\Omega)$ in (2.33), we obtain $A^*Au = f$. Hence $u \in H^{4m+r}(\omega)$ for any open bounded set ω such that $\bar\omega \subset \Omega$. Since $V^{2m}(\Omega)$ coincides locally with $H^{2m}(\Omega)$, we obtain (5.19) in [318], and we use the result of [318]. The proposition is proved. □

3 Fredholm theorems

We will use Green's formula (1.2) and Definition 1.2.

Proposition 3.1. *Let Conditions NS and NS* be satisfied, and $f_0 \in D(\bar\Omega)$, $f_j \in D(\Gamma)$, $j = 1, \ldots, m$. Then problem (1.1) has a solution $u \in H^{2m}(\Omega)$ if and only if*

$$\int_\Omega f_0\,\overline{v_0}\,dx + \sum_{j=1}^m \int_\Gamma f_j\,\overline{T_j v_0}\,ds = 0 \tag{3.1}$$

for any v_0 from the finite-dimensional space

$$N^* = \{v_0 : v_0 \in \mathcal{E}(\bar\Omega),\ A^*v_0 = 0,\ C_1 v_0 = 0, \ldots, C_m v_0 = 0\}. \tag{3.2}$$

Proof. Green's formula (1.2) implies that condition (3.1) is necessary. We will prove that it is sufficient.

1. We begin with the case where

$$f_0 \in D(\bar\Omega),\quad f_j = 0,\ j = 1, \ldots, m. \tag{3.3}$$

We introduce the space $V_*^{2m}(\Omega)$ similar to the space $V^{2m}(\Omega)$ in Section 2. Namely $V_*^{2m}(\Omega)$ is the collection of all $u \in H_\infty^{2m}(\Omega)$ such that

$$\int_\Omega |A^*u|^2 dx + \sum_{j=1}^m \langle C_j u, C_j u\rangle_j < \infty.$$

We introduce the inner product

$$[u,v] = \int_\Omega A^*u\overline{A^*v}dx + \sum_{j=1}^m \langle C_j u, C_j v\rangle_j + \int_\Omega Pu\overline{Pv}dx, \tag{3.4}$$

where P is the projector in $L^2(\Omega)$ on the space N^* defined by (3.2). More precisely, let $u_1, \ldots, u_k$ be a basis in N^* and $v_1, \ldots, v_k \in D(\Omega)$ be such that $(u_i, v_j) = \delta_{ij}$, $i,j = 1, \ldots, k$. Then

$$Pu = \sum_{i=1}^{k}(u, v_i)u_i.$$

The norm in $V_*^{2m}(\Omega)$ is given by the equality $\|u\|_{V_*^{2m}(\Omega)} = \sqrt{[u,u]}$. Similar to Section 2 we can prove that $V_*^{2m}(\Omega)$ is a Hilbert space and

$$\|u\|_{H_\infty^{2m}(\Omega)} \leq c\|u\|_{V_*^{2m}(\Omega)}, \quad \forall u \in V_*^{2m}(\Omega),$$
$$\|u\|_{V_*^{2m}(\Omega)} \leq c\|u\|_{H^{2m}(\Omega)}, \quad \forall u \in H^{2m}(\Omega),$$
$$H^{2m}(\Omega) \subset V_*^{2m}(\Omega) \subset H_\infty^{2m}(\Omega).$$

Set

$$[u,v]_0 = \int_\Omega A^*u\overline{A^*v}\,dx + \sum_{j=1}^{m}\langle C_j u, C_j v\rangle_j. \tag{3.5}$$

Then similar to Proposition 2.5 we have the following proposition.

Proposition 3.2. *Let $u \in V_*^{2m}(\Omega)$, $f \in L_1^2(\Omega) \cap H_{\mathrm{loc}}^r(\Omega)$, where r is a positive integer, and*

$$[u,v]_0 = \int_\Omega f\overline{v}\,dx, \quad \forall v \in V_*^{2m}(\Omega); \tag{3.6}$$

then $u \in H_{\mathrm{loc}}^{4m+r}(\Omega)$.

Since for $f_0 \in D(\bar\Omega)$, (v, f_0) is a continuous linear functional on $V_*^{2m}(\Omega)$ (see Proposition 2.4), then there exists $w \in V_*^{2m}(\Omega)$ such that

$$[w,v] = \int_\Omega f_0 \overline{v}\,dx, \quad \forall v \in V_*^{2m}(\Omega). \tag{3.7}$$

Write

$$w' = w - Pw. \tag{3.8}$$

Let $v' \in V_*^{2m}(\Omega)$ and $Pv' = 0$. Then from (3.7) we have

$$[w, v'] = \int_\Omega f_0 \,\overline{v'}\,dx. \tag{3.9}$$

But

$$[w, v'] = [w' + Pw, v'] = [w', v'] + [Pw, v']. \tag{3.10}$$

Obviously, $[Pw, v'] = (Pw, Pv')$ since $Pw \in N^*$. Hence $[Pw, v'] = 0$, and from (3.10) we get $[w, v'] = [w', v']$. Moreover, (3.9) implies

$$[w', v'] = (f_0, v'). \tag{3.11}$$

3. Fredholm theorems

Let $v \in V_*^{2m}(\Omega)$. Then we have
$$[w', v] = [w', v'] + [w', v''],$$
where $v'' = Pv$, $v' = (I - P)v$, and $Pv' = 0$. Further,
$$[w', v''] = [w', Pv] = (Pw', Pv) = 0$$
since $Pw' = 0$ by virtue of (3.8). It follows that $[w', v] = [w', v']$. From (3.11)
$$[w', v] = (f_0, v'). \tag{3.12}$$
From assumptions (3.3) and (3.1) we have $(f_0, v_0) = 0$ for all $v_0 \in N^*$. In particular, $(f_0, v'') = 0$ since $v'' = Pv \in N^*$. Hence
$$(f_0, v) = (f_0, v' + v'') = (f_0, v').$$
It follows from (3.12) that
$$[w', v] = (f_0, v), \quad \forall v \in V_*^{2m}(\Omega). \tag{3.13}$$
From (3.8) it follows that $Pw' = 0$. Therefore, by virtue of (3.5) $[w', v] = [w', v]_0$. From (3.13)
$$[w', v]_0 = (f_0, v), \quad \forall v \in V_*^{2m}(\Omega). \tag{3.14}$$
From Proposition 3.2 we conclude that
$$w' = H_{\text{loc}}^{4m+r}(\Omega) \tag{3.15}$$
for any integer r. Hence w' is a smooth function. Write
$$u = A^* w'. \tag{3.16}$$
From (3.5) and (3.14) we get
$$\int_\Omega u \overline{A^* v} \, dx + \sum_{j=1}^m \langle C_j w', C_j v \rangle_j = (f_0, v), \quad \forall v \in V_*^{2m}(\Omega). \tag{3.17}$$
In particular,
$$\int_\Omega u \overline{A^* v} \, dx = (f_0, v), \quad \forall v \in D(\Omega).$$
It follows that
$$Au = f_0. \tag{3.18}$$
Now, from (3.17) we obtain
$$(u, A^* v) + \sum_{i=1}^m \langle C_j w', C_j v \rangle_j = (Au, v), \quad \forall v \in V_*^{2m}(\Omega).$$

We take here $v \in D(\bar{\Omega})$ such that
$$C_j v = 0, \quad j = 1, \ldots, m. \tag{3.19}$$
Then $(u, A^*v) = (Au, v)$. We use Green's formula (1.2) for u and $v \in D(\bar{\Omega})$ satisfying (3.19). We get
$$\sum_{j=1}^{m} \int_{\Gamma} B_j u \, \overline{T_j v} \, ds = 0.$$
It follows that
$$B_j u = 0, \quad j = 1, \ldots, m \tag{3.20}$$
(see [318], Corollary 2.1). From (3.15), (3.16) it follows that $u \in H^{2m}_{loc}(\Omega)$. It remains to study the behavior of u at infinity. Since Condition NS is satisfied, and $\omega_\mu L u \in F_\infty$, we get $\mu u \in H^{2m}_\infty$ (Corollary 1.6). Therefore $u \in H^{2m}$. Proposition 3.1 is proved in the case (3.3).

2. Consider now the case
$$f_0 \in D(\bar{\Omega}), \quad f_j \in D(\Gamma), \quad j = 1, \ldots, m. \tag{3.21}$$
This case can be reduced to the previous one. Let $u_1 \in D(\bar{\Omega})$ be a function satisfying the boundary conditions in (1.1):
$$B_j u_1 = f_j \text{ on } \Gamma, \quad j = 1, \ldots, m.$$
Existence of such a function follows from [318], Lemma 2.2. Write $g_0 = f_0 - A u_1$. Then $g_0 \in D(\bar{\Omega})$. Consider the problem
$$A u_0 = g_0 \text{ in } \Omega, \quad B_j u_0 = 0 \text{ on } \Gamma, \quad j = 1, \ldots, m. \tag{3.22}$$
For any $v_0 \in N^*$ we have
$$\int_\Omega g_0 \overline{v_0} dx = \int_\Omega f_0 \overline{v_0} dx - \int_\Omega A u_1 \overline{v_0} dx. \tag{3.23}$$
Applying Green's formula to u_1 and v_0, we get
$$\int_\Omega A u_1 \overline{v_0} dx = -\sum_{j=1}^{m} \int_\Gamma B_j u_1 \overline{T_j v_0} ds = -\sum_{j=1}^{m} \int_\Gamma f_j \overline{T_j v_0} ds.$$
From (3.23) it follows that
$$\int_\Omega g_0 \overline{v_0} dx = \int_\Omega f_0 \overline{v_0} dx + \sum_{j=1}^{m} \int_\Gamma f_j \overline{T_j v_0} dx.$$
Therefore (3.1) implies
$$\int_\Omega g_0 \overline{v_0} dx = 0, \quad \forall v_0 \in N^*.$$

3. Fredholm theorems

It was proved in the previous section that there exists a solution $u_0 \in H^{2m}(\Omega)$ of problem (3.22). Obviously $u = u_0 + u_1$ is a solution of problem (1.1). Thus Proposition 3.1 is proved in the case (3.21). □

We introduce also the following duality: for any $f = (f_0, f_1, \ldots, f_m) \in L_\infty^{p_0}(\bar{\Omega}) \times L_\infty^{p_1}(\Gamma) \times \cdots \times L_\infty^{p_m}(\Gamma)$ and $v = (v_0, v_1, \ldots, v_m) \in \mathcal{E}(\bar{\Omega}) \times \mathcal{E}(\Gamma) \times \cdots \times \mathcal{E}(\Gamma)$ we write

$$\langle f, v \rangle = \int_\Omega f_0 \overline{v_0} dx + \sum_{j=1}^m \int_\Gamma f_j \overline{v_j} ds. \tag{3.24}$$

Here $p_i \geq 1$, $i = 0, 1, \ldots, m$. Consider the subspace

$$\hat{N} = \{v | v = (v_0, T_1 v_0, \ldots, T_m v_0), \ v_0 \in N^*\}$$

and denote by $\dot{W}_\infty^{s,p}(\mathbb{R}^n)$, $s \geq 0$, $1 < p < \infty$ the closure of D in $W_\infty^{s,p}(\mathbb{R}^n)$. As always, $\dot{W}_\infty^{s,p}(\Omega)$ is the restriction of $\dot{W}_\infty^{s,p}(\mathbb{R}^n)$ to Ω. The definition of the space $\dot{W}_\infty^{s,p}(\Gamma)$ is standard.

Theorem 3.3. *Let Conditions NS and NS* be satisfied. Then for any*

$$f = (f_0, \ldots, f_m), \ f_0 \in D(\bar{\Omega}), \ f_j \in D(\Gamma), \ i = 1, \ldots, m$$

such that $\langle f, v \rangle = 0$, $\forall v \in \hat{N}$ *the problem*

$$Lu = f \tag{3.25}$$

has a solution $u \in \mathcal{E}$.

Proof. It is proved above that problem (3.25) has a solution $u \in H^{2m}(\Omega)$. From a priori estimates it follows that $u \in H^{2m+r}(\Omega)$ for any integer $r \geq 0$. From Corollary 1.6 it follows that for any $l \geq 2m$ there exists $\epsilon > 0$ such that $\omega_\epsilon u \in H_\infty^l(\Omega)$, where $\omega_\epsilon(x) = e^{\epsilon \sqrt{1+|x|^2}}$. From embedding theorems it follows that for $|\alpha| < l - n/2$ we get

$$|D^\alpha(\omega_\epsilon u(x))| \leq c \|\omega_\epsilon u\|_{H^l(\Omega)}, \ \forall x \in \Omega.$$

Therefore for any α there exists $\epsilon > 0$ such that

$$|D^\alpha(\omega_\epsilon u(x))| \leq c_\alpha, \ \forall x \in \Omega. \tag{3.26}$$

Writing $v = \omega_\epsilon u$, we get

$$|D^\alpha u(x)| \leq |D^\alpha(\omega_{-\epsilon} v(x))| \leq \omega_{-\epsilon} c'_\alpha, \ \forall x \in \Omega.$$

It follows that $u \in \mathcal{E}$. The theorem is proved. □

Let $D(\Omega,\Gamma) = \{u | u = (u_0, u_1, \ldots, u_m), u_0 \in D(\bar{\Omega}), u_j \in D(\Gamma), j = 1, \ldots, m\}$. We take a basis of $\hat{N}$: $v^1, \ldots, v^l$. Let $e^1, \ldots, e^l \in D(\Omega,\Gamma)$ be a biorthogonal system:

$$\langle e^k, v^i \rangle = \delta_{ik}, \quad i, k = 1, \ldots, l.$$

It follows from Theorem 3.3 that the problem

$$Lu = f - \sum_{i=1}^{l} \langle f, v^i \rangle e^i \qquad (3.27)$$

has a solution $u \in \mathcal{E}$ for any $f \in D(\Omega,\Gamma)$. Let l be an integer, $l \geq 2m$.

Theorem 3.4. *Let Conditions NS and NS* be satisfied. Then for any*

$$f = (f_0, f_1, \ldots, f_m), \; f_0 \in \dot{W}_\infty^{l-2m,p}(\Omega), \; f_j \in \dot{W}_\infty^{l-s_j-1/p,p}(\Gamma), \; j = 1, \ldots, m,$$

such that $\langle f, v \rangle = 0$ *for any* $v \in \hat{N}$ *the problem*

$$Lu = f \qquad (3.28)$$

has a solution $u \in \dot{W}_\infty^{l,p}(\Omega)$.

Proof. Let $\{f^k\}, k = 1, 2, \ldots, f^k \in D(\Omega,\Gamma)$ be a sequence such that $f^k \to f$ in the space

$$F = W_\infty^{l-2m,p}(\Omega) \times \Pi_{j=1}^m W_\infty^{l-s_j-1/p,p}(\Gamma).$$

Consider a solution $u^k \in \mathcal{E}$ of problem (3.27) for $f = f^k$. From Lemma 3.5 below we have

$$\|w^k\|_{W_\infty^{l,p}(\Omega)} \leq c \| f^k - \sum_{i=1}^{l} <f^k, v^i> e^i \|_F,$$

where $w^k = (I-P)u^k$. It follows that $w^k \to u$ in $W_\infty^{l,p}(\Omega)$. Obviously u is a solution of (3.28). From Corollary 1.6 we conclude that $w^k \in \mathcal{E}$. Hence $u \in \dot{W}_\infty^{l,p}(\Omega)$. The theorem is proved. $\square$

Lemma 3.5. *Let an operator* $L : E \to F$ *have a closed image and a finite-dimensional kernel. Denote by* P *a projection operator of* E *on* $\ker L$: $P^2 u = Pu$, $Pu \in \ker L$, $\forall u \in E$. *Then for any* $u \in E$ *we have*

$$\|u - Pu\|_E \leq c\|Lu\|_F,$$

where c *is a constant independent of* u.

Proof. Set $E_0 = (I-P)E$, where I is the unit operator in E. Let L_0 be a restriction of L on E_0. Then L_0 is invertible. For any $u \in E$ write $u_0 = (E-P)u$. Then $L_0 u_0 = L u_0 = Lu$. Hence $\operatorname{Im} L_0 = \operatorname{Im} L$. By the Banach theorem L_0^{-1} is bounded. Therefore $u_0 = L_0^{-1} Lu$, $\|u_0\|_E \leq c\|Lu\|_F$. The lemma is proved. $\square$

4 Spaces E_q

In this section we will use again the notation

$$E = W^{l,p}_\infty(\Omega), \quad F = W^{l-2m,p}_\infty(\Omega) \times W^{l-s_1-1/p,p}_\infty(\Gamma) \times \cdots \times W^{l-s_m-1/p,p}_\infty(\Gamma).$$

We recall that $\dot{E}_\infty$ and $\dot{F}_\infty$ are the closures of D in the norms of the spaces E and F, respectively. In the previous section we proved the Fredholm property of the operator $L : \dot{E}_\infty \to \dot{F}_\infty$. In this section we will use this result to prove the Fredholm property of the operator $L_q : E_q \to F_q$, where we denote by L_q the same operator acting in the corresponding spaces. We begin with some auxiliary results.

Lemma 4.1. *Let an operator L acting from a Banach space E into another space F have a bounded inverse defined on its image $R(L) \subset F$. Suppose that the equation $L_\mu u = f$ has a solution, where $L_\mu = L + \mu K$, $K : E \to F$ is a bounded operator, $\|K\| \leq M$. Then for μ sufficiently small $\|u\|_E \leq C\|f\|_F$, where the constant C depends on μ and M but does not depend on the operator K.*

Proof. Since the equation $Lu + \mu Ku = f$ has a solution, then $f - \mu Ku \in R(L)$. Therefore $u = L^{-1}(f - \mu Ku)$. The assertion of the lemma follows from the estimate

$$\|u\|_E \leq \|L^{-1}\|\|f - \mu Ku\|_F \leq \|L^{-1}\|(\|f\|_F + |\mu|\|K\|\|u\|_E).$$

The lemma is proved. □

We generalize here the approach developed in [362] for the operators acting in $H^s(\mathbb{R}^n)$. As above we use the function $\omega_\delta(x) = \exp(\delta\sqrt{1+|x|^2})$ and a system of functions $e_i \in \dot{F}_\infty$ biorthogonal to the functions v_j that form a basis in $\hat{N}$ (Section 3). Since infinitely differentiable functions with bounded supports are dense in $\dot{F}_\infty$, we can choose functions e_i with bounded supports. We will assume this condition satisfied.

Lemma 4.2. *For any $f \in \dot{F}_\infty$ there exists a solution u_j of the equation*

$$Lu = \theta_j f - \sum_{i=1}^N \langle \theta_j f, v^i \rangle e^i, \tag{4.1}$$

and for δ sufficiently small the following estimate holds:

$$\|u_j(\cdot)\omega_\delta(\cdot - y_j)\|_{E_q} \leq C\|f\theta_j\|_F, \tag{4.2}$$

where B_j is a unit ball with its center at y_j, $\operatorname{supp}\theta_j \subset B_j$, and the constant C is independent of j.

Proof. Since the operator $L : \dot{E}_\infty \to \dot{F}_\infty$ satisfies the Fredholm property, then the equation

$$Lu = g - \sum_{i=1}^N \langle g, v^i \rangle e^i \tag{4.3}$$

is solvable for any $g \in \dot{F}_\infty$. Let $\operatorname{supp} g \in B_j$. Consider the function

$$\tilde{g}(x) = \left(g(x) - \sum_{i=1}^{N} \langle g, v_i \rangle e_i(x) \right) \omega_\delta(x - y_j).$$

We show that its norm in F_∞ is independent of j. We note first of all that $\omega_\delta(x-y_j)$ is bounded in B_j together with all derivatives independently of j. Therefore

$$\|g(x)\omega_\delta(x - y_j)\|_{F_\infty} \leq C\|g(x)\|_{F_\infty}$$

with a positive constant C independent of j. We use here that the norm of the multiplier in Ω and $\partial\Omega$ can be estimated by the norm in C^k. Taking (3.24) into account, we obtain

$$|\langle g, v^i \rangle| = |\langle g, \psi_j v^i \rangle| \leq C \sup_x |\omega_{-\mu}(x)\psi_j| \, \|g\|_{F_\infty},$$

where $\psi_j = \psi(x-y_j)$ is a function with a finite support equal to 1 in B_j, $v^i \in \mathcal{E}(\Omega)$. We have

$$\sup_x |\omega_{-\mu}(x)\psi_j| \leq \sup_x |\omega_{-\mu}(x + y_j)\psi(x)| \leq C\omega_{-\mu}(y_j)$$

with C independent of j. For $\delta \leq \mu$ the product $\omega_{-\mu}(y_j)\omega_\delta(x - y_j)$ is bounded independently of $y_j \in R^n$ and of $x \in \operatorname{supp} e^i(x)$. Hence $\|\tilde{g}\|_{F_\infty} \leq C\|g\|_{F_\infty}$, where the constant C depends on the diameter of the supports of e_i but is independent of j.

Since u is a solution of equation (4.3), then $\tilde{u} = S_\delta u$ is a solution of the equation

$$L_\delta \tilde{u} = \tilde{g}, \tag{4.4}$$

where $L_\delta = S_\delta L S_{-\delta}$, and S_δ is the operator of multiplication by $\omega_\delta(x - y_j)$. On the other hand, $L_\delta = L + \delta K$, where K is a bounded operator, $\|K\| \leq C$, where C does not depend on j and on δ for δ sufficiently small. By virtue of Lemma 4.1 the solution of (4.4), which belongs to the subspace supplementary to the kernel of the operator L, admits the estimate

$$\|\tilde{u}\|_{E_\infty} \leq C_1 \|\tilde{g}\|_{F_\infty} \leq C_2 \|g\|_{F_\infty}$$

independent of j. Let $\delta = \delta_1 + \delta_2$, where δ_1 and δ_2 are positive. Then

$$\|u(x)\omega_{\delta_1}(x - y_j)\|_{E_q} = \|\tilde{u}(x)\omega_{-\delta_2}(x - y_j)\|_{E_q} \leq C_3 \|\tilde{u}(x)\|_{E_\infty} \leq C_4 \|g\|_{F_\infty}.$$

Applying this estimate to equation (4.1), we obtain

$$\|u(x)\omega_{\delta_1}(x - y_j)\|_{E_q} \leq C_5 \|f\theta_j\|_{F_\infty} \leq C_6 \|f\theta_j\|_F.$$

The lemma is proved. □

4. Spaces E_q

Assumption 4.3. *Let $u_j \in E_q$, $j = 1, 2, \ldots,$ and*

$$\sum_{j=1}^{\infty} \|u_j\, \omega_\delta(x - y_j)\|_{E_q}^q < \infty.$$

Then the series $u = \sum_{j=1}^{\infty} u_j$ is convergent, and the following estimate holds:

$$\|u\|_{E_q}^q \leq C \sum_{j=1}^{\infty} \|u_j\, \omega_\delta(x - y_j)\|_{E_q}^q. \tag{4.5}$$

If this assumption is satisfied, then from the estimate in Lemma 4.2 we obtain $\|u\|_{E_q} \leq C\|f\|_{F_q}$. Therefore for any $f \in F_q (\subset F_\infty)$ there exists a solution $u \in E_q$ of the equation

$$Lu = f - \sum_{i=1}^{N} \langle f, v_i \rangle e_i. \tag{4.6}$$

From this follows that the operator L_q is normally solvable and the codimension of its image is finite. Its kernel is also finite dimensional since it is true for the operator L_∞. Hence L_q is a Fredholm operator. We note that estimate (4.5) characterizes the function spaces and is not related to the operators under consideration. It was proved in Section 5 of Chapter 5 that it is satisfied for Sobolev spaces.

Suppose that $l \geq 2m$ is an integer, and $1 < p < \infty$. Then the following theorem holds.

Theorem 4.4. *Let Conditions NS and NS* be satisfied. Then for $f \in W^{l-2m,p}(\Omega) \times \prod_{j=1}^{m} W^{l-s_j-1/p,p}(\Gamma)$ the problem $Lu = f$ has a solution $u \in W^{l,p}(\Omega)$ if and only if $\langle f, v \rangle = 0$ for any $v \in \hat{N}$.*

Let us now show that the Fredholm property holds not only for $q = p$ but also for $q \leq p$. We will verify Assumption 4.3. Let $E = L^p$, $E_q = (L^p)_q$. Then

$$\|u\|_{E_q}^q = \sum_j \|\phi_j u\|_{L^p}^q.$$

If $u = \sum_i u_i$, then from Lemma 5.7 of Chapter 5,

$$\|\phi_j u\|_{L^p}^p \leq C \sum_i \|\phi_j u_i \omega_\delta(x - y_i)\|_{L^p}^p.$$

From this estimate and the previous equality we have

$$\|u\|_{E_q}^q \leq C^{q/p} \sum_j \left(\sum_i \|\phi_j u_i \omega_\delta(x - y_i)\|_{L^p}^p \right)^{q/p}.$$

On the other hand,
$$\sum_i \|u_i\omega_\delta(x-y_i)\|_{E_q}^q = \sum_i \sum_j \|\phi_j u_i \omega_\delta(x-y_i)\|_{L^p}^q.$$

Therefore, to verify Assumption 4.3 it is sufficient to verify the estimate
$$\left(\sum_i \|\phi_j u_i \omega_\delta(x-y_i)\|_{L^p}^p\right)^{q/p} \leq \sum_i \|\phi_j u_i \omega_\delta(x-y_i)\|_{L^p}^q.$$

It is satisfied if $q \leq p$ (see Lemma 5.7 of Chapter 5). This proves Theorem 1.3. □

5 Smoothness of solutions

Denote by $C_B^\infty(\Omega)$ the class of infinitely differentiable functions defined in Ω and such that, for any multi-index α,
$$\sup_{x \in \Omega} |D^\alpha u(x)| < \infty.$$

Similarly $C_B^\infty(\Gamma)$ is the class of functions defined on the boundary Γ of Ω, all of whose derivatives are bounded. Consider elliptic problem (1.1). It is not supposed here to be regular.

Theorem 5.1. *If* $u \in W_q^{l,p}(\Omega)$ ($l \geq 2m, 1 < p < \infty, 1 \leq q \leq \infty$) *is a solution of problem* (1.1), *and* $f_0 \in C_B^\infty(\Omega)$, $f_j \in C_B^\infty(\Gamma)$ ($j = 1, \ldots, m$), *then* $u \in C_B^\infty(\Omega)$.

Proof. Since $W_q^{l,p}(\Omega) \subset W_\infty^{l,p}(\Omega)$, it suffices to consider $q = \infty$. The following estimate is proved in Chapter 3:

$$\|u\|_{W_\infty^{k,p}(\Omega)} \leq c_{kp} \left(\|f_0\|_{W_\infty^{k-2m,p}(\Omega)} + \sum_j \|f_j\|_{W_\infty^{k-s_j-1/p,p}(\Gamma)} + \|u\|_{L_\infty^p(\Omega)}\right) \quad (5.1)$$

for any $k \geq l$. We recall that the coefficients and the boundary of the domain are supposed to be infinitely differentiable. Since $f_0 \in C_B^\infty(\Omega)$ and $f_j \in C_B^\infty(\Gamma)$, we get
$$\|u\|_{W_\infty^{k,p}(\Omega)} \leq M_{kp}, \quad \forall k \geq l.$$

Let α be an arbitrary multi-index and $k > n/p$. Then by the Sobolev embedding theorem we have
$$\sup_{x \in \Omega} |D^\alpha v(x)| \leq c_{|\alpha|,k,p} \|v\|_{W^{|\alpha|+k,p}(\Omega)}, \quad \forall v \in W^{|\alpha|+k,p}(\Omega).$$

It follows from Definition 1.1 that
$$\|u\|_{W_\infty^{|\alpha|+k,p}(\Omega)} = \sup_{y \in \Omega} \|u(\cdot)\omega(\cdot - y)\|_{W^{|\alpha|+k,p}(\Omega)}.$$

Hence for any $y \in \Omega$ we have
$$\sup_{x \in \Omega} |D_x^\alpha(u(x)\omega(x-y))| \leq C_\alpha,$$
where the constant C_α does not depend on y. In particular, for $y = x$ we get
$$\sup_{x \in \Omega} |D_x^\alpha u(x)| \leq C_\alpha.$$
The theorem is proved. □

It follows from this theorem that Conditions NS and NS* can be formulated in the following equivalent form where the spaces are different in comparison with the definitions in Section 1.

Condition NS. Any limiting problem
$$\hat{L}u = 0, \quad x \in \Omega_*, \ u \in C_B^\infty(\Omega_*)$$
has only the zero solution.

Condition NS*. Any limiting problem
$$\hat{L}^* v = 0, \quad x \in \Omega_*, \ v \in C_B^\infty(\Omega_*)$$
has only the zero solution.

Corollary 5.2. *Let Condition NS be satisfied. Then any solution $u \in W_q^{l,p}(\Omega)$ of the problem $Lu = 0$ belongs to the space $\mathcal{E}(\bar{\Omega})$. Here $1 < p < \infty, 1 \leq q \leq \infty$, $\mathcal{E}(\bar{\Omega})$ is introduced in Section 1.*

The proof follows from Corollary 1.6 if we use the estimates from the proof of Theorem 5.1.

6 Examples

Scalar operators. Formally adjoint problems are introduced in this chapter for scalar elliptic operators by means of Green's formula. If the boundary operators B_j, $j = 1, \ldots, m$ form a normal system, then the formally adjoint problem can be defined. In some cases, boundary operators of the formally adjoint problem can be determined more explicitly. In the case of the Dirichlet boundary conditions,
$$B_j u = \frac{\partial^{j-1} u}{\partial n^{j-1}}, \quad j = 1, \ldots, m,$$
where n is an interior normal vector, there exists a formally adjoint operator also with the Dirichlet boundary conditions [318].

There are other examples and particular cases where formally adjoint operators can be easily determined. For example, the Laplace operator with the Neumann boundary condition is formally self-adjoint. In the case of the Laplace operator with oblique derivative, the formally adjoint operator may not be defined.

Vector operators in $\mathbb{R}^n$. Some results of this chapter can be generalized for some classes of vector operators, in particular, for operators in $\mathbb{R}^n$. Following [361], [362], consider uniformly elliptic operators in the sense of Petrovskii,

$$Lu = \sum_{|p|\leq r} A_p(x) D^p u,$$

where $A_p(x)$ are real matrix functions with their elements in $C^{r+\alpha}(\mathbb{R}^n)$, $0 < \alpha < 1$,

$$\inf_{x,\xi\in\mathbb{R}^n, |\xi|=} \left| \det \sum_{|p|=r} A_p(x)\xi^p \right| > 0.$$

The operator L acts from $H^r(\mathbb{R}^n)$ into $H^0(\mathbb{R}^n) = L^2(\mathbb{R}^n)$. The formally adjoint operator is given by the formula

$$L^*v = \sum_{|p|\leq r} (-1)^{|p|} D^p (A_p^T(x) v),$$

where the superscript T indicates the transposed matrices. This operator acts in the same spaces as the operator L.

Limiting operators $\hat{L}$ and $\hat{L}^*$ for the direct and for the formally adjoint operators are defined in the usual way. If all limiting equations $\hat{L}u = 0$ and $\hat{L}^*v = 0$ have only zero solutions in the space $H^0_\infty(\mathbb{R}^n)$, then the operator L satisfies the Fredholm property and the equation $Lu = f$ where $f \in H^0_\infty(\mathbb{R}^n)$ is solvable if and only if

$$\int_{\mathbb{R}^n} (f, v) dx = 0$$

for any solution $v \in H^r(\mathbb{R}^n)$ of the equation $L^*v = 0$.

Solvability conditions with formally adjoint operators. In the case where a formally adjoint operator can be defined, it allows a more explicit formulation of solvability conditions. It is discussed in this chapter for general scalar elliptic problems. A simpler approach can be developed under some additional conditions. We assume that an elliptic operator L has a formally adjoint operator L^*, that both of them satisfy the Fredholm property, and that the sum of their indices equals zero. Both operators L and L^* act from a space E into another space F. Here E and F are some Sobolev or Hölder spaces specific for elliptic problems. Denote by α the dimension of the kernel and by β the codimension of the image, $\kappa = \alpha - \beta$ is the index. For simplicity of presentation, we restrict ourselves to operators in $\mathbb{R}^n$ in order not to deal with boundary operators, and consider spaces of real-valued functions.

Lemma 6.1. $\beta(L) \geq \alpha(L^*)$.

6. Examples

Proof. The equation
$$Lu = f \tag{6.1}$$
is solvable for some $f \in F$ if and only if
$$\phi_i(f) = 0, \quad i = 1, \ldots, \beta(L), \tag{6.2}$$
where ϕ_i are some linearly independent functionals from the dual space F^*. Consider the functionals
$$\psi_j(f) = \int_\Omega f(x) v_j(x) dx, \quad j = 1, \ldots, \alpha(L^*),$$
where u_j are linearly independent solutions of the equation $L^* v = 0$. The integrals here are well defined since the functions v_j are exponentially decreasing at infinity. Obviously, $\psi_j \in F^*$. If $\beta(L) < \alpha(L^*)$, then among the functionals ψ_j there exists at least one, which is linearly independent with respect to the functional ϕ_i. Let it be for example ψ_1. Then there exists a function $f \in F$ such that conditions (6.2) are satisfied but $\psi_1(f) \neq 0$. Multiplying (6.1) by v_1 and integrating, we obtain a contradiction. The lemma is proved. □

We recall that L is a formally adjoint operator of the operator L^*. Hence, similarly to Lemma 6.1, it can be proved that $\beta(L^*) \geq \alpha(L)$. Therefore $\kappa(L) + \kappa(L^*) \leq 0$. If we suppose that
$$\kappa(L) + \kappa(L^*) = 0, \tag{6.3}$$
then $\beta(L) = \alpha(L^*)$, $\beta(L^*) = \alpha(L)$.

Theorem 6.2. *Suppose that condition (6.3) is satisfied. Then equation (6.1) is solvable if and only if*
$$\int_\Omega f(x) v_j(x) dx = 0, \quad j = 1, \ldots, \alpha(L^*),$$
where v_j are linearly independent solutions of the equation $L^* v = 0$.

Proof. Consider two subspaces of the space F^*. One of them is formed by the functionals ψ_j defined in the proof of Lemma 6.1, and another one by the functionals ϕ_j. From Lemma 6.1 it follows that their dimensions are equal to each other. It remains to conclude that they coincide. Indeed, otherwise we would obtain the same contradiction with the solvability of equation (6.1) as in the proof of the lemma. The theorem is proved. □

We illustrate this result and show how it can be adapted for problems in domains. Consider the scalar second-order operator
$$Lu = \Delta u + \sum_{i=1}^n b_i(x) \frac{\partial u}{\partial x_i} + c(x) u$$

with sufficiently smooth coefficients. Instead of the whole $\mathbb{R}^n$, we consider a domain Ω, with a sufficiently smooth boundary, and the Dirichlet boundary conditions. We do not consider the boundary operators but take into account the boundary condition in the definition of function spaces. In the case of Hölder spaces, the operator L can be considered as acting from the space

$$E = \{u \in C^{2+\delta}(\bar{\Omega}), \ u\,|_{\partial\Omega} = 0\}$$

into the space $F = C^{\delta}(\bar{\Omega})$. Denote by L^* the formally adjoint operator,

$$L^*u = \Delta u - \sum_{i=1}^{n} \frac{\partial(b_i(x)u)}{\partial x_i} + c(x)u$$

acting in the same spaces.

If the domain Ω is bounded, then both operators L and L^* satisfy the Fredholm property with zero index. The solvability conditions can be formulated in terms of orthogonality to solutions of the homogeneous formally adjoint equation. If the domain is unbounded, then we need to impose the additional condition on limiting operators. If we suppose for example that the function $c(x)$ is negative at infinity, that is $c(x) \leq c_0 < 0$ for $|x| \geq R$ and some positive R, then it can be verified that all limiting equations have only zero solutions. Therefore, the Fredholm property is satisfied. Moreover, since it is also satisfied for the operator $L - \lambda$ for any positive λ, and it is invertible for λ sufficiently large (Chapter 7), then the index of the operator L equals zero. The same is true for the operator L^*. The theorem formulated above is applicable.

Chapter 7

Elliptic Problems with a Parameter

In this chapter we study general elliptic problems with a parameter for mixed-order systems in unbounded domains $\Omega \subset \mathbb{R}^n$:

$$A(x, \lambda, D)u = f, \quad x \in \Omega, \tag{0.1}$$
$$B(x, \lambda, D)u = g, \quad x \in \partial\Omega. \tag{0.2}$$

Precise definitions are given below. We will obtain a priori estimates of solutions and will prove the existence and the uniqueness of solutions for λ in a given sector S, and $|\lambda| \geq \lambda_0 > 0$.

Elliptic problems with a parameter form an important class of elliptic problems because it is possible to prove for them the existence and the uniqueness of solutions. As it is well known, they are also used to study evolution problems. Moreover, it turns out that parameter-elliptic problems may be applied to study the Fredholm property of general elliptic problems in unbounded domains. It was proved in previous chapters that a general elliptic partial differential operator is Fredholm if its limiting operators at infinity are invertible. This is why we introduce here a new class of operators – elliptic operators with a parameter at infinity. They will be invertible for $\lambda \in S, |\lambda| \geq \lambda_0$ and some $\lambda_0 > 0$. It follows that elliptic operators in domain Ω, which are parameter-elliptic at infinity, are Fredholm for $\lambda \in S, |\lambda| \geq \lambda_0$. Obviously, any parameter-elliptic operator in the domain Ω is also parameter-elliptic at infinity but the corresponding values of λ_0 are different. This is essential in the analysis of the location of the Fredholm spectrum. We note also that the parameter-ellipticity at infinity plays an essential role in the index theory for elliptic operators in unbounded domains (see the next chapter).

We will use the Agranovich-Vishik method adapted for unbounded domains. We obtain these results not directly in Sobolev spaces but in the scale of spaces

$W_q^{l,p}$ ($1 < p < \infty$, $1 \leq q \leq \infty$). The space $W_q^{l,p}$ coincides with the Sobolev space $W^{l,p}$ for $q = p$ ($1 < p < \infty$) and with the Sobolev-Stepanov space for $q = \infty$, $1 < p < \infty$. For arbitrary domains $\Omega \subset \mathbb{R}^n$ with $C^{k+\alpha}$ boundary we obtain a priori estimates in the spaces $W_q^{l,p}$ ($1 \leq q \leq \infty$) and construct the inverse operator first for the case $q = \infty$. Then we use this result to prove an existence theorem for the case $q < \infty$. For this we consider functions f and g in (0.1), (0.2) with compact supports. The set of such functions is dense in the case $q < \infty$. It is known (see Chapters 4 and 5) that any solution $u \in W_\infty^{l,p}$ of (0.1), (0.2) with such f, g decays exponentially at infinity and hence belongs to $W_q^{l,p}$ ($q < \infty$). From this and a priori estimates we obtain the desired existence result.

1 Parameter-elliptic boundary value problems

Consider the matrix operator $A(x, \lambda, D)$ with elements

$$A_{ij}(x, \lambda, D) = \sum_{|\alpha|+\beta \leq \alpha_{ij}} a_{ij}^{\alpha\beta}(x) \lambda^\beta D^\alpha, \quad i, j = 1, \ldots, N$$

in an unbounded domain Ω, and the boundary operator $B(x, \lambda, D)$ with elements

$$B_{kj}(x, \lambda, D) = \sum_{|\alpha|+\beta \leq \beta_{kj}} b_{kj}^{\alpha\beta}(x) \lambda^\beta D^\alpha, \quad k = 1, \ldots, r, \ j = 1, \ldots, N.$$

Consider a sector S of the complex plane,

$$S = \{\lambda : \sigma_1 \leq \arg \lambda \leq \sigma_2\},$$

where we do not exclude the case $\sigma_1 = \sigma_2$. For each fixed $\lambda \in S$, let the operator $L = (A, B)$ be an elliptic operator in the Douglis-Nirenberg sense. We recall that this implies the existence of some integers s_i, t_j, σ_k such that

$$s_i + t_j = \alpha_{ij}, \quad \sigma_k + t_j = \beta_{kj}.$$

We suppose that the coefficients of the operator are defined for $x \in \mathbb{R}^n$ and

$$a_{ij}^{\alpha\beta}(x) \in C^{l-s_i+\theta}(\mathbb{R}^n), \quad b_{kj}^{\alpha\beta}(x) \in C^{l-\sigma_k+\theta}(\mathbb{R}^n), \quad 0 < \theta < 1.$$

We also assume that the domain Ω satisfies the following condition.

Condition D. For each $x_0 \in \partial\Omega$ there exists a neighborhood $U(x_0)$ such that:

1. $U(x_0)$ contains a sphere with radius δ and center x_0, where δ is independent of x_0.
2. There exists a homeomorphism $\psi(x; x_0)$ of the neighborhood $U(x_0)$ on the unit sphere $B = \{y : |y| < 1\}$ in $\mathbb{R}^n$ such that the images of $\Omega \bigcap U(x_0)$ and

1. Parameter-elliptic boundary value problems   253

$\partial\Omega \cap U(x_0)$ coincide with $B_+ = \{y : y_n > 0, |y| < 1\}$ and $B_0 = \{y : y_n = 0, |y| < 1\}$ respectively.

3. The function $\psi(x; x_0)$ and its inverse belong to the Hölder space $C^{m_0+\theta}$, $0 < \theta < 1$, $m_0 = \max_{i,j,k}(l + t_i, l - s_j, l - \sigma_k)$. Their $\|\cdot\|_{m_0+\theta}$-norms are bounded uniformly in x_0.

It can be proved that this condition is satisfied if and only if the domain is uniformly regular in the sense of [87]. The operator L is supposed to be uniformly elliptic with a parameter (see Definition 1.2 below). For any $x \in \bar{\Omega}$ consider the matrix $A(x, \lambda, \xi)$ with elements

$$A_{ij}(x, \lambda, \xi) = \sum_{|\alpha|+\beta \leq \alpha_{ij}} a_{ij}^{\alpha\beta}(x) \lambda^\beta \xi^\alpha, \quad i,j = 1,\ldots,N,$$

where $\xi = (\xi_1, \ldots, \xi_n)$. For any $x \in \partial\Omega$ consider the local coordinates (ξ', ν), where $\xi' = (\xi_1, \ldots, \xi_{n-1})$ are the coordinates in the tangential hyperspace, ν is the normal coordinate. Let $A(\lambda, \xi', \nu)$ and $B(\lambda, \xi', \nu)$ be the matrices with elements

$$A_{ij}(\lambda, \xi', \nu) = \sum_{|\alpha'|+\alpha_n+\beta=\alpha_{ij}} a_{ij}^{\alpha'\alpha_n\beta} \lambda^\beta \xi'^{\alpha'} \nu^{\alpha_n}, \quad i,j = 1,\ldots,N,$$

$$B_{ij}(\lambda, \xi', \nu) = \sum_{|\alpha'|+\alpha_n+\beta=\beta_{ij}} b_{ij}^{\alpha'\alpha_n\beta} \lambda^\beta \xi'^{\alpha'} \nu^{\alpha_n}, \quad i = 1,\ldots,r,\ j = 1,\ldots,N$$

(the dependence of the coefficients on x is not indicated). We recall that the Lopatinskii matrix is given by the equality

$$\Lambda(\lambda, \xi') = \int_{\gamma_+} B(\lambda, \xi', \mu) A^{-1}(\lambda, \xi', \mu) \Phi(\mu) d\mu,$$

where

$$\Phi(\mu) = (E, \mu E, \ldots, \mu^{s-1} E),$$

E is the identity matrix of the order N, $s = \max_{i,j} \alpha_{ij}$, γ_+ is a Jordan curve in the half-plane $\operatorname{Im}\mu > 0$ enclosing all the roots of $\det A(\lambda, \xi', \mu)$ with positive imaginary parts. By virtue of the condition of proper ellipticity there are r such roots, and there are no roots on the real axis. The Lopatinskii condition implies that the rank of the matrix $\Lambda(\lambda, \xi')$ equals r for all $|\xi'| \neq 0$.

We will use the following notation:

$$e_d^0 = \inf_{x \in \Omega, |\xi|=1} |\det A(x, 0, \xi)|,$$

$$e_d = \inf_{x \in \Omega, |\xi|+|\lambda|=1, \lambda \in S} |\det A(x, \lambda, \xi)|,$$

$$M_d = \max_{|\alpha|+\beta \leq \alpha_{ij}, i,j=1,\ldots,N} \|a_{ij}^{\alpha\beta}\|_{C^{l-s_i}(\Omega)},$$

$$M_\Gamma = \max_{|\alpha|+\beta \leq \beta_{ij}, i=1,\ldots,r, j=1,\ldots,N} \|b_{ij}^{\alpha\beta}\|_{C^{l-\sigma_i}(\Gamma)},$$

$$e_\Gamma^0 = \inf_{x \in \Gamma, |\xi'|=1} \sum_\alpha |\mu_\alpha(x, 0, \xi')|,$$

$$e_\Gamma = \inf_{x \in \Gamma, |\xi'|+|\lambda|=1, \lambda \in S} \sum_\alpha |\mu_\alpha(x, \lambda, \xi')|,$$

where $\mu_\alpha(x, \xi')$ are all r-minors of the Lopatinskii matrix in the local coordinates (ξ', ν) at the point x.

Definition 1.1. The operator $L(x, 0, D)$ is called uniformly elliptic if $e_d^0 > 0$, $e_\Gamma^0 > 0$, $M_d < \infty$, $M_\Gamma < \infty$.

Definition 1.2. The operator $L(x, \lambda, D)$ is called uniformly elliptic with a parameter if $e_d > 0$, $e_\Gamma > 0$, $M_d < \infty$, $M_\Gamma < \infty$.

2 Spaces

We consider the following function spaces:

$$E(\Omega) = \Pi_{i=1}^N W^{l+t_j, p}(\Omega),$$
$$F^d(\Omega) = \Pi_{j=1}^N W^{l-s_i, p}(\Omega),$$
$$F^b(\partial\Omega) = \Pi_{k=1}^r W^{l-\sigma_k - 1/p, p}(\partial\Omega)$$

and

$$F = F^d \times F^b.$$

The operator L can be considered as acting from E to F. In the case of unbounded domains it is convenient to use the spaces $E_q, 1 \leq q \leq \infty$ introduced in Chapter 2. Their norms are given by the equalities

$$\|u\|_{E_\infty(\Omega)} = \sup_i \|\phi_i u\|_{E(\Omega)},$$

$$\|u\|_{E_q(\Omega)} = \left(\sum_i \|\phi_i u\|_{E(\Omega)}^q\right)^{1/q}, \quad 1 \leq q < \infty.$$

We define similarly the spaces $F_q, 1 \leq q \leq \infty$. Here ϕ_i is a system of functions satisfying Condition 2.5 (Chapter 2) which we re-state here.

Condition 2.5 (Chapter 2). System of functions ϕ_i satisfies the following conditions:

1. $\phi_i(x) \geq 0$, $\phi_i \in D$.
2. For any i there exists no more than N functions ϕ_j such that $\operatorname{supp} \phi_j \cap \operatorname{supp} \phi_i \neq \emptyset$.
3. $\sup_i \|\phi_i\|_M < \infty$.

2. Spaces

4. $\phi(x) = \sum_{i=1}^{\infty} \phi_i(x) \geq m > 0$ for some constant m.
5. The following estimate holds: $\sup_x |D^\alpha \phi(x)| \leq M_\alpha$, where D^α denotes the operator of differentiation, and M_α are positive constants.

Here $\|\phi\|_M$ is the norm of a multiplier ϕ: $\|\phi u\|_E \leq \|\phi\|_M \|u\|_E$, $\forall u \in E$. For $E = W^{s,p}$ it is known that $\|\phi\|_M \leq \|\phi\|_{C^{[|s|]+1}}$, where K is a positive constant. For partitions of unity $\{\phi_i\}$ we always suppose that $\sup_i \|\phi_i\|_M < \infty$.

To study operators with a parameter, following [13] and [457] we introduce the norm
$$\|\|u\|\|_{W^{l,p}(\mathbb{R}^n)} = \|F^{-1}(1+|\xi|^2+|\lambda|^2)^{l/2}Fu\|_{L^p(\mathbb{R}^n)},$$
where F denotes the Fourier transform, $\xi = (\xi_1, \ldots, \xi_n)$. For $\lambda = 0$ this norm is the usual $W^{l,p}$ norm, for any λ fixed these norms are equivalent. On the other hand, the norm $\|\|\cdot\|\|_{W^{l,p}(\mathbb{R}^n)}$ is equivalent to the norm
$$[u]_{W^{l,p}(\mathbb{R}^n)} = \|u\|_{W^{l,p}(\mathbb{R}^n)} + |\lambda|^l \|u\|_{L^p(\mathbb{R}^n)}.$$

The proof uses Mikhlin's theorem. It is also used to prove the following assertion (cf. [13] and [457]).

Proposition 2.1. (Interpolation inequality.) *The estimate*
$$|\lambda|^{l-k}\|\|u\|\|_{W^{k,p}(\mathbb{R}^n)} \leq c_{kl}\|\|u\|\|_{W^{l,p}(\mathbb{R}^n)} \tag{2.1}$$
holds for any $u \in W^{l,p}(\mathbb{R}^n)$, $0 \leq k \leq l$ with some constants c_{kl} that depend on k, l, n, and p only.

To define the $\|\|\cdot\|\|$-norm in domains we put
$$\|\|u\|\|_{W^{l,p}(\Omega)} = \inf \|\|u^c\|\|_{W^{l,p}(\mathbb{R}^n)},$$
where the infimum is taken with respect to all $u^c \in W^{l,p}(\mathbb{R}^n)$ such that the restriction of $u^c(x)$ to Ω coincides with $u(x)$.

Proposition 2.2. (Interpolation inequality.) *The estimate*
$$|\lambda|^{l-k}\|\|u\|\|_{W^{k,p}(\Omega)} \leq c_{kl}\|\|u\|\|_{W^{l,p}(\Omega)} \tag{2.2}$$
holds for any $u \in W^{l,p}(\Omega)$, $0 \leq k \leq l$ with the same constants c_{kl} as in Proposition 2.1.

Proposition 2.3. *The estimate*
$$\|\|D^\alpha u\|\|_{W^{l,p}(\Omega)} \leq c\|\|u\|\|_{W^{l+|\alpha|,p}(\Omega)} \tag{2.3}$$
holds for any $u \in W^{l+|\alpha|,p}(\Omega)$ with a constant c independent of u and λ.

We consider next the spaces $W^{l-1/p,p}(\partial\Omega)$ with an integer $l \geq 1$, $1 < p < \infty$. We introduce the norm

$$\|\phi\|_{W^{l-1/p,p}(\partial\Omega)} = \inf \|u\|_{W^{l,p}(\Omega)},$$

where the infimum is taken over all $u \in W^{l,p}(\Omega)$ such that the restriction of u to $\partial\Omega$ coincides with ϕ.

Proposition 2.4. (Interpolation inequality.) *The estimate*

$$|\lambda|^{l-k}\|\phi\|_{W^{k-1/p,p}(\partial\Omega)} \leq c_{kl}\|\phi\|_{W^{l-1/p,p}(\partial\Omega)} \qquad (2.4)$$

holds for any $\phi \in W^{l-1/p,p}(\partial\Omega)$, $1 \leq k \leq l$ with the same constants c_{kl} as in Proposition 2.1.

We define next the $\|\|\cdot\|\|$-norms for the ∞-spaces. Let ϕ_i be a system of functions satisfying Condition 2.5 (Chapter 2). We put

$$\|\|u\|\|_{W^{l,p}_\infty(\Omega)} = \sup_i \|\phi_i u\|_{W^{l,p}(\Omega)}, \quad \forall u \in W^{l,p}_\infty(\Omega).$$

We note that $\|\|u\|\|_{W^{l,p}_\infty(\Omega)} < \infty$. It follows from the estimate

$$\|\phi_i u\|_{W^{l,p}(\Omega)} \leq c(\lambda)\|\phi_i u\|_{W^{l,p}(\Omega)}$$

that holds for any λ.

Proposition 2.5. *Let ϕ_i^1 and ϕ_i^2 be two equivalent systems of functions satisfying Condition* 2.5 *(Chapter* 2*). Then the norms $\|\|\cdot\|\|^1_{W^{l,p}_\infty(\Omega)}$ and $\|\|\cdot\|\|^2_{W^{l,p}_\infty(\Omega)}$ corresponding to these systems of functions are equivalent:*

$$c_1\|\|u\|\|^2_{W^{l,p}_\infty(\Omega)} \leq \|\|u\|\|^1_{W^{l,p}_\infty(\Omega)} \leq c_2\|\|u\|\|^2_{W^{l,p}_\infty(\Omega)}.$$

Here c_1 and c_2 are positive constants independent of λ.

This proposition remains similar for the space $W^{l-1/p,p}_\infty(\partial\Omega)$.

3 Model problem in $\mathbb{R}^n$

Consider the equation with constant coefficients $A(D,\lambda)u = f$ in $\mathbb{R}^n$. Applying the Fourier transform, we obtain $A(\xi,\lambda)\tilde{u}(\xi) = \tilde{f}(\xi)$. We use a tilde to denote a Fourier transform as well as the letter F. Hence

$$\tilde{u}(\xi) = A^{-1}(\xi,\lambda)\tilde{f}(\xi). \qquad (3.1)$$

3. Model problem in $\mathbb{R}^n$

In the notations of the previous section we have

$$\|u\|_{E(\mathbb{R}^n)} = \sum_{j=1}^{N} \|F^{-1}(1+|\xi|^2+|\lambda|^2)^{(l+t_j)/2} Fu_j\|_{L^p(\mathbb{R}^n)},$$

$$\|f\|_{F^d(\mathbb{R}^n)} = \sum_{j=i}^{N} \|F^{-1}(1+|\xi|^2+|\lambda|^2)^{(l-s_i)/2} Ff_i\|_{L^p(\mathbb{R}^n)}.$$

Set $\mu = \sqrt{1+|\xi|^2+|\lambda|^2}$. We have (cf. [542])

$$A(\mu\xi, \mu\lambda) = S(\mu) A(\xi, \lambda) T(\mu),$$

$$\|u\|_{E(\mathbb{R}^n)} = \|F^{-1}\mu^l T(\mu) Fu\|_{L^p(\mathbb{R}^n)}, \quad \|f\|_{F^d(\mathbb{R}^n)} = \|F^{-1}\mu^l S^{-1}(\mu) Fu\|_{L^p(\mathbb{R}^n)},$$

where

$$S(\rho) = (\delta_{ij}\rho^{s_i}), \quad T(\rho) = (\delta_{ij}\rho^{t_j})$$

are diagonal matrices. From (3.1)

$$F^{-1}\mu^l T(\mu) Fu = F^{-1} T(\mu) A^{-1}(\lambda,\xi) S(\mu) F F^{-1}\mu^l S^{-1}(\mu) F f. \tag{3.2}$$

With the notation

$$\Phi(\xi) = T(\mu) A^{-1}(\lambda,\xi) S(\mu), \quad v(x) = F^{-1}\mu^l S^{-1}(\mu) F f$$

we rewrite (3.2) as

$$F^{-1}\mu^l T(\mu) Fu = F^{-1}\Phi(\xi) Fv.$$

Hence

$$\|u\|_{E(\mathbb{R}^n)} = \|F^{-1}\Phi(\xi) Fv\|_{L^p(\mathbb{R}^n)}. \tag{3.3}$$

We note that

$$\Phi(\xi) = A^{-1}\left(\frac{\lambda}{\mu}, \frac{\xi}{\mu}\right)$$

and

$$\Delta(\xi) \equiv \det A\left(\frac{\lambda}{\mu}, \frac{\xi}{\mu}\right) \geq \left(\frac{|\lambda|}{\mu} + \frac{|\xi|}{\mu}\right)^{\sigma} e_d,$$

where $\sigma = s_1 + \cdots + s_N + t_1 + \cdots + t_N$. Taking into account that, for $\lambda \geq \lambda_0$,

$$\frac{|\lambda|+|\xi|}{\mu} \geq \lambda_1 \equiv \frac{\lambda_0}{\sqrt{1+|\lambda_0|^2}},$$

we obtain

$$\Delta(\xi) \geq \lambda_1^{\sigma} e_d.$$

It can be verified that

$$\xi^{\alpha} D^{\alpha} \Phi_{ik}(\xi) = P\left(\frac{\lambda}{\mu}, \frac{\xi}{\mu}\right) \Delta^{-m}(\xi),$$

where $\Phi_{ik}(\xi)$ are the elements of the matrix $\Phi(\xi)$, α is the multi-index from Mikhlin's theorem, P is a polynomial, m is a positive integer. Therefore $\Phi_{ik}(\xi)$ are Fourier multipliers and

$$|\xi^\alpha D^\alpha \Phi_{ik}(\xi)| \leq \frac{c_{ik}}{\lambda_1^\sigma e_d^m}$$

with some constants c_{ik}. We conclude from (3.3) that

$$|||u|||_{E(\mathbb{R}^n)} \leq \frac{c}{\lambda_1^\sigma e_d^m} |||f|||_{F^d(\mathbb{R}^n)}.$$

We have proved the following theorem.

Theorem 3.1. *The equation*

$$A(D,\lambda)u = f, \quad u \in E, \quad f \in F^d$$

has a unique solution u for any $\lambda \in S$, $\lambda \geq \lambda_0 > 0$. The estimate

$$|||u|||_{E(\mathbb{R}^n)} \leq \frac{c}{e_d^\kappa} |||f|||_{F^d(\mathbb{R}^n)}$$

holds with some constants c and κ independent of λ and u. The constant c depends on λ_0 and on M_d (see Section 1).

4 Model problem in a half-space

Consider the matrix operators $A(\lambda, D)$ and $B(\lambda, D)$ with elements

$$A_{ij}(\lambda, D) = \sum_{|\alpha|+\beta=\alpha_{ij}} a_{ij}^{\alpha\beta} \lambda^\beta D^\alpha, \quad i,j = 1,\ldots,N,$$

$$B_{kj}(\lambda, D) = \sum_{|\alpha|+\beta=\beta_{kj}} b_{kj}^{\alpha\beta} \lambda^\beta D^\alpha, \quad k = 1,\ldots,r, \quad j = 1,\ldots,N,$$

respectively, in the half-space $\mathbb{R}_+^n$ ($x_n \geq 0$). The coefficients of the operators are complex numbers. It is supposed that

$$A(\rho\xi, \rho\lambda) = S(\rho)A(\xi,\lambda)T(\rho), \quad B(\rho\xi,\rho\lambda) = M(\rho)B(\xi,\lambda)T(\rho), \qquad (4.1)$$

where

$$S(\rho) = (\delta_{ij}\rho^{s_i}), \quad T(\rho) = (\delta_{ij}\rho^{t_j}), \quad M(\rho) = (\delta_{ij}\rho^{\sigma_k})$$

are diagonal matrices (cf. [542]). We consider the operator $L = (A, B)$ from the space E into the space F with the $|||\cdot|||$-norms (see Section 2).

Proposition 4.1. *The operator $L : E \to F$ is bounded. The estimate*

$$|||Lu|||_F \leq c|||u|||_E, \quad \forall u \in E \qquad (4.2)$$

holds with a constant c independent of u and λ.

4. Model problem in a half-space

Proof. We have
$$\||Lu\||_F = \||Au\||_{F^d} + \||Bu\||_{F^b}.$$
Consider first the operator $Au = (A_1 u, \ldots, A_N u)$,
$$\||A_i u\||_{W^{l-s_i,p}(\Omega)} \leq M_d \sum_{j=1}^{N} \sum_{|\alpha|+\beta=\alpha_{ij}} |\lambda|^\beta \||D^\alpha u_j\||_{W^{l-s_i,p}(\Omega)},$$
where M_d is defined in Section 1. By virtue of Proposition 2.3,
$$\||A_i u\||_{W^{l-s_i,p}(\Omega)} \leq c M_d \sum_{j=1}^{N} \sum_{|\alpha|+\beta=\alpha_{ij}} |\lambda|^\beta \||u_j\||_{W^{l-s_i+|\alpha|,p}(\Omega)},$$
and by the interpolation inequality (Proposition 2.2),
$$\||A_i u\||_{W^{l-s_i,p}(\Omega)} \leq K_1 \sum_{j=1}^{N} \||u_j\||_{W^{l+t_j,p}(\Omega)}.$$
Hence
$$\||Au\||_{F^d} \leq K_2 \||u\||_E.$$
Here the constants c, K_1 and K_2 are independent of u and λ. Similarly,
$$\||Bu\||_{F^b} \leq K_3 \||u\||_E.$$
The last two estimates give (4.2). The proposition is proved. □

Proposition 4.2. *The operator $L : E_\infty \to F_\infty$ is bounded. The estimate*
$$\||Lu\||_{F_\infty} \leq c\||u\||_{E_\infty}, \quad \forall u \in F, \tag{4.3}$$
holds with a constant c independent of u and λ.

Proof. We have
$$\||Lu\||_{F_\infty} = \||Au\||_{F_\infty^d} + \||Bu\||_{F_\infty^b}.$$
Let us consider the first norm in the right-hand side,
$$\||Au\||_{F_\infty^d} = \sum_{i=1}^{N} \||A_i u\||_{W_\infty^{l-s_i,p}(\Omega)}.$$
Further, let ϕ_k be a partition of unity and $\psi_k(x) = 1$ for $x \in \operatorname{supp} \phi_k$. Then
$$\||\phi_k A_i u\||_{W^{l-s_i,p}(\Omega)} = \||\phi_k A_i(\psi_k u)\||_{W^{l-s_i,p}(\Omega)} \leq c_1 \||A_i(\psi_k u)\||_{W^{l-s_i,p}(\Omega)},$$

where $c_1 = \sup_k \|\phi_k\|_M$ is independent of λ. Here $\|\cdot\|_M$ denotes the norm of multipliers (see Lemma 4.3 below). From the boundedness of the operator A it follows that

$$\|\phi_k A_i \psi_k u\|_{W^{l-s_i,p}(\Omega)} \leq c_2 \|\psi_k u\|_E \leq c_3 \sum_{j=1}^N \|u_j\|_{W^{l+t_j,p}_\infty(\Omega)} = c_3 \|u\|_{E_\infty},$$

where c_3 does not depend on λ. Taking the supremum with respect to k in the last estimate and a sum with respect to i, we obtain

$$\|Au\|_{F^d_\infty} \leq c_3 \|u\|_{E_\infty}.$$

The estimates of the boundary operators are similar. The proposition is proved. □

Lemma 4.3. *For any $u \in W^{l,p}$,*

$$\|\phi u\|_{W^{l,p}} \leq c \|\phi\|_{C^l} \|u\|_{W^{l,p}},$$

where the constant c is independent of ϕ, u, and λ.

Proof. Consider first the norm $[\cdot]$. We have

$$[\phi u]_{W^{l,p}} = \|\phi u\|_{W^{l,p}} + |\lambda|^l \|\phi u\|_{L^p} \leq c_4 \|\phi\|_{C^l} \|u\|_{W^{l,p}} + |\lambda|^l \|\phi\|_C \|u\|_{L^p}$$
$$\leq (c_4 + 1) \|\phi\|_{C^l} [u]_{W^{l,p}},$$

where c_4 does not depend on ϕ, u, and λ. It remains to note that the norms $[\cdot]$ and $\|\|\cdot\|\|$ are equivalent. The lemma is proved. □

Consider the problem

$$Lu = f, \quad u \in E, \quad f \in F. \tag{4.4}$$

The remaining part of this section is devoted to the following theorem.

Theorem 4.4. *For any $\lambda \in S$, $\lambda \neq 0$ and for any $f \in F$ there exists a unique solution $u \in E$ of equation (4.4). If $\lambda \in S$ and $|\lambda| \geq \lambda_0$, where λ_0 is an arbitrary positive number, then the following estimate holds:*

$$\|u\|_E \leq c \|f\|_F \tag{4.5}$$

with a constant c that depends on the coefficients of the operator L only through $M_d, M_\Gamma, e_d, e_\Gamma, \lambda_0$ and does not depend on λ and f. Here S is a sector in the complex plane defined in Section 1.

4. Model problem in a half-space

Proof. We begin the proof of the theorem with some auxiliary results (cf. [542]). First of all, we write problem (4.4) in the form

$$Au = f^d, \quad u \in E, \quad f^d \in F^d, \tag{4.6}$$
$$Bu = f^b, \quad u \in E, \quad f^b \in F^b \tag{4.7}$$

and reduce problem (4.6), (4.7) to the case where $f^d = 0$. For this consider an extension $f_*(x)$ of the function $f^d(x)$ to $\mathbb{R}^n$ such that

$$f_* \in F^d(\mathbb{R}^n) = \Pi_{i=1}^N W^{l-s_i,p}(\mathbb{R}^n)$$

and

$$|||f_*|||_{F^d(\mathbb{R}^n)} \leq M |||f^d|||_{F^d(\Omega)}, \tag{4.8}$$

where M is a positive constant. Consider the equation

$$Au_* = f_* \quad \text{in } \mathbb{R}^n. \tag{4.9}$$

According to Theorem 3.1 there exists a solution $u_* \in \mathbb{R}^n$ of this equation, and the estimate

$$|||u_*|||_{E(\mathbb{R}^n)} \leq c_0 |||f_*|||_{F^d(\mathbb{R}^n)} \tag{4.10}$$

holds. Here c_0 depends on M_d, e_d (see Section 1), λ_0, and does not depend on λ and u.

Set

$$v(x) = u(x) - u_*(x), \quad x \in \Omega(= \mathbb{R}_+^n). \tag{4.11}$$

Then

$$\begin{aligned} Av &= 0, & x \in \Omega, \; v \in E(\Omega), \\ Bv &= g(x,\lambda), & x \in \partial\Omega, \end{aligned} \tag{4.12}$$

where

$$g(x,\lambda) = f^b(x) - B(D,\lambda)u_*(x)$$

We have

$$|||Bu_*|||_{F^b(\partial\Omega)} \leq c_1 |||u_*|||_{E(\Omega)} \leq c_2 |||f_*|||_{F^d(\mathbb{R}^n)} \leq c_3 |||f^d|||_{F^d(\Omega)}.$$

Hence $g \in F^b(\partial\Omega)$ and

$$|||g|||_{F^b(\partial\Omega)} \leq |||f^b|||_{F^b(\partial\Omega)} + c_3 |||f^d|||_{F^d(\Omega)}. \tag{4.13}$$

Thus we can consider problem (4.12) with $g \in F^b(\partial\Omega)$.

We will prove the estimate

$$|||v|||_E \leq c_4 |||Bv|||_{F^b(\partial\Omega)}, \quad \forall v \in E \tag{4.14}$$

with a constant c_4 independent of v and λ. Estimate (4.5) follows from it.

Proposition 4.5. *If the estimate*
$$|||u|||_E \leq c|||Lu|||_F \qquad (4.15)$$
holds for all $u \in S(\mathbb{R}^n_+)$, *then it is true for all* $u \in E$ *with the same constant* c.

The proof of this proposition is straightforward. It is based on the boundedness of the operator L. Therefore it is sufficient to prove estimate (4.14) for $v \in S(\mathbb{R}^n_+)$. We can assume that in (4.12) $g \in S(\mathbb{R}^{n-1})$. We fulfil the partial Fourier transform in (4.12) with respect to x' and use the notation $t = x_n$. We obtain

$$A(\lambda, \xi', D_t)\tilde{v}(\lambda, \xi', t) = 0, \quad t > 0, \qquad (4.16)$$
$$B(\lambda, \xi', D_t)\tilde{v}(\lambda, \xi', 0) = \tilde{g}(\lambda, \xi'), \qquad (4.17)$$

where $A(\lambda, \xi', D_t)$ and $B(\lambda, \xi', D_t)$ are the matrices with elements

$$A_{ij}(\lambda, \xi', D_t) = \sum_{|\alpha'|+\alpha_n+\beta=\alpha_{ij}} a_{ij}^{\alpha' \alpha_n \beta} \lambda^\beta \xi'^{\alpha'} D_t^{\alpha_n}, \quad i,j=1,\ldots,N,$$

$$B_{ij}(\lambda, \xi', D_t) = \sum_{|\alpha'|+\alpha_n+\beta=\beta_{ij}} b_{ij}^{\alpha' \alpha_n \beta} \lambda^\beta \xi'^{\alpha'} D_t^{\alpha_n}, \quad i=1,\ldots,r, \; j=1,\ldots,N.$$

Denote by $\omega(\lambda, \xi', t)$ the stable, that is decaying at $+\infty$, solution of the problem

$$A(\lambda, \xi', D_t)\omega(\lambda, \xi', t) = 0, \quad t > 0, \qquad (4.18)$$
$$B(\lambda, \xi', D_t)\omega(\lambda, \xi', 0) = I, \qquad (4.19)$$

where I is the identity matrix of order r.

Proposition 4.6. *Let the Lopatinskii condition be satisfied. Then the matrix* $\omega(\lambda, \xi', t)$ *can be represented in the form*

$$\omega(\lambda, \xi', t) = \int_{\gamma_+} e^{i\mu t} A^{-1}(\lambda, \xi', \mu) K(\lambda, \xi', \mu) d\mu,$$

where $K(\lambda, \xi', \mu)$ *is an* $N \times r$ *matrix polynomial with respect to* μ, *continuous with respect to* λ, ξ' *for* $|\lambda| + |\xi'| > 0$ *and such that the following homogeneity condition holds:*

$$K(\rho\lambda, \rho\xi', \rho\mu) = \rho^{-1} S(\rho) K(\lambda, \xi', \mu) M^{-1}(\rho).$$

Proof. We recall the definition of the Lopatinskii matrix,

$$\Lambda(\lambda, \xi') = \int_{\gamma_+} B(\lambda, \xi', \mu) A^{-1}(\lambda, \xi', \mu) \Phi(\mu) d\mu,$$

where
$$\Phi(\mu) = (E, \mu E, \ldots, \mu^{s-1} E)$$
(see Section 1). For some fixed (λ, ξ') such that $|\lambda| + \xi'| > 0$, $\lambda \in S$ we choose r columns of the matrix $\Phi(\mu)$,

$$H_0(\mu) = \Phi(\mu) \Psi_r$$

4. Model problem in a half-space

such that the matrix

$$\Lambda_0(\lambda, \xi') = \int_{\gamma_+} B(\lambda, \xi', \mu) A^{-1}(\lambda, \xi', \mu) H_0(\mu) d\mu$$

is invertible. Put

$$K(\lambda, \xi', \mu) = H_0(\mu)\Lambda_0^{-1}(\lambda, \xi').$$

Then

$$\omega(\lambda, \xi', t) = \int_{\gamma_+} e^{i\mu t} A^{-1}(\lambda, \xi', \mu) K(\lambda, \xi', \mu) d\mu$$

is a solution of (4.18), (4.19). Indeed,

$$A(\lambda, \xi', D_t)\omega(\lambda, \xi', t) = \int_{\gamma_+} e^{i\mu t} A(\lambda, \xi', \mu) A^{-1}(\lambda, \xi', \mu) K(\lambda, \xi', \mu) d\mu$$

$$= \int_{\gamma_+} e^{i\mu t} H_0(\mu) d\mu \Lambda_0^{-1}(\lambda, \xi') = 0,$$

$$B(\lambda, \xi', D_t)\omega(\lambda, \xi', t) = \int_{\gamma_+} e^{i\mu t} B(\lambda, \xi', \mu) A^{-1}(\lambda, \xi', \mu) K(\lambda, \xi', \mu) d\mu.$$

Hence

$$B(\lambda, \xi', D_t)\omega(\lambda, \xi', t)\big|_{t=0} = \int_{\gamma_+} B(\lambda, \xi', \mu) A^{-1}(\lambda, \xi', \mu) H_0(\mu) d\mu \Lambda_0^{-1}(\lambda, \xi') = I$$

by virtue of the definition of Λ_0.

It remains to verify the homogeneity of the matrix K. We have

$$\Lambda_0(\rho\lambda, \rho\xi') = \rho \int_{\tilde{\gamma}_+} B(\rho\lambda, \rho\xi', \rho\zeta) A^{-1}(\rho\lambda, \rho\xi', \rho\zeta) H_0(\rho\zeta) d\zeta$$

$$= \rho \int_{\tilde{\gamma}_+} M(\rho) B(\lambda, \xi', \zeta) T(\rho) T^{-1}(\rho) A^{-1}(\lambda, \xi', \zeta) S^{-1}(\rho) H_0(\rho\zeta) d\zeta \equiv P.$$

We use further that $H_0(\rho\zeta) = H_0(\zeta)h(\rho)$, $S^{-1}(\rho)H_0(\zeta) = H_0(\zeta)D(\rho)$, where $h(\rho)$ and $D(\rho)$ are some diagonal matrices,

$$P = \rho M(\rho)\Lambda_0(\lambda, \xi') D(\rho) h(\rho).$$

Hence

$$K(\rho\lambda, \rho\xi', \rho\mu) = H_0(\rho\mu)\Lambda_0^{-1}(\rho\lambda, \rho\xi')$$
$$= H_0(\mu)h(\rho)h^{-1}(\rho)D^{-1}(\rho)\Lambda_0^{-1}(\lambda, \xi')\rho^{-1}M^{-1}(\rho)$$
$$= H_0(\mu)(H_0(\mu)D(\rho))^{-1}H_0(\mu)\Lambda_0^{-1}(\lambda, \xi')\rho^{-1}M^{-1}(\rho)$$
$$= S(\rho)K(\lambda, \xi', \mu)\rho^{-1}M^{-1}(\rho).$$

The proposition is proved. □

The solution of problem (4.16), (4.17) has the form

$$\tilde{v}(\lambda, \xi', t) = \omega(\lambda, \xi', t)\tilde{g}(\lambda, \xi'). \tag{4.20}$$

We recall that the Lopatinskii matrix $\Lambda(\lambda, \xi')$ is introduced in Section 1. The Lopatinskii condition for problems with a parameter is formulated as

$$\operatorname{rank} \Lambda(\lambda, \xi') = r \ \ \forall \lambda, \xi', \lambda \in S, |\lambda| + |\xi'| > 0.$$

Proposition 4.7. *The Lopatinskii condition is equivalent to the condition $e_\Gamma > 0$.*

The proof of this proposition is based on the fact that any minor of $\Lambda(\lambda, \xi')$ of order r is a homogeneous function of (λ, ξ').

Let $\phi \in S(\mathbb{R}^{n-1}_+)$ be an extension of g:

$$g(\lambda, x') = \phi(\lambda, x', 0). \tag{4.21}$$

Then from (4.20)

$$\tilde{v}(\lambda, \xi', t) = \omega(\lambda, \xi', t)\tilde{\phi}(\lambda, \xi', 0).$$

Hence

$$v(\lambda, x) = (F')^{-1}(\omega(\lambda, \xi', t)\tilde{\phi}(\lambda, \xi', 0)), \ \ t = x_n. \tag{4.22}$$

Since the elements of the matrix $\omega(\lambda, \xi', t)$ exponentially decay as $t \to \infty$, then we have

$$\omega(\lambda, \xi', t)\tilde{\phi}(\lambda, \xi', 0) \tag{4.23}$$
$$= -\int_0^\infty D_\tau \omega(\lambda, \xi', t+\tau)\tilde{\phi}(\lambda, \xi', \tau)d\tau - \int_0^\infty \omega(\lambda, \xi', t+\tau)\tilde{\phi}(\lambda, \xi', \tau)d\tau D_\tau.$$

Let $M(\rho) = (\delta_{ij}\rho^{\sigma_i})$, $Q(\xi')$ be another diagonal matrix polynomial in ξ', which will be specified below. From (4.22), (4.23) we have

$$D_t^k Q(D')v(\lambda, x) = -(T_1 + T_2), \tag{4.24}$$

where

$$T_1 = D_t^k Q(D')(F')^{-1} \int_0^\infty D_\tau \omega(\lambda, \xi', t+\tau)\tilde{\phi}(\lambda, \xi', \tau)d\tau,$$

$$T_2 = D_t^k Q(D')(F')^{-1} \int_0^\infty \omega(\lambda, \xi', t+\tau)D_\tau\tilde{\phi}(\lambda, \xi', \tau)d\tau.$$

We can write these expressions in the form

$$T_1 = (F')^{-1} \int_0^\infty H(\lambda, \xi', t+\tau)\tilde{a}(\lambda, \xi', \tau)d\tau, \tag{4.25}$$

$$T_2 = (F')^{-1} \int_0^\infty G(\lambda, \xi', t+\tau)\tilde{b}(\lambda, \xi', \tau)d\tau, \tag{4.26}$$

4. Model problem in a half-space

where
$$H(\lambda, \xi', t) = \rho^{-l} Q(\xi') D_t^{k+1} w(\lambda, \xi', t) M(\rho),$$
$$G(\lambda, \xi', t) = \rho^{-l+1} Q(\xi') D_t^k w(\lambda, \xi', t) M(\rho),$$
(4.27)

$$a(\lambda, x) = (F')^{-1} \rho^l M^{-1}(\rho) F' \phi,$$
$$b(\lambda, x) = (F')^{-1} \rho^{l-1} M^{-1}(\rho) F' D_\tau \phi,$$
(4.28)

$\rho = \sqrt{|\lambda|^2 + |\xi'|^2}$.

We will prove that the matrix $H(\lambda, \xi', t)$ given by (4.27) is a Fourier multiplier in $L^p(\mathbb{R}^{n-1})$ with respect to ξ'. It follows from Proposition 4.6 that

$$D_t^{k+1} w(\lambda, \xi', t) = \int_{\gamma_+} \mu^{k+1} e^{i\mu t} \Phi(\lambda, \xi', \mu) d\mu,$$

where
$$\Phi(\lambda, \xi', \mu) = A^{-1}(\lambda, \xi', \mu) K(\lambda, \xi', \mu) d\mu.$$

Hence
$$H(\lambda, \xi', t) = \rho^{-l} Q(\xi') \int_{\gamma_+} \mu^{k+1} e^{i\mu t} \Phi(\lambda, \xi', \mu) d\mu M(\rho), \qquad (4.29)$$

where $Q(\xi') = (\delta_{ij} \xi'^{\gamma_i})$ is a diagonal matrix, γ_i are multi-indices.

Put
$$\lambda_0 = \frac{\lambda}{\rho}, \quad \xi_0' = \frac{\xi'}{\rho}, \quad \zeta = \frac{\mu}{\rho}.$$

Then $|\lambda_0|^2 + |\xi_0'|^2 = 1$ and

$$H(\lambda, \xi', t) = \rho^{-l} Q(\xi') \int_{\tilde{\gamma}_+} \rho^{k+1} \zeta^{k+1} e^{i\rho\zeta t} \Phi(\rho\lambda_0, \rho\xi_0', \rho\zeta) \rho d\zeta M(\rho), \qquad (4.30)$$

where $\tilde{\gamma}_+$ is a contour in the half-plane $\text{Im}\,\zeta > 0$ enclosing the zeros ζ of the polynomial $\det A(\lambda_0, \xi_0', \zeta)$ lying in this half-plane. We have

$$\Phi(\rho\lambda_0, \rho\xi_0', \rho\zeta) = A^{-1}(\rho\lambda_0, \rho\xi_0', \rho\zeta) K(\rho\lambda_0, \rho\xi_0', \rho\zeta)$$
$$= T^{-1}(\rho) A^{-1}(\lambda_0, \xi_0', \zeta) S^{-1}(\rho) \rho^{-1} S(\rho) K(\lambda_0, \xi_0', \zeta) M^{-1}(\rho)$$
$$= \rho^{-1} T^{-1}(\rho) \Phi(\lambda_0, \xi_0', \zeta) M^{-1}(\rho). \qquad (4.31)$$

Therefore
$$H(\lambda, \xi', t) = \rho^{-l} \hat{Q}(\rho) \rho^{k+1} T^{-1}(\rho) \Psi(\lambda_0, \xi_0'), \qquad (4.32)$$

where
$$\hat{Q}(\rho) = (\delta_{ij} \rho^{|\gamma_i|}),$$
$$\Psi(\lambda_0, \xi_0') = Q(\xi_0') \int_{\tilde{\gamma}_+} \zeta^{k+1} e^{i\rho\zeta t} \Phi(\lambda_0, \xi_0', \zeta) d\zeta.$$
(4.33)

Set
$$\omega = \inf \text{Im}\,\zeta, \quad \zeta \in \tilde{\gamma}_+(\lambda_0, \xi_0'), \quad |\lambda_0|^2 + |\xi_0'|^2 = 1, \quad \lambda_0 \in S. \qquad (4.34)$$

It will be proved below that $\omega > 0$. Hence
$$|e^{i\rho\zeta t}| \leq e^{-\rho\omega t}.$$

To estimate $\Psi(\lambda_0, \xi_0')$ we introduce the parametric representation of the contour $\tilde{\gamma}_+(\lambda_0, \xi_0') : \zeta = \zeta(s), \ 0 \leq s \leq l_\gamma$. Then
$$\Psi(\lambda_0, \xi_0') = Q(\xi_0') \int_0^{l_\gamma} (\zeta(s))^{k+1} e^{i\rho\zeta(s)t} \Phi(\lambda_0, \xi_0', \zeta(s)) \zeta'(s) ds.$$

Denote by Ψ_{jm} the elements of the matrix Ψ. We have
$$|\Psi_{jm}(\lambda_0, \xi_0')| \leq e^{-\rho\omega t} \int_0^{l_\gamma} |(\zeta(s))^{k+1}| |\Phi_{jm}(\lambda_0, \xi_0', \zeta(s))| |\zeta'(s)| ds.$$

Set
$$\psi = \sup_{|\lambda_0|^2 + |\xi_0'|^2 = 1, \lambda_0 \in S, j, m} \int_0^{l_\gamma} |(\zeta(s))^{k+1}| |\Phi_{jm}(\lambda_0, \xi_0', \zeta(s))| |\zeta'(s)| ds.$$

Then
$$|\Psi_{jm}(\lambda_0, \xi_0')| \leq \psi e^{-\rho\omega t}.$$

From (4.30)
$$|H_{jm}(\lambda, \xi', t)| \leq \rho^{-l+|\gamma_j|+k+1-t_j} \psi e^{-\rho\omega t} \leq \rho^{-l+|\gamma_j|+k-t_j} \frac{\psi}{\omega} \frac{1}{t}.$$

Therefore, if
$$|\lambda| \geq \lambda_0 > 0, \quad k + |\gamma_j| \leq l + t_j,$$
then
$$|H_{jm}(\lambda, \xi', t)| \leq \frac{1}{\lambda_0^\nu} \frac{\psi}{\omega} \frac{1}{t},$$
where $\nu = l - |\gamma_j| - k + t_j$.

Proposition 4.8. *The elements of the matrices $H(\lambda, \xi', t)$ and $G(\lambda, \xi', t)$ defined in (4.27) are Fourier multipliers in $L^p(\mathbb{R}^{n-1})$ with respect to ξ'. The norms of these multipliers admit the estimates*
$$\|H_{jm}(\lambda, \cdot, t)\|_M \leq \frac{K}{t}, \quad \|G_{jm}(\lambda, \cdot, t)\|_M \leq \frac{K}{t}, \tag{4.35}$$
where the constant K depends on M_d, M_Γ, e_d, and e_Γ only.

The proof of this proposition will be given below. We estimate now T_1 and T_2 given by (4.25), (4.26). Let
$$a = (a_1, \ldots, a_m), \quad \phi = (\phi_1, \ldots, \phi_m), \quad T_1 = (T_{11}, \ldots, T_{1m}),$$
$$A_{jk} = (F')^{-1} \int_0^\infty H_{jk}(\lambda, \xi', t+\tau) F' a_k(\lambda, \xi', \tau) d\tau.$$

4. Model problem in a half-space

Then

$$T_{1j} = \sum_{k=1}^{m} A_{jk}. \tag{4.36}$$

It follows from Proposition 1.9 (Chapter 3) (cf. [542], Proposition 5.2) that

$$\|A_{jk}\|_{L^p(\mathbb{R}^n_+)} \leq c_{jk} K \|a_k\|_{L^p(\mathbb{R}^n_+)}, \tag{4.37}$$

where K is the constant in (4.35), and c_{jk} are constants independent of the coefficients of the operators $A(D, \lambda)$ and $B(D, \lambda)$ under consideration (we will use the notation c with subscripts for such constants). It follows from (4.28) that

$$a_k(\lambda, x) = (F')^{-1}(\rho^{l-\sigma_k} F' \phi_k) = F^{-1}(\rho^{l-\sigma_k} F \phi_k). \tag{4.38}$$

Since we assume that $\phi_k \in S(\mathbb{R}^n_+)$, then it belongs to $W^{l-\sigma_k, p}(\mathbb{R}^n_+)$ and can be extended to $W^{l-\sigma_k, p}(\mathbb{R}^n)$ in such a way that

$$\|\|\phi_k\|\|_{W^{l-\sigma_k, p}(\mathbb{R}^n)} \leq 2\|\|\phi_k\|\|_{W^{l-\sigma_k, p}(\mathbb{R}^n_+)} \leq 4\|\|g_k\|\|_{W^{l-\sigma_k-1/p, p}(\Gamma)} \tag{4.39}$$

(see (4.21)). In view of (4.38), (4.39) we have for $|\lambda| \geq \lambda_0 > 0$:

$$\|a_k\|_{L^p(\mathbb{R}^n_+)} \leq \|a_k\|_{L^p(\mathbb{R}^n)} = \|F^{-1}(\rho^{l-\sigma_k} F \phi_k)\|_{L^p(\mathbb{R}^n)}$$
$$\leq c\|F^{-1}((1+|\xi|^2+|\lambda|^2)^{(l-\sigma_k)/2} F \phi_k)\|_{L^p(\mathbb{R}^n)} = c\|\|\phi_k\|\|_{W^{l-\sigma_k, p}(\mathbb{R}^n)}.$$

From (4.36), (4.37), (4.39) and the last estimate we obtain

$$\|T_{1j}\|_{L^p(\mathbb{R}^n_+)} \leq c_1 K \|\|g\|\|_{F^b(\Gamma)}, \quad j = 1, \ldots, N.$$

Similarly,

$$\|T_{2j}\|_{L^p(\mathbb{R}^n_+)} \leq c_2 K \|\|g\|\|_{F^b(\Gamma)}, \quad j = 1, \ldots, N.$$

By virtue of (4.24),

$$\|D_t^k (D')^{\gamma_j} v_j(\lambda, x)\|_{L^p(\mathbb{R}^n_+)} \leq c_3 K \|\|g\|\|_{F^b(\Gamma)}, \quad j = 1, \ldots, N.$$

We suppose that $|\gamma_j| + k \leq l + t_j$. Therefore

$$\|v_j\|_{W^{l+t_j, p}(\Omega)} \leq c_3 K \|\|g\|\|_{F^b(\Gamma)}, \quad j = 1, \ldots, N. \tag{4.40}$$

To estimate the norm $\|\|v_j\|\|_{W^{l+t_j, p}(\Omega)}$ it remains to estimate $|\lambda|^{l+t_j} \|v_j\|_{L^p(\Omega)}$. We can repeat the same construction as above replacing $Q(\xi')$ in (4.30) by $Q(\lambda) = (\delta_{ij}|\lambda|^{l+t_j})$. Then instead of (4.32) we get

$$H(\lambda, \xi', t) = \rho^{-l} Q(\lambda) \rho T^{-1}(\rho) \Psi(\lambda_0, \xi_0'), \tag{4.41}$$

where Ψ is the same as in (4.33) with $k = 0$ and $Q(\xi')$ is replaced by the identity matrix. It follows that

$$H_{jm}(\lambda, \xi', t) = \rho^{-l+1-t_j} \lambda^{l+t_j} \Psi_{jm}(\lambda_0, \xi_0').$$

Hence
$$|H_{jm}(\lambda,\xi',t)| \le \rho^{-l+1-t_j}\lambda^{l+t_j}\psi e^{-\rho\omega t} \le \frac{\psi}{\omega}\frac{1}{t}$$
since
$$\rho^{-l-t_j}\lambda^{l+t_j} \le 1, \quad \rho e^{-\rho\omega t} \le \frac{1}{\omega t}.$$

As before we get
$$|\lambda|^{l+t_j}\|v_j\|_{L^p(\Omega)} \le c_4 K \|\|g\|\|_{F^b(\Gamma)}, \quad j=1,\ldots,N.$$

Together with (4.40) this gives
$$\|\|v\|\|_{E(\Omega)} \le c_5 K \|\|g\|\|_{F^b(\Gamma)}. \tag{4.42}$$

Proof of Proposition 4.8. We consider the elements of the matrix $H(\lambda,\xi',t)$ in the form
$$H_{ij}(\lambda,\xi',t) = \rho^{-l+\sigma_j}|\lambda|^{\nu}(\xi')^{\gamma_i} D_t^{k+1}\psi_{ij}(\lambda,\xi',t), \tag{4.43}$$
which includes both (4.27) and (4.41). Here
$$\psi_{ij}(\lambda,\xi',t) = \int_{\gamma_+} e^{i\mu t}\Phi_{ij}(\lambda,\xi',\mu)d\mu, \tag{4.44}$$

$\Phi(\lambda,\xi',\mu)$ is the same matrix as in (4.29), and $\rho = \sqrt{|\xi'|^2 + |\lambda|^2}$. The proof of the proposition is based on the following lemma.

Lemma 4.9. *Consider the function*
$$f(\lambda,\xi',t) = \lambda^{\beta}(\xi')^{\alpha}(|\lambda|^2 + |\xi'|^2)^{s/2} D_t^k \psi(\lambda,\xi',t), \tag{4.45}$$
where
$$\psi(\lambda,\xi',t) = \int_{\gamma_+(\lambda,\xi')} e^{i\mu t}g(\lambda,\xi',\mu)d\mu, \tag{4.46}$$

$\gamma_+(\lambda,\xi')$ *is a contour lying in the half-plane* $\operatorname{Im}\mu > 0$ *and enclosing the zeros* μ *of the polynomial* $\det A(\lambda,\xi',\mu)$ *lying in this half-plane;* $g(\lambda,\xi',\mu)$ *is a function homogeneous with respect to* (λ,ξ',μ) *of degree* γ, *analytic in* μ *and infinitely differentiable in* ξ', *it is defined in a neighborhood of any point* (λ,ξ'); $\psi(\lambda,\xi',\mu)$ *is infinitely differentiable in* ξ'; α *is a multi-index,* β *is an integer,* $k \ge 0$ *is an integer.*

It is supposed that for any multi-index m,
$$|D_{\xi'}^m g(\lambda_0,(\xi_0)',\mu)| \le K_m, \quad \forall \mu \in \gamma_+(\lambda_0,\xi_0) \tag{4.47}$$
for $|(\xi_0)'|^2 + |\lambda_0|^2 = 1$, $\lambda_0 \in S$, *where* K_m *is a constant which depends on* $M_d, M_\Gamma, e_d, e_\Gamma$ *and does not depend directly on the coefficients of the operators* A *and* B *and on* λ. *If*
$$|\alpha| + \beta + s + k + \gamma \le 0, \quad |\lambda| \ge \lambda_0 > 0, \tag{4.48}$$

4. Model problem in a half-space

then $f(\lambda, \xi', \mu)$ is a Fourier multiplier in $L^p(\mathbb{R}^{n-1})$ with respect to ξ'. The norm of the multiplier admits the estimate

$$\|f(\lambda, \cdot, t)\|_M \leq \frac{K}{t}, \tag{4.49}$$

where K is a constant which depends on $M_d, M_\Gamma, e_d, e_\Gamma$ and does not depend directly on the coefficients of the operators A and B and on λ.

Proof. We fix a point $\lambda_0, (\xi_0)'$ such that $|(\xi_0)'|^2 + |\lambda_0|^2 = 1$, $\lambda_0 \in S$ and put

$$\xi' = \rho(\xi_0)', \quad \lambda = \rho\lambda_0, \quad \mu = \rho\zeta, \quad \rho = \sqrt{|\xi'|^2 + |\lambda|^2}.$$

From (4.46)

$$\psi(\lambda, \xi', t) = \rho \int_{\bar{\gamma}_+} e^{i\rho\zeta t} g(\rho\lambda_0, \rho(\xi_0)', \rho\zeta) d\zeta,$$

where $\bar{\gamma}_+$ is a contour, which encloses all zeros ζ of $\det A(\lambda_0, (\xi_0)', \zeta)$ lying in the half-plane $\operatorname{Im} \zeta > 0$. According to Lemma 4.10 below, the contour $\bar{\gamma}_+$ can be chosen in such a way that it lies in the set

$$|\zeta| \leq \frac{M_d}{e_d}, \quad \operatorname{Im} \zeta \geq \omega.$$

Since the function $g(\lambda, \xi', \mu)$ is homogeneous, we get

$$\psi(\lambda, \xi', t) = \rho^{\gamma+1} \int_{\bar{\gamma}_+} e^{i\rho\zeta t} g(\lambda_0, (\xi_0)', \zeta) d\zeta.$$

Hence in view of (4.45) and (4.47) with $m = 0$,

$$|f(\lambda, \xi', t)| \leq |\lambda|^\beta |\xi'|^{|\alpha|} \rho^{s+k+\gamma+1} e^{-\rho\omega t} \int_{\bar{\gamma}_+} |\zeta|^k |g(\lambda_0, (\xi_0)', \zeta)| |\zeta'(\sigma)| d\sigma$$

$$\leq \rho^{|\alpha|+\beta+s+k+\gamma} \frac{K_0}{\omega t} \int_{\bar{\gamma}_+} |\zeta|^k |\zeta'(\sigma)| d\sigma \leq M \rho^{|\alpha|+\beta+s+k+\gamma} \frac{1}{t},$$

where

$$M = \frac{K_0}{\omega} \int_{\bar{\gamma}_+} |\zeta|^k |\zeta'(\sigma)| d\sigma.$$

It follows from (4.48) that

$$|f(\lambda, \xi', t)| \leq M\lambda_0^\nu \frac{1}{t}, \tag{4.50}$$

where $\nu = |\alpha| + \beta + s + k + \gamma$. According to Lemma 4.10 below and condition (4.47) the constant M depends only on $M_d, M_\Gamma, e_d, e_\Gamma$.

Consider now

$$\xi_j \frac{\partial f(\lambda, \xi', t)}{\partial \xi_j} \tag{4.51}$$

$$= (\alpha_j \rho^2 + s\xi_j^2)\lambda^\beta (\xi')^\alpha \rho^{s-2} D_t^k \psi(\lambda, \xi', t) + \lambda^\beta (\xi')^\alpha \rho^s \xi_j \frac{\partial}{\partial \xi_j} D_t^k \psi(\lambda, \xi', t).$$

Obviously, the first term in the right-hand side of the last equality satisfies the conditions of the lemma. Consider the second term. Set

$$g_j(\lambda, \xi', \mu) = \xi_j \frac{\partial g(\lambda, \xi', \mu)}{\partial \xi_j}, \quad \psi_j(\lambda, \xi', t) = \int_{\gamma_+(\lambda, \xi')} e^{i\mu t} g_j(\lambda, \xi', \mu) d\mu,$$

$$f_j(\lambda, \xi', t) = \lambda^\beta (\xi')^\alpha \rho^s D_t^k \psi_j(\lambda, \xi', t).$$

Obviously, f_j coincides with the second term in (4.51) since taking the derivative $\partial \psi / \partial \xi_j$ we can retain the same contour γ_+. The function g_j is a homogeneous function of degree γ. It satisfies the estimate of the type (4.47). Therefore the function $f_j(\lambda, \xi', t)$ satisfies the conditions of the lemma. Consequently, the functions $\xi_j \frac{\partial f(\lambda, \xi', t)}{\partial \xi_j}$, $j = 1, \ldots, n-1$ also satisfy them. Hence f satisfies the conditions of Mikhlin's theorem, which implies that it is a Fourier multiplier and that estimate (4.49) holds. The lemma is proved. □

Lemma 4.10. *Let*

$$P(z) = a_0 z^m + \cdots + a_{m-1} z + a_m$$

be an arbitrary polynomial with complex coefficients. Suppose that

$$|a_0| \geq e_d, \quad \sum_{i=0}^{m} |a_i| \leq \mu_d, \tag{4.52}$$

where μ_d and e_d are given positive numbers. Then all roots of the equation $P(z) = 0$ admit the estimate

$$|z| \leq \frac{\mu_d}{e_d}. \tag{4.53}$$

If in addition $P(z) > 0$ for all real z, then

$$|\operatorname{Im} z| \geq \omega > 0 \tag{4.54}$$

for all roots of this polynomial. Here ω is a constant which depends only on μ_d and e_d.

Proof. Suppose that estimate (4.53) does not hold and

$$|z| > \frac{\mu_d}{e_d} \ (> 1). \tag{4.55}$$

4. Model problem in a half-space

From the equation we get

$$z = -\frac{1}{a_0 z^{m-1}}(a_1 z^{m-1} + \cdots + a_{m-1}z + a_m).$$

Therefore (since $|z| > 1$)

$$|z| \le \frac{1}{e_d}(|a_1| + \cdots + |a_m|) \le \frac{\mu_d}{e_d}.$$

This estimate contradicts (4.55).

Suppose next that (4.54) is not true. Then there is a sequence z_k such that $\operatorname{Im} z_k \to 0$ and $P_k(z_k) = 0$, where

$$P_k(z) \equiv a_0^k z^m + \cdots + a_{m-1}^k z + a_m^k$$

(the polynomial can depend on k). By virtue of (4.52), (4.53),

$$\sum_{i=0}^{m}|a_0^k| \le \mu_d, \quad |z_k| \le \frac{\mu_d}{e_d}.$$

Therefore we can choose convergent subsequences of the roots, $z_{k_i} \to z_0$, $\operatorname{Im} z_{k_i} \to 0$, and of the coefficients, $a_j^{k_i} \to a_j^0$, $j = 1, \ldots, m$. Passing to the limit in the equation, we get

$$a_0^0 z_0^m + \cdots + a_{m-1}^0 z_0 + a_m^0 = 0.$$

This equality contradicts the condition of the lemma that there are no real roots of the polynomials. The lemma is proved. □

We return to the proof of Proposition 4.8. We will verify that Lemma 4.9 is applicable to the functions H_{ij} given by (4.43). We begin with the properties of the contour $\gamma_+(\lambda, \xi')$. After the change of variables $\xi' = \rho(\xi_0)'$, $\lambda = \rho\lambda_0$, $\mu = \rho\zeta$, where $|(\xi_0)'|^2 + |\lambda_0|^2 = 1$, $\lambda_0 \in S$, it is reduced to the contour $\bar{\gamma}_+(\lambda_0, (\xi_0)')$. We need to verify that it is bounded and separated from the real axis uniformly in $(\lambda_0, (\xi_0)')$. For this we should verify the applicability of Lemma 4.10, that is of condition (4.52) for the polynomial $\det A(\lambda_0, (\xi_0)', \zeta)$ and that it does not have real roots. This follows from the conditions on the coefficients of the operator, from the ellipticity condition and the condition of proper ellipticity (see Section 1). Indeed, we can write

$$\det A(\lambda_0, (\xi_0)', z) = a_0 z^m + \cdots + a_{m-1}z + a_m.$$

Then $|a_0| = |\det A(0, 0, 1)| \ge e_d$, where e_d is determined in Section 1. Since $|\lambda_0| \le 1$ and $|(\xi_0)'| \le 1$, then the coefficients of this polynomial can be estimated by the coefficients of the operator A, that is by M_d.

We study next the properties of the function

$$g(\lambda, \xi', \mu) = \Phi_{ij}(\lambda, \xi', \mu).$$

We will verify that it is a homogeneous function of the order $\gamma = -1 - t_i - \sigma_j$. We recall that
$$\Phi(\lambda, \xi', \mu) = A^{-1}(\lambda, \xi', \mu) K(\lambda, \xi', \mu).$$

By virtue of (4.1) and Proposition 4.6 we have
$$\begin{aligned}\Phi(\rho\lambda, \rho\xi', \rho\mu) &= A^{-1}(\rho\lambda, \rho\xi', \rho\mu) K(\rho\lambda, \rho\xi', \rho\mu) \\ &= T^{-1}(\rho) A^{-1}(\xi, \lambda) S^{-1}(\rho) \rho^{-1} S(\rho) K(\lambda, \xi', \mu) M^{-1}(\rho) \\ &= T^{-1}(\rho) \Phi(\lambda, \xi', \mu) \rho^{-1} M^{-1}(\rho).\end{aligned}$$

This equality proves that $\Phi_{ij}(\lambda, \xi', \mu)$ is a homogeneous function of the order indicated above. Other conditions of Lemma 4.9 can be easily verified. Proposition 4.8 is proved. $\square$

We can now finish the proof of Theorem 4.4. We have proved estimate (4.14) for functions from $S(\mathbb{R}^n_+)$. Then we have estimate (4.5) first for such functions and then for all functions from E (Proposition 4.5). Existence of solutions for functions from $S(\mathbb{R}^n_+)$ can be easily obtained by the Fourier transform. From this and from the estimates of solutions we obtain the existence for all functions $f \in F$. The uniqueness of solutions follows from estimate (4.5). The theorem is proved. $\square$

5 Problem in $\mathbb{R}^n$

We consider the operators
$$A_{ij}(x, \lambda, D) = \sum_{|\alpha|+\beta=\alpha_{ij}} a_{ij}^{\alpha\beta}(x) \lambda^\beta D^\alpha, \quad i, j = 1, \cdots, N,$$

where $x \in \mathbb{R}^n$, and the matrix $A(x, \lambda, D)$ with the elements $A_{ij}(x, \lambda, D)$. Here
$$a_{ij}^{\alpha\beta}(x) \in C^{l-s_i+\theta}(\mathbb{R}^n), \quad 0 < \theta < 1.$$

Theorem 5.1. Suppose that the operator $A(0, \lambda, D)$ is elliptic. Then there exist constants $\epsilon > 0$, $\lambda_0 > 0$ and $K > 0$, which depend only on e_d and M_d, such that if
$$|a_{ij}^{\alpha\beta}(x) - a_{ij}^{\alpha\beta}(0)| < \epsilon, \quad i, j = 1, \ldots, N, \quad |\alpha| + \beta = \alpha_{ij}$$
for all x, then for all $\lambda \in S$, $|\lambda| \geq \lambda_0$ the operator $A(x, \lambda, D)$ has a right inverse R, for which the estimate
$$|||Rf|||_E \leq K |||f|||_{F^d}, \quad \forall f \in F^d$$
holds.

5. Problem in $\mathbb{R}^n$

Proof. According to Theorem 3.1 the operator $A(0, \lambda, D)$ has an inverse R_0 and

$$|||R_0 f|||_E \leq K_0 |||f|||_{F^d}, \tag{5.1}$$

where the constant $K_0 = |||R_0|||$ does not depend on f and λ. Consider the operator $A(x, \lambda, D) R_0$. We have

$$A(x, \lambda, D) R_0 = A(0, \lambda, D) R_0 + T = I + T, \tag{5.2}$$

where I is the identity operator in F^d and

$$T = A(x, \lambda, D) R_0 - A(0, \lambda, D) R_0.$$

Set $u = R_0 f$. The elements of the vector Tf have the form

$$T_i f = \sum_{j=1}^{N} \sum_{|\alpha|+\beta=\alpha_{ij}} (a_{ij}^{\alpha\beta}(x) - a_{ij}^{\alpha\beta}(0)) \lambda^\beta D^\alpha u_j. \tag{5.3}$$

We need to estimate the norm $|||T_i f|||_{W^{l-s_i,p}}$. Consider first $\|T_i f\|_{W^{l-s_i,p}}$. We have

$$D^\gamma (a_{ij}^{\alpha\beta}(x) - a_{ij}^{\alpha\beta}(0)) \lambda^\beta D^\alpha u_j = \lambda^\beta \sum_{\tau+\sigma=\gamma} c_{\tau\sigma} D^\tau (a_{ij}^{\alpha\beta}(x) - a_{ij}^{\alpha\beta}(0)) D^{\sigma+\alpha} u_j$$

$$= \lambda^\beta (a_{ij}^{\alpha\beta}(x) - a_{ij}^{\alpha\beta}(0)) D^{\gamma+\alpha} u_j + \lambda^\beta \sum_{\tau+\sigma=\gamma, |\tau|>0} c_{\tau\sigma} D^\tau a_{ij}^{\alpha\beta}(x) D^{\sigma+\alpha} u_j.$$

Hence

$$\|D^\gamma (a_{ij}^{\alpha\beta}(x) - a_{ij}^{\alpha\beta}(0)) \lambda^\beta D^\alpha u_j\|_{L^p} \tag{5.4}$$

$$\leq |\lambda^\beta| \epsilon \|D^{\gamma+\alpha} u_j\|_{L^p} + |\lambda^\beta| \sum_{\tau+\sigma=\gamma, |\tau|>0} |c_{\tau\sigma}| \|D^\tau a_{ij}^{\alpha\beta}(x) D^{\sigma+\alpha} u_j\|_{L^p} = S_1 + S_2.$$

Here $|\gamma| \leq l - s_i$. We estimate S_2:

$$S_2 \leq \|a_{ij}^{\alpha\beta}\|_{C^{l-s_i}} |\lambda^\beta| \sum_{|\sigma|<|\gamma|} \|D^{\sigma+\alpha} u_j\|_{L^p} \leq \|a_{ij}^{\alpha\beta}\|_{C^{l-s_i}} |\lambda^\beta| \sum_{|\sigma|<|\gamma|} \|u_j\|_{W^{|\sigma|+|\alpha|,p}}$$

$$\leq \|a_{ij}^{\alpha\beta}\|_{C^{l-s_i}} \frac{c_1}{|\lambda|} \sum_{|\sigma|<|\gamma|} |||u_j|||_{W^{\beta+|\sigma|+|\alpha|+1,p}}$$

by the interpolation inequality. Here and below, c with subscripts denotes constants independent of u, λ, and of the coefficients of the operator. We have

$$\beta + |\sigma| + |\alpha| + 1 \leq \alpha_{ij} + |\gamma| \leq \alpha_{ij} + l - s_i = l + t_j.$$

Therefore

$$S_2 \leq \|a_{ij}^{\alpha\beta}\|_{C^{l-s_i}} \frac{c_2}{|\lambda|} |||u|||_E.$$

Further,

$$S_1 \leq \epsilon|\lambda|^\beta \|u_j\|_{W^{|\gamma|+|\alpha|,p}} \leq \epsilon c_3 \|\|u_j\|\|_{W^{\beta+|\gamma|+|\alpha|,p}} \leq \epsilon c_3 \|\|u_j\|\|_{W^{l+t_j,p}} \leq \epsilon c_3 \|\|u\|\|_E.$$

From (5.4) we get

$$\|D^\gamma(a_{ij}^{\alpha\beta}(x) - a_{ij}^{\alpha\beta}(0))\lambda^\beta D^\alpha u_j\|_{L^p} \leq c_4 \left(\epsilon + \frac{1}{|\lambda|}\|a_{ij}^{\alpha\beta}\|_{C^{l-s_i}}\right) \|\|u\|\|_E. \tag{5.5}$$

From this estimate and (5.3) we obtain

$$\|T_i f\|_{W^{l-s_i,p}} \leq c_5 \left(\epsilon + \frac{1}{|\lambda|}M_d\right) \|\|u\|\|_E. \tag{5.6}$$

We now estimate the expression $|\lambda|^{l-s_i}\|T_i f\|_{L^p}$. We have

$$|\lambda|^{l-s_i}\|(a_{ij}^{\alpha\beta}(x) - a_{ij}^{\alpha\beta}(0))\lambda^\beta D^\alpha u_j\|_{L^p} \leq \epsilon|\lambda|^{\beta+l-s_i}\|\|D^\alpha u_j\|\|_{L^p}$$
$$\leq \epsilon|\lambda|^{\beta+l-s_i}\|\|u_j\|\|_{W^{|\alpha|,p}} \leq \epsilon\|\|u_j\|\|_{W^{|\alpha|+\beta+l-s_i,p}} = \epsilon\|\|u_j\|\|_{W^{l+t_j,p}}.$$

Therefore

$$|\lambda|^{l-s_i}\|T_i f\|_{L^p} \leq \epsilon c_6 \|\|u\|\|_E. \tag{5.7}$$

It follows from (5.6), (5.7) that

$$\|\|Tf\|\|_{W^{l-s_i,p}} \leq c_7 \left(\epsilon + \frac{1}{|\lambda|}M_d\right) \|\|u\|\|_E \tag{5.8}$$

and

$$\|\|Tf\|\|_{F^d} \leq c_8 \left(\epsilon + \frac{1}{|\lambda|}M_d\right) \|\|u\|\|_E \leq c_8 \left(\epsilon + \frac{1}{|\lambda|}M_d\right) K_0 \|\|f\|\|_{F^d}. \tag{5.9}$$

Thus, if

$$\left(\epsilon + \frac{1}{|\lambda|}M_d\right) K_0 \leq \frac{1}{2c_8}, \tag{5.10}$$

then $\|\|T\|\| \leq \frac{1}{2}$, the operator $I+T$ is invertible, and $\|\|(I+T)^{-1}\|\| \leq 2$. From (5.2),

$$A(x, \lambda, D) = (I+T)R_0^{-1}.$$

Hence, the operator $A(x, \lambda, D)$ is invertible,

$$R = (A(x, \lambda, D))^{-1} = R_0(I+T)^{-1}, \quad \text{and} \quad \|\|R\|\| \leq 2\|\|R_0\|\|.$$

Since $\|\|R_0\|\|$ depends only on e_d and M_d, then $\|\|R\|\|$ depends also only on them. The theorem is proved. □

6 Problem in $\mathbb{R}^n_+$

We consider the operators $A(x, \lambda, D)$ and $B(x, \lambda, D)$ in the half-space $\mathbb{R}^n_+ (x_n > 0)$. Here $A(x, \lambda, D)$ is the matrix with elements

$$A_{ij}(x, \lambda, D) = \sum_{|\alpha|+\beta=\alpha_{ij}} a_{ij}^{\alpha\beta}(x)\lambda^\beta D^\alpha, \quad i, j = 1, \ldots, N,$$

$B(x, \lambda, D)$ is a rectangular matrix with elements

$$B_{ij}(x, \lambda, D) = \sum_{|\alpha|+\beta=\beta_{ij}} b_{ij}^{\alpha\beta}(x)\lambda^\beta D^\alpha, \quad i = 1, \ldots, r, \ j = 1, \ldots, N,$$

$$a_{ij}^{\alpha\beta}(x) \in C^{l-s_i+\theta}(\mathbb{R}^n_+), \quad b_{ij}^{\alpha\beta}(x) \in C^{l-\sigma_i+\theta}(\mathbb{R}^n_+), \quad 0 < \theta < 1.$$

Theorem 6.1. *Suppose that the operator $L(0) = (A(0, \lambda, D), B(0, \lambda, D))$ is elliptic. Then there exist constants $\epsilon > 0$, $\lambda_0 > 0$ and $K > 0$, which depend only on e_d, e_Γ, M_d and M_Γ, such that if*

$$|a_{ij}^{\alpha\beta}(x) - a_{ij}^{\alpha\beta}(0)| < \epsilon, \quad x \in \mathbb{R}^n_+, \quad i, j = 1, \ldots, N, \quad |\alpha| + \beta = \alpha_{ij},$$

$$|b_{ij}^{\alpha\beta}(x) - b_{ij}^{\alpha\beta}(0)| < \epsilon, \quad x \in \mathbb{R}^n_+, \quad i = 1, \ldots, r, \ j = 1, \ldots, N, \quad |\alpha| + \beta = \beta_{ij},$$

then for all $\lambda \in S$, $|\lambda| \geq \lambda_0$ the operator $L(x) = (A(x, \lambda, D), B(x, \lambda, D))$ has a right inverse R, for which the estimate

$$|||Rf|||_E \leq K \, |||f|||_F, \quad \forall f \in F$$

holds.

Proof. According to Theorem 4.4 the operator $L(0)$ has inverse R_0 and

$$|||R_0 f|||_{E(\mathbb{R}^n_+)} \leq K_0 |||f|||_{F(\mathbb{R}^n_+)}. \tag{6.1}$$

Consider the operator

$$L(x)R_0 = L(0)R_0 + T = I + T, \tag{6.2}$$

where I is the identity operator in F and

$$T = (L(x) - L(0))R_0. \tag{6.3}$$

Set $u = R_0 f$. Then

$$Tf = (L(x) - L(0))u = (A(x, \lambda, D) - A(0, \lambda, D), B(x, \lambda, D) - B(0, \lambda, D))u.$$

Similarly as it is done in $\mathbb{R}^n$ we get the estimate

$$|||A(x, \lambda, D) - A(0, \lambda, D)u|||_{F^d} \leq c_1 \left(\epsilon + \frac{1}{|\lambda|} M_d\right) |||u|||_E, \tag{6.4}$$

where c_1 is a constant independent of the coefficients of the operator $L(x)$. In the same way we obtain a similar estimate for the boundary operator:

$$\||B(x,\lambda,D) - B(0,\lambda,D)u\||_{F^b} \leq c_2\left(\epsilon + \frac{1}{|\lambda|}M_b\right)\||u\||_E. \tag{6.5}$$

From (6.4), (6.5),

$$\||Tf\||_F \leq c_3\left(\epsilon + \frac{M_d + M_b}{|\lambda|}\right)\||f\||_F.$$

Here c_3 is a constant independent of the coefficients of the operator $L(x)$. We choose ϵ and λ such that

$$c_3\left(\epsilon + \frac{M_d + M_b}{|\lambda|}\right) \leq \frac{1}{2}.$$

Then $\||T\|| \leq \frac{1}{2}$ and the inverse operator $R = R_0(I+T)^{-1}$ admits the estimate $\||R\|| \leq 2\||R_0\||$. The theorem is proved. $\square$

7 Problem in Ω

Let Ω be an unbounded domain in $\mathbb{R}^n$. It is supposed that Condition D is satisfied. We consider the operators $A(x,\lambda,D)$ and $B(x,\lambda,D)$ in the domain Ω. Here $A(x,\lambda,D)$ is the matrix with elements

$$A_{ij}(x,\lambda,D) = \sum_{|\alpha|+\beta \leq \alpha_{ij}} a_{ij}^{\alpha\beta}(x)\lambda^\beta D^\alpha, \quad i,j = 1,\ldots,N,$$

$B(x,\lambda,D)$ is a rectangular matrix with elements

$$B_{ij}(x,\lambda,D) = \sum_{|\alpha|+\beta \leq \beta_{ij}} b_{ij}^{\alpha\beta}(x)\lambda^\beta D^\alpha, \quad i = 1,\ldots,r, \ j = 1,\ldots,N,$$

$a_{ij}^{\alpha\beta}(x) \in C^{l-s_i+\theta}(\mathbb{R}^n)$, $b_{ij}^{\alpha\beta}(x) \in C^{l-\sigma_i+\theta}(\mathbb{R}^n)$, $0 < \theta < 1$.

Write $L = (A, B)$. We will first consider this operator as acting in ∞-spaces introduced in Section 2 (see also Chapter 2).

Proposition 7.1. *The operator L is a bounded operator from E_∞ to F_∞. Moreover,*

$$\||Lu\||_{F_\infty} \leq K\||u\||_{E_\infty},$$

where K is a constant independent of u and λ. This is also true for the spaces E and F.

The proof of this proposition is similar to the proofs of Propositions 4.1, 4.2.

7. Problem in Ω

Let $\phi_i(x)$ be a partition of unity in $\mathbb{R}^n$, and $\psi_i(x)$ be a system of functions, $\psi_i(x) \in D$ such that $\psi_i(x) = 1$ in a neighborhood of $\operatorname{supp} \phi_i(x)$. We suppose that $\phi_i(x)$ and $\psi_i(x)$ satisfy the conditions which are specified for systems of functions in the construction of spaces E_q. Moreover, we suppose that $\operatorname{supp} \phi_i$ either do not intersect the boundary Γ or belong to a given covering of Γ. Similar assumptions are made for ψ_i (cf. [13], p. 86). It is easy to prove that for uniformly regular domains such systems of functions can be constructed. Moreover, we can suppose that the support of the function $\phi_i(x)$ is a ball $B_i(r)$ with its center at x_i and radius r, the support of the function $\psi_i(x)$ is the ball $B_i(2r)$ with the same center and radius $2r$.

Denote by X_i the space $E(\Omega \cap B_i(2r))$ and by L_i the restriction of the operator L to X_i. The operator L_i acts from X_i into $Y_i = F(\Omega \cap B_i(2r)) = F^d(\Omega \cap B_i(2r)) \times F^b(\Omega \cap B_i(2r))$. We have

$$Lu = \sum_{i=1}^{\infty} \phi_i L(\psi_i u). \tag{7.1}$$

By construction, $L(\psi_i u) = L_i(\psi_i u)$. If the support $\operatorname{supp} \psi_i$ is sufficiently small, then the coefficients of the operator L_i are close to constant (cf. [13], p. 86).

We denote next by L_{i0} the principal part of the operator L_i. According to Theorem 3.1 and Theorem 4.4 there exists λ_0 such that for $|\lambda| \geq \lambda_0$ the supports of the functions ψ_i can be chosen so small that the operators L_{i0} have locally right inverse operators $R_i : Y_i \to X_i$, that is

$$L_{i0} R_i f = f, \ \forall f \in Y_i.$$

The norms of the inverse operators $\|R_i\|_{F \to E}$ are bounded independently of i. It is supposed that the boundary of the domain is sufficiently smooth (see Condition D in Section 1) in order to map the problems under consideration into the problems in half-spaces.

For any $f \in F_\infty$, $R_i(\phi_i f)$ is defined. Therefore we can introduce the operator

$$Rf = \sum_{i=1}^{\infty} \psi_i R_i(\phi_i f), \quad f \in F_\infty. \tag{7.2}$$

This sum contains only a finite number of terms at any $x \in \bar{\Omega}$.

Proposition 7.2. *The operator R is a bounded operator from F_∞ to E_∞ with domain F_∞. Moreover,*

$$\|Rf\|_{E_\infty} \leq K \|f\|_{F_\infty},$$

where K is a constant independent of u and λ. It depends on M_d, M_b, e_d, e_Γ and the smoothness of the boundary.

Proof. Set
$$u_i = R_i(\phi_i f), \quad f \in F_\infty \tag{7.3}$$
and $u = Rf$. Then
$$u(x) = \sum_{i=1}^\infty \psi_i(x) u_i(x). \tag{7.4}$$
Here $u_i \in X_i$, $\psi_i(x)u_i(x)$ is defined in Ω and equals 0 outside the set $\Omega \cap B_i(2r)$. It follows from (7.4) that
$$\phi_k u = \sum_{i=1}^\infty \phi_k \psi_i u_i = \sum_{i'} \phi_k \psi_{i'} u_{i'},$$
where i' are all values of i such that $\phi_k \psi_i \neq 0$. By virtue of the condition on the systems of functions the number of such i' is limited by some number N_0. We have
$$\|\phi_k u\|_E \leq \sum_{i'}^\infty \|\phi_k \psi_{i'} u_{i'}\|_E \leq N_0 \sup_i \|\phi_k \psi_i u_i\|_E \leq N_0 c_1 \sup_i \|u_i\|_{X_i},$$
where the constant
$$c_1 = \sup_k \|\phi_k\|_M \sup_i \|\psi_i\|_M$$
does not depend on λ. From (7.3)
$$\|\phi_k u\|_E \leq N_0 c_1 K_1 \sup_i \|\phi_i f\|_{Y_i} \leq N_0 c_1 K_1 \|f\|_{F_\infty},$$
where
$$K_1 = \sup_i \|R_i\|_{Y_i \to X_i}.$$
Hence
$$\|u\|_{E_\infty} \leq N_0 c_1 K_1 \|f\|_{F_\infty}.$$
The proposition is proved. $\square$

It follows from Propositions 7.1, 7.2 that the operator LR is a bounded operator in F_∞ defined on all F_∞. Consider the operator T given by the expression
$$Tf = \sum_{j=1}^\infty \sum_{i=1}^\infty \phi_i((L_i \psi_j \psi_i u_i - \psi_j \psi_i L_i u_i) + \psi_j \psi_i (L_i u_i - L_{i0} u_i)),$$
where u_i is defined by (7.3). The expression in the right-hand side contains a finite number of terms for every $x \in \bar\Omega$.

Proposition 7.3.
$$LRf = f + Tf, \quad \forall f \in F_\infty. \tag{7.5}$$

7. Problem in Ω

Proof. Let $f \in F_\infty, u = Rf$. We have from (7.1), (7.4),

$$Lu = \sum_{j=1}^{\infty} \phi_j L(\psi_j u) = \sum_{j=1}^{\infty} \phi_j L_j(\psi_j u) = \sum_{j=1}^{\infty} \phi_j \sum_{i=1}^{\infty} L_j \psi_j \psi_i u_i.$$

The number of terms of this series is finite for each $x \in \bar{\Omega}$. We have

$$L_j \psi_j \psi_i u_i = L\psi_j \psi_i u_i = L_i \psi_j \psi_i u_i.$$

Hence

$$Lu = \sum_{j=1}^{\infty} \sum_{i=1}^{\infty} \phi_j L_i \psi_j \psi_i u_i.$$

Further

$$L\psi_j \psi_i u_i = L\psi_j \psi_i u_i - \psi_j \psi_i L_i u_i + \psi_j \psi_i (L_i u_i - L_{i0} u_i) + \psi_j \psi_i L_{i0} u_i.$$

Taking into account (7.3) we get

$$\psi_j \psi_i L_{i0} u_i = \psi_j \psi_i L_{i0} R_i \phi_i f = \psi_j \psi_i \phi_i f = \psi_j \phi_i f.$$

Therefore

$$Lu = \sum_{j=1}^{\infty} \sum_{i=1}^{\infty} \phi_j (L\psi_j \psi_i u_i - \psi_j \psi_i L_i u_i + \psi_j \psi_i (L_i u_i - L_{i0} u_i)) + \sum_{j=1}^{\infty} \sum_{i=1}^{\infty} \phi_j \phi_i f.$$

The last term in the right-hand side equals f. Thus,

$$Lu = Tu + f$$

almost everywhere. The proposition is proved. $\square$

Proposition 7.4. *For $\lambda \in S$, $|\lambda| > \lambda_0$, where λ_0 is sufficiently large, the following estimate holds:*

$$\|Tf\|_{F_\infty} \leq K|\lambda|^{-1} \|f\|_{F_\infty}, \quad \forall f \in F_\infty$$

with a constant K determined by M_d, M_b, e_d, e_Γ and the smoothness of the boundary (Condition D), and independent of f and λ.

Proof. Consider the operators

$$T_1 f = \sum_{j=1}^{\infty} \sum_{i=1}^{\infty} \phi_j (L_i \psi_j \psi_i u_i - \psi_j \psi_i L_i u_i),$$

$$T_2 f = \sum_{j=1}^{\infty} \sum_{i=1}^{\infty} \phi_j \psi_j \psi_i (L_i u_i - L_{i0} u_i),$$

where u_i is defined by (7.3). We get

$$Tf = T_1 f + T_2 f, \quad \forall f \in F_\infty.$$

Consider first $T_1 f$. We have

$$\phi_k T_1 f = \sum_{j'} \sum_{i'} \phi_k \phi_{j'} (L_{i'} \psi_{j'} \psi_{i'} u_{i'} - \psi_{j'} \psi_{i'} L_{i'} u_{i'}), \tag{7.6}$$

where i', j' are all those values of i, j for which $\phi_j \phi_k \neq 0$ and $\phi_i \phi_k \neq 0$.

Let A_i and B_i be the restrictions of the operators A and B to X_i. Then $L_i = (A_i, B_i)$. It follows from (7.6) that

$$\|\phi_k T_1 f\|_F \leq \sum_{j'} \sum_{i'} \|\phi_k \phi_{j'} (A_{i'} \psi_{j'} \psi_{i'} u_{i'} - \psi_{j'} \psi_{i'} A_{i'} u_{i'})\|_{F^d}$$
$$+ \sum_{j'} \sum_{i'} \|\phi_k \phi_{j'} (B_{i'} \psi_{j'} \psi_{i'} u_{i'} - \psi_{j'} \psi_{i'} B_{i'} u_{i'})\|_{F^b}. \tag{7.7}$$

Set $u_i = (v_1, \ldots, v_N)^T$ (the subscript i is omitted), $\omega = \psi_j \psi_i$ and consider the elements of the vector $(A_i \omega u_i - \omega A_i u_i)$:

$$\sum_{\tau=1}^{N} \sum_{|\alpha|+\beta \leq \alpha_{\sigma\tau}} \left(a_{\sigma\tau}^{\alpha\beta}(x) \lambda^\beta D^\alpha (\omega v_\tau) - a_{\sigma\tau}^{\alpha\beta}(x) \lambda^\beta \omega D^\alpha v_\tau \right).$$

We should estimate them in the norm $\|\|\cdot\|\|_{W^{l-s_\sigma,p}(\Omega_i)}$, where $\Omega_i = \Omega \cap B_i(2r)$.

We first estimate the $\|\cdot\|_{W^{l-s_\sigma,p}(\Omega_i)}$ norm:

$$\|a_{\sigma\tau}^{\alpha\beta}(x) \lambda^\beta D^\alpha (\omega v_\tau) - a_{\sigma\tau}^{\alpha\beta}(x) \lambda^\beta \omega D^\alpha v_\tau\|_{W^{l-s_\sigma,p}(\Omega_i)}$$
$$\leq \|a_{\sigma\tau}^{\alpha\beta}(x)\|_{C^{l-s_\sigma}} |\lambda|^\beta \|D^\alpha (\omega v_\tau) - \omega D^\alpha v_\tau\|_{W^{l-s_\sigma,p}(\Omega_i)}$$
$$\leq M_1 \|a_{\sigma\tau}^{\alpha\beta}(x)\|_{C^{l-s_\sigma}} |\lambda|^\beta \|v_\tau\|_{W^{l-s_\sigma+|\alpha|-1,p}(\Omega_i)}$$
$$\leq c_1 \frac{M_1}{|\lambda|} \|a_{\sigma\tau}^{\alpha\beta}(x)\|_{C^{l-s_\sigma}} \|v_\tau\|_{W^{l+t_\tau,p}(\Omega_i)} \leq \frac{M_2}{|\lambda|} \|u_i\|_{E(\Omega_i)}.$$

Here c_1 is a constant independent of the coefficients of the operator A and B, M_1 is a constant which depends on derivatives of ω. Hence it depends on ϵ in Theorems 5.1, 6.1 and, consequently, on M_d, M_b, e_d, e_Γ.

In a similar way we can estimate the norm $|\lambda|^{l-s_\sigma} \|\cdot\|_{L^p(\tilde{\Omega})}$. We get the same estimate as before. Hence

$$\sum_{j'} \sum_{i'} \|\phi_k \phi_{j'} (A_{i'} \psi_{j'} \psi_{i'} u_{i'} - \psi_{j'} \psi_{i'} A_{i'} u_{i'})\|_{F^d} \leq \sum_{j'} \sum_{i'} c_2 \frac{M_2}{|\lambda|} \|u_{i'}\|_{E(\Omega_{i'})}, \tag{7.8}$$

7. Problem in Ω

where c_2 is a constant independent of the coefficients of the operator A and B. We have

$$\|u_i\|_{X_i} = \|R_i(\phi_i f)\|_{X_i} \le K_1 \|\phi_i f\|_{Y_i} \le c_3 K_1 \|\phi_i f\|_F \le c_3 K_1 \|f\|_{F_\infty(\Omega)},$$

where $K_1 = \sup_i \|\|R_i\|\|_{Y_i \to X_i}$. Since the number of i' and j' is bounded by N, the last estimate together with (7.8) give

$$\sum_{j'} \sum_{i'} \|\|\phi_k \phi_{j'}(A_{i'} \psi_{j'} \psi_{i'} u_{i'} - \psi_{j'} \psi_{i'} A_{i'} u_{i'})\|\|_{F^d} \le c_4 K_1 \frac{M_2}{|\lambda|} \|f\|_{F_\infty(\Omega)}. \quad (7.9)$$

We next estimate the operator B. We have as above

$$\|b_{\nu\tau}^{\alpha\beta}(x)\lambda^\beta D^\alpha(\omega v_\tau) - b_{\nu\tau}^{\alpha\beta}(x)\lambda^\beta \omega D^\alpha v_\tau\|_{W^{l-s_\nu,p}(\Omega_i)} \le \frac{M_3}{|\lambda|} \|u_i\|_{E(\Omega_i)}.$$

Write $\Omega'_i = \partial\Omega \cap B_i(2r)$ assuming that this intersection is not empty. Since $\|\cdot\|_{W^{l-s_\nu-1/p,p}(\Omega'_i)} \le \|\cdot\|_{W^{l-s_\nu,p}(\Omega_i)}$, then

$$\|b_{\nu\tau}^{\alpha\beta}(x)\lambda^\beta D^\alpha(\omega v_\tau) - b_{\nu\tau}^{\alpha\beta}(x)\lambda^\beta \omega D^\alpha v_\tau\|_{W^{l-s_\nu-1/p,p}(\Omega'_i)} \le \frac{M_3}{|\lambda|} \|u_i\|_{E(\Omega_i)}.$$

Therefore

$$\sum_{j'} \sum_{i'} \|\|\phi_k \phi_{j'}(B_{i'} \psi_{j'} \psi_{i'} u_{i'} - \psi_{j'} \psi_{i'} B_{i'} u_{i'})\|\|_{F^b}$$

$$\le c_5 \frac{M_3}{|\lambda|} \|u_i\|_{E(\Omega_i)} \le c_6 K_1 \frac{M_3}{|\lambda|} \|f\|_{F_\infty(\Omega)},$$

where, as above, M_3 depends on M_d, M_b, e_d, e_Γ, c_6 is independent of the coefficients of the operator A and B (this convention about the constants is used also below). The last estimate together with (7.9) give

$$\|\phi_k T_1 f\|_{F(\Omega)} \le c_7 K_1 \frac{M_2 + M_3}{|\lambda|} \|f\|_{F_\infty(\Omega)}.$$

Hence

$$\|T_1 f\|_{F_\infty(\Omega)} \le c_7 K_1 \frac{M_2 + M_3}{|\lambda|} \|f\|_{F_\infty(\Omega)}.$$

Similarly we obtain the estimate

$$\|T_2 f\|_{F_\infty(\Omega)} \le c_8 K_1 \frac{M_2 + M_3}{|\lambda|} \|f\|_{E_\infty(\Omega)}.$$

Therefore,

$$\|Tf\|_{F_\infty(\Omega)} \le \frac{K}{|\lambda|} \|f\|_{F_\infty(\Omega)}.$$

The proposition is proved. $\square$

Theorem 7.5. *There exists $\lambda_0 > 0$ such that for any $\lambda \in S$, $|\lambda| \geq \lambda_0$ the equation*

$$Lu = f, \quad u \in E_\infty, \quad f \in F_\infty \tag{7.10}$$

is uniquely solvable for any $f \in F_\infty$. Moreover, the estimate

$$\|\|u\|\|_{E_\infty} \leq \kappa \|\|f\|\|_{F_\infty} \tag{7.11}$$

holds for the solution $u \in E_\infty$ of this problem. Here $\lambda_0 = 2K$, where K is the constant in Proposition 7.4, and hence depends on M_d, M_b, e_d, e_Γ and the smoothness of the boundary. The constant $\kappa = 2\|\|R\|\|$ depends on the same constants and is independent of λ and f.

Proof. According to (7.5) we have

$$LR = I + T, \tag{7.12}$$

where I is the identity operator in F_∞. From Proposition 7.4 we get $\|\|T\|\| \leq \frac{1}{2}$ for $\lambda_0 = 2k$. Hence the operator $I + T$ is invertible and

$$\|\|(I + T)^{-1}\|\| \leq 2. \tag{7.13}$$

From (7.12),

$$LR(I + T)^{-1}f = f, \quad \forall f \in F_\infty.$$

Write

$$u = R(I + T)^{-1}f. \tag{7.14}$$

Then $Lu = f$. From a priori estimates proved below it follows that the solution of this equation is unique. Hence any solution has the form (7.14). It follows that (7.11) holds with $\kappa = 2\|\|R\|\|$. The theorem is proved. $\square$

Lemma 7.6. *For any point $x_0 \in \bar{\Omega}$ there exists a neighborhood U of this point and a number $\lambda_0 > 0$ such that for any function $u \in E(\Omega)$ with the support in $U \cap \bar{\Omega}$ the following estimate holds:*

$$\|\|u\|\|_{E(\Omega)} \leq K \|\|Lu\|\|_{F(\Omega)}, \tag{7.15}$$

where $|\lambda| \geq \lambda_0$, $\lambda \in S$ and the constant K does not depend on u and λ.

Proof. The proof of this lemma follows from Theorems 5.1 and 6.1. For a function ϕ with its support in U we have, by virtue of these theorems, the estimate

$$\|\|\phi u\|\|_{E(\Omega)} \leq K \|\|L(\phi u)\|\|_{F(\Omega)}, \tag{7.16}$$

for any $u \in E_\infty(\Omega)$. The lemma is proved. $\square$

7. Problem in Ω

Lemma 7.7. *For a function ϕ with its support in U, the following estimate holds:*

$$|||\phi u|||_{E(\Omega)} \leq K_1 \left(|||\phi L u|||_{F(\Omega)} + \frac{1}{|\lambda|} |||\psi u|||_{E(\Omega)} \right), \quad \forall u \in E(\Omega), \qquad (7.17)$$

where K_1 is a constant independent of u and λ, $|\lambda| \geq \lambda_0$ and λ_0 is sufficiently large. Here $\psi \in D$ is such that $\psi(x) = 1$ for $x \in \operatorname{supp} \phi$.

Proof. We should estimate the norm $|||L(\phi u) - \phi L u|||_{F(\Omega)}$. We have

$$L(\phi u) - \phi L u = L(\phi \psi u) - \phi L(\psi u).$$

The required estimate can be obtained in the same way as the estimates in the proof of Proposition 7.4 (see (7.6)–(7.8)). The lemma is proved. $\square$

Let ϕ_i be a system of functions in the definition of the space E_q, $1 \leq q \leq \infty$, and $\psi_i(x) = 1$ for $x \in \operatorname{supp} \phi_i$. Then from (7.17),

$$|||\phi_i u|||_{E(\Omega)} \leq K_1 \left(|||\phi_i L u|||_{F(\Omega)} + \frac{1}{|\lambda|} |||\psi_i u|||_{E(\Omega)} \right), \quad \forall u \in E(\Omega). \qquad (7.18)$$

Let first $q = \infty$. Taking supremum with respect to i, we get

$$|||u|||_{E_\infty(\Omega)} \leq K_1 \left(|||L u|||_{F_\infty(\Omega)} + \frac{c}{|\lambda|} |||u|||_{E_\infty(\Omega)} \right).$$

The constant c appears as a result of equivalence of the norms E_∞ with the systems of functions ϕ_i and ψ_i. Let $|\lambda| \geq 2cK_1$. Then

$$|||u|||_{E_\infty(\Omega)} \leq 2K_1 |||L u|||_{F_\infty(\Omega)}. \qquad (7.19)$$

Consider now the case $1 \leq q < \infty$. From (7.18),

$$|||\phi_i u|||_{E(\Omega)}^q \leq 2^q K_1^q \left(|||\phi_i L u|||_{F(\Omega)}^q + \frac{1}{|\lambda|^q} |||\psi_i u|||_{E(\Omega)}^q \right).$$

Taking a sum with respect to i, we obtain

$$|||u|||_{E_q(\Omega)}^q \leq 2^q K_1^q \left(|||L u|||_{F_q(\Omega)}^q + \frac{c^q}{|\lambda|^q} |||u|||_{E_q(\Omega)}^q \right).$$

Let $|\lambda| \geq 4cK_1$. Then

$$\left(1 - \frac{1}{2^q} \right) |||u|||_{E_q(\Omega)}^q \leq 2^q K_1^q |||L u|||_{F_q(\Omega)}^q.$$

Hence

$$|||u|||_{E_q(\Omega)} \leq 4K_1 |||L u|||_{F_q(\Omega)}. \qquad (7.20)$$

We have proved the following theorem.

Theorem 7.8. *The following estimate holds:*

$$|||u|||_{E_q(\Omega)} \leq 4K_1|||Lu|||_{F_q(\Omega)}, \quad u \in E_q(\Omega), \ 1 \leq q \leq \infty \qquad (7.21)$$

with the constant K_1 from Lemma 7.7, $|\lambda| \geq \lambda_0$ and λ_0 is sufficiently large.

We prove next the solvability in the spaces E_q. Consider the equation

$$Lu = f, \quad u \in E_q(\Omega), \ f \in F_q(\Omega), \ 1 \leq q \leq \infty. \qquad (7.22)$$

Theorem 7.9. *If $\lambda \in S$, $|\lambda| \geq \lambda_0$, then equation (7.22) is uniquely solvable for any $f \in F_q$.*

Proof. From estimate (7.21) it follows that it is sufficient to prove the solvability of (7.22) for smooth f with compact support. Let $f \in D(\Omega) \times D(\Gamma)$. Then from Theorem 7.5 it follows that there exists a solution $u \in E_\infty(\Omega)$ of equation (7.22). Then (Theorem 2.12, Chapter 4)

$$u(x) \, e^{\mu\sqrt{1+|x|^2}} \in E_\infty(\Omega).$$

Hence $u \in E_q(\Omega)$. The theorem is proved. $\square$

8 Generation of analytic semigroups

We suppose that the operator A has homogeneous principal terms, that is $\alpha_{ij} = m$ for some m. Let

$$A(x, \lambda, D)u = \sum_{|\alpha| \leq m} a_\alpha(x) D^\alpha u - \lambda u, \qquad (8.1)$$

where $a_\alpha(x)$ are square $N \times N$ matrices and u is a vector. We suppose further that the boundary operator $B(x, D)$ does not contain the parameter λ. Let

$$E_q(\Omega) = W_q^{m,p}(\Omega), \quad 1 < p < \infty, \ 1 \leq q \leq \infty.$$

It is assumed that the domain of the operator A is

$$D(A) = \{u \in E_q(\Omega), \ B(x, D) = 0\}.$$

It is clear that $D(A)$ is dense in $L_q^p(\Omega)$. Indeed, the set of infinitely differentiable functions, which vanish at the boundary with their derivatives, is dense in $L_q^p(\Omega)$.

Assumption 8.1. Suppose that $\lambda = \mu^n$ in (8.1) and the operator $A(x, \lambda, D)$ with μ as a parameter is elliptic with respect to sector S such that the set $\{\mu : \text{Re } \mu^m \geq \omega\}$ belongs to S.

Theorem 8.2. *Consider the operator A acting in $L_q^p(\Omega)$ with the domain $D(A)$. If Assumption 8.1 is satisfied, then it is a generator of an analytic semigroup.*

Proof. It is known that an operator A is sectorial if the resolvent set $\rho(A)$ contains a half-plane
$$\{\lambda \in \mathbb{C} : \operatorname{Re} \lambda \geq \omega\} \tag{8.2}$$
and
$$\|\lambda R(\lambda, A)\| \leq M, \ \operatorname{Re} \lambda \geq \omega, \tag{8.3}$$
where $R(\lambda, A)$ is the resolvent of the operator A, M and ω are some constants. It follows from Theorem 7.9 that $\rho(A)$ contains the half-plane (8.2) for $|\mu|$ sufficiently large. Estimate (8.3) follows from Theorem 7.8. The theorem is proved. $\square$

9 Elliptic problems with a parameter at infinity

We consider the operators $A(x, \lambda, D)$ and $B(x, \lambda, D)$ defined in Section 4. We suppose that the sector S is given.

Definition 9.1. The operator $L(x, \lambda, D) = (A(x, \lambda, D), B(x, \lambda, D))$ is elliptic with a parameter at infinity if it is elliptic for $x \in \bar{\Omega}$, $\lambda = 0$ and elliptic with a parameter for $x \in \bar{\Omega}$, $|x| \geq R$ and $\lambda \in S$, where R is a sufficiently large number.

We will use the following notation:

$$e_d^0 = \inf_{x \in G, |\xi|=1} |\det A(X, 0, \xi)|,$$

$$e_d = \inf_{x \in G, |\xi|+|\lambda|=1, \lambda \in S} |\det A(X, \lambda, \xi)|,$$

$$M_d = \max_{|\alpha|+\beta \leq \alpha_{ij}, i,j=1,\ldots,N} \|a_{ij}^{\alpha\beta}\|_{C^{l-s_i}(\bar{G})},$$

$$M_\Gamma = \max_{|\alpha|+\beta \leq \beta_{ij}, i=1,\ldots,r, j=1,\ldots,N} \|b_{ij}^{\alpha\beta}\|_{C^{l-\sigma_i}(G \cap \Gamma)},$$

$$e_\Gamma^0 = \inf_{x \in G \cap \Gamma, |\xi'|=1} \sum_\alpha |\mu_\alpha(x, 0, \xi')|,$$

$$e_\Gamma = \inf_{x \in G \cap \Gamma, |\xi'|+|\lambda|=1, \lambda \in S} \sum_\alpha |\mu_\alpha(x, \lambda, \xi')|,$$

where $G \subset \bar{\Omega}$, $\mu_\alpha(x, \xi')$ are all r-minors of the Lopatinskii matrix in the local coordinates (ξ', ξ^n) at the point x.

Definition 9.2. The operator $L(x, 0, D)$ is called uniformly elliptic on the set $G \subset \bar{\Omega}$ if
$$e_d^0 > 0, \ e_\Gamma^0 > 0, \ M_d < \infty, \ M_\Gamma < \infty.$$

Definition 9.3. The operator $L(x, \lambda, D)$ is called uniformly elliptic on the set $G \subset \bar{\Omega}$ with a parameter if
$$e_d > 0, \ e_\Gamma > 0, \ M_d < \infty, \ M_\Gamma < \infty.$$

Definition 9.4. The operator $L(x, \lambda, D)$ is called uniformly elliptic with a parameter at infinity if the operator $L(x, 0, D)$ is uniformly elliptic in $\bar{\Omega}$ and $L(x, \lambda, D)$ is uniformly elliptic with a parameter in the domain $\{|x| \geq R\}$ for some R.

Proposition 9.5. *If the operator $L(x, \lambda, D)$ is uniformly elliptic with a parameter in the domain $\{x \in \Omega, |x| \geq R\}$, then all limiting operators are elliptic with a parameter with the same S and with the same values of the constants $e_d, e_\Gamma, M_d, M_\Gamma$ and with the same constants as in Condition D.*

The proof of the theorem follows from the definition of limiting operators.

Theorem 9.6. *If the operator $L(x, \lambda, D)$ is uniformly elliptic with a parameter at infinity, then there exist λ_0 such that for $|\lambda| \geq \lambda_0$, $\lambda \in S$ the operator $L(x, \lambda, D)$ is Fredholm as acting from the space E_∞ into F_∞.*

Proof. From Proposition 9.5 it follows that all limiting operators are elliptic operators with a parameter with the same S and constants $e_d, e_\Gamma, M_d, M_\Gamma$ and the same constants in the Condition D. According to Theorem 7.5 there exists a constant λ_0, which depends on the constants above, such that for $|\lambda| \geq \lambda_0$ all limiting operators are invertible. Hence according to the results of Chapter 5 the operator $L(x, \lambda, D)$ is Fredholm from E_∞ to F_∞. The theorem is proved. □

Remark 9.7. From the results of Chapter 5 it also follows that the operator $L(x, \lambda, D)$ is Fredholm from E_q into F_q for $q \leq p$ and in the properly chosen Hölder spaces if the coefficients and the boundary are sufficiently smooth. We recall that $E_q = E$ and $F_q = F$ for $q = p$. Therefore the Fredholm property is proved also for the usual Sobolev spaces.

10 Examples

A linear elliptic operator is invertible if it satisfies the Fredholm property, its index is zero, and the kernel is empty. Invertibility also follows from the unique solvability of the nonhomogeneous equation for any right-hand side. In some cases, in particular for scalar operators, the solvability conditions can be formulated with the help of formally adjoint operators.

10.1 Invertibility of second-order operators

One-dimensional case. Consider the operator

$$Lu = a(x)u'' + b(x)u' + c(x)u, \quad x \in \mathbb{R},$$

acting from $H^2(\mathbb{R})$ into $L^2()$. Assume that the coefficients are sufficiently smooth, $a(x) \geq a_0 > 0$ and that there exist the limits

$$a_\pm, b_\pm, c_\pm = \lim_{x \to \pm\infty} a(x), b(x), c(x).$$

10. Examples

The essential spectrum of this operator is given by the curves

$$\lambda_\pm(\xi) = -a_\pm \xi^2 + b_\pm i\xi + c_\pm, \ \xi \in \mathbb{R}.$$

If both of them lie in the left half-plane, then the index of the operator equals zero. It is invertible if and only if its kernel is empty, that is if the equation $Lu = 0$ has only the zero solution. In particular, this is the case if the principal eigenvalue lies in the left half-plane (see the next example).

Principal eigenvalue. Let us call the eigenvalue with the maximal real part the principal eigenvalue. Similar to matrices with positive off-diagonal elements, the principal eigenvalue of scalar elliptic problems is real, simple, and the corresponding eigenfunction is positive. Moreover, there are no positive eigenfunctions corresponding to other eigenvalues. In the case of bounded domains, where the inverse operator is compact, this result follows from the Krein-Rutman theorem. In the case of unbounded domains it is proved in [568], [563]. It remains also valid for some class of systems. Consider the operator

$$Lu = \Delta u$$

in a bounded domain $\Omega \subset \mathbb{R}^n$ with the Dirichlet boundary condition, $L : E \to F$, where

$$E = \{u \in H^2(\Omega), u|_{\partial\Omega} = 0\}, \ F = L^2(\Omega).$$

It follows from the maximum principle that the equation $Lu = \lambda u$ cannot have positive solutions for a real non-negative λ. Therefore, the principal eigenvalue is negative and the operator is invertible.

In the case of the Neumann boundary condition, $u = 1$ is a positive eigenfunction corresponding to the zero eigenvalue. Therefore, it is the principal eigenvalue. The operator $L-a$ is invertible for any positive a. For more general operators (10.1) the principal eigenvalue can be positive. However, it remains real and bounded. Therefore, this operator is invertible for sufficiently large ρ.

For unbounded domains, these results remain valid if the essential spectrum determined by limiting problems lies in the left half-plane.

Zero solution of the homogeneous equation. In this example we do not use the results on the principal eigenvalues and prove directly that the kernel of the operator is empty. Consider the operator

$$Lv = a(x)\Delta v + \sum_{k=1}^{n} b_k(x) \frac{\partial v}{\partial x_k} + c(x)v - \rho v \tag{10.1}$$

in a bounded domain $\Omega \subset \mathbb{R}^n$. Here the coefficients are sufficiently smooth real-valued functions, $a(x) \geq a_0 > 0$ for $x \in \Omega$, ρ is a real number.

We have

$$\int_\Omega (a(x)\Delta v, v)dx = -\sum_{k=1}^n \int_\Omega \left(a(x)\frac{\partial v}{\partial x_k}, \frac{\partial v}{\partial x_k}\right) dx$$
$$-\sum_{k=1}^n \int_\Omega \left(\frac{\partial a}{\partial x_k}\frac{\partial v}{\partial x_k}, v\right) dx + \int_{\partial\Omega}\left(a\frac{\partial v}{\partial \nu}, v\right) ds.$$

Using the ellipticity condition and the embedding of $H^1(G)$ into $L^2(\partial G)$, we obtain

$$\int_\Omega (a(x)\Delta v, v)dx \leq -\mu \int_\Omega \sum_{k=1}^n \left|\frac{\partial v}{\partial x_k}\right|^2 dx + M_1 \int_\Omega |v|^2 dx$$

for some positive constants μ and M_1. From this estimate and the inequality

$$\left|\int_\Omega \sum_{k=1}^n b_k(x)\frac{\partial v}{\partial x_k} + c(x)v, v)dx\right| \leq \epsilon\|\nabla v\|^2 + M_2\|v\|^2,$$

where $\epsilon > 0$ can be taken arbitrarily small, we finally obtain

$$(Lv, v) \leq (M_3 - \rho)\|v\|^2.$$

If $\rho \geq M_3 + 1$ and $Lv = 0$, then $v = 0$. If the homogeneous formally adjoint problem also has only the zero solution, then the equation $Lv = f$ is solvable for any right-hand side, and the operator L is invertible.

10.2 Operators with a parameter

Invertibility, sectorial operators. Consider the operator L given by (10.1) in a bounded or unbounded domain $\Omega \in \mathbb{R}^n$ with the homogeneous Dirichlet or Neumann boundary conditions. The usual conditions on the coefficients of the operator and on the boundary of the domain are imposed. The operator is considered as acting on functions from $H^2(\Omega)$, which satisfy the boundary conditions, into $L^2(\Omega)$. This is an operator with a parameter. Therefore, the equation $Lu = f$ is uniquely solvable for any f if ρ is real, positive and sufficiently large. Moreover,

$$\|u\|_{L^2(\Omega)} \leq K \frac{\|f\|_{L^2(\Omega)}}{|\rho|}$$

for some positive constant K independent of u, f, and ρ. In fact, invertibility and the last estimate hold for all ρ in some angle of the complex plane which contains some positive half-axis. Thus, the operator is sectorial and it generates an analytic semigroup. These results remain valid for more general elliptic operators and for other function spaces.

10. Examples

Nonhomogeneous boundary conditions. The estimate of the resolvent can depend on the boundary operator. Consider the operator

$$Au = \Delta u - \lambda^2 u \tag{10.2}$$

in a bounded domain $\Omega \subset \mathbb{R}^n$ with a sufficiently smooth boundary $\partial \Omega$. Here λ is a real positive number. Let the boundary operator be

$$Bu = \frac{\partial u}{\partial n}, \tag{10.3}$$

where n is the outer normal vector. Both operators act on functions from the Sobolev space $E = W^{2,p}(\Omega)$:

$$A: E \to F_d, \quad B: E \to F_b,$$

where

$$F_d = L^p(\Omega), \quad F_b = W^{1-1/p,p}(\partial \Omega), \quad p > 1.$$

Set $F = F_d \times F_b$. Then

$$L = (A, B): E \to F.$$

In the same spaces consider the norms

$$|||u|||_E = \|u\|_E + \lambda^2 \|u\|_{L^p(\Omega)}, \quad |||f|||_{F_d} = \|f\|_{F_d},$$
$$|||g|||_{F_b} = \inf_\phi |||\phi|||_{W^{1,p}(\Omega)}, \quad |||\phi|||_{W^{1,p}(\Omega)} = \|\phi\|_{W^{1,p}(\Omega)} + \lambda \|\phi\|_{L^p(\Omega)}$$

where the infimum is taken with respect to all functions $\phi \in W^{1,p}(\Omega)$, which coincide with g at $\partial \Omega$. From the results of this chapter, we can conclude that for λ sufficiently large, there exists a unique solution u of the problem

$$Au = f, \quad Bu = g \tag{10.4}$$

and

$$|||u|||_E \leq K(|||f|||_{F_d} + |||g|||_{F_b}). \tag{10.5}$$

From this estimate and the definition of the norms, we obtain

$$\lambda^2 \|u\|_{L^p(\Omega)} \leq \|u\|_E + \lambda^2 \|u\|_{L^p(\Omega)}$$
$$\leq K\left(\|f\|_{L^p(\Omega)} + \inf_\phi(\|\phi\|_{W^{1,p}(\Omega)} + \lambda \|\phi\|_{L^p(\Omega)})\right). \tag{10.6}$$

The dependence of the right-hand side of this estimate of λ is not explicit. It is possible to obtain a less precise estimate with an explicit dependence. We have for large λ:

$$\inf_\phi(\|\phi\|_{W^{1,p}(\Omega)} + \lambda \|\phi\|_{L^p(\Omega)}) \leq \lambda \inf_\phi(\|\phi\|_{W^{1,p}(\Omega)} + \|\phi\|_{L^p(\Omega)})$$
$$\leq 2\lambda \inf_\phi \|\phi\|_{W^{1,p}(\Omega)} = 2\lambda \|g\|_{F_b}.$$

Hence from (10.6)
$$\lambda^2 \|u\|_{L^p(\Omega)} \leq 2K \left(\|f\|_{L^p(\Omega)} + \lambda \|g\|_{W^{1-1/p,p}(\partial\Omega)} \right).$$

Thus,
$$\|u\|_{L^p(\Omega)} \leq \frac{K_1}{\lambda} \left(\|f\|_{L^p(\Omega)} + \|g\|_{W^{1-1/p,p}(\partial\Omega)} \right).$$

We note that the power of λ in the right-hand side of this estimate does not depend on p. This estimate may not be optimal.

Chapter 8

Index of Elliptic Operators

Ellipticity condition, proper ellipticity and Lopatinskii condition imply the Fredholm property of elliptic problems in bounded domains. In addition, invertibility of limiting problems determines the Fredholm property and solvability conditions of elliptic problems in unbounded domains. If this property is satisfied, then the index of the operator is defined. There is an extensive literature devoted to the index of elliptic operators in bounded domains and for some classes of operators in unbounded domains (see the bibliographical comments).

In this chapter we develop a new method which consists in approximation of problems in unbounded domains by problems in bounded domains. We will study convergence or stabilization of the index. We consider a sequence of elliptic operators L_k in bounded domains Ω_k assuming that the domains Ω_k converge to an unbounded domain Ω and that the operators converge to an operator L. Suppose that the indices of the operators L_k are known and equal to each other. Is it possible to conclude that the index of the operator L equals the index of the operators L_k? There are counterexamples that show that this may not be true. It appears that under some additional conditions formulated in terms of limiting operators the indices are equal to each other. This result provides a method to compute the index of elliptic operators in unbounded domains, approximating them by operators in bounded domains. We will prove general theorems about the stabilization of the index and will apply this approach to find the index of the Cauchy-Riemann system and of the Laplace operator with oblique derivative in unbounded domains.

We will introduce in this chapter a new notion of limiting domains, different in comparison with the definition given in Chapter 4. Limiting domains introduced before are constructed for an unbounded domain Ω through all possible sequences of shifted domains. Here we will consider sequences of domains Ω_k locally converging to the domain Ω and will define limiting domains for these sequences. It appears that for the same domain Ω, there can exist more limiting domains in the sense of the second definition than in the sense of the first one. Therefore, the condition about the invertibility of limiting problems is more restrictive in the sense of the second definition. The former provides the Fredholm property, while the latter the stabilization of the index.

The index equals the difference between the dimension of the kernel of the operator and the codimension of its image. The stabilization of the index does not necessarily signify that the dimensions of the kernels also stabilize. However, if it is the case, there is a convergence of solutions in bounded domains to solutions in unbounded domains.

For completeness of presentation, at the end of this chapter we will construct the index in two-dimensional domains. Multi-dimensional problems will be discussed in the bibliographical comments.

1 Definitions

As in previous chapters, we consider the operators

$$A_i u = \sum_{k=1}^{N} \sum_{|\alpha| \leq \alpha_{ik}} a_{ik}^{\alpha}(x) D^{\alpha} u_k, \quad i = 1, \ldots, N, \quad x \in \Omega, \tag{1.1}$$

$$B_j u = \sum_{k=1}^{N} \sum_{|\beta| \leq \beta_{jk}} b_{jk}^{\beta}(x) D^{\beta} u_k, \quad i = 1, \ldots, m, \quad x \in \partial\Omega, \tag{1.2}$$

where $u = (u_1, \ldots, u_N)$, $\Omega \subset R^n$ is an unbounded domain satisfying certain conditions given below. According to the definition of elliptic operators in the Douglis-Nirenberg sense [134] we suppose that

$$\alpha_{ik} \leq s_i + t_k, \ i, k = 1, \ldots, N, \quad \beta_{jk} \leq \sigma_j + t_k, \ j = 1, \ldots, m, \ k = 1, \ldots, N$$

for some integers s_i, t_k, σ_j such that $s_i \leq 0$, $\max s_i = 0$, $t_k \geq 0$.

Denote by E the space of vector-valued functions $u = (u_1, \ldots, u_N)$, where u_j belongs to the Sobolev space $W^{l+t_j,p}(\Omega)$, $j = 1, \ldots, N$, $1 < p < \infty$, l is an integer, $l \geq \max(0, \sigma_j + 1)$, $E = \Pi_{j=1}^{N} W^{l+t_j,p}(\Omega)$. The norm in this space is defined as

$$\|u\|_E = \sum_{j=1}^{N} \|u_j\|_{W^{l+t_j,p}(\Omega)}.$$

The operator A_i acts from E into $W^{l-s_i,p}(\Omega)$, the operator B_j from E into $W^{l-\sigma_j-1/p,p}(\partial\Omega)$. Set

$$\begin{aligned} L &= (A_1, \ldots, A_N, B_1, \ldots, B_m), \\ F &= \Pi_{i=1}^{N} W^{l-s_i,p}(\Omega) \times \Pi_{j=1}^{m} W^{l-\sigma_j-1/p,p}(\partial\Omega). \end{aligned} \tag{1.3}$$

1. Definitions

As before we will use the spaces E_q ($1 \leq q \leq \infty$) with the norms

$$\|u\|_{E_q} := \left(\int_{R^n} \|u(.)\omega(.-y)\|_E^q dy \right)^{1/q} < \infty, \quad 1 \leq q < \infty,$$

$$\|u\|_{E_\infty} := \sup_{y \in R^n} \|u(.)\omega(.-y)\|_E < \infty.$$

We will consider the operator L as acting from E_∞ into F_∞. We assume that the operator L is uniformly elliptic.

1.1 Limiting domains

Consider a sequence of domains $\Omega_k \subset \mathbb{R}^n$, $k = 1, 2, \ldots$ such that each of them contains the origin. Each domain Ω_k may be bounded or unbounded. We do not exclude the case where the domains are uniformly bounded, that is contained in a fixed ball independent of k. However we are interested in the case where the sequence of domains is not bounded.

In Chapter 4 the sequence of domains Ω_k is obtained as some shifts of an unbounded domain Ω. Here we consider an arbitrary sequence of domains satisfying the following condition. We keep for it the same name as before.

Condition D. For each $x_0 \in \partial \Omega$ there exists a neighborhood $U(x_0)$ such that:

1. $U(x_0)$ contains a sphere with radius δ and center x_0, where δ is independent of x_0.
2. There exists a homeomorphism $\psi(x; x_0)$ of the neighborhood $U(x_0)$ on the unit sphere $B = \{y : |y| < 1\}$ in $\mathbb{R}^n$ such that the images of $\Omega \cap U(x_0)$ and $\partial \Omega \cap U(x_0)$ coincide with $B_+ = \{y : y_n > 0, |y| < 1\}$ and $B_0 = \{y : y_n = 0, |y| < 1\}$ respectively.
3. The function $\psi(x; x_0)$ and its inverse belong to the Hölder space $C^{r+\theta}$, $0 < \theta < 1$. Their $\|\cdot\|_{r+\theta}$-norms are bounded uniformly in x_0.

For definiteness we suppose that $\delta < 1$. We also assume that

$$r \geq \max(l + t_i, l - s_i, l - \sigma_j + 1), \quad i = 1, \ldots, N, \ j = 1, \ldots, m.$$

The first expression in the maximum is used for a priori estimates of solutions, the second and the third will allow us to extend the coefficients of the operator.

We assume that domains Ω_k satisfy Condition D uniformly, that is δ and the norms of the functions $\psi(x; x_0)$ and of their inverse are independent of k. To define convergence of domains we use the following Hausdorff metric space. Let M and N denote two nonempty closed sets in $\mathbb{R}^n$. Set

$$\varsigma(M, N) = \sup_{a \in M} \rho(a, N), \quad \varsigma(N, M) = \sup_{b \in N} \rho(b, M),$$

where $\rho(a, N)$ denotes the distance from a point a to a set N, and let

$$\varrho(M, N) = \max(\varsigma(M, N), \varsigma(N, M)). \tag{1.4}$$

We denote by Ξ a metric space of bounded closed nonempty sets in R^n with the distance given by (1.4). We say that a sequence of domains Ω_k converges to a domain Ω in Ξ_{loc} if

$$\varrho(\bar{\Omega}_k \cap \bar{B}_R, \bar{\Omega} \cap \bar{B}_R) \to 0, \quad m \to \infty$$

for any $R > 0$ and $B_R = \{x : |x| < R\}$. Here the bar denotes the closure of domains.

Definition 1.1. Domain Ω_* is a limiting domain of a sequence Ω_k if $\Omega_k \to \Omega_*$ in Ξ_{loc} as $k \to \infty$.

Theorem 1.2. *Let a sequence Ω_k satisfy Condition D uniformly. Then there exists a subsequence Ω_{k_i} and a limiting domain Ω_* such that $\Omega_{k_i} \to \Omega_*$ in Ξ_{loc} as $k_i \to \infty$.*

The proof is the same as the proof of Theorem 1.6 in Chapter 4.

We recall that there are two types of limiting domains. The first one is defined for unbounded domains. If Ω is an unbounded domain, then we consider an unbounded sequence $x_n \in \Omega$ and a sequence of shifted domains Ω_n where x_n is translated to the origin. There exists a subsequence Ω_{n_k} locally convergent to a limiting domain $\hat{\Omega}$. Precise definitions are given in Chapter 4. We determine ensembles of limiting domains for all sequences x_n and all locally convergent subsequences of domains. Limiting domains that can be obtained from each other by translation are considered as the same domain.

The second type of limiting domains is defined for sequences of domains. Let Ω_n be a sequence of bounded or unbounded domains. We are particularly interested in the case where the domains Ω_n are bounded and locally converge to an unbounded domain Ω. As above, we take an unbounded sequence $x_n \in \Omega_n$ and consider the shifted domains $\tilde{\Omega}_n$ where x_n is translated to the origin. After that, we choose locally convergent subsequences of the shifted domains. Their limits are limiting domains of the second type.

Limiting domains of the first type and the corresponding limiting operators determine the Fredholm property of elliptic problems in unbounded domains. Limiting domains of the second type will determine the stabilization of the index. For the same unbounded domain Ω, there can exist more limiting domains of the second type than of the first type. This implies that conditions of the stabilization of the index are more restrictive than conditions that provide the Fredholm property.

We will restrict ourselves here to the simplest example of one-dimensional domains. Two other examples with a sequence of rectangles and with a sequence of circles are considered in the introduction. Let I_n denote the interval $[-n, n]$. Consider a sequence $x_n \in I_n$. Suppose that $x_n = -n + 1$ and consider the shifted

1. Definitions

intervals $\tilde{I}_n = [-1, 2n-1]$. Each such interval is obtained from I_n if we translate it in such a way that the point x_n moves to $x = 0$. The sequence $\tilde{I}_n$ converges to the right half-axis $I^+ = \{x \geq -1\}$. We can choose the sequence x_n in such a way that the limiting domain will be any other right half-axis. We identify them to each other. Other limiting domains are left half-axis and the whole axis. The corresponding sequences x_n can be easily constructed. Thus, there are three limiting domains of the second type. We note that for the unbounded interval $I = \mathbb{R}$, there exists only one limiting domain of the first type, the same unbounded interval I.

1.2 Limiting operators and problems

Consider a sequence of domains Ω_ν and the corresponding operators A_i^ν, B_j^ν with the coefficients, respectively, $a_{ik,\nu}^\alpha(x)$ and $b_{jk,\nu}^\beta(x)$. We suppose that

$$a_{ik,\nu}^\alpha(x) \in C^{l-s_i+\delta}(\bar{\Omega}_\nu), \quad b_{jk,\nu}^\beta(x) \in C^{l-\sigma_j+\delta}(\partial\Omega_\nu), \tag{1.5}$$

where $0 < \delta < 1$, and that these coefficients can be extended to R^n in such a way that

$$a_{ik}^\alpha(x) \in C^{l-s_i+\delta}(\mathbb{R}^n), \quad b_{jk}^\beta(x) \in C^{l-\sigma_j+\delta}(\mathbb{R}^n) \tag{1.6}$$

and

$$\|a_{ik,\nu}^\alpha(x)\|_{C^{l-s_i+\delta}(\mathbb{R}^n)} \leq M, \quad \|b_{jk,\nu}^\beta(x)\|_{C^{l-\sigma_j+\delta}(\mathbb{R}^n)} \leq M \tag{1.7}$$

with some constant M independent of ν. It follows from Theorem 1.2 that there exists a subsequence of the sequence Ω_ν, for which we keep the same notation, such that it converges to a limiting domain Ω_*. From (1.7) it follows that there exists a convergent subsequence of the coefficients:

$$a_{ik,\nu}^\alpha \to \hat{a}_{ik}^\alpha \text{ in } C^{l-s_i}(\mathbb{R}^n) \text{ locally}, \quad b_{jk,\nu}^\beta \to \hat{b}_{jk}^\beta \text{ in } C^{l-\sigma_j}(\mathbb{R}^n) \text{ locally}, \tag{1.8}$$

where $\hat{a}_{ik}^\alpha$ and $\hat{b}_{jk}^\beta$ are limiting coefficients,

$$\hat{a}_{ik}^\alpha \in C^{l-s_i+\delta}(\mathbb{R}^n), \quad \hat{b}_{jk}^\beta \in C^{l-\sigma_j+\delta}(\mathbb{R}^n).$$

We have constructed limiting operators:

$$\hat{A}_i u = \sum_{k=1}^N \sum_{|\alpha|\leq \alpha_{ik}} \hat{a}_{ik}^\alpha(x) D^\alpha u_k, \quad i = 1,\ldots,N, \quad x \in \Omega_*, \tag{1.9}$$

$$\hat{B}_j u = \sum_{k=1}^N \sum_{|\beta|\leq \beta_{jk}} \hat{b}_{jk}^\beta(x) D^\beta u_k, \quad i = 1,\ldots,m, \quad x \in \partial\Omega_*, \tag{1.10}$$

$$\hat{L} = (\hat{A}_1,\ldots,\hat{A}_N, \hat{B}_1,\ldots,\hat{B}_m). \tag{1.11}$$

We consider them as acting from $E_\infty(\Omega_*)$ into $F_\infty(\Omega_*)$. We assume that all limiting operators satisfy the condition of uniform ellipticity, that they are properly elliptic and satisfy the Lopatinskii condition.

We will deal with three different situations where a sequence of domains converges to some domain and a sequence of operators converges to some operator. First of all, we consider a sequence of bounded domains Ω_k that converge to some unbounded domain Ω and a sequence of operators $L_k : E(\Omega_k) \to F(\Omega_k)$ that converge to an operator $L : E_\infty(\Omega) \to F_\infty(\Omega)$. We will study the stabilization of the index of the operators L_k to the index of the operator L.

To study this question we introduce two types of limiting problems. Suppose that a sequence of domains Ω_k is not uniformly bounded. Then there exists a sequence of points $x_k \in \Omega_k$ such that the sequence $|x_k|$ is not bounded. From now on we will consider only such sequences that $|x_k| \to \infty$. Denote by $\tilde{\Omega}_k$ the shifted domains with the characteristic functions $\chi_k(x + x_k)$, where $\chi_k(x)$ is the characteristic function of the domain Ω_k. We consider all limiting domains for the sequence $\tilde{\Omega}_k$ and define the corresponding limiting operators $\hat{L}$ and the corresponding limiting problems. Limiting problems can depend on the choice of the sequence x_k and of converging subsequences of the domains and of the coefficients of the operator.

Limiting problems of the first type are the problems where $\Omega_k = \Omega$ and $L_k = L$ for all k. Here Ω is an unbounded domain. This means that we consider all shifts of the same domain and of the same operator and choose locally convergent subsequences. Limiting problems defined in this sense are introduced in Chapter 4.

Limiting problems of the second type are the problems for which a sequence of domains Ω_k (usually bounded though it is not necessary) converge to an unbounded domain Ω. We consider all shifts of the domains Ω_k and of the corresponding operators and choose locally convergent subsequences.

Limiting problems of the second type were not studied before. Let Ω_k be a sequence of domains locally convergent to an unbounded domain Ω. It appears that there can be more limiting problems of the second type than of the first type. Some properties of the operators should be formulated in terms of limiting domains of the first type (Fredholm property) and some others in terms of limiting problems of the second type (stabilization of the index).

1.3 Conditions NS and estimates

Let Ω be an unbounded domain. We will use the spaces E and F introduced above, the dual spaces E^* and F^*, and the corresponding ∞-spaces:

$$E_\infty(\Omega) = \Pi_{j=1}^N W_\infty^{l+t_j,p}(\Omega),$$
$$F_\infty(\Omega) = \Pi_{i=1}^N W_\infty^{l-s_i,p}(\Omega) \times \Pi_{j=1}^m W_\infty^{l-\sigma_j-1/p,p}(\partial\Omega),$$
$$(E^*(\Omega))_\infty = \Pi_{j=1}^N (\dot{W}^{-l-t_j,p'}(\Omega))_\infty,$$

1. Definitions

$$(F^*(\Omega))_\infty = \Pi_{i=1}^N (\dot{W}^{-l+s_i,p'}(\Omega))_\infty \times \Pi_{j=1}^m (W^{-l+\sigma_j+1/p,p'}(\partial\Omega))_\infty,$$
$$(F_{-1}^*(\Omega))_\infty = \Pi_{i=1}^N (\dot{W}^{-l+s_i-1,p'}(\Omega))_\infty \times \Pi_{j=1}^m (W^{-l+\sigma_j+1/p-1,p'}(\partial\Omega))_\infty,$$

where $\dot{W}^{-s,p'}(\Omega)$ is the closure in $W^{-s,p'}(\mathbb{R}^n)$ of infinitely differentiable functions with supports in Ω, $\dot{W}^{-s,p'}(\Omega) = (W^{s,p}(\Omega))^*$, $\frac{1}{p} + \frac{1}{p'} = 1$.

We recall Conditions NS and NS* introduced in Chapters 4 and 5 in order to study the Fredholm property.

Condition NS. Any limiting problem of the first type, $\hat{L}u = 0$, $u \in E_\infty(\hat{\Omega})$ has only the zero solution. Here $\hat{\Omega}$ is a limiting domain, $\hat{L}$ is a limiting operator.

Condition NS*. Any limiting problem of the first type, $\hat{L}^*v = 0$ for the adjoint operator $\hat{L}^*$ has only the zero solution in $(F^*(\hat{\Omega}))_\infty$.

It is proved that Condition NS is necessary and sufficient for normal solvability. If both conditions are satisfied, then the operator is Fredholm. We now modify Condition NS for sequences of domains. Let a sequence of domains Ω_k converge to an unbounded domain Ω.

Condition NS(seq). Any limiting problem of the second type, $\hat{L}u = 0$, $u \in E_\infty(\hat{\Omega})$ has only the zero solution.

Though the formulation is the same as before, the condition is different because limiting problems may not be the same.

Theorem 1.3. *Let a sequence Ω_k satisfy Condition D uniformly, $u \in E_\infty(\Omega_k)$. Then*
$$\|u\|_{E_\infty(\Omega_k)} \leq c \left(\|L_k u\|_{F_\infty(\Omega_k)} + \|u\|_{L_\infty^p(\Omega_k)} \right),$$
where the constant c does not depend on u and on k.

The proof of this theorem is similar to the proof of Theorem 2.2 in Chapter 3.

Theorem 1.4. *Let Condition NS(seq) be satisfied. Then there exist numbers M and R such that the following estimate holds:*
$$\|u\|_{E_\infty(\Omega_k)} \leq M \left(\|L_k u\|_{E_\infty(\Omega_k)} + \|u\|_{L^p(\Omega_R^k)} \right), \quad \forall u \in E_\infty(\Omega_k)$$
for any k. Here $\Omega_R^k = \Omega_k \cap \{|x| \leq R\}$, M and R do not depend on k.

The proof of this theorem is similar to the proof of Theorem 2.10 in Chapter 4.

Theorem 1.5. *Let a sequence Ω_k satisfy Condition D uniformly, $v \in (F^*(\Omega_k))_\infty$. Then*
$$\|v\|_{(F^*(\Omega_k))_\infty} \leq c \left(\|L_k^* v\|_{(E^*(\Omega_k))_\infty} + \|v\|_{(F_{-1}^*(\Omega_k))_\infty} \right),$$
where the constant c does not depend on v and on k.

The proof of this theorem is similar to the proof of Theorem 5.1 in Chapter 3.

We now introduce an analogue of Condition NS* for sequences of problems. Similar to Condition NS(seq), its formulation is the same as for a single unbounded domain but limiting problems can be different.

Condition NS*(seq). Any limiting homogeneous problem, $\hat{L}^* v = 0$ has only the zero solution in $(F^*(\hat{\Omega}))_\infty$, where $\hat{L}^*$ is the operator adjoint to the limiting operator $\hat{L}$, and $\hat{\Omega}$ is a limiting domain.

Theorem 1.6. *Let Condition NS*(seq) be satisfied. Then there exist positive numbers M and R such that for any $v \in (F^*(\Omega_k))_\infty$ and for any k the following estimate holds:*

$$\|v\|_{(F^*(\Omega_k))_\infty} \leq M \left(\|L_k^* v\|_{(E^*(\Omega_k))_\infty} + \|v\|_{F^*_{-1}(\Omega_R^k)} \right).$$

Here

$$F^*_{-1}(\Omega_R^k) = \Pi_{i=1}^N \dot{W}^{-l+s_i-1,p'}(\Omega_R^k) \times \Pi_{j=1}^m \dot{W}^{-l+\sigma_j+1/p-1,p'}(\Gamma_R^k),$$

Ω_R^k *and* Γ_R^k *are the intersections of* Ω_k *and* Γ_k *with the ball* $|x| < R$.

The proof of this theorem is similar to the proof of Theorem 1.1 in Chapter 5.

We finally present the theorem on uniform exponential estimates of solutions. Write $\omega_\mu = \exp(\mu \sqrt{1 + |x|^2})$, where μ is a real number.

Theorem 1.7. *Let Condition NS(seq) be satisfied. Then there exist positive constants M, R and μ_0 independent of k and such that for all μ, $0 < \mu < \mu_0$ the following estimate holds:*

$$\|\omega_\mu u\|_{E_\infty(\Omega_k)} \leq M \left(\|\omega_\mu L u\|_{F_\infty(\Omega_k)} + \|\omega_\mu u\|_{L^p(\Omega_R)} \right) \quad \text{if } \omega_\mu u \in E_\infty(\Omega_k). \quad (1.12)$$

This theorem follows from Theorem 1.4. Its proof is similar to the proof of Theorem 2.11 in Chapter 4. A similar theorem holds for the adjoint operator.

1.4 Results and examples

Consider a sequence of domains Ω_k and a sequence of elliptic operators L_k acting on functions defined in these domains. The corresponding function spaces will be introduced below. Suppose that the sequence of domains converges to a domain Ω in the sense specified in Section 1.1, and the sequence of operators converges to an operator L (Section 1.2). Suppose that all operators L_k and L satisfy the Fredholm property. Then the indices $\kappa(L_k)$ and $\kappa(L)$ are defined. Is it possible to affirm that $\kappa(L_k)$ converges to $\kappa(L)$ or, since the index is an integer, that for k sufficiently large, $\kappa(L_k) = \kappa(L)$? If this equality holds we say that the index stabilizes.

1. Definitions

We begin with an example that shows that the stabilization of the index may not occur. Consider the operator

$$Lu = u'' + cu' + b(x)u$$

acting from $C^{2+\delta}(\mathbb{R})$ into $C^\delta(\mathbb{R})$, $0 < \delta < 1$. Here c is a constant, $b(x)$ is a sufficiently smooth function with the limits $b_\pm$ at $\pm\infty$. If $c > 0$, $b_- < 0$, and $b_+ > 0$, then the index of this operator equals 1 (see the next chapter). Let L_n, $n = 1, 2, \ldots$ be a sequence of operators given by the same differential expression and acting from $C_0^{2+\delta}(I_n)$ into $C^\delta(I_n)$, where $I_n = \{-n \leq x \leq n\}$, and $C_0^{2+\delta}(I_n)$ denotes the space of functions defined in the interval I_n with the zero boundary conditions. It can be easily verified that the index of each operator L_n equals zero. Therefore $\text{ind}(L_n)$ does not converge to $\text{ind}(L)$.

It appears that, similar to the Fredholm property, the stabilization of the index is also determined by the invertibility of limiting problems. However the definition of limiting problems in this case is different in comparison with the previous one. The second type of limiting problems is defined not for a single unbounded domain through its shifts but for sequences of domains Ω_k and for the corresponding operators. Each of these domains can be bounded or unbounded but we will assume that the sequence is not uniformly bounded, that is there is no ball B_R of a given radius R which contains all domains.

Let $x_n \in \Omega_n$, $n = 1, 2, \ldots$ be a sequence of points such that $|x_n| \to \infty$. Denote by $\chi_n(x)$ the characteristic function of the domain Ω_n and consider the shifted domain $\tilde{\Omega}_n$ with the characteristic function $\chi_n(x + x_n)$. We start the procedure of construction of limiting domains corresponding to the sequence x_n. For a given ball B_r, we choose a subsequence from the sequence of domains $\tilde{\Omega}_n$ converging to some limiting domain inside B_r. Then we consider a ball of a larger radius and choose a subsequence of the previous subsequence converging to some limiting domain inside this new ball. Going on in the same manner, we will obtain a limiting domain $\hat{\Omega}$. It can depend on the sequence x_n and on the choice of converging subsequences of shifted domains.

Limiting operators are constructed here in the same way as above. We consider the shifted coefficients and choose convergent subsequences.

Consider the following example. Let $\Omega_n \subset \mathbb{R}^2$, $n = 1, 2, \ldots$ be rectangles

$$-n \leq x \leq n, \quad -1 \leq y \leq 1.$$

We take nonsmoooth domains for simplicity of presentation. In this case we can obtain three types of limiting domains: an infinite strip, a left half-strip, and a right half-strip. The domains Ω_n converge locally to the domain

$$\Omega = \{(x, y), -1 \leq y \leq 1\}.$$

There exists a unique (up to a shift) limiting domain for the domain Ω in the first sense. It is the domain Ω itself. However there are also two other limiting domains for the approximating sequence Ω_n.

This example shows that the definitions are different. There are more limiting domains in the sense of the second definition. This difference appears to be important for what follows. The invertibility of limiting operators in the sense of the first definition provides the Fredholm property for elliptic operators in unbounded domains. The invertibility of limiting operators in the sense of the second definition provides the stabilization of the index.

In Section 1.3 we have introduced Condition NS(seq) which says that any limiting problem in the sense of Definition 1.1 has only a trivial solution. The same condition will also be introduced for adjoint problems and for formally adjoint problems. We will prove in Section 2 that they are sufficient for the stabilization of the index. It is not clear whether they are also necessary. However, in the examples, for which the stabilization does not occur, Conditions NS(seq) or NS*(seq) are not satisfied either.

The result on stabilization of the index allows determination of the index of elliptic problems in unbounded domains. We apply it below for the Laplace operator with oblique derivative, for the Cauchy-Riemann system, and for some other problems.

One of the questions related to the stabilization of the index is convergence of solutions in bounded domain to solutions in unbounded domains. This convergence may or may not take place. If we consider solutions u_n of the problems

$$\Delta u - au = b, \quad u|_{\partial \Omega_n} = 0,$$

where a and b are some constants and Ω_n is a sequence of balls of radius n, then u_n converges to $-b/a$ if a is positive. This sequence of solutions is divergent if $a = 0$. Convergence of solutions for some particular elliptic problems is studied in [111]. Here we develop another method to study convergence of solutions which is applicable for general elliptic problems. Similar to the stabilization of the index it is based on limiting problems.

The main results of this chapter are given by Theorems 2.9 and 3.3. They concern stabilization of the index in bounded domains to the index in an unbounded domain. The convergence of solutions in bounded domains to solutions in unbounded domains is proved in Theorem 4.1. The convergence of the index and of the solutions occur under Conditions NS(seq) and NS*(seq). We essentially use here a priori estimates of solutions formulated in Section 1.3 and the results on ellipticity with a parameter in unbounded domains.

The results on the stabilization of the index are applied to find the index of various elliptic problems in unbounded domains. We consider several model problems, Cauchy-Riemann system, Laplace operator, canonical first-order systems. A specific example is given by the first-order system

$$\frac{\partial u}{\partial x} - \frac{\partial v}{\partial y} - \lambda u = 0, \quad \frac{\partial u}{\partial y} + \frac{\partial v}{\partial x} + \lambda v = 0 \qquad (1.13)$$

1. Definitions

in the half-plane $y \geq 0$ with the boundary condition

$$a(x)u + b(x)v = f(x), \tag{1.14}$$

where $a(x)$ and $b(x)$ are real functions, $a^2(x) + b^2(x) \neq 0$, λ is a positive number. If $\lambda = 0$, then it is the Hilbert problem for the Cauchy-Riemann system. In bounded domains its index is determined by the rotation of the vector $(a(x), b(x))$ along the boundary of the domain. In unbounded domains this problem does not satisfy the Fredholm property. For λ different from zero, it is satisfied under some conditions on the coefficients. If we assume that there exist limits $a_\pm$ and $b_\pm$ of the functions $a(x)$ and $b(x)$ at $\pm\infty$, then it is required that $a_\pm + b_\pm \neq 0$.

If the numbers $a_- + b_-$ and $a_+ + b_+$ have opposite signs, then we can construct a sequence of increasing bounded domains and of the operators in such a way that Conditions NS(seq) and NS*(seq) are satisfied. This allows us to find the index of the problem in the half-plane. Similar to the problem in bounded domains it equals $2N + 1$, where N is the rotation of the vector $(a(x), b(x))$ for x from $-\infty$ to $+\infty$ (Theorem 5.9). If $a_- + b_-$ and $a_+ + b_+$ have the same sign, the method of approximation by bounded domains cannot be directly used. It is quite possible that it is not a technical difficulty but a topological restriction.

We also consider the problem

$$\Delta u - \lambda^2 u = 0, \quad a(x)\frac{\partial u}{\partial x} - b(x)\frac{\partial u}{\partial y} = f(x). \tag{1.15}$$

If $\lambda = 0$, it is the classical problem for the Laplace operator with oblique derivative. In the case of unbounded domains it does not satisfy the Fredholm property. Problem (1.15) satisfies the Fredholm property in unbounded domains if $\lambda \neq 0$ and if the limits $b_\pm$ are different from zero. Its index can be found with the approximation method if b_- and b_+ have the same sign. In this case the index is even. The method cannot be directly used if the signs are opposite.

Thus, we find the index of problem (1.13), (1.14) when it is odd, and of problem (1.15) when it is even. It appears that these problems can be reduced to each other by a number of consecutive transformations. This allows us to find their indices in the cases not embraced by the approximation method: when $a_- + b_-$ and $a_+ + b_+$ have the same sign for problem (1.13), (1.14) and when b_- and b_+ have opposite signs for problem (1.15).

Finally, we study general first-order systems with complex coefficients on the plane. In the case where they can be reduced to canonical systems, their index in bounded domains can be found by reduction to singular integral equations. Then the approximation method can be used to find the index in unbounded domains.

2 Stabilization of the index

From now on we will consider the spaces E and F on the basis of the Hilbert space $W^{l,2}$ (see Section 1). Let Ω_k be a sequence of bounded domains that converge to an unbounded domain Ω, $L_k : E(\Omega_k) \to F(\Omega_k)$ be a sequence of operators acting on functions defined in Ω_k and converging to an operator $L : E_\infty(\Omega) \to F_\infty(\Omega)$. Suppose that the operators L_k satisfy the Fredholm property and

$$\dim \operatorname{Ker} L_k = m, \quad \operatorname{codim} \operatorname{Im} L_k = p.$$

We suppose also that Conditions NS(seq) and NS*(seq) are satisfied. It will be proved in the sequel that we can find a subsequence such that m and p do not depend on k. We will prove that the index of the operator L equals $m - p$.

Denote by $u_1^k, \ldots, u_m^k$ linearly independent solutions of the equation

$$L_k u = 0, \quad u \in E(\Omega_k) \tag{2.1}$$

and by $v_1^k, \ldots, v_p^k$ linearly independent solutions of the homogeneous adjoint equation

$$L_k^* v = 0, \quad v \in F^*(\Omega_k). \tag{2.2}$$

Then the equation

$$L_k u = f, \quad u \in E(\Omega_k), \ f \in F(\Omega_k) \tag{2.3}$$

is solvable if and only if

$$\langle f, v_j^k \rangle = 0, \quad j = 1, \ldots, p \tag{2.4}$$

and the equation

$$L_k u = f - \sum_{j=1}^{p} \langle f, v_j^k \rangle w_j^k \tag{2.5}$$

is solvable for any $f \in F(\Omega_k)$. Here $w_j^k \in F(\Omega_k), j = 1, \ldots, p$ are functions biorthogonal to the functionals $v_j \in F^*(\Omega_k)$:

$$\langle w_i^k, v_j^k \rangle = \delta_{ij}, \quad i, j = 1, \ldots, p. \tag{2.6}$$

Lemma 2.1. *There exist solutions $\hat{v}_j^k \in F^*(\Omega_k)$ of equation (2.2) and functions $w_i^k \in F(\Omega_k)$ with supports in a ball B_ρ with the radius $\rho > R$ independent of k and i such that*

$$\langle w_i^k, \hat{v}_j^k \rangle = \delta_{ij}, \quad i, j = 1, \ldots, p$$

and

$$\|w_i^k\|_{F(\Omega_k)} \leq C, \quad \|\hat{v}_i^k\|_{F^*(\Omega_k^h)} \leq 1, \quad i = 1, \ldots, p, \ k \geq k_0(\rho),$$

where C is a constant independent of i and k, R is the same as in Theorem 1.6.

2. Stabilization of the index

Proof. We recall that $F(\Omega_k)$ is a Hilbert space. Denote the corresponding inner product by $[\cdot,\cdot]$. Then for every $v_j^k \in F^*(\Omega_k)$ there exists $z_j^k \in F(\Omega_k)$ such that

$$\langle f, v_j^k \rangle = [f, z_j^k], \quad \forall f \in F(\Omega_k).$$

Without loss of generality we can assume that the domain Ω_R^k in Theorem 1.6 does not depend on k. We denote it by Ω_R. Let $\phi \in D$ and $\phi(x) = 1$ for $x \in \Omega_R$ and $\operatorname{supp}\phi(x) \subset B_\rho$ with some $\rho > R$. The functions $\tilde{z}_j^k = \phi z_j^k$, $j = 1,\ldots,p$ are linearly independent for every k fixed. Indeed, otherwise there exists a linear combination of these functions equal to zero, that is for some constants c_j not all equal to zero,

$$c_1 \phi v_1^k + \cdots + c_p \phi v_p^k = 0.$$

Then the restriction of the functional $\psi = c_1 v_1^k + \cdots + c_p v_p^k$ to Ω_R (that is its value on the functions with supports in Ω_R) equals zero. From Theorem 1.6 it follows that $\psi = 0$ in Ω_k, that is the functionals $v_j^k, j = 1,\ldots,p$ are linearly dependent in Ω_k. This contradiction proves that $\tilde{z}_j^k$ are linearly independent.

Denote by $\hat{z}_j^k$ the functions obtained as a linear combination of the functions $\tilde{z}_j^k$ (for a fixed k) and such that

$$[\hat{z}_i^k, \hat{z}_j^k] = \delta_{ij}, \quad i,j = 1,\ldots,p, \quad \forall k.$$

Such combinations exist by virtue of the linear independence of the functions $\tilde{z}_j^k$. Let $\hat{v}_j^k$ be obtained as linear combinations of v_j^k with the same coefficients. Then

$$\langle f, \phi \hat{v}_j^k \rangle = [f, \hat{z}_j^k], \quad \forall f \in F(\Omega_k).$$

Therefore

$$\langle \phi \hat{z}_j^k, \hat{v}_j^k \rangle = \delta_{ij}, \quad i,j = 1,\ldots,p, \quad \forall k.$$

We put $w_i^k = \phi \hat{z}_i^k$.

It remains to estimate the norm of the restriction of the functionals $\hat{v}_i^k$ to Ω_R. We have

$$\sup_{f, \operatorname{supp} f \subset \Omega_R} \frac{|\langle f, \hat{v}_i^k \rangle|}{\|f\|_{F(\Omega_k)}} = \sup_{f, \operatorname{supp} f \subset \Omega_R} \frac{|\langle f, \phi \hat{v}_i^k \rangle|}{\|f\|_{F(\Omega_k)}} = \sup_{f, \operatorname{supp} f \subset \Omega_R} \frac{|[f, \hat{z}_i^k]|}{\|f\|_{F(\Omega_k)}} \leq 1.$$

The lemma is proved. $\square$

Denote by $w_j^k \in E^*(\Omega_k), j = 1,\ldots,m$ functionals biorthogonal to the functions $u_j \in E(\Omega_k)$:

$$\langle u_i^k, w_j^k \rangle = \delta_{ij}, \quad i,j = 1,\ldots,m. \tag{2.7}$$

Lemma 2.2. *There exist solutions $\hat{u}_j^k \in E(\Omega_k)$ of equation (2.1) and functionals $w_j^k \in E^*(\Omega_k)$ with supports in a ball B_ρ with some radius $\rho > R$ independent of k and i such that*

$$\langle \hat{u}_i^k, w_j^k \rangle = \delta_{ij}, \quad i,j = 1,\ldots,m$$

and
$$\|\phi\hat{u}_i^k\|_{E(\Omega_k)} = 1, \quad \|\omega_i^k\|_{F^*(\Omega^k)} \leq C, \quad i = 1, \ldots, m, \quad k \geq k_0(R).$$
Here $\phi \in D$ is an arbitrary function equal to 1 for $x \in \Omega_R$.

Proof. The restriction of the functions $u_1^k, \ldots, u_m^k$ to Ω_R are linearly independent. Otherwise, from Theorem 1.4 it would follow that they are linearly dependent in Ω_k.

Consider the functions $\tilde{u}_i^k = \phi u_i^k$, where $\phi \in D$ equals 1 for $x \in \Omega_R$. They are linearly independent and can be orthonormalized in $E(\Omega_k)$. Denote by $\hat{u}_i^k$ a linear combination of u_i^k such that
$$[\phi\hat{u}_i^k, \phi\hat{u}_j^k] = \delta_{ij}.$$
Here $[\cdot, \cdot]$ denotes the inner product in $E(\Omega_k)$. We define the functionals ω_i^k by the equalities
$$\langle f, \omega_i^k \rangle = [f, \phi^2 \hat{u}_i^k], \quad \forall f \in E(\Omega_k).$$
We keep the same notation for the inner product in $E(\Omega_k)$ as for $F(\Omega_k)$. These functionals form a biorthogonal system to $\hat{u}_i^k$ and
$$\|\omega_i^k\|_{F^*(\Omega_k)} \leq C\|\phi\hat{u}_i^k\|_{E(\Omega_k)} = C.$$

The lemma is proved. □

Lemma 2.3. *The equation*
$$L_k^* v = g, \quad g \in E^*(\Omega_k) \tag{2.8}$$
is solvable in $F^*(\Omega_k)$ if and only if
$$\langle u_j^k, g \rangle = 0, \quad j = 1, \ldots, m. \tag{2.9}$$

For the proof see [256], Chapter IV, Theorem 5.13.

Lemma 2.4. *Let u_i^k and ω_i^k be the same as in Lemma 2.2, $i = 1, \ldots, m$ (we omit the hat here). Then there exist subsequences $u_i^{k_l}$, $i = 1, \ldots, m$ converging to some limiting functions u_i, $i = 1, \ldots, m$ locally in $E_\infty(\Omega)$, and subsequences $\omega_i^{k_l}$, $i = 1, \ldots, m$ converging to some functionals ω_i, $i = 1, \ldots, m$ in $E^*(\Omega_R)$ such that*
$$\langle u_i, \omega_j \rangle = \delta_{ij}, \quad i, j = 1, \ldots, m.$$
The functions $u_i, i = 1, \ldots, m$ are linearly independent.

Proof. Since
$$\|\phi u_i^k\|_{E(\Omega_k)} = 1, \quad i = 1, \ldots, m,$$
then there exist subsequences $u_i^{k_l}$ converging in $L^2(\Omega_R)$ to some limiting functions u_i. It follows from Theorem 1.4 that the same subsequences converge locally

2. Stabilization of the index

strongly in E_∞ to some limiting functions $u_i \in E_\infty(\Omega)$. This convergence occurs in the intersection of the domains Ω_k with any bounded set.

From the construction of the functionals $w_i^{k_l}$ it follows that they converge to the functionals w_i defined by the equalities

$$\langle f, w_i \rangle = [f, \phi^2 u_i], \quad i = 1, \ldots, m, \quad \forall f \in E(\Omega_R).$$

We have

$$\langle u_j, w_i \rangle = [\phi u_j, \phi u_i] = \lim_{k \to \infty} [\phi u_j^k, \phi u_i^k] = \delta_{ij}.$$

These relations imply linear independence of the functions u_i. Indeed, if for some linear combination $\sum_{i=1}^m c_i u_i = 0$ and $c_j \neq 0$ for some j, then

$$0 = \left\langle \sum_{i=1}^m c_i u_i, w_j \right\rangle = c_j \neq 0.$$

This contradiction shows that the linear combination cannot have nonzero coefficients. The lemma is proved. $\square$

Lemma 2.5. *Let v_i^k and w_i^k be the same as in Lemma 2.1, $i = 1, \ldots, p$ (the tilde is omitted). Then there exist subsequences $v_i^{k_l}$, $i = 1, \ldots, p$ converging to some limiting functionals $v_i \in (F^*(\Omega))_\infty$, $i = 1, \ldots, p$ locally in F^* and subsequences $w_i^{k_l}$, $i = 1, \ldots, p$ converging in $F(\Omega_\rho)$ to some functions w_i, $i = 1, \ldots, m$ with supports in Ω_ρ such that*

$$\langle w_i, v_j \rangle = \delta_{ij}, \quad i, j = 1, \ldots, p.$$

Proof. By virtue of Lemma 2.1 the restriction of the sequence v_i^k, $k = 1, 2, \ldots$ to Ω_R is bounded in $F^*(\Omega_R)$. Hence there is a subsequence converging in $F_{-1}^*(\Omega_R)$ to some limiting function. From Theorem 1.6 it follows that the same subsequence is fundamental in F^* on every bounded set. Thus there exists a limiting functional $v_i \in (F^*(\Omega))_\infty$.

From the convergence $v_i^{k_l} \to v_i$ it follows that

$$\phi v_i^{k_l} \to \phi v_i \text{ in } F^*(\Omega_\rho).$$

This convergence and the definition of the functionals z_i^k (see Lemma 2.1) imply the strong convergence of the sequence $z_i^{k_l}$ and consequently the convergence of the sequence $w_i^{k_l}$ to some limiting function $w_i \in F(\Omega)$. We have

$$\langle w_i, v_j \rangle = \lim_{k_l \to \infty} \langle w_i^{k_l}, v_j^{k_l} \rangle = \delta_{ij}, \quad i, j = 1, \ldots, p.$$

The lemma is proved. $\square$

Proposition 2.6. *The following inequalities hold:*

$$\dim \operatorname{Ker} L_k \leq \dim \operatorname{Ker} L, \quad \operatorname{codim} \operatorname{Im} L_k \leq \operatorname{codim} \operatorname{Im} L \quad (2.10)$$

for k sufficiently large.

Proof. Suppose that
$$\dim \operatorname{Ker} L_k > \dim \operatorname{Ker} L$$
for some sequence of k. Then there exists a number m, $m > r(= \dim \operatorname{Ker} L)$ such that equation (2.1) has m linearly independent solutions u_i^k, $i = 1, \ldots, m$. As in Lemma 2.4 we prove that there is a subsequence which converges locally in $E_\infty(\Omega)$ to linearly independent functions u_i, $i = 1, \ldots, m$. These functions are solutions of the equation $Lu = 0$. Hence $r \geq m$, which contradicts the assumption above. This contradiction proves the first inequality in (2.10). Using Lemma 2.5 we can similarly prove the second inequality in (2.10). The proposition is proved. $\square$

This proposition justifies the assumption in the beginning of this section that m and p can be chosen independent of k. We can pass to a subsequence if necessary. Consider now the operators $L : E(\Omega) \to F(\Omega)$ and $L^* : F^*(\Omega) \to E^*(\Omega)$. Let the equation
$$Lu = 0 \qquad (2.11)$$
have solutions $u_1, \ldots, u_r$ that form an orthonormal basis of its kernel. Then $r \geq m$. Suppose that $r = m + s$, where $s \geq 0$. Let further the equation
$$L^*v = 0 \qquad (2.12)$$
have solutions $v_1, \ldots, v_q$ that form an orthonormal basis of its kernel. Then $q \geq p$. Suppose that $q = p + s^*$, where $s^* \geq 0$. We will show that $s = s^*$.

Proposition 2.7. *The inequality $s^* \geq s$ holds.*

Proof. From the sequence u_j^k of solutions of the equations $L_k u = 0$, $j = 1, \ldots, m$, $k = 1, 2, \ldots$ we can choose a subsequence that converges locally in E_∞ to some limiting function u_j as $k \to \infty$. The limiting functions u_j, $j = 1, \ldots, m$ are linearly independent (Lemma 2.4) and belong to the kernel of the operator L. Denote by $u_{m+1}, \ldots, u_{m+s}$ some functions that complete the basis of the kernel.

From Theorem 1.4 it follows that solutions of the equation
$$Lu = 0, \quad u \in E_\infty(\Omega) \qquad (2.13)$$
admit the estimate
$$\|u\|_{E_\infty(\Omega)} \leq C \|u\|_{L^2(\Omega_R)}. \qquad (2.14)$$
Consider the functions
$$u_1, \ldots, u_m, u_{m+1}, \ldots, u_{m+s}, \qquad (2.15)$$
which form the basis of the kernel of the operator L. It follows from (2.14) that the restrictions of (2.15) to $E(\Omega_R)$ are linearly independent. Indeed, if a function $\tilde{u} = \sum_{i=1}^{m+s} c_i u_i$ equals zero in Ω_R, then it is also zero in Ω.

2. Stabilization of the index

Therefore there exists k_0 such that for $k > k_0$ the functions $u_1^k, \ldots, u_m^k$, $u_{m+1}, \ldots, u_{m+s}$ are linearly independent in $E(\Omega_R)$. It is convenient to introduce the notation
$$u_{m+j}^k = u_{m+j}, \quad j = 1, \ldots, s, \quad \forall k.$$

We recall the assumption that $\Omega_k = \Omega$, $k = 1, 2, \ldots$ inside a ball B_ρ with some $\rho > R$. Hence, the restrictions of the functions u_{m+j}^k to B_ρ are defined in Ω_k.

We construct a system of functionals w_j^k, $j = 1, \ldots, m+s$ biorthogonal to the functions $u_1^k, \ldots, u_m^k, u_{m+1}^k, \ldots, u_{m+s}^k$. We follow here the method of Lemma 2.2 with some specific details explained below. Let $\phi \in D$, $\phi(x) = 1$ in Ω_R, $\operatorname{supp}\phi \subset \Omega_\rho$ with the same ρ as above. Since the functions ϕu_i^k, $i = 1, \ldots, m$ are linearly independent, then there exists linear combinations
$$\tilde{u}_i^k = \sum_{i=1}^m c_{ij}^k u_j^k$$

such that the functions $\phi \tilde{u}_i^k$ are orthonormal in $E(\Omega_\rho)$. The functions $\tilde{u}_i^k$ are solutions of the equations $L_k u = 0$. As $k \to \infty$, they converge to some limiting functions $\tilde{u}_i$, $i = 1, \ldots, m$ that are solutions of the equation $Lu = 0$. The functions
$$\tilde{u}_1, \ldots, \tilde{u}_m, u_{m+1}, \ldots, u_{m+s}$$

form the basis of the kernel of the operator L. We can introduce their linear combinations
$$\tilde{u}_i = \sum_{i=1}^m c_{ij} \tilde{u}_j + \sum_{i=m+1}^{m+s} c_{ij} u_j, \quad i = m+1, \ldots, m+s$$

such that the functions $\phi \tilde{u}_i$, $i = 1, \ldots, m+s$ are orthonormal.

For the linear combinations
$$\tilde{u}_i^k = \sum_{i=1}^m c_{ij} \tilde{u}_j^k + \sum_{i=m+1}^{m+s} c_{ij} u_j^k, \quad i = m+1, \ldots, m+s$$

with the same coefficients, the functions $\phi \tilde{u}_i^k$ are "almost" orthonormal (the first m of them are orthonormal according to the construction) since $\tilde{u}_j^k \to \tilde{u}_j$, $j = 1, \ldots, m$ as $k \to \infty$ and $u_j^k = u_j$, $i = m+1, \ldots, m+s$ in Ω_ρ. Therefore with a small change of the coefficients we obtain an orthonormal system of functions,
$$\tilde{u}_i^k = \sum_{i=1}^m c_{ij}^k \tilde{u}_j^k + \sum_{i=m+1}^{m+s} c_{ij}^k u_j^k, \quad i = m+1, \ldots, m+s,$$
$$\tilde{u}_i^k \to \tilde{u}_i, \quad c_{ij}^k \to c_{ij} \text{ as } k \to \infty.$$

Similarly to Lemma 2.2, we now construct a system of functionals $w_i^k \in E^*(\Omega_k)$ with supports in Ω_ρ such that

$$\langle u_i^k, w_j^k \rangle = \delta_{ij}, \quad i,j = 1,\ldots, m+s,$$

$\|w_i^k\|_{E^*(\Omega_k)} \leq C$, where C is a constant independent of i and k. Moreover, $w_i^k \to w_i$ as $k \to \infty$ and

$$\langle u_i, w_j \rangle = \delta_{ij}, \quad i,j = 1,\ldots, m+s.$$

By virtue of the solvability conditions (Lemma 2.3) the equation

$$L_k^* v = w_{m+j}^k, \quad j = 1,\ldots, s \tag{2.16}$$

is solvable in $F^*(\Omega_k)$. It has a unique solution y_{p+j}^k, $j = 1,\ldots, s$, which satisfies the conditions

$$\langle w_i^k, y_{p+j}^k \rangle = 0, \quad j = 1,\ldots, s, \ i = 1,\ldots, p, \tag{2.17}$$

where w_i^k are the functions defined in Lemma 2.1.

Consider the equations

$$L^* v = w_{m+j}, \quad j = 1,\ldots, s. \tag{2.18}$$

Since the solvability conditions given by Lemma 2.3 are also applicable to the operator L^*, then the last equations are not solvable.

We explain the idea of the following proof. We consider solutions of equation (2.16). If they remain bounded as $k \to \infty$, then we pass to the limit in the equation and obtain bounded solutions of equation (2.18). This contradiction shows that the sequence of solutions of equation (2.16) is not bounded. We divide the equation by their norms and pass to the limit. This gives us additional solutions of equation (2.12).

From Theorem 1.6 it follows that the estimate

$$\|y_{p+j}^k\|_{(F^*(\Omega_k))_\infty} \leq C \left(\|w_{m+j}^k\|_{(E^*(\Omega_k))_\infty} + \|y_{p+j}^k\|_{F_{-1}^*(\Omega_R)} \right) \tag{2.19}$$

holds with a constant C independent of j and k. Consider first the case where

$$\|y_{p+j}^k\|_{F_{-1}^*(\Omega_R)} \leq M, \quad j = 1,\ldots, s, \ \forall k \tag{2.20}$$

with some positive constant M. Since the norms $\|w_{m+j}^k\|_{(E^*(\Omega_k))_\infty}$ are bounded, then from estimate (2.19) it follows that the norms $\|y_{p+j}^k\|_{(F^*(\Omega_k))_\infty}$ are also bounded. Therefore there exists a subsequence $y_{p+j}^{k_l}$ such that its restriction to Ω_R converges in $F_{-1}^*(\Omega_R)$ to some limiting element $y_{p+j} \in F_{-1}^*(\Omega_R)$. The same estimate and the convergence $w_{m+j}^{k_l} \to w_{m+j}$ imply that the sequence $y_{p+j}^{k_l}$ is fundamental in F^* on every bounded set. Therefore it converges locally to $y_{p+j} \in F^*(\Omega)$.

2. Stabilization of the index

We can pass to the limit in equation (2.16) and obtain solvability of equation (2.18). This contradiction shows that (2.20) cannot hold.

We note that we use the equality $\langle u, g \rangle = \langle L_k u, v \rangle$ to pass to the limit in the equation $L_k^* v = g$.

Suppose now that the sequence y_{p+j}^k is not bounded in the norm $F_{-1}^*(\Omega_R)$. Without loss of generality we can assume, passing to a subsequence if necessary, that the norms tend to infinity. Put

$$\tilde{y}_{p+j}^k = \frac{y_{p+j}^k}{\|y_{p+j}^k\|_{F_{-1}^*(\Omega_R)}}, \quad \tilde{\omega}_{m+j}^k = \frac{\omega_{m+j}^k}{\|y_{p+j}^k\|_{F_{-1}^*(\Omega_R)}}.$$

Then $\tilde{y}_{p+j}^k$ still satisfy equalities (2.17) and the equation

$$L_k^* \tilde{y}_{p+j}^k = \tilde{\omega}_{m+j}^k, \quad j = 1, \ldots, s. \tag{2.21}$$

For each k fixed, the restrictions of $\tilde{y}_{p+j}^k$, $j = 1, \ldots, s$ to Ω_R are linearly independent since $\tilde{\omega}_{m+j}^k$ are linearly independent on Ω_R. Therefore they can be orthonormalized in $F^*(\Omega_R)$. Denote by $\hat{y}_{p+j}^k$ the corresponding functionals given as linear combinations of $\tilde{y}_{p+j}^k$:

$$\hat{y}_{p+j}^k = c_{j1}^k \tilde{y}_{p+1}^k + \cdots + c_{js}^k \tilde{y}_{p+s}^k.$$

Then

$$L_k^* \hat{y}_{p+j}^k = c_{j1}^k \tilde{\omega}_{m+1}^k + \cdots + c_{js}^k \tilde{\omega}_{m+s}^k$$

or

$$L_k^* \hat{y}_{p+j}^k = d_{j1}^k \omega_{m+1}^k + \cdots + d_{js}^k \omega_{m+s}^k, \tag{2.22}$$

where

$$d_{ji}^k = \frac{c_{ji}^k}{\|y_{p+i}^k\|_{F_{-1}^*(\Omega_R)}}.$$

The coefficients d_{ji}^k in the right-hand side of (2.22) are uniformly bounded. Indeed, by construction

$$\langle u_{m+i}^k, \omega_{m+j}^k \rangle = \delta_{ij}, \quad i, j = 1, \ldots, s.$$

From (2.22),

$$\langle u_{m+i}^k, L_k^* \hat{y}_{p+j}^k \rangle = d_{ij}^k.$$

Hence

$$d_{ij}^k = \langle L_k u_{m+i}^k, \hat{y}_{p+j}^k \rangle.$$

The uniform boundedness of the coefficients d_{ji}^k follows from the uniform boundedness of the operators L_k.

Moreover, these coefficients converge to zero. Indeed, if they are bounded but do not converge to zero, then we can choose converging subsequences of the coefficients and pass to the limit in the equation as it was done above. It would give a contradiction with the solvability conditions.

Denote the right-hand side in (2.22) by g_j^k. We have the estimate

$$\|\hat{y}_{p+j}^k\|_{(F^*(\Omega_k))_\infty} \leq C \left(\|g_j^k\|_{(E^*(\Omega_k))_\infty} + \|\hat{y}_{p+j}^k\|_{F_{-1}^*(\Omega_R)} \right). \tag{2.23}$$

It allows us to conclude that there exists a subsequence of the sequence $\hat{y}_{p+j}^k$ converging locally to some y_{p+j}:

$$L^* y_{p+j} = 0, \quad j = 1, \ldots, s. \tag{2.24}$$

From the relations

$$\langle w_i^k, \hat{y}_{p+j}^k \rangle = 0, \quad j = 1, \ldots, s, \quad i = 1, \ldots, p$$

and

$$[\hat{y}_{p+i}^k, \hat{y}_{p+j}^k]_{F^*(\Omega_R)} = \delta_{ij}, \quad i, j = 1, \ldots, s$$

it follows that

$$\langle w_i, y_{p+j} \rangle = 0, \quad j = 1, \ldots, s, \quad i = 1, \ldots, p$$

and

$$[y_{p+i}, y_{p+j}]_{F^*(\Omega_R)} = \delta_{ij}, \quad i, j = 1, \ldots, s.$$

Therefore, $v_1, \ldots, v_p, y_{p+1}, \ldots y_{p+s}$ are linearly independent and belong to the kernel of the operator L^*. The proposition is proved. □

Proposition 2.8. *The inequality $s^* \leq s$ holds.*

Proof. From the sequence of functionals v_j^k we can choose a subsequence that converges locally in F^* to some functionals $v_j \in F^*(\Omega)$ (Lemma 2.5). The limiting functionals v_j, $j = 1, \ldots, p$ are linearly independent and belong to the kernel of the operator L^*. Let $v_{p+1}, \ldots, v_{p+s^*}$ be some functionals that complete the basis of the kernel. From the sequence w_j^k we can choose a subsequence that converges in $F(\Omega_R)$ to some limiting function w_j as $k \to \infty$. We have

$$\langle w_j, v_i \rangle = \delta_{ij}, \quad i, j = 1, \ldots, p. \tag{2.25}$$

Let the functions $w_{p+1}, \ldots, w_{p+s^*}$ complete them to a biorthogonal system, that is (2.25) holds for $i, j = 1, \ldots, p + s^*$. They can be chosen in such a way that they have supports in Ω_ρ. Let us construct the functions w_j^k and the functionals v_j^k in the same way as u_j^k and ω_j^k in the previous proposition and such that they satisfy the following properties:

$$\|w_j^k - w_j\|_{F(\Omega_\rho)} \to 0 \text{ as } k \to \infty, \quad j = 1, \ldots, p + s^*,$$
$$v_j^k \to v_j \text{ as } k \to \infty, \quad j = 1, \ldots, p + s^*$$

2. Stabilization of the index

locally in F^*,
$$\langle w_j^k, v_i^k \rangle = \delta_{ij}, \quad i,j = 1,\ldots,p+s^*.$$

Consider the equations
$$Lu = w_{p+j}, \quad j = 1,\ldots,s^*. \tag{2.26}$$

Since the right-hand sides do not satisfy the solvability conditions, they are not solvable. The equations
$$L_k u = w_{p+j}^k, \quad j = 1,\ldots,s^* \tag{2.27}$$

are solvable in $E_k(\Omega_k)$. We will repeat the same construction as in the previous proposition to prove that equation (2.11) has at least $m + s^*$ linearly independent solutions.

Denote by $u_1^k, \ldots, u_m^k$ linearly independent solutions of equation (2.1). By virtue of Theorem 1.4 their restrictions to Ω_R are also linearly independent. Without loss of generality we can assume that
$$[u_i^k, u_j^k]_{E(\Omega_R)} = \delta_{ij}, \quad i,j = 1,\ldots,m. \tag{2.28}$$

From the sequences u_i^k we can choose subsequences converging locally in E_∞ to some limiting functions $u_i \in E_\infty(\Omega)$, $i = 1,\ldots,m$. We have
$$[u_i, u_j]_{E(\Omega_R)} = \delta_{ij}, \quad i,j = 1,\ldots,m. \tag{2.29}$$

Denote by z_{m+j}^k, $j = 1,\ldots,s^*$ solutions of equation (2.27) such that
$$[u_i^k, z_{m+j}^k]_{E(\Omega_R)} = 0, \quad i = 1,\ldots,m, \ j = 1,\ldots,s^*. \tag{2.30}$$

We have the estimate
$$\|z_{m+j}^k\|_{E_\infty(\Omega_k)} \leq C \left(\|w_{p+j}^k\|_{F_\infty(\Omega_k)} + \|z_{m+j}^k\|_{L^2(\Omega_R)} \right). \tag{2.31}$$

We begin with the case where the norms $\|z_{m+j}^k\|_{L^2(\Omega_R)}$ are uniformly bounded. Since the norms $\|w_{p+j}^k\|_{F_\infty(\Omega_k)}$ are also uniformly bounded by construction, then by virtue of the previous estimate,
$$\|z_{m+j}^k\|_{E_\infty(\Omega_k)} \leq M \tag{2.32}$$

with some constant M independent of j and k. The restrictions of these functions to Ω_R are also uniformly bounded in $E(\Omega_R)$. Therefore we can choose subsequences converging in the norm $L^2(\Omega_R)$. We recall that the functions w_{p+j}^k have their supports in Ω_R and converge to the limiting functions w_{p+j} in $F(\Omega_R)$. Hence the sequences z_{m+j}^k converge locally in E_∞ to some limiting functions z_{m+s}. Passing to the limit in equation (2.27) we obtain solutions of equation (2.26). This contradiction shows that (2.32) cannot hold.

We now consider the case where the norms $\|z_{m+j}^k\|_{L^2(\Omega_R)}$ are not uniformly bounded. Without loss of generality we can assume that they monotonically increase with k. Put

$$\bar{z}_{m+j}^k = \frac{z_{m+j}^k}{\|z_{m+j}^k\|_{L^2(\Omega_R)}}, \quad \tilde{w}_{p+j}^k = \frac{w_{p+j}^k}{\|z_{m+j}^k\|_{L^2(\Omega_R)}}.$$

Then

$$L_k \bar{z}_{m+j}^k = \tilde{w}_{p+j}^k, \quad j = 1, \ldots, s^* \tag{2.33}$$

and

$$[u_i^k, \bar{z}_{m+j}^k]_{E(\Omega_R)} = 0, \quad i = 1, \ldots, m, \; j = 1, \ldots, s^*. \tag{2.34}$$

For every k fixed the functions w_{p+j}^k, $j=1,\ldots,s^*$ are linearly independent on Ω_R. Therefore the restrictions of the functions z_{m+j}^k to Ω_R are also linearly independent and can be orthonormalized. Denote by $\hat{z}_{m+j}^k$ their linear combinations

$$\hat{z}_{m+j}^k = c_{1j} \bar{z}_{m+1}^k + \cdots + c_{s^*j} \bar{z}_{m+s^*}^k$$

such that

$$[\hat{z}_{m+i}^k, \hat{z}_{m+j}^k]_{E(\Omega_R)} = \delta_{ij}, \quad i,j = 1, \ldots, s^*. \tag{2.35}$$

Then

$$L_k \hat{z}_{m+j}^k = d_{1j} w_{p+1}^k + \cdots + d_{s^*j} w_{p+s^*}^k, \quad j = 1, \ldots, s^*, \tag{2.36}$$

where

$$d_{ij}^k = \frac{c_{ij}^k}{\|z_{m+j}^k\|_{L^2(\Omega_R)}}$$

and

$$[u_i^k, \hat{z}_{m+j}^k]_{E(\Omega_R)} = 0, \quad i = 1, \ldots, m, \; j = 1, \ldots, s^*. \tag{2.37}$$

If the coefficients d_{ij} are not uniformly bounded, we obtain a contradiction in (2.36) with the uniform boundedness of the operators L_k. If these coefficients are uniformly bounded but do not converge to zero, then we can choose converging subsequences and pass to the limit in equation (2.36) to obtain a contradiction with the solvability conditions for equation (2.26). Hence the coefficients d_{ij}^k converge to zero. Denote the right-hand side in equation (2.36) by f_j^k. Then

$$\|\hat{z}_{m+j}^k\|_{E_\infty(\Omega_k)} \leq C \left(\|f_j^k\|_{F_\infty(\Omega_k)} + \|\hat{z}_{m+j}^k\|_{L^2(\Omega_R)} \right). \tag{2.38}$$

Since the first norm in the right-hand side of this estimate converges to zero, the second norm is bounded,

$$\|\hat{z}_{m+j}^k\|_{L^2(\Omega_R)} \leq \|\hat{z}_{m+j}^k\|_{E(\Omega_R)} = 1,$$

then the norms in the left-hand side are uniformly bounded. As above, we can conclude that there exist subsequences $\hat{z}_{m+j}^{k_l}$ converging locally in E_∞ to some limiting functions $\hat{z}_{m+j} \in E_\infty(\Omega)$. Therefore

$$L\hat{z}_{m+j} = 0, \quad j = 1, \ldots, s^*.$$

By virtue of the convergence

$$\|\hat{z}_{m+j}^k - \hat{z}_{m+j}\|_{E(\Omega_R)} \to 0, \quad k \to \infty,$$

(2.35) and (2.37) it follows that

$$[\hat{z}_{m+i}, \hat{z}_{m+j}]_{E(\Omega_R)} = \delta_{ij}, \quad i, j = 1, \ldots, s^*$$

and

$$[u_i, \hat{z}_{m+j}]_{E(\Omega_R)} = 0, \quad i = 1, \ldots, m, \ j = 1, \ldots, s^*.$$

Therefore the functions $u_1, \ldots, u_m, \hat{z}_{m+1}, \ldots, \hat{z}_{m+s^*}$ belong to the kernel of the operator L and they are linearly independent. The proposition is proved. □

We have proved the following theorem.

Theorem 2.9. *Let bounded domains Ω_n converge to an unbounded domain Ω and Fredholm operators $L_n : E(\Omega_n) \to F(\Omega_n)$ converge to an operator $L : E_\infty(\Omega) \to F_\infty(\Omega)$. If Conditions NS(seq) and NS*(seq) are satisfied, then the index of the operators L_n does not depend on n for n sufficiently large and equals the index of the operator L.*

Remark 2.10. Since solutions of the equations $Lu = 0$ and $L^*v = 0$ decay exponentially at infinity, the index of the operator $L : E_q(\Omega) \to F_q(\Omega)$ is independent of q, $1 \le q \le \infty$.

3 Formally adjoint problems

In this section we consider the scalar boundary value problem

$$Au = f_0 \text{ in } \Omega, \quad B_j u = f_j \text{ in } \Gamma, \ j = 1, \ldots, m. \tag{3.1}$$

Here

$$Au = \sum_{|\alpha| \le 2m} a^\alpha(x) D^\alpha u, \quad x \in \Omega,$$

$$B_j u = \sum_{|\alpha| \le s_j} b_j^\alpha(x) D^\alpha u, \quad x \in \Gamma, \ j = 1, \ldots, m, \ s_j < 2m,$$

$\alpha = (\alpha_1, \ldots, \alpha_n)$ is a multi-index, $|\alpha| = \alpha_1 + \cdots + \alpha_n$, $D^\alpha = D_1^{\alpha_1} \ldots D_n^{\alpha_n}$, $D_i^{\alpha_i} = \partial^{\alpha_i}/\partial x_i^{\alpha_i}$, $\Omega \subset R^n$ is an unbounded domain and Γ is its boundary. We suppose that problem (3.1) is uniformly elliptic. This means that (i) A is a uniformly elliptic operator, (ii) A is properly elliptic, (iii) the boundary operators satisfy uniformly the Lopatinskii condition. It is also supposed that B_j form a normal system: $0 \leq s_1 < \cdots < s_m$ and for each normal vector ν_x on Γ, where $x \in \Gamma$, it holds that

$$\sum_{|\alpha|=s_i} b_j^\alpha(x)\nu_x^\alpha \neq 0, \quad j=1,\ldots,m$$

(see [32], [318] and Chapter 6). Such problems are referred to as regular elliptic problems. The coefficients $a^\alpha(x)$ and $b_j^\alpha(x)$ in (3.1) are complex-valued functions given in $\bar\Omega$ which, for simplicity, are assumed to be infinitely differentiable with all bounded derivatives.

As it is known, Green's formula takes place for the operators under consideration:

$$\int_\Omega Au\,\bar v\,dx - \int_\Omega u\,\overline{A^*v}\,dx = \sum_{j=1}^m \int_\Gamma S_j u\,\overline{C_j v}\,ds - \sum_{j=1}^m \int_\Gamma B_j u\,\overline{T_j v}\,ds. \quad (3.2)$$

Here $u, v \in D(\bar\Omega)$, A^* is the formally adjoint operator,

$$A^*v = \sum_{|\alpha|\leq 2m} (-1)^{|\alpha|} D^\alpha(a^\alpha(x)v),$$

and the boundary operators B_j, S_j and C_j, T_j form Dirichlet systems (see [318] and Chapter 6). The boundary value problem

$$A^*v = f_0 \text{ in } \Omega, \quad C_j v = f_j \text{ in } \Gamma, \quad j=1,\ldots,m. \quad (3.3)$$

is called the formally adjoint problem with respect to (3.1).

Consider a sequence of bounded domains Ω_n that converge to an unbounded domain Ω and a sequence of operators $L_n : E(\Omega_n) \to F(\Omega_n)$ that converge to an operator $L : E(\Omega) \to F(\Omega)$. Let the operators L_n and L satisfy the Fredholm property. We suppose that the operators L and L_n have formally adjoint operators, L^* and L_n^*, respectively. Then the solvability conditions can be formulated in terms of orthogonality to solutions of the homogeneous formally adjoint problems (both for the direct and for the formally adjoint operators). In particular, problem (3.1) is solvable if and only if

$$\int_\Omega f_0\,\bar v\,dx + \sum_{j=1}^m \int_\Gamma f_j\,\overline{T_j v}\,ds = 0$$

for any solution v of the problem

$$A^*v = 0, \quad C_j v = 0, \quad j=1,\ldots,m.$$

3. Formally adjoint problems

Similar solvability conditions hold for the operators L^*, L_n, and L_n^* (see Chapter 6).

Condition NS(seq) is supposed to be satisfied for both sequences of operators, L_n and L_n^*. We restrict ourselves to the spaces of real-valued functions assuming that the coefficients of the operators are real.

Lemma 3.1. *Assume that the operators L_n are invertible and that* $\dim \operatorname{Ker}(L) = \mu \geq 0$. *Then* $\dim \operatorname{Ker}(L^*) \geq \mu$.

Proof. Let $u_1, \ldots, u_\mu$ be linearly independent solutions of the equation

$$Lu = 0, \quad u \in E(\Omega). \tag{3.4}$$

Since the operator L satisfies the Fredholm property, they decay exponentially at infinity (Chapters 4 and 5). We can choose a system of functions $f_0^j \in W^{l-2m}(\Omega)$ with a bounded support strictly inside Ω, biorthogonal to the functions u_i,

$$\int_\Omega u_i f_0^j dx = \delta_{ij}, \quad i,j = 1,\ldots,\mu. \tag{3.5}$$

Hence the problems

$$A^* v = f_0^j, \quad C_1 v = \cdots = C_m v = 0, \quad j = 1,\ldots,\mu \tag{3.6}$$

are not solvable because the solvability conditions

$$\int_\Omega u_i \overline{A^* v} \, dx + \sum_{j=1}^m \int_\Gamma S_j u_i \overline{C_j v} \, ds = 0, \quad i = 1,\ldots,\mu$$

are not satisfied.

Since the operators L_n are invertible, then the operators L_n^* are also invertible. Indeed, the homogeneous equation $L_n^* v = 0$ has only zero solution. Otherwise, it would be possible to choose such f that the equation $L_n u = f$ was not solvable. The nonhomogeneous equation $L_n^* v = f$ is solvable for any $f \in F(\Omega_n)$ since the equation $L_n u = 0$ has only the zero solution.

Put $f^i = (f_0^i, f_1^i, \ldots, f_m^i)$, where $f_j^i = 0, j = 1, \ldots, m$. The functions f_0^i have bounded supports in Ω. Without loss of generality we can assume that the domains Ω_n coincide with Ω for n sufficiently large inside some ball which includes the supports of the functions f_0^i. Therefore for n sufficiently large $f^i \in F(\Omega_n)$, and the problems

$$L_n^* v = f^i, \quad i = 1,\ldots,\mu \tag{3.7}$$

are uniquely solvable. Denote their solutions by v_i^n. For each n the functions $v_1^n, \ldots, v_\mu^n$ are linearly independent since f_0^j are linearly independent.

Thus, equations (3.7) are solvable, while equations (3.6) are not solvable. If there existed a convergent subsequence of the sequence of solutions of equations

(3.7), we would obtain a solution of equation (3.6). This contradiction shows that the sequence of norms of the solutions should be unbounded. This will allow us to prove the existence of solutions of the equation $L^*v = 0$.

Consider the estimate

$$\|v_i^n\|_{E_\infty(\Omega_n)} \leq M\left(\|L_n^* v_i^n\|_{E_\infty(\Omega_n)} + \|v_i^n\|_{L^2(\Omega_R^n)}\right), \quad i = 1,\ldots,\mu, \quad n = 1, 2, \ldots, \tag{3.8}$$

which follows from Theorem 1.4. Here $\Omega_R^n = \Omega_n \cap \{|x| \leq R\}$, M and R do not depend on i and n. Without loss of generality we can assume that the domains Ω_R^n do not depend on n and use the notation Ω_R. We recall that L_n^* are formally adjoint operators, which can be considered as acting from $E_\infty(\Omega_n)$ into $F_\infty(\Omega_n)$.

If the norms $\|v_i^n\|_{L^2(\Omega_R)}$ are uniformly bounded, then the norms $\|v_i^n\|_{E_\infty(\Omega_n)}$ are also bounded. Therefore we can pass to the limit in equations (3.7) and obtain that equation (3.6) is solvable. This contradicts the choice of the functions f^i. Therefore the sequences of norms are unbounded. Set

$$w_i^n = \frac{v_i^n}{\|v_i^n\|_{L^2(\Omega_R^n)}}, \quad g_i^n = \frac{f^i}{\|v_i^n\|_{L^2(\Omega_R^n)}}.$$

Then

$$L_n^* w_i^n = g_i^n, \quad i = 1, \ldots, \mu. \tag{3.9}$$

We can choose locally convergent subsequences of the functions w_i^n and pass to the limit in the equation to obtain nonzero solutions of the equation

$$L^* w = 0. \tag{3.10}$$

We will show that there exist at least μ linearly independent solutions of this equation.

The functions $w_1^n, \ldots, w_\mu^n$ are linearly independent. Indeed, if they were linearly dependent, then it would follow from equation (3.9) that the functions $g_1^n, \ldots, g_\mu^n$ were also linearly dependent. This contradicts the linear independence of the functions $f^1, \ldots, f^\mu$.

We construct the orthonormal system of functions:

$$\tilde{w}_1^n = \frac{w_1^n}{\int_\Omega |w_1^n|^2 dx}, \quad \tilde{w}_2^n = \frac{w_2^n - a_2^n \tilde{w}_1^n}{\int_\Omega |w_2^n - a_2^n \tilde{w}_1^n|^2 dx}, \quad a_2^n = \int_\Omega w_2^n \tilde{w}_1^n dx,$$

$$\tilde{w}_3^n = \frac{w_3^n - b_1^n \tilde{w}_1^n - b_2^n \tilde{w}_2^n}{\int_\Omega |w_3^n - b_1^n \tilde{w}_1^n - b_2^n \tilde{w}_2^n|^2 dx}, \quad b_1^n = \int_\Omega w_3^n \tilde{w}_1^n dx, \quad b_2^n = \int_\Omega w_3^n \tilde{w}_2^n dx, \ldots$$

The functions $\tilde{w}_1^n, \ldots, \tilde{w}_m^n$ satisfy the equations

$$L_n^* w = \tilde{g}_i^n, \quad i = 1, \ldots, \mu, \tag{3.11}$$

where $\tilde{g}_i^n$ are some linear combinations of the functions f^i:

$$\tilde{g}_1^n = k_{11}^n f^1 + \cdots + k_{1\mu}^n f^\mu, \quad \ldots, \quad \tilde{g}_\mu^n = k_{\mu1}^n f^1 + \cdots + k_{\mu\mu}^n f^\mu.$$

We note first of all that the constants k_{ij}^n are uniformly bounded. Indeed, otherwise we divide equations (3.11) by the maximal constant for each n and pass to the limit as $n \to \infty$. We obtain a contradiction because the L^2-norm of the solution would be necessarily zero while the right-hand side is different from zero.

Moreover, they converge to zero: if there was a sequence k_{ij}^n that did not converge to zero as $n \to \infty$, we could pass to the limit in the equation and would obtain a contradiction with equalities (3.5) and solvability conditions.

Since the functions $\tilde{g}_i^n$ have bounded supports and are uniformly bounded, then the functions $\tilde{w}_1^n, \ldots, \tilde{w}_\mu^n$ admit a uniform exponential estimate,

$$|\tilde{w}_i^n(x)| \leq \omega_\delta(x) \equiv K e^{-\delta \sqrt{1+|x|^2}}, \quad i=1,\ldots,\mu, \ n=1,2,\ldots, \tag{3.12}$$

where K and δ are some positive constants independent of i and n. This follows from Theorem 1.7. It can be applied since the domains Ω_n are bounded and, consequently, $\omega_\delta \tilde{w}_i^n \in E_\infty(\Omega_n)$.

From each sequence $\tilde{w}_i^n$, $n = 1, 2, \ldots$ we can choose a subsequence locally convergent to some limiting function $\tilde{w}_i^0$. It satisfies limiting equation (3.10). The functions $\tilde{w}_i^0$, $i = 1, 2, \ldots, \mu$ are orthonormal and, consequently, linearly independent. The lemma is proved. □

Corollary 3.2. *The index of the operator L equals zero.*

Proof. It sufficient to note that the operator L is formally adjoint to the operator L^*. Therefore $\dim \operatorname{Ker}(L) \geq \dim \operatorname{Ker}(L^*)$. The corollary is proved. □

The main result of this section is given by the following theorem.

Theorem 3.3. *Suppose that the dimension of the kernel of the operator L_n equals μ and the codimension of the image equals π for all n. Then the index of the operator L equals $\mu - \pi$.*

Proof. Denote by $u_1^n, \ldots, u_\mu^n$ linearly independent solutions of the equation $L_n u = 0$. Without loss of generality we can assume that they are orthonormal in $L^2(\Omega_n)$. Let $v_1^n, \ldots, v_\mu^n$ be a system of functions biorthogonal to $u_1^n, \ldots, u_\mu^n$. Since the functions $u_1^n, \ldots, u_\mu^n$ admit uniform exponential estimates (Theorem 1.7), then $v_1^n, \ldots, v_\mu^n$ can be chosen in such a way that they are uniformly bounded in the W^{l-2m} norm and that they have uniformly bounded supports strictly inside Ω.

The operator L_n^* formally adjoint to the operator L_n has a π-dimensional kernel. Denote by $w_1^n, \ldots, w_\pi^n$ its orthonormal basis. The problem

$$A_n^* v = f_0 - \sum_{i=1}^\mu v_i^n \int_{\Omega_n} f_0 u_i^n dx, \quad C_j^n v = 0, \ j = 1, \ldots, m \tag{3.13}$$

is solvable for any f_0 such that $f = (f_0, 0, \ldots, 0) \in F(\Omega_n)$.

Since the functions $u_1^n, \ldots, u_\mu^n$ admit uniform exponential estimates, then we can pass to the limit and obtain that the equation $Lu = 0$ has at least μ linearly independent solutions. Let the dimension of the kernel of the operator L be $\mu + s$, $s \geq 0$. Similarly, the dimension of the kernel of the operator L^* is $\pi + s^*$, $s^* \geq 0$. We will show that $s = s^*$.

Let $u_1^0, \ldots, u_\mu^0, u_{\mu+1}^0, \ldots, u_{\mu+s}^0$ be an orthonormal basis in $\operatorname{Ker} L$, where the first m functions are the limits of the sequences $u_1^n, \ldots, u_\mu^n$. Denote by $f_1^0, \ldots, f_{\mu+s}^0$ a system of functions with a finite support in Ω and biorthogonal to the functions $u_1^0, \ldots, u_{\mu+s}^0$:

$$\int_\Omega u_i^0 f_j^0 \, dx = \delta_{ij}, \quad i, j = 1, \ldots, \mu + s.$$

The problems

$$A^* v = f_j^0, \quad C_i v = 0, \quad i = 1, \ldots, m, \quad j = \mu + 1, \ldots, \mu + s \tag{3.14}$$

are not solvable since the solvability conditions for them are not satisfied. On the other hand, the problems

$$A_n^* v = f_j^0 - \sum_{i=1}^\mu v_i^n \int_{\Omega_n} u_i^n f_j^0 \, dx, \quad C_i^n v = 0, \quad i = 1, \ldots, m, \quad j = \mu + 1, \ldots, \mu + s \tag{3.15}$$

are solvable for n sufficiently large for which the support of f_j^0 belongs to Ω_n. There is a unique solution y_j^n of this equation such that

$$\int_{\Omega_n} y_j^n w_i^n \, dx = 0, \quad i = 1, \ldots, \pi. \tag{3.16}$$

Suppose that the sequence y_j^n, $n = 1, 2, \ldots$ is uniformly bounded in $L^2(\Omega_R)$, where Ω_R is the same as in Theorem 1.4. As before, we can assume that it does not depend on n. Then the norms $\|y_j^n\|_{E(\Omega_n)}$ are uniformly bounded.

Since

$$\int_{\Omega_n} u_i^n f_j^0 \, dx \to 0, \quad n \to \infty, \quad i = 1, \ldots, \mu, \quad j = \mu + 1, \ldots, \mu + s,$$

then we can pass to the limit in equation (3.15) and obtain that equation (3.14) is solvable. This contradiction shows that the sequence of norms $\|y_j^n\|_{L^2(\Omega_R)}$ is not bounded. Dividing equations (3.15) by these norms, passing to the limit in the equation and taking into account (3.16) we will obtain additional solutions of the equation $L^* v = 0$. We will show that there are at least s such solutions.

Denote by $z_1^n, \ldots, z_s^n$ an orthonormal linear combination of the functions $y_{\mu+1}^n, \ldots, y_{\mu+s}^n$. Then

$$\int_{\Omega_n} z_j^n w_i^n \, dx = 0, \quad i = 1, \ldots, \pi, \quad j = 1, \ldots, s.$$

The functions $z_1^n, \ldots, z_s^n$ admit a uniform exponential estimate at infinity. Similar to the proof of the previous lemma, we can show that from the sequences z_j^n, $n =$

1, 2, ... we can choose subsequences locally converging to some limiting functions $z_1^0, \ldots, z_s^0$ that satisfy the equation $L^*v = 0$. Moreover, these functions are linearly independent between them and with $w_i, i = 1, \ldots, \pi$.

Thus we have proved that the dimension of the kernel of the operator L^* is at least $\mu + s$, that is $s^* \geq s$. Since the operator L is formally adjoint to L^*, then we obtain similarly that $s \geq s^*$. The theorem is proved. □

4 Convergence of solutions

In this section we study convergence of solutions of elliptic problems in bounded domains to solutions in unbounded domains. This convergence may not occur. Some examples are discussed below. As before we assume that Conditions NS(seq) and NS*(seq) are satisfied. They provide the stabilization of the index. For the convergence of solutions we need an additional condition: not only the indices but also the dimensions of the kernels should stabilize.

Denote by $u_1^k, \ldots, u_m^k \in E(\Omega_k)$ all linearly independent solutions of the equation $L_k u = 0$ and by $v_1^k, \ldots, v_p^k \in F^*(\Omega_k)$ all linearly independent solutions of the equation $L_k^* v = 0$. We suppose that m and p are independent of k. We recall that E is a Hilbert space with the inner product $[\cdot, \cdot]$. As before, $\langle \cdot, \cdot \rangle$ denotes the duality in the corresponding spaces.

Theorem 4.1. *Let conditions NS(seq) and NS*(seq) be satisfied. Consider the equations*

$$L_k u = f_k, \tag{4.1}$$

where $f_k \in F_\infty(\Omega_k)$, $\|f_k\|_{F_\infty(\Omega_k)} \leq K$ and $f_k \to f \in F_\infty(\Omega)$ in F_{loc}. Suppose that the solvability conditions

$$\langle f_k, v_i^k \rangle = 0, \quad i = 1, \ldots, p \tag{4.2}$$

are satisfied and denote by u_k the solution of equation (4.1) such that

$$[u_k, u_i^k] = 0, \quad i = 1, \ldots, m. \tag{4.3}$$

If

$$\dim \operatorname{Ker} L = \dim \operatorname{Ker} L_k, \tag{4.4}$$

then u_k converges in E_{loc} to the unique solution u_0 of the equation

$$Lu = f \tag{4.5}$$

for which

$$[u_0, u_i] = 0, \quad i = 1, \ldots, m, \tag{4.6}$$

where u_i form a basis of the kernel of the operator L.

Proof. Consider the solutions $v_1^k, \ldots, v_p^k \in F^*(\Omega_k)$ of the homogeneous adjoint equations. From Lemma 2.5 it follows that there exist linearly independent solutions $v_1, \ldots, v_p \in F^*(\Omega)$ of the equation $L^* v = 0$ such that $v_j^k \to v_j$, $j = 1, \ldots, p$ locally in F_∞. The convergence can take place along a subsequence.

Since the index of the operator L equals the indices of the operators L_k for k sufficiently large and by virtue of (4.4) the dimensions of their kernels are also equal, then

$$\dim \operatorname{Ker} L^* = \dim \operatorname{Ker} L_k^*. \tag{4.7}$$

Hence $v_1, \ldots, v_p$ form a basis in the kernel of the operator L^*.

We will show that we can pass to the limit in (4.2) to obtain

$$\langle f, v_i \rangle = 0, \quad i = 1, \ldots, p. \tag{4.8}$$

Indeed, suppose that for some i_0 this equality does not hold. Then

$$|\langle f, v_{i_0} \rangle| = a \tag{4.9}$$

for some $a \neq 0$. Let a function $\phi_R(x) \in D$ be identically equal to 1 for $|x| \leq R$ and 0 for $|x| \geq R + 1$. By virtue of the local convergence of the solutions, we have

$$\langle f_k, \phi_R v_{i_0}^k \rangle \to \langle f, \phi_R v_{i_0} \rangle, \quad k \to \infty. \tag{4.10}$$

To continue the proof we need the following estimate. Let $f \in F(\Omega_k)$, $v \in F^*(\Omega_k)$, $\phi_i \in D$ be a partition of unity and $\psi_i \in D$ equal to 1 on the support of ϕ_i (D denotes infinitely differentiable functions with bounded supports). Then

$$|\langle f, v \rangle| \leq \sum_i |\langle \psi_i f, \phi_i v \rangle| \leq C_1 \sup_i \|\psi_i f\|_{F(\Omega_k)} \sum_i \|\phi_i v\|_{F^*(\Omega_k)}$$

$$\leq C_2 \|f\|_{F_\infty(\Omega_k)} \|v\|_{(F^*(\Omega_k))_1}.$$

Here C_1 and C_2 are some positive constants, $(F^*(\Omega_k))_1$ is a space of functions introduced in Section 2.1. We use here its equivalent norm defined through the partition of unity (Chapter 2). Using this estimate, we obtain

$$|\langle f_k, (1 - \phi_R) v_{i_0}^k \rangle| = |\langle \psi_R \omega^{-1} f_k, \omega(1 - \phi_R) v_{i_0}^k \rangle|$$

$$\leq \|\psi_R \omega^{-1} f_k\|_{F_\infty(\Omega_k)} \|\omega(1 - \phi_R) v_{i_0}^k\|_{(F^*(\Omega_k))_1}$$

$$\leq \|\psi_R\|_M e^{-\mu(R-1)} \|f_k\|_{F_\infty(\Omega_k)} \|1 - \phi_R\|_M \|\omega v_{i_0}^k\|_{(F^*(\Omega_k))_1}$$

$$\leq C e^{-\mu(R-1)}, \tag{4.11}$$

where $\psi_R(x)$ is an infinitely differentiable function equal to 1 for $|x| \geq R$, that is on the support of the function $1 - \phi_R(x)$, $\|\cdot\|_M$ denotes the norm of the multiplier in both spaces $F_\infty(\Omega_k)$ and $(F^*(\Omega_k))_\infty$. In the last estimate we take into account that the norms $\|\psi_R\|_M$ and $\|1 - \phi_R\|_M$ are bounded independently of R for R sufficiently large, as well as that the norms $\|f_k\|_{F_\infty(\Omega_k)}$ are bounded

4. Convergence of solutions

independently of k according to the condition of the theorem. Finally, the norms $\|\omega v_{i_0}^k\|_{(F^*(\Omega_k))_1}$ are bounded independently of k. This follows from the exponential estimates of solutions of the homogeneous adjoint equation (cf. Theorem 1.7).

We next obtain the estimate

$$|\langle f, (1-\phi_R)v_{i_0}\rangle| \leq Ce^{-\mu(R-1)}. \tag{4.12}$$

We use here the result on the exponential estimate of solutions of the homogeneous adjoint equation (see Chapter 5). From estimates (4.11), (4.12) it follows that we can choose R such that

$$|\langle f_k, (1-\phi_R)v_{i_0}^k\rangle| \leq \frac{a}{4}, \quad |\langle f, (1-\phi_R)v_{i_0}\rangle| \leq \frac{a}{4}. \tag{4.13}$$

From (4.2),
$$\langle f_k, (1-\phi_R)v_{i_0}^k\rangle + \langle f_k, \phi_R v_{i_0}^k\rangle = 0.$$

Hence $|\langle f_k, \phi_R v_{i_0}^k\rangle| \leq \frac{a}{4}$. Convergence (4.10) implies the estimate $|\langle f, \phi_R v_{i_0}\rangle| \leq \frac{a}{4}$. Together with the second estimate in (4.12) this gives $|\langle f, v_{i_0}\rangle| \leq \frac{a}{2}$. This estimate contradicts (4.9) and proves (4.8).

It follows from (4.8) that there exists a solution of equation (4.5). Denote by u_0 the unique solution of this equation, which satisfies conditions (4.6). We will show that solutions u_k of equations (4.1) converge to u_0. Consider first the case where

$$\|u_k\|_{E_\infty(\Omega_k)} \leq K, \quad \forall k \tag{4.14}$$

for some positive constant K. Then there exists a subsequence converging to some limiting function $u \in E_\infty(\Omega)$ locally in a weaker norm. The limiting function is a solution of equation (4.5). By virtue of Lemma 2.4 we can pass to the limit in (4.3) to obtain

$$[u, u_i] = 0, \quad i = 1, \ldots, m.$$

Since there exists a unique solution of equation (4.5), which satisfies these equalities, then $u = u_0$. Moreover, since the limit is unique the convergence occurs along the whole sequence u_k.

We next consider the case where (4.14) does not hold. Without loss of generality we can assume that this sequence tends to infinity. Put

$$w_k = \frac{u_k}{\|u_k\|_{E_\infty(\Omega_k)}}, \quad g_k = \frac{f_k}{\|u_k\|_{E_\infty(\Omega_k)}}.$$

Then
$$L_k w_k = g_k, \quad \|w_k\|_{E_\infty(\Omega_k)} = 1, \quad \|g_k\|_{E_\infty(\Omega_k)} \to 0, \ k \to \infty.$$

We note first of all that if $w_k \to 0$ locally, then it follows from Theorem 1.4 that $\|w_k\|_{E_\infty(\Omega_k)} \to 0$ as $k \to \infty$. This contradiction shows that the sequence w_k does not converge locally to zero. Since it is bounded, then there exists a subsequence that converges locally in a weaker norm to some $w_0 \in E_\infty(\Omega)$, $Lw_0 = 0$, $w_0 \neq 0$.

From (4.3) it follows that $[w_k, u_i^k] = 0$, $i = 1, \ldots, m$. The uniform exponential estimate (Theorem 1.7) allows the passage to the limit in these equalities:

$$[w_0, u_i] = 0, \quad i = 1, \ldots, m.$$

We obtain a contradiction with the assumption (4.4). The theorem is proved. $\square$

Remark 4.2. Conditions NS(seq) and NS*(seq) in the formulation of the theorem are related to the invertibility of limiting operators. If a limiting operator $\hat{L}$ is invertible as acting from E_1 into F_1, then the adjoint operator $\hat{L}^* : (F_1)^* \to (E_1)^*$ is also invertible. Therefore the equation $\hat{L}^* v = 0$ has only a trivial solution in the space $(F_1)^* = (F^*)_\infty$. This provides Condition NS*(seq). Condition NS(seq) implies the unique solvability of the equation $\hat{L} u = 0$ in the space E_∞. It is a stronger condition than the unique solvability of this equation in the space E_1 since $E_1 \subset E_\infty$. The difference between these conditions is essential. The equation $u'' = 0$ has a nonzero solution in $H^2_\infty(\mathbb{R}^1)$ but not in $H^2_1(\mathbb{R}^1)$. If the Laplace operator is among limiting operators for some operator L, the latter does not satisfy the Fredholm property since Condition NS is not satisfied. Stabilization of the index and convergence of solutions may not occur because Condition NS(seq) is not satisfied.

We consider some examples to illustrate the theorem on the convergence of solutions.

Convergence of solutions. Consider the problems

$$\Delta u - au = b, \quad u|_{\partial \Omega_k} = 0, \tag{4.15}$$

where a and b are some positive constants, Ω_k are bounded domains with uniformly $C^{2+\delta}$ boundaries. Each of them has a unique solution $u_k \in C^{2+\delta}(\Omega_k)$. It can be easily verified that

$$-\frac{b}{a} \leq u_k \leq 0, \quad x \in \Omega_k.$$

Therefore the norms $\|u_k\|_{C^{2+\delta}(\Omega_k)}$ are bounded by a constant independent of k (Theorem 1.4). Suppose that the sequence of domains Ω_k locally converges to an unbounded domain Ω. From the estimate of the norms of the solutions it follows that there exist a subsequence converging to a limiting function $u_0 \in C^{2+\delta}(\Omega)$, which is a solution of the same problem in the limiting domain Ω. This result is in agreement with Theorem 4.1.

Divergence of solutions. Consider the problems

$$\Delta u = b, \quad u|_{\partial \Omega_k} = 0, \tag{4.16}$$

where Ω_k is a ball of radius k with its center at the origin. Each problem has the unique solution $u_k = b(|x|^2 - k^2)/(2n)$. Obviously this sequence does not have a finite limit. In this case Condition NS(seq) is not satisfied. Indeed, one of the limiting problems, $\Delta u = 0$ in $\mathbb{R}^2$, has a nonzero bounded solution. Hence Theorem 4.1 is not applicable.

4. Convergence of solutions

Boundary layer. Consider the problem

$$\epsilon^2 u'' - a(x)u = b(x), \quad u(0) = u(1) = 0 \tag{4.17}$$

in the interval $0 \leq x \leq 1$. Put $x = \epsilon y$, $v_\epsilon(y) = u(x)$. Then $v_\epsilon(y)$ satisfies the problem

$$v_\epsilon'' - a(\epsilon y)v_\epsilon = b(\epsilon y), \quad v_\epsilon(0) = v_\epsilon\left(\frac{1}{\epsilon}\right) = 0. \tag{4.18}$$

According to Theorem 4.1 solution of (4.18) locally converges to the solution $v_0(y)$ of the problem

$$v'' - a(0)v = b(0), \quad v(0) = 0$$

on the half-axis $y \geq 0$. Therefore $u(\epsilon y) \to v_0(y)$ locally in H^2. In this and in the following examples we suppose that the function $a(x)$ is such that the conditions of the theorem are satisfied. For example, it can be positive in the interval $[0, 1]$.

Singular perturbations. If we put formally $\epsilon = 0$ in problem (4.17), then we find the so-called external solution $u_e(x) = -b(x)/a(x)$ ($a(x) \neq 0$). To justify this formal solution consider the change of variables $x = x_0 + \epsilon y$ for some $x_0 \in (0, 1)$. Put $v_\epsilon(y) = u(x)$. Then

$$v_\epsilon'' - a(x_0 + \epsilon y)v_\epsilon = b(x_0 + \epsilon y), \quad v_\epsilon\left(\frac{-x_0}{\epsilon}\right) = v_\epsilon\left(\frac{1 - x_0}{\epsilon}\right) = 0.$$

Assuming that we can use Theorem 4.1, we obtain the convergence of the solution $v_\epsilon(y)$ to the solution $v_e(y)$ of the limiting problem

$$v'' - a(x_0)v = b(x_0), \quad y \in \mathbb{R}^1.$$

If $a(x_0) > 0$, then this equation has a unique solution $v_e(y) \equiv -b(x_0)/a(x_0)$. If $a(x_0) \leq 0$, then Condition NS(seq) is not satisfied and the theorem cannot be applied. This is in agreement with the condition $a(x) \neq 0$ required to define the formal solution.

Taking into account the solution obtained in the previous example (internal solution), we can write an approximate solution of the singular perturbation problem in its conventional form

$$u(x) = v_0\left(\frac{x}{\epsilon}\right) - \frac{b(x)}{a(x)} + \frac{b(0)}{a(0)}.$$

The last term in the right-hand side is introduced because both, the internal solution at infinity and the external solution at zero, are equal to $-b(0)/a(0)$.

Oscillating coefficients. Consider the equation

$$\epsilon^2 u'' - a\left(\frac{x}{\epsilon}\right)u = b(x) \tag{4.19}$$

on the real axis. As above, we introduce the function $v_\epsilon(y) = u(x)$. It satisfies the equation
$$v'' - a(y)v = b(\epsilon y).$$
Under the conditions of Theorem 4.1 we obtain the convergence $u(\epsilon y) \to v_0(y)$, where $v_0(y)$ is the solution of the limiting equation
$$v'' - a(y)v = b(0), \quad y \in \mathbb{R}^1.$$
We note that the function $a(y)$ is not necessarily periodic as is often the case in homogenization problems. The boundary layer for oscillating coefficients can be considered in the same way as in the previous example.

Oscillating boundaries. Let $h(\xi) \in C^{2+\delta}(\mathbb{R})$. Put
$$\Omega_\epsilon = \left\{(x,y) : 0 \leq x \leq 1,\ 0 \leq y \leq h\left(\frac{x}{\epsilon}\right)\right\}.$$
We suppose that the domain is smoothed at the corners and keep for it the same notation. Consider the problem
$$\epsilon^2 \frac{\partial^2 u}{\partial x^2} + \frac{\partial^2 u}{\partial y^2} - au = b$$
in the domain Ω_ϵ with the zero boundary conditions. Put $x = \epsilon \xi$, $v_\epsilon(\xi, y) = u(x, y)$. Under the conditions of Theorem 4.1, $v_\epsilon(\xi, y)$ will converge to the solution $v_0(\xi, y)$ of the problem
$$\frac{\partial^2 v}{\partial \xi^2} + \frac{\partial^2 v}{\partial y^2} - au = b$$
in the domain
$$\Omega_0 = \{(\xi, y) : 0 \leq y \leq h(\xi)\}$$
with the zero boundary conditions.

5 Cauchy-Riemann system

Consider the Hilbert problem for the Cauchy-Riemann system:
$$\frac{\partial u}{\partial x} - \frac{\partial v}{\partial y} = 0, \quad \frac{\partial u}{\partial y} + \frac{\partial v}{\partial x} = 0, \tag{5.1}$$
$$a(s)u + b(s)v = f(s). \tag{5.2}$$

We suppose that $a^2(s) + b^2(s) = 1$. It is known that the index of the operator $L = (A, B)$ for a bounded domain Ω equals $2N + 1$, where N is the rotation of the vector (a, b) (see, e.g., [205] and Section 8 below). Instead of system (5.1) we will consider the system
$$\frac{\partial u}{\partial x} - \frac{\partial v}{\partial y} - \lambda u = 0, \quad \frac{\partial u}{\partial y} + \frac{\partial v}{\partial x} + \lambda v = 0 \tag{5.3}$$

5. Cauchy-Riemann system

where λ is a positive number. For bounded domains the corresponding operator satisfies the Fredholm property, and its index is the same as for $\lambda = 0$. We introduce the zero-order terms to ensure the Fredholm property in the case of unbounded domains.

Consider system (5.3) in the half-plane $\mathbb{R}_+^2$ with the boundary condition

$$au + bv = h, \tag{5.4}$$

where a and b are some constants, $h(x)$ is an infinitely differentiable function with a bounded support. Applying the partial Fourier transform with respect to x to system (5.3), we obtain the system

$$\begin{cases} \tilde{u}' + i\xi\tilde{v} + \lambda\tilde{v} = 0, \\ \tilde{v}' - i\xi\tilde{u} + \lambda\tilde{u} = 0 \end{cases}$$

on the half-axis $y > 0$. Here the tilde denotes the partial Fourier transform and prime denotes the derivative with respect to y. There exists a bounded solution for $y > 0$:

$$\begin{pmatrix} \tilde{u} \\ \tilde{v} \end{pmatrix} = \begin{pmatrix} p_1 \\ p_2 \end{pmatrix} e^{-\sqrt{\xi^2 + \lambda^2}\, y},$$

where

$$p_2 = \frac{\lambda - i\xi}{\sqrt{\xi^2 + \lambda^2}} \cdot p_1\,.$$

From the boundary condition (5.4) we have

$$\left(a + b \frac{\lambda - i\xi}{\sqrt{\xi^2 + \lambda^2}} \right) p_1 = \tilde{h}(\xi), \tag{5.5}$$

where $\tilde{h}$ is the Fourier transform of the function h. If $\lambda = 0$, $|\xi| = 1$, then this equation is solvable for any $\tilde{h}$. This means that the Lopatinskii condition is satisfied. Let $|\lambda| + |\xi| = 1$. Equation (5.5) is solvable for any such λ and ξ if $a \neq -b$. Hence the last condition provides the ellipticity with a parameter. Thus we have the following proposition.

Proposition 5.1. *Problem (5.3), (5.4) in the half-plane $\mathbb{R}_+^2$ satisfies the condition of ellipticity with a parameter if $a \neq -b$. The corresponding operator is invertible for real positive and sufficiently large λ.*

Corollary 5.2. *The operator corresponding to problem (5.3), (5.4) is invertible for any positive λ.*

Proof. The homogeneous problem has only the zero solution for any positive λ. Indeed, from (5.5) it follows that $p_1 = 0$. The Fourier transform should be understood here in the generalized sense since the corresponding operator acts from $E_\infty(\mathbb{R}_+^2)$ to $F_\infty(\mathbb{R}_+^2)$.

We verify that Condition NS is satisfied. Problem (5.3), (5.4) has constant coefficients. Therefore it has two limiting problems. One of them coincides with problem (5.3), (5.4) in the half-space. Another one is the problem in $\mathbb{R}^2$. Applying the Fourier transform to system (5.3) considered on the whole plane, we obtain

$$\begin{pmatrix} i\xi_1 - \lambda & , & -i\xi_2 \\ i\xi_2 & , & i\xi_1 + \lambda \end{pmatrix} \begin{pmatrix} \tilde{u} \\ \tilde{v} \end{pmatrix} = 0.$$

The determinant of the matrix in the left-hand side of this equality equals $-|\xi|^2 - \lambda^2$. Since it is different from zero for any positive λ, then the homogeneous problem in $\mathbb{R}^2$ has only the zero solution. Thus, both limiting problems have only trivial solutions, and Condition NS is satisfied for all positive λ. Therefore, the operator corresponding to problem (5.3), (5.4) is normally solvable for such λ.

By virtue of Proposition 5.1, the operator is invertible for large positive λ. From this and from its normal solvability for all $\lambda > 0$ it follows that it satisfies the Fredholm property and has the zero index for any positive λ. The unique solvability of the homogeneous equation implies its invertibility. The corollary is proved. □

We will verify the condition of ellipticity with a parameter at another point of the boundary. Consider system (5.3) in the half-plane $y_1 \geq 0$, where the coordinates (x_1, y_1) are obtained from (x, y) by rotation on the angle θ counterclockwise:

$$x = x_1 \cos\theta - y_1 \sin\theta, \quad y = x_1 \sin\theta + y_1 \cos\theta.$$

Let

$$\tilde{u}(x_1, y_1) = u(x, y), \quad \tilde{v}(x_1, y_1) = v(x, y).$$

Then

$$\frac{\partial \tilde{u}}{\partial x_1} = \cos\theta \frac{\partial u}{\partial x} + \sin\theta \frac{\partial u}{\partial y}, \quad \frac{\partial \tilde{u}}{\partial y_1} = -\sin\theta \frac{\partial u}{\partial x} + \cos\theta \frac{\partial u}{\partial y},$$

$$\frac{\partial \tilde{v}}{\partial x_1} = \cos\theta \frac{\partial v}{\partial x} + \sin\theta \frac{\partial v}{\partial y}, \quad \frac{\partial \tilde{v}}{\partial y_1} = -\sin\theta \frac{\partial v}{\partial x} + \cos\theta \frac{\partial v}{\partial y}.$$

Multiplying the first equation in (5.3) by $\cos\theta$, the second equation by $\sin\theta$ and taking their sum, and then multiplying the first equation in (5.3) by $-\sin\theta$, the second equation by $\cos\theta$ and taking their sum, we obtain the system

$$\frac{\partial \tilde{u}}{\partial x_1} - \frac{\partial \tilde{v}}{\partial y_1} + \lambda(-\tilde{u}\cos\theta + \tilde{v}\sin\theta) = 0, \tag{5.6}$$

$$\frac{\partial \tilde{u}}{\partial y_1} + \frac{\partial \tilde{v}}{\partial x_1} + \lambda(\tilde{u}\sin\theta + \tilde{v}\cos\theta) = 0. \tag{5.7}$$

Boundary condition (5.2) becomes

$$a\tilde{u} + b\tilde{v} = \tilde{f}, \tag{5.8}$$

where $\tilde{f}(x_1, y_1) = f(x, y)$.

5. Cauchy-Riemann system

Applying the Fourier transform to system (5.6), (5.7), we obtain

$$\begin{pmatrix} i\xi_1 - \lambda\cos\theta & , & -i\xi_2 + \lambda\sin\theta \\ i\xi_2 + \lambda\sin\theta & , & i\xi_1 + \lambda\cos\theta \end{pmatrix} \begin{pmatrix} \tilde{u} \\ \tilde{v} \end{pmatrix} = 0.$$

As before, the determinant of the matrix in the left-hand side of this equality equals $-|\xi|^2 - \lambda^2$. Therefore, the homogeneous system in $\mathbb{R}^2$ has only a trivial solution for all $\lambda \neq 0$.

We consider next problem (5.6)–(5.8) in the half-plane $y_1 \geq 0$. Applying the partial Fourier transform with respect to x_1, we obtain

$$\hat{u}' + \lambda\sin\theta\,\hat{u} + (i\xi + \lambda\cos\theta)\hat{v} = 0, \tag{5.9}$$
$$\hat{v}' + (-i\xi + \lambda\cos\theta)\hat{u} - \lambda\sin\theta\,\hat{v} = 0. \tag{5.10}$$

It has a bounded solution for $y_1 > 0$:

$$\begin{pmatrix} \hat{u} \\ \hat{v} \end{pmatrix} = \begin{pmatrix} p_1 \\ p_2 \end{pmatrix} e^{-\sqrt{\xi^2+\lambda^2}\,y_1},$$

where

$$p_1 = 1, \quad p_2 = \frac{\lambda\cos\theta - i\xi}{\sqrt{\xi^2+\lambda^2} + \lambda\sin\theta}, \quad |\xi| + |\theta - 3\pi/2| \neq 0,$$
$$p_1 = 0, \quad p_2 = 1, \quad |\xi| + |\theta - 3\pi/2| = 0 \quad (\lambda > 0).$$

Boundary condition (5.8) cannot be satisfied if

$$ap_1 + bp_2 = 0,$$

that is

$$a + b\frac{\cos\theta}{1+\sin\theta} = 0, \quad \theta \neq 3\pi/2, \tag{5.11}$$
$$a = 1, \quad b = 0, \quad \theta = 3\pi/2.$$

Consider the equation

$$\tan(\alpha(\theta)) = -\frac{\cos\theta}{1+\sin\theta}. \tag{5.12}$$

For each θ between 0 and 2π it has two solutions, $\alpha_1(\theta)$ and $\alpha_2(\theta)$ (Figure 6). The function $\alpha_1(\theta)$ is continuous, increasing, $\alpha_1(0) = 3\pi/4$, $\alpha_1(2\pi) = 7\pi/4$. The function $\alpha_2(\theta)$ is continuous and increasing for $\theta \neq \pi/2$. For this value of θ it jumps from 2π to 0, $\alpha_2(0) = 7\pi/4$, $\alpha_2(2\pi) = 3\pi/4$.

Let s denote the natural parameter along the boundary of the domain Ω, (x_1, y_1) be local coordinates at the boundary, and $\nu(s)$ be the inner normal vector to the boundary. Consider the vector (a, b) in the local coordinates and denote by

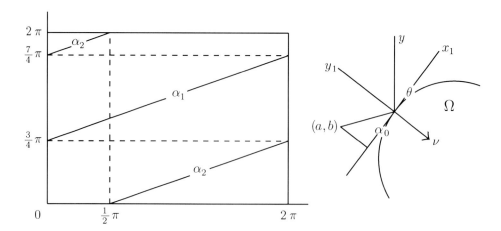

Figure 6: Schematic representation of the functions $\alpha_1(\theta)$ and $\alpha_2(\theta)$ (left); construction of the function $\alpha_0(s)$ (right).

$\alpha_0(s)$ the angle between this vector and the normal vector $\nu(s)$ (positive angle is counterclockwise). Then

$$\tan(\alpha_0(s)) = \frac{a}{b}. \tag{5.13}$$

Proposition 5.3. *For any bounded domain Ω there are points at its boundary where the condition of ellipticity with a parameter is not satisfied.*

Proof. If $\alpha_0(s) = \alpha_i(\theta(s))$, $i = 1, 2$, where $\theta(s)$ is the angle between $\nu(s)$ and the vertical direction, then (5.11) is satisfied. Therefore, problem (5.6)–(5.8) does not satisfy the condition of ellipticity with a parameter at this point of the boundary. Since the function $\alpha_0(\theta)$ is periodic and $\theta(s)$ takes all values between 0 and 2π, $\alpha_0(\theta)$ will necessarily intersect one of the functions $\alpha_i(\theta(s))$. The proposition is proved. $\square$

Remark 5.4. If the condition of ellipticity with a parameter were satisfied everywhere, the problem would have the zero index. However, the index should be odd. This is in agreement with the assertion of the previous proposition.

Definition 5.5. We say that a vector (a, b), $a^2 + b^2 \neq 0$ belongs to the half-plane Π_- if $a + b < 0$. It belongs to the half-plane Π_+ if $a + b > 0$.

Proposition 5.6. *If the vector $(a(s), b(s))$ belongs to the same half-plane for any $s \in \mathbb{R}$ and has limits $(a_\pm(s), b_\pm(s))$ as $s \to \pm\infty$, then problem (5.2), (5.3) in the half-plane $\mathbb{R}_+^2$ has the zero index.*

5. Cauchy-Riemann system

Proof. Consider the homotopy (continuous deformation)

$$\begin{pmatrix} a_\tau(s) \\ b_\tau(s) \end{pmatrix} = \begin{pmatrix} \cos\theta_\tau(s) & , & -\sin\theta_\tau(s) \\ \sin\theta_\tau(s) & , & \cos\theta_\tau(s) \end{pmatrix} \begin{pmatrix} a(s) \\ b(s) \end{pmatrix},$$

where $\theta_\tau(s)$ equals 0 for $\tau = 0$, and equals the angle between $(a(s), b(s))$ and $(a_-(s), b_-(s))$ for $\tau = 1$. Therefore this homotopy rotates the vector $(a(s), b(s))$ to the vector $(a_-(s), b_-(s))$. Moreover, $a_\tau(s) \neq -b_\tau(s)$. In particular, this is true for $s = +\infty$. Therefore Condition NS remains satisfied during the homotopy. For $\tau = 1$ we have the operator with constant coefficients in the boundary condition. It follows from Corollary 5.2 that for $\tau = 1$ the operator is invertible. Hence for $\tau = 0$ its index equals zero. The proposition is proved. □

Remark 5.7. If we do not assume that the vector $(a(s), b(s))$ has limits as $s \to \pm\infty$, then the operator is invertible for large positive λ. It follows from the ellipticity with a parameter.

Construction of domains 5.8. Consider a circle B_n with radius r_n and its center at $x_n = 0, y_n = r_n$. We translate the half-circle located at the half-plane $x \geq 0$ to the right at the distance $x = n$, and the half-circle located at the half-plane $x \leq 0$ to the left, also at the same distance $x = n$. We denote by Ω_n the domain bounded by the two half-circles and by the intervals $y = 0, -n \leq x \leq n$ and $y = 2r_n, -n \leq x \leq n$ (Figure 7).

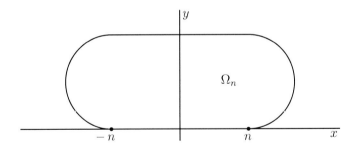

Figure 7: Domain Ω_n.

Theorem 5.9. *Consider problem (5.2), (5.3) in the half-plane $\mathbb{R}_+^2$. Suppose that there exist limits*

$$a_\pm = \lim_{s \to \pm\infty} a(s), \quad b_\pm = \lim_{s \to \pm\infty} b(s),$$

and $a_\pm \neq b_\pm$. If the vectors (a_-, b_-) and (a_+, b_+) belong to different half-planes in the sense of Definition 5.5, that is

$$(a_+ + b_+)(a_- + b_-) < 0,$$

then the index of the problem equals $2N+1$, where N is the number of rotations of the vector $(a(s), b(s))$ along the boundary from $-\infty$ to ∞ completed by the rotation from (a_+, b_+) to (a_-, b_-) counterclockwise.

Proof. Let us take s_0 with a sufficiently large modulus such that the vectors $(a(s_0), b(s_0))$ and $(a(-s_0), b(-s_0))$ belong to different half-planes. Let, for certainty,
$$(a(s_0), b(s_0)) \in \Pi_+, \quad (a(-s_0), b(-s_0)) \in \Pi_-.$$
Then there exists a function $\alpha_0(\theta)$ such that

$$\alpha_0(\theta) \neq \alpha_i(\theta), \quad 0 \leq \theta \leq 2\pi, \quad i = 1, 2, \tag{5.14}$$

$$\frac{3\pi}{4} < \alpha_0(0) < \frac{7\pi}{4}, \quad 0 \leq \alpha_0(2\pi) < \frac{3\pi}{4} \text{ or } \frac{7\pi}{4} < \alpha_0(2\pi) < 2\pi,$$

$$\tan(\alpha_0(0)) = \frac{a(s_0)}{b(s_0)}, \quad \tan(\alpha_0(2\pi)) = \frac{a(-s_0)}{b(-s_0)}. \tag{5.15}$$

Consider the domain Ω_{s_0} constructed in 5.8 with the lower boundary $y = 0, -s_0 \leq x \leq s_0$. Let z be a variable along the boundary. We define the functions $a(z), b(z)$ at the boundary $\partial \Omega_{s_0}$ as follows. At the lower part of the boundary they are the same as in the boundary conditions (5.3). At the remaining part of the boundary the vector $(a(z), b(z))$ forms the angle $\alpha_0(\theta(z))$ with the inner normal vector $\nu(z)$, where $\theta(z)$ is the angle between $\nu(z)$ and the vertical direction. Then

$$\tan \alpha_0(\theta(z)) = \frac{a}{b}.$$

Together with (5.15) this provides the continuity of the vector $(a(z), b(z))$ at the boundary of the domain Ω_{s_0}. On the other hand, from the last equality and (5.14) it follows that (5.11) does not hold.

Therefore, problem (5.2), (5.3) in the domain Ω_{s_0} with the constructed boundary condition satisfies the condition of ellipticity with a parameter at all points of the boundary $\partial \Omega_{s_0}$ for which $y > 0$. The rotation N of the vector $(a(z), b(z))$ along the boundary $\partial \Omega_{s_0}$ does not depend on s_0 if $|s_0|$ is sufficiently large. The index of the problem equals $2N + 1$.

It remains to verify that we can pass to the limit as s_0 goes to infinity and that the conditions of the stabilization of the index are satisfied. Consider the sequence of domains Ω_{s_j}, where $j \to \infty$. Let $x_j \in \Omega_{s_j}$ and $|x_j| \to \infty$. If the distance from the points x_j to the boundary $\partial \Omega_{s_j}$ tends to infinity, then the limiting domain is the whole $\mathbb{R}^2$. The corresponding limiting operator is invertible.

If for some sequence this distance remains bounded, then we denote by z_j the point of the boundary closest to x_j. If the points z_j belong to the upper part of the boundary, then the corresponding limiting domain is $\mathbb{R}^2_-$. The boundary conditions are constant, and the corresponding operator is invertible. If the points z_j belong to the right or to the left parts of the boundary (half-circles), we will denote by θ_j the corresponding angles. For each sequence θ_j converging to some limiting value θ we obtain a limiting problem in the half-plane $\mathbb{R}_\theta^2$ obtained from the half-plane $\mathbb{R}_-^2$ by rotation on the angle θ. The corresponding limiting problem has constant coefficients, and it is invertible. Therefore Conditions NS(seq) and NS*(seq) are satisfied and the results of Section 2 can be applied. The theorem is proved. □

6 Laplace operator with oblique derivative

In this section we consider the problem

$$\Delta u - \lambda^2 u = 0, \ \ a(x)\frac{\partial u}{\partial x} - b(x)\frac{\partial u}{\partial y} = h(x). \tag{6.1}$$

For $\lambda = 0$, it is a well-known problem for the Laplace operator with oblique derivative. In the case of bounded domains, its index is determined by the rotation of the vector $(c(x), d(x))$ along the boundary of the domain. It is independent of λ. For unbounded domains and $\lambda = 0$, the corresponding operator does not satisfy the Fredholm property because the limiting problems have a nonzero bounded solution u =const. We will consider problem (6.1) in unbounded domains with $\lambda \neq 0$ and will find its index.

Consider first problem (6.1) in the half-plane $\mathbb{R}_+^2 = \{(x, y) : -\infty < x < \infty, \ y \geq 0\}$ with constant a and b. We apply the partial Fourier transform with respect to x. For the homogeneous problem ($h = 0$) we have

$$\tilde{u}(\xi, y) = e^{-\sqrt{\xi^2 + \lambda^2}\, y}, \ \ y \geq 0.$$

It follows from the boundary condition that

$$\left(ai\xi + b\sqrt{\xi^2 + \lambda^2}\right)\tilde{u}(\xi, y) = 0. \tag{6.2}$$

We will suppose that $\lambda \neq 0$ and $b \neq 0$. Then the last equation has only a trivial solution for all real ξ. Therefore the corresponding operator is elliptic with a parameter.

Lemma 6.1. *Let a and b be some constants, $b \neq 0$. Then the operator $L = (A, B)$, where*

$$Au = \Delta u - \lambda^2 u = 0, \ \ Bu = a\frac{\partial u}{\partial x} - b\frac{\partial u}{\partial y},$$

$$L : H_\infty^2(\mathbb{R}_+^2) \to L_\infty^2(\mathbb{R}_+^2) \times H_\infty^{1/2}(\mathbb{R}^1)$$

is invertible for all real $\lambda \neq 0$.

Proof. We have verified that the operator L satisfies the condition of ellipticity with a parameter. Therefore it is invertible for large λ. On the other hand, it satisfies Condition NS for all $\lambda \neq 0$. Therefore, it is normally solvable with a finite-dimensional kernel. Hence, its index does not change when λ decreases from the large values for which the operator is invertible. Since the kernel of the operator is empty for all $\lambda \neq 0$, then it is invertible. The lemma is proved. □

Consider now the operator with variable coefficients and assume that they have limits at infinity,

$$a_\pm = \lim_{x \to \pm\infty} a(x), \quad b_\pm = \lim_{x \to \pm\infty} b(x).$$

If

$$b_\pm \neq 0 \tag{6.3}$$

then, as it is indicated in the proof of the previous lemma, the operator is normally solvable with a finite-dimensional kernel for all $\lambda \neq 0$. Therefore, its index is independent of λ. If $b(x) \neq 0$ for all $x \in \mathbb{R}^1$, then the operator satisfies the condition of ellipticity with a parameter. Hence it is invertible for large λ and, consequently, has the zero index for all $\lambda \neq 0$. However, if $b(x) = 0$ for some x, then the condition of ellipticity with a parameter is not satisfied. We will see that the index can be different from zero.

We will assume that

$$a^2(x) + b^2(x) \neq 0, \quad x \in \mathbb{R}^1. \tag{6.4}$$

If this condition is not satisfied at some $x = x_0$, then equation (6.2) with $a = a(x_0), b = b(x_0)$ has a nonzero solution, and the Lopatinskii condition is not satisfied.

We can consider homotopy classes of continuous curves $(a(x), b(x))$ under conditions (6.3), (6.4). Some examples are shown in Figure 8. Curves 1 and 3 belong to the same homotopy class. Curves 2 and 5 belong also to the same class, but to another one. All these curves can be retracted to a single point, that is to a problem with constant coefficients. Therefore the index of the corresponding

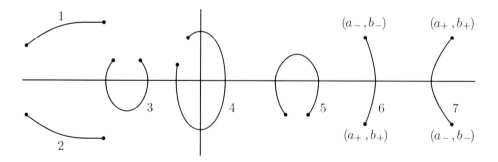

Figure 8: Examples of homotopy classes of curves $(a(x), b(x))$.

6. Laplace operator with oblique derivative

operator equals zero. Thus, there are at least two homotopy classes of operators with the zero index. All other curves belong to different homotopy classes. We will find the index of the corresponding operators approximating the half-plane by bounded domains.

Consider the problem

$$\Delta u - \lambda^2 u = 0, \quad a\frac{\partial u}{\partial \tau} - b\frac{\partial u}{\partial \nu} = 0 \tag{6.5}$$

in the half-plane $y_1 \geq 0$, where the coordinates (x_1, y_1) are obtained from (x, y) by rotation on the angle θ counterclockwise:

$$x = x_1 \cos\theta - y_1 \sin\theta, \quad y = x_1 \sin\theta + y_1 \cos\theta,$$

τ denotes the tangential vector to the boundary and ν the outer normal vector. Since the Laplace operator is invariant with respect to this change of variables, then problem (6.5) can be reduced to problem (6.1) (the y-direction in (6.1) corresponds to the inner normal vector). Therefore, the corresponding operator is invertible if $b \neq 0$.

Definition 6.2. Let $b_+ > 0, b_- > 0$ or $b_+ < 0, b_- < 0$ and I be the interval connecting the point (a_+, b_+) with the point (a_-, b_-). Denote by S the curve $(a(x), b(x))$, $x \in \mathbb{R}^1$ completed by the interval I and considered from (a_-, b_-) in the direction of growing values of x. Rotation κ of the vector $(a(x), b(x))$ is the number of rotations of the curve S around the origin with its direction taken into account. Positive direction is counterclockwise.

According to the definition, κ is an integer. It can be positive, negative or zero. In the examples in Figure 8, it is zero for the curves 1–3, 5; it can be $+1$ or -1 for the curve 4 depending on the direction along the curve. It is not defined for the curves 6 and 7.

Theorem 6.3. *Let conditions (6.3), (6.4) be satisfied and $b_+ > 0, b_- > 0$ or $b_+ < 0, b_- < 0$. Then the index of problem (6.1) equals 2κ.*

Proof. We note first of all that the index of problem (6.1) is understood as the difference between the number of linearly independent solutions of the homogeneous problem ($h = 0$) and the number of solvability conditions of the nonhomogeneous problem ($h \neq 0$). The index of the operator L introduced above (with variable coefficients) is not exactly the same because the right-hand side of the equation in (6.1) equals zero. However, the index of the operator and the index of the problem are equal to each other. This question is discussed in the next section in more detail.

We have discussed before that the operator L is normally solvable with a finite-dimensional kernel if $b_\pm \neq 0$. If we change the values of the coefficients at infinity in a continuous way with condition (6.3) being satisfied, then the index of the operator does not change. Hence we can reduce problem (6.1) to the problem for which $a_- = a_+ = 0, b_- = b_+$. Moreover, we can assume that $b(x) = b_+$ for $x \geq x_+$, $b(x) = b_-$ for $x \leq x_-$ for some $x_\pm$.

Consider the domains Ω_n, introduced in the previous section, assuming that $n \geq |x_\pm|$ and the problems

$$\Delta u - \lambda^2 u = 0, \quad a(s)\frac{\partial u}{\partial x} - b(s)\frac{\partial u}{\partial y} = h(s), \quad (x,y) \in \Omega_n \tag{6.6}$$

where s is a parameter along the boundary $\partial \Omega$. The coefficients $a(s), b(s)$ at the boundary are defined as follows. At the lower part of the boundary, where $y = 0$, they are equal to $a(x), b(x)$, respectively, for the corresponding values of x. At all other parts of the boundary we put

$$a(s) = b_\pm \cos\alpha(s), \quad b(s) = -b_\pm \sin\alpha(s),$$

where $\alpha(s)$ is the angle between the outer normal vector and the x-axis. This means that the derivative in the boundary condition is taken along the direction normal to the boundary of the domain. It is the outer normal derivative if $b_\pm > 0$ and the inner normal derivative if $b_\pm < 0$.

We verify that problems (6.6) satisfy Conditions NS(seq) and NS*(seq). All limiting operators have the form

$$\hat{L} = (\hat{A}, \hat{B}), \quad \hat{A}u = \Delta u - \lambda^2 u, \quad \hat{B}u = \frac{\partial u}{\partial \nu}$$

in half-planes $\mathbb{R}^2_\theta$, where $\theta \in [0, 2\pi]$ is the angle between the normal vector to the boundary and the vertical direction. Since $\lambda \neq 0$, then the equation $\hat{L}u = 0$ has only a trivial solution in the space $H^2_\infty(\mathbb{R}^2_\theta)$. Therefore Condition NS(seq) is satisfied.

The operator $\hat{L} : E_1 \to F_1$, where

$$E_1 = H^2_1(\mathbb{R}^2_\theta), \quad F_1 = L^2_1(\mathbb{R}^2_\theta) \times H^{1/2}_1(\mathbb{R}^2_\theta)$$

is invertible. Hence the adjoint operator $\hat{L}^* : (F_1)^* \to (E_1)^*$ is also invertible. From this it follows that the equation $\hat{L}^*v = 0$ has only a trivial solution in $(F_1)^*$. However, $(F_1)^* = (F^*)_\infty$. Thus, Condition NS*(seq) is satisfied.

From the results of Section 2 it follows that the index of problem (6.1) in the half-plane $\mathbb{R}^2_+$ equals the index of problems (6.6) for n sufficiently large. It remains to find the index of these problems in bounded domains. We note that the index of the problems in the bounded domains does not depend on λ. We put $\lambda = 0$ and obtain the Laplace operator with oblique derivative. Let γ be the rotation of the vector $(a(s), b(s))$ along the boundary. Then the index of problem (6.6) equals $2\gamma + 2$. On the other hand, $\gamma = \kappa - 1$. Therefore, the index of problem (6.6) equals 2κ. The lemma is proved. □

6. Laplace operator with oblique derivative

Examples 6.4. If the coefficients in the boundary condition in (6.1) are constant, then $\kappa = 0$, and the index of the problem equals zero. This follows from the invertibility of the corresponding operator.

For the curve 4 in Figure 8, $\kappa = 1$ or $\kappa = -1$ depending on the direction along the curve. Therefore the index of the problem equals ± 2. Theorem 6.3 is not applicable for the curves 6 and 7. In the next section we will show that in this case the index can be found by reduction of the Laplace operator to the Cauchy-Riemann system.

In the remaining part of this section we study the problem in an unbounded strip. Let
$$\Omega = \{(x, y),\ x \in \mathbb{R}^1,\ 0 \leq y \leq 1\}.$$
Consider the problem
$$\Delta u - \lambda^2 u = 0, \quad y = 0 : a(x)\frac{\partial u}{\partial x} - b(x)\frac{\partial u}{\partial y} = h(x),$$
$$y = 1 : c(x)\frac{\partial u}{\partial x} - d(x)\frac{\partial u}{\partial y} = g(x). \tag{6.7}$$

We begin with the case where $a, b, c,$ and d are constants. Applying the partial Fourier transform with respect to x, it can be easily verified that the homogeneous problem does not have nonzero bounded solutions if $b \neq 0$, $d \neq 0$.

In the case of the problem with variable coefficients we assume that there exist limits
$$a_\pm = \lim_{x \to \pm\infty} a(x),\quad b_\pm = \lim_{x \to \pm\infty} b(x),\quad c_\pm = \lim_{x \to \pm\infty} c(x),\quad d_\pm = \lim_{x \to \pm\infty} d(x)$$
and
$$a^2(x) + b^2(x) \neq 0,\quad c^2(x) + d^2(x) \neq 0,\quad b_\pm \neq 0,\quad d_\pm \neq 0. \tag{6.8}$$
Then the limiting operators are invertible, Conditions NS and NS* are satisfied, and the operator corresponding to problem (6.7) satisfies the Fredholm property. Moreover, the index of the problem does not change under a continuous deformation of the coefficients of the operator in such a way that (6.8) is satisfied. Therefore, we can reduce problem (6.7) to the case where
$$a_\pm = 0,\quad c_\pm = 0,\quad |b_-| = |d_-|,\quad |b_+| = |d_+|.$$

Theorem 6.5. *Suppose that*
$$b_- = -d_-,\quad b_+ = -d_+. \tag{6.9}$$
Then the index of problem (6.7) equals $2(\kappa_+ - \kappa_-)$, where κ_+ is the rotation of the vector $(a(x), b(x))$ and κ_- is the rotation of the vector $(c(x), d(x))$.

Remark 6.6. We have defined the rotation of the vector $(a(x), b(x))$ in the case where b_+ and b_- have the same sign (Definition 6.2). We do not impose this condition in this theorem. If b_- and b_+ have opposite signs, then the rotation is not integer. In this case we put $\kappa_+ = \kappa_0 \pm 1/2$, where κ_0 is the rotation of the vector $(a(x), b(x))$ from $x = -\infty$ to the maximal value $x = x_0$ for which $b(x)$ has the same sign as b_-. We add or subtract $1/2$ depending on the direction of rotation from x_0 to $+\infty$ ($+$ if the rotation is counterclockwise).

The rotation of the vector $(c(x), d(x))$ is defined similarly. If b_- and b_+ have opposite signs, then, by virtue of condition (6.9), it is also true for d_- and d_+. Hence, κ_+ and κ_- are both integer or both non-integer. Their difference is always integer, and the index is even. We can conjecture that if (6.9) is not satisfied, then the index may be also odd.

Proof. We approximate the infinite strip by the domains Ω_n similar to those constructed in the previous section. The difference is that in this case their height is constant and equals 1. Thus, the lower part of the boundary of the domain Ω_n is located at the straight line $y = 0$, the upper part of the boundary at $y = 1$. The left and the right parts of the boundary are half-circles.

The boundary condition at the boundary of the domain Ω_n is defined as follows. At the lower and upper parts of the boundary the coefficients in the boundary condition are the same as for problem (6.7). At the left and at the right parts of the boundary the derivative in the boundary condition is taken in the direction normal to the boundary, inner or outer depending on the sign of the coefficients. We use here condition (6.9) without which the vectors at the lower and at the upper parts of the boundary could not be connected without passing through the tangential direction.

We should verify Conditions NS(seq) and NS*(seq). There are three limiting domains: the whole cylinder, the left half-cylinder and the right half-cylinder. In all cases the derivative in the boundary condition is in the normal direction. Therefore, the corresponding operators are invertible ($\lambda \neq 0$), and the required conditions are satisfied.

Thus, we can apply the theorem on the stabilization of the index. The index of problem (6.7) equals the index of the problem in the domain Ω_n for n sufficiently large. The latter equals $2(\kappa_+ - \kappa_-)$. The theorem is proved. □

7 Cauchy-Riemann system and Laplace operator

7.1 Reduction of the problems to each other

In Section 5 we found the index of the Cauchy-Riemann problem in the case where the limiting values (a_-, b_-) and (a_+, b_+) of the coefficients in the boundary condition at $\pm\infty$ belong to different half-planes in the sense of Definition 5.5. On the other hand, in the previous section we found the index for the Laplace operator

7. Cauchy-Riemann system and Laplace operator

with oblique derivative in the case where (a_-, b_-) and (a_+, b_+) belong to the same half-plane. In this section we will show how the indices of these two problems are related to each other. It will allow us to find the index for both problems for all values of vectors (a_-, b_-) and (a_+, b_+) for which the index is defined.

In this section we establish an inter-connection between two problems:

$$\frac{\partial u}{\partial x} - \frac{\partial v}{\partial y} - \lambda u = 0, \quad \frac{\partial u}{\partial y} + \frac{\partial v}{\partial x} + \lambda v = 0, \tag{7.1}$$

$$r(x)u + s(x)v = h(x) \tag{7.2}$$

and

$$\Delta w - \lambda^2 w = 0, \quad r(x)\frac{\partial w}{\partial x} - s(x)\frac{\partial w}{\partial y} = h(x) \tag{7.3}$$

considered in the half-plane $\mathbb{R}_+^2 = \{(x,y) : -\infty < x < \infty, \ y \geq 0\}$. If $\lambda = 0$, the first one is the Hilbert problem for the Cauchy-Riemann system, the second one is the Laplace operator with oblique derivative. If we put $u = \partial w/\partial x$, $v = -\partial w/\partial y$, then we obtain (7.1), (7.2) from (7.3). However, if $\lambda \neq 0$, this transition cannot be made directly and requires several intermediate steps.

We begin with the operator $L_1 = (A_1^1, A_1^2, B_1)$, where

$$A_1^1(u,v) = \frac{\partial u}{\partial x} - \frac{\partial v}{\partial y} - \lambda u, \quad A_1^2(u,v) = \frac{\partial u}{\partial y} + \frac{\partial v}{\partial x} + \lambda v, \quad B_1(u,v) = r(x)u + s(x)v,$$

$$L_1 : (E^1)_\infty = (W_\infty^{1,2}(\mathbb{R}_+^2))^2 \to (F^1)_\infty = (L_\infty^2(\mathbb{R}_+^2))^2 \times W_\infty^{1/2,2}(\mathbb{R}_+^2).$$

Assume that the coefficients of the operator have limits at infinity,

$$r(x) \to r_\pm, \quad s(x) \to s_\pm, \quad x \to \pm\infty,$$

and $r_\pm + s_\pm \neq 0$. Then for positive λ the limiting problems $L_1^\pm(u,v) = 0$ have only trivial solutions in the space $(E^1)_\infty$. Therefore, Condition NS is satisfied and the operator L_1 is normally solvable. By virtue of the ellipticity with a parameter, for large λ the limiting operators $L_1^\pm$ are invertible as acting from $(E^1)_1$ into $(F^1)_1$. From normal solvability it follows that they are invertible for all $\lambda > 0$. Hence, the adjoint operator $(L_1^\pm)^* : ((F^1)_1)^* \to ((E^1)_1)^*$ is also invertible. Consequently, the homogeneous adjoint equation $(L_1^\pm)^*(\cdot) = 0$ has only a trivial solution in the space $((F^1)_1)^* = ((F^1)^*)_\infty$. Thus, Condition NS* is satisfied, and the operator L_1 satisfies the Fredholm property. Its index is defined and the solvability conditions of the equation

$$A_1^1(u,v) = p, \quad A_1^2(u,v) = q, \quad B_1(u,v) = h \tag{7.4}$$

are given by the equalities $\phi_j(p,q,h) = 0$, where $\phi_j \in ((F^1)_\infty)^*$ are linearly independent solutions of the equation $(L_1)^*\phi = 0$.

Consider the functionals $\phi_j = (\phi_j^1, \phi_j^2, \phi_j^3)$ and the functionals $\psi_j = (0, 0, \phi_j^3)$. Suppose that there are N linearly independent functionals ϕ_j and M linearly independent functionals ψ_j. We show that $M = N$. Indeed, the inequality $N \geq M$ is obvious. If $N > M$, then there exists a linear combination

$$\Psi = \sum_{j=1}^{N} c_j \phi_j$$

such that $\Psi = (\Psi^1, \Psi^2, 0)$, where at least one of the functionals Ψ^1 and Ψ^2 is different from zero. Hence, there exist functions p and q such that $\Psi(p, q, h) \neq 0$ for any h. System (7.4) is not solvable for such p, q and for any h. We will show that this conclusion cannot hold. Consider some extensions $\tilde{p}$ and $\tilde{q}$, respectively, of the functions p and q to $L^2(\mathbb{R}^2)$. The system

$$A_1^1(u,v) = \tilde{p}, \quad A_1^2(u,v) = \tilde{q}$$

is solvable in $W_\infty^{1,2}(\mathbb{R}^2)$. Denote by $(\tilde{u}, \tilde{v})$ its solution and put $\hat{u} = u - \tilde{u}, \hat{v} = v - \tilde{v}$. Then $(\hat{u}, \hat{v})$ satisfies the problem

$$A_1^1(\hat{u}, \hat{v}) = 0, \quad A_1^2(\hat{u}, \hat{v}) = 0, \quad B_1(\hat{u}, \hat{v}) = h - ru_0 - sv_0,$$

where u_0 is the trace of the function $\tilde{u}$ and v_0 the trace of $\tilde{v}$. There exists some h such that the last problem is solvable. This contradiction shows that $M = N$.

Lemma 7.1. *The problem*

$$\frac{\partial u}{\partial x} - \frac{\partial v}{\partial y} - \lambda u = 0, \quad \frac{\partial u}{\partial y} + \frac{\partial v}{\partial x} + \lambda v = 0, \quad (7.5)$$

$$r(x)u + s(x)v = h, \quad (7.6)$$

is solvable if and only if $\psi_j(h) = 0$, $j = 1, \ldots, N$, where ψ_j are linearly independent functionals.

Proof. We have proved before the lemma that the functionals ψ_j are linearly independent. Since problem (7.4) is solvable if and only if $\phi_j(p, q, h) = 0$, then problem (7.5), (7.6) is solvable if and only if $\psi_j(h) = 0$. The lemma is proved. □

Corollary 7.2. *The index of the operator L_1 equals the index of problem (7.5), (7.6), that is the difference between the number of linearly independent solutions of the homogeneous problems and the number of solvability conditions of the non-homogeneous problem.*

Next, we consider the problem

$$\frac{\partial f}{\partial x} + \frac{\partial g}{\partial y} + \lambda f = 0, \quad -\frac{\partial f}{\partial y} + \frac{\partial g}{\partial x} - \lambda g = 0, \quad (7.7)$$

$$c(x)f + d(x)g = h. \quad (7.8)$$

It can be obtained from problem (7.5), (7.6) if we replace x by $-x$ and put $c(x) = r(-x)$, $d(x) = s(-x)$.

7. Cauchy-Riemann system and Laplace operator

Lemma 7.3. *Consider the problem*

$$\frac{\partial u}{\partial x} - \frac{\partial v}{\partial y} - \lambda u = f, \quad \frac{\partial u}{\partial y} + \frac{\partial v}{\partial x} + \lambda v = g, \tag{7.9}$$

$$au + bv = 0, \tag{7.10}$$

in the half-plane $\mathbb{R}_+^2$. *Here* a *and* b *are constants,* $a + b \neq 0$, (f, g) *is a solution of problem* (7.7), (7.8). *There exists a one-to-one correspondence between solutions of problem* (7.7)–(7.10) *and of problem* (7.7), (7.8).

Proof. It is sufficient to note that problem (7.9), (7.10) is uniquely solvable for any f and g. The lemma is proved. □

Substituting f and g from (7.9) into (7.7), (7.8) we obtain:

$$\Delta u - \lambda^2 u = 0, \tag{7.11}$$

$$\Delta v - \lambda^2 v = 0, \tag{7.12}$$

$$c\left(\frac{\partial u}{\partial x} - \frac{\partial v}{\partial y} - \lambda u\right) + d\left(\frac{\partial u}{\partial y} + \frac{\partial v}{\partial x} + \lambda v\right) = h. \tag{7.13}$$

Corollary 7.4. *The index of problem* (7.7)–(7.10) *equals the index of problem* (7.7), (7.8). *The index of problem* (7.7)–(7.10) *is understood as the difference between the number of linearly independent solutions of the problem with $h = 0$ and the number of solvability conditions on the function h.*

Lemma 7.5. *There exists a one-to-one correspondence between solutions of problem* (7.7)–(7.10) *and of problem* (7.10)–(7.13).

Proof. Let (u, v, f, g) be a solution of problem (7.7)–(7.10). Substituting f and g from (7.9) into (7.7), (7.8), we obtain (7.11)–(7.13). We note that f and g belong to $W_\infty^{1,2}(\mathbb{R}_+^2)$ as a solution of (7.7), (7.8). Therefore, u and v belong to $W_\infty^{2,2}(\mathbb{R}_+^2)$ as a solution of (7.9), (7.10). Hence, expressions (7.11)–(7.13) are well defined.

Let (u, v) be a solution of problem (7.10)–(7.13). Put

$$f = \frac{\partial u}{\partial x} - \frac{\partial v}{\partial y} - \lambda u, \quad g = \frac{\partial u}{\partial y} + \frac{\partial v}{\partial x} + \lambda v.$$

Then

$$\frac{\partial f}{\partial x} + \frac{\partial g}{\partial y} + \lambda f = \Delta u - \lambda^2 u = 0,$$

$$-\frac{\partial f}{\partial y} + \frac{\partial g}{\partial x} - \lambda g = \Delta v - \lambda^2 v = 0,$$

$$cf + dg = c\left(\frac{\partial u}{\partial x} - \frac{\partial v}{\partial y} - \lambda u\right) + d\left(\frac{\partial u}{\partial y} + \frac{\partial v}{\partial x} + \lambda v\right) = h.$$

The lemma is proved. □

Corollary 7.6. *The indices of problems (7.7)–(7.10) and (7.10)–(7.13) are equal to each other.*

We introduce the operator L_2 corresponding to problem (7.10)–(7.13):

$$L_2 = (A_2^1, A_2^2, B_2^1, B_2^2),$$

$$L_2 : (E^2)_\infty = (W_\infty^{2,2}(\mathbb{R}_+^2))^2 \to (F^2)_\infty$$

$$= (L_\infty^2(\mathbb{R}_+^2))^2 \times W_\infty^{3/2,2}(\mathbb{R}_+^2) \times W_\infty^{1/2,2}(\mathbb{R}_+^2),$$

$$A_2^1(u,v) = \Delta u - \lambda^2 u, \quad A_2^2(u,v) = \Delta v - \lambda^2 v,$$

$$B_2^1(u,v) = au + bv,$$

$$B_2^2(u,v) = c\left(\frac{\partial u}{\partial x} - \frac{\partial v}{\partial y} - \lambda u\right) + d\left(\frac{\partial u}{\partial y} + \frac{\partial v}{\partial x} + \lambda v\right).$$

As above, we assume that a and b are some constants, $a + b \neq 0$, $c(x)$ and $d(x)$ have limits at infinity, $c_\pm + d_\pm \neq 0$.

We prove first of all that the operator L_2 satisfies the Fredholm property. We use the same approach as for the operator L_1. Consider the limiting operators $L_2^\pm$. Since the limiting problems $L_2^\pm(u,v) = 0$ are equivalent to the limiting problems for problem (7.7)–(7.10) with $h = 0$, then they have only a trivial solution for all positive λ. Therefore, the operator L_2 satisfies Condition NS for all $\lambda > 0$.

Next, we verify that the limiting operators satisfy the condition of ellipticity with a parameter. Consider the problem

$$\Delta u - \lambda^2 u = p, \quad \Delta v - \lambda^2 v = q$$

in $\mathbb{R}^2$. Applying the Fourier transform we obtain the unique solvability of this system for any ξ and λ, $|\xi| + |\lambda| = 1$. For the problem

$$A_2^1(u,v) = 0, \quad A_2^2(u,v) = 0, \quad B_2^1(u,v) = h^1, \quad B_2^2(u,v) = h^2$$

in the half-plane $\mathbb{R}_+^2$ we apply the partial Fourier transform with respect to x. We have

$$\tilde{u} = k_1 e^{-\sqrt{\xi^2 + \lambda^2}\, y}, \quad \tilde{v} = k_2 e^{-\sqrt{\xi^2 + \lambda^2}\, y}.$$

We substitute these expressions into the boundary conditions:

$$\begin{pmatrix} a & b \\ ci\xi + d\sqrt{\xi^2 + \lambda^2} - c\lambda & di\xi + c\sqrt{\xi^2 + \lambda^2} + d\lambda \end{pmatrix} \begin{pmatrix} \tilde{u} \\ \tilde{v} \end{pmatrix} = \begin{pmatrix} \tilde{h}^1 \\ \tilde{h}^2 \end{pmatrix}. \quad (7.14)$$

The determinant of the matrix in the left-hand side satisfies

$$\det = ad(i\xi + \lambda) - bc(i\xi - \lambda) + (bd + ac)\sqrt{\xi^2 + \lambda^2} \neq 0, \quad \forall \xi, \ \lambda > 0.$$

Therefore, the condition of ellipticity with a parameter is satisfied, and the limiting operators $L_2^\pm$ are invertible for large positive λ. From this and from Condition NS

7. Cauchy-Riemann system and Laplace operator

it follows that it is a Fredholm operator with the zero index for all positive λ. Since its kernel is empty, then it is invertible for all $\lambda > 0$. Hence, this is also true for the adjoint operator, which implies Condition NS*. From Conditions NS and NS* it follows that the operator L_2 is Fredholm.

Lemma 7.7. *The index of the operator L_2 equals the index of problem (7.10)–(7.13).*

Proof. The homogeneous equation $L_2(u,v) = 0$ coincides with problem (7.10)–(7.13) with $h = 0$. Therefore the dimensions of their kernels are the same.

It remains to verify that the codimensions of the images are also the same. We prove this in the same way as for the operator L_1. Let ϕ_j, $j = 1, \ldots, N$ be linearly independent solutions of the homogeneous adjoint equation $L_2^* \phi = 0$. They can be represented in the form $\phi_j = (\phi_j^1, \phi_j^2, \phi_j^3, \phi_j^4)$. Consider the functionals $\psi_j = (0, 0, 0, \phi_j^4)$. Let M be the number of them that are linearly independent. Then $N \geq M$.

We will show that $N = M$. Indeed, if $N > M$, then there exists a linear combination

$$\Psi = \sum_{j=1}^{N} c_j \phi_j$$

such that $\Psi = (\phi_j^1, \phi_j^2, \phi_j^3, 0)$, where at least one of the functionals $\phi_j^1, \phi_j^2, \phi_j^3$ is different from zero. Therefore, there exist some p, q, h^1 such that the problem

$$A_2^1(u,v) = p, \quad A_2^2(u,v) = q, \quad B_2^1(u,v) = h^1, \quad B_2^2(u,v) = h^2 \qquad (7.15)$$

is not solvable for any h^2.

Denote by (u_0, v_0) a solution of the problem

$$\Delta u - \lambda^2 u = p, \quad \Delta v - \lambda^2 v = q, \quad au + bv = h^1.$$

It can be constructed as follows. We extend the function $p(x)$ to $L^2(\mathbb{R}^2)$ and denote by u_0 the solution of the equation

$$\Delta u - \lambda^2 u = p, \quad x \in \mathbb{R}^2.$$

Then v_0 is a solution of the problem

$$\Delta v - \lambda^2 v = q, \quad bv = h^1 - a\hat{u},$$

where $\hat{u}$ is the trace of the function u_0.

The functions $\tilde{u} = u - u_0$, $\tilde{v} = v - v_0$ satisfy the problem

$$A_2^1(\tilde{u}, \tilde{v}) = 0, \quad A_2^2(\tilde{u}, \tilde{v}) = 0, \quad B_2^1(\tilde{u}, \tilde{v}) = 0, \quad B_2^2(\tilde{u}, \tilde{v}) = h^2 - B_2^2(u_0, v_0).$$

The function h^2 can be chosen in such a way that this problem has a solution. Hence $u = u_0 + \tilde{u}$, $v = v_0 + \tilde{v}$ is a solution of problem (7.15). We obtain a contradiction with the assumption that it is not solvable for any h^2. This contradiction proves that $M = N$.

Thus, there are N linearly independent functionals ψ_j. Since problem (7.15) is solvable if and only if $\phi_j(p, q, h^1, h^2) = 0, j = 1, \ldots, N$ and

$$\phi_j(0, 0, 0, h^2) = \psi_j(h^2),$$

then problem (7.10)–(7.13) is solvable if and only if $\psi_j(h^2) = 0$. The lemma is proved. □

We introduce an operator L^τ which depends on the parameter $\tau \in [0, 1]$. It acts in the same spaces as the operator L_2 and is given by the following relations:

$$L^\tau = (A_2^1, A_2^2, B_2^1, \hat{B}_\tau),$$
$$A_2^1(u, v) = \Delta u - \lambda^2 u, \quad A_2^2(u, v) = \Delta v - \lambda^2 v, \quad B_2^1(u, v) = au + bv,$$
$$\hat{B}_\tau(u, v) = c\left(\frac{\partial u}{\partial x} - \left(\tau \frac{\partial v}{\partial y} - (1-\tau)\frac{a}{b}\frac{\partial u}{\partial y}\right) - \lambda u\right) + d\left(\frac{\partial u}{\partial y} + \frac{\partial v}{\partial x} + \lambda v\right).$$

It coincides with the operator L_2 for $\tau = 1$ and with the operator

$$\hat{L}_2 = (A_2^1, A_2^2, B_2^1, \hat{B}_2)$$

for $\tau = 0$. Here

$$\hat{B}_2 = c\left(\frac{\partial u}{\partial x} + \frac{a}{b}\frac{\partial u}{\partial y} - \lambda u\right) + d\left(\frac{\partial u}{\partial y} + \frac{\partial v}{\partial x} + \lambda v\right).$$

Similar to the operator L_2, the operator L^τ satisfies Condition NS. Instead of the matrix in (7.14), here we have another matrix but with the same determinant:

$$\det \begin{pmatrix} a & b \\ ci\xi - ((1-\tau)c\frac{a}{b} + d)\sqrt{\xi^2 + \lambda^2} - c\lambda & di\xi + \tau c\sqrt{\xi^2 + \lambda^2} + d\lambda \end{pmatrix}$$
$$= ad(i\xi + \lambda) - bc(i\xi - \lambda) + (bd + ac)\sqrt{\xi^2 + \lambda^2} \neq 0, \quad \forall \xi, \; \lambda > 0.$$

Since the operator L_2 is Fredholm, the operator L^τ is normally solvable with a finite-dimensional kernel, then the operator $\hat{L}_2$ is also Fredholm. Its index equals the index of the operator L_2.

Lemma 7.8. *The problem*

$$\Delta u - \lambda^2 u = 0, \quad \Delta v - \lambda^2 v = 0, \tag{7.16}$$
$$au + bv = 0, \tag{7.17}$$
$$c\left(\frac{\partial u}{\partial x} - \mu\frac{\partial u}{\partial y} - \lambda u\right) + d\left(\frac{\partial u}{\partial y} + \mu\frac{\partial u}{\partial x} + \mu\lambda u\right) = h, \tag{7.18}$$

where $\mu = -a/b$, has the same index as the operator $\hat{L}_2$.

Proof. The proof is similar to the proof of the previous lemma. We use the relation $v = \mu u$ in the boundary condition. The lemma is proved. □

7. Cauchy-Riemann system and Laplace operator

Lemma 7.9. *There exists a one-to-one correspondence between solutions of problem* (7.16)–(7.18) *and of the problem*

$$\Delta u - \lambda^2 u = 0, \tag{7.19}$$

$$(c + d\mu)\frac{\partial u}{\partial x} + (d - c\mu)\frac{\partial u}{\partial y} + \lambda(-c + d\mu)u = h. \tag{7.20}$$

Proof. Obviously, any solution of problem (7.16)–(7.18) provides a solution of problem (7.19), (7.20). Inversely, let u be a solution of problem (7.19), (7.20). Then we can find a unique v that satisfies the problem

$$\Delta v - \lambda^2 v = 0, \quad v = -\frac{a}{b}u.$$

Then (u, v) is a solution of problem (7.16)–(7.18). The lemma is proved. $\square$

Along with problem (7.19), (7.20) we consider the problem

$$\Delta u - \lambda^2 u = 0, \tag{7.21}$$

$$(c + d\mu)\frac{\partial u}{\partial x} + (d - c\mu)\frac{\partial u}{\partial y} + \tau\lambda(-c + d\mu)u = h, \tag{7.22}$$

where $\tau \in [0, 1]$. For $\tau = 1$ we have problem (7.19), (7.20). For $\tau = 0$ we obtain the problem with oblique derivative. We will find conditions on c and d such that the problem satisfies Condition NS during this deformation. In fact, we need to introduce the operator that corresponds to problem (7.19), (7.20) and to show that it has the same index. Then we construct a continuous deformation of this operator in such a way that Condition NS is satisfied for all values of the parameter τ. Then we can conclude that the index is preserved. After that, instead of the operator with $\tau = 0$ we consider the corresponding problem (7.21), (7.22) and show that it has the same index. Since these intermediate steps are the same as before for the operator L_2^τ, we will work directly with problem (7.21), (7.22).

We consider the limiting values of c and d at $\pm\infty$ and will keep for them the same notation. We should verify that problem (7.21), (7.22) with constant coefficients and with $h = 0$ does not have nonzero solutions. We apply the partial Fourier transform with respect to x. We find from (7.21):

$$\tilde{u}(\xi, y) = k e^{-\sqrt{\xi^2 + \lambda^2}\, y}.$$

The problem will have a nonzero bounded solution if for some real ξ,

$$(c + d\mu)\, i\xi - \sqrt{\xi^2 + \lambda^2}(d - c\mu) + \tau\lambda(-c + d\mu) = 0. \tag{7.23}$$

We will find the value of μ for which this equality does not hold for any ξ, $\lambda > 0$, and $\tau \in [0, 1]$. We recall that $\mu = -a/b$, where a and b some constants.

If $\xi \neq 0$, then $c = -d\mu$. Therefore, since $d \neq 0$ (otherwise, $d = c = 0$), then

$$\frac{\sqrt{\xi^2 + \lambda^2}}{\lambda} = \frac{2\tau\mu}{1+\mu^2}.$$

The left-hand side of this equality is greater than or equal to 1. The right-hand side is less than or equal to 1. The equality can take place only for $\tau = 1$ and $\mu = 1$. We suppose however that $\mu \neq 1$. Therefore it cannot hold for any real ξ and positive λ.

Consider now the case $\xi = 0$. Then from (7.23)

$$-(d - c\mu) + \tau(-c + d\mu) = 0. \tag{7.24}$$

If we take for example $\mu = -1$, then this equality cannot hold for any $\tau \in [0, 1]$ since $c + d \neq 0$. We will use this example below.

We have proved the following lemma.

Lemma 7.10. *For any limits $c_\pm, d_\pm$ of the coefficients $c(x)$ and $d(x)$ at infinity, $c_\pm + d_\pm \neq 0$, there exists $\mu \neq 1$ such that $d_\pm - c_\pm\mu \neq 0$ and problem (7.19), (7.20) has the same index as the problem*

$$\Delta u - \lambda^2 u = 0, \tag{7.25}$$

$$(c + d\mu)\frac{\partial u}{\partial x} + (d - c\mu)\frac{\partial u}{\partial y} = h. \tag{7.26}$$

We note that the vector

$$(c + d\mu, d - c\mu) = (c, d) + \mu(d, -c)$$

is obtained from the vector (c, d) by rotation on a constant angle. Therefore the rotation of these two vectors along the boundary is the same. Hence the index of problem (7.25), (7.26) is the same as for the problem

$$\Delta u - \lambda^2 u = 0, \quad c(x)\frac{\partial u}{\partial x} + d(x)\frac{\partial u}{\partial y} = h. \tag{7.27}$$

In this case we should assume additionally that $d_\pm \neq 0$.

We have proved the following theorem.

Theorem 7.11. *Suppose that there exist the limits*

$$c_\pm = \lim_{x \pm \infty} c(x), \quad d_\pm = \lim_{x \pm \infty} d(x) \quad \text{and} \quad c_\pm + d_\pm \neq 0, \quad d_\pm \neq 0.$$

Then the index of problem (7.7), (7.8) equals the index of problem (7.27).

Corollary 7.12. *The indices of problems (7.1), (7.2) and (7.3) are equal to each other.*

Proof. It is sufficient to replace x by $-x$ in both problems and put $r(x) = c(-x)$, $s(x) = d(-x)$. □

7.2 Formula for index

In Section 5 we have found the index of problem (5.2), (5.3) in the half-plane assuming that the quantities $(a_- + b_-)$ and $(a_+ + b_+)$ are different from zero and have opposite signs. Its index equals $2N + 1$, where N is the number of rotations of the vector $(a(x), b(x))$ for x from $-\infty$ to ∞ completed by the rotation from $(a_+ + b_+)$ to $(a_- + b_-)$ counterclockwise. Some examples are shown in Figure 9: $N = 0$ for the curve 1 and $N = 1$ for the curve 2.

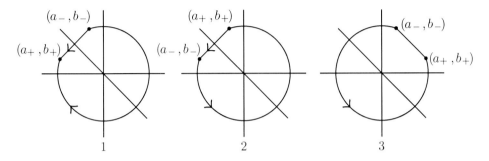

Figure 9: Rotation of the vector $(a(x), b(x))$ in the boundary condition of problem (5.2), (5.3).

The results of Section 5 are not applicable for the third example in the same figure since $(a_- + b_-)$ and $(a_+ + b_+)$ have the same sign. In this case we use the results of the present section that allow us to reduce problem (5.2), (5.3) to problem (6.1) with the same rotation of the vector in the boundary condition. They have the same index. Therefore it is sufficient to find the index of problem (6.1). For this we need to verify that conditions of Theorem 6.3 are satisfied: the limits $b_\pm$ of the function $b(x)$ in (6.1) should have the same signs.

Therefore we have two conditions: $(a_- + b_-)$ and $(a_+ + b_+)$ have the same sign; b_- and b_+ have the same sign. The second condition does not follow of course from the first one. It is satisfied for the example in Figure 9 but it is not necessarily the case. Hence the reduction to problem (6.1) can be done but Theorem 6.3 may not be directly applicable.

We recall that when we reduce problem (7.7), (7.8) to problem (7.25), (7.26), there is an additional parameter μ that appears in the latter problem. It should satisfy certain conditions in order for the problems to have the same index. In particular, it was indicated above that these conditions were satisfied if $\mu = -1$. Introduction of this value of μ in the boundary condition (7.26) signifies that the vector $(c(x), d(x))$ is turned through an angle $\pi/4$ counterclockwise. This means that the third curve in Figure 9 will also be turned through the same angle. Since $(a_- + b_-)$ and $(a_+ + b_+)$ are different from zero and have the same sign, then after this rotation we will obtain that b_- and b_+ are also different from zero and have the same sign. Hence we can apply Theorem 6.3. For the example in Figure 9, the index equals 2.

Thus, we can find the index of the Cauchy-Riemann system directly (approximating by bounded domains) if the index is odd. In the case of an even index we can do it by reduction to the Laplace operator. Similarly, the index of problem (6.1) can be found directly if it is even. If it is odd, it can be found by reduction to the Cauchy-Riemann system (examples 6 and 7 in Figure 8).

8 General first-order systems on the plane

8.1 Reduction to canonical systems

Consider the problem

$$Au \equiv a(x)\frac{\partial u}{\partial x_1} - \frac{\partial u}{\partial x_2} + c(x)u = f, \quad x \in \Omega, \tag{8.1}$$

$$Bu \equiv b(x)u = g, \quad x \in \Gamma. \tag{8.2}$$

Here Ω is a bounded simply connected domain with a sufficiently smooth boundary Γ, $a(x)$ is a real square matrix of the order $2r$, $a \in C^l(\bar{\Omega})$, $b(x)$ is a real $r \times 2r$ matrix. Problem (8.1), (8.2) is supposed to be elliptic.

Theorem 8.1. *There exists a square matrix $P(x)$ of the order $2r$, $P(x) \in C^l(\bar{\Omega})$ invertible for all $x \in \bar{\Omega}$ and such that its last r rows are complex conjugate for the first r rows and the equality*

$$P(x)a(x)P^{-1}(x) = \begin{pmatrix} a_0(x) & 0 \\ 0 & \overline{a_0(x)} \end{pmatrix} \tag{8.3}$$

holds, where $a_0(x)$ is a square complex matrix, for which all eigenvalues for any $x \in \bar{\Omega}$ have positive imaginary parts.

The proof of this theorem is given in [547].

Put

$$\tilde{a}(x) = \begin{pmatrix} a_0(x) & 0 \\ 0 & \overline{a_0(x)} \end{pmatrix}.$$

We set $v = Pu$. Then problem (8.1), (8.2) takes the form

$$\tilde{a}\frac{\partial v}{\partial x_1} - \frac{\partial v}{\partial x_2} + \tilde{a}_0 v = \tilde{f}, \quad x \in \Omega, \tag{8.4}$$

$$\tilde{b}(x)u = g, \quad x \in \Gamma, \tag{8.5}$$

where

$$\tilde{a}(x) = P(x)a(x)P^{-1}(x), \quad \tilde{b}(x) = b(x)P^{-1}(x), \quad \tilde{f} = Pf,$$

$$\tilde{a}_0(x) = P\left(a\frac{\partial P^{-1}}{\partial x_1} - \frac{\partial P^{-1}}{\partial x_2} + cP^{-1}\right).$$

8. General first-order systems on the plane

We will call problem (8.4), (8.5) canonical.

Proposition 8.2. *Canonical problem* (8.4), (8.5) *remains canonical after an orthogonal change of the independent variables.*

Proof. Consider the change of variables

$$y_1 = \alpha_1 x_1 + \alpha_2 x_2, \quad y_2 = -\alpha_2 x_1 + \alpha_1 x_2,$$

where $\alpha_1 = \cos\theta$, $\alpha_2 = \sin\theta$. Then system (8.4) takes the form

$$\hat{a}(y)\frac{\partial v}{\partial x_1} - \frac{\partial v}{\partial x_2} + \cdots,$$

where

$$\hat{a}(y) = \begin{pmatrix} c & 0 \\ 0 & \bar{c} \end{pmatrix}, \quad c = (a_0\alpha_2 + \alpha_1)^{-1}(a_0\alpha_1 - \alpha_2).$$

We have to prove that all eigenvalues of the matrix c have positive imaginary parts. We suppose that x and θ are fixed. Suppose, first, that all eigenvalues of the matrix a_0 are different from each other. Let T be such that

$$Ta_0 T^{-1} = \begin{pmatrix} \lambda_1 & \cdots & 0 \\ \cdots & \cdots & \cdots \\ 0 & \cdots & \lambda_r \end{pmatrix}.$$

We have $\operatorname{Im}\lambda_k \geq \rho > 0$, $k = 1, \ldots, r$ for some ρ,

$$TcT^{-1} = \begin{pmatrix} \phi(\lambda_1) & \cdots & 0 \\ \cdots & \cdots & \cdots \\ 0 & \cdots & \phi(\lambda_r) \end{pmatrix}, \quad \phi(\lambda) = \frac{\lambda\alpha_1 - \alpha_2}{\lambda\alpha_2 + \alpha_1}.$$

Since $\alpha_1^2 + \alpha_2^2 = 1$, then

$$\operatorname{Im}\phi(\lambda_k) = \frac{\operatorname{Im}\lambda_k}{|\lambda_k\alpha_2 + \alpha_1|^2} > \frac{\rho}{|\lambda_k\alpha_2 + \alpha_1|^2} > 0.$$

Hence the proposition is proved in the case where all eigenvalues of the matrix a_0 are different from each other. The general case can be reduced to this one by a small perturbation. The proposition is proved. □

Theorem 8.3. *The Lopatinskii condition for canonical system* (8.4), (8.5) *does not depend on the matrix $\tilde{a}$ and on the normal vector to the boundary Γ. It has the form*

$$\det \tilde{b}_1(x) \neq 0, \quad \det \bar{\tilde{b}}_1(x) \neq 0 \quad \forall x \in \Gamma, \tag{8.6}$$

where $\tilde{b}(x) = (\tilde{b}_1(x), \tilde{b}_2(x))$.

Proof. Consider, first, the case where Ω is the half-plane $x_2 > 0$, and the coefficients are constant. After the Fourier transform with respect to x_1 we obtain, assuming that $f = 0, \tilde{a}_0 = 0, x_2 = t$:

$$\tilde{a} \, i \, \xi \, \tilde{v} - \frac{d\tilde{v}}{dt} = 0, \quad t > 0, \tag{8.7}$$

$$\tilde{b}\tilde{v} = \tilde{g}, \quad t = 0. \tag{8.8}$$

Let $\tilde{v} = (v_1, v_2)^T$. Then from (8.7)

$$\frac{d\tilde{v}_1}{dt} = a_0 \, i \, \xi \, v_1, \quad \frac{d\tilde{v}_2}{dt} = \overline{a_0} \, i \, \xi \, v_2. \tag{8.9}$$

Suppose, first, that $\xi > 0$. Since all eigenvalues of the matrix a_0 have positive imaginary parts, we conclude that the stable (decaying at infinity) solution of system (8.5) is

$$v_1 = e^{a_0 i \xi t}, \quad v_2 = 0.$$

Substituting it into (8.8), we get

$$\tilde{b}_1 c = \tilde{g}.$$

Here $\tilde{b} = (\tilde{b}_1, \tilde{b}_2)$. Therefore in the case $\xi > 0$ the Lopatinskii condition has the form $\det \tilde{b}_1 \neq 0$. Similarly, for $\xi < 0$, it is $\det \tilde{b}_2 \neq 0$.

Consider now the general case of an arbitrary domain Ω with a smooth boundary. Let $x \in \Gamma$, and ν be the inward normal vector to the boundary at the point x. We consider the tangent half-plane and fulfil the corresponding orthogonal change of the independent variables. According to Proposition 8.2, the problem (8.4), (8.5) remains canonical in the new coordinates. Moreover, the boundary condition (8.5) does not change. Therefore, the Lopatinskii condition is given by (8.6). The theorem is proved. □

We now return to problem (8.1), (8.2) and will obtain for it the Lopatinskii condition. Since the matrix $P(x)$ in Theorem 8.1 has the property that its last r rows are complex conjugate to the first ones, it is easy to see that

$$P^{-1}(x) = (T(x), \overline{T(x)}), \tag{8.10}$$

where $T(x)$ is a $2r \times r$ matrix continuous in x, and $\overline{T(x)}$ is its complex conjugate. From the definition of the matrix $\tilde{b}(x)$ and of the matrices $\tilde{b}_i(x)$ (see Theorem 8.3) it follows that

$$\tilde{b}_1(x) = b(x)T(x), \quad \tilde{b}_2(x) = b(x)\overline{T(x)}.$$

The first condition in (8.7) has the form

$$\det(b(x), T(x)) \neq 0 \quad \forall x \in \Gamma. \tag{8.11}$$

Since $b(x)$ is a real matrix, then the second inequality in (8.7) follows from this one. We have proved the following theorem.

Theorem 8.4. *The Lopatinskii condition for problem (8.1), (8.2) does not depend on the direction of the normal to the boundary and has the form (8.11).*

To give an interpretation of the matrix $T(x)$ we note that from (8.3) it follows that
$$a(x)T(x) = T(x)a_0(x). \tag{8.12}$$
Suppose for simplicity that all eigenvalues of the matrix $a(x)$ at the point x are different from each other. Let $\lambda_1, \ldots, \lambda_r$ be its eigenvalues with positive imaginary parts. Then they constitute all eigenvalues of the matrix a_0. Therefore it can be reduced to the diagonal form,
$$a_0(x) = R\Lambda R^{-1}, \tag{8.13}$$
where R is an invertible matrix of order r, and Λ is a diagonal matrix with diagonal elements $\lambda_1, \ldots, \lambda_r$. From (8.12), (8.13) we obtain
$$aTR = TR\Lambda.$$
Hence TR is a matrix whose columns are eigenvectors of the matrix $a(x)$ corresponding to all its eigenvalues with positive imaginary parts. Therefore the columns of the matrix $T(x)$ are linear combinations of these eigenvectors.

Up to now we have considered problem (8.1), (8.2) with real coefficients. Consider now the case where the matrices $a(x), b(x)$, and $c(x)$ are complex. Suppose that there exists a matrix $P(x)$ of the order $2r$, $P(x) \in C^l(\bar{\Omega})$, invertible for all $x \in \bar{\Omega}$, and such that
$$P(x)a(x)P^{-1}(x) = \begin{pmatrix} a_+(x) & 0 \\ 0 & a_-(x) \end{pmatrix}, \tag{8.14}$$
where $a_+(x)$ ($a_-(x)$) are square matrices of order r whose eigenvalues for all $x \in \bar{\Omega}$ have positive (negative) imaginary parts. In the case of a real matrix $a(x)$ existence of such a matrix is given by Theorem 8.1.

We set $v = Pu$. Then we obtain problem (8.1), (8.2) with matrix (8.14). This problem is canonical. Hence the Lopatinskii condition is given by (8.6), where $\tilde{b}(x) = b(x)P^{-1}(x)$.

Let $P^{-1}(x) = (T_1(x), T_2(x))$, where T_1 and T_2 are some $2r \times r$ matrices. Then the Lopatinskii condition for problem (8.1), (8.2) is
$$\det(b(x)T_1(x)) \neq 0, \quad \det(b(x)T_2(x)) \neq 0 \quad \forall x \in \Gamma. \tag{8.15}$$

We have proved the following theorem.

Theorem 8.5. *If there exists such a matrix $P(x)$ that condition (8.14) is satisfied, then the Lopatinskii condition for problem (8.1), (8.2) with complex coefficients has the form (8.15).*

8.2 Index in bounded domains

Consider problem (8.1), (8.2) with complex coefficients. We begin with canonical systems, that is we assume that

$$a(x) = \begin{pmatrix} a_+(x) & 0 \\ 0 & a_-(x) \end{pmatrix},$$

where $a_+(x)(a_-(x))$ are square matrices of order r whose eigenvalues for any $x \in \bar{\Omega}$ have positive (negative) imaginary part. Let $b(x) = (b_1(x), b_2(x))$. It follows from Theorem 8.3 that the Lopatinskii condition does not depend on the matrix a(x) and on the normal direction to the boundary. It has the form

$$\det b_1(x) \neq 0, \quad \det b_2(x) \neq 0 \quad \forall x \in \Gamma. \tag{8.16}$$

We assume that this condition is satisfied.

We can fulfil a continuous deformation of system (8.1) to the system with

$$a(x) = \begin{pmatrix} iE & 0 \\ 0 & -iE \end{pmatrix}, \quad c(x) = 0,$$

where E is the identity matrix of order r. Put $u = (u_1, u_2)$. Then we obtain the following problem (for $f = 0$):

$$i\frac{\partial u_1}{\partial x_1} - \frac{\partial u_1}{\partial x_2} = 0, \quad -i\frac{\partial u_2}{\partial x_1} - \frac{\partial u_2}{\partial x_2} = 0, \tag{8.17}$$

$$b_1(x)u_1 + b_2(x)u_2 = g. \tag{8.18}$$

Since the Lopatinskii condition is satisfied in the process of this deformation, then the index of the corresponding operator does not change.

To find the index of the corresponding operator we suppose that Ω is a unit disk with its center at the origin. The problem above can be reduced to this case by a conformal mapping. Put $z = x_1 + ix_2$ and write $u_1(z)$ instead of $u_1(x_1, x_2)$. Then $u_1(z)$ is an analytic function. Further, if $u_2(z) = u_2(x_1, x_2)$, then its complex conjugate $\bar{u}_2(z)$ is analytic.

Denote by Ω_- the domain $|z| > 1$. Let $v_2(z) = u_2(1/\bar{z})$. The function $v_2(z)$ is analytic in Ω_- and bounded (c.f. [532], p. 171). If $z \in \Gamma$, then

$$v_2(z) = u_2\left(\frac{1}{\bar{z}}\right) = u_2(z).$$

Therefore problem (8.17), (8.18) can be reduced to the problem to find an analytic function $u_1(z)$ in Ω and another analytic function $v_2(z)$ in Ω_- such that

$$b_1(z)u_1^+(z) + b_2(z)v_2^-(z) = g(z), \quad z \in \Gamma.$$

8. General first-order systems on the plane

Here the superscripts $\pm$ denote the values of the corresponding functions at the boundary Γ. Setting

$$u(z) = \begin{cases} u_1(z), & z \in \Omega \\ u_2\left(\frac{1}{z}\right), & z \in \Omega_- \end{cases},$$

we obtain the following Hilbert problem: to find a piecewise holomorphic function $u(z)$ satisfying the condition

$$u^+(z) = G(z)u^-(z) + h(z), \quad z \in \Gamma, \tag{8.19}$$

where

$$G(z) = -b_1^{-1}(z)b_2(z), \quad h(z) = b_1^{-1}(z)h(z).$$

Consider first the case

$$u_2(0) = 0. \tag{8.20}$$

Then

$$\lim_{z \to \infty} u(z) = 0.$$

In this case the index κ of problem (8.19) is given by the formula

$$\kappa = \frac{1}{2\pi}[\arg\, \det\, G(z)]_\Gamma \tag{8.21}$$

(see [532], pp. 41, 44).

It is easy to see that without condition (8.20) we obtain the following formula for index:

$$\kappa = \frac{1}{2\pi}[\arg\, \det\, G(z)]_\Gamma + r$$

or

$$\kappa = \frac{1}{2\pi}[\arg\, \det\, b_2(x)]_\Gamma - \frac{1}{2\pi}[\arg\, \det\, b_1(x)]_\Gamma + r. \tag{8.22}$$

We have proved the following theorem.

Theorem 8.6. *If problem (8.1), (8.2) is canonical and the Lopatinskii condition (8.16) is satisfied, then the index of this problem is given by formula (8.22).*

If the problem is not canonical, then we can use the result about the reduction to the canonical form. In conditions of Theorem 8.5 the index is given by the formula

$$\kappa = \frac{1}{2\pi}[\arg\, \det\, (b(x)T_2(x))]_\Gamma - \frac{1}{2\pi}[\arg\, \det\, (b(x)T_1(x))]_\Gamma + r.$$

8.3 Index in unbounded domains

In this section we consider the problem

$$i\frac{\partial u_1}{\partial x_1} - \frac{\partial u_1}{\partial x_2} + \lambda u_2 = 0, \quad -i\frac{\partial u_2}{\partial x_1} - \frac{\partial u_2}{\partial x_2} + \lambda u_1 = 0, \tag{8.23}$$

$$b_1(x_1)u_1 + b_2(x_1)u_2 = g \tag{8.24}$$

in the half-plane $x_2 \geq 0$. The functions u_1, u_2, and the coefficients $b_1(x_1), b_2(x_1)$ are complex valued, λ is a real constant, $\lambda \geq 0$. We suppose that $b_i(x_1) \neq 0$, $i = 1, 2$, $x_1 \in \mathbb{R}^1$. It is shown in Section 8.1 that in this case the Lopatinskii condition is satisfied.

Next, we verify that Conditions NS and NS* are also verified. For this we consider limiting problems corresponding to problem (8.23), (8.24). We assume that there exist limits $b_i^{\pm} = \lim_{x_1 \to \pm\infty} b(x_1)$, $i = 1, 2$. There are two types of limiting problems: in $\mathbb{R}^2$ and in $\mathbb{R}^2_+$. Applying the Fourier transform, it is easy to verify that the operator corresponding to the limiting problem in $\mathbb{R}^2$ is invertible in the corresponding spaces (see Section 7). Consider the limiting problem in the half-plane. Since there are limits of the coefficients at infinity, we obtain two problems (at $\pm\infty$) with constant coefficients.

Applying the partial Fourier transform with respect to x_1, we obtain from (8.23)

$$\tilde{u}_1' = -\xi\tilde{u}_1 + \lambda\tilde{u}_2, \quad \tilde{u}_2' = \xi\tilde{u}_2 + \lambda\tilde{u}_1, \quad x_2 \geq 0.$$

We look for bounded solutions of this system in the form

$$\tilde{u}_1(\xi, x_2) = k_1(\xi)e^{\mu x_2}, \quad \tilde{u}_2(\xi, x_2) = k_2(\xi)e^{\mu x_2}.$$

We obtain

$$\mu = -\sqrt{\xi^2 + \lambda^2}, \quad k_2 = \frac{\xi - \sqrt{\xi^2 + \lambda^2}}{\lambda}k_1.$$

From the boundary condition

$$\left(b_1^{\pm} + \frac{\xi - \sqrt{\xi^2 + \lambda^2}}{\lambda}b_2^{\pm}\right)k_1(\xi) = 0.$$

Suppose that

$$b_1^{\pm} + \frac{\xi - \sqrt{\xi^2 + \lambda^2}}{\lambda}b_2^{\pm} \neq 0, \quad \forall \xi \in \mathbb{R}^1, \lambda \geq 0, |\xi| + |\lambda| \neq 0. \tag{8.25}$$

Then the limiting problems have only the zero solution and Condition NS is satisfied for all positive λ. It is not satisfied for $\lambda = 0$ because the limiting problems have some constant nonzero solutions. From (8.25) it follows that the operators corresponding to the limiting problems are elliptic with a parameter as acting from $E_1(\mathbb{R}^2_+)$ into $F_1(\mathbb{R}^2_+)$ (see Section 1 for the definition of the spaces). Hence they are

8. General first-order systems on the plane

invertible for large λ. From Condition NS it follows that they are normally solvable with a finite-dimensional kernel, moreover their kernels are empty. Therefore, they are invertible for all positive λ. Then Condition NS* is also satisfied. From Conditions NS and NS* we conclude that the operator corresponding to problem (8.23), (8.24) satisfies the Fredholm property. We will find its index approximating the half-plane by bounded domains.

Since $b_i(x) \neq 0$, $i = 1, 2$, then, without loss of generality, we can assume that $b_2(x) \equiv 1$. In this case condition (8.25) means that $b_1^\pm$ are not real nonnegative numbers. The limits b_1^+ and b_1^- are not necessarily equal to each other. We can use a continuous deformation of the problem, that is of the function $b_1(x_1)$, in such a way that it remains different from 0 for all $x_1 \in \mathbb{R}^1$ and that condition (8.25) is satisfied. Figure 10 shows two examples. The first curve can be retracted

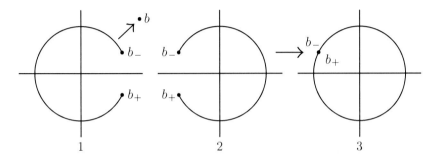

Figure 10: Examples of homotopy classes for the function $b_1(x_1)$.

to a point, that is to the case where $b_1(x_1) \equiv const$. The second curve cannot be reduced to a point. It can be deformed to a closed curve where the limits at $\pm\infty$ coincide.

Consider the domains Ω_n constructed in Section 6. The problems in the bounded domains are defined as follows. We consider the same equations and boundary conditions (8.23), (8.24). The function $b_2(x)$ in the boundary condition is supposed to be identically 1 on the whole boundary. The function $b_1(x)$ on the lower part of the boundary of the domain Ω_n is the same as for the problem in the half-plane. We should define it on the remaining part of the boundary in such a way that the sequence of the problems satisfies Conditions NS(seq) and NS*(seq).

Consider, first, an auxiliary problem for system (8.23), (8.24) with constant coefficients b_1 and b_2 in the half-plane

$$R_\theta^2 = \{(x_1, x_2) : -x_1 \sin\theta + x_2 \cos\theta \geq 0\}$$

obtained by rotation on the angle θ counterclockwise. Consider the change of variables

$$y_1 = x_1 \cos\theta + x_2 \sin\theta, \quad y_2 = -x_1 \sin\theta + x_2 \cos\theta$$

and put $v_i(y_1, y_2) = u_i(x_1, x_2)$, $i = 1, 2$. We obtain

$$i\frac{\partial v_1}{\partial y_1} - \frac{\partial v_1}{\partial y_2} + \frac{\lambda}{a_1} v_2 = 0, \quad -i\frac{\partial v_2}{\partial y_1} - \frac{\partial v_2}{\partial y_2} + \frac{\lambda}{a_2} v_1 = 0, \tag{8.26}$$

$$b_1 v_1 + b_2 v_2 = g, \tag{8.27}$$

where

$$a_1 = \cos\theta + i\sin\theta, \quad a_2 = \cos\theta - i\sin\theta.$$

Instead of condition (8.25) we obtain

$$b_1 + a_1 \frac{\xi - \sqrt{\xi^2 + \lambda^2}}{\lambda} b_2 \neq 0, \quad \forall \xi \in \mathbb{R}^1, \; \lambda \geq 0, \; |\xi| + |\lambda| \neq 0. \tag{8.28}$$

For $b_2 = 1$, this condition means that b_1 does not belong to the half-ray $\mathbb{R}_\theta^1$ obtained from the half-axis $\mathbb{R}_+^1$ by rotation on the angle θ counterclockwise.

We now return to problem (8.23), (8.24) in the domain Ω_n. The function $b_1(x)$ is defined as follows. At the lower part of the boundary, as indicated above, it is the same as for the problem in the half-plane. Without loss of generality we can suppose that $b_1^+ = b_1^-$ and $b_1(x_1) = b_1^\pm$ for $|x_1| \geq N$ for N sufficiently large. The general case can be reduced to this one by a continuous deformation.

We recall that the right and the left parts of the boundary $\partial\Omega_n$ are half-circles. Let x be at the right half-circle. Consider the tangent to the boundary at this point and denote by θ its angle with the x_1-axis. It varies from 0 to π at the right part of the boundary and from π to 2π at the left part. We denote by $\tilde{b}_1$ the coefficient in the boundary condition at the part of the boundary where $x_2 > 0$ and define it as a function of the angle θ. It should be continuous (its values for $\theta = 0$ and $\theta = 2\pi$ are equal to each other and coincide with $b_1^\pm$) and satisfy condition (8.28). We can put for example $\tilde{b}_1(\theta) = a_1 \tilde{b}_1(0)$, where a_1 is the same as in (8.28). Then this condition is satisfied and

$$\frac{1}{2\pi} \arg \tilde{b}_1(x)|_0^{2\pi} = 1. \tag{8.29}$$

This equality is independent of the explicit form of the function $\tilde{b}_1(\theta)$.

From Theorem 8.6 it follows that the index κ of problem (8.23), (8.24) in the half-plane is determined by the argument of the function $b_1(x_1)$:

$$\kappa = -\frac{1}{2\pi} \arg b_1(x)|_{-\infty}^\infty.$$

If the values b_1^+ and b_1^- are not equal to each other, then this formula cannot be directly applied. In this case we reduce the problem to the case where the limits at infinity are the same.

We have proved the following theorem.

Theorem 8.7. *Let the function* $b(x_1) = b_1(x_1)/b_2(x_1)$ *intersect the real positive half-axis in a finite number of points,* n_+ *be the number of such points when its argument increases and* n_- *where it decreases. Then the index* κ *of problem* (8.23), (8.24) *is given by the formula*

$$\kappa = n_- - n_+.$$

9 Examples

Hilbert problem for the Cauchy-Riemann system. Consider the Cauchy-Riemann system

$$\frac{\partial u}{\partial x} - \frac{\partial v}{\partial y} = 0, \quad \frac{\partial u}{\partial y} + \frac{\partial v}{\partial x} = 0 \tag{9.1}$$

in the unit circle $x^2 + y^2 \leq 1$ with the boundary condition

$$a(s)u + b(s)v = f(s). \tag{9.2}$$

Here s is the length of the arc of the circle, $a(s), b(s)$, and $f(s)$ are sufficiently smooth functions, $a^2(s) + b^2(s) > 0$. The solution is supposed to be continuous in the closed domains.

Consider the vector $(a(s), b(s))$ and denote by N the number of rotations of this vector around the origin when s changes from 0 to 2π counterclockwise. Positive values of N correspond to the counterclockwise rotation of the vector, negative values to the clockwise rotation.

Theorem 9.1. *If* $N \geq 0$, *then problem* (9.1), (9.2) *has* $2N+1$ *linearly independent solutions. If* $N < 0$, *then it is solvable if and only if* $-(2N+1)$ *conditions are satisfied:*

$$\int_0^{2\pi} f(s)\phi_k(s)ds = 0, \quad k = 1, 2, \ldots, 2|N| - 1.$$

Here ϕ_k *are linearly independent functions defined on the circle.*

Thus, the difference between the number of solutions of the homogeneous problem and the number of solvability conditions of the nonhomogeneous problem, that is the index, equals $2N + 1$. Direct proof of this theorem can be found in [205]. More general problems on the plane can be studied by reduction to singular integral equations (Section 8).

Laplace operator with oblique derivative. We consider the problem

$$\Delta w = 0, \tag{9.3}$$

in the unit circle $x^2 + y^2 \leq 1$ with the boundary condition

$$a(s)\frac{\partial w}{\partial x} - b(s)\frac{\partial w}{\partial y} = f(s). \tag{9.4}$$

Similar to the previous example, we follow here the presentation in [205]. If we put
$$u = \frac{\partial w}{\partial x}, \quad v = -\frac{\partial w}{\partial y},$$
then we reduce this problem to (9.1), (9.2). Conversely, from the solution of the Hilbert problem we can determine the solution w of (9.3), (9.4) up to an arbitrary constant. Therefore, if $N \geq 0$, then problem (9.3), (9.4) is solvable and has $2N + 2$ linearly independent solutions. If $N < 0$, then it is solvable under $2|N| - 1$ conditions and it is determined up to an arbitrary constant. Therefore, in both cases the index of the problem equals $2N + 2$.

In the case of the Neumann boundary condition, $a(s) = \cos s, b(s) = -\sin s$. Hence $N = -1$ and the index of problem (9.3), (9.4) equals 0. There is one solvability condition and the solution is determined up to an arbitrary constant.

Poincaré problem. We follow the presentation in [532] which uses the results of the paper [71]. Let Ω be a bounded simply connected domain in $\mathbb{R}^2$ with the boundary $\partial \Omega$ which is a simple bounded contour with a finite curvature. Consider the elliptic system

$$\Delta u + A(x,y)\frac{\partial u}{\partial x} + B(x,y)\frac{\partial u}{\partial y} + C(x,y)u = 0 \tag{9.5}$$

with the boundary condition

$$D(s)\frac{\partial u}{\partial x} + E(s)\frac{\partial u}{\partial y} + F(s)u = G(s), \tag{9.6}$$

where A, B, C, E, F, G are $n \times n$ matrices, G is a vector, $s \in \partial\Omega$. All of them are supposed to be sufficiently smooth real-valued functions of their arguments. We look for solutions of this problem continuous together with the second derivatives inside the domain and together with the first derivatives in the closure of the domain.

It is shown that if

$$\det(D(s) + iE(s)) \neq 0, \quad s \in \partial\Omega, \tag{9.7}$$

then the index κ of problem (9.5), (9.6) equals

$$\kappa = \frac{1}{\pi}[\arg \det(D - iE)]_{\partial\Omega} + 2n.$$

If $n = 1$, we obtain the formula from the previous example.

The proof is based on the reduction of problem (9.5), (9.6) to singular integral equations at the boundary for which the index is known. If condition (9.7) is satisfied, then problem (9.5), (9.6) is equivalent to the equation

$$M(t)\mu(t) + \int_{\partial\Omega} \frac{K(t,t_1)\mu(t_1)dt_1}{t_1 - t} = G(t),$$

where M and K are some matrices expressed through the coefficients of the equation and of the boundary condition.

First-order systems in the canonical form. Consider the system of equations

$$A(z)\frac{\partial u}{\partial x} - \frac{\partial u}{\partial y} + B(z)u = f(z) \tag{9.8}$$

in a bounded simply connected domain $\Omega \in \mathbb{R}^2$ with a sufficiently smooth boundary, with the boundary condition

$$a(t)u(t) = 0, \quad t \in \partial\Omega. \tag{9.9}$$

Here $z = x + iy$, the matrix A is Hölder continuous in the closure of the domain together with its first derivatives with respect to x and y, the matrix B and the vector $f(z)$ are Hölder continuous in $\overline{\Omega}$; A and B are square matrices of order $2r$, the order of the matrix a is $r \times 2r$. It is supposed that the matrix $A(z)$ can be represented in the form

$$A(z) = \begin{pmatrix} A^{(1)}(z) & -A^{(2)}(z) \\ A^{(2)}(z) & A^{(1)}(z) \end{pmatrix},$$

where the matrix $A^{(1)}(z) + iA^{(2)}(z)$ is such that all its eigenvalues have positive imaginary parts. In this case system (9.8) is called canonical (cf. Section 8.1).

Let $a = (a_1, a_2)$, where a_i, $i = 1, 2$ are square matrices of order r. Then the condition of normal solvability of problem (9.8), (9.9) is given by

$$\det(a_1(t) + ia_2(t)) \neq 0, \quad \forall t \in \partial\Omega, \tag{9.10}$$

and its index

$$\kappa = \frac{1}{\pi}[\arg\det(a_1 + ia_2)]_{\partial\Omega} + r$$

[547]. Condition (9.10) and the index formula are clearly related to those in the previous example: second-order operator can be reduced to the first-order operator. General first-order systems can be reduced, under some conditions, to the canonical systems.

Chapter 9

Problems in Cylinders

In the case of cylindrical domains, conditions which provide normal solvability and the Fredholm property can be formulated in a more simple and explicit form. Moreover, there are more direct methods to compute the index. In particular, we will use spectral decomposition and reduction to one-dimensional problems for which we can use fundamental solutions. Let us illustrate it with the following example. Consider the following equation on the real axis:

$$u'(t) = r(t)u(t) + f(t).$$

Let us assume, for simplicity, that $r(t) = r_+$ for t sufficiently large, $r(t) = r_-$ for $-t$ sufficiently large. The solution of this equation can be written explicitly:

$$u(t) = e^{\int_0^t r(s)ds} \left(\int_0^t e^{-\int_0^y r(s)ds} f(y)dy + u(0) \right).$$

If $r_\pm \neq 0$, then there are four cases depending on the signs of r_+ and r_-. For each of them we can easily find the number of linearly independent bounded solutions of the homogeneous equation and the number of solvability conditions of the nonhomogeneous equation. In particular, if $r_+ > 0$ and $r_- < 0$, then the solvability condition is

$$\int_{-\infty}^{\infty} e^{-\int_0^y r(s)ds} f(y)dy = 0.$$

The index κ of the problem is given by the following formula: $\kappa = s_+ + s_- - 1$, where s_+ is the number of solutions of the equation $du/dt = r_+ u$ which are bounded near $+\infty$, s_- is the number of solutions of the equation $du/dt = r_- u$ which are bounded near $-\infty$. Obviously, s_+ and s_- are determined by the signs of r_+ and r_-. A similar formula remains valid for systems of equations: $\kappa = s_+ + s_- - p$. Here s_+ and s_- are the numbers of bounded solutions of the corresponding limiting equations at $+\infty$ and $-\infty$, and p is the dimension of the system. In the other words, s_+ is the number of eigenvalues of the matrix r_+ with negative real parts, s_- is the number of eigenvalues of the matrix r_- with positive real parts. We then use the index of one-dimensional equations to determine the index of elliptic problems in cylinders [114]. This method can be applicable in the cases where the method of approximation by bounded domains, developed in the previous chapter, fails.

We finish this chapter with lower estimates of elliptic operators in cylinders. These estimates determine their Fredholm property and allow construction of the topological degree for nonlinear problems [561], [562].

1 Ordinary differential operators on the real line

1.1 Representation of solutions

Hereafter, for any integer l the notation $C_b^l(\mathbb{R}, \mathbb{R}^p)$ denotes the space of functions which are defined on $\mathbb{R}$, take their values in $\mathbb{R}^p$, and whose derivatives of all orders up to l are continuous and bounded. In particular, the space $C_b^0(\mathbb{R}, \mathbb{R}^p)$ is a Banach space for the uniform norm. Let M be a continuous matrix-valued function defined on $\mathbb{R}$, having limits at infinity:

$$M(t) \to A_\pm \quad \text{as} \quad t \to \pm\infty.$$

On $C_b^0(\mathbb{R}, \mathbb{R}^p)$ we consider the operator T defined by

$$u \mapsto u' + Mu,$$

with domain $D(T) = C_b^1(\mathbb{R}, \mathbb{R}^p)$. Together with T we consider the operator $\tilde{T}$ which has the same domain, and is associated to some continuous matrix-valued function A (in place of M) such that

$$A(t) = A_- \quad \text{for} \quad t \leq -1, \quad A_+ \quad \text{for} \quad 1 \leq t. \tag{1.1}$$

We will denote by $E_\pm$ the spectral projector associated to the set of all eigenvalues of $A_\pm$ having positive real part. We shall repeatedly use the fact that $E_\pm$ commutes with any function of $A_\pm$ whenever the latter can be defined.

We will assume the following condition on the spectrum of $A_\pm$:

$$\sigma(A_\pm) \cap i\mathbb{R} = \emptyset. \tag{1.2}$$

We wish to show that under this assumption the operator $\tilde{T}$ is Fredholm, and to compute its index in terms of the projectors $E_\pm$ (see [206, 487, 231] for related results on the half-line). The main tool we use is the variation of constants formula.

Let $f \in C_b^0(\mathbb{R}, \mathbb{R}^p)$ be given, and let $u \in D(\tilde{T})$ be a solution to the equation

$$\frac{du}{dt} + Au = f. \tag{1.3}$$

1. Ordinary differential operators on the real line

Let us denote by $Y = Y(t)$ the fundamental matrix associated to $\tilde{T}$, defined by

$$\frac{dY}{dt} + A(t)Y = 0, \quad Y(-1) = I.$$

We remark that Y can be written explicitly:

$$\text{for } t \leq -1: \quad Y(t) = e^{-(t+1)A_-}, \tag{1.4}$$

$$\text{for } t \geq 1: \quad Y(t) = e^{-(t-1)A_+}\Lambda, \tag{1.5}$$

where $\Lambda = Y(1)$ is an invertible $p \times p$ matrix. Therefore, in the range $|t| \geq 1$, the solution u is given by

$$\text{for } t \leq -1: \quad u(t) = e^{-(t+1)A_-}u(-1) - \int_t^{-1} e^{-(t-s)A_-}f(s)\,ds,$$

$$\text{for } 1 \leq t: \quad u(t) = e^{-(t-1)A_+}\Lambda u(-1) + e^{-(t-1)A_+}\Lambda \int_{-1}^{1} Y(s)^{-1}f(s)\,ds$$

$$+ e^{-(t-1)A_+}\int_1^t e^{(s-1)A_+}f(s)\,ds. \tag{1.6}$$

Let us first consider the case $t > 1$. The formula defining $u(t)$ yields

$$e^{(t-1)A_+}(I - E_+)u(t) = (I - E_+)\Lambda u(-1) + (I - E_+)\Lambda \int_{-1}^{1} Y(s)^{-1}f(s)\,ds$$

$$+ \int_1^t e^{(s-1)A_+}(I - E_+)f(s)\,ds. \tag{1.7}$$

The left-hand side of (1.7) tends to 0 as $t \to +\infty$, thus we get a first compatibility relation:

$$(I - E_+)\Lambda u(-1) = -(I - E_+)\Lambda \int_{-1}^{1} Y(s)^{-1}f(s)\,ds$$

$$- \int_1^\infty e^{(s-1)A_+}(I - E_+)f(s)\,ds. \tag{1.8}$$

Using this relation we can now express $u(t)$ for $t > 1$ as follows:

$$u(t) = e^{-(t-1)A_+}E_+\Lambda[u(-1) + \int_{-1}^{1} Y(s)^{-1}f(s)\,ds]$$

$$+ \int_1^t e^{-(t-s)A_+}E_+f(s)\,ds - \int_t^\infty e^{-(t-s)A_+}(I - E_+)f(s)\,ds. \tag{1.9}$$

Note that all integrals appearing in this expression are automatically finite. Similarly for $t < -1$ we have the relation

$$e^{(t+1)A_-}E_-u(t) = E_-u(-1) - \int_t^{-1} e^{(s+1)A_-}E_-f(s)\,ds, \tag{1.10}$$

which yields
$$E_-u(-1) = \int_{-\infty}^{-1} e^{(s+1)A_-} E_-f(s)\,ds. \tag{1.11}$$

This in turn gives the following expression of $u(t)$ for $t < -1$:
$$u(t) = e^{-(t+1)A_-}(I - E_-)u(-1) - \int_t^{-1} e^{-(t-s)A_-}(I - E_-)f(s)\,ds$$
$$+ \int_{-\infty}^t e^{-(t-s)A_-} E_-f(s)\,ds. \tag{1.12}$$

To summarize, we have established the following proposition.

Proposition 1.1. *Assume that (1.2) is satisfied. If the function $f \in C_b^0(\mathbb{R}, \mathbb{R}^p)$ is such that $f = \tilde{T}u$ for some $u \in D(\tilde{T})$, then any such u is given by (1.9) for $t > 1$ and (1.12) for $t < -1$, for some $u(-1) \in \mathbb{R}^p$ satisfying (1.8) and (1.11).*

1.2 Calculation of the index

In what follows, for any operator L we denote the null-space (kernel) by $N(L)$, the range by $R(L)$, the α-characteristic of L is the dimension of $N(L)$ and the β-characteristic is the codimension of $R(L)$.

If $\tilde{T}u = 0$, then by setting $f = 0$ in (1.8) and (1.11) we obtain
$$u(-1) \in N(E_-), \quad \Lambda u(-1) \in N(I - E_+).$$

Noting that Λ is an isomorphism, we may conclude that
$$\alpha(\tilde{T}) = \dim[N(E_-) \cap \Lambda^{-1} N(I - E_+)].$$

For the sake of convenience, let us introduce the following notation for $f \in C_b^0(\mathbb{R}, \mathbb{R}^p)$:
$$I_1(f) = (I - E_+)\Lambda \int_{-1}^{1} Y(s)^{-1} f(s)\,ds + \int_1^\infty e^{(s-1)A_+}(I - E_+)f(s)\,ds,$$
$$I_2(f) = \int_{-\infty}^{-1} e^{(s+1)A_-} E_-f(s)\,ds.$$

Note that $(I - E_+)I_1(f) = I_1(f)$, and that $E_-I_2(f) = I_2(f)$. If $f \in C_b^0(\mathbb{R}, \mathbb{R}^p)$ is in the range of $\tilde{T}$, then from (1.8), (1.11) we obtain for some $u(-1) \in \mathbb{R}^p$ the condition
$$E_-u(-1) = I_2(f), \quad (I - E_+)\Lambda u(-1) = -I_1(f). \tag{1.13}$$

We first give an equivalent form of this condition.

Lemma 1.2. *The last condition is equivalent to the requirement*
$$\Lambda I_2(f) + I_1(f) \in N(I - E_+) + \Lambda N(E_-). \tag{1.14}$$

1. Ordinary differential operators on the real line

Proof. To prove necessity, assuming (1.13), we note that

$$(I - E_+)(I_1(f) + \Lambda u(-1)) = 0, \quad E_-(I_2(f) - u(-1)) = 0,$$

thus for some v_1, v_2 we must have

$$u(-1) = I_2(f) + (I - E_-)v_2, \quad \Lambda u(-1) = -I_1(f) - E_+ v_1.$$

Multiplying the first equality by Λ and subtracting, we obtain (1.14). To prove sufficiency, suppose that

$$I_1(f) + \Lambda I_2(f) = -\Lambda(I - E_-)v_2 - E_+ v_1.$$

Define $u(-1)$ by

$$u(-1) = I_2(f) + (I - E_-)v_2. \tag{1.15}$$

Then we have

$$\Lambda u(-1) = \Lambda I_2(f) + \Lambda(I - E_-)v_2$$
$$= -I_1(f) - E_+ v_1. \tag{1.16}$$

Thus (1.15) and (1.16) respectively yield

$$E_-(u(-1) - I_2(f)) = 0,$$
$$(I - E_+)(\Lambda u(-1) + I_1(f)) = 0.$$

Then (1.13) is satisfied, and this proves the lemma. □

We have seen that if $f \in C_b^0(\mathbb{R}, \mathbb{R}^p)$ is in the range of $\tilde{T}$, then (1.14) is satisfied. Conversely if (1.14) is satisfied, then for some $u(-1) \in \mathbb{R}^p$, (1.13) is satisfied. We may then define u by (1.9) for $t > 1$ and (1.12) for $t < -1$. To show that $f \in R(\tilde{T})$, it remains to verify that we can define u in the interval $[-1, 1]$ so as to obtain a global solution. The Cauchy problem with initial data at $t = -1$ has a unique solution which satisfies

$$u(1) = \Lambda[u(-1) + \int_{-1}^{1} Y(s)^{-1} f(s)\, ds].$$

Thus from (1.9), we see that we can obtain a global solution if we have continuity at $t = 1$, i.e., if the relation

$$E_+ \Lambda[u(-1) + \int_{-1}^{1} Y(s)^{-1} f(s)\, ds] - \int_{1}^{\infty} e^{(s-1)A_+}(I - E_+) f(s)\, ds$$
$$= \Lambda\left[u(-1) + \int_{-1}^{1} Y(s)^{-1} f(s)\, ds\right]$$

holds true. This relation is an immediate consequence of the second equality in (1.13), thus we obtain that $f \in R(\tilde{T})$.

Using condition (1.14) we can now compute the β-characteristic of $\tilde{T}$. Define a linear map
$$\Phi : C_b^0(\mathbb{R}, \mathbb{R}^p) \longrightarrow \frac{\mathbb{R}^p}{N(I - E_+) + \Lambda N(E_-)}$$
by assigning to each $f \in C_b^0(\mathbb{R}, \mathbb{R}^p)$ the coset of $I_1(f) + \Lambda I_2(f)$, denoted by $[I_1(f) + \Lambda I_2(f)]$. From what we have seen, we have
$$R(\tilde{T}) = N(\Phi).$$
Thus
$$\beta(\tilde{T}) = \operatorname{codim} R(\tilde{T}) = \dim R(\Phi).$$
We now show that Φ is surjective. Pick $x \in \mathbb{R}^p$, we need to find $f \in C_b^0(\mathbb{R}, \mathbb{R}^p)$ such that
$$I_1(f) + \Lambda I_2(f) - x \in N(I - E_+) + \Lambda N(E_-).$$

If we were working with the space of (not necessarily continuous) bounded functions we could take
$$f(t) = -A_+ x \quad \forall t > 1, \quad f(t) = 0 \quad \forall t < 1.$$
This function f satisfies
$$I_1(f) + \Lambda I_2(f) - x = -E_+ x,$$
and therefore $\Phi(f) = [x]$. However this f is not continuous, thus using this idea we set
$$f_k(t) = 0 \quad \text{for} \quad t \leq 1,$$
$$k(1-t)A_+ x \quad \text{for} \quad 1 \leq t \leq 1 + \frac{1}{k},$$
$$-A_+ x \quad \text{for} \quad t \geq 1 + \frac{1}{k}.$$
It is easy to compute that
$$I_1(f_k) + \Lambda I_2(f_k) - x = -E_+ x + x_k,$$
where
$$x_k \to 0 \quad \text{as} \quad k \to \infty.$$
Since the projection map $[\cdot]$ is continuous we have
$$\Phi(f_k) \to [x],$$

and since the image of Φ (being finite-dimensional) is closed we conclude again that it contains the point $[x]$.

Summarizing, we have shown the following:

$$\beta(\tilde{T}) = p - \dim[N(I - E_+) + \Lambda N(E_-)]$$
$$= p - \dim[\Lambda^{-1}N(I - E_+) + N(E_-)].$$

Note that to obtain the last equality we used the fact that Λ is an isomorphism.

Obviously, the operator T is Fredholm if and only if the operator $\tilde{T}$ is Fredholm. They have the same index since they can be reduced to each other by a continuous deformation in the class of Fredholm operators. Finally we obtain the index of the Fredholm operator T:

$$\text{ind}(T) = \dim \Lambda^{-1} N(I - E_+) + \dim N(E_-) - p$$
$$= \dim N(I - E_+) + \dim N(E_-) - p.$$

Therefore we have shown the following theorem.

Theorem 1.3. *If condition (1.2) is satisfied then T is a Fredholm operator, and its index is given by*

$$\text{ind}(T) = \dim N(I - E_+) - \dim N(I - E_-).$$

In particular, we see that the index is zero if $E_- = E_+$.

We note that the formula for the index can be also formulated in terms of the number of linearly independent bounded solutions at infinity. Namely, the index is equal to the sum of the numbers of bounded linearly independent solutions at minus and plus infinity, minus the dimension of the system p.

2 Second-order equations

2.1 Reduction to first-order equations

We consider the operator

$$L : C^{2+\delta}(\mathbb{R}, \mathbb{R}^p) \longrightarrow C^\delta(\mathbb{R}, \mathbb{R}^p) \tag{2.1}$$

defined by the expression

$$Lu = a(x)u'' + b(x)u' + c(x)u.$$

Here $a(x)$, $b(x)$, and $c(x)$ are smooth $p \times p$ matrices having respectively the limits $a^\pm, b^\pm, c^\pm$ as $x \to \pm\infty$. We will use the results of the previous subsection to show that under some appropriate condition L is a Fredholm operator, and to compute its index. We begin with the following lemma.

Lemma 2.1. *Assume that*

$$\forall \xi \in \mathbb{R}: \quad T^\pm(\xi) = -a^\pm \xi^2 + b^\pm i\xi + c^\pm \quad \text{is an invertible matrix.} \quad (2.2)$$

Then the operator L is normally solvable with a finite-dimensional kernel.

Proof. If the condition of the lemma is satisfied, then the limiting equation

$$a^\pm u'' + b^\pm u' + c^\pm u = 0$$

does not have nonzero bounded solutions. The lemma is proved. □

We first rewrite the system $Lu = f$ as a first-order system, and consider the first-order ordinary differential operators M and T:

$$M : C^{2+\delta}(\mathbb{R}, \mathbb{R}^p) \times C^{1+\delta}(\mathbb{R}, \mathbb{R}^p) \longrightarrow C^{1+\delta}(\mathbb{R}, \mathbb{R}^p) \times C^{\delta}(\mathbb{R}, \mathbb{R}^p), \quad (2.3)$$

$$T : C_b^1(\mathbb{R}, \mathbb{R}^{2p}) \longrightarrow C_b^0(\mathbb{R}, \mathbb{R}^{2p}), \quad (2.4)$$

associated to the expression

$$(u, q) \longmapsto (u, q)' + A(u, q).$$

Here A denotes the matrix

$$A(x) = \begin{pmatrix} 0 & -I_p \\ a^{-1}c & a^{-1}b \end{pmatrix},$$

and I_p denotes the $p \times p$ identity matrix. The result in Section 1.2 (Theorem 1.3) shows that if condition (1.2) is satisfied, then T is a Fredholm operator, and we have a formula for its index in terms of the limit matrices $A^\pm$. We remark that condition (1.2) is equivalent to condition (2.2). We now proceed to show that under this condition, the operators M and L are also Fredholm, and have the same index as T.

Lemma 2.2. *If T is Fredholm, then M also is, and has the same index.*

Lemma 2.3. *Assume that M is Fredholm, and that L is normally solvable. Then L also is Fredholm, and has the same index as M.*

Combining these three lemmas we immediately obtain the desired result.

Theorem 2.4. *Assume that condition (2.2) is satisfied. Then L is a Fredholm operator, and its index is given by*

$$\mathrm{ind}(L) = \kappa^+ - \kappa^-,$$

where κ^+ and κ^- are, respectively, the number of eigenvalues of the matrices A^+ and A^- with positive real part.

2. Second-order equations

Proof of Lemma 2.2. Let us consider the system of equations

$$u' = p + f_1, \quad p' = a^{-1}(bp + cu + f_2).$$

If the solution of the homogeneous problem belongs to $\mathcal{C}^1(\mathbb{R}, \mathbb{R}^p)$, then $u' = p \in \mathcal{C}^{1+\delta}(\mathbb{R}, \mathbb{R}^p)$ (since the coefficients are Hölder continuous), and $u(x) \in \mathcal{C}^{2+\delta}(\mathbb{R}, \mathbb{R}^p)$. Hence the dimension of the kernel remains the same. Now let $f_1 \in \mathcal{C}^{1+\delta}(\mathbb{R}, \mathbb{R}^p)$, $f_2 \in \mathcal{C}^\delta(\mathbb{R}, \mathbb{R}^p)$. If (f_1, f_2) satisfies the solvability conditions in C^1, then there is a solution $u \in \mathcal{C}^1(\mathbb{R}, \mathbb{R}^p), p \in \mathcal{C}^1(\mathbb{R}, \mathbb{R}^p)$. As above, it follows that $u \in \mathcal{C}^{2+\delta}(\mathbb{R}, \mathbb{R}^p)$ and $p \in \mathcal{C}^{1+\delta}(\mathbb{R}, \mathbb{R}^p)$. If (f_1, f_2) does not satisfy the solvability conditions in C^1, then obviously there are no solutions of this problem with $u \in \mathcal{C}^{2+\delta}(\mathbb{R}, \mathbb{R}^p)$ and $p \in \mathcal{C}^{1+\delta}(\mathbb{R}, \mathbb{R}^p)$. Hence the solvability conditions and their number remain the same. □

Proof of Lemma 2.3. We first consider the kernel of L. If $u \in N(L)$, then we have $(u, u') \in N(M)$. Conversely if $(u, q) \in N(M)$, then $q = u'$, and $u \in N(L)$. This shows that the map $u \mapsto (u, u')$ realizes an isomorphism between $N(L)$ and $N(M)$, thus we have $\alpha(L) = \alpha(M)$. Let us now consider the range of L. For fixed $(f_1, f_2) \in \mathcal{C}^{1+\delta}(\mathbb{R}, \mathbb{R}^p) \times \mathcal{C}^\delta(\mathbb{R}, \mathbb{R}^p)$, from $M(u, q) = (f_1, f_2)$ it follows that $Lu = af_1' + bf_1 + af_2$. For convenience we define the map

$$\pi : \mathcal{C}^{1+\delta}(\mathbb{R}, \mathbb{R}^p) \times \mathcal{C}^\delta(\mathbb{R}, \mathbb{R}^p) \longrightarrow \mathcal{C}^\delta(\mathbb{R}, \mathbb{R}^p), \tag{2.5}$$

$$(f_1, f_2) \longmapsto af_1' + bf_1 + af_2. \tag{2.6}$$

Then clearly we have

$$\pi(R(M)) = R(L).$$

By assumption $R(M)$ has finite codimension in $\mathcal{C}^{1+\delta}(\mathbb{R}, \mathbb{R}^p) \times \mathcal{C}^\delta(\mathbb{R}, \mathbb{R}^p)$, and $R(L)$ is closed in $\mathcal{C}^\delta(\mathbb{R}, \mathbb{R}^p)$. Our aim is to show that the codimensions of these two spaces are equal. Define $N_1 = \beta(M) < \infty$, and $N_2 = \beta(L) \leq \infty$, and denote by $\{\Phi_i, i = 1, \ldots, N_1\}$ a basis of the annihilator of $R(M)$. Each Φ_i is a bounded linear functional on $\mathcal{C}^{1+\delta}(\mathbb{R}, \mathbb{R}^p) \times \mathcal{C}^\delta(\mathbb{R}, \mathbb{R}^p)$. Similarly, denote by $\{F_k, k = 1, \ldots, N_2\}$ a basis of the annihilator of $R(L)$. Then for fixed $f = (f_1, f_2) \in \mathcal{C}^{1+\delta}(\mathbb{R}, \mathbb{R}^p) \times \mathcal{C}^\delta(\mathbb{R}, \mathbb{R}^p)$ we have the following equivalence:

$$f \in R(M) \Leftrightarrow \Phi_i(f) = 0, \quad i = 1, \ldots, N_1,$$
$$\Leftrightarrow \pi f \in R(L) \Leftrightarrow F_k(\pi f) = 0, \quad k = 1, \ldots, N_2.$$

Writing $G_k = F_k \pi$, we thus have:

$$\{f : \Phi_i(f) = 0, i = 1, \ldots, N_1\} = \{f : G_k(f) = 0, k = 1, \ldots, N_2\}.$$

The functionals G_k form a family of linearly independent bounded linear functionals on the space $\mathcal{C}^{1+\delta}(\mathbb{R}, \mathbb{R}^p) \times \mathcal{C}^\delta(\mathbb{R}, \mathbb{R}^p)$. Thus it follows that $N_1 = N_2$, i.e., the β-characteristics of the operators L and M are the same. □

Remarks 2.5.

- It is easy to verify that $\kappa^\pm$ is the number of solutions to the equation

$$det(a^\pm\lambda^2 - b^\pm\lambda + c^\pm) = 0$$

which have positive real part. As for the case of first-order systems of equations, we can say that the index of the operator L is equal to the sum of the numbers of linearly independent bounded solutions of the equation $Lu = 0$ at plus and minus infinity, minus $2p$.

- $\alpha(L)$ and $\beta(L)$ are determined by the projectors E^+ and E^-. These projectors are determined by the matrices A^+ and A^- and remain the same if we consider the operator M as acting from $C^{1+\delta}(\mathbb{R}, \mathbb{R}^p) \times C^\delta(\mathbb{R}, \mathbb{R}^p)$ or from $C^\delta(\mathbb{R}, \mathbb{R}^p) \times C^\delta(\mathbb{R}, \mathbb{R}^p)$.

As an example we consider the case $p = 1$ where $u(x)$ is a scalar-valued function and suppose that $a(x) \equiv 1$, b is a positive constant. If $c^+ \neq 0$, $c^- \neq 0$, then the operator is Fredholm. Its index equals 0 if c^+ and c^- have the same sign, it is 1 for $c^+ > 0$, $c^- < 0$, and -1 for $c^+ < 0$, $c^- > 0$. The essential spectrum of the operators consists of two parabolas on the complex plane. In the case $c^+ < 0$, $c^- < 0$ both of them are completely in the left half-plane. In the case $c^+ > 0$, $c^- > 0$ both of them are partially in the right half-plane having the origin inside the region where the large negative numbers are. Obviously, these two curves cannot be moved to the left half-plane by a continuous deformation such that they do not intersect zero. Hence two elliptic operators with the same index are not necessarily homotopic in the class of Fredholm operators of the same form.

2.2 Solvability conditions

2.2.1. Scalar equations. In Chapter 4 conditions of normal solvability are formulated in terms of limiting problems. In the one-dimensional case it can be also useful to apply another approach based on the integral representation of solutions. Consider first the scalar equation

$$u'' + b(x)u' + c(x) = f(x) \tag{2.7}$$

assuming that the coefficients belong to the Hölder space $C^{1+\delta}(\mathbb{R})$ with some $0 < \delta < 1$. Denote by $u_1(x)$ and $u_2(x)$ two linearly independent solutions of the homogeneous equation

$$u'' + b(x)u' + c(x)u = 0 \tag{2.8}$$

and by $v_1(x)$ and $v_2(x)$ two linearly independent solutions of the adjoint homogeneous equation

$$v'' - (b(x)v)' + c(x)v = 0. \tag{2.9}$$

2. Second-order equations

The solutions $u_i(x)$, $v_i(x)$, $i = 1, 2$ are not supposed to be bounded. It can be directly verified that the functions

$$v_1(x) = \frac{-u_2}{u_1 u_2' - u_1' u_2}, \quad v_2(x) = \frac{u_1}{u_1 u_2' - u_1' u_2} \tag{2.10}$$

are linearly independent and satisfy equation (2.9). We have

$$u_1 v_1 + u_2 v_2 = 0, \quad u_1' v_1 + u_2' v_2 = 1. \tag{2.11}$$

The solution of equation (2.7) can be represented in the form

$$u(x) = u_1(x) \int_{x_1}^{x} v_1(y) f(y) + u_2(x) \int_{x_2}^{x} v_2(y) f(y), \tag{2.12}$$

where x_1 and x_2 can be bounded or unbounded. This can be verified substituting (2.12) to (2.7) with the use of (2.11).

Suppose that the solutions $u_1(x)$ and $u_2(x)$ behave exponentially at infinity,

$$u_i(x) \sim a_i^{\pm} e^{\lambda_i^{\pm} x}, \quad x \to \pm\infty, \ i = 1, 2,$$

$a_i^{\pm} \neq 0$, $\lambda_i^{\pm} \neq 0$. We assume here for simplicity that $\lambda_1^{\pm} \neq \lambda_2^{\pm}$. Then from (2.10)

$$v_i(x) \sim b_i^{\pm} e^{-\lambda_i^{\pm} x}, \quad x \to \pm\infty, \ i = 1, 2$$

for some $b_i^{\pm} \neq 0$. Depending on the signs of $\lambda_i^{\pm}$ there are the following cases:

- $u_1(x)$ decays at $\pm\infty$, $u_2(x)$ grows at $\pm\infty$. Then $v_1(x)$ grows at $\pm\infty$, $v_2(x)$ decays at $\pm\infty$. We put

$$u(x) = u_1(x) \int_{x_1}^{x} v_1(y) f(y) + u_2(x) \int_{-\infty}^{x} v_2(y) f(y), \tag{2.13}$$

where x_1 is bounded. For any bounded f, the first term in the right-hand side of (2.13) is bounded. Consider the second term. If the lower limit in the integral was different from $-\infty$, then the integral did not necessarily converge to 0 as $x \to -\infty$, and this term would not be bounded since $u_2(x)$ grows at $-\infty$. Similarly, the integral should converge to 0 at $+\infty$. Therefore we need the condition

$$\int_{-\infty}^{\infty} v_2(y) f(y) dy = 0 \tag{2.14}$$

in order for equation (2.7) to have a bounded solution.

If $f(x) \in C^0(\mathbb{R})$, then $u(x) \in C^2(\mathbb{R})$. If $f(x) \in C^{\delta}(\mathbb{R})$, then $u(x) \in C^{2+\delta}(\mathbb{R})$. The operator L,

$$Lu = u'' + b(x) u' + c(x) u$$

considered as acting in these spaces has zero index.

- $u_1(x)$ decays at $+\infty$, and grows at $-\infty$, $u_2(x)$ grows at $\pm\infty$. Then $v_1(x)$ grows at $+\infty$, and decays at $-\infty$, $v_2(x)$ decays at $\pm\infty$. We put

$$u(x) = u_1(x) \int_{-\infty}^{x} v_1(y)f(y) + u_2(x) \int_{-\infty}^{x} v_2(y)f(y). \tag{2.15}$$

Solvability condition (2.14) should be satisfied. The index of L equals -1.

- $u_1(x)$ grows at $+\infty$, and decays at $-\infty$, $u_2(x)$ grows at $\pm\infty$. Then $v_1(x)$ decays at $+\infty$, and grows at $-\infty$, $v_2(x)$ decays at $\pm\infty$. We put

$$u(x) = -u_1(x) \int_{x}^{\infty} v_1(y)f(y) + u_2(x) \int_{-\infty}^{x} v_2(y)f(y). \tag{2.16}$$

Solvability condition (2.14) should be satisfied. The index of L equals -1.

- $u_1(x)$ grows at $\pm\infty$, $u_2(x)$ grows at $\pm\infty$. Then $v_1(x)$ decays at $\pm\infty$, $v_2(x)$ decays at $\pm\infty$. We put

$$u(x) = u_1(x) \int_{-\infty}^{x} v_1(y)f(y) + u_2(x) \int_{-\infty}^{x} v_2(y)f(y). \tag{2.17}$$

Solvability condition

$$\int_{-\infty}^{\infty} v_i(y)f(y)dy = 0, \quad i = 1, 2 \tag{2.18}$$

should be satisfied. The index of L equals -2.

- $u_1(x)$ decays at $\pm\infty$, $u_2(x)$ grows at $-\infty$, decays at ∞. Then $v_1(x)$ grows at $\pm\infty$, $v_2(x)$ decays at $-\infty$, grows at ∞. We put

$$u(x) = u_1(x) \int_{x_1}^{x} v_1(y)f(y) + u_2(x) \int_{-\infty}^{x} v_2(y)f(y), \tag{2.19}$$

where x_1 is bounded. Here we do not need solvability conditions. The index of L equals 1.

- $u_1(x)$ decays at $\pm\infty$, $u_2(x)$ decays at $\pm\infty$. Then $v_1(x)$ grows at $\pm\infty$, $v_2(x)$ grows at $\pm\infty$. We put

$$u(x) = u_1(x) \int_{x_1}^{x} v_1(y)f(y) + u_2(x) \int_{x_2}^{x} v_2(y)f(y), \tag{2.20}$$

where x_1 and x_2 is bounded. Here we do not need solvability conditions. The index of L equals 2.

Similarly other cases can be considered. We summarize them in the following theorem.

2. Second-order equations

Theorem 2.6. *Suppose that solutions $u_1(x)$ and $u_2(x)$ of the homogeneous equation (2.8) behave exponentially at infinity with different exponents. Then the operator L satisfies the Fredholm property and*

$$\text{ind } L = n_- + n_+ - 2,$$

where n_- is the number of bounded solutions (among u_1 and u_2) at $-\infty$ and n_+ at ∞.

In the next theorem we do not explicitly assume the exponential behavior of solutions of the homogeneous and of the homogeneous adjoint equation at infinity. We will use the following notation:

$$\Phi_0^i(x) = |u_i(x)| \int_0^x |v_i(y)| dy, \quad \Phi_\infty^i(x) = |u_i(x)| \int_x^\infty |v_i(y)| dy,$$

$$\Psi_0^i(x) = |u_i(x)| \int_x^0 |v_i(y)| dy, \quad \Psi_\infty^i(x) = |u_i(x)| \int_{-\infty}^x |v_i(y)| dy, \quad i = 1, 2.$$

Theorem 2.7. *Suppose that $\Phi_0^i(x)$, $i = 1, 2$ is uniformly bounded for $x \geq 0$ or, if it is not uniformly bounded, then $\Phi_\infty^i(x)$ is defined and is uniformly bounded for $x \geq 0$. Similarly, suppose that $\Psi_0^i(x)$, $i = 1, 2$ is uniformly bounded for $x \leq 0$ or, if it is not uniformly bounded, then $\Psi_\infty^i(x)$ is defined and is uniformly bounded for $x \leq 0$. Then the operator L satisfies the Fredholm property. Equation (2.7) is solvable if and only if conditions (2.18) are satisfied for those i for which each component of the couple (Φ_0^i, Ψ_0^i) is unbounded.*

Proof. Suppose that Φ_0^i and Ψ_0^i are bounded for $i = 1, 2$. Then the solution

$$u(x) = u_1(x) \int_0^x v_1(y) f(y) dy + u_2(x) \int_0^x v_2(y) f(y) dy$$

is uniformly bounded for all $x \in \mathbb{R}$. If Φ_0^1 is not uniformly bounded, and all other functions are, then we put

$$u(x) = u_1(x) \int_x^\infty v_1(y) f(y) dy + u_2(x) \int_0^x v_2(y) f(y) dy. \tag{2.21}$$

This function is uniformly bounded for any bounded f and for $x \geq 0$ by virtue of the assumptions on Φ_∞^1 and Φ_0^2. For $x \leq 0$ we have

$$u(x) = u_1(x) \int_0^\infty v_1(y) f(y) dy + u_1(x) \int_x^0 v_1(y) f(y) dy + u_2(x) \int_x^0 v_2(y) f(y) dy.$$

Since Ψ_0^i is bounded, then the last terms in the right-hand side are bounded. Moreover, $u_1(x)$ is bounded for all $x \leq 0$. Therefore the first term in the right-hand side is also bounded.

Let us next consider the case where Φ_0^1 and Ψ_0^1 are not bounded. We consider $u(x)$ given by (2.21). As above, it is uniformly bounded for $x \geq 0$:. Since

$$\int_{-\infty}^{\infty} v_1(y) f(y) dy = 0, \tag{2.22}$$

then along with (2.21) we can use the representation

$$u(x) = u_1(x) \int_{-\infty}^{x} v_1(y) f(y) dy + u_2(x) \int_{0}^{x} v_2(y) f(y) dy.$$

Hence $u(x)$ is bounded for $x \leq 0$.

We have verified that condition (2.22) is sufficient for the solvability of equation (2.7). Its necessity can be easily obtained by multiplication of equation (2.7) by $v_1(x)$ and integration over $\mathbb{R}$. Similarly we can consider the case where Φ_0^2 or Ψ_0^2 are unbounded. Solvability condition (2.22) appears because both Φ_0^1 and Ψ_0^1 are not bounded. Therefore the codimension of the image equals the number of unbounded couples.

Let us determine the dimension of the kernel. If Φ_0^1 is bounded, then $u_1(x)$ is bounded for $x \geq 0$. If Ψ_0^1 is bounded, then $u_1(x)$ is bounded for $x \leq 0$. Therefore each bounded couple provides a bounded solution of the homogeneous equation. On the other hand, only bounded couples can provide a bounded solution. Indeed, if Φ_0^1 is not bounded, then Φ_∞^1 is defined and bounded. This means that the integral $\int_x^\infty |v_1(y)| dy$ is defined, that is the integral $\int_0^x |v_1(y)| dy$ has limit as $x \to \infty$. Thus the assumption that Φ_0^1 is not bounded implies that $|u_1(x)|$ is not bounded for $x \geq 0$. The theorem is proved. □

2.2.2. Systems of equations. The integral representation of solutions for systems of equations and the solvability conditions can be obtained similarly to the scalar equation. Consider the system of equations

$$a(x) u'' + b(x) u' + c(x) u = f, \tag{2.23}$$

where a, b, c are $C^{1+\delta}(\mathbb{R})$, $n \times n$ matrices, u and f are n-vectors. Assuming that the matrix $a(x)$ is invertible for each x, and that its inverse is bounded uniformly in x, we can reduce this system to the case where $a(x) \equiv E$, where E is the identity matrix.

We will obtain first of all the representation of the solution through Green's function. Consider the homogeneous equation

$$u'' + b(x) u' + c(x) u = 0 \tag{2.24}$$

and denote by $u^1, \ldots, u^{2n}$ its linearly independent solutions. Let U^1 be an $n \times n$ matrix whose columns are the first n solution, and U^2 the matrix composed from

2. Second-order equations

the last n solutions. We will use the notation $P^i = (U^i)'$, $i = 1, 2$. System (2.24) can be written as

$$\begin{pmatrix} u \\ p \end{pmatrix}' = \begin{pmatrix} 0 & E \\ -c(x) & -b(x) \end{pmatrix} \begin{pmatrix} u \\ p \end{pmatrix}. \tag{2.25}$$

Its fundamental solution is

$$\Phi(x) = \begin{pmatrix} U^1 & U^2 \\ P^1 & P^2 \end{pmatrix}.$$

Consider next the system adjoint to (2.25):

$$\begin{pmatrix} q \\ v \end{pmatrix}' = -\begin{pmatrix} 0 & -(c(x))^T \\ E & -(b(x))^T \end{pmatrix} \begin{pmatrix} q \\ v \end{pmatrix}. \tag{2.26}$$

Here the superscript T denotes the transposed matrix. Then

$$q' = (c(x))^T v, \quad v' = -q + (b(x))^T v.$$

Differentiating the second equation, we obtain

$$v'' - ((b(x))^T v)' + (c(x))^T v = 0, \tag{2.27}$$

that is the system adjoint to (2.24).

Let us represent the inverse matrix $(\Phi(x))^{-1}$ as a block matrix

$$(\Phi(x))^{-1} = \begin{pmatrix} W^1 & V^1 \\ W^2 & V^2 \end{pmatrix}.$$

Then the fundamental matrix of system (2.26) is

$$\Psi(x) = ((\Phi(x))^{-1})^T = \begin{pmatrix} (W^1)^T & (W^2)^T \\ (V^1)^T & (V^2)^T \end{pmatrix}.$$

Thus the columns of the matrices $(V^1)^T$ and $(V^2)^T$ provide $2n$ linearly independent solutions of system (2.27). On the other hand,

$$U^1 V^1 + U^2 V^2 = 0, \quad P^1 V^1 + P^2 V^2 = E.$$

Applying these equalities we can represent the solution $u(x)$ of equation (2.23) in the form

$$u(x) = U^1(x) \int^x V^1(y) f(y) dy + U^2(x) \int^x V^2(y) f(y) dy. \tag{2.28}$$

The lower limits in the integrals will be specified below.

We recall that the lines of the matrices V^1 and V^2 are solutions of system (2.27). Therefore we can rewrite (2.28) as

$$u(x) = \begin{pmatrix} u_1^1 & & u_1^n \\ \vdots & \cdots & \vdots \\ u_n^1 & & u_n^n \end{pmatrix} \begin{pmatrix} F_1(x) \\ \vdots \\ F_n(x) \end{pmatrix} + \begin{pmatrix} u_1^{n+1} & & u_1^{2n} \\ \vdots & \cdots & \vdots \\ u_n^{n+1} & & u_n^{2n} \end{pmatrix} \begin{pmatrix} F_{n+1}(x) \\ \vdots \\ F_{2n}(x) \end{pmatrix} \quad (2.29)$$

where

$$F_i(x) = \int_{x_i}^x (v^i(y), f(y)) dy, \quad i = 1, \ldots, 2n, \quad v^i(y) = (V^i(y))^T.$$

The lower limits x_i can be finite or infinite and can be different from each other.

We will use the following notation:

$$\Phi_0^{ijk}(x) = |u_i^j(x)| \int_0^x |v_k^j(y)| dy, \quad \Phi_\infty^{ijk}(x) = |u_i^j(x)| \int_x^\infty |v_k^j(y)| dy,$$

$$\Psi_0^{ijk}(x) = |u_i^j(x)| \int_x^0 |v_k^j(y)| dy, \quad \Psi_\infty^{ijk}(x) = |u_i^j(x)| \int_{-\infty}^x |v_k^j(y)| dy,$$

where $i, k = 1, \ldots, n$, $j = 1, \ldots, 2n$.

Theorem 2.8. *Suppose that for each j fixed either all Φ_0^{ijk} are bounded, that is there exists a constant M such that*

$$\sup_{x \geq 0} \Phi_0^{ijk}(x) \leq M, \quad i, k = 1, \ldots, n,$$

or all of them tend to infinity as $x \to +\infty$. In this case all Φ_∞^{ijk} are assumed to be defined and

$$\sup_{x \geq 0} \Phi_\infty^{ijk}(x) \leq M, \quad i, k = 1, \ldots, n.$$

Suppose next that for each j fixed either all Ψ_0^{ijk} are bounded, that is there exists a constant M such that

$$\sup_{x \leq 0} \Psi_0^{ijk}(x) \leq M, \quad i, k = 1, \ldots, n,$$

or all of them tend to infinity as $x \to -\infty$. In this case all Ψ_∞^{ijk} are assumed to be defined and

$$\sup_{x \leq 0} \Psi_\infty^{ijk}(x) \leq M, \quad i, k = 1, \ldots, n.$$

Then equation (2.23) has a bounded solution for a bounded function f if and only if

$$\int_{-\infty}^\infty (v^j(y), f(y)) dy = 0, \quad (2.30)$$

3. Second-order problems in cylinders   375

for all j for which both elements of the couple $(\Phi_0^{ijk}, \Psi_0^{ijk})$ are unbounded. The number of linearly independent solutions of this equation equals the number of couples with both bounded components.

The proof of this theorem is similar to the proof of Theorem 2.7. We note that the solvability conditions appear in the case where u^i is unbounded at both infinities and v^i is bounded. We can use this theorem to compute the index.

3 Second-order problems in cylinders

3.1 Normal solvability

We consider the operator

$$Lu = a(x)\Delta u + \sum_{i=1}^{m} b_i(x)\frac{\partial u}{\partial x_i} + c(x)u, \qquad (3.1)$$

acting from the space of functions $u \in C^{2+\delta}(\bar{\Omega})$ satisfying the boundary condition

$$\Lambda u \equiv \alpha \frac{\partial u}{\partial \nu} + \beta(x')u \mid_S = 0 \qquad (3.2)$$

into the space $C^\delta(\bar{\Omega})$. We suppose that the boundary of the cylinder belongs to the class $C^{2+\delta}$ and the coefficients of the operator to $C^{1+\delta}(\bar{\Omega})$. We assume also that the ellipticity condition $(a(x)q, q) \geq \sigma \mid q \mid^2$, $x \in \bar{\Omega}$ is satisfied. Here q is a constant vector, $\sigma > 0$, $(\cdot, \cdot)$ denotes the inner product in R^p. We assume that there exist the limits

$$a^\pm(x') = \lim_{x_1 \to \pm\infty} a(x), \quad b^\pm(x') = \lim_{x_1 \to \pm\infty} b(x), \quad c^\pm(x') = \lim_{x_1 \to \pm\infty} c(x).$$

We consider the limiting operators

$$L^\pm u = a^\pm(x')\Delta u + \sum_{i=1}^{n} b_i^\pm(x')\frac{\partial u}{\partial x_i} + c^\pm(x')u$$

with the coefficients independent of x_1.

Condition 3.1. The equation

$$L^\pm u = 0, \quad \Lambda u = 0 \qquad (3.3)$$

has no nontrivial solution in $C^{2+\delta}(\bar{\Omega})$.

The operators $L^\pm$ are limiting operators for the operator L, and Condition 3.1 is a particular case of Condition NS (Chapter 4). We know that it is necessary and sufficient for the operator L to be normally solvable and for the dimension of its kernel $\alpha(L)$ to be finite. Condition 1 can be written in a different form. If we apply a formal Fourier transform with respect to the variable x_1 to the corresponding equations, we obtain the following condition:

Condition 3.1'. The equation

$$\tilde{L}_\xi^\pm \tilde{u} = 0, \quad \Lambda \tilde{u} = 0$$

has no nontrivial solution in $C^{2+\delta}(G)$ for any real ξ. Here

$$\tilde{L}_\xi^\pm \tilde{u} = -\xi^2 a^\pm(x')\tilde{u} + a^\pm(x')\Delta'\tilde{u} + i\xi b_1^\pm(x')\tilde{u} + \sum_{k=2}^{m} b_k^\pm(x')\frac{\partial \tilde{u}}{\partial x_k} + c^\pm(x')\tilde{u},$$

$$\Delta'\tilde{u} = \sum_{k=2}^{m} \frac{\partial^2 \tilde{u}}{\partial x_k^2}.$$

We now prove that Conditions 3.1 and 3.1' are equivalent. We consider the problem

$$a(x')\Delta u + \sum_{j=1}^{m} b_j(x')\frac{\partial u}{\partial x_j} + c(x')u = 0, \tag{3.4}$$

$$\alpha \frac{\partial u}{\partial \nu} + \beta(x')u \mid_{\partial \Omega} = 0. \tag{3.5}$$

We suppose that the matrices a, b_j, and c belong to $C^\delta(\bar{G})$ and β to $C^{1+\delta}(\bar{G})$. If we consider $u(x)$ in the form

$$u(x) = e^{\lambda x_1} v(x'), \tag{3.6}$$

we obtain

$$a(x')\Delta'v + \sum_{j=2}^{m} b_j(x')\frac{\partial v}{\partial x_j} + (a(x')\lambda^2 + b_1(x')\lambda + c(x'))v = 0, \tag{3.7}$$

$$\alpha \frac{\partial v}{\partial \nu} + \beta(x')v \mid_{\partial G} = 0. \tag{3.8}$$

If problem (3.7), (3.8) has a solution for some $\lambda = i\xi$ with a real ξ, then (3.6) is a bounded solution of (3.4), (3.5). The following theorem establishes the equivalence between existence of solutions of problems (3.4), (3.5) and (3.7), (3.8) for $\lambda = i\xi$. In L^2 spaces the Fourier transform may be used to study this connection. If we consider the problem (3.4), (3.5) in Hölder spaces, then we cannot apply the classical Fourier transform directly.

Theorem 3.2. *The problem (3.4), (3.5) has a nontrivial bounded solution if and only if the problem (3.7), (3.8) has a nontrivial solution for some $\lambda = i\xi$.*

3. Second-order problems in cylinders

Proof. Let $u(x)$ be a bounded solution of (3.4), (3.5). Consider the Laplace transform of the function $u(x)$ in the x_1 variable,

$$v(x') = \int_0^\infty e^{-px_1} u(x) dx_1.$$

Here Re $p > 0$. Then

$$a(x')\Delta' v + \sum_{j=2}^m b_j(x') \frac{\partial v}{\partial x_j} + (a(x')p^2 + b_1(x')p + c(x'))v \quad (3.9)$$
$$= a(x')(u(0, x')p + u'(0, x')) + b_1(x')u(0, x').$$

Here $u' = \partial u/\partial x_1$. We consider the operator

$$A_p v = a(x')\Delta' v + \sum_{j=2}^m b_j(x') \frac{\partial v}{\partial x_j} + (a(x')p^2 + b_1(x')p + c(x'))v$$

acting from $C^{(2+\delta)}(\bar{G})$ with boundary conditions (3.8) into $C^{(\delta)}(\bar{G})$ and suppose that for any $p = i\xi$, $-\infty < \xi < \infty$ it does not have a zero eigenvalue. Let $p_0 = i\xi_0$. For each $p = p_0 + \delta$ with small nonnegative δ, the equation

$$A_p v = f_p, \quad (3.10)$$

where

$$f_p(x') = a(x')(u(0, x')p + u'(0, x')) + b_1(x')u(0, x'),$$

is uniquely solvable, and its solution depends continuously on p. Denote it by v_p. Then

$$v_p(x') = \int_0^\infty e^{-px_1} u(x) dx_1, \quad \delta > 0$$

and

$$\int_0^\infty e^{-px_1} u(x) dx_1 \to v_{i\xi_0}(x'), \quad \delta \to 0. \quad (3.11)$$

Similarly we put

$$w(x') = \int_{-\infty}^0 e^{qx_1} u(x) dx_1, \quad q = -i\xi_0 + \delta,$$

$$B_q w = a(x')\Delta' w + \sum_{j=2}^m b_j(x') \frac{\partial w}{\partial x_j} + (a(x')q^2 - b_1(x')q + c(x'))w,$$

$$g_q(x') = a(x')(u(0, x')q - u'(0, x')) - b_1(x')u(0, x').$$

The equation

$$B_q w = g_q \quad (3.12)$$

has a unique solution for small δ,
$$w_q(x') = \int_{-\infty}^{0} e^{qx_1} u(x) dx_1, \quad \delta > 0$$

and
$$\int_{-\infty}^{0} e^{qx_1} u(x) dx_1 \to w_{-i\xi_0}(x'), \quad \delta \to 0. \tag{3.13}$$

We note that $v_{i\xi_0}(x') = -w_{-i\xi_0}(x')$. Then from (3.11), (3.13) we obtain
$$\int_{-\infty}^{0} e^{(-i\xi_0+\delta)x_1} u(x) dx_1 + \int_{0}^{\infty} e^{(-i\xi_0-\delta)x_1} u(x) dx_1 \to 0, \quad \delta \to 0. \tag{3.14}$$

If we pass formally to the limit in the left-hand side of (3.14), we obtain that the Fourier transform of the function $u(x)$ in the x_1 variable is zero since ξ_0 is arbitrary. It would contradict the assumption that $u(x)$ is a nontrivial solution of (3.4), (3.5) and prove that the operator $A_{i\xi}$ (or the same $B_{i\xi}$) has a zero eigenvalue for some ξ. We now justify this passage to the limit.

For some bounded ϕ, and $x' \in G$,
$$\int_{-\infty}^{+\infty} u(x)\phi(x_1) dx_1 \neq 0. \tag{3.15}$$

Then
$$\int_{G} dx' \left| \int_{-\infty}^{+\infty} u(x)\phi(x_1) dx_1 \right| \neq 0. \tag{3.16}$$

Set
$$u_\delta(x) = \begin{cases} u(x) e^{-\delta x_1}, & x_1 > 0, \\ u(x) e^{\delta x_1}, & x_1 < 0. \end{cases}$$

Then
$$2\pi \int_{-\infty}^{+\infty} u_\delta(x)\phi(x_1) dx_1 = \int_{-\infty}^{+\infty} \tilde{u}_\delta(\xi, x')\psi(\xi) d\xi, \tag{3.17}$$

where
$$\tilde{u}_\delta(\xi, x') = 2\pi \int_{-\infty}^{+\infty} e^{-i\xi} u_\delta(x) dx_1,$$

and ψ is the Fourier transform of ϕ. Then
$$\int_{G} dx' \left| \int_{-\infty}^{+\infty} u_\delta(x)\phi(x_1) dx_1 \right| = \frac{1}{2\pi} \int_{G} dx' \left| \int_{-\infty}^{+\infty} \tilde{u}_\delta(\xi, x')\psi(\xi) d\xi \right|. \tag{3.18}$$

The left-hand side in (3.18) tends to the integral in (3.16) as $\delta \to 0$. We now show that the right-hand side in (3.18) tends to zero. From (3.14) it follows that $\tilde{u}_\delta(\xi, x') \to 0$, $\delta \to 0$ for each fixed ξ. Moreover this convergence is uniform in ξ

3. Second-order problems in cylinders

on every bounded interval because the solutions v_p and w_q of the equations (3.10) and (3.12) depend continuously on p and q, respectively. It remains to estimate the behavior of the function $\tilde{u}_\delta(\xi, x')$ for large ξ.

We have $\tilde{u}_\delta(\xi, x') = 2\pi(v_p(x') + w_q(x'))$. So we should estimate the solutions v_p and w_q, $p = i\xi + \delta$, $q = -i\xi + \delta$ for large ξ. Substituting $p = i\xi + \delta$ in (3.9), we obtain

$$\Delta' v + \sum_{j=2}^{m} a(x')^{-1} b_j(x') \frac{\partial v}{\partial x_j} - \xi^2 v + B(x', \xi) v \qquad (3.19)$$
$$= (u(0, x')(i\xi + \delta) + u'(0, x')) + a(x')^{-1} b_1(x') u(0, x'),$$

where

$$B(x', \xi) = (2i\xi\delta + \delta^2) + a(x')^{-1} b_1(x')(i\xi + \delta) + a(x')^{-1} c(x').$$

Consider the problem

$$\Delta' v + \sum_{j=2}^{m} a(x')^{-1} b_j(x') \frac{\partial v}{\partial x_j} - \xi^2 v = f(x), \quad \alpha \frac{\partial v}{\partial \nu} + \beta(x') v \mid_{\partial G} = 0.$$

Its solution satisfies the estimate

$$\|v(x')\|_{L^2(G)} \le K \frac{\|f(x')\|_{L^2(G)}}{\xi^2}$$

for some constant K (for large ξ) because the operator which corresponds to the problem (3.1), is sectorial in L^2. Then for the solution of (3.19)

$$\|v_p(x')\|_{L^2(G)} \le K \left(\frac{\|f_p(x')\|_{L^2(G)}}{\xi^2} + \frac{\|B(\xi, x') v\|_{L^2(G)}}{\xi^2} \right)$$

uniformly in δ, $0 \le \delta \le \delta_0$. Since the norm of the matrix $B(\xi, x')/\xi^2$ becomes less than 1 for large ξ, then

$$\|v_p(x')\|_{L^2(G)} \le \frac{c}{|\xi|}$$

for some positive c. We obtain a similar estimate for w_p. Hence we have, uniformly in δ,

$$\|\tilde{u}_\delta(\xi, x')\|_{L^2(G)} \le \frac{c}{|\xi|}.$$

Here we use the same notation c for different constants. We have thus

$$\int_G dx' \left| \int_{-\infty}^{+\infty} \tilde{u}_\delta(\xi, x') \psi(\xi) d\xi \right| \le \int_G dx' \left| \int_{|\xi| \le N} \tilde{u}_\delta(\xi, x') \psi(\xi) d\xi \right|$$
$$+ \int_G dx' \left| \int_{|\xi| \ge N} \tilde{u}_\delta(\xi, x') \psi(\xi) d\xi \right|.$$

It remains to note that the second integral in the right-hand side of the last inequality tends to zero uniformly in δ as N increases, the first integral goes to zero as $\delta \to 0$ for any given N. It proves that the right-hand side of (3.18) goes to zero and contradicts (3.16). This contradiction shows that the operator $A_{i\xi}$ has a zero eigenvalue for some ξ. Thus, we have proved necessity. Sufficiency is obvious, and this completes the proof of the theorem. $\square$

3.2 Calculation of the index

1. Projection on the eigenvalues of the transversal Laplacian. We consider a linear elliptic operator of the form

$$Lu = a(x)\Delta u + \sum_{i=1}^{n} b_i(x)\frac{\partial u}{\partial x_i} + c(x)u. \tag{3.20}$$

Here $x = (x_1, \ldots, x_n) \in \Omega$, where $\Omega \subset \mathbb{R}^n$ is an unbounded cylinder: $\Omega = \mathbb{R} \times \Omega'$, and Ω' is a bounded domain in $\mathbb{R}^{n-1}$. Thus the axis of the cylinder is parallel to the x_1-direction, and we denote by $x' = (x_2, \ldots, x_n) \in \Omega'$ the transversal variable. The unknown function is vector valued: $u = (u_1, \ldots, u_p)$, and $a(x)$, $b_i(x)$, $c(x)$ are $p \times p$ matrices. The matrix $a(x)$ is supposed to be positive definite,

$$(a(x)u, u) \geq k(u, u)$$

for some $k > 0$, and any $u \in \mathbb{R}^p$, $x \in \bar{\Omega}$. The boundary S of the cylinder Ω is supposed to be of class $C^{2+\delta}$ for some fixed $0 < \delta < 1$. We will use homogeneous Dirichlet boundary conditions

$$u = 0 \quad \text{on} \quad S = \mathbb{R} \times \partial\Omega'.$$

For any integer l, by $\mathcal{C}^{l+\delta}(\bar{\Omega}, \mathbb{R}^p)$ we denote the Banach space of functions which are bounded and continuous in $\bar{\Omega}$, together with all partial derivatives up to order l, and such that the partial derivatives of order l are uniformly Hölder continuous with exponent δ.

We suppose that all the coefficients of the operator L belong to $\mathcal{C}^\delta(\bar{\Omega}, \mathbb{R})$, and have limits as $x_1 \to \pm\infty$:

$$a^{\pm} = \lim_{x_1 \to \pm\infty} a(x), \quad b_i^{\pm} = \lim_{x_1 \to \pm\infty} b_i(x), \quad c^{\pm} = \lim_{x_1 \to \pm\infty} c(x).$$

Here $a^{\pm}$, $b_i^{\pm}$, and $c^{\pm}$ are constant matrices.

In this section we compute the index of the operator defined by (3.20). We assume that $b_k \to 0$, $k = 2, \ldots, n$ as $x_1 \to \pm\infty$. We consider L as acting between the space

$$E := \{u \in \mathcal{C}^{2+\delta}(\bar{\Omega}, \mathbb{R}^p) : \quad u|_S = 0\}$$

and the space $\mathcal{C}^\delta(\bar{\Omega}, \mathbb{R}^p)$. The corresponding norms will be denoted by $|\cdot|_{2+\delta,\Omega}$ and $|\cdot|_{\delta,\Omega}$ respectively. However to compute this index it will be more convenient to view

3. Second-order problems in cylinders

L as an unbounded operator acting in $\mathcal{C}^\delta(\bar{\Omega}, \mathbb{R}^p)$, with domain $D(L) = E$. We do so from now on. The index is the same, since the kernel and range are unchanged. To compute $\mathrm{ind}(L)$ we connect L to the operator $\tilde{L}$ acting in $\mathcal{C}^\delta(\bar{\Omega}, \mathbb{R}^p)$, with the same domain E, defined by

$$\tilde{L}u = \tilde{a}(x_1)\Delta u + \tilde{b}(x_1)\frac{\partial u}{\partial x_1} + \tilde{c}(x_1)u. \tag{3.21}$$

For the operator $\tilde{L}$ we take coefficients which only depend on the variable x_1:

$$\tilde{a}(x_1) = a^+\psi(x_1) + a^-(1 - \psi(x_1)), \quad \tilde{b}(x_1) = b_1^+\psi(x_1) + b_1^-(1 - \psi(x_1)), \tag{3.22}$$
$$\tilde{c}(x_1) = c^+\psi(x_1) + c^-(1 - \psi(x_1)), \tag{3.23}$$

where $\psi(x_1)$ is a sufficiently smooth function, equal to 1 for $x_1 \geq 1$ and to 0 for $x_1 \leq 0$.

We note that during the deformation

$$\tau\tilde{L} + (1 - \tau)L, \quad \tau \in [0, 1],$$

Condition 3.1 is satisfied and the index of the operator does not change. Thus instead of the operator (3.20) we will consider the operator (3.21). From now on we omit the tilde and it is this operator which we denote by L.

Consider the (scalar) Laplace operator in the transverse domain:

$$\Delta': \quad \mathcal{C}^\delta(\bar{\Omega}', \mathbb{R}) \longrightarrow \mathcal{C}^\delta(\bar{\Omega}', \mathbb{R}),$$

with domain

$$D(\Delta') := \{u \in \mathcal{C}^{2+\delta}(\bar{\Omega}', \mathbb{R}) : u = 0 \text{ on } \partial\Omega'\}.$$

Since Δ' is formally self-adjoint and Ω' is bounded, the spectrum of this operator consists of a countable sequence of eigenvalues. Let us consider these eigenvalues in the following order (without repetition):

$$\cdots \omega_k < \cdots < \omega_2 < \omega_1 < 0.$$

For each k, the multiplicity m_k of ω_k is finite, and we will denote by $\phi_i^k, i = 1, \ldots, m_k$ an $L^2(\Omega')$ orthonormal basis of the associated eigenspace:

$$i = 1, \ldots, m_k: \quad \phi_i^k(x') \in \mathbb{R}, \, \Delta'\phi_i^k = \omega_k\phi_i^k \quad \text{in } \Omega', \, \phi_i^k = 0 \quad \text{on } \partial\Omega'.$$

Now consider the (vector) Laplacian in the transverse domain:

$$\underline{\Delta}': \quad \mathcal{C}^\delta(\bar{\Omega}', \mathbb{R}^p) \longrightarrow \mathcal{C}^\delta(\bar{\Omega}', \mathbb{R}^p),$$
$$D(\underline{\Delta}') := E' := \{u \in \mathcal{C}^{2+\delta}(\bar{\Omega}', \mathbb{R}^p) : u = 0 \text{ on } \partial\Omega'\}. \tag{3.24}$$

The spectrum of this self-adjoint operator is the same as for the scalar Laplacian, but the multiplicity of ω_k is now pm_k. Denote by π'_k the spectral projector associated to the eigenvalue ω_k of the vector Laplacian. Then π'_k is a bounded operator acting in $\mathcal{C}^\delta(\bar{\Omega}', \mathbb{R}^p)$, and takes its values in E'. We will use the fact that π'_k is continuous for the $\mathcal{C}^0(\bar{\Omega}', \mathbb{R}^p)$ norm. This is an immediate consequence of the representation of π'_k as a Dunford integral over a bounded contour, see for instance [206]. If $u \in R(\pi'_k)$, then each component of u is a linear combination of the functions $\phi^k_i, i = 1, \ldots, m_k$. Thus we obtain

$$u(x') = \sum_{i=1}^{m_k} p_i \phi^k_i(x'), \quad (3.25)$$

with $p_i \in \mathbb{R}^p$ for each i. From now on let s be a fixed integer. We set

$$P'_s = \sum_{k=1}^{s} \pi'_k.$$

In particular, if Γ is a closed contour in the complex plane containing the first s eigenvalues, for any $v \in \mathcal{C}^\delta(\bar{\Omega}', \mathbb{R}^p)$ we have

$$P'_s v = \frac{1}{2i\pi} \int_\Gamma (\Delta' - \lambda)^{-1} v \, d\lambda,$$

and the range of P'_s is equal to the corresponding eigenspace which we denote by E'_s. Finally, we define Q'_s as a bounded operator acting in $\mathcal{C}^\delta(\bar{\Omega}', \mathbb{R}^p)$ by

$$Q'_s u = u - P'_s u.$$

We have

$$E'_s = P'_s[\mathcal{C}^\delta(\bar{\Omega}', \mathbb{R}^p)], \quad \mathcal{C}^\delta(\bar{\Omega}', \mathbb{R}^p) = E'_s \oplus \tilde{E}'_s, \quad \text{where} \quad \tilde{E}'_s := Q'_s[\mathcal{C}^\delta(\bar{\Omega}', \mathbb{R}^p)].$$

Remark 3.3. As we did for π'_k, we note that P'_s and Q'_s are continuous for the $\mathcal{C}^0(\bar{\Omega}', \mathbb{R}^p)$ norm.

Let us set

$$E_s = \{u \in \mathcal{C}^\delta(\bar{\Omega}, \mathbb{R}^p) : \forall x_1 \in \mathbb{R} : \quad u(x_1, \cdot) \in E'_s\},$$
$$\tilde{E}_s = \{u \in \mathcal{C}^\delta(\bar{\Omega}, \mathbb{R}^p) : \forall x_1 \in \mathbb{R} : \quad u(x_1, \cdot) \in \tilde{E}'_s\}.$$

We now define two operators P_s, Q_s on $\mathcal{C}^\delta(\bar{\Omega}, \mathbb{R}^p)$ by

$$(P_s u)(x_1, \cdot) = P'_s(u(x_1, \cdot)), \quad (Q_s u)(x_1, \cdot) = Q'_s(u(x_1, \cdot)). \quad (3.26)$$

We first show a preliminary result:

Lemma 3.4. P_s and Q_s are bounded projectors acting in $\mathcal{C}^\delta(\bar{\Omega}, \mathbb{R}^p)$. Moreover for any $u \in E$ we have

$$P_s u, Q_s u \in E, \quad \text{and} \quad P_s L u = L P_s u, \quad Q_s L u = L Q_s u. \quad (3.27)$$

3. Second-order problems in cylinders

Proof. We first show that P_s (and therefore Q_s) takes its values in $\mathcal{C}^\delta(\bar{\Omega}, \mathbb{R}^p)$. This amounts to checking the regularity of $P_s u$ with respect to x_1, for $u \in \mathcal{C}^\delta(\bar{\Omega}, \mathbb{R}^p)$. For fixed $x_1 \in \mathbb{R}$, $x' \in \bar{\Omega}'$, using Remark 3.3 we have

$$\begin{aligned}|(P_s u)(x_1 + h, x') - (P_s u)(x_1, x')| &\leq ||P_s'(u(x_1 + h, \cdot)) - P_s'(u(x_1, \cdot))||_{0,\Omega'} \\ &\leq K ||u(x_1 + h, \cdot) - u(x_1, \cdot)||_{0,\Omega'} \quad (3.28) \\ &\leq K ||u||_{\delta,\Omega} h^\delta.\end{aligned}$$

Inequality (3.28) shows that P_s, Q_s are bounded operators acting in $\mathcal{C}^\delta(\bar{\Omega}, \mathbb{R}^p)$. Clearly (from the analogous relations for P_s', Q_s') we have

$$(P_s)^2 = P_s, \quad (Q_s)^2 = Q_s, \quad P_s + Q_s = I.$$

Now since P_s' takes its values in E', it follows that for $u \in \mathcal{C}^{2+\delta}(\bar{\Omega}, \mathbb{R}^p)$, the function $P_s u$ satisfies the boundary condition $P_s u = 0$ on S. To show that $P_s u \in E$, again we only need to check regularity with respect to x_1. If $u \in \mathcal{C}^{1+\delta}(\bar{\Omega}, \mathbb{R}^p)$ we have, as $h \to 0$:

$$\frac{u(x_1 + h, \cdot) - u(x_1, \cdot)}{h} \to \frac{\partial u}{\partial x_1}(x_1, \cdot) \quad \text{in} \quad \mathcal{C}^0(\bar{\Omega}', \mathbb{R}^p).$$

Thus, using Remark 3.3:

$$\frac{(P_s u)(x_1 + h, \cdot) - (P_s u)(x_1, \cdot)}{h} = P_s' \frac{u(x_1 + h, \cdot) - u(x_1, \cdot)}{h}$$

$$\to P_s' \frac{\partial u}{\partial x_1}(x_1, \cdot) = \left(P_s \frac{\partial u}{\partial x_1}\right)(x_1, \cdot) \quad \text{in} \quad \mathcal{C}^0(\bar{\Omega}', \mathbb{R}^p).$$

We obtain for any $u \in \mathcal{C}^{1+\delta}(\bar{\Omega}, \mathbb{R}^p)$:

$$\frac{\partial}{\partial x_1}(P_s u) = P_s \left(\frac{\partial u}{\partial x_1}\right),$$

thus for $u \in \mathcal{C}^{2+\delta}(\bar{\Omega}, \mathbb{R}^p)$:

$$\frac{\partial^2}{\partial x_1^2}(P_s u) = P_s \left(\frac{\partial^2 u}{\partial x_1^2}\right). \quad (3.29)$$

This shows that E is invariant under P_s and Q_s. Clearly P_s' (as a spectral projector of Δ') commutes with Δ', and with multiplication by functions of x_1. Thus (3.27) follows from (3.29), and this completes the proof of the lemma. $\square$

Remark 3.5. $R(Q_s)$ is closed for the topology of uniform convergence on compact subsets of $\bar{\Omega}$. Indeed if $f_n \in R(Q_s)$ converges to f uniformly on compact subsets of $\bar{\Omega}$, then for all x_1, the function $f_n(x_1, \cdot)$ converges to $f(x_1, \cdot)$ uniformly on $\bar{\Omega}'$. Then it follows from Remark 3.3 that $(P_s' f)(x_1, \cdot) = 0$, which means that $f \in R(Q_s)$.

To compute the index of L we now define two operators L_1, L_2 by restricting L to $R(P_s)$ and $R(Q_s)$ respectively, and with domains

$$D(L_1) = D(L) \cap R(P_s), \quad D(L_2) = D(L) \cap R(Q_s).$$

It follows from (3.27) that their ranges are given by

$$R(L_1) = R(L) \cap R(P_s), \quad R(L_2) = R(L) \cap R(Q_s).$$

This proves that if Condition 3.1 holds true, then the unbounded operators

$$L_1, L_2 : \mathcal{C}^\delta(\bar{\Omega}, \mathbb{R}^p) \longrightarrow \mathcal{C}^\delta(\bar{\Omega}, \mathbb{R}^p)$$

are normally solvable. We will use the following lemma.

Lemma on the index of a direct sum. *Let X be a Banach space, and $L : X \to X$ an unbounded normally solvable operator with domain $D(L) = E$. Let P, Q be two bounded projectors acting in X, with $P + Q = I_X$. Assume that for any $x \in E$*

$$Px \in E, \quad \text{and} \quad LPx = PLx.$$

Define two (unbounded) operators L_1, L_2 by taking restrictions of L:

$$L_1 : R(P) \to R(P), \quad L_2 : R(Q) \to R(Q),$$
$$D(L_1) = R(P) \cap E, \quad D(L_2) = R(Q) \cap E.$$

Then L_1 and L_2 are normally solvable, and $\mathrm{ind}(L) = \mathrm{ind}(L_1) + \mathrm{ind}(L_2)$.

Thus we obtain the following result.

Theorem 3.6. *Assume the operator L satisfies Condition 3.1, and define the operators L_1, L_2 as above. If these two operators are Fredholm then L also is, and*

$$\mathrm{ind}(L) = \mathrm{ind}(L_1) + \mathrm{ind}(L_2).$$

2. Computation of $\mathrm{ind}(L_1)$. Defining a map π_k by

$$(\pi_k u)(x_1, \cdot) = \pi'_k(u(x_1, \cdot)),$$

we obtain (as we did with P_s, Q_s) a bounded projector acting in $\mathcal{C}^\delta(\bar{\Omega}, \mathbb{R}^p)$. Also, E is invariant under π_k, and on E, the operators L and π_k commute. We can now define an unbounded operator L^k by restricting L to $R(\pi_k)$, with domain $D(L^k) = E \cap R(\pi_k)$. By the same argument as above we have

$$\mathrm{ind}(L_1) = \sum_{k=1}^{s} \mathrm{ind}(L^k).$$

3. Second-order problems in cylinders

If $u \in R(\pi_k)$ we have $u(x_1, \cdot) \in R(\pi'_k)$ for each x_1, thus in view of (3.25) we obtain

$$u(x_1, x') = \sum_{i=1}^{m_k} p_i(x_1)\phi_i^k(x'),$$

with $p_i \in C^\delta(\mathbb{R}, \mathbb{R}^p)$ for each i. Then for a given $f \in R(\pi_k)$ written in the form

$$f(x_1, x') = \sum_{i=1}^{m_k} g_i(x_1)\phi_i^k(x'),$$

the equation $L^k u = f$ may be rewritten as

$$\sum_{i=1}^{m_k} (a(x_1)p_i'' + b(x_1)p_i' + (c(x_1) + \omega_k a(x_1))p_i - g_i(x_1))\phi_i^k(x') = 0.$$

Multiplying by $\phi_i^k(x')$ and integrating over Ω' we obtain the ordinary differential system

$$ap_i'' + bp_i' + (c + \omega_k a)p_i = g_i.$$

The index of this one-dimensional problem (which is given by the results in the previous section), which we denote by ind_k^0 does not depend on i, and we obtain

$$\text{ind}(L^k) = m_k \, \text{ind}_k^0.$$

3. Computation of $\text{ind}(L_2)$. We now show that if s is sufficiently large, then $\text{ind}(L_2) = 0$. We note that for μ large enough, the operator

$$L - \mu: \quad C^\delta(\bar{\Omega}, \mathbb{R}^p) \longrightarrow C^\delta(\bar{\Omega}, \mathbb{R}^p)$$

with domain E has a bounded inverse defined on all of $C^\delta(\bar{\Omega}, \mathbb{R}^p)$, thus has index 0. Therefore it suffices to show that for all $\lambda \geq 0$ the operator

$$L_2 - \lambda: R(Q_s) \to R(Q_s)$$

is normally solvable and has finite α characteristic (for then we may use again the same method, with $\lambda = \tau\mu$, $\tau \in [0, 1]$). We first prove a preliminary lemma:

Lemma 3.7. *If s is large enough, then for any λ with $\text{Re}(\lambda) \geq 0$, any $j \geq s+1$ and $\xi \in \mathbb{R}$:*

$$\det(-a^\pm \xi^2 + ib^\pm \xi + c^\pm + a^\pm \omega_j - \lambda) \neq 0.$$

Proof. We use the fact that $\omega_j \to -\infty$ as $j \to \infty$. Consider an eigenvalue λ of the matrix

$$(\xi^2 - \omega_j)\left(-a^\pm + \frac{1}{\xi^2 - \omega_j}(ib^\pm \xi + c^\pm)\right).$$

Noting that
$$\left\|\frac{1}{\xi^2 - \omega_j}(ib^\pm \xi + c^\pm)\right\| \leq \frac{1}{2\sqrt{-\omega_j}}\|b^\pm\| + \frac{\|c^\pm\|}{-\omega_j},$$
we obtain that any such λ (recall $a^\pm$ is positive definite) has negative real part, provided j is large enough. This proves the claim. □

Therefore it only remains to prove the following two propositions.

Proposition 3.8. *For all j, ξ, λ, define*
$$E_j^\pm(\xi, \lambda) = -a^\pm \xi^2 + ib^\pm \xi + c^\pm + a^\pm \omega_j - \lambda.$$

If $\lambda \in \mathbb{C}$ is such that
$$\forall \xi \in \mathbb{R},\, j \geq s+1: \quad \det(E_j^\pm(\xi, \lambda)) \neq 0, \tag{3.30}$$
then the operator
$$L_2 - \lambda : R(Q_s) \to R(Q_s)$$
has finite α characteristic.

Proposition 3.9. *Under the same assumption on λ, the operator $L_2 - \lambda$ is normally solvable.*

Before proving these results we show the following lemma.

Lemma 3.10. *Under the same assumption on λ, the problem*
$$a^\pm \Delta u + b^\pm \frac{\partial u}{\partial x_1} + c^\pm u = \lambda u, \quad u_{|S} = 0 \tag{3.31}$$
has no nontrivial solution in $R(Q_s)$.

Proof. Proceeding by contradiction, let us assume that (3.30) holds true, and let $u \in R(Q_s)$ be a nontrivial solution to (3.31). For fixed $k > s$ and $1 \leq i \leq m_k$ we multiply (3.31) by $\phi_i^k(x')$ and integrate over Ω'. Defining p_i^k by
$$p_i^k(x_1) = \int_{\Omega'} u(x_1, x')\phi_i^k(x')\, dx' \in \mathbb{R}^p$$
we obtain after two integration by parts,
$$a^\pm p_i^{k''} + b^\pm p_i^{k'} + (a^\pm \omega_k + c^\pm - \lambda)p_i^k = 0.$$

Here the prime denotes derivation with respect to x_1. Since p_i^k is bounded we may view it as a tempered distribution, thus we may apply the Fourier transform with respect to x_1. Since $E_k^\pm(\xi, \lambda)$ is invertible we obtain that $\hat{p}_i^k$ is identically zero, thus by the inversion theorem p_i^k also is, a contradiction. □

3. Second-order problems in cylinders

Proof of Proposition 3.8. We prove that $\alpha(L_2 - \lambda)$ is finite by showing that the unit ball in the null space $N(L_2 - \lambda)$ is compact in $\tilde{E}_s$. Consider a sequence $u_k \in N(L_2-\lambda)$, with $|u_k|_{2+\delta,\Omega} = 1$. Then we can find a subsequence (still denoted by u_k) which converges to some function u_0 in the space $C^2(A)^p$ for any compact subset A of $\bar{\Omega}$. Note that $u_0 \in C^{2+\delta}(\bar{\Omega}, \mathbb{R}^p)$. We are going to show that the convergence is uniform on $\bar{\Omega}$. If this were not the case, we would have (again for some subsequence)

$$|u_k(x^k) - u_0(x^k)| \geq \delta > 0$$

for some $x^k \in \Omega$. Because of the uniform convergence on compact sets we have $|x^k| \to \infty$. Consider now the following shifted functions:

$$w_k(x) = u_k(x_1 + x_1^k, x') - u_0(x_1 + x_1^k, x').$$

Note that the sequence w_k is bounded in $C^{2+\delta}(\bar{\Omega}, \mathbb{R}^p)$, and that

$$|w_k(0, x_2^k, \cdots, x_n^k)| \geq \delta.$$

Thus we can find a subsequence which converges to some $w_0 \in C^{2+\delta}(\bar{\Omega}, \mathbb{R}^p)$, in $C^2(A)^p$ for any compact subset A of $\bar{\Omega}$. Passing to the limit $k \to \infty$ in the relations

$$a(x_1 + x_1^k)\Delta w_k(x) + b(x_1 + x_1^k)\frac{\partial w_k}{\partial x_1}(x) + c(x_1 + x_1^k)w_k(x) = \lambda w_k(x) \quad x \in \Omega,$$

$$w_k(x) = 0 \quad x \in \partial\Omega,$$

we obtain that w_0 is a nontrivial solution to the problem (3.31). We now show that $w_0 \in R(Q_s)$. Applying Remark 3.5 to u_k we see that u_0, and therefore $u_k - u_0$, belong to $R(Q_s)$. Clearly any shift of an element of $R(Q_s)$ remains in $R(Q_s)$, thus $w_k, w_0 \in R(Q_s)$. This contradicts the conclusion of Lemma 3.10, thus we have shown that u_k converges to u_0 uniformly on $\bar{\Omega}$. It then follows from Schauder's estimates

$$\|u\|^{(2+\delta)} \leq k_1 \|Lu\|^{(\delta)} + k_2 \max_x |u|$$

(see Chapter 1) that the convergence also holds in $C^{2+\delta}(\bar{\Omega}, \mathbb{R}^p)$, and this completes the proof of Proposition 3.8. □

Proof of Proposition 3.9. Pick a sequence $f_n = (L_2 - \lambda)u_n$ in $R(L_2 - \lambda) = R(L) \cap R(Q_s)$, converging to f in $C^\delta(\bar{\Omega}, \mathbb{R}^p)$. Because Q_s is a bounded operator we have $f \in R(Q_s)$. From Proposition 3.8 it follows that $N(L_2-\lambda)$ has a direct complement W in $R(Q_s)$, thus we may write

$$u_n = v_n + w_n, \quad v_n \in N(L_2 - \lambda), w_n \in W.$$

If w_n is bounded in $C^{2+\delta}(\bar{\Omega}, \mathbb{R}^p)$, then as in the previous proof we can obtain a subsequence (still denoted w_n) which converges to some $w_0 \in E$ in $C^2(A, \mathbb{R}^p)$ for any compact subset A of $\bar{\Omega}$. Thus $f_n = (L_2 - \lambda)w_n$ converges to $(L_2 - \lambda)w_0$, and $f \in R(L_2 - \lambda)$. If w_n is not bounded in $\mathcal{C}^{2+\delta}(\bar{\Omega}, \mathbb{R}^p)$, then let us set:

$$\tilde{w}_n = \frac{w_n}{|w_n|_{2+\delta,\Omega}}, \quad \tilde{f}_n = \frac{f_n}{|w_n|_{2+\delta,\Omega}}.$$

As above we extract a subsequence of $\tilde{w}_n$ converging in C^2 on compact sets to some $w_0 \in W$. Then $(L_2 - \lambda)\tilde{w}_n = \tilde{f}_n$ converges to 0, thus $(L_2 - \lambda)w_0 = 0$, a contradiction. This completes the proof of the proposition. □

Thus we have proved the following theorem.

Theorem 3.11. *Consider the operator*

$$L: \quad \mathcal{C}^{\delta}(\bar{\Omega}, \mathbb{R}^p) \longrightarrow \mathcal{C}^{\delta}(\bar{\Omega}, \mathbb{R}^p), \quad D(L) = \{u \in \mathcal{C}^{2+\delta}(\bar{\Omega}, \mathbb{R}^p) : u_{|S} = 0\}$$

defined by

$$Lu = a(x)\Delta u + \sum_{j=1}^{n} b_j(x)\frac{\partial u}{\partial x_j} + c(x)u.$$

Assume that for any $j \geq 2$ we have $b_j(x) \to 0$ as $x_1 \to \pm\infty$, and assume Condition 3.1. Define the coefficients $\tilde{a}(x_1)$, $\tilde{b}(x_1)$, and $\tilde{c}(x_1)$ by (3.22) (3.23), and let ω_k denote the eigenvalues of the Laplace operator in the section of the cylinder, with the boundary condition $u = 0$ on $\partial\Omega'$.

Then the operator L is Fredholm if and only if the operators

$$L^k: \quad \mathcal{C}^{2+\delta}(\mathbb{R}, \mathbb{R}^p) \longrightarrow \mathcal{C}^{\delta}(\mathbb{R}, \mathbb{R}^p)$$

defined by

$$L^k v = \tilde{a}(x_1)v'' + \tilde{b}(x_1)v' + (\tilde{c}(x_1) + \tilde{a}(x_1)\omega_k)v$$

are Fredholm for all k. This occurs if and only if for any integer k and all $\xi \in \mathbb{R}$, the matrices

$$E_k(\xi) := -a^{\pm}\xi^2 + ib_1^{\pm}\xi + c^{\pm} + a^{\pm}\omega_k$$

are invertible. In this case only finitely many of the indices $\mathrm{ind}(L^k)$ are different from zero and

$$\mathrm{ind}(L) = \sum_{k=1}^{\infty} m_k \, \mathrm{ind}(L^k),$$

where m_k is the multiplicity of the eigenvalue ω_k.

4 Lower estimates of elliptic operators

In this section we present another approach to study elliptic problems in cylinders. It is based on lower estimates of elliptic operators. These estimates imply the Fredholm property of linear operators and properness of nonlinear operators. They also allow the construction of topological degree.

4.1 Operators with constant coefficients

In this section we consider the operator

$$\langle Lu, v\rangle = \int_\Omega \left(\sum_{k=1}^m \left(a\frac{\partial u}{\partial x_k}, \frac{\partial v}{\partial x_k}\right) - (bu, v)\right) dx$$

acting from $W_2^1(\Omega)$ into the dual space $(W_2^1(\Omega))^*$. Here Ω is a cylindrical domain with its axis along the x_1-direction and with a bounded smooth cross-section D, $x = (x_1, x')$, $x' \in D$, a and b are constant matrices, a is positive definite, $\langle Lu, v\rangle$ denotes the action of the functional Lu on the element $v \in W_2^1(\Omega)$. We will denote by $[\cdot, \cdot]$ the inner product in $W_2^1(\Omega)$,

$$[u, v] = \int_\Omega \left(\sum_{k=1}^m \left(\frac{\partial u}{\partial x_k}, \frac{\partial v}{\partial x_k}\right) + (u, v)\right) dx$$

and by $\|\cdot\|$ the norm in this space.

We suppose that the following condition is satisfied. It is related to Condition NS and to Conditions 3.1 and 3.1' of the previous section.

Condition 4.1. All eigenvalues of the matrix $b - a\xi^2$ are in the left half-plane for all real ξ.

Theorem 4.2. *There exists a linear, bounded, symmetric, positive definite operator T acting in $W_2^1(\Omega)$, such that for any $u \in W_2^1(\Omega)$,*

$$\langle Lu, Tu\rangle = \|u\|^2. \tag{4.1}$$

To prove the theorem we begin with some auxiliary results. Consider the eigenvalue problem in D:

$$-\Delta' g = \lambda g, \quad \frac{\partial v}{\partial n}\bigg|_{\partial D} = 0, \tag{4.2}$$

where Δ' is the Laplace operator in the section of the cylinder, and n is the outer normal vector. Let g_i be a complete orthonormal system of eigenfunctions of this eigenvalue problem. Then for any $u \in W_2^1(\Omega)$ we have

$$u(x) = \sum_{i=1}^\infty v_i(x_1) g_i(x'), \tag{4.3}$$

where v_i are the coefficients of the expansion

$$v_i(x_1) = \int_D u(x) g_i(x') dx'.$$

The functions v_i belong to $L^2(\mathbb{R})$. Indeed,

$$\int_{-\infty}^\infty v_i^2 dx_1 \leq \int_{-\infty}^\infty dx_1 \left(\int_D |u|^2 dx'\right)\left(\int_D |g_i|^2 dx'\right) \leq \|u\|^2.$$

Similarly,
$$\|v'_i\|_{L^2(\mathbb{R})} \leq \|u\|.$$

Consequently, $v_i \in W_2^1(\mathbb{R})$.

We introduce linear operators T_i, acting in $W_2^1(\Omega)$ by the formula

$$\widetilde{T_i w} = R_i(\xi)\tilde{w}(\xi). \tag{4.4}$$

Here tilde denotes the Fourier transform, $R_i(\xi)$ is a symmetric positive definite matrix, which satisfies the equality

$$((a(\xi^2 + \lambda_i) - b)p, R_i(\xi)p) = ((1 + \xi^2 + \lambda_i)p, p) \tag{4.5}$$

for any vector p.

To construct the matrix $R_i(\xi)$, put

$$c_i(\xi) = \frac{1}{1 + \xi^2 + \lambda_i}(-a(\xi^2 + \lambda_i) + b).$$

Since $\lambda_i \geq 0$, then this matrix is well defined. Thus

$$(c_i(\xi)p, R_i(\xi)p) = -(p, p). \tag{4.6}$$

We put

$$R_i(\xi) = 2\int_0^\infty e^{c_i^*(\xi)s} e^{c_i(\xi)s} ds, \tag{4.7}$$

where c_i^* is the matrix conjugate to c_i. The existence of the integral in the right-hand side of (4.7) follows from Condition 4.1. Indeed, it implies that there exists a contour Γ such that it contains the spectrum of the matrices $c_i(\xi)$ inside it for all real ξ and all i, and it lies in the half-plane Re $\lambda < -\omega$, where ω is a positive number. Then

$$e^{c(\xi)s} = \frac{1}{2\pi i}\int_\gamma e^{s\lambda}(\lambda I - c(\xi))^{-1} d\lambda,$$

where I is the identity matrix. Hence

$$\|e^{c_i(\xi)s}\| \leq Me^{-\omega s}$$

with some positive constant M. Similarly,

$$\|e^{c_i^*(\xi)s}\| \leq Me^{-\omega s}.$$

Therefore the integral in (4.7) exists, and $R_i(\xi)$ is a bounded symmetric matrix. Since

$$R_i c_i + c_i^* R_i = 2\int_0^\infty \frac{d}{ds}\left(e^{c_i^*(\xi)s} e^{c_i(\xi)s}\right) ds = -2I,$$

then (4.5) holds.

4. Lower estimates of elliptic operators

We define now the operator T:

$$Tu = \sum_{i=1}^{\infty} T_i(v_i)g_i(x'), \qquad (4.8)$$

where u is given by (4.3).

Lemma 4.3. *The operator T is a linear, bounded, symmetric and positive definite operator in $W_2^1(\Omega)$.*

Proof. The linearity of T is obvious. We will show that it is bounded. From (4.3) we have

$$\int_\Omega |u|^2 dx = \sum_{i=1}^{\infty} \int_{-\infty}^{\infty} |v_i(x_1)|^2 dx_1,$$

$$\int_\Omega \left|\frac{\partial u}{\partial x_1}\right|^2 dx = \sum_{i=1}^{\infty} \int_{-\infty}^{\infty} |v_i'(x_1)|^2 dx_1.$$

Since

$$\sum_{j=2}^{m} \int_\Omega \frac{\partial g_i}{\partial x_j}\frac{\partial g_k}{\partial x_j} dx = -\int_D \Delta' g_i g_k dx' = \lambda_i \delta_i^k,$$

where δ_i^k is the Kronecker symbol, then

$$\sum_{j=2}^{m} \int_\Omega \left|\frac{\partial u}{\partial x_j}\right|^2 dx = \sum_{i=1}^{\infty} \lambda_i \int_{-\infty}^{\infty} |v_i(x_1)|^2 dx_1.$$

From these equalities and the estimate $\sup_{\xi,i} \|R_i(\xi)\| \leq K$ with some positive constant K we obtain

$$\int_\Omega |Tu|^2 dx = \frac{1}{2\pi}\sum_{i=1}^{\infty} \int_{-\infty}^{\infty} |R_i(\xi)\tilde{v}_i(x_1)|^2 d\xi \leq K^2 \sum_{i=1}^{\infty} \int_{-\infty}^{\infty} |v_i(x_1)|^2 dx_1,$$

$$\int_\Omega \left|\frac{\partial(Tu)}{\partial x_1}\right|^2 dx = \frac{1}{2\pi}\sum_{k=1}^{\infty} \int_{-\infty}^{\infty} |i\xi R_k(\xi)\tilde{v}_k(x_1)|^2 d\xi \leq K^2 \sum_{i=1}^{\infty} \int_{-\infty}^{\infty} |v_i'(x_1)|^2 dx_1,$$

$$\sum_{j=2}^{m} \int_\Omega \left|\frac{\partial(Tu)}{\partial x_j}\right|^2 dx \leq K^2 \sum_{i=1}^{\infty} \lambda_i \int_{-\infty}^{\infty} |v_i(x_1)|^2 dx_1.$$

Thus $\|Tu\| \leq K\|u\|$.

We now show that the operator T is symmetric. Let $w \in W_2^1(\Omega)$ and

$$w = \sum_{i=1}^{\infty} w_i g_i, \quad Tw = \sum_{i=1}^{\infty} T_i(w_i)g_i.$$

We have

$$\int_\Omega (Tu, w)dx = \frac{1}{2\pi} \sum_{i=1}^{\infty} \int_{-\infty}^{+\infty} (R_i\tilde{v}_i, \overline{\tilde{\omega}_i})d\xi,$$

$$\int_\Omega (u, Tw)dx = \frac{1}{2\pi} \sum_{i=1}^{\infty} \int_{-\infty}^{+\infty} (\tilde{v}_i, \overline{R_i\tilde{\omega}_i})d\xi,$$

$$\int_\Omega \left(\frac{\partial(Tu)}{\partial x_1}, \frac{\partial w}{\partial x_1}\right)dx = \frac{1}{2\pi} \sum_{k=1}^{\infty} \int_{-\infty}^{+\infty} (i\xi R_k\tilde{v}_k, \overline{i\xi\tilde{\omega}_k})d\xi,$$

$$\int_\Omega \left(\frac{\partial u}{\partial x_1}, \frac{\partial(Tw)}{\partial x_1}\right)dx = \frac{1}{2\pi} \sum_{k=1}^{\infty} \int_{-\infty}^{+\infty} (i\xi\tilde{v}_k, \overline{i\xi\tilde{R}_k\omega_k})d\xi,$$

$$\sum_{j=2}^{m} \int_\Omega \left(\frac{\partial(Tu)}{\partial x_j}, \frac{\partial w}{\partial x_j}\right)dx = \sum_{i=1}^{\infty} \lambda_i \int_{-\infty}^{+\infty} (T_i(v_i), \omega_i)dx_1,$$

$$\sum_{j=2}^{m} \int_\Omega \left(\frac{\partial u}{\partial x_j}, \frac{\partial(Tw)}{\partial x_j}\right)dx = \sum_{i=1}^{\infty} \lambda_i \int_{-\infty}^{+\infty} (v_i, T(\omega_i))dx_1.$$

Since the matrices R_i are symmetric and real, we obtain

$$[Tu, w] = [u, Tw],$$

that is the operator T is symmetric in $W_2^1(\Omega)$.

To prove that it is positive definite we note that all eigenvalues λ of the matrices $R_i(\xi)$ satisfy the inequality $\lambda \geq \mu$ for some $\mu > 0$ and for all ξ and i. Therefore

$$(R_i(\xi)p, p) \geq \mu(p, p) \quad \forall i, \xi.$$

Then

$$\int_\Omega (Tu, u)dx = \frac{1}{2\pi} \sum_{i=1}^{\infty} \int_{-\infty}^{+\infty} (R_i\tilde{v}_i, \tilde{v}_i)d\xi \geq \mu \int_\Omega |u|^2 dx,$$

$$\sum_{i=1}^{m} \int_\Omega \left(\frac{\partial(Tu)}{\partial x_j}\right)dx \geq \mu \sum_{j=1}^{m} \int_\Omega \left|\frac{\partial u}{\partial x_j}\right|^2 dx.$$

Thus

$$[Tu, u] \geq \mu[u, u].$$

The lemma is proved. □

We note that the operator T is bounded, symmetric, and positive definite in $L^2(\Omega)$ also.

4. Lower estimates of elliptic operators

Proof of Theorem 4.2. To prove the theorem it remains to verify (4.1). We have

$$\langle Lu, Tu\rangle = \int_\Omega \left(\sum_{k=1}^m \left(a\frac{\partial u}{\partial x_k}, \frac{\partial(Tu)}{\partial x_k}\right) - (bu, Tu)\right) dx$$

$$= \frac{1}{2\pi}\sum_{k=1}^\infty \int_{-\infty}^{+\infty} ((ai\xi\tilde{v}_k, \overline{i\xi R_k\tilde{v}_k}) + \lambda_k(a\tilde{v}_k, \overline{R_k\tilde{v}_k}) - (b\tilde{v}_k, \overline{R_k\tilde{v}_k}))d\xi$$

$$= \frac{1}{2\pi}\sum_{k=1}^\infty \int_{-\infty}^{+\infty} ((1+\xi^2+\lambda_i)\tilde{v}_i, \tilde{v}_i)d\xi = \|u\|^2.$$

The theorem is proved. □

4.2 Variable coefficients

Consider the operator

$$\langle Lu, v\rangle = \int_\Omega \left(\sum_{k=1}^m \left(a\frac{\partial u}{\partial x_k}, \frac{\partial v}{\partial x_k}\right) - (b(x)u, v)\right) dx$$

acting in the same spaces as in the previous section. Here a is a constant, symmetric, positive definite matrix, $b(x)$ is a continuous matrix having limits at infinity,

$$b_1 = \lim_{x_1\to-\infty} b(x), \quad b_2 = \lim_{x_1\to\infty} b(x)$$

uniformly with respect to x'. The constant matrices b_1 and b_2 are supposed to satisfy Condition 4.1.

Theorem 4.4. *There exists a linear, bounded, symmetric, positive definite operator S_0 acting in $W_2^1(\Omega)$ such that the estimate*

$$\langle Lu, S_0 u\rangle \geq \|u\|^2 + \theta(u)$$

holds. Here $\theta(u)$ is a functional defined on $W_2^1(\Omega)$ and such that $\theta(u_n) \to 0$ as $u_n \to 0$ weakly.

Proof. We first consider the case where $b(x) = b_0(x)$ and

$$b_0(x) = \phi_1(x)b_1 + \phi_2(x_1)b_2,$$

$\phi_i(x)$, $i=1,2$ are smooth functions, $0 \leq \phi_i(x_1) \leq 1$, $\phi_1(x_1) + \phi_2(x_1) \equiv 1$, $\phi_1(x_1) = 0$ for $x_1 > 1$, $\phi_2(x_1) = 0$ for $x_1 < -1$.

Denote by $T^{(i)}$ the operator which is defined for the matrix b_i as it is done in Theorem 4.2, and

$$T^{(0)} = \phi_1 T^{(1)}\phi_1 + \phi_2 T^{(2)}\phi_2.$$

We have

$$\int_\Omega \sum_{k=1}^m \left(a\frac{\partial u}{\partial x_k}, \frac{\partial(T^{(0)}u)}{\partial x_k}\right) dx = \sum_{i=1}^2 \int_\Omega \sum_{k=1}^m \left(a\frac{\partial(\phi_i u)}{\partial x_k}, \frac{\partial(T^{(i)}\phi_i u)}{\partial x_k}\right) dx$$
$$+ \sum_{i=1}^2 \int_\Omega \sum_{k=1}^m \left(\left(a\frac{\partial u}{\partial x_1}, \phi_i' T^{(i)}\phi_i u\right) - \left(au\phi_i', \frac{\partial(T^{(i)}\phi_i u)}{\partial x_1}\right)\right) dx. \quad (4.9)$$

The second summand in the right-hand side of (4.9) tends to zero as $u \to 0$ weakly (see Lemma 4.5 below). We have further

$$\int_\Omega (bu, T^{(0)}u) dx = \sum_{i=1}^2 \int_\Omega (\phi_i b_i u, \phi_i T^{(i)} \phi_i u) dx + \theta(u)$$
$$= \sum_{i=1}^2 \int_\Omega (\phi_i b_i u, T^{(i)} \phi_i u) dx + \theta(u).$$

Here $\theta(u)$ denotes all functionals that satisfy the condition of the theorem. Thus

$$\langle Lu, T^{(0)}\rangle = \sum_{i=1}^2 \int_\Omega \left(\sum_{k=1}^m \left(a\frac{\partial(\phi_i u)}{\partial x_k}, \frac{\partial(T^{(i)}\phi_i u)}{\partial x_k}\right) - (b_i \phi_i u, T^{(i)}\phi_i u)\right) dx + \theta(u).$$

From Theorem 4.2 it follows that

$$\langle Lu, T^{(0)}u\rangle = \|\phi_1 u\|^2 + \|\phi_2 u\|^2 + \theta(u) \geq \frac{1}{2}\|u\|^2 + \theta(u). \quad (4.10)$$

Let now $b(x)$ be an arbitrary matrix satisfying the conditions above. Then from Lemma 4.5 below it follows that

$$\int_\Omega ((b - b_0))u, u) dx = \theta(u).$$

To complete the proof of the theorem we should show that

$$T^{(0)} = \frac{1}{2}S^{(0)} + K, \quad (4.11)$$

where $S^{(0)}$ is a symmetric, positive definite operator, K is a compact operator in $W_2^1(\Omega)$.

We note that the operator T_i defined by (4.4) satisfies the equality

$$\frac{\partial(T_i u)}{\partial x_1} = T_i \frac{\partial u}{\partial x_1}$$

4. Lower estimates of elliptic operators

and, consequently, a similar equality is valid for the operator T and for the operators $T^{(i)}$. Since the operators $T^{(i)}$ are symmetric in L^2 we have

$$\int_\Omega (\phi_i T^{(i)} \phi_i u, v) dx = \int_\Omega (u, \phi_i T^{(i)} \phi_i v) dx,$$

$$\int_\Omega \left(\phi_i T^{(i)} \phi_i \frac{\partial u}{\partial x_1}, \frac{\partial v}{\partial x_1} \right) dx = \int_\Omega \left(\frac{\partial u}{\partial x_1}, \phi_i T^{(i)} \phi_i \frac{\partial v}{\partial x_1} \right) dx,$$

$$\sum_{k=2}^m \int_\Omega \left(\frac{\partial}{\partial x_k}(\phi_i T^{(i)} \phi_i u), \frac{\partial v}{\partial x_k} \right) dx = \sum_{k=2}^m \int_\Omega \left(\frac{\partial u}{\partial x_k}, \frac{\partial}{\partial x_k}(\phi_i T^{(i)} \phi_i v) \right) dx.$$

It is easy to verify now that

$$[T^{(0)} u, v] - [u, T^{(0)} v]$$
$$= \sum_{i=1}^2 \int_\Omega \left(\left(\phi_i' T^{(i)} \phi_i u + \phi_i T^{(i)} \phi_i' u, \frac{\partial v}{\partial x_1} \right) - \left(\frac{\partial u}{\partial x_1}, \phi_i' T^{(i)} \phi_i v + \phi_i T^{(i)} \phi_i' v \right) \right) dx.$$

Denote the right-hand side of this equality by $\Phi(u, v)$. This is a bilinear bounded functional in $W_2^1(\Omega)$. Therefore

$$\Phi(u, v) = [u, K^{(0)} v], \quad u, v \in W_2^1(\Omega),$$

where K^0 is a linear bounded operator. We will prove that it is compact. Indeed, let $v_n \to 0$ weakly in $W_2^1(\Omega)$. Denote $y_n = K^0 v_n$. Then

$$\|y_n\|^2 = [y_n, K^0 v_n] = \Phi(y_n, v_n).$$

From Lemma 4.5 below it follows that $\Phi(y_n, v_n) \to 0$ since $v_n, y_n \to 0$ weakly in $W_2^1(\Omega)$ and the derivatives $\partial v_n / \partial x_1$, $\partial y_n / \partial x_1$ are uniformly bounded in $L^2(\Omega)$. This proves the compactness of the operator K^0 and, consequently, the compactness of the operator

$$(T^{(0)})^* - T^{(0)} = K^0. \tag{4.12}$$

We have further

$$[T^{(0)} u, v] + [u, T^{(0)} v] = \Phi_1(u, v) + \Phi_2(u, v),$$

where

$$\Phi_1(u, v) = \sum_{i=1}^2 \int_\Omega \sum_{k=1}^m \left(\left(\frac{\partial}{\partial x_k}(T^{(i)} \phi_i u), \frac{\partial}{\partial x_k}(\phi_i v) \right) \right.$$
$$\left. + \left(\frac{\partial}{\partial x_k}(\phi_i u), \frac{\partial}{\partial x_k}(T^{(i)} \phi_i v) \right) \right) dx$$
$$+ 2 \sum_{i=1}^2 \int_\Omega (T^{(i)} \phi_i u, \phi_i v) dx,$$

$$\Phi_2(u,v) = \sum_{i=1}^{2} \int_\Omega \left(\left(\phi_i' T^{(i)} \phi_i u, \frac{\partial v}{\partial x_1} \right) \right.$$
$$\left. - \left(\frac{\partial}{\partial x_1}(T^{(i)}\phi_i u), \phi_i' v \right) + \left(\frac{\partial u}{\partial x_1}, \phi_i' T^{(i)} \phi_i v \right) \right) dx$$
$$- \sum_{i=1}^{2} \int_\Omega \left(\phi_i' u, \frac{\partial}{\partial x_1}(T^{(i)}\phi_i v) \right) dx.$$

Since Φ_1 and Φ_2 are bounded bilinear functionals in $W_2^1(\Omega)$ then
$$\Phi_1(u,v) = [S_0 u, v], \quad \Phi_2(u,v) = [Bu, v],$$
where S_0 and B are bounded linear operators. The operator S_0 is symmetric and positive definite since the functional $\Phi_1(u,v)$ is symmetric and
$$[S_0 u, u] = \Phi_1(u,u) = \sum_{i=1}^{2}([T^{(i)}\phi_i u, \phi_i u] + [\phi_i u, T^{(i)}\phi_i u]) \geq 2\mu \sum_{i=1}^{2} \|\phi_i u\|^2 \geq \mu \|u\|^2.$$

From Lemma 4.5 it follows that the operator B is compact.

Equality (4.11) with $K = (B-K^0)/2$ follows now from (4.12) and the equality
$$T^0 + (T^0)^* = S_0 + B.$$

The theorem is proved. □

We used the following lemma.

Lemma 4.5. *Let a sequence of functions f_n be uniformly bounded in $L^2(\Omega)$ and a sequence converge weakly to zero in $W_2^1(\Omega)$. Let, further, $\psi(x)$ be a bounded smooth function which tends to zero as $x_1 \to \pm\infty$ uniformly with respect to x'. Then*
$$\int_\Omega \psi(x)(f_n(x), g_n(x)) dx \to 0, \quad n \to \infty.$$

Proof. We have
$$\left| \int_\Omega \psi(f_n, g_n) dx \right| \leq \left(\int_\Omega |f_n|^2 dx \right)^{1/2} \left(\int_\Omega \psi^2 |g_n|^2 dx \right)^{1/2},$$
and it is sufficient to show that the second integral in the right-hand side of this inequality tends to zero. We represent it in the form
$$\int_\Omega \psi^2 |g_n|^2 dx = \int_{|x_1| \geq R} \int_D \psi^2 |g_n|^2 dx_1 dx' + \int_{|x_1| < R} \int_D \psi^2 |g_n|^2 dx_1 dx'. \quad (4.13)$$

For any given $\epsilon > 0$ we can choose R such that the first integral in the right-hand side of (4.13) is less than $\epsilon/2$ since $\psi \to 0$ as $x_1 \to \pm\infty$. For R fixed the second

4. Lower estimates of elliptic operators

integral tends to zero as $n \to \infty$ since weak convergence of g_n in $W_2^1(\Omega)$ implies strong convergence of ϕg_n in $L^2(\Omega)$ for any smooth function ϕ with a bounded support, and hence strong convergence of g_n in $L^2(\Omega_R)$. Here $\Omega_R = D \times \{x_1 < R\}$. Thus there is an integer number N such that $\|\psi g_n\|_{L^2(\Omega)}^2 \leq \epsilon$ for $n \geq N$. The lemma is proved. $\square$

4.3 Weighted spaces

We now consider the weighted space $W_{2,\mu}^1(\Omega)$ with the inner product

$$[u,v]_\mu = \int_\Omega \left(\sum_{k=1}^m \left(\frac{\partial u}{\partial x_k}, \frac{\partial v}{\partial x_k} \right) + (u,v) \right) \mu(x_1) dx.$$

The weight function μ depends on x_1 only, and it is supposed to satisfy the following conditions:

1. $\mu(x_1) \geq 1$, $\mu(x_1) \to \infty$ as $|x_1| \to \infty$,
2. μ'/μ and μ''/μ are continuous functions that tend to zero as $|x_1| \to \infty$.

For example we can take $\mu(x_1) = 1 + x_1^2$.

We consider the operator $L: W_{2,\mu}^1(\Omega) \to (W_{2,\mu}^1(\Omega))^*$ given by

$$\langle Lu, v \rangle = \int_\Omega \left(\sum_{k=1}^m \left(a \frac{\partial u}{\partial x_k}, \frac{\partial v}{\partial x_k} \right) - (bu, v) \right) \mu dx,$$

where $u, v \in W_{2,\mu}^1(\Omega)$, the matrices a and b satisfy the same conditions as in the previous section.

Theorem 4.6. *Let the matrices b_1 and b_2 satisfy Condition 4.1. Then there exists a linear, bounded, symmetric, positive definite operator S acting in $W_{2,\mu}^1(\Omega)$ such that for any $u \in W_{2,\mu}^1(\Omega)$,*

$$\langle Lu, Su \rangle \geq \|u\|_\mu^2 + \theta_\mu(u),$$

where $\|\cdot\|_\mu$ is the norm in $W_{2,\mu}^1(\Omega)$, $\theta_\mu(u)$ is a functional defined on $W_{2,\mu}^1(\Omega)$ and such that $\theta_\mu(u_n) \to 0$ as $u_n \to 0$ weakly.

Proof. We put
$$T = \omega^{-1} T^{(0)} \omega,$$
where $\omega = \sqrt{\mu}$ and $T^{(0)}$ is defined in the previous section. Introducing the notation $w = \omega u$, we obtain

$$\langle Lu, Tu \rangle = \int_\Omega \left(\sum_{k=1}^m \left(a \frac{\partial w}{\partial x_k}, \frac{\partial}{\partial x_k}(T^{(0)} w) \right) - (bw, T^{(0)} w) \right) dx + I,$$

where

$$I = \int_\Omega \left(\left(a\omega^{-1}(\omega^{-1})' \frac{\partial w}{\partial x_1}, T^{(0)}w \right) + (a(\omega^{-1})'w, (\omega^{-1})'T^{(0)}w) \right) \omega^2 dx$$

$$+ \int_\Omega \left(a(\omega^{-1})'w, \omega^{-1} \frac{\partial}{\partial x_1}(T^{(0)}w) \right) \omega^2 dx \ .$$

It is easy to verify that the operator of multiplication by ω is a bounded operator from $W^1_{2,\mu}(\Omega)$ into $W^1_2(\Omega)$, and the operator of multiplication by ω^{-1} is a bounded operator from $W^1_2(\Omega)$ into $W^1_{2,\mu}(\Omega)$. Hence $w \in W^1_2(\Omega)$, and from (4.10) we have

$$\int_\Omega \left(\sum_{k=1}^m \left(a \frac{\partial w}{\partial x_k}, \frac{\partial}{\partial x_k}(T^{(0)}w) \right) - (bw, T^{(0)}w) \right) dx$$

$$\geq \frac{1}{2} \|w\|^2 + \theta(w) \geq c\|u\|_\mu^2 + \theta_\mu(u),$$

where $c = 1/(2N^2)$, N is the norm of the multiplication operator ω^{-1}.

From Lemma 4.5 it follows that $I \to 0$ as $w \to 0$ weakly in $W^1_2(\Omega)$. Thus we have shown that

$$\langle Lu, Tu \rangle \geq c\|u\|_\mu^2 + \theta_\mu(u).$$

We use the notation θ_μ here for all functionals which satisfy the condition of the theorem.

To complete the proof of the theorem it is sufficient to show that

$$T = cS + K, \qquad (4.14)$$

where S is a symmetric positive definite operator, K is a compact operator in $W^1_{2,\mu}(\Omega)$. For this we construct the operators acting in $W^1_{2,\mu}(\Omega)$ from the operators acting in $W^1_2(\Omega)$ in the following way. To each linear bounded operator A acting in $W^1_2(\Omega)$ we assign a linear operator A_μ in $W^1_{2,\mu}(\Omega)$ by means of the equality

$$[u, A_\mu v]_\mu = [\omega u, A\omega v], \quad u, v \in W^1_{2,\mu}(\Omega), \qquad (4.15)$$

where, as above, $[\cdot, \cdot]_\mu$ and $[\cdot, \cdot]$ are the inner products in $W^1_{2,\mu}(\Omega)$ and in $W^1_2(\Omega)$, respectively. Going over in (4.11) to operators acting in $W^1_{2,\mu}(\Omega)$ by this rule, we obtain $T^{(0)}_\mu = \frac{1}{2} S^{(0)}_\mu + K_\mu$. From Lemma 4.7 below it follows that $S^{(0)}_\mu$ is a bounded symmetric, positive definite operator, K_μ is a compact operator. By the same lemma and equation $T = \omega^{-1}T^{(0)}\omega$ we have $T^{(0)}_\mu = T + B$, where B is a compact operator in $W^1_{2,\mu}(\Omega)$. Hence we obtain (4.14), where $S = \frac{1}{2c}S^{(0)}_\mu$, $K = K_\mu - B$. The theorem is proved. □

Lemma 4.7. *Let an operator A_μ acting in $W^1_{2,\mu}(\Omega)$ be defined according to the operator A acting in $W^1_2(\Omega)$ by equality (4.15).*

1. *If A is a linear, bounded operator, then A_μ is also linear and bounded.*
2. *If A is compact, then A_μ is also compact.*

4. Lower estimates of elliptic operators

3. If A is symmetric, positive definite, then A_μ is also symmetric and positive definite.
4. $A_\mu = \omega^{-1} A \omega + B$, where B is a linear compact operator in $W_{2,\mu}^1(\Omega)$.

Proof. Assertions 1–3 can be verified directly. We shall prove assertion 4. Set $\tilde{A} = \omega^{-1} A \omega$. We have for $u, v \in W_{2,\mu}^1(\Omega)$:

$$[u, \tilde{A}v]_\mu = \int_\Omega \left(\sum_{k=1}^m \left(\frac{\partial u}{\partial x_k}, \frac{\partial (\tilde{A}v)}{\partial x_k} \right) + (u, \tilde{A}v) \right) \mu dx.$$

Put $y = \omega u$, $z = \omega v$. Then we obtain

$$[u, \tilde{A}v]_\mu = [y, Az] + \Phi(y, z),$$

where

$$\Phi(y, z) = \int_\Omega \left(((\omega^{-1})'\omega y, (\omega^{-1})' \omega A z) + \left(\frac{\partial y}{\partial x_1}, (\omega^{-1})' \omega A z \right) \right.$$
$$\left. + \left((\omega^{-1})' \omega y, \frac{\partial (Az)}{\partial x_1} \right) \right) dx.$$

We note that $\Phi(y, z)$ is a bilinear bounded form in $W_2^1(\Omega)$. Therefore

$$\Phi(y, z) = [y, Kz],$$

where K is a linear bounded operator in $W_2^1(\Omega)$. From Lemma 4.5 it follows, as above, that K is compact. Hence (see assertion 2) K_μ is compact in $W_{2,\mu}^1(\Omega)$. We have further

$$[u, \tilde{A}v]_\mu = [u, A_\mu v]_\mu + [u, K_\mu v]_\mu.$$

This means that $\tilde{A} = A_\mu + K_\mu$. The lemma is proved. $\square$

4.4 Fredholm property and applications

The lower estimates of elliptic operators obtained above allow us to prove their Fredholm property and the unique solvability for operators with a parameter. Similar to Section 4.2, we consider the operator $L: W_2^1(\Omega) \to (W_2^1(\Omega))^*$ defined by the equality

$$\langle Lu, v \rangle = \int_\Omega \left(\sum_{k=1}^m \left(a \frac{\partial u}{\partial x_k}, \frac{\partial v}{\partial x_k} \right) - (b(x)u, v) \right) dx,$$

and also the operator J,

$$\langle Ju, v \rangle = \int_\Omega (u, v) dx$$

acting in the same spaces.

Theorem 4.8. *Let Condition 4.1 be satisfied. Then for all $\lambda \geq 0$ the operator $L+\lambda J$ is Fredholm. For all such λ, except possibly for a finite number, it has a bounded inverse defined on the whole $(W_2^1(\Omega))^*$.*

The proof of this theorem is similar to the proof of Theorem 1.4 of Chapter 2 in [568] and we will not present it here. The estimates of this type were first obtained in [561] in the one-dimensional case. The theorem above remains valid for the operators acting in weighted spaces (Section 4.3) and for more general elliptic problems. Compared with the method of limiting operators developed in Chapters 4 and 5, the method of lower estimates is more explicit but it uses the specific form of the operators and of the domains and it cannot be used for general elliptic problems.

The lower estimates of the same type hold for nonlinear elliptic problems [562]. They provide Condition α introduced by Skrypnik [492], [493], which allows construction of a topological degree. In the case of unbounded domains, we need to use weighted spaces. Otherwise, the operators may not be proper and the degree may not be defined (Chapter 11).

Chapter 10

Non-Fredholm Operators

The theory of elliptic problems is essentially based on their Fredholm property which determines solvability conditions and a well-defined index. The Fredholm property and index are preserved under small perturbations of the operators. The situation is quite different if the Fredholm property is not satisfied. A general theory of such problems does not exist, solvability conditions are generally not known, and properties of such problems may not be preserved under small perturbations of the operators.

In this chapter we introduce weakly non-Fredholm operators (Section 1). It is a class of operators which satisfy the Fredholm property in spaces with a small exponential weight. In this case we can formulate solvability conditions. They can be similar to those for Fredholm operators or different, depending on function spaces and on the right-hand side of the equation.

As we know, the Fredholm property is determined by Conditions NS and NS* for limiting operators. If the set of limiting operators is finite, then "good" limiting operators, which satisfy these conditions, preserve this property under small perturbations. In this sense we can say that the structure of weakly non-Fredholm operators is stable under small perturbations. In some cases, we can use this spatial structuring and separate Fredholm and non-Fredholm parts of the operator (Section 3).

In Section 2 we study ordinary differential equations on the real axis and introduce some other solvability conditions. They involve solutions of homogeneous adjoint equations but cannot be written as functionals from the dual space. For some classes of strongly non-Fredholm operators, briefly discussed at the end of this chapter, solvability conditions can be obtained by the spectral theory of self-adjoint operators.

1 Weakly non-Fredholm operators

It was shown in Chapter 5 that Conditions NS and NS* provide the Fredholm property of elliptic operators in unbounded domains. Condition NS is necessary and sufficient for normal solvability with a finite-dimensional kernel. Condition NS* implies a finite codimension of the image. In the same chapter we have introduced the operator $L_\mu = S_\mu L S_{-\mu}$, where S_μ is the operator of multiplication by an exponential weight. If L is a Fredholm operator, then the operator L_μ satisfies the Fredholm property for the values of μ in some domain G_0 of the complex plane which contains the origin. At the boundary of this domain, the operator L_μ is not Fredholm. Thus, the complex plane can be represented as a union of a finite or countable number of domains G_i, where the operator L_μ is Fredholm, and of a complementary closed set, where the Fredholm property is not satisfied.

In this section we will study the operators which belong to the boundaries of the domains G_i. More precisely, suppose that the operator L does not satisfy the Fredholm property and the operator L_μ satisfies it for all μ sufficiently small. We will call such operators weakly non-Fredholm operators. It appears that solvability conditions for such operators can be determined. There are several different situations. In the simplest case, the solvability conditions are the same as for Fredholm operators: the equation $Lu = f$ is solvable if and only if $(f, v) = 0$ for all solutions v of the homogeneous adjoint equation $L^*v = 0$. It is also possible that this condition provides the existence of solutions in a weaker sense. We call them solutions in the sense of sequence. This means that there exists a sequence u_n such that $Lu_n = f_n \to f$ as $n \to \infty$ but this sequence is not necessarily convergent to a solution u of the equation $Lu = f$. Finally, the expression (f, v) may not be defined. We recall that v is exponentially decaying at infinity in the case of Fredholm operators. However, this may not be the case for non-Fredholm operators resulting in the fact that v does not belong to the dual space. Then we define solvability conditions in the sense of sequence: $(f_n, v) = 0$, $n = 1, 2, \ldots$, where the sequence f_n is such that the last expressions are well defined and $f_n \to f$. This convergence can be strong or local depending on the function spaces where the operator acts. We illustrate these situations in Section 1.3 with the example of one-dimensional operators.

1.1 Some properties of operators

Suppose that the convergence $Lu_k \to f$ takes place for some sequence of functions u_k. Is it possible to conclude that the equation $Lu = f$ has a solution? This question is directly related to normal solvability of the operator L. Furthermore, we can use this convergence to generalize the notion of solution. We will call a sequence $u_1, u_2, \ldots$ the solution in the sense of sequence of the equation $Lu = f$ (or the sequence solution) if $Lu_k \to f$. The sequence $u_1, u_2, \ldots$ provides approximation of the usual solution in the case of Fredholm operators. For non-Fredholm operators, sequence solutions may not converge to usual solutions.

1. Weakly non-Fredholm operators

We briefly recall here the notion of G-convergence (see, for example, [593], [594]) which is related to the notion of sequence solutions but also differs from it. A sequence of operators A_k G-converges to an operator A if for any right-hand side f a sequence u_k of solutions of the equations $A_k u = f$ converges to a solution of the equation $Au = f$. Function spaces, and in what sense the sequence u_k converges, should be specified. Thus, G-convergence implies convergence of solutions for any right-hand side. Convergence of sequence solutions depends on the right-hand side and may not occur.

1.1.1. Solutions in the sense of sequence. Consider a bounded operator $L : E \to F$, where E and F are Banach spaces, and the equation

$$Lu = f. \tag{1.1}$$

Definition 1.1. A sequence $u = (u_1, u_2, \ldots)$ such that $u_k \in E$ is called a solution of equation (1.1) in the sense of sequence or a sequence solution if $\|Lu_k - f\|_F \to 0$ as $k \to \infty$. We say that this solution belongs to the space E if $\|u_k\|_E \leq M$ for some constant M and for all k. A sequence solution $u = (u_1, u_2, \ldots)$ is strongly convergent if there exists $u_0 \in E$ such that $\|u_k - u_0\|_E \to 0$ as $k \to \infty$. Two sequence solutions $u = (u_1, u_2, \ldots)$ and $v = (v_1, v_2, \ldots)$ are equivalent if $\|u_k - v_k\|_E \to 0$ as $k \to \infty$.

If a sequence solution is strongly convergent, then u_0 is a solution of equation (1.1) in the usual sense. Therefore, we can identify strongly convergent sequence solutions with usual solutions. In this case, we will use the same notation u for the function and for the sequence. If two sequence solutions are strongly convergent and equivalent, then they correspond to the same usual solution.

A subsequence $\hat{u} = (u_{k_1}, u_{k_2}, \ldots)$ of a sequence solution $u = (u_1, u_2, \ldots)$ will be called a sub-solution of the solution u.

Proposition 1.2. *Let $u \in E$ be a sequence solution of equation (1.1). If L is a Fredholm operator, then f belongs to its image and there exists a strongly convergent sub-solution $\hat{u}$ of the solution u, that is $\hat{u}$ is the usual solution of equation (1.1).*

We recall that the image of the Fredholm operator L is closed, its kernel has a finite dimension, the operator is invertible from the complement to its kernel into its image. This provides the assertion of the proposition.

In what follows we consider elliptic operators and the corresponding function spaces. In the case of bounded domains, sequence solutions approximate usual solutions.

Example 1.3. Consider the Poisson equation with the Dirichlet boundary condition

$$\Delta u = f, \quad u|_{\partial \Omega} = 0, \tag{1.2}$$

where $f \in L^2(\Omega)$, Ω is a bounded domain with the $C^{2+\alpha}$ boundary. Let f_k be infinitely differentiable functions converging to f in $L^2(\Omega)$, and u_k such that $\Delta u_k =$

f_k. Then $u = (u_1, u_2, \dots)$ is a sequence solution of problem (1.2). It converges in $H^2(\Omega)$ to the usual solution u of this problem.

In the case of bounded domains with sufficiently smooth boundaries, sequence solutions are equivalent to usual solutions, and their introduction does not have much sense. The difference between sequence solutions and usual solutions becomes clear in the case of unbounded domains.

Theorem 1.4. *Suppose that Condition NS is satisfied. If $u \in E_\infty$ is a sequence solution of equation (1.1), then $f \in F_\infty$ belongs to the image of the operator, u has a strongly convergent sub-solution and equation (1.1) is solvable in the usual sense.*

Proof. We know that Condition NS provides normal solvability of the operator and a finite dimension of its kernel. The assertion of the theorem follows from this. It can be also proved directly. Consider for certainty Sobolev spaces. Since $u \in E_\infty$, then the sequence u_k is uniformly bounded in the norm of this space. Therefore, there exists a subsequence of this sequence which converges to some function $u_0 \in E_\infty$ locally in L^2. Moreover, $Lu_0 = f$. From Theorem 2.10 (Chapter 4), it follows that this convergence is strong. The theorem is proved. □

Remark 1.5. To prove the existence of a usual solution, we do not need to assume that the sequence solution belongs to the space E_∞. If $u = (u_1, u_2, \dots)$, $u_k = v_k + w_k$, where w_k belongs to the kernel of the operator and v_k to its supplement, then $Lv_k \to f$. Hence the sequence v_k is uniformly bounded and converges to some $v_0 \in E_\infty$, $Lv_0 = f$.

If Condition NS is not satisfied, that is one of the limiting problems has a nonzero bounded solution in the corresponding space, then there exist a sequence solution $u \in E_\infty$ of the equation $Lu = 0$ such that it does not have a strongly convergent sub-solution. Such a sequence can be constructed as in the proof of Theorem 2.15 (Chapter 4). Thus, any solution in the sense of sequence corresponds to a usual solution if and only if Condition NS is satisfied.

The notion of solutions in the sense of sequence is related to approximation of solutions. If we consider an operator equation $Lu = f$, then we conventionally understand approximation of its solution u by a sequence of functions u_k if this sequence converges to u in some norm. We can also understand it in a different way. Namely, if Lu_k converges to f. The discussion above shows that they may not be equivalent.

1.1.2. Exponential dichotomy. We proved in Chapter 4 that all bounded solutions of the equation $Lu = 0$ exponentially decay at infinity if the operator L is normally solvable with a finite-dimensional kernel. In the previous chapter, we discussed the converse assertion for one-dimensional problems: exponential decay or growth of solutions at infinity ensure the Fredholm property. We will introduce the notion of exponential dichotomy for solutions in the sense of sequence and will discuss its relation with normal solvability.

1. Weakly non-Fredholm operators

We recall the definition of the norm in the space E_∞ for an unbounded domain Ω:

$$\|u\|_{E_\infty(\Omega)} = \sup_i \|\phi_i u\|_{E(\Omega)},$$

where ϕ_i is a partition of unity. We will suppose that supports of the functions ϕ_i are unit balls B_i with their centers at some points x_i.

Definition 1.6. A solution $u = (u_1, u_2, \ldots)$ of the equation $Lu = 0$ in the sense of sequence, where the operator L acts from $E_\infty(\Omega)$ into $F_\infty(\Omega)$, satisfies the condition of exponential dichotomy in some sequence of domains $\Omega_k \subset \Omega$ if there exist positive constants M and ϵ such that one of the following estimates

$$\|u_k \phi_i\|_E \leq M e^{\epsilon(|x_j|-|x_i|)} \|u_k \phi_j\|_E, \quad \forall x_i, x_j \in \Omega_k, \quad |x_i| > |x_j|, \quad (1.3)$$

$$\|u_k \phi_j\|_E \leq M e^{\epsilon(|x_j|-|x_i|)} \|u_k \phi_i\|_E, \quad \forall x_i, x_j \in \Omega_k, \quad |x_i| > |x_j| \quad (1.4)$$

holds for all k. Here M and ϵ are independent of the solution and of k.

This condition provides an exponential decay or growth of sequence solutions with respect to local norms. It generalizes the conventional notion of exponential dichotomy for solutions of ordinary differential systems of equations. We note that the exponential estimate in (1.3) and (1.4) depends on the difference of the modulus x_i and x_j and not on the distance between these points. This definition is more convenient for what follows. Another definition is also possible.

We next define sectors $Q(\xi, \delta)$ in $\mathbb{R}^n$. Let S be the unit sphere, $S = \{x \in \mathbb{R}^n, |x| = 1\}$. We put $Q(\xi, \delta) = \{x \in \mathbb{R}^n, x = \zeta t\}$, for all $t > 0$ and $\zeta \in S$, $|\zeta - \xi| < \delta$. Let

$$\Omega(\xi, \delta, \rho) = \Omega \cap Q(\xi, \delta) \cap \{x \in \mathbb{R}^n, |x| > \rho\}.$$

Definition 1.7. The operator $L : E_\infty(\Omega) \to F_\infty(\Omega)$ satisfies the condition of exponential dichotomy at infinity if, for any solution $u = (u_1, u_2, \ldots)$ of the equation $Lu = 0$ in the sense of sequence, there exist δ and ρ such that, for any ξ and any sequence of domains $\Omega_k \subset \Omega(\xi, \delta, \rho)$, the sequence solution u satisfies the condition of exponential dichotomy in these domains.

Theorem 1.8. *If the operator $L : E_\infty(\Omega) \to F_\infty(\Omega)$ satisfies the condition of exponential dichotomy at infinity, then it is normally solvable with a finite-dimensional kernel.*

Proof. We will prove that the operator L satisfies Condition NS. Then the assertion of the theorem will follow from Theorem 2.13 of Chapter 4. Suppose that this is not the case and there exists a limiting operator $\hat{L} : E_\infty(\hat{\Omega}) \to F_\infty(\hat{\Omega})$ such that the equation $\hat{L}u = 0$ has a nonzero solution $\hat{u} \in E_\infty(\hat{\Omega})$. We then construct a solution of the equation $Lu = 0$ in the sense of sequence, which does not satisfy the condition of exponential dichotomy at infinity.

Similar to Theorem 2.15 of Chapter 4, we can construct a sequence $u_k(x) = \psi_k(x - h_k)\hat{u}(x - h_k)$ such that $Lu_k \to 0$ as $k \to \infty$. Here h_k is a sequence of constant vectors for which $|h_k| \to \infty$ as $k \to \infty$, $\psi_x(x)$ is a C^∞ function equal 1 for $|x| \le r_k$ and 0 for $|x| \ge R_k$, where $r_k, R_k \to \infty$ as $k \to \infty$.

We will verify that the sequence solution $u = (u_1, u_2, \ldots)$ does not satisfy the condition of exponential dichotomy at infinity. In order to do this, for any δ and ρ fixed, we need to construct a sequence of domains $\Omega_k \subset \Omega(\xi, \delta, \rho)$ such that the sequence u_k does not satisfy either (1.3) or (1.4).

First of all, we choose the value of ξ. Let us take projections ξ_k of the points h_k at the unit sphere S, and choose a convergent subsequence, for which we keep the same notation, $\xi_k \to \xi_0$ as $k \to \infty$. Consider the domain $\Omega(\xi_0, \delta_0, \rho_0)$ for some δ_0 and ρ_0 and put $\Omega_k = \Omega \cap B(h_k, \hat{r}_k)$, where $B(h_k, \hat{r}_k)$ is the ball with center at h_k and radius $\hat{r}_k$. The values $\hat{r}_k$ can be chosen in such a way that $\hat{r}_k < r_k$, $\hat{r}_k \to \infty$ as $k \to \infty$ and $\Omega_k \subset \Omega(\xi_0, \delta_0, \rho_0)$.

Suppose that the sequence u_k satisfies inequality (1.3). Put $x_i^k = h_k$. We can choose x_j^k in such a way that $|x_j^k| = |x_i^k| - \hat{r}_k$ and $x_j^k \in \Omega_k$. Then

$$\|u_k \phi_{i_k}\|_E \le M e^{-\hat{r}_k} \|u_k \phi_{j_k}\|_E \le MCe^{-\hat{r}_k} \|\hat{u}\|_{E_\infty(\Omega)}.$$

Since u_k is obtained by means of a shift of the function $\hat{u}$, then the last inequality and the choice of x_i^k imply

$$\|\hat{u}\phi_0\|_E \le MCe^{-\hat{r}_k} \|\hat{u}\|_{E_\infty(\Omega)},$$

where ϕ_0 is the function from the partition of unity in the definition of the E_∞-norm such that its support is the unit ball with center at the origin. Since $\hat{r}_k \to \infty$, then $\|\hat{u}\phi_0\|_E = 0$. Similarly, it can be shown that $\|\hat{u}\phi_i\|_E = 0$ for any i. Therefore, $\hat{u} = 0$. This contradicts the assumption that this is a nonzero solution. The proof remains similar if the sequence u_k satisfies inequality (1.4). The theorem is proved. □

The condition of exponential dichotomy provides an exponential decay or growth of solutions at infinity. In this case, limiting problems do not have nonzero bounded solutions and the operator is normally solvable. Suppose that exponential dichotomy does not hold and there is a limiting problem which has a nonzero bounded solution. This solution can be isolated in the sense that all other solutions decay or grow at infinity with the exponents separated from zero. Then the limiting problem will not have nonzero bounded solutions in a weighted space, and the operator will be normally solvable in this space. This is related to the structure of non-Fredholm operators discussed in the next section and to solvability conditions discussed in Section 1.2.

1.1.3. Structure of non-Fredholm operators. Let us consider general elliptic operators $L : E_\infty(\Omega) \to F_\infty(\Omega)$, where Ω is an unbounded domain. Denote by $\mathcal{L}(L)$ the set of its limiting operators acting on functions defined in the corresponding

1. Weakly non-Fredholm operators

limiting domains. We recall that if Condition NS is satisfied, that is if all limiting problems $\hat{L}u = 0$ have only zero solutions in the spaces $E_\infty(\Omega)$, then the operator L is normally solvable with a finite-dimensional kernel. If, moreover, Condition NS* is satisfied, then it satisfies the Fredholm property.

If one of the limiting problems has a nonzero solution, then the Fredholm property is not satisfied. In other words, the origin belongs to the essential spectrum of the operator L. In some cases, a small exponential weight can move the spectrum and the Fredholm property can be satisfied in the weighted space.

Denote by S the operator of multiplication by the function

$$\omega_\mu(x) = \exp(\mu\sqrt{1 + |x|^2}),$$

and consider the operator $L_\mu = S_\mu L S_\mu^{-1}$ acting from $E_\infty(\Omega)$ into $F_\infty(\Omega)$. Then $L_\mu = L + \mu K(\mu)$, where $K(\mu)$ is a bounded operator polynomial with respect to μ. We consider here real exponents μ though they can also be complex.

Suppose for simplicity that the set $\mathcal{L}(L)$ contains a finite number of limiting operators. We split this set into two subsets, $\mathcal{L}^+(L)$ and $\mathcal{L}^-(L)$. The operators from the subset $\mathcal{L}^+(L)$ satisfy Conditions NS and NS*, and the operators from the subset $\mathcal{L}^-(L)$ do not satisfy at least one of them. Then for all μ sufficiently small the limiting operators $\hat{L}_\mu = \hat{L} + \mu \hat{K}(\mu)$, where $\hat{L} \in \mathcal{L}^+(L)$ will also satisfy Conditions NS and NS* (see the proposition below). If the operators $\hat{L}_\mu = \hat{L} + \mu \hat{K}(\mu)$ with $\hat{L} \in \mathcal{L}^-(L)$ also satisfy these conditions, then the operator L_μ is Fredholm.

We note that the set of limiting operators $\mathcal{L}(\hat{L})$ of a limiting operator $\hat{L}$ belong to the set of limiting operators $\mathcal{L}(L)$ of the operator L.

Proposition 1.9. *Suppose that the set $\mathcal{L}(L)$ is finite, and for any $\hat{L} \in \mathcal{L}^+(L)$ the following inclusion takes place:*

$$\mathcal{L}(\hat{L}) \subset \mathcal{L}^+(L), \tag{1.5}$$

that is for each limiting operator $\hat{L}$ satisfying Conditions NS and NS, all its limiting operators also satisfy them. Then for all μ sufficiently small, the corresponding operators $\hat{L}_\mu$ satisfy these conditions, $\hat{L}_\mu \in \mathcal{L}^+(L_\mu)$.*

Proof. By virtue of (1.5), each operator $\hat{L} \in \mathcal{L}^+(L)$ satisfies the Fredholm property. Moreover, the operator $\hat{L}$ is invertible. Indeed, its kernel is empty by virtue of Condition NS. The equation $\hat{L}u = f$ is solvable if and only if $\phi(f) = 0$ for any solution $\phi \in (F_\infty(\Omega))^*$ of the equation $L^*\phi = 0$. If there is a nonzero solution of this equation, then it belongs to the space $(F^*(\Omega))_\infty$ (it follows from Lemma 2.8 of Chapter 5, Lemma 10.5 of Chapter 2 and from the inclusion $(F^*)_1 \subset (F^*)_\infty$). This contradicts Condition NS*. Hence, the homogeneous adjoint equation has only the zero solution and the equation $\hat{L}u = f$ is solvable for any right-hand side. Thus, the operator $\hat{L}$ is invertible.

Therefore, the operator $\hat{L}_\mu = \hat{L} + \mu \hat{K}(\mu)$ is invertible for all μ sufficiently small. Since there is a finite number of limiting operators, then we can choose μ_0 such that the invertibility of all operators occurs for $0 < \mu < \mu_0$. Hence the operators $\hat{L}_\mu$ satisfy Condition NS for all $0 < \mu < \mu_0$.

We next verify that they satisfy Condition NS*. Since the operator $\hat{L} : E_\infty(\hat{\Omega}) \to F_\infty(\hat{\Omega})$ satisfies the Fredholm property, then the same operator considered in other spaces, namely, $\hat{L} : E_1(\hat{\Omega}) \to F_1(\hat{\Omega})$ also satisfies it (see the end of Section 6, Chapter 5). Moreover, this operator is invertible. Indeed, its kernel is empty since $E_1 \subset E_\infty$. Its image coincides with F_1 because $(F_1(\hat{\Omega}))^* = (F^*(\hat{\Omega}))_\infty$ (Section 2, Chapter 5), and the equation $\hat{L}^*\phi = 0$ does not have nonzero solutions in $(F^*(\hat{\Omega}))_\infty$ according to Condition NS*.

From the invertibility of the operator $\hat{L} : E_1(\hat{\Omega}) \to F_1(\hat{\Omega})$ it follows that the operator $\hat{L}_\mu : E_1(\hat{\Omega}) \to F_1(\hat{\Omega})$ is also invertible for all μ sufficiently small. Hence, the equation $\hat{L}^*_\mu \phi = 0$ does not have nonzero solutions in $(F^*(\hat{\Omega}))_\infty$. The proposition is proved. □

In the proof of the proposition we do not use the particular structure of the operator L_μ related to the introduction of weighted spaces. It remains valid for any operator sufficiently close to the operator L. The assertion of the proposition gives, in a certain sense, stability of the structure of non-Fredholm operators: the limiting operators satisfying Conditions NS and NS* preserve these properties under small perturbations of the operator. In particular, if all limiting operators satisfy Conditions NS and NS*, then we obtain the well-known result that the Fredholm property is preserved under small perturbations of the operator.

1.2 Solvability conditions

In this section we discuss solvability conditions for weakly non-Fredholm operators. We consider the operator $L_\mu = S_\mu L S_\mu^{-1}$, where S_μ is the operator of multiplication by the weight function $\omega_\mu(x) = \exp(\mu\sqrt{1 + |x|^2})$.

Definition 1.10. An operator L is called a weakly non-Fredholm operator if it does not satisfy at least one of the Conditions NS and NS*, and if these conditions are satisfied for the operator L_μ for all positive μ sufficiently small.

For simplicity of presentation we suppose that μ is positive. This definition implies that the operator L does not satisfy the Fredholm property while the operator L_μ satisfies it for all positive μ sufficiently small. To verify the conditions of the definition, we can use exponential dichotomy (Section 1.1.2) or consider directly limiting problems. Everywhere below in this section we will assume that the operator L is weakly non-Fredholm. Consider the equation

$$Lu = f, \tag{1.6}$$

1. Weakly non-Fredholm operators

where $L : E_q(\Omega) \to F_q(\Omega)$, $1 \leq q \leq \infty$. We also consider the adjoint operator $L^* : (F_q(\Omega))^* \to (E_q(\Omega))^*$ and the homogeneous adjoint equation

$$L^*v = 0. \tag{1.7}$$

Let v_i, $i = 1, \ldots, k$ be its linearly independent solutions. Applying the functionals v_i to both sides of equality (1.6), we obtain necessary solvability conditions in their conventional form. These solvability conditions remain valid for solutions in the sense of sequence.

Proposition 1.11. *If there exists a solution $u = (u_1, u_2, \ldots)$ of equation (1.6) in the sense of sequence, then*

$$(f, v) = 0 \tag{1.8}$$

for any solution $v \in (F_q(\Omega))^*$ of equation (1.7).

Proof. By definition of the solution in the sense of sequence, $Lu_i = f_i$, where $u_i \in E_q(\Omega)$, $f_i \in F_q(\Omega)$ and $f_i \to f$ in the norm of the space $F_q(\Omega)$. We have

$$(f_i, v) = (Lu_i, v) = (u_i, L^*v) = 0.$$

Equality (1.8) follows from the convergence $f_i \to f$. The proposition is proved. □

In the case of Fredholm operators, solutions of equation (1.7) exponentially decay at infinity. Consequently, they belong to the spaces $(F_q(\Omega))^*$ with any q, $1 \leq q \leq \infty$. In the case of non-Fredholm operators, this may not be the case. If there is a solution of equation (1.7), which belongs to the space $F_\infty^*(\Omega)$ but not to the dual space $(F_q(\Omega))^*$, then the left-hand side of equality (1.8) may not be well defined.

Set

$$w = S_\mu u, \quad g = S_\mu f.$$

We assume that $S_{\mu_0} f \in F_\infty(\Omega)$ for some $\mu_0 > 0$. If $\mu < \mu_0$, then $g \in F_q(\Omega)$ for all q, $1 \leq q \leq \infty$. Equation (1.6) can be written as

$$L_\mu w = g. \tag{1.9}$$

Since the operator L_μ satisfies Conditions NS and NS*, then it satisfies the Fredholm property. Hence, equation (1.9) is solvable if and only if

$$(g, z) = 0 \tag{1.10}$$

for all solutions z of the homogeneous adjoint equation

$$L_\mu^* z = 0, \quad z \in (F_q(\Omega))^*. \tag{1.11}$$

The last equation can be also written in the form

$$S_\mu^{-1} L^* S_\mu z = 0.$$

If we compare it with equation (1.7), we can conclude that $v = S_\mu z$. Therefore, from (1.10),
$$0 = (g, z) = (S_\mu f, S_\mu^{-1} v) = (f, v).$$
Thus, condition (1.8) is sufficient for solvability of equation (1.9) in the space $E_q(\Omega)$. It can be easily verified that $u = S_\mu^{-1} w$ is a solution of equation (1.6) (see the proposition below).

Definition 1.12. A function f is said to be exponentially decaying at infinity if there exists a positive μ such that $S_\mu f \in F_\infty(\Omega)$. A functional v is not exponentially growing at infinity if $S_{-\mu} v \in F_\infty^*(\Omega)$ for any positive μ.

In what follows, we consider functionals from the space $F_\infty^*(\Omega)$ and functionals that are not exponentially growing at infinity. The difference between them is that there can exist functionals growing at infinity with the rate of growth weaker than exponential. A priori, we cannot ignore them because, multiplied by a decaying exponential weight function, they belong to the function spaces under consideration and determine the solvability conditions. This is related to exponential dichotomy.

Proposition 1.13. *Let f be exponentially decaying at infinity. Equation (1.6) is solvable with a solution u exponentially decaying at infinity if and only if condition (1.8) is satisfied for any solution v of equation (1.7), which is not exponentially growing at infinity.*

Proof. Let $S_{\mu_0} f \in F_\infty(\Omega)$ for some $\mu_0 > 0$. Then $S_\mu f \in F_1(\Omega)$ for any μ such that $0 < \mu < \mu_0$. Since $S_{-\mu} v \in F_\infty^*(\Omega) = (F_1(\Omega))^*$ (Theorem 10.2, Chapter 2), then $(f, v) = (S_\mu f, S_{-\mu} v)$ is well defined.

On the other hand, if u is exponentially decaying at infinity, then taking into account that L is a differential operator, direct calculations show that Lu is also exponentially decaying at infinity. Then there exists $\mu' > 0$ such that $S_{\mu'} Lu \in E_1(\Omega)$, $S_{-\mu'} v \in F_\infty^*(\Omega) = (F_1(\Omega))^*$. Hence
$$(Lu, v) = (S_{\mu'} Lu, S_{-\mu'} v) = (u, (S_{\mu'} L)^* S_{-\mu'} v) = (u, L^* v) = 0.$$
Thus, we can apply the functional v to both sides of the equality (1.6). We obtain $(f, v) = 0$. Necessity is proved.

To prove sufficiency, we show first of all that if v is not exponentially growing at infinity, then $S_{-\mu} v \in (F_q(\Omega))^*$ for any $\mu > 0$ and any q, $1 \leq q \leq \infty$. Let $0 < \mu_1 < \mu$. By virtue of Definition 1.12, $S_{-\mu_1} v \in F_\infty^*(\Omega) (= (F_1(\Omega))^*)$. Let $g \in F_q(\Omega)$. Then
$$(S_{-\mu} v, g) = (S_{-\mu_1} v, S_{-\nu} g),$$
where $\nu = \mu - \mu_1 > 0$. Since $S_{-\nu} g \in F_1(\Omega)$, then the right-hand side of the last equality is well defined. Therefore, $S_{-\mu} v$ is a linear bounded functional over $F_q(\Omega)$.

Thus, any solution v of equation (1.7), which is not exponentially growing at infinity, determines a solution $z = S_{-\mu} v \in (F_q(\Omega))^*$ of equation (1.11). Let us

1. Weakly non-Fredholm operators

now show that the number of linearly independent solutions $v_i, i = 1, \ldots, N$ of equation (1.7), which are not exponentially growing at infinity, and of solutions $z_i, i = 1, \ldots, M$ of equation (1.11) is the same. It remains to verify that $N \geq M$. Suppose that this is not true and $N < M$. It follows from Theorem 4.5 (Chapter 5) that M is independent of μ. Consider the functionals $v_i = S_\mu z_i, i = 1, \ldots, M$ for some given μ. Since they are linearly independent and $M > N$, then some of them do not satisfy conditions of Definition 1.12 being exponentially growing at infinity. Let it be, for example, v_1. Then there exists $\mu_2 > 0$ such that $S_{-\mu} v_1$ remains exponentially growing at infinity for $0 < \mu < \mu_2$ and, consequently, does not belong to $(F_q(\Omega))^*$. Hence, the linearly independent solutions of equation (1.11) for such values of μ become $w_1, z_2, \ldots, z_M$, that is z_1 is replaced by some other solution w_1. Obviously, w_1 is linearly independent with respect to $z_2, \ldots, z_M$ and with respect to $S_{-\mu} v_1$. On the other hand, each of the functionals $S_{-\mu} v_1, w_1, z_2, \ldots, z_M$ determines a solution of equation (1.11) for $\mu > \mu_2$. There are $M + 1$ of them and they are linearly independent. This contradiction proves the one-to-one correspondence between solutions of equation (1.7), which are not growing at infinity, and of equation (1.11).

Let f be exponentially decaying at infinity. According to the definition, there exists $\mu_0 > 0$ such that $S_{\mu_0} f \in F_\infty(\Omega)$. Write $g = S_\mu f$ for $0 < \mu < \mu_0$. Then $g \in F_q(\Omega)$ for any $q, 1 \leq q \leq \infty$. Since the operator L_μ satisfies the Fredholm property and conditions (1.10) are satisfied for all solutions of equation (1.11), then equation (1.9) has a solution $w \in E_q(\Omega)$. Hence, $u = S_{-\mu} w$ is a solution of equation (1.6). It decays exponentially at infinity. The proposition is proved. □

Theorem 1.14. *Let L be a weakly non-Fredholm operator. Suppose that all solutions v of equation (1.7), which are not exponentially growing at infinity, belong to the dual space $(F_q(\Omega))^*$. Then equation (1.6) with $f \in F_q(\Omega), 1 \leq q < \infty$ is solvable in $E_q(\Omega)$ in the sense of sequence if and only if equality (1.8) is satisfied for all such v.*

Proof. Necessity follows from Proposition 1.11. To prove sufficiency, put $f_n = S_{-\mu_n} f$, where $\mu_n > 0, \mu_n \to 0$ as $n \to \infty$, and consider the equation

$$Lu = f_n. \qquad (1.12)$$

The functions f_n are exponentially decaying at infinity and $f_n \to f$ in the norm of the space $F_q(\Omega)$ as $n \to \infty$. Here we use the assumption that $q < \infty$. Therefore, $(f_n, v) \to 0$.

If $(f_n, v) = 0$, then equation (1.12) is solvable by virtue of the previous proposition. If this equality is not satisfied, then we will modify the right-hand side. Assume, for simplicity, that there is only one linearly independent solution v of equation (1.7). Let $\psi \in F_q(\Omega)$ be a function with a bounded support and such that $(\psi, v) = 1$. Put

$$\tilde{f}_n = f_n - (f_n, v)\psi.$$

Then $(\tilde{f}_n, v) = 0$, $\tilde{f}_n$ is exponentially decaying at infinity and $\tilde{f}_n \to f$ since $(f_n, v) \to 0$. Denote by u_n the solution of the equation $Lu = \tilde{f}_n$. Then $u =$

$(u_1, u_2, \ldots)$ is a solution of equation (1.6) in the sense of sequence. We do not affirm here that the norms of the functions u_i are uniformly bounded. The theorem is proved. □

If $f \in F_q(\Omega)$ and $v \notin (F_q(\Omega))^*$, then, generally speaking, the expression (f, v) is not defined. However, it may be possible to approximate f by a sequence f_n in such a way that all (f_n, v) are defined. Instead of solvability conditions (1.8), we will formulate solvability condition in the sense of sequence.

Proposition 1.15. *Suppose that there is a solution v of equation (1.7), which is not exponentially growing at infinity and which does not belong to the dual space $(F_q(\Omega))^*$. If equation (1.6) with $f \in F_q(\Omega)$, $1 \le q < \infty$ is solvable in $E_q(\Omega)$, then there exists a sequence f_n of functions exponentially decaying at infinity such that $f_n \to f$ in $F_q(\Omega)$ and $(f_n, v) = 0$.*

Proof. Let $u \in E_q(\Omega)$ be a solution of equation (1.6). Put $u_n = S_{-\mu_n} u$, where μ_n is a sequence of positive numbers converging to 0. Then $u_n \to u$ in the norm of the space $E_q(\Omega)$. Set $f_n = L u_n$. Then

$$f_n - f = L(u_n - u) \to 0$$

in the norm of the space $F_q(\Omega)$ since L is a bounded operator.

Taking into account that L is a differential operator, we can easily verify that $L(S_{-\mu} u) = S_{-\mu} T u$, where $T : E_q(\Omega) \to F_q(\Omega)$ is a bounded operator. Hence f_n is exponentially decaying at infinity and (f_n, v) is well defined. Applying v to both sides of the equality $L u_n = f_n$, we obtain $(f_n, v) = 0$ (cf. the proof of necessity in Proposition 1.13). The proposition is proved. □

Theorem 1.16. *Let L be a weakly non-Fredholm operator. Then equation (1.6) with $f \in F_q(\Omega)$, $1 \le q < \infty$ is solvable in $E_q(\Omega)$ in the sense of sequence if and only if there exists a sequence f_n of functions exponentially decaying at infinity such that $f_n \to f$ in $F_q(\Omega)$ and $(f_n, v) = 0$ for all solutions v of equation (1.7) which are not exponentially growing at infinity.*

Proof. Suppose that there exists a solution of equation (1.6) in the sense of sequence, that is a sequence $u_n \in E_q(\Omega)$, $1 \le q < \infty$ such that the sequence $f_n = L u_n$ converges to f in the norm of the space $F_q(\Omega)$. The functions f_n are not necessarily exponentially decaying at infinity. We will modify them in order to satisfy this and other conditions. We put $\tilde{f}_n = L(S_{-\mu_n} u_n)$, where $\mu_n > 0$ is sufficiently small. As in the proof of the previous proposition, we conclude that $\tilde{f}_n \in F_q(\Omega)$, $\tilde{f}_n$ decay exponentially at infinity and $\tilde{f}_n \to f$ in the norm of the space $F_q(\Omega)$ under the proper choice of the sequence μ_n converging to zero. Finally, since the equation $Lu = \tilde{f}_n$ has a solution $S_{-\mu_n} u_n$ exponentially decaying at infinity, then we can apply the functional v to obtain $(\tilde{f}_n, v) = 0$.

If there exists a sequence f_n of functions exponentially decaying at infinity such that $f_n \to f$ in $F_q(\Omega)$ and $(f_n, v) = 0$, then the equation $Lu = f_n$ is solvable

1. Weakly non-Fredholm operators

by virtue of Proposition 1.13. Its solutions $u_1, u_2, \ldots$ form a solution in the sense of sequence. The theorem is proved. □

Remark 1.17. We do not assume in this theorem that solutions v of equation (1.7), which are not exponentially growing at infinity, belong to the space $(F_q(\Omega))^*$. The assumption that L is a weakly non-Fredholm operator, that is the operator L_μ satisfies the Fredholm property, ensures that the number of linearly independent solutions of equation (1.7), which are not growing at infinity, is finite.

We considered above the right-hand side f of equation (1.6) either assuming that it was exponentially decaying at infinity or that it could be approximated by functions exponentially decaying at infinity in the norm $F_q(\Omega)$. This assumption imposes a restriction on the function spaces: $q < \infty$. Functions from the space $F_\infty(\Omega)$ may not be approximated by exponentially decaying functions in the $F_\infty(\Omega)$-norm.

In the remaining part of this section, we discuss solvability conditions for the operator L acting from $E_\infty(\Omega)$ into $F_\infty(\Omega)$. As before, we will approximate functions f from this space by sequences f_n of exponentially decaying at infinity functions. However, the convergence $f_n \to f$ in this case is local in the $F_\infty(\Omega)$-norm. This means that $\|(f_n - f)\phi\|_{F_\infty(\Omega)} \to 0$ for any C^∞ function ϕ with a bounded support.

Let us consider a partition of unity ϕ_i, $i = 1, 2, \ldots$ such that $\sum_{i=1}^\infty \phi_1 \equiv 1$, ϕ_i are C^∞ functions, $0 \leq \phi(x) \leq 1$ for all $x \in \mathbb{R}^n$, $\|\phi_i\|_{C^k} \leq M_k$ for all i and any k, and

$$\phi_i(x) \leq M e^{-\sigma\sqrt{1+|x-x_i|^2}}, \quad x \in \mathbb{R}^n, \ i = 1, 2, \ldots$$

for some positive constants M, σ and some x_i. We will call such partition of unity exponential. A particular case of an exponential partition of unity is the case where the functions ϕ_i have bounded supports. We will suppose that the distance between any two points x_i is greater than some given positive number independent of i. Then

$$\sup_{x \in \mathbb{R}^n} \sum_{i=1}^\infty e^{-\sigma\sqrt{1+|x-x_i|^2}} < \infty. \tag{1.13}$$

Proposition 1.18. *Suppose that there exists a solution $u \in E_\infty(\Omega)$ of equation (1.6) with $f \in F_\infty(\Omega)$. Then there exists a sequence $f_n \in F_\infty(\Omega)$ of exponentially decaying at infinity functions such that $f_n \to f$ locally in $F_\infty(\Omega)$ and $(f_n, v) = 0$ for any solution v of equation (1.7), which is not exponentially growing at infinity.*

Proof. Consider an exponential partition of unity ϕ_i. Set

$$u_n = \sum_{i=1}^n \phi_i u, \quad f_n = L u_n.$$

The functions u_n are exponentially decaying at infinity, and $u_n \to u$ locally in $E_\infty(\Omega)$. Therefore, f_n are also exponentially decaying at infinity and $f_n \to f$

locally in $F_\infty(\Omega)$ since L is a bounded local operator. Applying v to both sides of the equality $Lu_n = f_n$ we conclude that $(f_n, v) = 0$. The proposition is proved. □

This proposition gives a necessary solvability condition. An important question is whether it is also sufficient. By virtue of Proposition 1.13, we can define the sequence u_n of solutions of the equations $Lu = f_n$. However, we do not know whether this sequence is convergent. We discuss this question in the remaining part of this section. Denote $g_i = \phi_i f$, where ϕ_i is a function from the exponential partition of unity, and consider the equation $Lu = g_i$ assuming that $(g_i, v) = 0$ for all solutions v of equation (1.7), which are not exponentially growing at infinity. According to Proposition 1.13, this equation has a solution. Let us denote it by w_i. We know that it decays exponentially at infinity. This allows us to expect that the sum $u = \sum_{i=1}^\infty w_i$ determines a function from $E_\infty(\Omega)$. In order to prove this assertion, we need some auxiliary results.

Denote by $S_\mu(h)$ the operator of multiplication by the function $\omega(x) = e^{\mu\sqrt{1+|x-h|^2}}$ and
$$L_\mu(h) = S_\mu(h) L S_\mu^{-1}(h).$$
Thus, we consider a family of operators depending on the parameter $h \in \Omega$. We will define limiting operators with respect to both, the independent variable x and the parameter h. We recall that the coefficients $a_{ik}^\alpha(x)$, $b_{jk}^\beta(x)$ of the elliptic operator L belong to some Hölder spaces (Section 2.2.2, Chapter 4). The coefficients $a_{ik,h}^\alpha(x)$, $b_{jk,h}^\beta(x)$ of the operators $L_\mu(h)$ belong to the same spaces with an estimate of the norm independent of h.

Definition 1.19. Let $h_m, x_m \in \Omega$ and $|h_m| + |x_m| \to \infty$. Consider the shifted coefficients
$$\tilde{a}_{ik,m}^\alpha(x) = a_{ik,h_m}^\alpha(x+x_m), \quad \tilde{b}_{jk,m}^\beta(x) = b_{jk,h_m}^\beta(x+x_m)$$
and a subsequence of these coefficients locally convergent to some limiting functions $\hat{a}_{ik}^\alpha(x)$, $\hat{b}_{jk}^\beta(x)$. Limiting operator $\hat{L}_\mu$ is an operator with the limiting coefficients in the limiting domain $\hat{\Omega}$ (Section 1, Chapter 4).

We note that for any fixed h, limiting operators of the operator $L_\mu(h)$ are the same as for the operator $L_\mu = L_\mu(0)$. Therefore, if the operator L_μ satisfies Conditions NS and NS*, then this is also true for the operators $L_\mu(h)$.

Lemma 1.20. *Suppose that the equations*
$$L_\mu(h)u = 0, \quad u \in E_\infty(\Omega)$$
have only zero solutions for all finite h, and all limiting equations (in the sense of Definition 1.19)
$$\hat{L}_\mu u = 0, \quad u \in E_\infty(\hat{\Omega})$$

1. Weakly non-Fredholm operators

also have only zero solutions. If a family of functions g_h, $h \in \mathbb{R}$ satisfies the conditions

$$\|g_h\|_{F_\infty(\Omega)} \leq M, \quad \forall h \tag{1.14}$$

and

$$(g_h, z_h) = 0 \tag{1.15}$$

for all solutions z_h of the equation

$$L_\mu^*(h)z = 0, \quad z \in (F_\infty(\Omega))^*, \tag{1.16}$$

then the equation

$$L_\mu(h)w = g_h, \quad w \in E_\infty(\Omega) \tag{1.17}$$

has a solution w_h and the norm $\|w_h\|_{E_\infty(\Omega)}$ is independent of h.

Proof. Since the operator $L_\mu(h)$ satisfies Conditions NS and NS* for any h fixed, then it is a Fredholm operator. Equation (1.17) is solvable due to solvability conditions (1.15). Its solution is unique by virtue of the assumption that the homogeneous equation has only the zero solution.

We need to obtain an estimate of the solutions w_h independent of h. Suppose that the uniform estimate does not hold and $\|w_{h_i}\|_{E_\infty(\Omega)} \to \infty$ for some sequence h_i. Put

$$\tilde{w}_i = \frac{w_{h_i}}{\|w_{h_i}\|_{E_\infty(\Omega)}}, \quad \tilde{g}_i = \frac{g_{h_i}}{\|w_{h_i}\|_{F_\infty(\Omega)}}.$$

Then $\|\tilde{w}_i\|_{E_\infty(\Omega)} = 1$, $\|\tilde{g}_i\|_{F_\infty(\Omega)} \to 0$ and $L_\mu(h_i)\tilde{w}_i = \tilde{g}_i$. By our definition of the norm (Chapter 2),

$$\|\tilde{w}_i\|_{E_\infty(\Omega)} = \sup_j \|\phi_j \tilde{w}_i\|_E,$$

where ϕ_j is a partition of unity. Therefore, for some j which may depend on i, $\|\phi_j \tilde{w}_i\|_E \geq 1/2$. Without loss of generality, we can assume that the supports of the functions ϕ_j are unit balls B_j with centers at some points x_j. We shift the domain Ω and the coefficients of the operator in such a way that the point x_j is translated to the origin. Denote the corresponding domain by Ω_i and the corresponding operator by L_i. Then

$$L_i \hat{w}_i = \hat{g}_i, \tag{1.18}$$

where $\hat{w}_i(x) = \tilde{w}_i(x + x_{j(i)})$, $\hat{g}_i(x) = \tilde{g}_i(x + x_{j(i)})$. It can be easily seen that

$$\|\phi_0 \hat{w}_i\|_E \geq \frac{1}{2}, \quad \|\hat{g}_i\|_{F_\infty(\Omega_i)} \to 0. \tag{1.19}$$

Here ϕ_0 is a function of the partition of unity with its support at the ball with center at the origin. We consider the coefficients of the operators L_i and choose their locally convergent subsequences. Denote the operator with the limiting coefficients by $\hat{L}$ and the corresponding limiting domain by $\hat{\Omega}$. The sequence $\hat{w}_i$

contains a subsequence locally convergent to some limiting function $w^0 \in E_\infty(\hat{\Omega})$, the convergence being in a weaker norm (Theorem 2.7, Chapter 4). It satisfies the equation $\hat{L}w = 0$ and $\|\phi_0 w^0\|_E \geq 1/2$ by virtue of (1.19). If $|h_i| + |x_{j(i)}|$ are uniformly bounded, then we obtain a nonzero solution of the equation $L_\mu(h)u = 0$ for some h. If this sequence tends to infinity, then we obtain a nonzero solution of the equation $\hat{L}_\mu u = 0$, where $\hat{L}_\mu$ is a limiting operator in the sense of Definition 1.19. In both cases we obtain a contradiction with the assumptions of the lemma. The lemma is proved. □

Proposition 1.21. *Suppose that L is a weakly non-Fredholm operator, and for all positive μ sufficiently small, the operator $L_\mu(h)$ satisfies the assumptions of Lemma 1.20. Let ϕ_i be an exponential partition of unity, and for some $f \in F_\infty(\Omega)$, the functions $f_i = \phi_i f$ satisfy the conditions*

$$(f_i, v) = 0, \quad i = 1, 2, \ldots \tag{1.20}$$

for all solutions v of equation (1.7), which are not exponentially growing at infinity. Then there exists a solution $u \in E_\infty(\Omega)$ of equation (1.6).

Proof. Consider the equations

$$Lu = f_i, \quad i = 1, 2, \ldots. \tag{1.21}$$

Set $g_i = S_\mu(x_i)f_i$, where $0 < \mu < \sigma$, σ is the exponent in the definition of exponential partition of unity. Then $\|g_i\|_{E_\infty(\Omega)} \leq M$ for some M and for any i.

Consider the equations

$$L_\mu(x_i)w = g_i, \quad i = 1, 2, \ldots. \tag{1.22}$$

The operators $L_\mu(x_i)$ satisfy the Fredholm property since L is a weakly non-Fredholm operator: the limiting operators do not depend on x_i and satisfy Conditions NS and NS*. Therefore, these equations are solvable if and only if

$$(g_i, z) = 0 \tag{1.23}$$

for all solutions z of the equations

$$L_\mu^*(x_i)z = 0, \quad z \in (F_\infty(\Omega))^*.$$

All solutions of these equations have the form $z = S_{-\mu}(x_i)v$, where v is a solution of equation (1.7), which is not exponentially growing at infinity. Therefore,

$$(g_i, z) = (S_\mu(x_i)f_i, S_{-\mu}(x_i)v) = (f_i, v),$$

that is solvability conditions (1.23) have the form (1.20). The right-hand side of the last equality is well defined since f_i is exponentially decaying at infinity.

1. Weakly non-Fredholm operators

Thus, there exists a solution w_i of equation (1.22). By virtue of Lemma 1.20, $\|w_i\|_{E_\infty(\Omega)} \leq M$ for all i and for some M independent of i. The functions $u_i = S_{-\mu}(x_i)w_i$ are solutions of equations (1.21). If the series $u = \sum_{i=1}^\infty u_i$ converges, then u is a solution of equation (1.6). To prove the convergence, put $u_n = \sum_{i=1}^n u_i$. Let ψ_j be a function from the partition of unity in the definition of the $E_\infty(\Omega)$-norm. Without loss of generality, we can assume that their supports are unit balls B_j with the centers at some points x^j. Then

$$\|u_n\|_{E_\infty(\Omega)} = \sup_j \|\psi_j u_n\|_E,$$

$$\|\psi_j u_n\|_E \leq \sum_{i=1}^n \|\psi_j u_i\|_E = \sum_{i=1}^n \|\psi_j S_{-\mu}(x_i) w_i\|_E$$

$$\leq \sum_{i=1}^n K e^{-\mu|x_i - x^j|} \|\psi_j w_i\|_E \leq KM \sum_{i=1}^\infty e^{-\mu|x_i - x^j|}.$$

By virtue of (1.13), the right-hand side of the last inequality is bounded independently of j. Hence, $\|\psi_j u_n - \psi_j u\|_E \to 0$ and $u \in E_\infty(\Omega)$. The proposition is proved. □

We summarize solvability conditions obtained above in the following theorem.

Theorem 1.22. *Let L be a weakly non-Fredholm operator. Then:*

- *If $f \in F_\infty(\Omega)$ is exponentially decaying at infinity, then equation (1.6) is solvable with solutions $u \in E_\infty(\Omega)$ exponentially decaying at infinity if and only if conditions (1.8) are satisfied for all solutions v of equation (1.7), which are not exponentially growing at infinity,*
- *If $f \in F_q(\Omega)$, $1 \leq q < \infty$, then equation (1.6) is solvable in $E_q(\Omega)$ in the sense of sequence if and only if there exists a sequence f_n of functions exponentially decaying at infinity such that $f_n \to f$ in $F_q(\Omega)$ and $(f_n, v) = 0$ for all solutions v of equation (1.7) which are not exponentially growing at infinity,*
- *Let $f \in F_\infty(\Omega)$. If there exists a solution $u \in E_\infty(\Omega)$ of equation (1.6), then there exists a sequence $f_n \in F_\infty(\Omega)$ of exponentially decaying at infinity functions such that $f_n \to f$ locally in $F_\infty(\Omega)$ and $(f_n, v) = 0$ for any solution v of equation (1.7), which is not exponentially growing at infinity.*
If assumptions of Lemma 1.20 are satisfied and $(f_i, v) = 0$, $i = 1, 2, \ldots$, where $f_i = \phi_i f$, ϕ_i is an exponential partition of unity, and v are solutions of equation (1.7), which are not exponentially growing at infinity, then there exists a solution $u \in E_\infty(\Omega)$ of equation (1.6).

1.3 Examples of weakly non-Fredholm operators

1.3.1. One-dimensional equations. Consider the equation

$$u'' + cu = f, \quad -\infty < x < \infty, \qquad (1.24)$$

where c is a positive constant. Let us verify that

$$Lu = u'' + cu$$

is a weakly non-Fredholm operator (Definition 1.10). We will consider it as acting from $E_q(\mathbb{R})$ into $F_q(\mathbb{R})$, where $E = H^2$ and $F = L^2$. It is also possible to consider Hölder spaces. The operator L coincides with its limiting operators. The corresponding limiting equations have two linearly independent bounded solutions, $u_1(x) = \sin(\sqrt{c}x)$, $u_2(x) = \cos(\sqrt{c}x)$. Therefore, it does not satisfy the Fredholm property.

Consider now weighted spaces. Set

$$v = ue^{\mu\sqrt{1+x^2}}, \quad g = fe^{\mu\sqrt{1+x^2}},$$

where $\mu > 0$. Then v satisfies the equation

$$v'' - 2\mu \frac{x}{\sqrt{1+x^2}} v' + \left(\mu^2 \frac{x^2}{1+x^2} - \mu \frac{1}{(1+x^2)^{3/2}} + c\right) v = g. \qquad (1.25)$$

The operator $L_\mu = S_\mu L S_\mu^{-1}$, where S_μ is the operator of multiplication by the weight function $w_\mu(x) = e^{\mu\sqrt{1+x^2}}$, is given by the left-hand side of the previous equation,

$$L_\mu v = v'' - 2\mu \frac{x}{\sqrt{1+x^2}} v' + \left(\mu^2 \frac{x^2}{1+x^2} - \mu \frac{1}{(1+x^2)^{3/2}} + c\right) v.$$

Its limiting operators are

$$L_\mu^\pm v = v'' \mp 2\mu v' + (\mu^2 + c)v.$$

The limiting problems $L_\mu^\pm v = 0$ do not have nonzero bounded solutions for $\mu \neq 0$. Therefore the operator L_μ is normally solvable with a finite-dimensional kernel. From the results of Chapter 9 it follows that it satisfies the Fredholm property and its index equals -2. Indeed, $\text{ind}(L_\mu) = n_+ + n_- - 2$, where $n_\pm$ is the number of bounded solutions of the equation $L_\mu^\pm u = 0$ at $\pm\infty$. It can be easily verified that $n_+ = n_- = 0$.

The dimension of its kernel is zero. Indeed, if there exists a bounded solution v_0 of the equation $L_\mu v = 0$, then there is the exponentially decaying solution $u_0 = v_0 \exp(-\mu\sqrt{1+x^2})$ of the equation $u'' + cu = 0$, which is not possible.

1. Weakly non-Fredholm operators

Thus, the codimension of the image equals 2, and there are two solvability conditions. In order to formulate them, consider the formally adjoint operator $L_\mu^* : E_q(\mathbb{R}) \to F_q(\mathbb{R})$,

$$L_\mu^* v = v'' + 2\mu \left(\frac{x}{\sqrt{1+x^2}} v\right)' + \left(\mu^2 \frac{x^2}{1+x^2} - \mu \frac{1}{(1+x^2)^{3/2}} + c\right) v.$$

It can be written in a form similar to the operator L_μ where μ is replaced by $-\mu$:

$$L_\mu^* v = v'' + 2\mu \frac{x}{\sqrt{1+x^2}} v' + \left(\mu^2 \frac{x^2}{1+x^2} + \mu \frac{1}{(1+x^2)^{3/2}} + c\right) v.$$

The equation

$$L_\mu^* v = 0$$

has two linearly independent solutions

$$v_1(x) = \sin(\sqrt{c}x)\, e^{-\mu\sqrt{1+x^2}}, \quad v_2(x) = \cos(\sqrt{c}x)\, e^{-\mu\sqrt{1+x^2}}.$$

Therefore, equation (1.25) is solvable if and only if

$$\int_{-\infty}^{\infty} g(x) v_i(x)\, dx = 0, \quad i = 1, 2 \tag{1.26}$$

or

$$\int_{-\infty}^{\infty} f(x) \sin(\sqrt{c}x)\, dx = \int_{-\infty}^{\infty} f(x) \cos(\sqrt{c}x)\, dx = 0. \tag{1.27}$$

Thus, the operator L is weakly non-Fredholm. Equation (1.24) with $f \in F_\mu$ is solvable in E_μ, where $E_\mu = \{u : ue^{\mu\sqrt{1+x^2}} \in E_\infty(\mathbb{R})\}$, $F_\mu = \{f : fe^{\mu\sqrt{1+x^2}} \in F_\infty(\mathbb{R})\}$, if and only if conditions (1.27) are satisfied. The solvability conditions are formulated here in terms of formally adjoint operators and not of adjoint operators as in the previous section. We use here the results of Chapter 6.

We can also use the explicit representation of solutions

$$u(x) = u_1^0(x) \int_{-\infty}^{x} f(y) v_1^0(y)\, dy + u_2^0(x) \int_{-\infty}^{x} f(y) v_2^0(y)\, dy, \tag{1.28}$$

where

$$u_1^0(x) = \frac{1}{\sqrt{c}} \sin(\sqrt{c}x), \quad u_2^0(x) = \frac{1}{\sqrt{c}} \cos(\sqrt{c}x),$$
$$v_1^0(y) = \cos(\sqrt{c}y), \quad v_2^0(y) = -\sin(\sqrt{c}y)$$

(Section 2.2, Chapter 9). If f is exponentially decaying at infinity then the integrals above are well defined and $u(x)$ is exponentially decaying at $-\infty$. Solvability conditions (1.27) allow us to replace the integrals $\int_{-\infty}^{x} f(y) v_i^0(y)\, dy$ by the integrals $\int_{x}^{\infty} f(y) v_i^0(y)\, dy$ and to conclude that $u(x)$ decays exponentially also at $+\infty$.

We next discuss solvability of equation (1.24) with a bounded f not necessarily decaying exponentially at infinity. If $f \in F_q(\mathbb{R})$, $1 \leq q < \infty$, then we can approximate f by a sequence f_k of exponentially decaying functions and consider the sequence solution

$$u_k(x) = u_1^0(x) \int_{-\infty}^x f_k(y) v_1^0(y) dy + u_2^0(x) \int_{-\infty}^x f_k(y) v_2^0(y) dy. \qquad (1.29)$$

Without additional conditions on f we cannot conclude that the sequence u_k converges. We recall that the homogeneous equations $L_\mu v = 0$ and $L_\mu^\pm v = 0$ do not have nonzero bounded solutions. Therefore, we can apply Proposition 1.21. Consider equation (1.24) with $f(x) = \cos(px) \in F_\infty(\mathbb{R})$, where p is an integer, $p \neq 1$ and $c = 1$. Put

$$f_k(x) = \begin{cases} \cos(px) &, \quad 2\pi k < x < 2\pi(k+1), \\ 0 &, \quad x \leq 2\pi k, \; x \geq 2\pi(k+1). \end{cases}$$

Then $f(x) = \sum_{k=-\infty}^\infty f_k(x)$. Since f_k satisfies solvability conditions (1.27), then the equation

$$u'' + cu = f_k$$

has a solution. Explicit calculations with formula (1.29) give

$$u_k(x) = \begin{cases} (\cos(px) - \cos(x))/(1 - p^2) &, \quad 2\pi k < x < 2\pi(k+1), \\ 0 &, \quad x \leq 2\pi k, \; x \geq 2\pi(k+1). \end{cases}$$

We note that u_k is continuous with its first derivative. The second derivative is discontinuous because the function f_k is discontinuous. For simplicity of calculations, we use here discontinuous functions ϕ_k in the partition of unity, which are characteristic functions of the corresponding intervals, $f_k = \phi_k f$.

Thus, equation (1.24) is solvable in $E_\infty(\mathbb{R})$, and

$$u(x) = \sum_{k=-\infty}^\infty u_k(x) = \frac{1}{1-p^2}(\cos(px) - \cos(x)).$$

Obviously, other solutions of this equation can be obtained as linear combinations of $u(x)$ with the solutions of the homogeneous equation.

Next, let us consider the equation

$$u'' + bu' = f, \quad -\infty < x < \infty. \qquad (1.30)$$

If

$$v = u e^{\mu \sqrt{1+x^2}}, \quad g = f e^{\mu \sqrt{1+x^2}},$$

then

$$v'' + \left(b - 2\mu \frac{x}{\sqrt{1+x^2}}\right) v' + \left(\mu^2 \frac{x^2}{1+x^2} - \mu \frac{1}{(1+x^2)^{3/2}} - b\mu \frac{x}{\sqrt{1+x^2}}\right) v = g. \qquad (1.31)$$

1. Weakly non-Fredholm operators

Consider the operator

$$L_\mu v = v'' + \left(b - 2\mu \frac{x}{\sqrt{1+x^2}}\right) v' + \left(\mu^2 \frac{x^2}{1+x^2} - \mu \frac{1}{(1+x^2)^{3/2}} - b\mu \frac{x}{\sqrt{1+x^2}}\right) v$$

and the corresponding limiting operators

$$L_\mu^\pm v = v'' + (b \mp 2\mu)v' + (\mu^2 \mp b\mu)v.$$

The operator L_μ satisfies the Fredholm property for sufficiently small $\mu \neq 0$. The formally adjoint operator is

$$L_\mu^* v = v'' - \left(b - 2\mu \frac{x}{\sqrt{1+x^2}}\right) v' + \left(\mu^2 \frac{x^2}{1+x^2} + \mu \frac{1}{(1+x^2)^{3/2}} - b\mu \frac{x}{\sqrt{1+x^2}}\right) v.$$

If $b \neq 0$, then the homogeneous formally adjoint equation $L_\mu^* v = 0$ has a unique, up to a constant factor, bounded solution

$$v_1(x) = e^{-\mu\sqrt{1+x^2}}.$$

The index of the operator $L_\mu : C^{2+\alpha}(\mathbb{R}) \to C^\alpha(\mathbb{R})$ equals -1, equation (1.31) is solvable if and only if

$$\int_{-\infty}^{\infty} g(x) v_1(x) dx = \int_{-\infty}^{\infty} f(x) dx = 0.$$

If $b = 0$, then there are two solutions

$$v_1(x) = e^{-\mu\sqrt{1+x^2}}, \quad v_2(x) = xe^{-\mu\sqrt{1+x^2}},$$

two solvability conditions

$$\int_{-\infty}^{\infty} g(x) v_i(x) dx = 0, \quad i = 1, 2,$$

and the index of the operator equals -2.

In the case of variable coefficients with limits at infinity, limiting operators can be easily found, solvability conditions can be obtained in the same way as above. If the coefficients do not have limits at infinity, construction of limiting operators and of solvability conditions is less explicit.

1.3.2. Cylindrical domains.
In this section we consider the equation

$$\Delta u + cu = f \tag{1.32}$$

in an unbounded cylinder $\Omega \subset \mathbb{R}^n$ with axis x_1 and orthogonal variables $x' = (x_2, \ldots, x_n)$. Here $c \geq 0$ is some constant. Consider the homogeneous Neumann boundary condition

$$\frac{\partial u}{\partial n} = 0, \tag{1.33}$$

where n is an outer normal vector. The corresponding operator L, acting from the space

$$E = \left\{W^{2,2}(\Omega), \; \frac{\partial u}{\partial n} = 0\right\}$$

into the space $F = L^2(\Omega)$, does not satisfy the Fredholm property since the limiting problem

$$\Delta u + cu = 0, \quad \frac{\partial u}{\partial n} = 0 \tag{1.34}$$

has a nonzero solution.

In order to solve this problem, let us put $v(x) = p(x_1)\phi(x')$, where $\phi(x')$ is an eigenfunction of the Laplace operator in the section of the cylinder with the homogeneous Neumann boundary condition corresponding to an eigenvalue σ. Then

$$p'' + (c + \sigma)p = 0. \tag{1.35}$$

This equation has a bounded for all $x_1 \in \mathbb{R}$ solution if $c + \sigma \geq 0$. Denote the eigenvalues of the Laplace operator in the section of the cylinder by σ_i, $i = 0, 1, \ldots$, assuming for simplicity that they are simple, and suppose that

$$c + \sigma_i > 0, \; i = 1, \ldots, k, \quad c + \sigma_i < 0, \; i = k+1, \ldots.$$

Then bounded solutions of problem (1.34) are

$$u_i^1(x) = \cos(a_i x_1)\phi_i(x'), \quad u_i^2(x) = \sin(a_i x_1)\phi_i(x'), \quad i = 1, \ldots, k, \quad a_i = \sqrt{c + \sigma_i}.$$

Let us introduce the weighted spaces

$$E_\mu = \{u(x) : u(x)\mu(x_1) \in E\}, \quad F_\mu = \{u(x) : u(x)\mu(x_1) \in F\},$$

where

$$\mu(x_1) = e^{\mu\sqrt{1+x_1^2}}.$$

Set

$$v(x) = u(x)e^{\mu\sqrt{1+x_1^2}}, \quad g(x) = f(x)e^{\mu\sqrt{1+x_1^2}}.$$

Then v satisfies the equation

$$\Delta v - 2\mu \frac{x_1}{\sqrt{1+x_1^2}} \frac{\partial v}{\partial x_1} + \left(\mu^2 \frac{x_1^2}{1+x_1^2} - \mu \frac{1}{(1+x_1^2)^{3/2}} + c\right) v = g. \tag{1.36}$$

Consider the operator

$$L_\mu v = \Delta v - 2\mu \frac{x_1}{\sqrt{1+x_1^2}} \frac{\partial v}{\partial x_1} + \left(\mu^2 \frac{x_1^2}{1+x_1^2} - \mu \frac{1}{(1+x_1^2)^{3/2}} + c\right) v,$$

acting from E into F, and the corresponding limiting problems

$$L_\mu^\pm v = \Delta v \mp 2\mu \frac{\partial v}{\partial x_1} + (\mu^2 + c)v, \quad \frac{\partial v}{\partial n} = 0.$$

1. Weakly non-Fredholm operators

Set
$$p(x_1) = \int_G v(x)\phi(x')dx',$$
where G is the cross-section of the cylinder. Multiplying the limiting equation $L_\mu^\pm v = 0$ by $\phi(x')$ and integrating over G, we obtain the equation
$$p'' \mp 2\mu p' + (\sigma + c + \mu^2)p = 0.$$
Here σ is the eigenvalue corresponding to the eigenfunction $\phi(x')$. Since the eigenvalues of the Laplace operator in the section of the cylinder form a discrete set, then $\sigma + c + \mu^2 \neq 0$ for all $\mu \neq 0$ sufficiently small. Hence the last equation cannot have bounded solutions for such μ, and Condition NS is satisfied.

The same remains valid for the formally adjoint operator. Consequently, we can use the results of Chapter 6 in order to conclude that the operator $L_\mu : E \to F$ satisfies the Fredholm property. Equation (1.36) is solvable if and only if the solvability conditions
$$\int_\Omega g(y)v(y)dy = 0 \qquad (1.37)$$
are verified for any solution v of the formally adjoint equation
$$\Delta v + 2\mu \frac{\partial}{\partial x_1}\left(\frac{x_1}{\sqrt{1+x_1^2}}v\right) + \left(\mu^2 \frac{x_1^2}{1+x_1^2} - \mu \frac{1}{(1+x_1^2)^{3/2}} + c\right)v = 0.$$
If we write it in the form
$$\Delta v + 2\mu \frac{x_1}{\sqrt{1+x_1^2}}\frac{\partial v}{\partial x_1} + \left(\mu^2 \frac{x_1^2}{1+x_1^2} + \mu \frac{1}{(1+x_1^2)^{3/2}} + c\right)v = 0, \qquad (1.38)$$
we can easily see that it can be obtained from (1.36) replacing μ by $-\mu$.

Any solution $u \in E_\mu$ of equation (1.32) has the form
$$u = \exp(-\mu\sqrt{1+x_1^2})\, v,$$
where $v \in E$ is a solution of equation (1.36). Any solution $u \in E_{-\mu}$ of the equation formally adjoint to (1.32) (it coincides with (1.32)) has the form
$$u = \exp(\mu\sqrt{1+x_1^2})\, v,$$
where $v \in E$ is a solution of equation (1.38). We have proved the following theorem.

Theorem 1.23. *Problem (1.32), (1.33), where $u \in E_\mu$, $f \in F_\mu$ satisfies the Fredholm property for any real $\mu \neq 0$ such that $\mu^2 + c + \sigma_i \neq 0$ for all eigenvalues σ_i of the Laplace operator in the section of the cylinder. This problem is solvable in E_μ for $f \in F_\mu$ if and only if*
$$\int_\Omega u_0(x)\, f(x)\, dx = 0$$
for any solution $u_0 \in E_{-\mu}$ of the homogeneous problem (1.34).

1.3.3. Problems in $\mathbb{R}^2$.
Consider the operator

$$Lu = \Delta u + c\frac{\partial u}{\partial x_2} + b(x_2)u$$

in $\mathbb{R}^2$. Here c is some constant, the function b depends only on the variable x_2, $b(\pm\infty) < 0$. Suppose that the operator

$$L_0 u = u'' + cu' + b(x_2)u$$

has a zero eigenvalue, which is simple, with the corresponding eigenfunction $u_0(x_2)$, while all other points of its spectrum are in the left half-plane. Here ' denotes the differentiation with respect to x_2.

There are three limiting problems,

$$\Delta u + c\frac{\partial u}{\partial x_2} + b(\pm\infty)u = 0$$

and

$$\Delta u + c\frac{\partial u}{\partial x_2} + b(x_2)u = 0.$$

The last one has a nonzero bounded solution $u(x) = u_0(x_2)$. Therefore, the operator L is not Fredholm. Consider the equation

$$Lu = f.$$

We will use the same notations v and g as in the previous example. Then

$$\Delta v + c\frac{\partial v}{\partial x_2} - 2\mu\frac{x_1}{\sqrt{1+x_1^2}}\frac{\partial v}{\partial x_1} + \left(\mu^2\frac{x_1^2}{1+x_1^2} - \mu\frac{1}{(1+x_1^2)^{3/2}} + b(x_2)\right)v = g. \tag{1.39}$$

The limiting problems

$$\Delta v + c\frac{\partial v}{\partial x_2} \mp 2\mu\frac{\partial v}{\partial x_1} + \left(\mu^2 + b(\pm\infty)\right)v = 0,$$

obtained as $|x_1|, |x_2| \to \infty$, have constant coefficients. For sufficiently small μ they have only zero solutions. Consider next the limiting problems ($|x_1| \to \infty$),

$$\Delta v + c\frac{\partial v}{\partial x_2} \mp 2\mu\frac{\partial v}{\partial x_1} + \left(\mu^2 + b(x_2)\right)v = 0.$$

We use the partial Fourier transform with respect to x_1:

$$\tilde{v}'' + c\tilde{v}' + \left(\mu^2 - \xi^2 \mp 2\mu i\xi + b(x_2)\right)\tilde{v} = 0.$$

This equation can be written in the form

$$L_0\tilde{v} = -\left(\mu^2 - \xi^2 \mp 2\mu i\xi\right)\tilde{v}.$$

1. Weakly non-Fredholm operators

Since zero is a simple eigenvalue of the operator L_0 and all other spectrum is in the left half-plane, then this equation does not have nonzero solutions for small positive μ and any real ξ.

For a fixed x_1 and $|x_2| \to \infty$ we obtain another type of limiting problems (x_1 can be replaced here by $x_1 + h$):

$$\Delta v + c\frac{\partial v}{\partial x_2} - 2\mu \frac{x_1}{\sqrt{1+x_1^2}}\frac{\partial v}{\partial x_1} + \left(\mu^2 \frac{x_1^2}{1+x_1^2} - \mu \frac{1}{(1+x_1^2)^{3/2}} + b(\pm\infty)\right)v = 0.$$

Since $b(\pm\infty) < 0$, then for sufficiently small μ this equation has only the zero solution in the class of bounded functions.

Thus, Condition NS is satisfied. The formally adjoint operator can be studied in the same way. The operator L considered in the weighted space satisfies the Fredholm property.

1.3.4. Linearly dependent systems. Systems of equations can have some additional features in comparison with the scalar equation. Consider the operator

$$L\begin{pmatrix}u\\v\end{pmatrix} = \begin{cases} u'' + cu' + a(x)u + b(x)v \\ dv'' + cv' + a(x)u + b(x)v \end{cases}$$

on the whole axis. Here d and c are some constants, d is positive, the functions $a(x)$ and $b(x)$ belong to $C^\delta(\mathbb{R})$. Suppose that they have limits $a_\pm$, $b_\pm$ at $\pm\infty$. Since the zero-order terms are the same in the first and in the second expressions, then the limiting problems

$$\begin{cases} u'' + cu' + a_\pm u + b_\pm v = 0, \\ dv'' + cv' + a_\pm u + b_\pm v = 0 \end{cases}$$

have nonzero bounded solutions. Indeed, the constant vector $u \equiv 1, v \equiv -a_+/b_+$ satisfies this system. Therefore, the operator does not satisfy the Fredholm property. Such operators arise in some applications to reaction-diffusion problems [137], [138]. We will study them in the second volume.

Limiting equations for the weighted operator L_μ has the form

$$\begin{cases} u'' + (c+2\mu)u' + (a_\pm + c\mu + \mu^2)u + b_\pm v = 0, \\ dv'' + (c+2d\mu)v' + a_\pm u + (b_\pm + c\mu + d\mu^2)v = 0. \end{cases}$$

It does not have nonzero bounded solutions for small $\mu \neq 0$ if $(a_\pm + b_\pm)c \neq 0$. Therefore Condition NS is satisfied. Similarly, Condition NS* can be verified for the formally adjoint system. Hence the operator L_μ is Fredholm. Its index can be computed using the results of Chapter 9.

2 Non-Fredholm solvability conditions

Solvability conditions discussed in the previous section are similar to the usual Fredholm type solvability conditions. The right-hand side f should be orthogonal to the solutions v of the homogeneous adjoint equation, $(f, v) = 0$. In the cases where this duality is not defined, we can approximate f by a sequence f_n and consider solvability conditions in the sense of sequence, $(f_n, v) = 0$. In the case of one-dimensional problems, a more explicit form of the solutions allows us to introduce solvability conditions in a different form. These results can be generalized for some problems in cylinders.

2.1 Example of first-order ODE

We begin with a simple example that illustrates the classical Fredholm type solvability conditions and other type solvability conditions when the Fredholm property is not satisfied. Consider the scalar equation

$$\frac{du}{dt} = a(t)u + f(t), \quad t \in \mathbb{R}. \tag{2.1}$$

A solution of (2.1) can be written as

$$u(t) = u_0(t) \int_0^t v_0(\tau) f(\tau) d\tau, \tag{2.2}$$

where

$$u_0(t) = e^{\int_0^t a(\tau) d\tau}, \quad v_0(t) = e^{-\int_0^t a(\tau) d\tau} = \frac{1}{u_0(t)},$$

$u_0(t)$ is a solution of the homogeneous equation, and $v_0(t)$ is a solution of the homogeneous adjoint equation

$$\frac{du_0}{dt} = a(t)u_0, \quad \frac{dv_0}{dt} = -a(t)v_0.$$

Let us introduce the functions

$$\Phi^+(t) = |u_0(t)| \int_0^t |v_0(\tau)| d\tau, \quad \Psi^+(t) = |u_0(t)| \int_t^\infty |v_0(\tau)| d\tau, \quad t > 0,$$

$$\Phi^-(t) = |u_0(t)| \int_t^0 |v_0(\tau)| d\tau, \quad \Psi^-(t) = |u_0(t)| \int_{-\infty}^t |v_0(\tau)| d\tau, \quad t < 0.$$

Condition 2.1. There exists a positive constant M such that:
- either $\Phi^+(t) \leq M$ for all $t \geq 0$ or the integral in the expressions for $\Psi^+(t)$ is defined and $\Psi^+(t) \leq M$ for all $t \geq 0$,
- either $\Phi^-(t) \leq M$ for all $t \leq 0$ or the integral in the expressions for $\Psi^-(t)$ is defined and $\Psi^-(t) \leq M$ for all $t \leq 0$.

2. Non-Fredholm solvability conditions

Proposition 2.2. *Let Condition 2.1 be satisfied. If at least one of the functions $\Phi^+(t)$ and $\Phi^-(t)$ is bounded, then equation (2.1) has a bounded solution for any bounded function f. If both of them are not bounded, then a bounded solution exists if and only if*

$$\int_{-\infty}^{\infty} v_0(t)f(t)dt = 0. \tag{2.3}$$

Proof. Suppose that both functions $\Phi^+(t)$ and $\Phi^-(t)$ are bounded. Then the solution of equation (2.1) is given by expression (2.2), and it is obviously bounded.

Suppose next that $\Phi^+(t)$ is bounded and $\Phi^-(t)$ is not bounded. Then $\Psi^-(t)$ is defined. Put

$$u^-(t) = u_0(t) \int_{-\infty}^{t} v_0(\tau)f(\tau)d\tau. \tag{2.4}$$

It is easy to verify that $u^-(t)$ is bounded on the whole axis for any bounded f. Moreover, since the integral in $\Psi^-(t)$ is bounded and the function $\Phi^-(t)$ is not bounded at $-\infty$, then $u_0(t)$ is not bounded as $t \to -\infty$. Hence the function $u^-(t)$ is the only solution of (2.1) bounded as $t \to -\infty$ for any bounded f. Indeed, the integral should converge to zero as $t \to -\infty$. Therefore its lower limit is $-\infty$.

The case where $\Phi^-(t)$ is bounded and $\Phi^+(t)$ is not, is similar. The bounded solution is given by the formula

$$u^+(t) = -u_0(t) \int_{t}^{\infty} v_0(\tau)f(\tau)d\tau. \tag{2.5}$$

This is the only solution bounded as $t \to +\infty$.

If both functions $\Phi^+(t)$ and $\Phi^-(t)$ are not bounded but $\Psi^+(t)$ and $\Psi^-(t)$ are bounded, then $u_0(t)$ is not bounded as $t \to \pm\infty$. Therefore the functions u^- and u^+ defined by (2.4) and (2.5) are the only solutions bounded as $t \to -\infty$ and $t \to +\infty$ respectively. The solution bounded on the whole axis exists if and only if $u^+(0) = u^-(0)$. This gives us equality (2.3) The proposition is proved. □

Example 2.3. Suppose that $a(t) = a^+$ for t sufficiently large, and $a(t) = a^-$ for $-t$ sufficiently large. If $a^\pm \neq 0$, then $u_0(t)$ and $v_0(t)$ behave exponentially at infinity. Then Condition 2.1is satisfied.

Condition (2.3) is a typical Fredholm type solvability condition. It may not be sufficient for solvability of equation (2.1) if Condition 2.1 is not satisfied. Indeed, suppose that $v_0(t)$ is integrable. We can choose such t_0 that for the function

$$f(t) = \begin{cases} 1, & t \geq t_0, \\ -1, & t < t_0 \end{cases}$$

equality (2.3) is satisfied. From integrability of $v_0(t)$ it follows that $u_0(t)$ is not bounded as $t \to \pm\infty$. Therefore, the functions $\Phi^+(t)$ and $\Phi^-(t)$ are not bounded either. If Condition 2.1 is not satisfied, then at least one of the functions $\Psi^+(t)$

and $\Psi^-(t)$ is not bounded. Hence there is no bounded solution of equation (2.1) with such f. Thus, (2.3) does not provide solvability of equation (2.1). Condition 2.1 is not satisfied for polynomial u and v: $u(t) \sim |t|^k, v(t) \sim 1/|t|^k$, $k > 1$ as $t \to \pm\infty$.

To illustrate another type of solvability conditions suppose that the function

$$b(t) = \int_0^t a(s)ds$$

is uniformly bounded. Then $v_0(t)$ is bounded and $|u_0(t)| \geq \varepsilon > 0$ for some ε. Therefore the solution given by (2.2) is bounded if and only if

$$\sup_t \left| \int_0^t v_0(s)f(s)ds \right| < \infty. \tag{2.6}$$

As above, the solvability condition is given in terms of bounded solutions of the homogeneous adjoint equation. The principal difference is that condition (2.6), contrary to Fredholm type solvability conditions, cannot be formulated in the form $\phi(f) = 0$, where ϕ is a functional from the dual space.

2.2 Ordinary differential systems on the real line

In this section we discuss solvability conditions for non-Fredholm operators in the case of ordinary differential systems of equations. We will obtain solvability conditions similar to those in the previous section. We follow here the presentation in [279] where these results are also used to study elliptic operators in cylinders.

Let $u \in \mathbb{R}^n$. Denote by $|\cdot|$ the Euclidian vector norm in $\mathbb{R}^n$ and the corresponding matrix norm and by $\langle\cdot,\cdot\rangle$ the scalar product in $\mathbb{R}^n$. Consider the linear system

$$u' = P(x)u \tag{2.7}$$

where the matrix $P(x)$ is defined, bounded and continuous on the interval $(a, b) \subset \mathbb{R}$. Here a is a real number or $-\infty$ and b is a real number or $+\infty$. Let $\Phi(x,t)$ be the Cauchy matrix of system (2.7).

Definition 2.4. ([417]) Let I be a closed convex subset of $\mathbb{R}$. Consider an $n \times n$ matrix $P(x)$, continuous and bounded on I. System (2.7) is *dichotomic* on I if there exist positive constants c and λ, and subspaces $U^s(x)$ and $U^u(x)$ of $\mathbb{R}^n$, defined for all $x \in I$ and such that

1. $\Phi(x,\xi)U^{s,u}(\xi) = U^{s,u}(x)$ for all $x, \xi \in I$;
2. $U^s(x) \oplus U^u(x) = \mathbb{R}^n$ for every $x \in I$;
3. $|\Phi(x,\xi)u_0| \leq c\exp(-\lambda(x-\xi))|u_0|$ for all $x,\xi \in I$: $x \geq \xi$, $u_0 \in U^s(\xi)$;
4. $|\Phi(x,\xi)u_0| \leq c\exp(\lambda(x-\xi))|u_0|$, if $x,\xi \in I$: $x \leq \xi$, $u_0 \in U^u(\xi)$.

This property is also called hyperbolicity and the corresponding system is called hyperbolic. Nevertheless, we shall always call it dichotomic in order not to

2. Non-Fredholm solvability conditions

confuse this notion with hyperbolicity of partial differential equations. Note that Definition 2.4 coincides with the definition of exponential dichotomy given by Coppel [115, p. 10] with the additional assumption of the boundedness of the matrix P.

Definition 2.5. System (2.7) is *almost dichotomic* on (a,b) with positive constants c and λ if for every $x \in (a,b)$ there exist three spaces $M_S(x)$ (stable space), $M_U(x)$ (unstable space) and $M_B(x)$ (zero space), satisfying the following conditions:

1. $M_S(x) \oplus M_U(x) \oplus M_B(x) = \mathbb{R}^n$ for all $x \in (a,b)$;
2. $\Phi(x,t)M_\sigma(t) = M_\sigma(x)$ for all $\sigma \in \{S,U,B\}$, $x,t \in (a,b)$;
3. $|\Phi(x,t)u_0| \leq c\exp(-\lambda(x-t))|u_0|$ for all $x \geq t$, $x,t \in (a,b)$, $u_0 \in M_S(t)$;
4. $|\Phi(x,t)u_0| \leq c\exp(\lambda(x-t))|u_0|$ for all $x \leq t$, $x,t \in (a,b)$, $u_0 \in M_U(t)$;
5. $|\Phi(x,t)u_0| \leq c|u_0|$ for all $x,t \in (a,b)$, $u_0 \in M_B(t)$.

The following statement is evident.

Lemma 2.6. *Let matrix $P(x)$ be constant, i.e., $P(x) \equiv P$. System (2.7) is almost dichotomic if and only if for every purely imaginary eigenvalue λ of the matrix P the number of linearly independent eigenvectors corresponding to λ is equal to the multiplicity of λ.*

Remark 2.7. In other words, the condition is the following: for every $\lambda \in i\mathbb{R}$ every block in the Jordan form of the matrix A corresponding to λ is simple. The statement of the lemma holds true if the matrix P does not have purely imaginary eigenvalues at all. In this case the space M_B is trivial and system (2.7) is dichotomic.

Definition 2.8. ([2]) Consider the change of variables

$$u = L(x)v, \quad x \in \mathbb{R}. \tag{2.8}$$

It is called a *Lyapunov transform* if the matrix $L(x)$ is C^1-smooth invertible and all matrices $L(x)$, $L^{-1}(x)$ and $L'(x)$ are bounded.

Lemma 2.9. *Let system (2.7) be almost dichotomic and let the dimensions of the corresponding spaces $M_S(x)$, $M_U(x)$ and $M_B(x)$ be n_S, n_U and n_B, respectively. Then for every x there exist continuous projectors $\Pi_S(x)$, $\Pi_U(x)$ and $\Pi_B(x)$ on the spaces $M_S(x)$, $M_U(x)$ and $M_B(x)$ respectively, such that $\Pi_S(x) + \Pi_U(x) + \Pi_B(x) \equiv$ id. These projectors are uniformly bounded. Also, there exists a Lyapunov transform (2.8), which reduces system (2.7) to the form*

$$v' = \widetilde{P}(x)v, \tag{2.9}$$

where $v = (v_S, v_U, v_B)$, $\widetilde{P}(x) = \text{diag}(P_S(x), P_U(x), P_B(x))$, and system (2.9) splits into three subsystems:

$$v_S' = P_S(x)v_S, \tag{2.10}$$
$$v_U' = P_U(x)v_U, \tag{2.11}$$
$$v_B' = P_B(x)v_B. \tag{2.12}$$

Systems (2.10)–(2.12) *satisfy the following properties:*

1. *System* (2.10) *is steadily dichotomic, i.e., it is dichotomic and the corresponding stable space coincides with the space* $\mathbb{R}^{n_S}$ *for all* x.
2. *System* (2.11) *is unsteadily dichotomic, i.e., it is dichotomic and the corresponding unstable space coincides with the space* $\mathbb{R}^{n_U}$ *for all* x.
3. *Every solution of system* (2.12) *is bounded.*

Remark 2.10. The matrix $\tilde{P}(x)$ can be found by the formula

$$\tilde{P}(x) = L^{-1}(x)P(x)L(x) - L^{-1}(x)L'(x). \tag{2.13}$$

Since the matrix $P(x)$ is bounded, the matrix $\tilde{P}(x)$ is also bounded. If for a certain $\delta \geq 0$, $P(x) \in C^{\delta}$ and $L(x) \in C^{1+\delta}$, then $\tilde{P}(x) \in C^{\delta}$.

The proof of Lemma 2.9 is the same as the proof for dichotomic (hyperbolic) ordinary differential systems [115, Lemma 3, p. 41], [418, Theorem 0.1, p. 14].

Lemma 2.11. *If system* (2.7) *is steadily dichotomic, the dual system*

$$u' = -P^T(x)u, \tag{2.14}$$

is unsteadily dichotomic. If (2.7) *is an unsteadily dichotomic system, then system* (2.14) *is steadily dichotomic. If system* (2.7) *is almost dichotomic with all solutions bounded, the dual system also is.*

The lemma above follows from the fact that for every fundamental matrix $\Phi(x)$ of system (2.7), the matrix $(\Phi^{-1})^T(x)$ is fundamental for system (2.14). The following statement is evident.

Lemma 2.12. *Any system* (2.7), *which splits into almost dichotomic blocks, is almost dichotomic. The stable, unstable and bounded spaces are direct products of the corresponding spaces for blocks.*

Having fixed a number $\delta \geq 0$, define spaces $X = C^{\delta}(\mathbb{R} \to \mathbb{R}^n)$, $Y = C^{1+\delta}(\mathbb{R} \to \mathbb{R}^n)$ and consider a function $f \in X$.

Theorem 2.13. *Let system* (2.7) *be almost dichotomic on* $\mathbb{R}$, *and the matrix* $P(x)$ *be bounded in* $C^{\delta}(\mathbb{R} \to \mathbb{R}^{n^2})$. *Then for any* $f \in X$ *the system*

$$u' = P(x)u + f(x) \tag{2.15}$$

has a solution $v(x) \in Y$ *if and only if*

$$\sup_{x \in \mathbb{R}} \left| \int_0^x \langle \varphi(s), f(s) \rangle \, ds \right| < +\infty \tag{2.16}$$

for every bounded solution $\varphi(s)$ *of system* (2.14).

2. Non-Fredholm solvability conditions

Proof. Transformation (2.8), which exists due to Lemma 2.9, reduces system (2.15) to the form
$$v' = \widetilde{P}(x)v + g(x) \tag{2.17}$$
where $\widetilde{P}(x)$ satisfies (2.13), and $g(x) = L^{-1}(x)f(x)$. If $f(x) \in X$, then $g(x) \in X$ and vice versa. System (2.17) splits into three subsystems:
$$v_S' = P_S(x)v_S + g_S(x), \tag{2.18}$$
$$v_U' = P_U(x)v_U + g_U(x), \tag{2.19}$$
$$v_B' = P_B(x)v_B + g_B(x). \tag{2.20}$$

Here $g(x) = (g_S(x), g_U(x), g_B(x))$. Systems (2.15) and (2.17) have bounded solutions if and only if each system (2.18), (2.19), and (2.20) has a bounded solution.

Let $\Psi(x,t)$ be the Cauchy matrix of system (2.9). It can be written in the form
$$\Psi(x,t) = \mathrm{diag}\,(\Psi_S(x,t), \Psi_U(x,t), \Psi_B(x,t))$$
where $\Psi_S(x,t)$, $\Psi_U(x,t)$ and $\Psi_B(x,t)$ are the Cauchy matrices for systems (2.10), (2.11) and (2.12), respectively. Since systems (2.10) and (2.11) are dichotomic, the nonhomogeneous systems (2.18) and (2.19) have, for every g, bounded solutions of the form
$$v_S(x) = \int_{-\infty}^{x} \Psi_S(x,t)g_S(t)\,dt; \quad v_U(x) = -\int_{x}^{\infty} \Psi_U(x,t)g_U(t)\,dt.$$

All solutions of the system (2.20) have the form
$$\Psi_B(x)C + \int_0^x \Psi_B(x,t)g_B(t)\,dt.$$

Here $\Psi_B(x) = \Psi_B(x,0)$. Every solution of system (2.12) is bounded. Therefore the matrix $\Psi_B(x)$ is also bounded. Hence, it is sufficient to verify that the solution
$$v_B(x) = \int_0^x \Psi_B(x,t)g_B(t)\,dt = \Psi_B(x)\int_0^x \Psi_B^{-1}(t)g_B(t)\,dt$$
is bounded. Let c be the constant from Definition 2.5 for system (2.7), and $K > 0$ be such that $\max(\|L(x)\|_{C^{1+\delta}}, \|L^{-1}(x)\|_{C^{1+\delta}}) < K$. Then every column of the matrices $\Psi_B(x)$ and $\Psi_B^{-1}(x)$ is bounded by cK. Hence
$$\max(\|\Psi_B(x)\|_{C^{1+\delta}}, \|\Psi_B^{-1}(x)\|_{C^{1+\delta}}) \leq \sqrt{n}cK.$$

Thus, $v_B(x)$ is bounded if and only if the integral
$$I(x) = \int_0^x \Psi_B^{-1}(t)g_B(t)\,dt$$

is bounded. Consider the matrix $\Xi(x)$ which is obtained from Ψ_B^{-1} by adding $n_U + n_S$ zero rows. It follows from Lemmas 2.11 and 2.12 that every bounded solution of the system

$$v' = -\widetilde{P}^T(x)v \qquad (2.21)$$

is a linear combination of columns of $\Xi(x)$. Hence $I(x)$ is bounded if and only if the condition

$$\sup_{x \in \mathbb{R}} \left| \int_0^x \langle \eta(t), g(t) \rangle \, dt \right| < +\infty \qquad (2.22)$$

is satisfied for every bounded solution $\eta(x)$ of (2.21).

On the other hand, $\Phi(x) = L(x)\Psi(x)$ is a fundamental matrix of system (2.7). Then $\Psi^{-1}(x) = \Phi^{-1}(x)L(x)$. Hence every bounded solution $\eta(x)$ of system (2.21) can be written in the form $\eta(x) = L^T(x)\varphi(x)$, where $\varphi(x)$ is a bounded solution of (2.14). It is easy to see that this correspondence is one-to-one. Consequently, we can rewrite the integral in (2.22) in the form

$$\int_0^x \langle L^T(t)\varphi(t), L^{-1}(t)f(t) \rangle \, dt = \int_0^x \langle \varphi(t), f(t) \rangle \, dt. \qquad (2.23)$$

Thus, there exists a bounded solution of system (2.15) if and only if expression (2.23) is uniformly bounded. The theorem is proved. $\square$

2.3 Second-order equations

Consider the scalar equation

$$u'' + b(x)u' + c(x)u = f(x) \qquad (2.24)$$

assuming that the coefficients belong to the Hölder space $C^{1+\delta}(\mathbb{R})$ with some δ, $0 < \delta < 1$. Denote by $u_1(x)$ and $u_2(x)$ two linearly independent solutions of the homogeneous equation

$$u'' + b(x)u' + c(x)u = 0 \qquad (2.25)$$

and by $v_1(x)$ and $v_2(x)$ two linearly independent solutions of the homogeneous adjoint equation

$$v'' - (b(x)v)' + c(x)v = 0 \qquad (2.26)$$

such that

$$u_1v_1 + u_2v_2 = 0, \quad u_1'v_1 + u_2'v_2 = 1. \qquad (2.27)$$

The solution of equation (2.24) can be represented in the form

$$u(x) = u_1(x) \int_0^x v_1(y)f(y) + u_2(x) \int_0^x v_2(y)f(y) \qquad (2.28)$$

(cf. Section 2.2 of Chapter 9). If the functions u_i and v_i are exponentially decaying or growing at infinity, then such behavior is specific for Fredholm operators. Let

2. Non-Fredholm solvability conditions

us suppose that the functions u_i and v_i, $i = 1, 2$ are bounded on $\mathbb{R}$. It follows from (2.27) that not all of them can be exponentially decaying. Therefore, the operator is not Fredholm.

Theorem 2.14. *Suppose that the solutions u_i and v_i, $i = 1, 2$ of equations (2.25) and (2.26) are bounded. Let $f \in C^\delta(\mathbb{R})$. Then equation (2.24) is solvable in $C^{2+\delta}(\mathbb{R})$ if and only if the following solvability conditions are satisfied:*

$$\sup_x \left| \int_0^x f(y) v_i(y) dy \right| < \infty, \quad i = 1, 2. \tag{2.29}$$

Proof. Existence of a bounded solution follows from representation (2.28), condition (2.29) and from the assumption that the functions u_i are bounded. From the equalities

$$u'(x) = u_1'(x) \int_0^x v_1(y) f(y) + u_2'(x) \int_0^x v_2(y) f(y),$$

$$u''(x) = u_1''(x) \int_0^x v_1(y) f(y) + u_2''(x) \int_0^x v_2(y) f(y) + f(x)$$

and regularity of solutions of the homogeneous equation, $u_i \in C^{2+\delta}(\mathbb{R})$, we conclude that $u \in C^{2+\delta}(\mathbb{R})$.

We next prove necessity. Suppose that equation (2.24) has a solution $u \in C^{2+\delta}(\mathbb{R})$. We multiply this equation by v_i and integrate. Taking into account that v_i is a solution of the homogeneous adjoint equation, we obtain

$$\int_0^x f(y) v_i(y) dy = u' v_i |_0^x - u v_i' |_0^x + b u v_i |_0^x.$$

It remains to note that the right-hand side of this equality is uniformly bounded. The theorem is proved. $\square$

Corollary 2.15. *If u_i and v_i, $i = 1, 2$ are bounded, then equation (2.24) has a solution $u \in C^{2+\delta}(\mathbb{R})$ for any $f \in C^\delta(\mathbb{R}) \cap L^1(\mathbb{R})$.*

Example 2.16. Consider the equation

$$u'' + c(x) u = f(x), \tag{2.30}$$

where

$$c(x) = \begin{cases} c_+ , & x \geq N, \\ c_- , & x \leq -N, \end{cases}$$

for some N and for some positive constants $c_\pm$.

All solutions of the homogeneous equation

$$u'' + c(x) u = 0$$

for $|x| \geq N$ have the form

$$u(x) = k_1 \sin(\sqrt{c_\pm} x) + k_2 \cos(\sqrt{c_\pm} x),$$

where k_1 and k_2 are some constants. Since the equation is self-adjoint, then the solutions u_i and v_i are bounded. For any bounded function f, the integrals

$$I_i(x) = \int_0^x f(y) v_i(y) dy, \quad i = 1, 2$$

are bounded for $|x| \leq N$. Condition (2.29) is satisfied if the integrals

$$\int_N^x f(y) \sin(\sqrt{c_+} y) dy, \quad \int_N^x f(y) \cos(\sqrt{c_+} y) dy, \quad x \geq N,$$

$$\int_x^{-N} f(y) \sin(\sqrt{c_-} y) dy, \quad \int_x^{-N} f(y) \cos(\sqrt{c_-} y) dy, \quad x \leq -N$$

are bounded uniformly in x. If $c(x) \equiv c$, where c is a positive constant, then conditions (2.29) become

$$\sup_x \left| \int_0^x f(y) \sin(\sqrt{c} y) dy \right| < \infty, \quad \sup_x \left| \int_0^x f(y) \cos(\sqrt{c} y) dy \right| < \infty.$$

These solvability conditions are satisfied for the function $f(x) = \sin(kx)$, where $k \neq \sqrt{c_\pm}$. (cf. Section 1.3.1).

It can be verified that the image of the operator is not closed. Indeed, let $c = 1$ and $v(x) = \cos x$, $f_k(x) = (\phi_k(x) \cos x)/\sqrt{1+x^2}$, where $\phi_k(x)$ is a smooth function equal to 1 for $|x| \leq k$ and 0 for $|x| \geq k+1$, $f(x) = (\cos x)/\sqrt{1+x^2}$. Then $f_k \to f$ in $C^\delta(\mathbb{R})$ and

$$\sup_x \left| \int_0^x f_k(y) v(y) dy \right| < \infty, \quad \sup_x \left| \int_0^x f(y) v(y) dy \right| = \infty.$$

The integrals with $v(x) = \sin x$ have the same properties. Thus, the functions f_k satisfy the solvability conditions and belong to the image of the operator while f does not belong to it.

The results of this section can be generalized for one-dimensional systems of equations and for some special classes of multi-dimensional problems, e.g., in unbounded cylinders. Applicability of such solvability conditions for general multi-dimensional problems is not clear.

3 Space decomposition of operators

In the previous section, solvability conditions for non-Fredholm operators were formulated in the one-dimensional case. In this section, we will consider multi-dimensional elliptic operators in the whole space and will use some space decomposition in order to reduce non-Fredholm properties to some one-dimensional

3. Space decomposition of operators

operator. This is related to the structure of non-Fredholm operators discussed in Section 1. Consider the operator

$$Lu = a(x)\Delta u + \sum_{j=1}^{n} b_j(x)\frac{\partial u}{\partial x_j} + c(x)u \tag{3.1}$$

acting from $C^{2+\alpha}(\mathbb{R}^n)$ into $C^\alpha(\mathbb{R}^n)$. Here $u = (u_1, \ldots, u_p)$, $a(x), b_j(x), c(x)$ are real-valued $p \times p$ matrices with $C^\alpha(\mathbb{R}^n)$ entries, and $a(x)$ is symmetric positive definite,

$$(a(x)\xi, \xi) \geq a_0|\xi|^2$$

for any vector $\xi \in \mathbb{R}^p$, $x \in \mathbb{R}^n$ with a constant $a_0 > 0$. To simplify the presentation, we consider the case $n = 2$ with the independent variables x and y.

In order to determine explicitly the location of the essential spectrum, we make some simplifying assumptions. We assume existence of the limits

$$a(x,y) \to a^\pm(y), \ b_j(x,y) \to b_j^\pm(y), \ c(x,y) \to c^\pm(y)$$

as $x \to \pm\infty$. This convergence is uniform with respect to y on every bounded set in $\mathbb{R}^1$. If $y \to \pm\infty$, then

$$a(x,y) \to a_\pm^0, \ b_j(x,y) \to b_{j\pm}^0, \ c(x,y) \to c_\pm^0$$

uniformly on every bounded set. Here $a_\pm^0, b_{j\pm}^0$, and $c_\pm^0$ are constant matrices,

$$a_\pm^0 = \lim_{y \to \pm\infty} a^\pm(y), \ b_{j\pm}^0 = \lim_{y \to \pm\infty} b_j^\pm(y), \ c_\pm^0 = \lim_{y \to \pm\infty} c^\pm(y).$$

These assumptions allow us to define the limiting operators

$$L^\pm u = a^\pm(y)\Delta u + b_1^\pm(y)\frac{\partial u}{\partial x} + b_2^\pm(y)\frac{\partial u}{\partial y} + c^\pm(y)u, \tag{3.2}$$

$$L_\pm^0 u = a_\pm^0 \Delta u + b_{1\pm}^0 \frac{\partial u}{\partial x} + b_{2\pm}^0 \frac{\partial u}{\partial y} + c_\pm^0 u. \tag{3.3}$$

Consider the problems

$$L^\pm u = \lambda u, \ L_\pm^0 u = \lambda u. \tag{3.4}$$

If one of them has a nonzero solution in $C^{2+\alpha}(\mathbb{R}^2)$, then the corresponding value of λ belongs to the essential spectrum of the operator L, i.e., the operator $L - \lambda I$ is not Fredholm.

We suppose that the last problem in (3.4) does not have nonzero solutions for any λ with nonnegative real part and that there exists a nonzero solution of at least one of the limiting problems

$$L^+ u = 0, \tag{3.5}$$

$$L^- u = 0, \tag{3.6}$$

in $C^{2+\alpha}(\mathbb{R}^2)$. Then the operator L is not Fredholm. We will reduce the operator to a subspace where its image is closed and the kernel is finite-dimensional. This allows us to localize the non-Fredholm properties of the operator in a complementary subspace.

3.1 Normal solvability on a subspace

Suppose for certainty that both problems (3.5) and (3.6) have nonzero solutions and impose the following conditions.

Condition 3.1. Problems (3.5) and (3.6) have unique nonzero solutions $v^+ \in C^{2+\alpha}(\mathbb{R}^1)$ and $v^- \in C^{2+\alpha}(\mathbb{R}^1)$, respectively, which are functions of the variable y.

Condition 3.2. The coefficients of the limiting operators are sufficiently smooth, and the formally adjoint problems

$$\hat{L}^\pm u \equiv \Delta(\hat{a}^\pm(y)u) - \frac{\partial(\hat{b}_1^\pm(y)u)}{\partial x} - \frac{\partial(\hat{b}_2^\pm(y)u)}{\partial y} + \hat{c}^\pm(y)u = 0 \tag{3.7}$$

have nonzero solutions $\hat{v}^\pm \in C^{2+\alpha}(\mathbb{R}^1) \cap L^1(\mathbb{R}^1)$ which depend only on y and such that

$$\int_{\mathbb{R}^1} (v^+(y), \hat{v}^+(y)) \, dy \neq 0, \quad \int_{\mathbb{R}^1} (v^-(y), \hat{v}^-(y)) \, dy \neq 0.$$

Here $\hat{a}^\pm, \hat{b}_j^\pm, \hat{c}^\pm$ are the matrices transposed to $a^\pm, b_j^\pm, c^\pm$, respectively and $(\,,\,)$ denotes the inner product in $\mathbb{R}^2$.

Condition 3.3. Let

$$a_{11} = \int_{\mathbb{R}^1} (v^+(y), \hat{v}^+(y)) dy, \quad a_{12} = \int_{\mathbb{R}^1} (v^-(y), \hat{v}^+(y)) dy,$$

$$a_{21} = \int_{\mathbb{R}^1} (v^+(y), \hat{v}^-(y)) dy, \quad a_{22} = \int_{\mathbb{R}^1} (v^-(y), \hat{v}^-(y)) dy.$$

Then $a_{11} a_{22} \neq a_{12} a_{21}$.

Condition 3.4. The limiting problems

$$L_0 u \equiv a_\pm^0 \Delta u + b_{1\pm}^0 \frac{\partial u}{\partial x} + b_{2\pm}^0 \frac{\partial u}{\partial y} + c_\pm^0 u = 0$$

do not have nonzero solutions in $C^{2+\alpha}(\mathbb{R}^2)$.

Let us introduce one-dimensional operators

$$L_1^\pm u = a^\pm(y) \frac{\partial^2 u}{\partial y^2} + b_2^\pm(y) \frac{\partial u}{\partial y} + c^\pm(y) u$$

3. Space decomposition of operators

acting from $C^{2+\alpha}(\mathbb{R}^1)$ into $C^{\alpha}(\mathbb{R}^1)$. Due to the assumptions above, $L_1^{\pm}v^{\pm} = 0$, $(L_1^{\pm})^*\hat{v}^{\pm} = 0$. The essential spectrum $\lambda(\xi)$ of these operators, which is also a part of the essential spectrum of the operator L, is a set of all complex numbers λ such that
$$\det(-a_{\pm}^0 \xi^2 + ib_{2\pm}^0 \xi + c_{\pm}^0 - \lambda) = 0, \quad \xi \in \mathbb{R}^1.$$
If it lies in the left half-plane, then the operators $L_1^{\pm}$ are Fredholm with zero index (Chapter 9). According to Conditions 3.1 and 3.2, the operators $L_1^{\pm}$ have a simple zero eigenvalue.

Set $E = C^{2+\alpha}(\mathbb{R}^2)$, $E' = C^{\alpha}(\mathbb{R}^2)$ and
$$E_0 = \left\{ u \in E : \int_{\mathbb{R}^1} u(x,y)\hat{v}^+(y)dy = \int_{\mathbb{R}^1} u(x,y)\hat{v}^-(y)dy = 0, \ \forall x \in \mathbb{R}^1 \right\}.$$

Lemma 3.5. *For any $u \in E$ the following representation holds:*
$$u(x,y) = u_0(x,y) + c^+(x)v^+(y) + c^-(x)v^-(y),$$
where $u_0 \in E_0$, $c^{\pm} \in C^{2+\alpha}(\mathbb{R}^1)$.

Proof. Consider the system
$$a_{11}c^+(x) + a_{12}c^-(x) = \int_{\mathbb{R}^1} u(x,y)\hat{v}^+(y)dy, \qquad (3.8)$$
$$a_{21}c^+(x) + a_{22}c^-(x) = \int_{\mathbb{R}^1} u(x,y)\hat{v}^-(y)dy. \qquad (3.9)$$

The integrals in the right-hand sides of (3.8), (3.9) are well defined. They belong to $C^{2+\alpha}(\mathbb{R}^1)$ as functions of x. By virtue of Condition 3.3 we can find a solution of this system, $c^{\pm} \in C^{2+\alpha}(\mathbb{R}^1)$. The function $u_0(x,y) = u(x,y) - c^+(x)v^+(y) - c^-(x)v^-(y)$ belongs to E_0. The lemma is proved. $\square$

Thus we can represent the space E as a direct sum of E_0 and of the complementary subspace
$$\hat{E} = \left\{ u \in E : u = c^+(x)v^+(y) + c^-(x)v^-(y), \ c^{\pm} \in C^{2+\alpha}(\mathbb{R}^1) \right\}.$$

Remark 3.6. If $v^+(y) \equiv v^-(y)$, then Condition 3.3 is not satisfied. Instead of the representation in Lemma 3.5, in this case we have $u(x,y) = u_0(x,y) + c(x)v(y)$.

Lemma 3.7. *Let a sequence $u_k \in E_0$ be bounded in the norm of E. If $u_k \to u_0$ uniformly on every bounded set, then $u_0 \in E_0$.*

The proof of the lemma is obvious.

Lemma 3.8. *The kernel of the operator $L : E_0 \to E'$ is finite dimensional.*

Proof. Consider a sequence $u_k \in E_0$, $\|u_k\| \leq 1$ in the kernel of the operator L,

$$Lu_k = 0. \tag{3.10}$$

We will show that it has a converging subsequence. Then we will conclude that the unit sphere in the kernel of the operator is compact and the kernel is finite dimensional. Since u_k is bounded in $C^{2+\alpha}(\mathbb{R}^2)$, there exists a subsequence, still denoted by u_k, which converges to some $u_0 \in E$ in C^2 uniformly on every bounded set. Passing to the limit in (3.10), we obtain $Lu_0 = 0$. Set $v_k = u_k - u_0$. Then $Lv_k = 0$.

We show that the convergence $v_k \to 0$ is uniform in $\mathbb{R}^2$. Suppose that it is not and there exists a sequence (x_k, y_k) such that $|v_k(x_k, y_k)| \geq \varepsilon > 0$. By virtue of the local convergence of this sequence of functions to zero, we have $x_k^2 + y_k^2 \to \infty$.

Consider first the case where the values y_k are uniformly bounded. Without loss of generality, we can assume that $y_k \to y_0$ and that x_k converges to $+\infty$. Put $w_k(x, y) = v_k(x + x_k, y + y_k)$. Then

$$a(x + x_k, y + y_k)\Delta w_k + b_1(x + x_k, y + y_k)\frac{\partial w_k}{\partial x}$$
$$+ b_2(x + x_k, y + y_k)\frac{\partial w_k}{\partial y} + c(x + x_k)w_k = 0,$$

$|w_k(0)| \geq \varepsilon$, and w_k converges to some w_0 in C^2 uniformly on every bounded set. Therefore

$$a^+(y + y_0)\Delta w_0 + b_1^+(y + y_0)\frac{\partial w_0}{\partial x} + b_2^+(y + y_0)\frac{\partial w_0}{\partial y} + c^+(y + y_0)w_0 = 0.$$

By virtue of Condition 3.1, $w_0(x, y) \equiv v^+(y + y_0)$. On the other hand

$$\int_{\mathbb{R}^1} v_k(x + x_k, y + y_k)\hat{v}^+(y + y_k)\,dy$$
$$= \int_{\mathbb{R}^1} (u_k(x + x_k, y + y_k) - u_0(x + x_k, y + y_k))\hat{v}^+(y + y_k)dy = 0, \quad \forall x \in \mathbb{R}^1,$$

for all k, because u_k and u_0 belong to E_0, and

$$\int_{\mathbb{R}^1} w_0(x, y)\hat{v}^+(y + y_0)\,dy = \lim_{k \to \infty} \int_{\mathbb{R}^1} v_k(x + x_k, y + y_k)\hat{v}^+(y + y_k)\,dy.$$

We obtain a contradiction with Condition 3.2. Therefore, $v_k \to 0$ uniformly in $\mathbb{R}^2$. The Schauder estimate implies the convergence $v_k \to 0$ in $C^{2+\alpha}(\mathbb{R}^2)$. Hence the unit sphere in the kernel of the operator is compact.

Suppose now that $|y_k|$ is unbounded. As above, we obtain a nonzero solution of one of the limiting problems

$$a_\pm^0 \Delta u + b_{1\pm}^0 \frac{\partial u}{\partial x} + b_{2\pm}^0 \frac{\partial u}{\partial y} + +c_\pm^0 u = 0.$$

This contradicts Condition 3.4. The lemma is proved. □

3. Space decomposition of operators

Lemma 3.9. *The image of the operator* $L : E_0 \to E'$ *is closed.*

Proof. Let
$$Lu_k = f_k, \qquad (3.11)$$
$f_k \in E'$, $f_k \to f_0$, $u_k \in E_0$. We will show that there exists $u_0 \in E_0$ such that $Lu_0 = f_0$. Consider first the case where the sequence u_k is bounded in E. Since the functions u_k are uniformly bounded in the norm $C^{2+\alpha}(\mathbb{R})$, then we can choose a subsequence converging to some $u_0 \in E$ in C^2 uniformly on every bounded set. Therefore $u_0 \in E_0$. Passing to the limit in (3.11), we obtain $Lu_0 = f_0$.

Suppose now that the sequence u_k is unbounded. Since the kernel of the operator L in E_0 is finite dimensional, we can represent E_0 as a direct sum of $\mathrm{Ker}\, L$ and a complementary subspace $\hat{E}_0$. Then $u_k = \hat{u}_k + u_k^0$, where $\hat{u}_k \in \hat{E}_0$, $u_k^0 \in \mathrm{Ker}\, L$. Then $L\hat{u}_k = f_k$. If the sequence $\hat{u}_k$ is bounded, we can proceed as above to obtain a function $\hat{u}_0 \in E_0$ such that $L\hat{u}_0 = f_0$.

If $\hat{u}_k$ is not bounded, then we write $v_k = \hat{u}_k / \|\hat{u}_k\|_E$, $g_k = f_k/\|\hat{u}_k\|_E$. Hence
$$Lv_k = g_k. \qquad (3.12)$$
We will show that there exists a subsequence of v_k converging to some $v_0 \in \hat{E}_0$ in E and such that $Lv_0 = 0$. This will contradict the definition of $\hat{E}_0$. Since v_k is bounded, there exists a subsequence, denoted again by v_k, converging to some $v_0 \in E$ in C^2 uniformly on every bounded set. Let us show that this convergence is uniform in $\mathbb{R}^2$. Passing to the limit in (3.12), we obtain $Lv_0 = 0$ and, for $w_k = v_k - v_0$, $Lw_k = g_k$. The sequence w_k converges to 0 uniformly on every bounded set. If this convergence is not uniform in $\mathbb{R}^2$ then there exists a sequence (x_k, y_k) such that $|w_k(x_k, y_k)| \geq \varepsilon$ and $x_k^2 + y_k^2 \to \infty$. If y_k is unbounded, then we obtain a nonzero solution of one of the limiting problems
$$u_\pm^0 \Delta u + b_{1\pm}^0 \frac{\partial u}{\partial x} + b_{2\pm}^0 \frac{\partial u}{\partial y} + c_\pm^0 u - 0.$$

This contradicts Condition 3.4. If y_k is bounded, then we can assume that $y_k \to y_0$, $x_k \to +\infty$. Put $w_k(x, y) = w_k(x + x_k, y + y_k)$. We can choose a subsequence w_k converging to some w_0 in C^2 uniformly on every bounded set. From the equation
$$a(x + x_k, y + y_k)\Delta w_k + b_1(x + x_k, y + y_k) \frac{\partial w_k}{\partial x}$$
$$+ b_2(x + x_k, y + y_k) \frac{\partial w_k}{\partial y} + c(x + x_k, y + y_k)w_k = g_k(x + x_k, y + y_k)$$

we obtain
$$a^+(y + y_0)\Delta w_0 + b_1^+(y + y_0) \frac{\partial w_0}{\partial x} + b_2^+(y + y_0) \frac{\partial w_0}{\partial y} + c^+(y + y_0)w_0 = 0.$$

Hence $w_0(x, y) = v^+(y + y_0)$. As in the previous lemma we have

$$\int_{R^1} w_0(x, y) \hat{v}^+(y + y_0)\, dy$$
$$= \lim_{k \to \infty} \int_{R^1} (v_k(x + x_k, y + y_k) - v_0(x + x_k, y + y_k))\hat{v}^+(y + y_k)\, dy = 0.$$

This contradicts Condition 3.2. Thus we have proved that $v_k \to v_0$ uniformly on R^2. From the Schauder estimate we obtain the convergence in E. Therefore $v_0 \in \hat{E}_0$ and $Lv_0 = 0$. This contradiction proves the lemma. □

These results can be generalized for the subspaces

$$E_{r,s} = \left\{ u \in E : \int_{R^1} u(x, y)\hat{v}^+(y)\, dy = 0,\ \forall x \geq r,\right.$$
$$\left.\int_{R^1} u(x, y)\hat{v}^-(y)\, dy = 0,\ \forall x \leq s \right\}.$$

Theorem 3.10. *For any r and s the operator $L : E_{r,s} \to E'$ has a finite-dimensional kernel and a closed image.*

3.2 Scalar equation

In this section we consider a scalar equation in R^2. We will obtain more complete solvability conditions. Consider the operator

$$Lu = \Delta u + b(y)u \tag{3.13}$$

acting from $C^{2+\delta}(R^2)$ into $C^\delta(R^2)$. According to Condition 3.1, the equation

$$u''(y) + b(y)u(y) = 0 \tag{3.14}$$

has a solution $v(y)$. We assume that $b(\pm\infty) < 0$ and that the principal eigenvalue of the operator L is zero. We will use here that the principal eigenfunction is positive, that is $v(y) > 0$, $y \in R^1$. For scalar second-order elliptic operators in bounded domains this follows from the Krein-Rutman type theorems. They are not directly applicable for operators in unbounded domains. However, this property remains valid under the assumption that the essential spectrum of the operator is located in the left half-plane [563], [568]. Consider the equation

$$Lu = g,\quad g \in C^\delta(R^2), \tag{3.15}$$

and write

$$k(x) = \int_{-\infty}^{\infty} g(x, y)v(y)\, dy. \tag{3.16}$$

3. Space decomposition of operators

This integral is well defined since $v(y)$ decays exponentially as $|y| \to \infty$. Moreover, $k \in C^\delta(\mathbb{R}^2)$. We can represent $g(x,y)$ in the form

$$g(x,y) = k(x)v(y) + g_0(x,y). \tag{3.17}$$

Without loss of generality we can assume that $\int_{-\infty}^{\infty} v^2(y)\, dy = 1$. Then

$$\int_{-\infty}^{\infty} g_0(x,y)v(y)\, dy = 0, \quad \forall x \in \mathbb{R}^1. \tag{3.18}$$

Thus we can represent the space $E = C^{2+\delta}(\mathbb{R}^2)$ as a direct sum $E = E_0 + E_1$, where E_0 is the subspace of functions satisfying (3.18) and E_1 is the subspace of functions of the form $k(x)v(y)$. We will now consider the restriction of the operator L to E_0. Let $u \in E_0$. It is easy to note that

$$\int_{-\infty}^{\infty} Lu(x,y)v(y)dy = 0.$$

Indeed,

$$\int_{-\infty}^{\infty} \frac{\partial^2 u}{\partial x^2} v(y)\, dy = \frac{\partial^2}{\partial x^2} \int_{-\infty}^{\infty} u(x,y)v(y)\, dy = 0,$$

$$\int_{-\infty}^{\infty} \left(\frac{\partial^2 u}{\partial y^2} + b(y) \right) v(y)\, dy = \int_{-\infty}^{\infty} u(x,y) \left(\frac{\partial^2}{\partial y^2} + b(y) \right) v(y)\, dy = 0.$$

Hence we can consider L as acting from E_0 into

$$\hat{E}_0 = \left\{ g \in C^\delta(\mathbb{R}^2) : \int_{-\infty}^{\infty} g(x,y)v(y)\, dy = 0 \right\}.$$

Obviously, L is a bounded operator. Solvability of equation (3.15) is given by the following theorem. Its proof can be found in [559].

Theorem 3.11. *The operator $L : E_0 \to \hat{E}_0$ has a bounded inverse. Equation (3.15) is solvable in $C^{2+\delta}(\mathbb{R}^2)$ if and only if the equation $\phi'' = k$, where $k(x)$ is given by (3.16), is solvable in $C^{2+\delta}(\mathbb{R}^1)$.*

The simple model example considered in this section shows how we can reduce the dimension of the problem. The invertibility of the operator L on the subspace allows us to reduce equation (3.15) to the one-dimensional equation with a non-Fredholm operator. Solvability of the equation $\phi'' = k$ was discussed in the examples above (Section 1.3). We will finish this section with one more example. Let $k(x)$ be given as a Fourier series, $k(x) = \sum_{j=1}^{\infty} a(\xi_j) \cos(\xi_j x)$. Then $\phi(x) = - \sum_{j=1}^{\infty} (a(\xi_j) \cos(\xi_j x))/\xi_j^2$. If $\xi_j \to 0$, then the first series can be convergent while the second divergent. In this case, partial sums correspond to a solution in the form of sequence (Section 1.1), which is not a solution in the usual sense.

4 Strongly non-Fredholm operators

Let us call non-Fredholm operators that do not satisfy Definition 1.10 strongly non-Fredholm operators. A simple example is given by the Laplace operator. Consider the equation
$$\Delta u = f \tag{4.1}$$
in $\mathbb{R}^2$. Let
$$v = e^{\mu\sqrt{1+x_1^2+x_2^2}} u, \quad g = e^{\mu\sqrt{1+x_1^2+x_2^2}} f.$$
Then
$$\Delta v - 2\mu \left(\frac{x_1}{r} \frac{\partial v}{\partial x_1} + \frac{x_2}{r} \frac{\partial v}{\partial x_2} \right) + \left(-\mu \frac{1+r^2}{r^{5/2}} + \mu^2 \frac{r^2-1}{r^2} \right) v = g, \tag{4.2}$$
where $r = \sqrt{1 + x_1^2 + x_2^2}$. The operator $L_\mu = S_\mu L S_\mu^{-1}$ corresponds to the left-hand side of (4.2), L is the Laplacian. There is a family of limiting operators of the form
$$\hat{L}_\mu v = \Delta v - 2\mu \left(a \frac{\partial v}{\partial x_1} + b \frac{\partial v}{\partial x_2} \right) + \mu^2 v,$$
where a and b are arbitrary real numbers such that $a^2 + b^2 = 1$. Equation $\hat{L}_\mu v = 0$ has nonzero bounded solutions for any positive μ. Indeed, substituting $v(x_1, x_2) = \exp(i(\xi_1 x_1 + \xi_2 x_2))$, we obtain
$$a\xi_1 + b\xi_2 = 0, \quad \xi_1^2 + \xi_2^2 = \mu^2.$$
This system has a solution for any a, b, and μ. Therefore, L_μ does not satisfy the Fredholm property.

The operator $Lu = \Delta u + au$ considered in $\mathbb{R}^n$, where a is a positive constant, is also a strongly non-Fredholm operator. Since the equation $\Delta u + au = f$ has constant coefficients, we can apply the Fourier transform. It has a solution $u \in L^2(\mathbb{R}^n)$ if and only if $\hat{f}(\xi)/(a - \xi^2) \in L^2(\mathbb{R}^n)$, where $\hat{}$ denotes the Fourier transform. The solvability conditions are given by the equality
$$\int_{\mathbb{R}^n} e^{-i\xi x} f(x) dx = 0 \tag{4.3}$$
for any $\xi \in \mathbb{R}^n$ such that $|\xi|^2 = a$. Formally, they are similar to solvability conditions for Fredholm operators: the right-hand side is orthogonal to all solutions of the homogeneous formally adjoint problem.

It should be noted that the left-hand side in (4.3) is not a bounded functional over $L^2(\mathbb{R}^n)$. Therefore, these orthogonality conditions do not imply the closeness of the range of the operator. Indeed, we can construct a sequence $f_n \in L^2(\mathbb{R}^n)$ such that it converges in $L^2(\mathbb{R}^n)$ to some f_0; then all functions f_n satisfy the solvability conditions while f_0 does not satisfy them. In order to construct such a sequence, we consider the Fourier transforms $\hat{f}_n(\xi)$ and assume that they vanish at $|\xi|^2 = a$.

4. Strongly non-Fredholm operators

These functions can converge in $L^2(\mathbb{R}^n)$ to a function $\hat{f}_0$ which does not vanish at $|\xi|^2 = a$. Hence, f_0 does not belong to the range of the operator. Thus, the range of the operator is not closed and similarity with Fredholm solvability conditions is only formal.

Schrödinger equation. In the case of the Schrödinger equation

$$\Delta u + V(x)u + au = f, \tag{4.4}$$

instead of the usual Fourier transform in the example above, we can apply a generalized Fourier transform. It allows us to prove the following theorem [573].

Theorem 4.1. *Suppose that*

$$|V(x)| \leq C/(1+|x|^{3.5+\epsilon}), \quad x \in \mathbb{R}^3$$

for some positive C and ϵ, and the norms $\|V\|_{L^\infty(\mathbb{R}^3)}$, $\|V\|_{L^{3/2}(\mathbb{R}^3)}$, $\|V\|_{L^{4/3}(\mathbb{R}^3)}$ are sufficiently small[1]. Assume further that $f(x) \in L^2(\mathbb{R}^3)$ and $|x|f(x) \in L^1(\mathbb{R}^3)$. Then equation (4.4) with $a \geq 0$ is solvable in $L^2(\mathbb{R}^3)$ if and only if

$$\int_{\mathbb{R}^3} f(x)\phi(x)dx = 0 \tag{4.5}$$

for all bounded solutions ϕ of the equation

$$\Delta u + V(x)u + au = 0. \tag{4.6}$$

Equation (4.6) is self-adjoint. Therefore solvability conditions (4.5) represent orthogonality to solutions of the homogeneous adjoint equation. As before, this similarity to Fredholm solvability conditions is only formal. Some generalizations of these results are given in [574].

Diffusion-convection equation. The diffusion equation with convective terms

$$\Delta u + v.\nabla u + c(x)u = f, \tag{4.7}$$

where $v = -\nabla p$ is the velocity vector and p is the pressure (Darcy's law), can be reduced to the Schrödinger equation by the change of variables $z = e^{-p/2}u$. We have

$$\Delta z + W(x)z = g, \tag{4.8}$$

where

$$W(x) = c(x) + \frac{1}{2}\Delta p - \frac{1}{4}|\nabla p|^2, \quad g(x) = f(x)\,e^{-p(x)/2}.$$

Assuming that W and g satisfy conditions of Theorem 4.1, we obtain the solvability conditions

$$\int_{\mathbb{R}^n} g(x)\psi(x)dx = 0 \tag{4.9}$$

---
[1] Exact bounds for the norms are given in [573].

for all bounded solutions ψ of the homogeneous equation

$$\Delta z + W(x)z = 0. \tag{4.10}$$

Set $\phi = \psi e^{p/2}$. If ϕ is a solution of the equation

$$\Delta u + v.\nabla u + c(x)u = 0, \tag{4.11}$$

then ψ is a solution of (4.10). Condition (4.9) can be written as

$$\int_{\mathbb{R}^n} f(x)\phi(x)\, e^{-p(x)} dx = 0. \tag{4.12}$$

Set $\Phi(x) = \phi(x)\, e^{-p(x)}$. Then $\Phi(x)$ satisfies the equation

$$\Delta y - \nabla.(vy) + c(x)y = 0 \tag{4.13}$$

adjoint to (4.11). Thus solvability conditions for equation (4.7) are given by the equality

$$\int_{\mathbb{R}^n} f(x)\Phi(x)dx = 0 \tag{4.14}$$

for all bounded solutions $\Phi(x)$ of equation (4.13) adjoint to (4.11). If the pressure is bounded, then there is one-to-one correspondence between bounded solutions $\Phi(x)$ of equation (4.13) and bounded solutions $\psi(x)$ of equation (4.10).

Chapter 11

Nonlinear Fredholm Operators

The theory of linear Fredholm operators will be used in this chapter to study nonlinear elliptic problems. Nonlinear operators are called Fredholm operators if the corresponding linearized operators satisfy this property.

We will introduce general nonlinear elliptic problems first in Hölder spaces and then in Sobolev spaces. In Section 2 we will study properness of elliptic operators. It signifies that the inverse image of a compact set is compact in any bounded closed set. In particular, this implies that the set of solutions of operator equations is compact in bounded closed sets. In the case of unbounded domains, we need to introduce special weighted spaces and to impose Condition NS which provides normal solvability of elliptic problems. Otherwise, the operators may not be proper.

We construct the topological degree for Fredholm and proper operators. This construction is devised for abstract operators but keeping in mind elliptic problems in bounded or unbounded domains. As before, in the case of unbounded domains we need to introduce weighted spaces. Otherwise, the degree with the required properties may not exist.

Topological degree is a powerful tool of nonlinear analysis. We will discuss some of its applications to existence and bifurcation of solutions of operator equations.

1 Nonlinear elliptic problems

Let $\beta = (\beta_1, \ldots, \beta_n)$ be a multi-index, β_i nonnegative integers, $|\beta| = \beta_1 + \cdots + \beta_n$, $D^\beta = D_1^{\beta_1} \ldots D_n^{\beta_n}$, $D_i = \partial/\partial x_i$. We consider the operators

$$A_i u = \sum_{k=1}^{p} \sum_{|\beta| \leq \beta_{ik}} a_{ik}^\beta(x) D^\beta u_k \quad (i = 1, \ldots, p), \quad x \in \Omega, \tag{1.1}$$

$$B_i u = \sum_{k=1}^{p} \sum_{|\beta| \leq \gamma_{ik}} b_{ik}^\beta(x) D^\beta u_k \quad (i = 1, \ldots, r), \quad x \in \partial\Omega. \tag{1.2}$$

According to the definition of elliptic problems in the Douglas-Nirenberg sense, there are some integers $s_1, \ldots, s_p; t_1, \ldots, t_p; \sigma_1, \ldots, \sigma_r$ such that

$$\beta_{ij} \leq s_i + t_j, \; i,j = i, \ldots, p; \quad \gamma_{ij} \leq \sigma_i + t_j, \; i = 1, \ldots, r, j = 1, \ldots, p, \; s_i \leq 0.$$

We suppose that the number $m = \sum_{i=1}^{p}(s_i + t_i)$ is even and put $r = m/2$.

We assume that the problem is elliptic, i.e., the ellipticity condition

$$\det \left(\sum_{|\beta| = \beta_{ik}} a_{ik}^\beta(x) \xi^\beta \right)_{ik=1}^{p} \neq 0, \quad \beta_{ik} = s_i + t_k$$

is satisfied for any $\xi \in \mathbb{R}^n$, $\xi \neq 0$, $x \in \bar{\Omega}$, as well as the condition of proper ellipticity and the Lopatinskii conditions. Here $\xi = (\xi, \ldots, \xi_n)$, $\xi^\beta = \xi_1^\beta \ldots \xi_n^\beta$. The system is uniformly elliptic if the last determinant is bounded from below by a positive constant for all $|\xi| = 1$ and $x \in \bar{\Omega}$. Exact definitions are given in the introduction (Section 2.2).

In this section we work with Hölder spaces. Nonlinear elliptic problems in Sobolev spaces will be discussed in Section 2.5. Everywhere below $C^{k+\alpha}(\Omega)$ denotes the standard Hölder space of functions bounded in Ω together with their derivatives up to order k, and the latter satisfies the Hölder condition uniformly in x.

Denote by E_0 a space of vector-valued functions $u(x) = (u_1(x), \ldots, u_p(x))$, $u_j \in C^{l+t_j+\alpha}(\Omega)$, $j = 1, \ldots, p$, where l and α are given numbers, $l \geq \max(0, \sigma_i)$, $0 < \alpha < 1$. Therefore

$$E_0 = C^{l+t_1+\alpha}(\Omega) \times \cdots \times C^{l+t_p+\alpha}(\Omega).$$

The domain Ω can be bounded or unbounded. To avoid uncertainty, from now on we will consider only unbounded domains unless it is explicitly indicated. The results about degree construction and its applications remains valid, and usually much simpler for bounded domains. The boundary $\partial\Omega$ of the domain Ω is supposed to be of the class $C^{l+\lambda+\alpha}$, where $\lambda = \max(-s_i, -\sigma_i, t_j)$, and to satisfy

1. Nonlinear elliptic problems

Condition D (Chapter 4, Section 1). The coefficients of the operator satisfy the following regularity conditions:

$$a_{ij}^{\beta} \in C^{l-s_i+\alpha}(\Omega), \quad b_{ij}^{\beta} \in C^{l-\sigma_i+\alpha}(\partial\Omega).$$

The operator A_i acts from E_0 into $C^{l-s_i+\alpha}(\Omega)$, and B_i from E_0 into $C^{l-\sigma_i+\alpha}(\partial\Omega)$. Write $A = (A_1, \ldots, A_p)$, $B = (B_1, \ldots, B_r)$. Then

$$A : E_0 \to E_1, \quad B : E_0 \to E_2, \quad (A, B) : E_0 \to E,$$

where $E = E_1 \times E_2$,

$$E_1 = C^{l-s_1+\alpha}(\Omega) \times \cdots \times C^{l-s_p+\alpha}(\Omega), \quad E_2 = C^{l-\sigma_1+\alpha}(\partial\Omega) \times \cdots \times C^{l-\sigma_r+\alpha}(\partial\Omega).$$

We will consider weighted Hölder spaces $E_{0,\mu}$ and E_μ with the norms

$$\|u\|_{E_{0,\mu}} = \|u\mu\|_{E_0}, \quad \|u\|_{E_\mu} = \|u\mu\|_E.$$

We use also the notation $C_\mu^{k+\alpha}$ for a weighted Hölder space with norm $\|u\|_{C_\mu^{k+\alpha}} = \|u\mu\|_{C^{k+\alpha}}$.

We suppose that the weight function μ is a positive infinitely differentiable function defined for all $x \in \mathbb{R}^n$, $\mu(x) \to \infty$ as $|x| \to \infty$, $x \in \Omega$, and

$$\left|\frac{1}{\mu(x)} D^\beta \mu(x)\right| \to 0, \quad |x| \to \infty, \quad x \in \Omega \tag{1.3}$$

for any multi-index β, $|\beta| > 0$. In fact, we will use its derivative only up to a certain order.

Operator (A, B), considered in weighted Hölder spaces, acts from $E_{0,\mu}$ into E_μ.

We consider general nonlinear elliptic operators

$$F_i(x, D^{\beta_{i1}} u_1, \ldots, D^{\beta_{ip}} u_p) = 0, \quad i = 1, \ldots, p, \quad x \in \Omega \tag{1.4}$$

with nonlinear boundary operators

$$G_j(x, D^{\gamma_{j1}} u_1, \ldots, D^{\gamma_{jp}} u_p) = 0, \quad j = 1, \ldots, r, \quad x \in \partial\Omega \tag{1.5}$$

in $\Omega \subset \mathbb{R}^n$.

Here $D^{\beta_{ik}} u_k$ is a vector with components $D^\alpha u_k = \partial^{|\alpha|} u_k / \partial x_1^{\alpha_1} \ldots \partial x_n^{\alpha_n}$ where the multi-index $\alpha = (\alpha_1, \ldots, \alpha_n)$ takes all values such that $0 \leq |\alpha| = \alpha_1 + \cdots + \alpha_n \leq \beta_{ik}$, β_{ik} are given integers. The vectors $D^{\gamma_{jk}} u_k$ are defined similarly. The regularity of the real-valued functions F_i, G_i, $u = (u_1, \ldots, u_p)$, and of the domain Ω is determined by β_{ik}, γ_{jk}, $i, k = 1, \ldots, p$, $j = 1, \ldots, r$ (see below).

In what follows we will also use the notation
$$F_i(x, \mathcal{D}_i u) = F_i(x, D^{\beta_{i1}} u_1, \ldots, D^{\beta_{ip}} u_p),$$
$$G_j(x, \mathcal{D}_j^b u) = G_j(x, D^{\gamma_{j1}} u_1, \ldots, D^{\gamma_{jp}} u_p).$$

The corresponding linear operators are

$$A_i(v, \eta_i) = \sum_{k=1}^{p} \sum_{|\alpha| \leq \beta_{ik}} a_{ik}^{\alpha}(x, \eta_i) D^{\alpha} v_k, \quad i = 1, \ldots, p, \quad x \in \Omega, \tag{1.6}$$

$$B_j(v, \zeta_i) = \sum_{k=1}^{p} \sum_{|\alpha| \leq \gamma_{jk}} b_{jk}^{\alpha}(x, \zeta_i) D^{\alpha} v_k, \quad j = 1, \ldots, r, \quad x \in \partial\Omega, \tag{1.7}$$

where
$$a_{ik}^{\alpha}(x, \eta_i) = \frac{\partial F_i(x, \eta_i)}{\partial \eta_{ik}^{\alpha}}, \quad b_{jk}^{\alpha}(x, \zeta_i) = \frac{\partial G_j(x, \zeta_j)}{\partial \zeta_{jk}^{\alpha}},$$

$\eta_i \in \mathbb{R}^{n_i}$ and $\zeta_j \in \mathbb{R}^{m_j}$ are vectors with components η_{ik}^{α} and ζ_{jk}^{α}, respectively, ordered in the same way as the derivatives in (1.4), (1.5).

The system (1.4), (1.5) is called *elliptic* if the corresponding system (1.6), (1.7) is elliptic for all values of parameters η_i, ζ_j. When we mention the Lopatinskii condition for operators (1.4), (1.5) we mean the corresponding condition for operators (1.6), (1.7) for any $\eta_i \in \mathbb{R}^{n_i}$ and $\zeta_j \in \mathbb{R}^{m_j}$.

We suppose that the functions F_i (G_i) satisfy the following conditions: for any positive number M and for all multi-indices β and γ: $|\beta + \gamma| \leq l - s_i + 2$ ($|\beta + \gamma| \leq l - \sigma_i + 2$), $|\beta| \leq l - s_i$ ($|\beta| \leq l - \sigma_i$) the derivatives $D_x^{\beta} D_{\eta}^{\gamma} F_i(x, \eta)$ ($D_x^{\beta} D_{\zeta}^{\gamma} G_i(x, \zeta)$) as functions of $x \in \Omega$, $\eta \in \mathbb{R}^{n_i}$, $|\eta| \leq M$ ($x \in V$, $\zeta \in \mathbb{R}^{m_i}$, $|\zeta| \leq M$) satisfy the Hölder condition in x uniformly in η (ζ) and the Lipschitz condition in η (ζ) uniformly in x (with constants possibly depending on M). Write $F = (F_1, \ldots, F_p)$, $G = (G_1, \ldots, G_r)$. Then (F, G) acts from $E_{0,\mu}$ into E_μ.

2 Properness

We recall that an operator A acting from a Banach space E_0 into another Banach space E is called proper on bounded closed sets if for any bounded closed set $D \subset E_0$ the intersection of the inverse image of any compact set in E with D is compact. For the sake of brevity, we will call such operators proper.

2.1 Lemma on properness of operators in Banach spaces

Let E_0 and E be two Banach spaces. Suppose that a topology is introduced in E_0 such that the convergence in this topology, which we denote by $\rightharpoonup$, has the following property: for any sequence $\{u_n\}$, $u_n \in E_0$, bounded in the E_0-norm there is a subsequence $\{u_{n_k}\} : u_{n_k} \rightharpoonup u_0 \in E_0$.

2. Properness

We consider an operator $T(u) : D \to E$, where $D \subset E_0$. Suppose that this operator is *closed* with respect to the convergence $\rightharpoonup$ in the following sense: if $T(u_k) = f_k$, $u_k \in D$, $f_k \in E$ and $u_k \rightharpoonup u_0 \in E_0$, $f_k \to f_0$ in E, then $u_0 \in D$ and $T(u_0) = f_0$.

Lemma 2.1. *Suppose that D is a bounded closed set in E_0, the operator $T(u)$ is closed with respect to the convergence $\rightharpoonup$ and for any $u_0 \in D$ there exists a linear bounded operator $S(u_0) : E_0 \to E$, which has a closed range and finite-dimensional kernel, such that for any sequence $\{v_k\}$, $v_k \in D$, $v_k \rightharpoonup u_0 \in D$ we have*

$$\|T(u_0) - T(v_k) - S(u_0)(u_0 - v_k)\|_E \to 0.$$

Then $T(u)$ is a proper operator.

Proof. Consider a sequence $\{u_n\}$ in D such that $f_n = T(u_n) \to f_0$ in E. We have to prove that there exists a subsequence of $\{u_n\}$ which is convergent in E_0. Consider a subsequence $\{u_{n_i}\}$ such that $u_{n_i} \rightharpoonup u_0 \in E_0$. Then since $T(u)$ is closed, we have $u_0 \in D$ and $T(u_0) = f_0$. Set $v_i = u_{n_i} - u_0$ and $h_i = S(u_0)v_i$. Then $h_i = [S(u_0)(u_{n_i} - u_0) - (T(u_{n_i}) - T(u_0))] + (T(u_{n_i}) - T(u_0)) \to 0$ in E. Suppose that $w_1, \ldots, w_k$ is a basis of $\ker S(u_0)$ and $\{\varphi_i\}$ is a biorthogonal sequence of functionals in the dual to E_0 space. Write $E_1 = \{u \in E_0, <\varphi_i, u> = 0, i = 1, \ldots, k\}$. Then we have

$$v_i = \sum_{j=1}^{k} <\varphi_j, v_i> w_j + v_i^1, \quad v_i^1 \in E_1. \tag{2.1}$$

Denote by S_1 the restriction of $S(u_0)$ on E_1. Then $S_1 v_i^1 = h_i$. By the Banach theorem, S_1 has a bounded inverse. So v_i^1 is a convergent in E_0 sequence. Since $u_{n_i} \in D$ and so v_i is a bounded sequence in E_0, it follows from (2.1) that we can find a convergent subsequence of v_i. The lemma is proved. $\square$

2.2 Properness of elliptic operators in Hölder spaces

We will prove properness of elliptic problems in unbounded domains. The proof remains valid and becomes even simpler for bounded domains. In this case we do not need to introduce weighted spaces and to impose additional conditions which provide normal solvability of linearized operators.

We will show that the operator $T = (F, G) : E_{0,\mu} \to E_\mu$ defined above satisfies conditions of Lemma 2.1 under the assumptions formulated below. The convergence $\rightharpoonup$ is convergence in the space $E_{0,\mu}(\Omega_R)$ for $\alpha = 0$ and any $R > 0$. Here Ω_R is the intersection of Ω with a ball B_R in $\mathbb{R}^n$ with radius R and center at 0. It is clear that any bounded in $E_{0,\mu}$ sequence has a $\rightharpoonup$ convergent subsequence.

As a domain D we take a closed ball in $E_{0,\mu}$ with its center at zero. Obviously the operator $T = (F, G)$ is closed with respect to the convergence $\rightharpoonup$.

We construct below the operator S introduced in Lemma 2.1. Let $F = (F_1, \ldots, F_p)$, where F_i is the operator (1.4),

$$\eta_i = (\eta_{i1}, \ldots, \eta_{in_i}), \quad \text{and} \quad \eta_i^0 = (\eta_{i1}^0, \ldots, \eta_{in_i}^0)$$

are two vectors in $\mathbb{R}^{n_i}$. Then by Taylor's formula we can write

$$F_i(x, \eta_i) = F_i(x, \eta_i^0) + \sum_{j=1}^{n_i} F'_{i\eta_{ij}}(x, \eta_i^0)(\eta_{ij} - \eta_{ij}^0)$$

$$+ \int_0^1 (1-s) \sum_{j,k=1}^{n_i} F''_{i\eta_{ij}\eta_{ik}}(x, \eta_i^0 + s(\eta_i - \eta_i^0)) ds \, (\eta_{ij} - \eta_{ij}^0)(\eta_{ik} - \eta_{ik}^0).$$

Therefore for any $u, u^0 \in E_{0,\mu}$ we have

$$F_i(x, \mathcal{D}_i u) - F_i(x, \mathcal{D}_i u^0) = A_i(u - u^0, \mathcal{D}_i u^0) + \Phi_i(u, u^0),$$

where A_i is given by (1.6) and

$$\Phi_i(u, u^0) = \int_0^1 (1-s) \sum_{j,k=1}^{n_i} F''_{iv_j v_k}(x, v^0 + s(v - v^0)) ds \, (v_j - v_j^0)(v_k - v_k^0),$$

$v(x) = \mathcal{D}_i u(x)$, $v^0(x) = \mathcal{D}_i u^0(x)$.

Lemma 2.2. $\|\Phi_i(u^m, u^0)\|_{C_\mu^{l-s_i+\alpha}(\Omega)} \to 0$ if $u^m \rightharpoonup u^0$ and $\|u^m\|_{E_{0,\mu}}$ is bounded.

Proof. It is sufficient to prove that

$$\|D^\beta (F''_{iv_j v_k}(x, v^0 + s(v^m - v^0))(v_j^m - v_j^0)(v_k^m - v_k^0)\mu)\|_{C^\alpha(\Omega)} \to 0$$

for $|\beta| \leq l - s_i$. Here $v^m(x) = \mathcal{D}_i u^m(x)$. We will prove that

$$\|D^\beta F''_{iv_j v_k}(x, v^0 + s(v^m - v^0))\|_{C^\alpha(\Omega)} \leq M, \quad |\beta| \leq l - s_i, \quad (2.2)$$

where M is a constant and

$$\|D^\beta((v_j^m - v_j^0)(v_k^m - v_k^0)\mu)\|_{C^\alpha(\Omega)} \to 0, \quad |\beta| \leq l - s_i \quad \text{as} \quad m \to \infty. \quad (2.3)$$

We begin with (2.2). Let $u^m = (u_1^m, \ldots, u_p^m)$. By assumption

$$\|u_k^m\|_{C_\mu^{l+t_k+\alpha}(\Omega)} \leq M_1 \quad (k = 1, \ldots, p).$$

Here and below M with subscripts denotes constants independent of u and v. It follows that

$$\|u_k^m\|_{C^{l+t_k+\alpha}(\Omega)} \leq M_2. \quad (2.4)$$

2. Properness

Indeed, write $w = \mu u_k^m$. Then $\|w\|_{C^{l+t_k+\alpha}(\Omega)} \leq M_1$, $u_k^m = \frac{1}{\mu}w$, and (2.4) follows easily from the properties of the function $\mu(x)$ since by (1.3), $D^\beta \frac{1}{\mu}$ is bounded for any multi-index β.

Obviously, (2.4) implies

$$\|v^m\|_{C^{l-s_i+\alpha}(\Omega)} \leq M_3. \tag{2.5}$$

Inequality (2.2) follows from this inequality and from the conditions of smoothness of the functions F_i.

We now prove (2.3). Set $w_j^m = v_j^m - v_j^0$. Obviously $D^\beta(w_j^m \, w_k^m \, \mu)$ is a sum of expressions of the form

$$D^\gamma w_j^m \, D^\tau w_k^m \, D^\sigma \mu = [\mu D^\gamma w_j^m]\, [\mu D^\tau w_k^m]\, \frac{1}{\mu}\left(\frac{1}{\mu} D^\sigma \mu\right) \tag{2.6}$$

with constant coefficients, where γ, τ, σ are multi-indices, $\gamma + \tau + \sigma \leq \beta$. The last factor in (2.6) is bounded by virtue of (1.3). From the properties of the function μ we conclude that $\frac{1}{\mu}$ and $D_i \frac{1}{\mu}$ ($i = 1, \ldots, n$) tend to 0 as $|x| \to \infty, x \in \Omega$. So

$$\left\|\frac{1}{\mu}\right\|_{C^\alpha(\Omega_R^-)} \to 0 \text{ as } R \to \infty. \tag{2.7}$$

Here Ω_R^- is the intersection of Ω with the ball $|x| > R$.

We prove next that

$$\|\mu \, D^\gamma w_j^m\|_{C^\alpha(\Omega)} \leq M_4, \tag{2.8}$$

where $|\gamma| \leq l - s_i$. Set $y_k^m = u_k^m - u_k^0$. Then w_j^m has the form $D^\sigma y_k^m$ with $|\sigma| \leq s_i + t_k$. So $D^\gamma w_j^m$ has the form $D^{\sigma+\gamma} y_k^m$ with $|\sigma + \gamma| \leq l + t_k$. By conditions of the lemma $\|y_k^m\|_{C_\mu^{l+t_k+\alpha}(\Omega)} \leq M_5$, and (2.8) follows from this.

From (2.7) and (2.8) we obtain the convergence $\|D^\beta (w_j^m w_k^m \mu)\|_{C^\alpha(\Omega_R^-)} \to 0$ as $R \to \infty$. So to prove (2.3) it is sufficient to verify that $\|D^\beta (w_j^m w_k^m \mu)\|_{C^\alpha(\Omega_R)} \to 0$ for any R as $m \to \infty$. This follows from (2.6) and the fact that $\|\mu D^\beta w_j^m\|_{C^\alpha(\Omega_R)}$ is bounded and $\|\mu D^\beta w_j^m\|_{C(\Omega_R)} \to 0$ as $m \to \infty$ for $|\beta| \leq l - s_i$ since $u^m \rightharpoonup u^0$. Therefore the Hölder norm of the product of the first two factors in the right-hand side of (2.6) converges to zero. The lemma is proved. $\square$

Lemma 2.2 implies the convergence

$$\|F_i(x, \mathcal{D}_i u^m) - F_i(x, \mathcal{D}_i u^0) - A_i(u^m - u^0, \mathcal{D}_i u^0)\|_{C_\mu^{l-s_i+\alpha}(\Omega)} \to 0 \tag{2.9}$$

if $u^m \rightharpoonup u^0$ and $\|u^m\|_{E_{0,\mu}}$ is bounded.

Similarly we have for the operators $G_j(x, \mathcal{D}_j^b u(x))$ ($j = 1, \ldots, r$):

$$\|G_j(x, \mathcal{D}_j^b u^m) - G_j(x, \mathcal{D}_j^b u^0) - B_j(u^m - u^0, \mathcal{D}_j^b u^0)\|_{C_\mu^{l-\sigma_j+\alpha}(\partial\Omega)} \to 0 \tag{2.10}$$

if $u^m \rightharpoonup u^0$ and $\|u^m\|_{E_{0,\mu}}$ is bounded.

Consider the operator

$$S(u_0)u = (A_1(u, \mathcal{D}_1 u_0), \ldots, A_p(u, \mathcal{D}_p u_0),$$
$$B_1(u, \mathcal{D}_1^b u_0), \ldots, B_r(u, \mathcal{D}_r^b u_0)) : E_{0,\mu} \to E_\mu.$$

We are interested in limiting operators for $S(u_0)$ in the sense of the previous section. We consider also the operator

$$S_0 u = (A_1(u, 0), \ldots, A_p(u, 0), B_1(u, 0), \ldots, B_r(u, 0)) : E_{0,\mu} \to E_\mu,$$

which does not depend on u_0.

Lemma 2.3. *For any $u_0 \in E_{0,\mu}$ the limiting operators for $S(u_0)$ and S_0 coincide.*

Proof. Consider first the operator $A_i(u; \eta_i)$ defined by (1.6). Since $u_0 \in E_{0,\mu}$, then $\mu D^\beta u_{0k}(x) \in C^\alpha(\Omega)$ for $|\beta| \leq l + t_k$. So $\mu \mathcal{D}_i u_0 \in C^\alpha(\Omega)$ and therefore

$$|\mathcal{D}_i u_0(x)| \leq \frac{M}{\mu(x)} \to 0. \tag{2.11}$$

Let $|x^m| \to \infty, x^m \in \Omega$. Then $|x + x^m| \to \infty$ for all $x \in B_R$. So there exists m_0 such that for all $m > m_0$ and all $x \in \Omega_* \cap B_R$ the inequality $|\mathcal{D}_i u_0(x + x^m)| \leq 1$ holds. Here Ω_* is a limiting domain which corresponds to the sequence x^m.

Set $f_{ik}^\beta(x, \eta_i) = \frac{\partial F_i(x, \eta_i)}{\partial \eta_k^\beta}$. It follows from the properties of the function F_i that for $m > m_0$ we have

$$|f_{ik}^\beta(x + x^m, 0) - f_{ik}^\beta(x + x^m, \mathcal{D}_i u_0(x + x^m))| \leq K |\mathcal{D}_i u_0(x + x^m)| \leq \frac{K M}{\mu(x + x^m)} \to 0$$

as $|x^m| \to \infty, x \in B_R$. Therefore if one of the functions

$$f_{ik}^\beta(x + x^m, 0), \ f_{ik}^\beta(x + x^m, \mathcal{D}_i u_0(x + x^m))$$

has a limit as $|x^m| \to \infty$, then the same is true for another one and the limits coincide. Thus the lemma is proved for the operator (1.6). The proof is similar for the operator (1.7). The lemma is proved. □

Theorem 2.4. *Suppose that the system of operators (1.4) is uniformly elliptic and for the system of operators (1.4), (1.5) Lopatinskii conditions are satisfied. Assume further that all limiting operators for the operator S_0 satisfy Condition NS. Then the operator $(F, G) : E_{0,\mu} \to E_\mu$ is proper.*

Proof. We use Lemma 2.1 for the operator $T = (F, G)$. For any $u_0 \in E_{0,\mu}$ we take

$$S(u_0) = (A_1(u_0, \mathcal{D}_1 u_0), \ldots, (A_p(u_0, \mathcal{D}_p u_0),$$
$$B_1(u_0, \mathcal{D}_1^b u_0), \ldots, B_r(u_0, \mathcal{D}_r^b u_0)) : E_{0,\mu} \to E_\mu.$$

2. Properness

From (2.9) and (2.10) we obtain

$$\|T(u_0) - T(u^m) - S(u_0)(u_0 - u^m)\|_{E_\mu} \to 0$$

if $u^m \rightharpoonup u_0$ and $\|u^m\|_{E_{0,\mu}}$ is bounded. If all limiting operators for S_0 satisfy Condition NS, then according to Lemma 2.3 the same is true for all limiting operators for $S(u_0)$ for any $u_0 \in E_{0,\mu}$. The results of the previous section imply that $S(u_0)$ has a closed range and a finite-dimensional kernel. The theorem is proved. □

Remark 2.5. Functions from the weighted space $E_{0,\mu}$ tend to zero at infinity. If we look for solutions, which are not zero at infinity, we can represent them in the form $u + \psi$, where ψ is a given function with a needed behavior at infinity, and u belongs to $E_{0,\mu}$.

2.3 Operators depending on a parameter

Consider an operator $T(u,t) : D \times [0,1] \to E$, $D \subset E_0$ depending on parameter $t \in [0,1]$. We suppose here as in Section 2.1 that E_0 and E are arbitrary Banach spaces. We will obtain conditions of its properness with respect to both variables u and t. First of all, we modify the definition of closed operators given in Section 2.1:

Let $T(u_k, t_k) = f_k$, $t_k \to t_0$, $u_k \in D$, $f_k \in E$, $u_k \rightharpoonup u_0 \in E_0$, $f_k \to f_0$ in E, then $u_0 \in D$ and $T(u_0, t_0) = f_0$.

Lemma 2.1'. *Suppose that D is a bounded set in E_0, the operator $T(u,t)$ is closed, and for any $u_0 \in D$ there exists a linear bounded operator $S(u_0) : E_0 \to E$, which has a closed range and a finite-dimensional kernel, such that for any sequence $\{v_k\}$, $v_k \in D$, $v_k \rightharpoonup u_0 \in D$ and $t_k \to t_0$ we have*

$$\|T(u_0, t_0) - T(v_k, t_k) - S(u_0)(u_0 - v_k)\|_E \to 0.$$

Then $T(u,t)$ is a proper operator.

The proof of the lemma remains the same as above. Suppose now that the operator $T(u,t)$ satisfies the conditions of Lemma 2.1' for any $t \in [0,1]$ fixed, and it depends on t continuously in the operator norm, i.e.,

$$\|T(u,t) - T(u,t_0)\|_E \leq c(t,t_0), \quad \forall u \in D,$$

where $c(t,t_0) \to 0$ as $t \to t_0$. Then

$$\|T(u_0, t_0) - T(v_k, t_k) - S(u_0)(u_0 - v_k)\|_E$$
$$\leq \|T(u_0, t_0) - T(v_k, t_0) - S(u_0)(u_0 - v_k)\|_E + \|T(v_k, t_0) - T(v_k, t_k)\|_E.$$

Therefore, if the conditions of Lemma 2.1 are satisfied for each t fixed and the operator depends continuously on a parameter, then Lemma 2.1' holds.

On the other hand, if the operator $T(u,t)$ is closed in the sense of Section 2.1 for each t fixed, and if it depends continuously on a parameter, then it is also closed in the sense of the definition given in this section. Thus, under conditions of Section 2.2 elliptic operators depending continuously on a parameter are proper with respect to two variables.

2.4 Example of non-proper operators

Properness of an operator $A : E_0 \to E$ implies that the set of solutions of the equation $A(u) = 0$ is compact in any bounded closed set $D \subset E_0$. In the proof of properness of elliptic operators in unbounded domains, we used weighted spaces. In the spaces without weight this property may not be satisfied.

Consider the operator

$$A_\tau(u) = u'' + F_\tau(u)$$

acting from $C^{2+\alpha}(\mathbb{R})$ into $C^\alpha(\mathbb{R})$, $0 < \alpha < 1$. Here

$$F(u) = u(u - \tau)(1 - u),$$

where $0 < \tau < 1/2$. We look for a positive solution of the equation $A(u) = 0$ with zero limits at infinity, $u(\pm\infty) = 0$. It can be found explicitly by reduction of the second-order equation to a system of two first-order equations:

$$u' = p, \quad p' = -F(u) \tag{2.12}$$

and by integration of this system. Denote this solution by $u_\tau(x)$. We note that it is invariant with respect to translation in space, that is $u_\tau(x + h)$ is also a solution for any real h. The set of solutions $u_\tau(x + h)$, $h \in \mathbb{R}$ is not compact in $C^{2+\alpha}(\mathbb{R})$. Hence the operator A is not proper. However, the invariance of the solution with respect to translation is mostly a technical difficulty. There are various methods to get rid of it. In particular, we can consider the problem on the half-axis $x \geq 0$ with the boundary condition $u'(0) = 0$. It is equivalent to choose such a shift that the solution attains its maximum at $x = 0$.

The absence of properness of such operators is not a technical but a principal problem related to unbounded domains. In order to show this, we will construct another set of solutions, which is uniformly bounded but not compact. We will consider that such solutions $u_\tau(x)$ have their maxima at $x = 0$. Write $\mu_\tau = u_\tau(0)$. It can be verified that

$$\int_0^{\mu_\tau} F(u)du = 0.$$

Since $\tau < 1/2$, then $\tau < \mu_\tau < 1$.

The phase plane of system (2.12) is shown in Figure 11. The homoclinic trajectory corresponds to the solution $u_\tau(x)$. There are also two heteroclinic trajectories going from the point $(0,0)$ to $(1,0)$ and from $(1,0)$ to $(0,0)$.

2. Properness

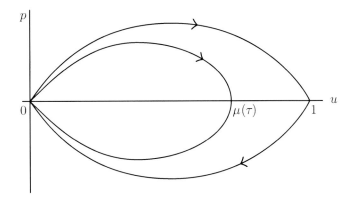

Figure 11: Trajectories of system (2.12).

Consider now a sequence $\tau_n \uparrow 1/2$ and the corresponding solutions $u_{\tau_n}(x)$. Then $\mu_{\tau_n} \to 1$, and the homoclinic trajectory tends to the heteroclinic ones. The sequence of functions $u_{\tau_n}(x)$ converges to 1 uniformly in any bounded interval. On the other hand, they tend to zero at infinity. Hence, this sequence is not compact in $C^{2+\alpha}(\mathbb{R})$, and the operator $A(u)$ is not proper.

Suppose now that the operator $A(u)$ acts in weighted spaces, from the space $C_\mu^{2+\alpha}(\mathbb{R})$ into the space $C_\mu^{\alpha}(\mathbb{R})$. The norms in these spaces are defined by the equality

$$\|u\|_{C_\mu^{k+\alpha}(\mathbb{R})} = \|u\mu\|_{C^{k+\alpha}(\mathbb{R})}, \ k = 0, 2.$$

The weight function $\mu(x) = \sqrt{1+x^2}$ grows at infinity. The sequence of functions $u_{\tau_n}(x)$ is not uniformly bounded in the norm of the space $C_\mu^{2+\alpha}(\mathbb{R})$. Therefore, there can exist only a finite number of elements of this sequence in any bounded closed set $D \subset C_\mu^{2+\alpha}(\mathbb{R})$. The operator is proper in the weighted spaces.

2.5 Properness in Sobolev spaces

The proof of properness of elliptic problems in Sobolev spaces is close to that for Hölder spaces. However, we need to specify the function spaces and the conditions on the operators. For simplicity of presentation, we restrict ourselves here to scalar elliptic problems. As before, we carry out the proof for the case of unbounded domains. We consider nonlinear elliptic operators

$$F(x, D^{2m}u), \ x \in \Omega, \tag{2.13}$$

where D^{2m} is a vector with the components D^α, $0 \leq |\alpha| \leq 2m$, and nonlinear boundary operators

$$G_j(x, D^{m_j}u), \ x \in \partial\Omega, \ j = 1, \ldots, m, \tag{2.14}$$

where D^{m_j} is a vector with the components D^α, $0 \leq |\alpha| \leq m_j$. The regularity conditions of the functions F, G_j and of the domain Ω is determined by m and m_j (see below).

The corresponding linear operators are

$$A(v, \eta) = \sum_{|\alpha| \leq 2m} a^\alpha(x, \eta) D^\alpha v, \quad x \in \Omega, \tag{2.15}$$

$$B_j(v, \xi) = \sum_{|\alpha| \leq m_j} b_j^\alpha(x, \xi_j) D^\alpha v, \quad x \in \partial\Omega, \ j = 1, \ldots, m, \tag{2.16}$$

where

$$a^\alpha(x, \eta) = \frac{\partial F(x, \eta)}{\partial \eta^\alpha}, \quad b_j^\alpha(x, \xi_j) = \frac{\partial G_j(x, \xi_j)}{\partial \xi_j^\alpha},$$

η and ξ_j are vectors with the components η^α, ξ_j^α, respectively, ordered in the same way as the derivatives in (2.13), (2.14).

Operators (2.13), (2.14) are called *elliptic* if the corresponding operators (2.15), (2.16) are elliptic for all values of the parameters η, ξ_j. When we mention the Lopatinskii condition for operators (2.13), (2.14), we mean the corresponding condition for operators (2.15), (2.16) for any η and ξ_j.

Let l be an integer, $l > max(2m, m_j + 1)$ and p be a real number, $p > n$. We introduce the space $E_0 = E_0(\Omega) = W_\infty^{l,p}(\Omega)$ and the corresponding weighted space $E_{0,\mu} = W_{\infty,\mu}^{l,p}(\Omega)$ with the norm $\|u\|_{E_{0,\mu}} = \|u\mu\|_{E_0}$. The weight function μ is supposed to satisfy the conditions of Section 1.

We impose the following smoothness conditions on the functions F and G_j:

– The derivatives

$$D_x^\alpha D_\eta^\beta F(x, \eta), \quad |\alpha| \leq l - 2m, \ |\beta| \leq l - 2m + 2 \tag{2.17}$$

exist, they are bounded for $x \in \Omega$, $|\eta| \leq R$ for any $R > 0$ (with constants depending on R), and they satisfy the Hölder condition.

– The functions G_j are supposed to be extended on the domain $\bar\Omega$ in such a way that the derivatives

$$D_x^\alpha D_\xi^\beta G_j(x, \xi), \quad |\alpha| \leq l - m_j, \ |\beta| \leq l - m_j + 2 \tag{2.18}$$

exist, they are bounded for $x \in \Omega$, $|\xi| \leq R$ for any $R > 0$ (with constants depending on R), and they satisfy the Hölder condition.

– Moreover it is supposed that $\mu(x) D_x^\alpha F(x, 0)$, $|\alpha| \leq l - 2m$ and $\mu(x) D_x^\alpha G_j(x, 0)$, $|\alpha| \leq l - m_j$, $j = 1, \ldots, m$ are bounded in Ω.

– The domain Ω is supposed to be of the class $C^{l+\theta}$ and to satisfy Condition D (Chapter 4).

2. Properness

Lemma 2.6. *The operator $F(x, D^{2m}u)$ acts from the space $E_{0,\mu}(\Omega)$ into*
$$W_{\infty,\mu}^{l-2m,p}(\Omega).$$

Proof. Let $v = D^{2m}u$, $v_\alpha = D^\alpha u$, $0 \leq |\alpha| \leq 2m$. We will prove that
$$F(x, v) \in W_{\infty,\mu}^{l-2m,p}(\Omega),$$
or
$$D^\alpha(\mu F(x, v)) \in L_\infty^p(\Omega), \quad 0 \leq |\alpha| \leq l - 2m. \tag{2.19}$$

We have for $0 \leq |\alpha| \leq l - 2m$:
$$D^\alpha(\mu F(x, v)) = \sum_{\beta+\gamma \leq \alpha} c_{\beta\gamma} D^\beta \mu D^\gamma F(x, v),$$

where $c_{\beta\gamma}$ are constants. Then for $y \in \Omega$,
$$\|D^\alpha(\mu F(x,v))\|_{L^p(\Omega \cap Q_y)} \leq \sum_{\beta+\gamma \leq \alpha} c_{\beta\gamma} \left\|\frac{D^\beta \mu}{\mu}\right\| \mu D^\gamma F(x,v)\|_{L^p(\Omega \cap Q_y)}$$
$$\leq M \sum_{\gamma \leq l-2m} \|\mu D^\gamma F(x,v)\|_{L^p(\Omega \cap Q_y)}$$

since $|D^\beta \mu|/\mu$ is bounded.

We use the following formula of differentiation of a superposition:
$$\bar{D}_x^\alpha F(x, v(x)) = D_x^\alpha F(x, v(x)) + \sum_{i=1}^{s} \frac{\partial F(x, v(x))}{\partial v_i} D_x^\alpha v_i(x) + T_{|\alpha|}(v; F), \tag{2.20}$$

where
$$T_{|\alpha|}(v; F) = \sum c_{\beta\sigma\gamma_1\ldots\gamma_s} D_x^\sigma D_v^\beta F(x, v(x)) \Pi_{\tau=1}^s D_x^{\gamma_\tau} v_\tau(x)$$

and the sum is taken over all $0 \leq |\beta| \leq |\alpha|$, $0 \leq |\sigma| < |\alpha|$, $0 \leq |\gamma_1| < |\alpha|$, $\ldots$, $0 \leq |\gamma_s| < |\alpha|$. Here $\bar{D}_x^\alpha F(x, v(x))$ is the full derivative in x, $D_x^\alpha F(x, v(x))$ is a partial derivative in x (for fixed v), $D_v^\beta F(x, v)$ is a partial derivative in v, $v = (v_1, \ldots, v_s)$; $\alpha = (\alpha_1, \ldots, \alpha_n)$, $\beta = (\beta_1, \ldots, \beta_s)$, $\gamma_\tau = (\gamma_{\tau_1}, \ldots, \gamma_{\tau_s})$ are multi-indices, $c_{\beta\sigma\gamma_1\ldots\gamma_s}$ are constants.

According to the assumptions on F, all derivatives in the right-hand side of (2.20) exist. Formula (2.20) can be proved directly by induction in $|\alpha|$.

If $u \in W_{\infty,\mu}^{l,p}(\Omega)$ and $v = D^{2m}u$, then
$$D^\alpha v \in L_{\infty,\mu}^p(\Omega) \quad \text{for} \quad |\alpha| \leq l - 2m. \tag{2.21}$$

Indeed, denote $w = \mu u$. For $|\gamma| \leq 2m$ we have

$$D^\gamma u = \sum_{\sigma+\tau \leq \gamma} c_{\sigma\tau} D^\sigma \frac{1}{\mu} D^\tau w,$$

$$\mu D^{\alpha+\gamma} u = \sum_{\sigma+\tau \leq \gamma} c_{\sigma\tau} \mu D^\alpha \left(D^\sigma \frac{1}{\mu} D^\tau w \right)$$

$$= \sum_{\sigma+\tau \leq \gamma} c_{\sigma\tau} \mu \sum_{\rho+\nu \leq \alpha} c_{\rho\nu} \left(D^{\rho+\sigma} \frac{1}{\mu} D^{\tau+\nu} w \right).$$

From the properties of the weight function μ we obtain that $\mu D^{\rho+\sigma} \frac{1}{\mu}$ is bounded. Since $w \in W_\infty^{l,p}(\Omega)$, we get (2.21).

It follows from (2.21) that

$$D^\alpha(\mu v) \in L_\infty^p(\Omega), \quad |\alpha| \leq l - 2m,$$

and the embedding theorems imply

$$\mu v \in C^{l-2m-1}(\Omega). \tag{2.22}$$

To prove (2.19) we multiply (2.20) by μ and estimate the right-hand side.

From (2.22) and (2.17) we conclude that $D_x^\alpha F(x, v)$ satisfies the Lipschitz condition in v. Hence

$$|\mu D_x^\alpha F(x, v(x))| \leq \mu |D_x^\alpha F(x, 0)| + \mu K |v(x)|$$

with some constant K. Therefore $\mu D_x^\alpha F(x, v(x))$ is bounded and, consequently, belongs to $L_\infty^p(\Omega)$.

For the second term in the right-hand side of (2.20) we get

$$\mu \sum_{i=1}^s \frac{\partial F(x, v(x))}{\partial v_i} D_x^\alpha v_i(x) \in L_\infty^p(\Omega)$$

since $\frac{\partial F(x,v(x))}{\partial v_i}$ is bounded and $\mu D_x^\alpha v_i(x) \in L_\infty^p(\Omega)$ by virtue of (2.21). Finally, $\mu T_{|\alpha|}(v; F)$ is bounded and, consequently, belongs to $L_\infty^p(\Omega)$. Thus (2.19) is proved. The lemma is proved. □

Lemma 2.7. *The operator $G_j(x, D^{m_j} u)$, $j = 1, \ldots, m$ acts from the space $E_{0,\mu}(\Omega)$ into the space $W_{\infty,\mu}^{l-m_j-1/p,p}(\partial\Omega)$.*

Proof. Similarly to the proof of the previous lemma, we obtain that $G_j(x, D^{m_j} u)$ acts from $E_{0,\mu}$ into $W_{\infty,\mu}^{l-m_j+p}(\Omega)$. It is sufficient to consider the trace of G_j on $\partial\Omega$. The lemma is proved. □

2. Properness

We consider the operator $T = (F, G)$, where $G = (G_1, \ldots, G_m)$. It acts from $E_{0,\mu}$ into E_μ, where

$$E_\mu = W^{l-2m,p}_{\infty,\mu}(\Omega) \times W^{l-m_1-1p,p}_{\infty,\mu}(\partial\Omega) \times \cdots \times W^{l-m_m-1/p,p}_{\infty,\mu}(\partial\Omega).$$

We will prove in this section that the operator $T = (F, F) : E_{0,\mu} \to E_\mu$ defined above satisfies the conditions of Lemma 2.1 under the assumptions formulated below.

The convergence $\rightharpoonup$ is a local weak convergence in the space $E_{0,\mu}$. More precisely, we say that a sequence u_k, $u_k \in E_{0,\mu}$ converges *locally weakly* to an element $u_0 \in E_{0,\mu}$ and write $u_k \rightharpoonup u_0$ if for any $R > 0$, $u_k \to u_0$ in $E_{0,\mu}(\Omega_R)$ weakly.

Here we use the same notation for a function from $E_{0,\mu}(\Omega)$ and for its restriction to $E_{0,\mu}(\Omega_R)$, $\Omega_R = \Omega \cap \{|x| < R\}$.

Lemma 2.8. *Let $u_k \in E_{0,\mu}(\Omega)$ be a bounded sequence, $\|u_k\|_{E_{0,\mu}(\Omega)} \leq M$, where M does not depend on k. Then there exists a subsequence u_{k_i} and a function $u_0 \in E_{0,\mu}(\Omega)$ such that $u_{k_i} \rightharpoonup u_0$.*

Proof. It is sufficient to prove the lemma for the space $E_0(\Omega)$ without weight. Indeed, if $\|u_k\|_{E_{0,\mu}(\Omega)} \leq M$, then $\|v_k\|_{E_0(\Omega)} \leq M$, where $v_k = \mu u_k$. For any functional $f(v) \in (E_{0,\mu}(\Omega_R))^*$ we conclude that $f(\frac{v}{\mu}) \in (E_0(\Omega_R))^*$. Hence if v_{k_i} converges locally weakly in E_0 to $v_0 \in E_0$, then $u_{k_i} = v_{k_i}/\mu$ converges locally weakly in $E_{0,\mu}$ to $u_0 = v_0/\mu \in E_{0,\mu}$.

Therefore we will prove the lemma for the space $E_0(\Omega)$. Let $\|v_k\|_{E_0(\Omega)} \leq M$. We can extend the functions $v_k(x)$ on $\mathbb{R}^n$ in such a way that $\|v_k\|_{E_0(\mathbb{R}^n)} \leq M_1$. We will prove that there exists $v_0 \in E_0(\mathbb{R}^n)$ such that a subsequence v_{k_i} of v_k tends to v_0 locally weakly in $E_0(\mathbb{R}^n)$. Then it will follow that v_{k_i} tends to the restriction of v_0 on $E_0(\Omega)$ locally weakly.

Let $N = 1, 2, \ldots$. Denote by B_N the ball $\{x : |x| < N\}$. For $N = 1$ we can find a subsequence v_k^1 of v_k such that $v_k^1 \to v^1$ in $E_0(B_1)$ weakly. For $N = 2$ we find a subsequence v_k^2 of v_k^1 such that $v_k^2 \to v^2$ in $E_0(B_2)$ weakly. Proceeding in this manner we construct a sequence v_k^N for any N, and then by the diagonal process we find a subsequence w_k of v_k and a function $v_0(x)$, which belongs to $E_0(B_N)$ for all N, such that $w_k \to v_0$ weakly in $E_0(B_N)$ for any N.

We now prove that $v_0 \in E_0(\mathbb{R}^n)$ and that w_k tends to v_0 locally weakly in $E_0(\mathbb{R}^n)$. Let B be an arbitrary ball in $\mathbb{R}^n$. Then $w_k \to v_0$ weakly in $E_0(B)$. Indeed, suppose that it is not so. Then there exists $\epsilon > 0$, a functional $f \in (E_0(B))^*$ and a subsequence w_{k_i} of w_k such that

$$|f(w_{k_i}) - f(v_0)| > \epsilon. \tag{2.23}$$

Since $\|w_{k_i}\|_{E_0(B)} \leq M_1$, there exists a subsequence $\tilde{w}_k$ of w_{k_i} such that $\tilde{w}_k \to \tilde{w}$ weakly in $E_0(B)$.

Let ϕ belong to the space D of infinitely differentiable functions with compact support, $\operatorname{supp} \phi \subset B$. Then $g(u) = \int u(x)\phi(x)dx$ is a bounded functional in $E_0(B)$. Therefore

$$g(\tilde{w}_k) \to g(\tilde{w}). \tag{2.24}$$

If N is so large that $B \subset B_N$, then g is also a bounded functional in $E_0(B_N)$. Hence

$$g(\tilde{w}_k) \to g(v_0). \tag{2.25}$$

Taking into account that ϕ can be arbitrary, we obtain from (2.24) and (2.25) that $\tilde{w} = v_0$ as elements of $E_0(B)$. Therefore $\tilde{w}_k \to v_0$ weakly in $E_0(B)$. This contradicts (2.23). This contradiction proves that $w_k \to v_0$ weakly in $E_0(B)$.

It remains to prove that $v_0 \in E_0(\mathbb{R}^n)$. For any ball $Q_a = \{x : |x - a| < 1\}$ we consider the shifted functions $w_k^a(x) = w_k(x + a)$ in the ball Q_0. We have

$$w_k^a \to w_0^a \in E_0(Q_0) \text{ weakly.} \tag{2.26}$$

We prove next that the set of all w_0^a, $a \in \mathbb{R}^n$ is bounded in $E_0(Q_0)$. Suppose that it is not so. Then since any weakly bounded set is strongly bounded, there exists $f \in (E_0(Q_0))^*$ and a sequence $w_0^{a_i}$ such that

$$f(w_0^{a_i}) \to \infty \text{ as } i \to \infty. \tag{2.27}$$

From (2.26) it follows that $f(w_k^{a_i}) \to f(w_0^{a_i})$ as $k \to \infty$. This and (2.27) imply that there exists a subsequence $w_{k_i}^{a_i}$ of $w_k^{a_i}$ such that

$$f(w_{k_i}^{a_i}) \to \infty \text{ as } i \to \infty. \tag{2.28}$$

On the other hand, since $w_{k_i}^{a_i}(x) = w_{k_i}(x + a_i)$, we have from the estimate

$$\|v_k\|_{E_0(\mathbb{R}^n)} \le M_1$$

that $\|w_{k_i}^{a_i}\|_{E_0(Q_0)} \le M_1$, which contradicts (2.28). This contradiction proves that $v_0 \in E_0(\mathbb{R}^n)$. The lemma is proved. □

We now return to the operator $T = (F, G) : E_{0,\mu} \to E_\mu$.

Lemma 2.9. *The operator T is closed with respect to the convergence $\rightharpoonup$ in the following sense: if $T(u_k) = f_k$, $u_k \in E_{0,\mu}$, $f_k \in E_\mu$, and $u_k \rightharpoonup u_0 \in E_{0,\mu}$, $f_k \to f_0$ in E_μ, then $T(u_0) = f_0$.*

Proof. Let $u_k \rightharpoonup u_0$. This means that $u_k \to u_0$ in $E_{0,\mu}(\Omega_R)$ weakly for any $R > 0$. Write $v_k = \mu u_k$, $v_0 = \mu u_0$. Then obviously $v_k \to v_0$ weakly in $E_0(\Omega_R) = W_\infty^{l,p}(\Omega_R)$. Since the embedding operator $W^{l,p}(\Omega_R) \to C^{l-1}(\Omega_R)$ is compact, it follows that $v_k \to v_0$ strongly in $C^{l-1}(\Omega_R)$. From this we obtain the strong convergence $u_k \to u_0$ in $C^{l-1}(\Omega_R)$. Consider the operator (2.13). Since $l - 1 \ge 2m$, it follows that

$$F(x, D^{2m}u_k(x)) \to F(x, D^{2m}u_0(x)) \text{ as } k \to \infty$$

uniformly in Ω_R. This convergence proves the lemma for operator (2.13). Similarly it can be proved for the operator (2.14). The lemma is proved. □

2. Properness

We begin construction of the operator S introduced in Lemma 2.1. Consider the function $F(x, \eta)$, which corresponds to the operator (2.13). Here η is a vector, $\eta \in \mathbb{R}^s$, where s is the number of components η^α. Let $\eta^0 \in \mathbb{R}^s$ be another vector. By Taylor's formula we can write

$$F(x,\eta) = F(x,\eta^0) + \sum_{j=1}^s F'_{\eta_j}(x,\eta^0)(\eta_j - \eta_j^0)$$

$$+ \int_0^1 (1-\tau) \sum_{j,k=1}^s F''_{\eta_j \eta_k}(x, \eta^0 + \tau(\eta - \eta^0))d\tau \ (\eta_j - \eta_j^0)(\eta_k - \eta_k^0).$$

Therefore for any $u, u^0 \in E_{0,\mu}$ we have

$$F(x, D^{2m}u) - F(x, D^{2m}u^0) = A(u - u^0, D^{2m}u^0) + \Phi(u, u^0),$$

where A is given by (2.15) and

$$\Phi(u, u^0) = \int_0^1 (1-\tau) \sum_{j,k=1}^s F''_{v_j v_k}(x, v^0 + \tau(v - v^0))d\tau \ (v_j - v_j^0)(v_k - v_k^0),$$

$v(x) = D^{2m}u(x), \ v^0(x) = D^{2m}u^0(x)$.

Lemma 2.10. *Let $u^i \in E_{0,\mu}$ be a bounded sequence, $u^i \rightharpoonup u^0$. Then*

$$\|\Phi(u^i, u^0)\|_{W^{l-2m,p}_{\infty,\mu}(\Omega)} \to 0$$

as $i \to \infty$.

Proof. We will prove that

$$\|D^\alpha(\mu\Phi(u^i, u^0))\|_{L^p_\infty(\Omega)} \to 0 \ \text{ as } \ i \to \infty, \ |\alpha| \leq l - 2m.$$

Set $v^i(x) = D^{2m}u^i(x), v^0(x) = D^{2m}u^0(x)$,

$$y^i(x) = F''_{v_j v_k}(x, v^0 + \tau(v^i(x) - v^0(x)), \quad z^i(x) = (v_j^i(x) - v_j^0(x))(v_k^i(x) - v_k^0(x)). \tag{2.29}$$

We have to prove that

$$\|D^\alpha(\mu y^i z^i)\|_{L^p_\infty(\Omega)} \to 0, \ i \to \infty. \tag{2.30}$$

We have

$$D^\alpha(\mu y^i z^i) = \sum_{\beta+\gamma \leq \alpha} c_{\beta\gamma} D^\beta y^i D^\gamma(\mu z^i). \tag{2.31}$$

We estimate the terms in the right-hand side. Consider first the case $|\beta| \leq l - 2m - 1$. We prove that

$$\|D^\beta y^i\|_{C(\Omega)} \leq M, \tag{2.32}$$

$$\|D^\gamma \mu z^i\|_{L^p_\infty(\Omega)} \to 0 \ \text{ as } \ i \to \infty. \tag{2.33}$$

By the assumption of the lemma, $\|u^i\|_{\hat{W}^{l,p}_\mu(\Omega)} \leq M_1$. Here and in what follows M with subscripts denote constants independent of i. Then

$$\|v^i\|_{W^{l-2m,p}_{\infty,\mu}(\Omega)} \leq M_2. \tag{2.34}$$

Using the embedding theorem and the properties of the weight function μ we get

$$\|v^i\|_{C^{l-2m-1}(\Omega)} \leq M_3, \quad v^0 \in C^{l-2m-1}(\Omega). \tag{2.35}$$

From the smoothness assumptions imposed above on the function F and from (2.35) we obtain (2.32).

To prove (2.33) write $w^i_j(x) = v^i_j(x) - v^0_j(x)$. We show that

$$\|D^\gamma(\mu w^i_j w^i_k)\|_{L^p_\infty(\Omega)} \to 0 \text{ as } i \to \infty, \tag{2.36}$$

where $|\gamma| \leq l - 2m$. We have

$$D^\gamma(\mu w^i_j w^i_k) = \sum_{\rho+\sigma+\tau \leq \gamma} c_{\rho\sigma\tau} D^\rho \mu D^\sigma w^i_j D^\tau w^i_k, \tag{2.37}$$

where $c_{\rho\sigma\tau}$ are constants. Hence we should estimate

$$D^\rho \mu D^\sigma w^i_j D^\tau w^i_k = \frac{1}{\mu}\left(\frac{1}{\mu} D^\rho \mu\right)\left(\mu D^\sigma w^i_j\right)\left(\mu D^\tau w^i_k\right). \tag{2.38}$$

Since $(D^\rho \mu)/\mu$ is bounded, we have

$$\|D^\rho \mu D^\sigma w^i_j D^\tau w^i_k\|_{L^p(\Omega \cap Q_y)} \leq M_4 \sup_{x \in \Omega \cap Q_y} \frac{1}{\mu}\|\left(\mu D^\sigma w^i_j\right)\left(\mu D^\tau w^i_k\right)\|_{L^p(\Omega \cap Q_y)}. \tag{2.39}$$

Here $|\sigma| + |\tau| \leq l - 2m$. Hence at least one of the numbers $|\sigma|$ and $|\tau|$ is less than or equal to $l - 2m - 1$. Suppose for definiteness that $|\sigma| \leq l - 2m - 1$. Since

$$\|\mu u^i\|_{W^{l,p}_\infty(\Omega)} \leq M_5, \tag{2.40}$$

it follows from the embedding theorem that

$$\|\mu u^i\|_{C^{l-1}(\Omega)} \leq M_6.$$

Then

$$\|\mu D^\sigma w^i_j\|_{C(\Omega)} \leq M_7.$$

Therefore (2.39) implies

$$\|D^\rho \mu D^\sigma w^i_j D^\tau w^i_k\|_{L^p(\Omega \cap Q_y)} \leq M_8 \sup_{x \in \Omega \cap Q_y} \frac{1}{\mu}\|\mu D^\tau w^i_k\|_{L^p(\Omega \cap Q_y)}, \tag{2.41}$$

2. Properness

where $|\tau| \leq l - 2m$. By (2.40) we get

$$\|\mu D^\tau w_k^i\|_{L^p(\Omega \cap Q_y)} \leq M_9. \tag{2.42}$$

From (2.37), (2.41), (2.42) we obtain

$$\|D^\gamma(\mu w_j^i w_k^i)\|_{L^p(\Omega \cap Q_y)} \leq M_{10} \sup_{x \in \Omega \cap Q_y} \frac{1}{\mu}.$$

To prove (2.36) it is sufficient to show that

$$\sup_{y \in \Omega_R} \|D^\gamma(\mu w_j^i w_k^i)\|_{L^p(\Omega \cap Q_y)} \to 0 \text{ as } i \to \infty \tag{2.43}$$

for any $R > 0$, $|\gamma| \leq l - 2m$.

By the assumption of the lemma, $u^i \to u^0$. This means that

$$u^i \to u^0 \text{ weakly in } W_\mu^{l,p}(\Omega_R) \tag{2.44}$$

for any $R > 0$. Since for $y \in \Omega(R)$, $\|\cdot\|_{L^p(\Omega \cap Q_y)} \leq \|\cdot\|_{L^p(\Omega_{R+1})}$, then (2.43) will follow from the convergence

$$\|D^\gamma(\mu w_j^i w_k^i)\|_{L^p(\Omega_R)} \to 0 \text{ as } i \to \infty$$

for any $R > 0$, or, according to (2.37), (2.38) from the convergence

$$\|(\mu D^\sigma w_j^i)(\mu D^\tau w_k^i)\|_{L^p(\Omega_R)} \to 0 \text{ as } i \to \infty, \tag{2.45}$$

where $|\sigma| + |\tau| \leq l - 2m$. Let for definiteness $|\sigma| \leq l - 2m - 1$. From (2.42) it follows that $\|\mu D^\tau w_k^i\|_{L^p(\Omega_R)} \leq M_{11}$. Therefore to prove (2.45) we should verify that

$$\|\mu D^\sigma w_j^i\|_{C(\Omega_R)} \to 0 \text{ as } i \to \infty. \tag{2.46}$$

From (2.44) we conclude that $\mu u^i \to \mu u^0$ weakly in $W^{l,p}(\Omega_R)$ and, consequently, this convergence is strong in $C^{l-1}(\Omega_R)$. Then $\|\mu w_j^i\|_{C^{l-2m-1}(\Omega_R)} \to 0$ as $i \to \infty$, and (2.46) follows from the assumption $|\sigma| \leq l - 2m - 1$. Thus (2.43), (2.36), and (2.33) are proved.

Let us return to (2.30). We have studied the terms in (2.31) with $|\beta| \leq l - 2m - 1$. Since

$$|\beta| + |\gamma| \leq |\alpha| \leq l - 2m,$$

we have to consider $|\beta| = l - 2m$, $|\gamma| = 0$. We will prove that

$$\|D^\beta y^i\|_{\hat{L}^p(\Omega)} \leq M_{12}, \tag{2.47}$$

$$\|D^\gamma(\mu z^i)\|_{C(\Omega)} \to 0 \text{ as } i \to \infty. \tag{2.48}$$

We begin with (2.47). According to (2.34) we have for $|\beta| \leq l - 2m$,

$$\lambda \|D^\beta v^i\|_{\hat{L}^p(\Omega)} \leq \|\mu D^\beta v^i\|_{\hat{L}^p(\Omega)} \leq M_{13}. \tag{2.49}$$

Here $0 < \lambda \leq \mu(x)$. We recall that y^i is defined by (2.29). We use the formula (2.20) of differentiation of superposition to the function $F''_{v_j v_k}$. We have

$$D^\beta y_i = D^\beta_x F''_{v_j v_k}(x, v^0(x) + \tau(v^i(x) - v^0(x)))$$
$$+ \sum_{s=1}^{\kappa} \frac{\partial F''_{v_j v_k}(x, v^0(x) + \tau(v^i(x) - v^0(x)))}{\partial v_s} \qquad (2.50)$$
$$\times D^\beta_x(v^0_s(x) + \tau(v^i_s(x) - v^0_s(x))) + T_{|\beta|}(v; F''_{v_j v_k}).$$

Here κ is the number of components of the vector v. From (2.49) by the embedding theorem we get

$$\|v^i\|_{C^{l-2m-1}(\Omega)} \leq M_{14}. \qquad (2.51)$$

From the smoothness conditions imposed above on the function F we conclude that the first term in the right-hand side of (2.50) is bounded in the C-norm and, consequently, in the $L^p_\infty(\Omega)$-norm.

Consider the second term in the right-hand side of (2.50).

The factor $\partial F''_{v_j v_k}/\partial v_s$ is bounded in the C-norm, which follows from (2.51) and from the smoothness conditions on F. Hence from (2.49) it follows that the second term in the right-hand side of (2.50) is bounded in the $L^p_\infty(\Omega)$-norm. It is easy to verify that the last term on the right in (2.50) is bounded in the C-norm and so in the $L^p_\infty(\Omega)$-norm. Therefore (2.47) is proved.

Consider now (2.48). Since $|\gamma| = 0$ we have to prove the convergence

$$\|\mu z^i\|_{C(\Omega)} \to 0 \text{ as } i \to \infty$$

or

$$\|\mu w^i_j w^i_k\|_{C(\Omega)} \to 0 \text{ as } i \to \infty. \qquad (2.52)$$

This can be written in the form

$$\left\|\frac{1}{\mu}(\mu w^i_j)(\mu w^i_k)\right\|_{C(\Omega)} \to 0 \text{ as } i \to \infty.$$

We have $w^i_j(x) = v^i_j(x) - v^0_j(x)$, and from (2.49) it follows that

$$\|D^\beta(\mu w^i_j)\|_{L^p_\infty(\Omega)} \leq M_{15}.$$

From the embedding theorem we have

$$\|\mu w^i_j\|_{C(\Omega)} \leq M_{16}. \qquad (2.53)$$

To prove (2.52) it is sufficient to verify that

$$\|w^i_k\|_{C(\Omega)} \to 0 \text{ as } i \to \infty. \qquad (2.54)$$

By the assumptions of the lemma, $u^i \rightharpoonup u^0$ as $i \to \infty$. This means that $u^i \to u^0$ in $W^{l,p}_{\infty,\mu}(\Omega_R)$ weakly. From the embedding theorem it follows that $\mu u^i \to \mu u^0$

2. Properness

strongly in $C^{l-1}(\Omega_R)$. Therefore $u^i \to u^0$ in $C^{l-1}(\Omega_R)$. Since $v^i(x) = D^{2m}u^i(x)$, $v^0(x) = D^{2m}u^0(x)$, $w_k^i = v_k^i - v_k^0$, then

$$\|w_k^i\|_{C^{l-2m-1}(\Omega_R)} = \|v_k^i - v_k^0\|_{C^{l-2m-1}(\Omega_R)} \to 0 \text{ as } i \to \infty.$$

Together with the estimate

$$|w_k^i(x)| \leq \frac{M_{17}}{\mu(x)} \to 0 \text{ as } |x| \to \infty,$$

which follows from (2.53), it proves (2.54). Therefore (2.48) is proved. From (2.31), (2.47), (2.48) we obtain (2.30). The lemma is proved. □

Consider now the boundary operators (2.14). As it was supposed above, the function $G_j(x,\xi)$ was given for $x \in \bar{\Omega}$, $\xi \in \mathbb{R}^{s_j}$, where s_j was the number of components of the vector ξ. We study $G_j(x,\xi)$ similarly to $F(x,\eta)$. Let ξ, ξ^0 be vectors in $\mathbb{R}^{s_j}$. By Taylor's formula

$$G_j(x,\xi) = G_j(x,\xi^0) + \sum_{k=1}^{s_j} G'_{j,\xi_k}(x,\xi^0)(\xi_k - \xi_k^0)$$

$$+ \int_0^1 (1-\tau) \sum_{k,l=1}^{s_j} G''_{j,\xi_k \xi_l}(x,\xi^0 + \tau(\xi - \xi^0))d\tau (\xi_k - \xi_k^0)(\xi_l - \xi_l^0),$$

where ξ_k and ξ_l are the components of the vector ξ.

Therefore for any $u, u^0 \in E_{0,\mu}$ we have

$$G_j(x, D^{m_j}u) - G_j(x, D^{m_j}u^0) = B_j(u - u^0, D^{m_j}u_0) + \Phi_j(u, u_0), \quad (2.55)$$

where B_j is given by (2.16) and

$$\Phi_j(u, u^0) = \int_0^1 (1-\tau) \sum_{k,l=1}^{s_j} G''_{j,\xi_k \xi_l}(x, v^0 + \tau(v - v^0))d\tau (v_j - v_j^0)(v_k - v_k^0),$$

$v(x) = D^{m_j}u(x)$, $v^0(x) = D^{m_j}u^0(x)$.

Lemma 2.11. *Let u^i be a sequence, $u^i \in E_{0,\mu}$, such that $\|u^i\|_{E_{0,\mu}}$ is bounded and $u^i \to u^0$. Then*

$$\|\Phi_j(u^i, u^0)\|_{W^{l-m_j,p}_{\infty,\mu}(\Omega)} \to 0$$

as $i \to \infty$.

The proof of the lemma is the same as the proof of the previous lemma. It follows from Lemma 2.10 that

$$\|F(x, D^{2m}u^i) - F(x, D^{2m}u^0) - A(u^i - u^0, D^{2m}u^0)\|_{W^{l-2m,p}_{\infty,\mu}(\Omega)} \to 0 \quad (2.56)$$

if $u^i \rightharpoonup u^0$ and $\|u^i\|_{W^{l,p}_{\infty,\mu}}$ is bounded. Similarly Lemma 2.11 implies that

$$\|G_j(x, D^{m_j}u^i) - G_j(x, D^{m_j}u^0) - B_j(u^i - u^0, D^{m_j}u^0)\|_{W^{l-m_j-1/p,p}_{\infty,\mu}(\partial\Omega)} \to 0 \quad (2.57)$$

if $u^i \rightharpoonup u^0$ and $\|u^i\|_{W^{l,p}_{\infty,\mu}}$ is bounded. Indeed, from Lemma 2.11 we have this convergence in the norm $\|\cdot\|_{W^{l-m_j,p}_{\infty,\mu}(\Omega)}$. Convergence (2.57) follows from the embedding theorem.

Consider the operators

$$S(u_0)u = (A(u, D^{2m}u_0), B_1(u, D^{m_1}u_0), \ldots, B_m(u, D^{m_m}u_0)) : E_{0,\mu} \to E_\mu$$

and

$$S_0 u = (A(u, 0), B_1(u, 0), \ldots, B_m(u, 0)) : E_{0,\mu} \to E_\mu.$$

Lemma 2.12. *For any $u_0 \in E_{0,\mu}$ the limiting operators for $S(u_0)$ and S_0 coincide.*

Proof. Consider first the operator $F(x, D^{2m}u_0)$. We will prove that

$$|D^{2m}u_0(x)| \leq \frac{M}{\mu(x)}, \quad x \in \Omega, \quad (2.58)$$

where M does not depend on x. Indeed, we have $u_0 \in E_{0,\mu} = W^{l,p}_{\infty,\mu}(\Omega)$. It follows that $\mu u_0 \in W^{l,p}_\infty(\Omega)$. By the embedding theorem we have $\mu u_0 \in C^{l-1}(\Omega)$. Then $D^{2m}(\mu u_0) \in C^{l-2m-1}$. Since $l \geq 2m+1$, we have $D^{2m}(\mu u_0) \in C(\Omega)$. Write $v_0 = \mu u_0$. We have for $|\alpha| \leq 2m$:

$$|D^\alpha v_0| \leq M_1, \quad (2.59)$$

$$D^\alpha u_0 = D^\alpha(\frac{1}{\mu}v_0) = \sum_{\beta+\gamma\leq\alpha} c_{\beta\gamma}D^\beta\frac{1}{\mu}D^\gamma v_0 = \sum_{\beta+\gamma\leq\alpha} c_{\beta\gamma}\frac{1}{\mu}\left(\mu D^\beta\frac{1}{\mu}\right)D^\gamma v_0.$$

The estimate (2.58) follows from (2.59).

Let $|x_k| \to \infty$, $x_k \in \Omega$. Then $|y + x_k| \to \infty$ for all $y \in \Omega_k \cap B_R$, where Ω_k is the shifted domain. Then $x = y + x_k \in \Omega$. It follows from (2.58) that

$$|D^{2m}u_0(y + x_k)| \leq \frac{M}{\mu(y + x_k)} \to 0 \quad \text{as} \quad |x_k| \to \infty$$

uniformly in $y \in \Omega_k \cap B_R$. Then for $k > k_0$ we have $|D^{2m}u_0(y+x_k)| \leq 1$ for all $y \in \Omega_k \cap B_R$. By the smoothness conditions on F we have

$$|a^\alpha(x, D^{2m}u_0(x)) - a^\alpha(x, 0)| \leq K|D^{2m}u_0(x)|$$

for $x \in \Omega$, $|D^{2m}u_0(x)| \leq 1$. Hence for $y \in \Omega_k \cap B_R$ we get

$$|a^\alpha(y+x_k, D^{2m}u_0(y+x_k)) - a^\alpha(y+x_k, 0)| \leq K|D^{2m}u_0(y+x_k)|$$

$$\leq \frac{KM}{\mu(y+x_k)} \to 0 \quad \text{as} \quad |x_k| \to \infty, \ x_k \in \Omega \quad (2.60)$$

uniformly in $y \in \Omega_k \cap B_R$.

2. Properness

Consider now the limiting coefficients for the operators $S(u_0)$ and S_0. Denote them by $\hat{a}^\alpha(x)$ and $a^{0,\alpha}(x)$, respectively. By definition, for any $y \in \bar{\Omega}_* \cap B_R$ we have

$$\hat{a}^\alpha(y) = \lim_{k \to \infty} a_k^\alpha(y), \quad a^{0,\alpha}(y) = \lim_{k \to \infty} a_k^{0,\alpha}(y), \tag{2.61}$$

where

$$a_k^\alpha = a^\alpha(y + x_k, D^{2m} u_0(y + x_k)), \quad y \in B_R,$$
$$a_k^{0,\alpha}(y) = a^\alpha(y + x_k, 0), \quad y \in B_R$$

and $a_k^\alpha(y)$ and $a_k^{0,\alpha}(y)$ are supposed to be extended from Ω_k to $\mathbb{R}^n$. If the limit in (2.61) exists, it is uniform in $y \in B_R$.

Now, we show that

$$\hat{a}^\alpha(y) = a^{0,\alpha}(y), \quad \forall y \in \bar{\Omega}_* \cap B_R. \tag{2.62}$$

Indeed, according to (2.60) we have

$$|a_k^\alpha(y) - a_k^{0,\alpha}(y)| \leq \frac{KM}{\mu(y + x_k)} \to 0 \text{ as } |x_k| \to \infty, \; x_k \in \Omega \tag{2.63}$$

uniformly in $y \in \Omega_k \cap B_R$. For any $y_0 \in \bar{\Omega}_* \cap B_R$ we can find $y_k \in \bar{\Omega}_k \cap B_R$ in such a way that $y_k \to y_0$. We have

$$\begin{aligned}&|\hat{a}^\alpha(y_0) - a^{0,\alpha}(y_0)| \\ &\leq |\hat{a}^\alpha(y_0) - a_k^\alpha(y_k)| + |a_k^{0,\alpha}(y_k) - a^{0,\alpha}(y_0)| + |a_k^\alpha(y_k) - a_k^{0,\alpha}(y_k)|.\end{aligned} \tag{2.64}$$

For the first term in the right-hand side we can write

$$|\hat{a}^\alpha(y_0) - a_k^\alpha(y_k)| \leq |\hat{a}^\alpha(y_0) - \hat{a}^\alpha(y_k)| + |\hat{a}^\alpha(y_k) - a_k^\alpha(y_k)|.$$

The first term on the right in this inequality tends to 0 since $\hat{a}^\alpha(y)$ is continuous. The second term converges to 0 since the limit in (2.61) is uniform. Therefore the first term in the right-hand side in (2.64) converges to 0 as $k \to \infty$. The same is true for the second term. Finally, for the last term we have from (2.60)

$$|a_k^\alpha(y_k) - a_k^{0,\alpha}(y_k)| \leq \frac{KM}{\mu(y_k + x_k)} \to 0 \text{ as } |x_k| \to \infty$$

since the limit in (2.60) is uniform in y. Thus (2.62) is proved.

It is also clear that if one of the limits in (2.61) exists, then the second limit also exists and coincides with the first one. Indeed, suppose that there exists the first limit in (2.61). Then

$$|\hat{a}^\alpha(y) - a_k^{0,\alpha}(y)| \leq |\hat{a}^\alpha(y) - a_k^\alpha(y)| + |a_k^\alpha(y) - a_k^{0,\alpha}(y)|, \; y \in \Omega_*.$$

The first term in the right-hand side of this inequality tends to 0 as $k \to \infty$ by virtue of the first equality in (2.61); the second term converges to a limit by virtue of (2.63) since we can approximate y by $y_k \in \Omega_k \cap B_R$.

The operator $G_j(x, D^{m_j}u)$ can be studied similarly. The lemma is proved. □

The main result of this section is given by the following theorem.

Theorem 2.13. *Suppose that the operator $F(x, D^{2m})$ is uniformly elliptic and that the Lopatinskii condition is satisfied for the operators (2.13), (2.14). Assume further that all limiting operators for the operator S_0 satisfy Condition NS. Then the operator $(F, G) : E_{0,\mu} \to E_\mu$ is proper.*

3 Topological degree

Topological degree for infinite-dimensional operators was introduced by Leray and Schauder in the case of compact perturbations of the identity operator. It can be generalized for Fredholm and proper operators.

Topological degree is an integer which depends on the operator and on the domain in the Banach space. This integer is supposed to satisfy three conditions: homotopy invariance, additivity, normalization (see below). The existence of such characteristics is not a priori known and should be verified. If the degree can be defined, it provides a powerful tool to study solutions of operator equations.

The degree constructed in this section is adapted for elliptic problems. In particular, even in the abstract setting we will work with pairs of operators. In the context of elliptic problems, one of them corresponds to the operators inside the domains while another one to the boundary operators. Another way to do it is to include boundary conditions in function spaces (see [567] and the bibliographical comments).

In Section 3.2 we introduce the notion of orientation of linear operators used for construction of the topological degree. In Section 3.3 a topological degree is constructed for Fredholm and proper operators with the zero index and with some additional conditions on the essential spectrum. All these conditions can be verified for elliptic problems in bounded or unbounded domains. We will prove that the degree, which satisfies the three conditions mentioned above, is unique. It appears that the topological degree may not exist for certain operators and spaces. An example is presented in Section 3.6.

3.1 Definition and main properties of the degree

We recall the definition of a topological degree. Let E_0 and E be two Banach spaces. Suppose we are given a class F of operators acting from E_0 into E and a

3. Topological degree

class H of homotopies, i.e., mappings

$$A_\tau(u) : E_0 \times [0,1] \to E, \quad \tau \in [0,1], \; u \in E_0$$

such that $A_\tau(u) \in F$ for any $\tau \in [0,1]$. Assume moreover that for any bounded open set $D \subset E_0$ and any operator $A \in F$ such that

$$A(u) \neq 0, \quad u \in \partial D$$

(∂D denotes the boundary of D) there is an integer $\gamma(A, D)$ satisfying the following conditions:

(i) *Homotopy invariance.* Let $A_\tau(u) \in H$ and

$$A_\tau(u) \neq 0, \quad u \in \partial D, \tau \in [0,1].$$

Then

$$\gamma(A_0, D) = \gamma(A_1, D).$$

(ii) *Additivity.* Let $D \subset E_0$ be an arbitrary bounded open set in E_0, and $D_1, D_2 \subset D$ be open sets such that $D_1 \cap D_2 = \emptyset$. Suppose that $A \in F$ and

$$A(u) \neq 0, \quad u \in \bar{D} \setminus (D_1 \cup D_2).$$

Then

$$\gamma(A, D) = \gamma(A, D_1) + \gamma(A, D_2).$$

(iii) *Normalization.* There exists a bounded linear operator $J : E_0 \to E$ with a bounded inverse defined on all of E such that for any bounded open set $D \subset E_0$ with $0 \in D$,

$$\gamma(J, D) = 1.$$

The integer $\gamma(A, D)$ is called a *topological degree*.

Some of the properties of the degree easily follow from its definition. In particular, if there are no solutions of the equation $A(u) = 0$ in D, then $\gamma(A, D) = 0$. Hence, if $\gamma(A, D) \neq 0$, then there are some solutions $u \in D$ of this equation. This argument is used to prove existence of solutions (Section 4).

Another property of the degree provides the persistence (structural stability) of solutions. If u_0 is an isolated solution, that is there are no other solutions in its small neighborhood U, and $\gamma(A, U) \neq 0$, then the solution will persist under small perturbations of the operator. Some other properties and applications will be discussed below in Section 4.

3.2 Orientation of operators

Let E_0, E_1 and E_2 be Banach spaces. We suppose that $E_0 \subset E_1$. This means that if $u \in E_0$, then $u \in E_1$ and $\|u\|_{E_1} \leq K\|u\|_{E_0}$, where K does not depend on u. Write $E = E_1 \times E_2$. We consider linear operators $A_1 : E_0 \to E_1$, $A_2 : E_0 \to E_2$, $A = (A_1, A_2) : E_0 \to E$, and the following class of operators.

Class O is a class of bounded operators $A : E_0 \to E$ satisfying the following conditions:

(i) Operator $(A_1 + \lambda I, A_2) : E_0 \to E$ is Fredholm with index 0 for all $\lambda \geq 0$.
(ii) Equation $A_1 u = 0$, $A_2 u = 0$ $(u \in E_0)$ has only the zero solution.
(iii) There exists $\lambda_0 = \lambda_0(A)$ such that the equation
$$(A_1 + \lambda I)u = 0, \quad A_2 u = 0 \quad (u \in E_0)$$
has only the zero solution for all $\lambda > \lambda_0$. Here I is the identity operator in E_0.

Proposition 3.1. *Let operator $A = (A_1, A_2)$ belong to class O. Then the eigenvalue problem*
$$A_1 u + \lambda u = 0, \quad A_2 u = 0 \quad (u \in E_0) \tag{3.1}$$
has only a finite number of positive eigenvalues λ. Each of them has a finite multiplicity.

Remark 3.2. Instead of the eigenvalue problem (3.1) we can consider the eigenvalue problem
$$A_{1,2} u + \lambda u = 0, \quad u \in E_{0,2}, \tag{3.2}$$
where $E_{0,2}$ is the space of all $u \in E_0$ such that $A_2 u = 0$, and $A_{1,2}$ is the restriction of A_1 on the space $E_{0,2}$. By multiplicity of λ in (3.1) we mean the multiplicity of λ in (3.2).

Proof of Proposition 3.1. Since $A \in O$, the operator $A_{1,2} + \lambda I$ is Fredholm with index 0 for all $\lambda \geq 0$ and invertible for $\lambda = 0$ and $\lambda > \lambda_0$. The proposition follows from known properties of Fredholm operators. □

Definition 3.3. The number
$$o(A) = (-1)^\nu,$$
where ν is the sum of multiplicities of all positive eigenvalues of problem (3.1), is called *orientation* of the operator A. Operators A belonging to the class O are called *orientable*.

Definition 3.4. Operators $A^0 \in O$ and $A^1 \in O$ are said to be *homotopic* if there exists an operator $A(\tau) : E_0 \times [0, 1] \to E$ such that $A(\tau) \in O$ for all $\tau \in [0, 1]$, $A(\tau)$ is continuous in the operator norm with respect to τ, $\lambda_0(A(\tau))$ is bounded, and
$$A(0) = A^0, \quad A(1) = A^1.$$

3. Topological degree

Theorem 3.5. *If A^0 and A^1 are homotopic, then*
$$o(A^0) = o(A^1). \tag{3.3}$$

Proof. Let $\tau_0 \in [0,1]$. It is sufficient to prove that
$$o(A(\tau)) = o(A(\tau_0)) \tag{3.4}$$
for τ in some neighborhood of τ_0. Indeed, covering the interval $[0,1]$ by such neighborhoods and taking a finite subcovering we get (3.3).

To prove (3.4) consider the eigenvalue problems
$$A_1(\tau_0)u + \lambda u = 0, \quad A_2(\tau_0)u = 0, \quad u \in E_0 \tag{3.5}$$
and
$$A_1(\tau)u + \lambda u = 0, \quad A_2(\tau)u = 0, \quad u \in E_0. \tag{3.6}$$

We should prove that for τ close to τ_0 the sum of multiplicities of positive eigenvalues λ of problems (3.5) and (3.6) coincide modulo 2. It is convenient to consider the problem
$$A_1(\tau_0)u + \lambda u = 0, \quad A_2(\tau)u = 0, \quad u \in E_0 \tag{3.7}$$
and to compare (3.5) and (3.6) with (3.7).

Consider first problems (3.5) and (3.7). Consider also operators $A_{1,2}(\tau_0)$ and $A_{1,2}(\tau)$, the restrictions of $A_1(\tau_0)$ on the spaces
$$E_{0,2}(\tau_0) = \{u : u \in E_0, \; A_2(\tau_0)u = 0\}$$
and
$$E_{0,2}(\tau) = \{u : u \in E_0, \; A_2(\tau)u = 0\}, \tag{3.8}$$
respectively. By (i) and (ii) of Definition of Class O, $A_{1,2}(\tau_0)$ is invertible.

It is easy to see that for τ sufficiently close to τ_0 the operator $A_{1,2}(\tau)$ is also invertible and has a uniformly bounded inverse. Indeed, denote $K(\tau) = (A_1(\tau_0), A_2(\tau)) : E_0 \to E$. Obviously $\|K(\tau) - K(\tau_0)\| \le \|A_2(\tau) - A_2(\tau_0)\|$. Since $K(\tau_0)$ is invertible, we conclude that if τ is sufficiently close to τ_0, then $K(\tau)$ has uniformly bounded inverse. Consider the equation $A_{1,2}(\tau)u = f$, $u \in E_{0,2}(\tau)$ or $A_1(\tau_0)u = f$, $A_2(\tau)u = 0$, $u \in E_0, f \in E_1$. Since $K(\tau)$ is invertible, this equation has a unique solution for any $f \in E_1$. So $A_{1,2}(\tau)$ is invertible and $\|A_{1,2}^{-1}(\tau)\| \le \|K^{-1}(\tau)\|$.

Set
$$J = A_{1,2}^{-1}(\tau)A_{1,2}(\tau_0) : E_{0,2}(\tau_0) \to E_{0,2}(\tau).$$
The problems (3.5) and (3.7) can be written as
$$A_{1,2}(\tau_0)v + \lambda v = 0, \quad v \in E_{0,2}(\tau_0) \tag{3.9}$$

and
$$A_{1,2}(\tau)u + \lambda u = 0, \quad u \in E_{0,2}(\tau). \qquad (3.10)$$

Let $u = Jv$, $v \in E_{0,2}(\tau_0)$, $u \in E_{0,2}(\tau)$. Then from (3.10)
$$\frac{1}{\lambda}v + S_1 v = 0, \quad v \in E_{0,2}(\tau_0), \qquad (3.11)$$

where $S_1 = J^{-1} A_{1,2}^{-1}(\tau) J = A_{1,2}^{-1}(\tau_0) A_{1,2}^{-1}(\tau) A_{1,2}(\tau_0)$. We have from (3.9)
$$\frac{1}{\lambda}v + S_0 v = 0, \quad v \in E_{0,2}(\tau_0), \qquad (3.12)$$

where $S_0 = A_{1,2}^{-1}(\tau_0)$.

We will prove that for any $\epsilon > 0$ there exists $\delta > 0$ such that
$$||S_1 - S_0|| < \epsilon \quad \text{if} \quad |\tau - \tau_0| < \delta. \qquad (3.13)$$

Consider the problems
$$A_1(\tau_0)u = f, \quad A_2(\tau_0)u = 0, \quad u \in E_0, \quad f \in E_1 \qquad (3.14)$$
and
$$A_1(\tau_0)u_1 = f, \quad A_2(\tau)u_1 = 0, \quad u_1 \in E_0, \quad f \in E_1 \qquad (3.15)$$
or
$$A_{1,2}(\tau_0)u = f, \quad A_{1,2}(\tau)u_1 = f, \quad u \in E_{0,2}(\tau_0), \quad u_1 \in E_{0,2}(\tau). \qquad (3.16)$$

Let $B = A_2(\tau_0) - A_2(\tau)$. Write $w = u - u_1$. Then from (3.14) and (3.15)
$$A_1(\tau_0)w = 0, \quad A_2(\tau_0)w = -Bu_1.$$

We have from (3.16)
$$A_1(\tau_0)w = 0, \quad A_2(\tau_0)w = -BA_{1,2}^{-1}(\tau)f. \qquad (3.17)$$

Set
$$L = (A_1(\tau_0), A_2(\tau_0)) : E_0 \to E.$$

Then (3.17) implies
$$||w||_{E_0} \leq ||L^{-1}|| \, ||B|| \, ||A_{1,2}^{-1}|| \, ||f||_{E_1}. \qquad (3.18)$$

By (3.16) we have
$$||A_{1,2}^{-1}(\tau_0)f - A_{1,2}^{-1}(\tau)f||_{E_0} \leq ||L^{-1}|| \, ||B|| \, ||A_{1,2}^{-1}|| \, ||f||_{E_1}.$$

Therefore
$$||A_{1,2}^{-1}(\tau_0) - A_{1,2}^{-1}(\tau)|| \leq ||L^{-1}|| \, ||B|| \, ||A_{1,2}^{-1}(\tau)||. \qquad (3.19)$$

Since $A_2(\tau) \to A_2(\tau_0)$ as $\tau \to \tau_0$ in the operator norm, we get $||B|| \to 0$ as $\tau \to \tau_0$, and from (3.19) we obtain (3.13).

Using (3.13) we will prove that if τ is sufficiently close to τ_0, then the sum of multiplicities of the negative eigenvalues of the operator S_1 coincides modulo 2 with the sum of multiplicities of the negative eigenvalues of the operator S_0. Indeed, taking into account that $||S_1||$ and $\lambda_0(A(\tau))$ are uniformly bounded, we conclude that there exists an interval $[\alpha, \beta]$, $\alpha < \beta < 0$ such that all negative eigenvalues of the operators S_1 and S_0 lie in this interval. Let Γ be a rectifiable contour in the λ-plane which contains the interval $[\alpha, \beta]$ and such that all points inside this contour, except for negative eigenvalues of the operator S_0, are regular points of this operator. From the known results on root spaces (see [207]) it follows that the sum of multiplicities of all eigenvalues of S_1 lying inside Γ coincides with the sum of the multiplicities of the negative eigenvalues of S_0 if δ in (3.13) is sufficiently small. Therefore the sum of multiplicities of negative eigenvalues of S_0 and S_1 coincide modulo 2. It follows that the sum of the multiplicities of positive eigenvalues of the problems (3.9) and (3.10), and consequently of the problems (3.5) and (3.7) coincide modulo 2.

We obtain now the same results for problems (3.6) and (3.7). Denote by $B(\tau_0)$ and $B(\tau)$ the restrictions of $A_1(\tau_0)$ and $A_1(\tau)$ on the space $E_{0,2}(\tau)$ (see (3.8)), respectively. Then obviously

$$||B(\tau) - B(\tau_0)|| \le ||A(\tau) - A(\tau_0)|| \to 0$$

as $\tau \to \tau_0$. By the same arguments that we used for the operators S_0 and S_1 above we prove that the sum of multiplicities of the negative eigenvalues of the operators $B(\tau)$ and $B(\tau_0)$ coincide modulo 2. The theorem is proved. $\square$

Remark 3.6. The requirement that $\lambda_0(A(\tau))$ is bounded in Definition 3.4 can be omitted if we replace (iii) in class O by the following:

(iii*)   There exists $\lambda_0 = \lambda_0(A)$ such that the operator $(A_1 + I\lambda, A_2) : E_0 \to E$ has an inverse for $\lambda > \lambda_0$ which is uniformly bounded.

Indeed, let $A(\tau, \lambda) = (A_1(\tau) + I\lambda, A_2(\tau))$. Let $\tau_0 \in [0, 1]$. Then $A(\tau, \lambda) = A(\tau_0, \lambda) + B(\tau)$, where $B(\tau) = A(\tau) - A(\tau_0)$. For $\lambda > \lambda_0(A(\tau_0))$ we have

$$A(\tau, \lambda) = A(\tau_0, \lambda)[I + A^{-1}(\tau_0, \lambda)B(\tau)].$$

Since $||B(\tau)|| \to 0$ as $\tau \to \tau_0$, we can take $\delta(\tau_0) > 0$ such that $||A^{-1}(\tau_0, \lambda)B(\tau)|| \le 1/2$ for all $\lambda > \lambda_0(A(\tau_0))$, $|\tau - \tau_0| < \delta(\tau_0)$. So for these values of τ and λ the operator $A(\tau, \lambda)$ has a uniformly bounded inverse. Taking the corresponding covering of the interval $[0, 1]$ and choosing a finite subcovering, we obtain that $\lambda_0(A(\tau))$ is bounded for $\tau \in [0, 1]$.

Class O with the property (iii*) instead of (iii) will be used in construction of the topological degree.

3.3 Degree for Fredholm and proper operators

Let E_0, E_1, E_2 and $E = E_1 \times E_2$ be the same spaces as in Section 3.2, $G \subset E_0$ be an open bounded set. We consider the following classes of linear (Φ) and nonlinear (F) operators.

Class Φ is a class of bounded linear operators $A = (A_1, A_2) : E_0 \to E$ satisfying the following conditions:

(i) Operator $(A_1 + I\lambda, A_2) : E_0 \to E$ is Fredholm for all $\lambda \geq 0$.
(ii) There exists $\lambda_0 = \lambda_0(A)$ such that operators $(A_1 + I\lambda, A_2) : E_0 \to E$ have inverses which are uniformly bounded for all $\lambda > \lambda_0$.

Class F is a class of proper operators $f \in C^1(G, E)$ such that for any $x \in G$ the Fréchet derivative $f'(x)$ belongs to Φ.

We introduce also the following class of homotopies.

Class H is a class of proper operators $f(x, t) \in C^1(G \times [0, 1], E)$ which for any $t \in [0, 1]$ belong to class F.

Two operators $f_0(x) : G \to E$ and $f_1(x) : G \to E$ are said to be *homotopic* if there exists $f(x, t) \in H$ such that

$$f_0(x) = f(x, 0), \quad f_1(x) = f(x, 1). \tag{3.20}$$

In this section we will construct a topological degree for the classes F and H. In what follows D will denote an open set such that $\overline{D} \subset G$.

Let $a \in E$, $f \in C^1(G, E)$,

$$f(x) \neq a \quad (x \in \partial D), \tag{3.21}$$

where ∂D is the boundary of D. Suppose that the equation

$$f(x) = a \quad (x \in D) \tag{3.22}$$

has a finite number of solutions $x_1, \ldots, x_m$ and $f'(x_k)$ ($k = 1, \ldots, m$) are invertible operators belonging to the class Φ. Then the orientation of these operators is defined. We shall use the notation

$$\gamma(f, D; a) = \sum_{k=1}^{m} o(f'(x_k)). \tag{3.23}$$

If equation (3.22) does not have solutions, it is supposed that $\gamma(f, D; a) = 0$.

Lemma 3.7. *Let $f(x, t) \in H$, $a \in E$ be a regular value of $f(., 0)$ and $f(., 1)$. Suppose that*

$$f(x, t) \neq a \quad (x \in \partial D, \ t \in [0, 1]). \tag{3.24}$$

Then

$$\gamma(f(., 0), D; a) = \gamma(f(., 1), D; a). \tag{3.25}$$

3. Topological degree

Proof. The main part of the proof of the lemma is done under the assumption that a is a regular value of the homotopy under consideration. Since this is not supposed in the formulation of the lemma, we replace $f(x,t)$ by a close function $g(x,t)$ for which a is a regular value and

$$\gamma(g(.,0), D; a) = \gamma(f(.,0), D; a), \tag{3.26}$$
$$\gamma(g(.,1), D; a) = \gamma(f(.,1), D; a) \tag{3.27}$$

(see [404]). Then we prove that

$$\gamma(g(.,0), D; a) = \gamma(g(.,1), D; a). \tag{3.28}$$

To construct the function $g(x,t)$ we use the following result (see [404]). For any $\eta > 0$ an operator $h \in C^1(G \times [0,1] \times [0,1], E)$ with the following properties can be constructed:

(i) $||h(.,\tau) - f||_{1, G \times [0,1]} < \eta$ for any $\tau \in [0,1]$.
(ii) h is proper.
(iii) For $\tau \in [0,1]$, $h(.,\tau)$ is Fredholm of index 1.
(iv) $h(.,0) = f$ and a is a regular value of $h(.,1)$.

Here we use the notation $||f||_{1, G \times [0,1]} = \sup ||f(x,t)|| + \sup ||f'(x,t)||$ for $f \in C^1(G \times [0,1], E)$ (the supremum is taken over $(x,t) \in G \times [0,1]$ and f' is the Fréchet derivative of f).

We can now put $g(x,t) = h(x,t,1)$, $x \in G, t \in [0,1]$. From (3.24) it follows that $\eta > 0$ can be taken such that

$$g(x,t) \neq a \quad (x \in \partial D, \ t \in [0,1]). \tag{3.29}$$

We will prove that for a proper choice of $\eta > 0$ the equality (3.26) holds. Since a is a regular value of $f(x,0)$, $f(x,0) \neq a$, $x \in \partial D$ and $f(x,0)$ is a proper operator, it follows that the equation

$$f(x,0) = a, \ x \in D \tag{3.30}$$

has a finite number of solutions.

If equation (3.30) does not have solutions, then taking η sufficiently small we conclude that the equation

$$g(x,0) = a, \ x \in D \tag{3.31}$$

does not have solutions either. In this case both parts of the equality (3.26) equal 0.

Suppose that equation (3.30) has solutions. We denote them by $x_1, \ldots, x_m$. Let B_k ($k = 1, \ldots, m$) be open balls with centers at x_k and radius r. We suppose that r is taken such that the closures of the balls are disjoint and belong to D. If $\eta > 0$ is taken sufficiently small, then equation (3.31) has exactly m solutions and moreover the equation $g(x,0) = a$, $x \in B_k$ has one and only one solution ($k = 1, \ldots, m$) (see [404]). Denote this solution by ξ_k.

Taking into account that $f'_x(x_k, 0)$ belongs to Φ and that it is invertible, it is easy to prove for a proper choice of r and η that $g'_x(\xi_k, 0)$ also belongs to Φ and is invertible. So the orientation of this operator is defined. Moreover applying Theorem 3.5 to the homotopy $(1 - \tau)f'_x(x_k, 0) + \tau g'_x(\xi_k, 0)$, $\tau \in [0, 1]$ we obtain

$$o(g'_x(\xi_k, 0)) = o(f'_x(x_k, 0))$$

and (3.26) follows from this. Decreasing η, if necessary, we obtain (3.27) in the same way.

We prove now (3.28). If both of the equations

$$g(x,0) = a, \quad g(x,1) = a \quad (x \in D) \qquad (3.32)$$

have no solutions, then (3.28) is true: both parts of the equality are equal to 0.

Suppose that at least one of the equations (3.32) has a solution. Then the set $S = g^{-1}(a) \cap \overline{D} \times [0, 1]$ is not empty. Since a is a regular value of g, $g^{-1}(a)$ is a one-dimensional submanifold of $\overline{D} \times [0, 1]$. The set S is compact since the map is proper. Because of (3.29) the set S cannot have joint points with the set $\partial D \times [0, 1]$. Suppose that the equation $g(x, 0) = a$ has m solutions ($m > 0$): $\xi_1, \ldots, \xi_m$,

$$g(\xi_k, 0) = a \quad (k = 1, \ldots, m). \qquad (3.33)$$

We denote by l_k the connected component of S which contains the point $(\xi_k, 0)$. The set l_k is homeomorphic to a closed interval $\Delta = [0, 1]$. We denote the endpoints of l_k by $P_0 = (\xi_k, 0)$ and P_1 and suppose that P_0 corresponds to the point 0 in Δ and P_1 to 1.

Write $y = (x, t)$ ($x \in G$, $t \in [0, 1]$). We introduce local coordinates on l_k by a finite number of sets $\{U_i\}$ such that each of them is homeomorphic to an open or half-open interval Δ_i. Moreover we can suppose that U_i is given by the equation

$$y = y(s) \quad (s \in \Delta_i) \qquad (3.34)$$

and that there exists a derivative in the norm $||y|| = ||x|| + |t|$. We have $g(y(s)) = a$ and therefore

$$g'(y(s))y'(s) = 0. \qquad (3.35)$$

Since a is a regular value, then the range of the operator $g'(y(s))$ coincides with E. Moreover, the index of $g'(y(s))$ is 1. So $y'(s)$ is the only (up to a real factor) solution of equation (3.35). We have $y(s) = (x(s), t(s))$, where $x(s) \in E_0$, $t(s)$ is a real-valued function. It is easy to see that we can construct a functional $\phi(s) \in E_0^*$ which is continuous with respect to $s \in \Delta_i$ and

$$\langle \phi(s), x'(s) \rangle > 0 \quad \text{if} \quad ||x'(s)|| > 0, \qquad (3.36)$$

where $\langle\,,\,\rangle$ denotes the action of a functional.

3. Topological degree

We can find η in (i) such that for all y satisfying the equation $g(y) = a$, the operators $g'_x(y)$ belong to Φ, and $\lambda_0(g'_x(y))$ are uniformly bounded. Indeed, denote by T the set of all solutions of the equation $f(y) = a$. From (i) and properness of f it follows that for any $\varepsilon > 0$ we can find $\eta > 0$ such that all solutions of the equation $g(y) = a$ belong to an ε-neighborhood of T. Since T is compact, ε and η can be found such that $g'_x(y)$ has the mentioned property.

We represent g in the form $g = (g_1, g_2)$, where $g_1 : G \times [0,1] \to E_1$, $g_2 : G \times [0,1] \to E_2$. Denote by $g'_{ix}(x,t)$ and $g'_{it}(x,t)$ $(i = 1, 2)$ the partial derivatives in x and t, respectively.

Consider the operators

$$A_1(s) = \begin{bmatrix} g'_{1x}(y(s)) & g'_{1t}(y(s)) \\ \phi(s) & t'(s) \end{bmatrix}, \quad A_2(s) = (g'_{2x}(y(s)), g'_{2t}(y(s)),$$

where $A_1(s) : E_0 \times \mathbb{R} \to E_1 \times \mathbb{R}$, $A_2(s) : E_0 \times \mathbb{R} \to E_2$, $\mathbb{R}$ is the space of real numbers.

Set $A(s) = (A_1(s), A_2(s)) : E_0 \times \mathbb{R} \to (E_1 \times \mathbb{R}) \times E_2$. It is easy to see that A is a Fredholm operator of index 0.

The equation $A(s)w = 0$, $w \in E_0 \times \mathbb{R}$ has only a zero solution. Indeed, let $w = (u, v)$, $u \in E_0, v \in \mathbb{R}$. Then

$$g'(y(s))w = 0, \quad \langle \phi(s), u \rangle + t'(s)v = 0.$$

It follows that

$$w = \alpha(s)y'(s) \; : \; u = \alpha(s)x'(s), \quad v = \alpha(s)t'(s).$$

So

$$\langle \phi(s), u \rangle + t'(s)v = \alpha(s)\left(\langle \phi(s), x'(s) \rangle + t'^2(s)\right).$$

Since $y'(s) \neq 0$, then $\langle \phi(s), x'(s) \rangle + t'^2(s) \neq 0$, and therefore $\alpha(s) = 0$.

Let J be the identity operator in $E_1 \times \mathbb{R}$. Then the operator

$$(A_1(s) + \lambda J, A_2(s)) : E_0 \times \mathbb{R} \to (E_1 \times \mathbb{R}) \times E_2$$

is a Fredholm operator of index 0 for $\lambda \geq 0$.

Let $s \in \Delta_i$. We will prove that there exists $\lambda_0 > 0$ such that for $\lambda > \lambda_0$ the operator $(A_1(s) + \lambda J, A_2(s))$ has a uniformly in λ bounded inverse. Indeed, consider the equation

$$(A_1(s) + \lambda J, A_2(s))w = \psi, \quad w \in E_0 \times \mathbb{R}, \; \psi \in (E_1 \times \mathbb{R}) \times E_2.$$

Let $w = (w_1, w_2)$, $\psi = (\psi_1, \psi_2, \psi_3)$, $w_1 \in E_0$, $w_2 \in \mathbb{R}$, $\psi_1 \in E_1$, $\psi_2 \in \mathbb{R}$, $\psi_3 \in E_2$. We have

$$(g'_{1x} + \lambda I)w_1 + g'_{1t}w_2 = \psi_1, \tag{3.37}$$

$$(\phi, w_1) + (t' + \lambda)w_2 = \psi_2, \tag{3.38}$$

$$g'_{2x}w_1 + g'_{2t}w_2 = \psi_3. \tag{3.39}$$

We can find w_1 from (3.37) and (3.39) for $\lambda > \lambda_0$ since $(g'_{1x} + \lambda I, g'_{2x})$ has a uniformly bounded inverse, and substitute in (3.38). Obviously the equation so obtained for w_2 can be solved for $\lambda > \lambda_0$ if λ_0 is sufficiently large. It is clear that the solution w_1, w_2 of (3.37)–(3.39) is unique and can be estimated by a constant independent of λ. So we have proved that $(A_1(s) + \lambda J, A_2(s))$ has an inverse uniformly bounded for $\lambda > \lambda_0$.

Operator $A(s)$ satisfies conditions formulated in the previous subsection. So the orientation $o(A(s))$ of operator $A(s)$ can be constructed, and it does not depend on s. By standard arguments we can prove that the orientation does not depend on the choice of covering of l_k.

Suppose now that for some s the operator $g'_x(y(s)) : E_0 \to E$ is invertible and $t'(s) \neq 0$. We will prove the formula

$$o(A(s)) = o(g'_x(y(s))) \operatorname{sgn} t'(s). \qquad (3.40)$$

Consider the operator $A(s;\tau) = (A_1(s;\tau), A_2(s;\tau))$, $0 \leq \tau \leq 1$,

$$A_1(s;\tau) = \begin{bmatrix} g'_{1x}(y(s)) & \tau g'_{1t}(y(s)) \\ \tau \phi(s) & t'(s) \end{bmatrix}, \quad A_2(s;\tau) = (g'_{2x}(y(s)), \tau g'_{2t}(y(s))).$$

As before we prove that this operator satisfies conditions of the previous subsection and, consequently,

$$o(A(s)) = o(A(s;0)).$$

The equality (3.40) easily follows from the definition of orientation.

Consider now the operator $A(s)$ at the endpoints of the line $l_k : P_0 = (\xi_k, 0)$ and P_1. We begin with the point P_0. The operator $g'_x(\xi_k, 0)$ is invertible. For small t we can take $s = t$. Then $t'(s) = 1$.

There are two possibilities for the point P_1:

$$P_1 = (\xi_l, 0) \quad (l \neq k), \qquad (3.41)$$

and

$$P_1 = (\bar{x}, 1), \qquad (3.42)$$

where $(\bar{x}, 1)$ is a solution of the equation

$$g(\bar{x}, 1) = a. \qquad (3.43)$$

Consider first the case (3.41). We can take $s = 1 - t$ in the neighborhood of the point P_1 (this corresponds to the positive orientation), and so $t'(s) = -1$. From (3.40) it follows that

$$o(g'_x(P_0)) = -o(g'_x(P_1)).$$

In the case (3.42) by the same reasoning we have

$$o(g'_x(P_0)) = o(g'_x(P_1)).$$

The proof of (3.28) follows directly from these equalities. The lemma is proved. □

3. Topological degree

Remark 3.8. According to Remark 3.6, we have replaced condition (iii) by condition (iii*) (Section 3.2). In the proof of Lemma 3.7, it is verified that the operator $(A_1(s) + \lambda J, A_2(s))$ has a uniformly bounded inverse. If we return to condition (iii), then we should verify that the corresponding homogeneous equation has only a zero solution for all λ sufficiently large. In this case, we put $\psi = 0$ in equations (3.37)–(3.39). In order to prove that this system has only a zero solution, we can impose a condition, which is slightly weaker than condition (iii*) (and a similar condition (ii) in the definition of the class Φ in the beginning of this section).

Suppose that there exist Banach spaces E_0' and E' such that $E_0 \subset E_0'$, $E \subset E'$, the operator $(A_1 + \lambda I, A_2)$ can be considered as acting from E_0' into E', and it has a uniformly bounded inverse in these spaces for all λ sufficiently large. This means that we replace condition (iii*) by a similar condition in weaker spaces. Then, as in the proof of Lemma 3.7, we can express $w_1 \in E_0'$ from (3.37), (3.39) and substitute into (3.38). The term $\langle \phi, w_1 \rangle$ has an estimate independent of λ for λ sufficiently large. Therefore, equation (3.38) has only a zero solution if λ is large enough. Thus, equation $(A_1(s) + \lambda J, A_2(s)) = 0$ has only a zero solution in E_0' and, consequently, in E_0. The remaining part of the proof of Lemma 3.7 does not change.

This generalization, though it is not very significant, will be used to construct a topological degree for elliptic operators in Hölder spaces.

Theorem 3.9. *Let $f \in F$ and B be a ball $\|a\| < r, a \in E$ such that $f(x) \neq a$ $(x \in \partial D)$ for all $a \in B$. Then for all regular values $a \in B$, $\gamma(f, D; a)$ does not depend on a.*

Proof. Let a_0 and a_1 be two regular values belonging to B. Set $a_t = a_0(1-t) + a_1 t$, $t \in [0, 1]$ and consider the operator $f(x, t) = f(x) - a_t$. It is easy to see that all conditions of Lemma 3.7 are satisfied for this operator if we set $a = 0$ in this lemma. So the equality (3.25) is valid. From (3.23) we get $\gamma(f, D; a_0) = \gamma(f, D; a_1)$. The theorem is proved. $\square$

Using this theorem we can give the following definition of the topological degree $\gamma(f, D)$.

Definition 3.10. Let $f \in F$ and $f(x) \neq 0$ $(x \in \partial D)$. Let B be a ball $\|a\| < r$ in E such that $f(x) \neq a$ $(x \in \partial D)$ for all $a \in B$. Then

$$\gamma(f, D) = \gamma(f, D; a) \qquad (3.44)$$

for any regular value $a \in B$.

Existence of regular values $a \in B$ of f follows from the Sard-Smale theorem ([496], [432]).

Theorem 3.11. (Homotopy invariance). *Let $f(x, t) \in H$ and (3.20) hold. Suppose that*

$$f(x, t) \neq 0 \quad (x \in \partial D, t \in [0, 1]) \qquad (3.45)$$

for an open set D, $\overline{D} \subset G$. Then
$$\gamma(f_0, D) = \gamma(f_1, D). \tag{3.46}$$

Proof. We take a number $\varepsilon > 0$ so small that
$$f(x, t) \neq a \quad (x \in \partial D, t \in [0, 1])$$
for all a such that $\|a\| < \varepsilon$. Let a be a regular value for both $f_0(x)$ and $f_1(x)$. Consider the function $\tilde{f}(x, t) = f(x, t) - a$. This function satisfies the conditions of Lemma 3.7 if we set $a = 0$ in this lemma. So
$$\gamma(\tilde{f}(., 0), D; 0) = \gamma(\tilde{f}(., 1), D; 0)$$
and therefore
$$\gamma(f_0, D; a) = \gamma(f_1, D; a).$$
This implies (3.46). The theorem is proved. $\square$

Additivity of the topological degree follows from (3.23). We suppose that the class F is not empty. Let $f \in F$, $x \in G$, $f'(x) = (A_1, A_2)$, where $A_1 : E_0 \to E_1$, $A_2 : E_0 \to E_2$. Suppose that $\lambda > 0$ is so large that operator $J = (A_1 + \lambda I, A_2) : E_0 \to E$ is invertible. Then the operator J can be taken as a *normalization* operator. Thus the topological degree for the class F of operators and class H of homotopies is constructed.

3.4 Uniqueness of the degree

The uniqueness of the degree will be proved here for more general classes of operators compared with the existence proved in the previous section.

Let Ψ be a set of linear Fredholm operators $L : E_1 \to E_2$. We suppose that Ψ is connected. This means that for any two operators $L_0 \in \Psi$ and $L_1 \in \Psi$ there exists a homotopy $L_\tau : E_1 \times [0, 1] \to E_2$ which connects them. We suppose that L_τ is continuous with respect to τ in the operator norm. We also assume that the class Ψ is complete with respect to finite-dimensional linear operators, i.e., for any $L \in \Psi$ we have $L + K \in \Psi$, where K is an arbitrary finite-dimensional operator from E_1 into E_2. We consider the following classes of nonlinear operators and homotopies.

Class F' is the set of all proper operators $A(u)$ acting from E_0 into E_2 which are continuous, have Fréchet derivative $A'(u)$ at any point $u \in E_0$ and $A'(u) \in \Psi$. Here E_0 is a given open bounded set in E_1.

Class H' is the set of all proper operators $A_\tau(u) : E_0 \times [0, 1] \to E_2$ which are continuous with respect to u and τ, and belong to F' for any $\tau \in [0, 1]$.

Theorem 3.12. *For the classes F' and H' and a given normalization operator J the topological degree satisfying conditions* (i)–(iii) *is unique.*

3. Topological degree

We begin with some auxiliary results. Denote by Ψ_0 the set of all invertible operators $A \in \Psi$. We use also the following notation. E_1^* is a space dual to E_1, $<u, \phi>$ is the value of the functional $\phi \in E_1^*$ at the element $u \in E_1$.

As usual, the index of an isolated stationary point $u_0 \in E_1$ is defined as the degree of a small ball centered at this point, $\mathrm{ind}(A, u_0) = \gamma(A, B)$, where $B = \{u : \|u - u_0\|_1 < r\}$ for r sufficiently small. Here $\|\cdot\|_1$ is the norm in E_1.

Lemma 3.13. *Let $A \in \Psi_0$, $A - K \in \Psi_0$, where*

$$Ku = \langle u, \phi \rangle e, \quad u \in E_1, \; \phi \in E_1^*, \; e \in E_1.$$

Let $f(u) = A(u - u_0)$, $\tilde{f}(u) = (A - K)(u - u_0)$, where $u_0 \in D \subset E_1$. Then $\langle A^{-1}e, \phi \rangle \neq 1$ and for any topological degree γ,

$$\gamma(\tilde{f}, D) = \gamma(f, D) \quad \text{if} \quad \langle A^{-1}e, \phi \rangle < 1, \tag{3.47}$$

$$\gamma(\tilde{f}, D) = -\gamma(f, D) \quad \text{if} \quad \langle A^{-1}e, \phi \rangle > 1. \tag{3.48}$$

Proof. Consider first the case $\langle A^{-1}e, \phi \rangle = 0$. Let

$$f_\tau(u) = A(u - u_0) - \tau \langle u - u_0, \phi \rangle e, \quad 0 \leq \tau \leq 1.$$

For all $\tau \in [0, 1]$ the equation $f_\tau(u) = 0$ has only one solution $u = u_0$. So $\gamma(f_\tau, D)$ does not depend on τ and consequently $\gamma(f, D) = \gamma(\tilde{f}, D)$. Hence (3.47) is proved.

Suppose now that $\langle A^{-1}e, \phi \rangle \neq 0$. Define $\beta = \langle A^{-1}e, \phi \rangle$, $\psi = \phi/\beta$. Then

$$\langle A^{-1}e, \psi \rangle = 1, \quad \langle u - u_0, \phi \rangle = \beta \langle u - u_0, \psi \rangle.$$

So $\tilde{f}(u) = A(u - u_0) - \beta \langle u - u_0, \psi \rangle e$. Consider the function $g(u) = A(u - u_0) - \hat{\mu}(\langle u - u_0, \psi \rangle)e$, where $\hat{\mu}(\xi)$ is a smooth function of the real variable ξ. It is easy to see that u is a solution of the equation

$$g(u) = 0 \tag{3.49}$$

if and only if $c = \hat{\mu}(<u - u_0, \psi>)$ is a solution of the equation

$$c = \hat{\mu}(c). \tag{3.50}$$

For any solution c of (3.50), $u = u_0 + cA^{-1}e$, is a solution of (3.49). Suppose that $\hat{\mu}(\xi) = \xi^2 + \xi - \alpha$, where α is a real number. Then equation (3.50) has the form $c^2 = \alpha$. If $\alpha > 0$, then equation (3.49) has two solutions $u = u_0 \pm \sqrt{\alpha}A^{-1}e$. Denote by B the ball $\|u - u_0\|_1 < r$, where $r = \sqrt{\alpha}\|A^{-1}e\|_1 + \delta$, $\delta > 0$ is a given number. Then

$$\gamma(g, B) = 0. \tag{3.51}$$

We suppose that α and δ are so small that $B \subset D$. To prove (3.51) consider the function

$$\hat{\mu}_\tau(\xi) = \xi^2 + \xi + (2\tau - 1)\alpha, \quad \tau \in [0, 1].$$

Define
$$g_\tau(u) = A(u - u_0) - \hat{\mu}_\tau(\langle u - u_0, \psi\rangle)e.$$

The equation $g_\tau(u) = 0$ has the solution

$$u = u_0 \pm \sqrt{(1 - 2\tau)\alpha} A^{-1} e \qquad (3.52)$$

for $0 \leq \tau \leq 1/2$, and does not have any solution for $1/2 < \tau \leq 1$. For the solutions (3.52) we have $\|u - u_0\|_1 < r$. Therefore, $\gamma(g_\tau, D)$ does not depend on τ, $\gamma(g, B) = \gamma(g_1, B) = 0$ and (3.51) is proved. Let $u_\pm = u_0 \pm \sqrt{\alpha} A^{-1} e$. From (3.51)

$$\mathrm{ind}(g, u_+) = -\mathrm{ind}(g, u_-). \qquad (3.53)$$

For the derivative $g'(u)$ of g we have

$$g'(u_\pm)u = Au - (1 \pm 2\sqrt{\alpha})\langle u, \psi\rangle e.$$

Define

$$\hat{f}_\sigma(u) = Au - \sigma\langle u, \psi\rangle e.$$

Let T be the translation operator, $T(y)\hat{f}_\sigma(x) = \hat{f}_\sigma(x - y)$. Then we have for the ball $B_+ = \{u : \|u - u_+\|_1 < \rho\}$, where ρ is sufficiently small,

$$\begin{aligned}\mathrm{ind}(g, u_+) &= \gamma(g, B_+) = \gamma(T(u_+)\hat{f}_{2\sqrt{\alpha}+1}, B_+) \\ &= \gamma(T(u_+)\hat{f}_{2\sqrt{\alpha}+1}, D) = \gamma(T(u_0)\hat{f}_{2\sqrt{\alpha}+1}, D) = \gamma(\tilde{f}, D)\end{aligned} \qquad (3.54)$$

if $\beta > 1$. Similarly,

$$\mathrm{ind}(g, u_-) = \gamma(T(u_0)\hat{f}_{-2\sqrt{\alpha}+1}, D) = \gamma(f, D). \qquad (3.55)$$

We have used the fact that $\gamma(T(u_0)\hat{f}_\sigma, D)$ does not depend on σ for all $\sigma > 1$ and for all $\sigma < 1$. From (3.54), (3.55), and (3.53) we obtain (3.48). The equality (3.47) follows from the fact that $\gamma(T(u_0)\hat{f}_\sigma, D)$ does not depend on σ for all $\sigma < 1$. The lemma is proved. $\square$

Lemma 3.14. *Let $A \in \Psi_0$, $A - K \in \Psi_0$, where K is a finite-dimensional operator acting from E_1 into E_2. Define*

$$f(u) = A(u - u_0), \quad \tilde{f}(u) = (A - K)(u - u_0),$$

where $u_0 \in D \subset E_1$. Then

$$\gamma(\tilde{f}, D) = \pm\gamma(f, D), \qquad (3.56)$$

where the sign $+$ or $-$ does not depend on choice of the topological degree γ.

3. Topological degree

Proof. If K is one-dimensional, then the assertion of the lemma follows from the previous lemma. We can use induction with respect to the dimension n of K. If K is n-dimensional, we can represent it in the form $K = K_0 + K_1$, where K_0 is $(n-1)$-dimensional and

$$K_1 u = \langle u, \phi \rangle e, \quad u \in E_1, \ \phi \in E_1^*, \ e \in E_1.$$

Consider the operator $A_\rho = A - \rho K_0 - K_1$, where ρ is a real number. The operator $A - \rho K_0$ is invertible for all $\rho \in (1 - \delta, 1 + \delta)$ if δ is positive and sufficiently small, with the possible exception $\rho = 1$. Indeed, $A - \rho K_0 = \rho A(\frac{1}{\rho} I - A^{-1} K_0)$ and the operator $\frac{1}{\rho} I - A^{-1} K_0$ is not invertible only for a discrete set of ρ.

Let $f_\rho(u) = A_\rho(u - u_0)$, $\rho \in (1 - \delta, 1 + \delta)$. Obviously $f_1 = \tilde{f}$ and if δ is sufficiently small, we have

$$\gamma(f_\rho, D) = \gamma(\tilde{f}, D). \tag{3.57}$$

If we choose ρ such that $A - \rho K_0$ is invertible, then Lemma 3.13 implies

$$\gamma(f_\rho, D) = \epsilon \gamma(g_\rho, D), \tag{3.58}$$

where $g_\rho(u) = (A - \rho K_0)(u - u_0)$ and $\epsilon = \pm 1$ does not depend on the choice of γ. By the induction assumption

$$\gamma(g_\rho, D) = \epsilon_1 \gamma(f, D), \tag{3.59}$$

where $\epsilon_1 = \pm 1$ does not depend on the choice of γ. From (3.57)–(3.59) follows (3.56). The lemma is proved. $\square$

Lemma 3.15. *Let* $A \in \Psi_0$. *Write* $f(u) = A(u - u_0)$, $u_0 \in E_0$. *Then*

$$\mathrm{ind}(f, u_0) = \pm 1, \tag{3.60}$$

where the sign $+$ *or* $-$ *does not depend on the choice of the topological degree* γ *and the point* u_0.

Proof. Since Ψ is a connected set, there exists an operator $A_\tau : E_1 \times [0, 1] \to E_2$ such that $A_\tau \in \Psi$ for any $\tau \in [0, 1]$, A_τ is continuous in τ in the operator norm, and $A_0 = J$, $A_1 = A$, where J is the normalization operator. For any $\tau_0 \in [0, 1]$ there exists a finite-dimensional operator $K_{\tau_0} : E_1 \to E_2$ such that $A_{\tau_0} + K_{\tau_0}$ is invertible. Consequently, there exists a neighborhood $\Delta(\tau_0)$ of the point τ_0 in $[0, 1]$ such that the operator $A_\tau + K_{\tau_0}$ is invertible for $\tau \in \Delta(\tau_0)$. We can cover the interval $[0, 1]$ with such neighborhoods and choose a finite subcovering. Define $f_{\tau_0, \tau} u = (A_\tau + K_{\tau_0})(u - u_0)$. Obviously $\mathrm{ind}(f_{\tau_0, \tau}, u_0)$ does not depend on $\tau \in \Delta(\tau_0)$. If $\Delta(\tau_0)$ and $\Delta(\tau_1)$ have a common point $\bar{\tau}$, then $A_{\bar{\tau}} + K_{\tau_0}$ and $A_{\bar{\tau}} + K_{\tau_1}$ differ only by a finite-dimensional operator. By Lemma 3.13,

$$\mathrm{ind}(f_{\tau_0, \bar{\tau}}, u_0) = \epsilon \ \mathrm{ind}(f_{\tau_1, \bar{\tau}}, u_0),$$

where $\epsilon = \pm 1$ does not depend on the choice of γ. Therefore,
$$\text{ind}(f_{\tau_0,\tau_0}, u_0) = \epsilon \ \text{ind}(f_{\tau_1,\tau_1}, u_0).$$

It follows that
$$\text{ind}(A(u - u_0), u_0) = \epsilon \ \text{ind}(J(u - u_0), u_0) = \epsilon,$$

where $\epsilon = \pm 1$ does not depend on the choice of γ. The fact that (3.60) does not depend on u_0 follows from the construction. The lemma is proved. □

Lemma 3.16. *If $A \in F'$ and u_0 is a regular point of A, $A(u_0) = 0$, then*
$$\text{ind}(A, u_0) = \pm 1,$$

where the sign + or − does not depend on the choice of the topological degree.

The proof is obvious.

Proof of Theorem 3.12. Let $D \subset E$ be an open set, $A \in F'$,
$$A(u) \neq 0, \quad u \in \partial D.$$

Since A is a proper operator, we have
$$A(u) \neq a, \quad u \in \partial D$$

for $\|a\| < \epsilon$, if ϵ is sufficiently small. We can suppose that a is a regular value of A (see [22]). Then the equation
$$A(u) = a$$

has only finitely many solutions $u = u_k$, $k = 1, \ldots, n$ in D. Set $\tilde{A}(u) = A(u) - a$, $A_\tau(u) = A(u) - \tau a$, $\tau \in [0, 1]$. Then $A_\tau(u) \neq 0$, $u \in \partial D$, $\tau \in [0, 1]$. Then
$$\gamma(A, D) = \gamma(\tilde{A}, D) = \sum_{k=1}^{n} \text{ind}(\tilde{A}, u_k)$$

does not depend on the choice of γ by virtue of Lemma 3.16. The theorem is proved. □

3.5 Degree for elliptic operators

A topological degree is constructed above under the conditions presented in the beginning of Section 3.3. The nonlinear operator is supposed to be of the class C^1 and to be proper. This imposes certain conditions on the function spaces. In particular, in the case of unbounded domains they should be weighted spaces (Section 2).

3. Topological degree

The linearized operator $A = (A_1, A_2)$ is such that $A_\lambda = (A_1 + \lambda I, A_2)$ is Fredholm for all $\lambda \geq 0$ and has a uniformly bounded inverse operator for λ sufficiently large. The invertibility for large λ and the estimate of the inverse operator should be verified using the results on an elliptic operator with a parameter (Chapter 8).

The condition that the operator A_λ is Fredholm can be replaced by the condition of normal solvability with a finite-dimensional kernel. Indeed, if this is the case for all $\lambda \geq 0$ and if the operator is invertible for large λ, then it satisfies the Fredholm property and has the zero index for all non-negative λ. Thus, we should require that the operator A_λ satisfies Condition NS for all non-negative λ. Normal solvability and the Fredholm property in weighted spaces is proved in Section 3.4 (Chapter 4) and in Section 7 (Chapter 5).

Consider as an example a nonlinear second-order operator,

$$F(D^2 u, Du, u, x, t), \quad x \in \Omega, \quad G(Du, u, x, t), \quad x \in \partial\Omega$$

depending on parameter $t \in [0, 1]$ and acting from a domain $D \subset E_0$ into E. In the case of Hölder spaces, the conditions on the functions $F(\eta, x, t)$ and $G(\xi, x, t)$ are given in Section 1, and for Sobolev spaces in Section 2.5. Here the vector η (ξ) corresponds to the variables $D^2 u, Du, u$ (Du, u) of the function F (G). We also consider the corresponding linear operators:

$$A_1(v, \eta, t) = -\sum_{i,j=i}^{n} a_{ij}(\eta, x, t) \frac{\partial^2 v}{\partial x_i \partial x_j} + \sum_{i=i}^{n} a_i(\eta, x, t) \frac{\partial v}{\partial x_i} + a(\eta, x, t) v, \quad x \in \Omega,$$

$$A_2(v, \xi, t) = \sum_{i=1}^{n} b_i(\xi, x, t) \frac{\partial v}{\partial x_i} + b(\xi, x, t) v, \quad x \in \partial\Omega,$$

where the coefficients a_{ij}, a_i, a are partial derivatives of the function F with respect to the corresponding variables, and the coefficients b_i, b are partial derivatives of the function G. The operator A is supposed to be uniformly elliptic:

$$\sum_{ij=1}^{n} a_{ij}(\eta, x, t) \tau_i \tau_j \geq a_0 |\tau|^2, \quad |\eta| \leq R, \; x \in \bar{\Omega}, \; t \in [0, 1],$$

where $\tau = (\tau_1, \ldots, \tau_n)$, $a_0 > 0$. The constant R depends on the domain D. It is such that for any $u \in D$, the estimate

$$\sup_{x \in \bar{\Omega}} (|u|, |Du|, |D^2 u|) \leq R$$

holds. In the case of Sobolev spaces, we use here their embedding into C.

In the case of the oblique derivative in the boundary condition, it is supposed that

$$\sum_{i=1}^{n} b_i(\xi, x, t) \nu_i(x) \neq 0, \quad |\eta| \leq R, \; x \in \partial\Omega, \; t \in [0, 1],$$

where $\nu = (\nu_1, \ldots, \nu_n)$ is the normal vector to the boundary. In the case of the Dirichlet boundary condition, $A_2(v, \xi, t) = v$.

Consider first the case of a bounded domain Ω and, for certainty, the Dirichlet boundary condition. As function spaces, we can take

$$E_0 = C^{2+\alpha}(\bar{\Omega}), \quad E_1 = C^\alpha(\bar{\Omega}), \quad E_2 = C^{2+\alpha}(\partial\Omega), \tag{3.61}$$

above $0 < \alpha < 1$, or

$$E_0 = W^{3,p}(\Omega), \quad E_1 = W^{1,p}(\Omega), \quad E_2 = W^{3-1/p,p}(\partial\Omega), \tag{3.62}$$

where $p > n$.

Under the conditions imposed on the functions F and G, the nonlinear operator (F, G) is proper. The linear operator $(A_1 + \lambda I, A_2)$ satisfies the Fredholm property for all real $\lambda \geq 0$. Instead of the condition that the operator $(A_1 + \lambda I, A_2)$ has a uniformly bounded inverse for all $\lambda \geq \lambda_0$ in spaces (3.61), we impose the same condition in the spaces

$$E_0' = W^{2,2}(\Omega), \quad E_1' = L^2(\Omega), \quad E_2' = W^{3/2,2}(\partial\Omega) \tag{3.63}$$

(Remark 3.8). The value of λ_0 and the estimate of the norm of the inverse operator should be independent of η, ξ, t. For some fixed values of the parameters η_0, ξ_0, t_0, ellipticity with a parameter follows from some algebraic conditions. They will be also satisfied in some neighborhood of η_0, ξ_0, t_0. It remains to note that η, ξ, t vary on a compact set.

In the case of spaces (3.62), the invertibility of the operator $(A_1 + \lambda I, A_2)$ can be established directly or in some weaker spaces similar to (3.63).

Suppose now that the domain Ω is unbounded. We consider weighted spaces $E_{0,\mu}$ and E_μ with a polynomial weight. It can be for example $\mu(x) = \sqrt{1 + |x|^2}$. If $u \in C_\mu^{2+\alpha}(\Omega)$, then it converges to zero at infinity. If we need to study a problem with nonhomogeneous limits at infinity, then subtracting the corresponding function we can reduce this case to the homogeneous case. In the case of Sobolev spaces, $E_{0,\mu} = W_q^{l,p}(\Omega)$, where $q = \infty$ or $1 < q < \infty$.

In the case of unbounded domains, normal solvability of the operator $(A_1 + \lambda I, A_2)$ is provided by Condition NS, that is all limiting problems are supposed to have only zero solutions in the corresponding spaces.

When the degree is defined, we may need to find its value in order to study the existence of solutions (Section 4). The most natural way to do it is to use the definition of the degree through orientation. Consider a nonlinear operator $T : E_0 \to E$. Suppose that the equation $T(u) = 0$ has a finite number of isolated solutions u_k, $k = 1, \ldots, m$ in some domain $D \subset E_0$, and that the linearized operators $B'(u_k)$ does not have zero eigenvalues. Then the degree can be determined by the formula

$$\gamma(T, D) = \sum_{k=1}^m (-1)^{\nu_k}, \tag{3.64}$$

where ν_k is the number of positive eigenvalues of the operator $T'(u_k)$.

3. Topological degree 487

We will finish this section with some remarks about the relation of the index of Fredholm operators and the topological degree. The topological degree discussed in this section is constructed for a Fredholm and proper operator with the zero index. It is based on Smale's generalization of Sard's lemma about critical values. It says that if $f : M \to V$ is a C^q Fredholm map with $q > \max(\text{index } f, 0)$, then the regular values of f are almost all of V [496]. Almost all here means "except for a set of first category".

It follows that for Fredholm maps with the zero index, equation $f = a$ has a discrete set of solutions for almost all a, which is crucial for degree construction. If the index is negative, then the image of the map contains no interior points. Hence, even if the degree can be defined, it does not make sense because its value should be zero. Finally, if the index is positive, then for almost all a, $f^{-1}(a)$ has a dimension equal to the index or is empty. Therefore, the manifold of solutions has a positive dimension, the corresponding linearized operators have zero eigenvalues, and the orientation in the sense of Section 3.2 is not defined. We will discuss some approaches to prove existence of solutions in the case of a nonzero index in Section 4.2.3.

3.6 Non-existence of the degree

Similar to Section 2.4, consider the operator $T_\tau(u) = u'' + F(u)$ and suppose that it acts from the space

$$E_0 = \{u \in C^{2+\alpha}(\mathbb{R}_+),\ u'(0) = 0,\ u(\infty) = 0\}$$

into the space $E = C^\alpha(\mathbb{R}_+)$. Here $\mathbb{R}_+$ denotes the half-axis $x \geq 0$, $F(u) = u(u - \tau)(1 - u)$, $0 < \tau < 1/2$. It is shown in Section 2.4 that the equation $T(u) = 0$ has a unique nonzero solution $u_\tau(x)$. It can be constructed explicitly. We will show that the operator linearized about this solution does not have zero eigenvalues.

Lemma 3.17. *The linearized equation*

$$v'' + F'(u_\tau(x))v = 0 \tag{3.65}$$

does not have nonzero solutions in E_0.

Proof. The function $v_0(x) = u'_\tau(x)$ satisfies equation (3.65) but not the boundary condition $v'_0(0) = 0$. Indeed, if $v'_0(0) = 0$, then $u''_\tau(0) = 0$ and, by virtue of the equation, $F(u_\tau(0)) = 0$, which is not true since the value $u_\tau(0)$ satisfies the equality $\int_0^{u_\tau(0)} F(u)du = 0$. Hence, $v_0 \notin E_0$.

Suppose that equation (3.65) has a nonzero solution $v_1 \in E_0$. We note that $v_1(0) > 0$. Indeed, if $v_1(0) = 0$, then $v'_1(0) \neq 0$ (see, e.g., [200], Lemma 3.4), which contradicts the boundary condition.

Consider first the case where $v_1(x) > 0$ for $0 \leq x < \infty$. Put $w_s = sv_1$ with some positive constant s. We can choose s sufficiently large such that $w_s(x_0) >$

$v_0(x_0)$, where x_0 is an arbitrary value for which $F'(u_\tau(x_0)) < 0$. Then $w_s(x) \geq v_0(x)$ for all $x \geq x_0$. Indeed, suppose that this is not the case and $w_s(x_1) < v_0(x_1)$ for some $x_1 > x_0$. Set $z = w_s - v_0$. Then

$$z'' + F'(u_\tau(x))z = 0, \quad z(x_0) > 0, \quad z(\infty) = 0. \tag{3.66}$$

Since $z(z_1) < 0$, then it has a negative minimum at some point $x_2 > x_0$. This gives a contradiction in signs in the equation at $x = x_2$.

Hence, for s sufficiently large, $w_s(x) > v_0(x)$ for all x, $0 \leq x < \infty$. Let us now decrease s till its first value s_0 such that

$$w_{s_0}(x) \geq v_0(x), \quad 0 \leq x < \infty, \quad \exists \, x_2 > 0 \text{ such that } w_{s_0}(x_2) = v_0(x_2).$$

The existence of such x_2 contradicts the positiveness theorem. Indeed, the difference $z_1 = w_{s_0} - v_0$ satisfies the equation in (3.66), it is non-negative and not identically zero. According to the positiveness theorem, it should be strictly positive. This contradiction shows that there are no positive solutions of equation (3.65) in E_0.

Suppose now that there exists a solution of this equation which belongs to E_0 and which has a variable sign. As above, denote it by $v_1(x)$. Since $v_1(0) \neq 0$, then, multiplying this function by -1 if necessary, we can assume that $v_1(0) < 0$. Then for some $x > 0$, $v_1(x)$ is positive. We consider the function $w_s = sv_0$, where s is some positive constant. As above, we can prove that for s sufficiently large, $w_s(x) > v_1(x)$ for all x. Decreasing s, we can choose a value s_0 such that $w_{s_0}(x) \geq v_1(x)$ for all x and that this inequality is not strict. As before, we obtain a contradiction with the positiveness theorem. The lemma is proved. □

Let B_R be a ball in E_0 of the radius R. If R is sufficiently large, then

$$\|u_\tau\|_{C^{2+\alpha}(\mathbb{R})} < R. \tag{3.67}$$

Indeed, it can be easily verified that $\sup_x |u(x)| < 1$ (Figure 11). Then the second derivative is bounded by virtue of the equation. The estimate of the Hölder norm of the second derivative can also be obtained from the equation.

We can now compute the degree by formula (3.64). If $0 < \tau < 1/2$, then there are two solutions of equation (3.65) in B_R, $u = u_\tau$ and $u = 0$. For both of them, the corresponding linearized operator does not have zero eigenvalues. Therefore, the right-hand side in (3.64) is well defined. The value of the degree $\gamma(A, B_R)$ is an even number.

We recall that topological degree is homotopy invariant. This means that if the operator depends continuously on the parameter and $A(u) \neq 0$ at the boundary ∂B_R, that is (3.67) holds, then the degree does not change.

On the other hand, for $\tau = 1/2$, solution $u_\tau(x)$ with zero limits at infinity does not exist. The corresponding homoclinic trajectory tends to two heteroclinic trajectories as $\tau \to 1/2$ (Figure 11). In the limit, the homoclinic trajectory does not exist. Hence, there is only one solution of equation (3.65) in D for $\tau = 1/2$, $u = 0$. The value of the degree becomes odd. Homotopy invariance of the degree is not preserved.

This example shows that the degree with the required properties may not exist. This is related to the absence of properness of the corresponding operators. The degree is defined in the appropriate weighted spaces where the operators are proper.

4 Existence and bifurcations of solutions

4.1 Methods of nonlinear analysis

In this section we recall some results of nonlinear analysis which will be used to study nonlinear elliptic problems. They are closely related to the properties of linear operators (Fredholm property, spectrum, solvability conditions, invertibility) and to some properties of nonlinear operators discussed in this chapter (properness, topological degree). Detailed presentation of these results can be found elsewhere.

4.1.1. Fixed point theorems. Let E be a complete metric space and $D \subset E$ be a closed set. Suppose that there is a map A from D into itself. A point $u_0 \in D$ is a fixed point of the map A if $A(u_0) = u_0$. Therefore, it is a solution of the equation

$$A(u) = u. \tag{4.1}$$

The mapping is contracting if there is a number $\alpha < 1$ such that

$$\rho(A(u_1), A(u_2)) \leq \alpha \rho(u_1, u_2)$$

for any $u_1, u_2 \in D$. Here ρ denotes the distance between two points.

Theorem 4.1 (Banach [47]). *If an operator A is contracting, then there exists a unique solution u_0 of equation (4.1) in D.*

We suppose now that E is a Banach space and formulate another classical fixed point theorem. Various generalizations are known (see, e.g., [276]).

Theorem 4.2. (Schauder [463]). *Let a continuous operator A map a convex closed set D of a Banach space E into a compact subset of D. Then A has at least one fixed point in D, that is a solution $u \in D$ of the equation $A(u) = u$.*

4.1.2. Implicit function theorem. The presentation of this section and of Section 4.1.3 follows the book by Kantorovich, Akilov [254]. Let E_1, E_2 and F be Banach spaces. Consider an operator $A(u,v) : E_1 \times E_2 \to F$, $u \in E_1, v \in E_2$ and the operator equation

$$A(u,v) = 0. \tag{4.2}$$

Theorem 4.3. *Suppose that the operator A is given in a neighborhood D of a point $(u_0, v_0) \in E_1 \times E_2$, it maps it into the space F, and it is continuous at (u_0, v_0). If, moreover, the following conditions are satisfied:*

1. *(u_0, v_0) satisfies equation (4.2).*
2. *The partial derivative A'_v exists in D, and it is continuous at (u_0, v_0).*
3. *$A'_v(u_0, v_0)$ is a bounded operator from E_2 into F, and it has a bounded inverse.*

Then there exists an operator Φ given in some neighborhood $G \subset E_1$ of the point u_0 such that it maps this neighborhood into the space E_2 and satisfies the following properties:

1. *$(u, \Phi(u))$ satisfies equation (4.2) in G.*
2. *$\Phi(u_0) = v_0$.*
3. *Φ is continuous at u_0.*

The operator Φ is uniquely determined by these properties.

Under the conditions of the theorem, if A is continuous everywhere in D, then the operator Φ is continuous in some neighborhood of the point u_0. If we assume, moreover, that the partial derivative A'_u exists in D and is continuous at (u_0, v_0), then the operator Φ is differentiable at u_0 and

$$\Phi'(u_0) = -(A'_v(u_0, v_0))^{-1} A'_u(u_0, v_0).$$

4.1.3. Approximate methods. In this section we briefly recall some approximate methods of solutions of operator equations (see, e.g., [254], [275], [276]).

1. *Projection methods.* Consider Banach spaces E and F and the equation

$$Lu = f, \tag{4.3}$$

where L is a linear operator (bounded or unbounded) with the domain $\text{Dom}(L) \subset E$ and the image $R(L) \subset F$. Two sequences of subspaces E_n and F_n are considered such that

$$E_n \subset \text{Dom}(L) \subset E, \quad F_n \subset F, \quad n = 1, 2, \ldots.$$

Let $P_n : F \to F$ be projectors: $P_n^2 = P_n$, $P_n F = F_n$, $n = 1, 2, \ldots$. Instead of equation (4.3) we consider the equation

$$P_n(Lu_n - f) = 0 \tag{4.4}$$

where its solution u_n is sought in E_n. If $E = F$ and $E_n = F_n$ this projection method is called the Galerkin method.

In the case of Hilbert spaces, consider two complete sequences $\phi_j \in \text{Dom}(L) \subset E$ and $\psi_j \in F$, $j = 1, 2, \ldots$ and look for an approximate solution in the form $u_n = \sum_{j=1}^n c_j \phi_j$. We assume that $Lu_n - f \in F/F_n$, that is

$$\sum_{j=1}^n (L\phi_j, \psi_i) c_j = (f, \psi_i), \quad i = 1, \ldots, n.$$

4. Existence and bifurcations of solutions

These conditions allow us to find the coefficients c_j and the approximate solution u_n.

Consider next the Galerkin method in the case of nonlinear operators. Let A be a completely continuous operator acting in a Banach space E, D be a bounded domain, $D \subset E$. We look for solutions of the equation

$$u = A(u), \quad u \in D. \tag{4.5}$$

Along with the operator A, consider a sequence of completely continuous operators A_n acting in E and the approximate equations

$$u = A_n(u), \quad n = 1, 2, \ldots. \tag{4.6}$$

The following theorem provides convergence of the approximations [276].

Theorem 4.4. *Suppose that equation (4.5) does not have solutions at the boundary ∂D of the domain Ω and the degree $\gamma(I - A, D)$ is different from 0. If*

$$\lim_{n \to \infty} \sup_{u \in \bar{D}} \|A_n(u) - A(u)\| = 0,$$

then equations (4.6) have solutions in D for all n sufficiently large and these solutions converge to some solutions of equation (4.5).

The proof of this theorem is based on properties of the topological degree. It can be also proved for more general operators for which the degree is defined [492].

2. *Newton's method.* Consider the operator $A : E \to F$ and the equation

$$A(u) = 0. \tag{4.7}$$

Newton's method consists in the approximation of its solution by the iterations

$$u_{n+1} = u_n - (A'(u_n))^{-1} A(u_n), \quad n = 0, 1, \ldots.$$

It is assumed here that the linearized operator is invertible. In the modified Newton's method, the inverse operator in the right-hand side is taken at the initial approximation $u = u_0$:

$$u_{n+1} = u_n - (A'(u_0))^{-1} A(u_n), \quad n = 0, 1, \ldots.$$

The convergence of these methods is given by the following theorem [254]. We will use the notations:

$$B_R(u_0) = \{u \in F, \|u - u_0\| < R\}, \quad B_r(u_0) = \{u \in F, \|u - u_0\| \leq r\}, \quad r < R.$$

Theorem 4.5. *Suppose that the operator $A(u)$ is defined in $B_R(u_0)$ and has a continuous second derivative in $B_r(u_0)$. Assume that:*

1. There exists a continuous inverse operator $(A'(u_0))^{-1}$.
2. $\|(A'(u_0))^{-1} A(u_0)\| \leq \eta$.
3. $\|(A'(u_0))^{-1} A''(u)\| \leq K$, $u \in B_r(u_0)$.

If
$$h = K\eta \leq \frac{1}{2}, \quad r \geq r_0 = \frac{1 - \sqrt{1-2h}}{h}\, \eta,$$
then equation (4.7) has a solution u^*; Newton's method and modified Newton's method converge to it. Moreover, $\|u^* - u_0\| \leq r_0$.

If
$$h < \frac{1}{2} \; \left(h = \frac{1}{2}\right), \quad r < r_1 = \frac{1 + \sqrt{1-2h}}{h}\, \eta \quad (r \leq r_1),$$
then the solution u^* is unique in $B_r(u_0)$.

4.1.4. Leray-Schauder method. Consider a bounded and continuous nonlinear operator $A : E \to F$. We are interested in the existence of solutions of the operator equation (4.7) in some domain $D \subset E$. Let us assume that there exists an operator $A_\tau(u)$ which depends continuously on the parameter $\tau \in [0,1]$ and which satisfies the following conditions:

1. Topological degree $\gamma(A_\tau, D)$ is defined.
2. There are no solutions at the boundary of the domain:
$$A_\tau(u) \neq 0, \quad u \in \partial D, \ \tau \in [0,1]. \tag{4.8}$$

3. The homotopy A_τ connects the operator $A = A_1$ with an operator A_0 for which the degree is different from zero, $\gamma(A_0, D) \neq 0$.

Then there exists a solution u of equation (4.7) in D. Indeed, from the properties of the degree and by virtue of (4.8), it follows that $\gamma(A, D) = \gamma(A_0, D) \neq 0$. Since the degree is different from zero, then (4.7) has a solution in D (principle of nonzero rotation).

Suppose that solutions of the equation $A_\tau(u) = 0$ admit *a priori estimates*, that is for any such solution $\|u\|_E < R$, where a positive constant R does not depend on τ and on solution. If we put $D = B_{R_1}$ where B_{R_1} is a ball with radius $R_1 > R$, then condition (4.8) will be satisfied.

Thus, if we can construct a continuous deformation of the operator to a model operator with nonzero degree and obtain a priori estimates of solutions, then the existence of solutions will be proved. This is the Leray-Schauder method [303] widely used for elliptic problems and for other operator equations. This is how it was formulated by Leray and Schauder:

Théorème. Soit l'équation
$$x - \mathcal{F}(x, k) = 0. \tag{1}$$
Faisons les trois séries d'hypothèses:

(H_1) L'inconnue x et toutes les valeurs de $\mathcal{F}$ appartiennent à un espace linéaire, normé et complet, $\mathcal{E}$.
L'ensemble de valeur du paramètre k constitue en segment K de l'axe des nombres réels.
$\mathcal{F}(x,k)$ est définie pour tous les couples (x,k) où x est un élément quelconque de $\mathcal{E}$, k est un élément quelconque de K.
En chaque point k de K, $\mathcal{F}(x,k)$ est *complétement continue*[1]; ceci signifie que $\mathcal{F}(x,k)$ transforme tout ensemble borné de points x de $\mathcal{E}$ en un ensemble compact.
Sur tout sous-ensemble de $\mathcal{E}$ borné, $\mathcal{F}(x,k)$ est *uniformément continue par rapport à k*.

(H_2) En un point particulier k_* de K toutes les solutions sont connues et l'on peut étudier leurs indice par l'intermédiaire du Chapitre II; nous supposons *la somme de ces indices non nulle*.

(H_3) Enfin nous supposons démontré par un procédé quelconque que les solutions de (1) sont bornées dans leurs ensemble. (*Limitation a priori* indépendante de k.)

Conclusion. Alors il existe sûrement dans l'espace $[\mathcal{E} \times K]$ un continu de solutions le long duquel k prend toutes les valeurs de K.

4.2 Existence of solutions

Existence of solutions of nonlinear problems is one of the main topics in the theory of elliptic equations. We can refer to the monographs by Miranda [352], Ladyzhenskaya and Uraltseva [285], Gilbarg and Trudinger [200], Temam [514], Skrypnik [493], Krylov [278], Caffarelli and Cabre [99], Gasinski and Papageorgiou [192], Ambrosetti and Malchiodi [18]. In this section we illustrate application of the methods of nonlinear analysis to study existence of solutions of elliptic problems. The proof of the existence of solutions is often based on a priori estimates. This method was first suggested in the works by Bernstein (see Historical notes) and was further developed by Leray and Schauder. Nowadays, this is one of the most widely used methods to prove existence of solutions of operator equations.

4.2.1. Second-order equations

Semi-linear equations. Consider the linear second-order operator

$$Lu = -\sum_{i,j=i}^{n} a_{ij}(x)\frac{\partial^2 u}{\partial x_i \partial x_j} + \sum_{i=i}^{n} a_i(x)\frac{\partial u}{\partial x_i} + a(x)u$$

in a bounded domain $\Omega \subset \mathbb{R}^n$ and the semi-linear operator

$$T(u) = Lu + F(Du, u, x)$$

---
[1] Here and below italic by the authors.

with the Dirichlet boundary condition $u|_{\partial\Omega} = 0$. Here Du denotes the first derivatives of u. We suppose that the operators L and T act from $C^{2+\alpha}(\bar{\Omega})$ into $C^\alpha(\bar{\Omega})$. The coefficients of the operator L belong to $C^\alpha(\bar{\Omega})$, the function $F(\eta_1, \eta_2, x)$ (η_1 here is a vector) has second derivatives with respect to η_1, η_2 which satisfy the Hölder condition in x uniformly in $\eta = (\eta_1, \eta_2)$ and the Lipschitz condition in η uniformly in x. Then the topological degree can be defined. In order to obtain a priori estimates of solutions we assume that the operator L is invertible and

$$|F(\eta_1, \eta_2, x)| \leq M$$

for all real η_1, η_2, and $x \in \bar{\Omega}$ and for some positive constant M. This condition is restrictive but we will use it for the sake of simplicity. Other examples will be considered below.

Suppose that the problem

$$T(u) = 0, \quad u|_{\partial\Omega} = 0 \tag{4.9}$$

has a solution in $C^{2+\alpha}(\bar{\Omega})$. Denote $f(x) = F(Du, u, x)$. Then $f \in L^p(\Omega)$ for any p and from the equation $Lu = -f$ we obtain the estimate

$$\|u\|_{W^{2,p}(\Omega)} \leq \|L^{-1}\|_p \|f\|_{L^p(\Omega)} \leq M \|L^{-1}\|_p |\Omega|^{1/p}. \tag{4.10}$$

Here $\|L^{-1}\|_p$ is the norm of the operator inverse to the operator $L : W^{2,p}(\Omega) \to L^p(\Omega)$, $|\Omega|$ is the measure of the domain. For p sufficiently large, from the last estimate and from embedding theorems, we obtain an estimate of the norm $\|u\|_{C^{1+\alpha}(\bar{\Omega})}$ and, consequently, of the norm $\|f\|_{C^\alpha(\bar{\Omega})}$. Since the operator L is invertible from $C^{2+\alpha}(\bar{\Omega})$ into $C^\alpha(\bar{\Omega})$, we obtain an estimate of the norm $\|u\|_{C^{2+\alpha}(\bar{\Omega})}$.

Consider now the operator depending on a parameter,

$$T_\tau(u) = Lu + \tau F(Du, u, x).$$

A priori estimates of solutions of the operator equation $T_\tau(u) = 0$ can be obtained as before. If $\tau = 0$, then $T(u) = Lu$. From the invertibility of the operator it follows that the equation $Lu = 0$ has a unique solution $u_0 = 0$ and $\gamma(T_0, u_0) = \pm 1$. Here $\gamma(T_0, u_0)$ is the index of the stationary point u_0, that is the topological degree taken with respect to a small neighborhood of u_0 in the function space.

Thus, all conditions of the Leray-Schauder method are satisfied. This proves the existence of solutions of problem (4.9). This approach can be applied for other semi-linear equations and systems.

Let us now discuss the case where the domain Ω is unbounded. The approach described above is not applicable because of estimate (4.10). We will use here the method of approximation of unbounded domains by a sequence of bounded domains. For linear problems, it was discussed in Chapter 8. Let Ω_k be a sequence of bounded domains with uniformly $C^{2+\alpha}$ boundaries, converging to the domain Ω.

4. Existence and bifurcations of solutions

This is a local convergence defined in Section 1 of Chapter 4. For each of the domains Ω_k we can use the existence result obtained above. Consider the operators

$$L^k u = -\sum_{i,j=i}^{n} a_{ij}^k(x) \frac{\partial^2 u}{\partial x_i \partial x_j} + \sum_{i=i}^{n} a_i^k(x) \frac{\partial u}{\partial x_i} + a^k(x) u,$$

$$T^k(u) = L^k u + F^k(Du, u, x)$$

and denote by u_k a solution of the problem

$$T^k(u) = 0, \quad u|_{\partial \Omega_k} = 0. \tag{4.11}$$

It exists if the operators L^k are invertible. If we obtain uniform estimates of the functions u_k in an appropriate norm, then we can choose a convergent subsequence and obtain a solution of problem (4.9).

Let us extend the coefficients of the operator L to the whole $\mathbb{R}^n$ in such a way that their norms are preserved. As the coefficients of the operator L^k, we can take their restrictions to the domain Ω_k. We define the functions F^k in a similar way and suppose that

$$\sup_{\eta, x} |F^k(\eta_1, \eta_2, x)| \leq M \quad \forall k. \tag{4.12}$$

Consider the equations $L^k u = f_k$, where $f_k(x) = F^k(Du_k, u_k, x)$. From Theorem 1.3 (Section 1.3, Chapter 8) we obtain the estimate

$$\|u_k\|_{W^{2,p}_\infty(\Omega_k)} \leq c_1 \left(\|f_k\|_{L^p_\infty(\Omega)} + \|u_k\|_{L^p_\infty(\Omega)} \right), \tag{4.13}$$

where the constant c_1 is independent of u_k and Ω_k. Suppose that there exists a constant m such that

$$\sup_x |u_k| \leq m \quad \forall k. \tag{4.14}$$

This assumption together with (4.12) provides an estimate of the norm $\|u_k\|_{W^{2,p}_\infty(\Omega_k)}$ independently of k. From this estimate and embedding theorems we obtain an estimate of the norm $\|u\|_{C^{1+\alpha}(\bar{\Omega})}$ and, consequently, of the norm $\|f\|_{C^\alpha(\bar{\Omega})}$. Then we can use the estimate

$$\|u_k\|_{C^{2+\alpha}(\bar{\Omega}_k)} \leq c_2 \left(\|f_k\|_{\alpha(\bar{\Omega}_k)} + \|u_k\|_{C^0(\bar{\Omega})} \right), \tag{4.15}$$

where the constant c_2 is independent of k. From the sequence u_k, we can choose a subsequence locally convergent in C^2 to a limiting function u. It belongs to $C^{2+\alpha}(\bar{\Omega})$ and satisfies (4.9). We will consider some other examples in Section 4.3.

Quasi-linear equations. Consider the quasi-linear equation

$$\sum_{i,j=1}^{n} a_{ij}(Du, u, x) \frac{\partial^2 u}{\partial x_i \partial x_j} = F(Du, u, x) \tag{4.16}$$

in a bounded domain $\Omega \subset \mathbb{R}^n$ with the Dirichlet boundary conditions. Here Du is the vector of the first partial derivatives; the coefficients of the operator and the boundary of the domain are sufficiently smooth.

Let $v \in C^{2+\alpha}(\bar\Omega)$. Following [303], consider the linear equation

$$\sum_{i,j=1}^n a_{ij}(Dv, v, x) \frac{\partial^2 u}{\partial x_i \partial x_j} = F(Dv, v, x)$$

with respect to u. Its solvability is known. Therefore, we obtain an operator A which puts in correspondence a solution u of this equation to each function v. A fixed point of this operator, that is a solution of the equation $A(u) = u$, gives a solution of equation (4.16). The existence of solutions of this equation can be studied by fixed point theorems or with the use of the topological degree. The main difficulty here is to obtain a priori estimates of solutions. They were obtained by Schauder [470] who generalized the results of Bernstein previously obtained in the case of analytical solutions. The authors prove the existence theorem in the 2D case which is formulated, in the original version, as follows.

Toute équation

$$a\left(x_1, x_2; z; \frac{\partial z}{\partial x_1}, \frac{\partial z}{\partial x_1}\right) \frac{\partial^2 z}{\partial x_1^2} + 2b\left(x_1, x_2; z; \frac{\partial z}{\partial x_1}, \frac{\partial z}{\partial x_1}\right) \frac{\partial^2 z}{\partial x_1 \partial x_2}$$
$$+ c\left(x_1, x_2; z; \frac{\partial z}{\partial x_1}, \frac{\partial z}{\partial x_1}\right) \frac{\partial^2 z}{\partial x_2^2} = 0$$

du type elliptique admet au moins une solution qui soit définie dans un domaine convexe donné, Δ, et qui prenne des valeurs données sur sa frontière Δ'. (Rappelons que nous avons dû faire des hypothèses concernant la régularité de la courbe Δ', des valeurs frontières et des fonctions a, b, c.)

This result is generalized in [285] in the first theorem for uniformly, and in the second theorem for non-uniformly, elliptic equations.

Theorem 4.6 (Ladyzhenskaya, Uraltseva [285]). *Let all solutions $u(x,\tau) \in C^{2+\alpha}(\bar\Omega)$ of the problems*

$$\sum_{i,j=1}^2 a_{ij}(Du, u, x) \frac{\partial^2 u}{\partial x_i \partial x_j} + \tau a(Du, u, x) = 0, \quad u|_{\partial\Omega} = 0, \quad \tau \in [0,1] \qquad (4.17)$$

in a bounded domain $\Omega \subset \mathbb{R}^2$ satisfy the estimate $\max_\Omega |u(x,\tau)| \le M$, and $\partial\Omega \in C^{2+\alpha}$. Suppose that the functions $a_{ij}(p, u, x)$ and $a(p, u, x)$ belong to C^α for $x \in \bar\Omega$, $|u| \le M$, $|p| \le M_1$ and satisfy the following conditions:

$$\nu(|u|)|\xi|^2 \le a_{ij}(p, u, x)\xi_i \xi_j \le \mu(|u|)|\xi|^2,$$
$$|a(p, u, x)| \le \delta(M)|p|^2 + \mu_1(M)(1 + |p|^{2-\delta_1}).$$

4. Existence and bifurcations of solutions

Then problem (4.17) with $\tau = 1$ has at least one solution in $C^{2+\alpha}(\bar{\Omega})$. Here $\nu(t)$ and $\mu(t)$ are positive continuous functions defined for $t \geq 0$, ν non-increasing, μ non-decreasing; $\delta_1 > 0$, $M\delta(M)$ is sufficiently small, $(2+\sqrt{2})cM\delta(M) < 1$, $c = 2\mu(M)\nu^{-2}(M)$.

Theorem 4.7 (Ladyzhenskaya, Uraltseva [285]). *The problem*

$$\sum_{i,j=1}^{2} a_{ij}(Du, u, x)\frac{\partial^2 u}{\partial x_i \partial x_j} = 0, \quad u|_{\partial\Omega} = \phi|_{\partial\Omega} \tag{4.18}$$

has at least one solution $u \in C^{2+\alpha}(\bar{\Omega})$ if Ω is a strictly convex domain, $\phi \in C^{2+\alpha}(\bar{\Omega})$, $\partial\Omega \in C^{2+\alpha}$,

$$a_{ij}(p, u, x)\xi_i\xi_j > 0,$$

and $a_{ij}(p, u, x)$ satisfies the Hölder condition with $\alpha > 0$ with respect to (p, u, x) for $x \in \bar{\Omega}$, $|u| \leq M$, $|p| \leq M_1$, where $M = \max_{\partial\Omega}|\phi|$, M_1 some constant determined only by ϕ and $\partial\Omega$.

The existence results for quasi-linear equations in the case $n \geq 2$, for other boundary value problems and for quasi-linear systems can be found in [285], [200].

Fully non-linear equations. Following [200], consider the equation

$$F[u] \equiv F(D^2u, Du, u, x) = 0, \tag{4.19}$$

where $F(r, p, z, x)$ is a real-valued function, Du denotes the vector of the first partial derivatives of the function u, and D^2 of the second derivatives, $r \in \mathbb{R}^{n \times n}, p \in \mathbb{R}^n, z \in \mathbb{R}, x \in \Omega$. Here $\mathbb{R}^{n \times n}$ denotes real symmetric square matrices of the order n. The set of all such (r, p, z, x) is denoted by Γ.

If F is linear with respect to the variables r then the operator is quasi-linear, otherwise it is fully nonlinear. The operator F is called elliptic if the matrix with the elements $F_{ij} = \frac{\partial F}{\partial r_{ij}}$ is positive definite. Suppose that the following conditions are satisfied:

$$0 < \lambda|\xi|^2 \leq F_{ij}\xi_i\xi_j \leq \Lambda|\xi|^2, \tag{4.20}$$

$$|F_p|, |F_z|, |F_{rx}|, |F_{px}|, |F_{zx}| \leq \mu\lambda, \tag{4.21}$$

$$|F_x|, |F_{xx}| \leq \mu\lambda(1 + |p| + |r|) \tag{4.22}$$

for all nonzero $\xi \in \mathbb{R}^n$, $(r, p, z, x) \in \Gamma$, λ is a non-increasing and Λ, μ are non-decreasing functions of $|z|$. We recall that the condition of *external sphere* implies that for any point $x \in \partial\Omega$ there exists a ball B such that $\bar{B} \cap \bar{\Omega} = \{x\}$.

Theorem 4.8 (Gilbarg, Trudinger [200]). *Let $\Omega \subset \mathbb{R}^n$ be a bounded domain satisfying the condition of an external sphere at each boundary point. Suppose that the function $F \in C^2(\Gamma)$ is concave (or convex) with respect to all variables z, p, r, it is non-increasing in z and satisfies conditions (4.20)–(4.22). Then the classical*

Dirichlet problem $F[u] = 0$ in Ω, $u = \phi$ on $\partial\Omega$ is uniquely solvable in $C^2(\Omega) \cap C^0(\bar{\Omega})$ for any $\phi \in C^0(\partial\Omega)$.

Solvability of nonlinear problems has been discussed by Leray and Schauder [303], and later in [352] and [493]. The method of viscosity solutions is used in [99]. Other references can be found in the cited monographs.

4.2.2. Navier-Stokes equations. There is an extensive literature devoted to existence of solutions of the Navier-Stokes equations. In this section we follow the presentation in [284]. Consider the Navier-Stokes equations

$$-\nu \Delta v_i + \sum_{k=1}^{3} v_k \frac{\partial v_i}{\partial x_k} = -\frac{\partial p}{\partial x_i} + f_i(x), \quad i = 1, 2, 3 \tag{4.23}$$

for the incompressible fluid,

$$\operatorname{div} v \equiv \sum_{k=1}^{3} \frac{\partial v_k}{\partial x_k} = 0, \tag{4.24}$$

in a domain $\Omega \subset \mathbb{R}^3$, which can be bounded or unbounded. Here $v = (v_1, v_2, v_3)$ is the velocity, p is the pressure, ν the viscosity. We consider the homogeneous Dirichlet boundary condition

$$v|_{\partial\Omega} = 0. \tag{4.25}$$

If the domain is unbounded, the same condition is imposed at infinity

$$v|_{x=\infty} = 0. \tag{4.26}$$

Denote by $H(\Omega)$ the Hilbert space of functions v obtained as a closure of the set $J(\Omega)$ of all solenoidal (that is div $v = 0$) smooth vector-functions with support in Ω in the norm corresponding to the scalar product

$$[u, v] = \int_{\Omega} \sum_{k=1}^{3} \frac{\partial u}{\partial x_k} \frac{\partial u}{\partial x_k} dx.$$

We define a generalized solution of problem (4.23)–(4.26) as a function $v \in H(\Omega)$ which satisfies the equality

$$\int_{\Omega} \left(\nu \frac{\partial v}{\partial x_k} \frac{\partial \Phi}{\partial x_k} + v_k \frac{\partial v}{\partial x_k} \Phi \right) dx = \int_{\Omega} f \Phi dx \tag{4.27}$$

for any $\Phi \in J(\Omega)$. Here the repeated subscript k implies summation, the product of two vectors signifies the inner product in $\mathbb{R}^3$.

Theorem 4.9 ([284]). *Problem (4.23)–(4.25) in a bounded domain Ω has at least one generalized solution for any f for which the integral $\int_{\Omega} f \Phi dx$ determines a linear functional in $H(\Omega)$.*

4. Existence and bifurcations of solutions 499

The proof of this theorem is based on the Leray-Schauder method. The problem is represented as an equation with respect to v:

$$v - \frac{1}{\nu}(Av + F) = 0,$$

where A is a nonlinear compact operator acting in $H(\Omega)$. A priori estimates of solutions allow one to prove the existence of solutions of this equation.

If ν is sufficiently large, which corresponds to small Reynolds numbers, then the solution is unique. The existence can also be proved in the case of unbounded domains. In this case, we consider a sequence of bounded domains converging to the unbounded domain Ω and the corresponding sequence of solutions. Its weak limit is a generalized solution of problem (4.23)–(4.26). The existence theorem remains valid in the case of inhomogeneous Dirichlet boundary condition $v|_{\partial\Omega} = a|_{\partial\Omega}$ under some conditions on the function a and on the smoothness of the domain. We note that the regularity of the solution is determined by the regularity of the data. In particular, if f in (4.23)–(4.25) satisfies the Hölder condition, then the solution also satisfies it together with its second derivatives.

4.2.3. Operators with nonzero index. The topological degree constructed in this chapter is applicable for Fredholm and proper operators with zero index. Degree can also be constructed in the case of positive index. However, its application to elliptic problems in this case is rather complex and only a few examples have been studied [383], [448]. In this section we discuss another approach to study existence of solutions of some classes of elliptic problems with nonzero index.

Positive index. Let $A(u) = Lu - B(u)$, where $L : E \to F$ is a linear Fredholm operator and $B(u) : E \to F$ is a nonlinear continuous compact operator. Suppose that the dimension of the kernel E_0 of the operator L equals n, and its image coincides with the space F. Let E_1 be a complementary subspace, $E = E_0 \oplus E_1$. The restriction $\hat{L}$ of the operator L to E_1 has a bounded inverse. Put $u = v + w$, where $v \in E_0$ and $w \in E_1$. The equation $A(u) = 0$ can be written as $w = \hat{L}^{-1}B(u)$. We can now use a fixed point theorem or the Leray-Schauder degree to find its solution.

Let us illustrate the application of this approach to the two-dimensional problem with oblique derivative:

$$\Delta u + F(u, x, y) = 0, \quad (x, y) \in \Omega, \qquad (4.28)$$

$$a(x, y)\frac{\partial u}{\partial x} - b(x, y)\frac{\partial u}{\partial y} = 0, \quad (x, y) \in \partial\Omega. \qquad (4.29)$$

Here Ω is a bounded domain, $a^2 + b^2 = 1$. The coefficients, the nonlinearity and the boundary of the domain satisfy conditions of Sections 1 or 2.5. Consider also the linear equation

$$\Delta u = f(x) \qquad (4.30)$$

and the operator $L : W^{2,2}(\Omega) \to L^2(\Omega) \times W^{1/2,2}(\partial\Omega)$ corresponding to problem (4.29), (4.30). The index κ of this operator equals $2N + 2$, where N is the number of rotations of the vector (a, b) counterclockwise (Chapter 8). If $\kappa > 0$, then the dimension of the kernel E_0 of the operator L equals κ, the image of the operator is the whole space F. The operator $\hat{L} : E_1 \to F$ has a bounded inverse.

Denote by T the operator which puts in correspondence a solution $w \in E_1$ of problem (4.29), (4.30) with $f(x) = -F(z, x, y)$ to a function z, $T(z) = -\hat{L}^{-1} F(z, \cdot)$. A fixed point of the mapping $w = T(z)$ gives a solution of problem (4.28), (4.29). An appropriate choice of function spaces and of conditions on F allows us to use the Schauder fixed point theorem. Let, for example, F be a bounded function,

$$|F(u, x, y)| \leq M, \quad \forall u \in \mathbb{R}, \ (x, y) \in \bar{\Omega}.$$

Then $\|w\|_{W^{2,2}(\Omega)} \leq KM|\Omega|$, where K is the norm of the inverse operator $\hat{L}^{-1}$, $|\Omega|$ is the measure of the domain Ω. From the embedding of the space $C^\alpha(\bar{\Omega})$ into $W^{2,2}(\Omega)$ for some $\alpha, 0 < \alpha < 1$ we conclude that $\|w\|_{C^\alpha(\bar{\Omega})} \leq K_1$, where the constant K_1 is independent of the function z. Consider the ball $B_r = \{\|z\|_{C^\beta(\bar{\Omega})} \leq r\}$ in the Hölder space $C^\beta(\bar{\Omega})$ with some positive $\beta < \alpha$. If $r > K_1$, then $T(z)$ maps the ball B_r in its compact part. Hence there exists a fixed point of this mapping.

We note that each solution of problem (4.28), (4.29) originates a family of solutions with the dimension equal to the index (cf. [496]). The approach discussed above is applicable for other problems and in the case of less restrictive conditions on the function F.

Negative index. In the case of a negative index, the operator equation may not have solutions since the image of the linear part of the operator is not the whole space. Let us consider the operator depending on the vector parameter τ, $A_\tau(u) = Lu - B(u, \tau)$. We should choose the value of the parameter in such a way that the equation $A_\tau(u) = 0$ has a solution.

We suppose that the index κ of the operator L is negative and that its kernel is empty. The linear equation $Lu = f$ is solvable if and only if $(\phi_i, f) = 0$ for some functionals $\phi_i \in F^*$, $i = 1, \ldots, m$, where $m = |\kappa|$. Consider the equation

$$Lu = B(z, \tau) \tag{4.31}$$

for a given function z. In order to satisfy the solvability conditions

$$(\phi_i, B(z, \tau)) = 0, \quad i = 1, \ldots, m,$$

we introduce some functionals $\tau(z) = (\tau_1(z), \ldots, \tau_m(z))$ instead of the given real-valued constant vector $\tau = (\tau_1, \ldots, \tau_m)$. Its dimension is supposed to be the same as the number of solvability conditions. This is the so-called functionalization of the parameters (see, e.g., [276]). Let us assume, for the sake of simplicity, that the dependence on the parameters is linear:

$$B(z, \tau) = B_0(z) + \tau_1(z) B_1(z) + \cdots + \tau_m(z) B_m(z).$$

4. Existence and bifurcations of solutions

From the solvability conditions we obtain the linear algebraic system of equations with respect to τ_i:

$$a_{i1}\tau_1 + \cdots + a_{im}\tau_m = b_i, \quad i = 1, \ldots, m,$$

where

$$a_{ij} = (\phi_i, B_j(z)), \quad b_i = -(\phi_i, B_0(z)), \quad i, j = 1, \ldots, m.$$

It has a solution if the determinant of the matrix $(a_{ij})_{i,j=1}^m$ is different from zero. In this case we can determine the functionals $\tau_i(z)$ in such a way that the solvability conditions are satisfied.

Consider a domain $D \subset E$. If

$$\det((\phi_i, B_j(z)))_{i,j=1}^m \neq 0, \quad \forall z \in D,$$

then we can introduce such functionals $\tau_i(z)$ that equation (4.31) is solvable for any $z \in D$. Since the operator L is an isomorphism between the space E and its image, then we can consider the mapping $u = L^{-1}B(z, \tau(z))$ defined on the domain D. Similar to the case of positive index, a fixed point of this mapping is a solution of the equation $A_\tau(u) = 0$, where $\tau = \tau(u)$.

4.3 Elliptic problems in unbounded domains

Most of the results of this book, a priori estimates, Fredholm property, index, operators with a parameter for linear elliptic problems, properness and topological degree for nonlinear problems are applicable both for bounded and unbounded domains. In the case of unbounded domains, there can be some additional conditions on the operators. In particular, the Fredholm property requires the invertibility of limiting operators while properness and topological degree demand the introduction of weighted spaces.

Existence of solutions of nonlinear elliptic problems in unbounded domains can be studied either by approximation of the unbounded domain by a sequence of bounded domains or directly in the unbounded domain. There is a vast literature devoted to the existence of positive solutions (see, e.g., [319] and the references therein), to various properties of solutions such as monotonicity and symmetry (see [58]–[61]), bifurcations, symmetry breaking, decay and growth (see the bibliographical comments), to some specific solutions including travelling waves (see [568] and the references therein).

4.3.1. Approximation by bounded domains. We have already discussed this approach in Section 4.2.1 and will briefly describe here the general framework. Consider a sequence of bounded domains Ω_k locally convergent to an unbounded domain Ω in the sense of Section 1.1 of Chapter 8. For each domain Ω_k, we define function spaces $E(\Omega_k)$ and $F(\Omega_k)$ and the operators $A_k : E(\Omega_k) \to F(\Omega_k)$. The operator A acts from $E(\Omega)$ into $F(\Omega)$. The coefficients of the operators A_k and A

are extended to the whole $\mathbb{R}^n$, both for the operators in domains and for boundary operators. We assume that the coefficients of the operators A_k locally converge to the coefficients of the operator A in $C^\alpha(\mathbb{R}^n)$ or in another norm determined by the operators and function spaces.

Suppose that each equation $A_k(u) = 0$ has a solution u_k and the uniform estimate

$$\|u_k\|_{E_\infty(\Omega_k)} \leq M, \quad k = 1, 2, \ldots \tag{4.32}$$

holds. Then from the properties of the operators and function spaces it may be possible to verify that a subsequence of the sequence u_k is locally convergent in some weaker norm to a solution u of the equation $A(u) = 0$. For example, if $E = C^{s+\alpha}$, then there exists a subsequence of the sequence u_k locally convergent in C^s to some limiting function $u \in C^{s+\alpha}(\bar{\Omega})$ which satisfies the equation $A(u) = 0$.

The key point here is to prove estimates (4.32). If these are a priori estimates, then it can be possible to prove existence of solutions of the equations $A_k(u) = 0$ by the Leray-Schauder method. In this case, we may not assume the existence of solutions in bounded domains.

A priori estimates of solutions of linear elliptic problems independent of the domains (Chapter 8) can be used to obtain estimates (4.32). In particular, in the case where $A_k(u) = L_k u + B_k(u)$, L_k are linear operators and B_k are nonlinear operators such that $\|B_k(u)\|_{F_\infty(\Omega_k)} \leq K$ for all $u \in E_\infty(\Omega_k)$ and with a constant K independent of k (cf. Section 4.2.1). If the operators $L_k : E(\Omega_k) \to F(\Omega_k)$ are invertible with the norms of the inverse operators bounded independently of k, then we have a uniform estimate of u_k:

$$\|u_k\|_{E_\infty(\Omega_k)} \leq \|L^{-1}\| \, \|B_k(u)\|_{F_\infty(\Omega_k)}.$$

If the operators L_k do not have uniformly bounded inverse operators but the functions u_k are uniformly bounded in a weaker norm, for example if

$$\sup_{k, x \in \bar{\Omega}_k} |u_k(x)| \leq m,$$

then we can use the estimate

$$\|u_k\|_{E_\infty(\Omega_k)} \leq c \left(\|B_k(u)\|_{F_\infty(\Omega_k)} + \|u_k\|_{E^0_\infty(\Omega_k)} \right),$$

where E^0_∞ can be L^p_∞ or C^0 depending on the spaces E and F.

Laplace operator with nonlinear terms. Consider the problem

$$-\Delta u + F(Du, u, x) = 0, \quad u|_{\partial\Omega} = 0, \tag{4.33}$$

which is a particular case of the semi-linear problem discussed in Section 4.2.1. It has been studied by Picard and Bernstein, Schauder [463] used the Brower fixed point theorem and Nemytiskii [374] proved the existence of solutions for domains with a sufficiently small measure. The review of old works can be found

4. Existence and bifurcations of solutions

in Lichtenshtein [315], Miranda [352]. More recent results and literature reviews are presented in the monographs [450], [192].

We will not assume here, as in Section 4.2.1, that the function F is bounded. The principle of contracting mapping, fixed point theorems or the topological degree can be used to study the existence of solutions. We will present one of the possible approaches. Put

$$F(Du, u, x) = cu + F_0(Du, u, x).$$

We will suppose that c is a positive constant, $F_0(\eta_1, \eta_2, x)$ is a sufficiently smooth function of its arguments (see Section 4.2.1), and

$$|F_0(\eta_1, \eta_2, x)| \leq a|\eta_1| + b|\eta_2| + d(x), \tag{4.34}$$

$d(x)$ is a non-negative function from $L^2(\Omega)$.

<u>Bounded domains.</u> We begin with the case of a bounded domain Ω with a $C^{2+\alpha}$ boundary. Suppose that $u \in C^{2+\alpha}(\bar{\Omega})$. Multiplying equation (4.33) by u, integrating over Ω, and taking into account (4.34), we obtain

$$\int_\Omega |\nabla|^2 dx + c\int_\Omega u^2 dx \leq a\int_\Omega |\nabla u||u|dx + b\int_\Omega u^2 dx + \int_\Omega |u|d(x)dx$$

$$\leq \frac{a}{2}\int_\Omega |\nabla|^2 dx + \left(\frac{a}{2}+b\right)\int_\Omega u^2 dx + \left(\int_\Omega u^2 dx\right)^{1/2}\left(\int_\Omega d^2 dx\right)^{1/2}.$$

Hence

$$\left(1 - \frac{a}{2}\right)\int_\Omega |\nabla|^2 dx + \left(c - \frac{a}{2} - b\right)\int_\Omega u^2 dx \leq \left(\int_\Omega u^2 dx\right)^{1/2}\left(\int_\Omega d^2 dx\right)^{1/2}.$$

If $a < 2$ and $b + 1 < c$, then this inequality allows us to estimate, first, the norm $\|u\|_{L^2(\Omega)}$ and then $\|u\|_{H^1(\Omega)}$. Hence, using (4.34), we can estimate the norm $\|F_0(Du, u, x)\|_{L^2(\Omega)}$. Put $f(x) = -F_0(Du, u, x)$. Then from a priori estimates of solutions of the linear equation

$$-\Delta u + cu = f, \tag{4.35}$$

where $f \in L^2(\Omega)$, we obtain an estimate of the norm $\|u\|_{H^2(\Omega)}$. From the equalities

$$\frac{\partial f}{\partial x_i} = -\sum_{j=1}^n \frac{\partial F_0}{\partial \eta_{1j}}\frac{\partial^2 u}{\partial x_i \partial x_j} - \frac{\partial F_0}{\partial \eta_2}\frac{\partial u}{\partial x_i} - \frac{\partial F_0}{\partial x_i}$$

we can now estimate the norm $\|f\|_{H^1(\Omega)}$ assuming that the first partial derivatives of the function F_0 are bounded. Then from equation (4.35) we obtain an estimate of the norm $\|u\|_{H^3(\Omega)}$. If the dimension of the space $n \leq 3$, then from the embedding theorems follows an estimate of the norm $\|u\|_{C^{1+\alpha}(\bar{\Omega})}$ for some $\alpha \in (0, 1)$ and,

504　　　　　　　　　　　　　　　　　Chapter 11. Nonlinear Fredholm Operators

consequently, of the norm $\|f\|_{C^\alpha(\bar\Omega)}$. From equation (4.35), we estimate the norm $\|u\|_{C^{2+\alpha}(\bar\Omega)}$. If $n = 4$, we should continue with similar estimates,

$$\|u\|_{H^3(\Omega)} \to \|f\|_{H^2(\Omega)} \to \|u\|_{H^4(\Omega)} \to \|u\|_{C^{1+\alpha}(\bar\Omega)} \to \|f\|_{C^\alpha(\bar\Omega)} \to \|u\|_{C^{2+\alpha}(\bar\Omega)},$$

which require additional conditions on the derivatives of the function F_0.

Thus, we finally obtain an a priori estimate of the norm $\|u\|_{C^{2+\alpha}(\bar\Omega)}$. We can now apply the topological degree and the Leray-Schauder method. In order to do this, consider the operator

$$A_\tau(u) = -\Delta u + cu + \tau F_0(Du, u, x)$$

acting from the space E_0 into E,

$$E_0 = \{u \in C^{2+\alpha}(\bar\Omega),\ u|_{\partial\Omega} = 0\}, \quad E = C^\alpha(\bar\Omega).$$

A priori estimates of solutions for $\tau \in [0,1]$ can be obtained in a similar way. For $\tau = 0$, the equation $A_\tau(u) = 0$ has a unique solution $u = 0$. The index of this stationary point equals 1, since the corresponding linearized operator has only positive eigenvalues. Hence, $\gamma(A_0, B) = 1$, where B is a ball in E_0 of a sufficiently big radius determined by a priori estimates. Therefore, $\gamma(A_1, B) = 1$, and there exists at least one solution of problem (4.33) in the ball B.

<u>Unbounded domains.</u> Suppose that the domain Ω is unbounded with the uniformly $C^{2+\alpha}$ boundary satisfying Condition D (Chapter 4). Let ϕ_i, $i = 1, 2, \ldots$ be a partition of unity which satisfies the conditions in the definition of ∞-spaces (Chapter 2) and

$$0 \leq \phi_i \leq 1, \quad \sup_{x,i,j}\left|\frac{\partial \phi_i}{\partial x_j}\right| \leq m$$

for some positive constant m. Suppose that the norm $\|d\|_{L^2_\infty(\Omega)} = \sup_i \|d\phi_i\|_{L^2(\Omega)}$ is bounded. Let us multiply equation (4.33) by $u\phi_i^2$ and integrate over Ω:

$$\int_\Omega |\nabla u|^2 \phi_i^2 dx + c\int_\Omega u^2 \phi_i^2 dx + \int_\Omega u\nabla u \cdot \nabla(\phi_i^2) dx = -\int_\Omega F_0(Du, u, x)\phi_i^2 dx. \quad (4.36)$$

We have, further,

$$\int_\Omega |u\nabla u \cdot \nabla(\phi_i^2)| dx \leq Km\int_\Omega |\nabla u||u|\phi_i dx \leq Km\int_\Omega |\nabla u||u|\phi_i \sum_{j=1}^N \phi_j dx,$$

where we denote by K all constants which depend only on the dimension n of the space, the sum in the right-hand side is taken with respect to all functions ϕ_j from the partition of unity such that $\sum_{j=1}^N \phi_j(x) = 1$ at the support of the function ϕ_i. This sum does not include the functions equal to zero at the support

4. Existence and bifurcations of solutions

of ϕ_i. According to the assumptions on the partition of unity, the number of such functions is bounded by a constant N independent of i. Then

$$\int_\Omega |u \nabla u \cdot \nabla(\phi_i^2)| dx \leq KmN\epsilon \int_\Omega |\nabla u|^2 \phi_i^2 dx + Km \frac{1}{\epsilon} \sum_{j=1}^N \int_\Omega |u|^2 \phi_j^2 dx$$

$$\leq KmN \left(\epsilon \sup_i \int_\Omega |\nabla u|^2 \phi_i^2 dx + \frac{1}{\epsilon} \sup_i \int_\Omega |u|^2 \phi_i^2 dx \right).$$

Here ϵ is some positive constant which will be specified below. From (4.36) and (4.34),

$$\int_\Omega |\nabla u|^2 \phi_i^2 dx + c \int_\Omega u^2 \phi_i^2 dx - KmN \left(\epsilon \sup_i \int_\Omega |\nabla u|^2 \phi_i^2 dx + \frac{1}{\epsilon} \sup_i \int_\Omega |u|^2 \phi_i^2 dx \right)$$

$$\leq a \int_\Omega |\nabla u||u|\phi_i^2 dx + b \int_\Omega u^2 \phi_i^2 dx + \int_\Omega |u|d(x)\phi_i^2 dx.$$

Hence

$$\left(1 - \frac{a}{2}\right) \int_\Omega |\nabla u|^2 \phi_i^2 dx + \left(c - \frac{a}{2} - b\right) \int_\Omega u^2 \phi_i^2 dx$$

$$\leq \|u\|_{L^2_\infty(\Omega)} \|d\|_{L^2_\infty(\Omega)} + KmN \left(\epsilon \sup_i \int_\Omega |\nabla u|^2 \phi_i^2 dx + \frac{1}{\epsilon} \|u\|^2_{L^2_\infty(\Omega)} \right) \equiv T.$$

Suppose that $a < 2$ and $c > b + 1$. Since the last inequality holds for any i, then

$$\left(c - \frac{a}{2} - b\right) \|u\|^2_{L^2_\infty(\Omega)} \leq T, \quad \left(1 - \frac{a}{2}\right) \sup_i \int_\Omega |\nabla u|^2 \phi_i^2 dx \leq T.$$

Therefore

$$\left(\epsilon \sup_i \int_\Omega |\nabla u|^2 \phi_i^2 dx + \frac{1}{\epsilon} \|u\|^2_{L^2_\infty(\Omega)} \right) \leq RT,$$

where

$$R = \frac{\epsilon}{1 - a/2} + \frac{1}{\epsilon(c - a/2 - b)}.$$

From the last estimate and the definition of T it follows that

$$(1 - KmNR) \left(\epsilon \sup_i \int_\Omega |\nabla u|^2 \phi_i^2 dx + \frac{1}{\epsilon} \|u\|^2_{L^2_\infty(\Omega)} \right) \leq R \|u\|_{L^2_\infty(\Omega)} \|d\|_{L^2_\infty(\Omega)}.$$

If $KmNR < 1$, then we obtain an estimate of the norm $\|u\|_{L^2_\infty(\Omega)}$ and then of the norm $\|u\|_{H^1_\infty(\Omega)}$. Next, we use the estimate

$$\|u\|_{H^3_\infty(\Omega)} \leq k \left(\|f\|_{H^1_\infty(\Omega)} + \|u\|_{L^2_\infty(\Omega)} \right)$$

of the solution of equation (4.35) (Theorem 2.2, Chapter 3). We proceed as in the case of bounded domains and obtain an estimate of the norm $\|u\|_{C^{2+\alpha}(\bar{\Omega})}$. Similar estimates can be obtained for other boundary value problems for scalar and vector operators.

Contrary to the case of bounded domains, for unbounded domains such estimates are not sufficient to conclude the existence of solutions. We can use the method of approximation by a sequence of bounded domains. In order to do this, we should consider a sequence of bounded domains Ω_k locally convergent to the domain Ω. If the domains satisfy Condition D uniformly and if there exists a solution u_k for the problem in each domain, then it can be possible to obtain an estimate of the norm $\|u_k\|_{C^{2+\alpha}(\bar{\Omega}_k)}$ independent of k. This is the case for the example under consideration. Then we can chose a locally convergent in C^2 subsequence from the sequence u_k. The limiting function $u \in C^{2+\alpha}(\bar{\Omega})$ is a solution of the problem in the domain Ω.

Another approach is to use the topological degree and the Leray-Schauder method. It requires the introduction of weighted spaces (Section 3). In this case, we need to obtain an a priori estimate of solutions in the norm $\|u\|_{C^{2+\alpha}_\mu(\bar{\Omega})}$ with a polynomially growing at infinity weight function μ. We will return to this question in Section 4.3.3.

4.3.2. Solutions in the sense of sequence. Consider a bounded and continuous nonlinear operator A acting from a Banach space E into another Banach space F. Similar to the case of linear non-Fredholm operators, we can introduce here the notion of solutions in the sense of sequence.

Definition 4.10. *A sequence of functions $u_k \in E$ is called a solution in the sense of sequence if $\|A(u_k)\|_F \to 0$ as $k \to \infty$. This solution is uniformly bounded if $\|u_k\| \leq M$ for some positive constant M independent of k.*

If the operator A is proper on closed bounded sets, then from the existence of a uniformly bounded solution in the sense of sequence, it follows that there exists a solution in the usual sense, that is a function $u \in E$ such that $A(u) = 0$. Indeed, in this case the sequence u_k is compact since it is an inverse image of the compact set composed by the functions $f_k = A(u_k)$. Therefore, there exists a subsequence u_{n_k} convergent to some $u_0 \in E$, $A(u_0) = 0$.

If the operator A is not proper, then, generally speaking, the existence of a solution in the sense of sequence does not necessarily imply the existence of the usual solution (see the example in Section 2.4).

In the case of elliptic operators in unbounded domains, the operators are proper in weighted spaces and may not satisfy this property in the spaces without weight (Section 2). Specific features of the operators and spaces allow us to affirm the existence of usual solutions even in the case where the operator is not proper.

Theorem 4.11. *Suppose that nonlinear elliptic operator $A(u) : E(\Omega) \to F(\Omega)$ acts in Hölder spaces (see Section 1 for the definition of spaces and operators). If there exists a uniformly bounded solution $u_k \in E(\Omega)$ of the equation $A(u) = 0$ in the sense of sequence, then there exists a solution $u_0 \in E(\Omega)$ in the usual sense.*

4. Existence and bifurcations of solutions 507

To prove the theorem, it is sufficient to note that if a sequence of functions u_k is bounded in a Hölder norm $C^{s+\alpha}(\bar\Omega)$, then there exists a subsequence locally convergent in C^s to some limiting function $u_0 \in C^{s+\alpha}(\bar\Omega)$ which satisfies the equation. Similar results can be obtained for Sobolev spaces.

We should emphasize that in the case of proper operators, the convergence u_{n_k} is strong, that is in the norm of the space $E(\Omega)$. If the operator is not proper, then this convergence is local and in a weaker norm. This difference can be essential for the behavior of solutions at infinity. For example, all functions u_k can converge to zero at infinity while the limiting function may not converge to zero.

Solutions in the sense of sequence can be naturally introduced in the case of a sequence of bounded domains locally convergent to an unbounded domain (see the previous section) and for small perturbations of the operator. If the equations $A(u) + \tau_k B(u) = 0$, where $\tau_k \to 0$ as $k \to 0$, have solutions for all k such that their norms are uniformly bounded, and the operator B is bounded, then u_k is a uniformly bounded solution of the equation $A(u) = 0$ in the sense of sequence.

The elements u_k of the solution in the sense of sequence can be considered as approximate solutions. This does not necessarily mean that they approximate the exact solution but that the values of the operator on such functions are close to zero.

4.3.3. Weighted spaces. In order to use properness and the topological degree constructed in Section 3, we need to introduce weighted spaces. As we discussed above, elliptic operators in unbounded domains are not generally proper in spaces without weight, and the topological degree may not exist. The weight function $\mu(x)$ is polynomially growing at infinity. The fact that growth is polynomial allows us to preserve solutions exponentially decaying at infinity, which is a specific behavior in the case of Fredholm operators.

Elliptic problems in unbounded domains may possess families of solutions uniformly bounded in Sobolev or Hölder spaces but not being compact. One of the simplest examples is provided by the problem

$$u'' + F(u) = 0, \quad u(\pm\infty) = 0$$

in $\mathbb{R}$. Under certain conditions on the function F (Section 2.4) it has a bounded nontrivial solution $u_0(x)$ exponentially decaying at infinity. We can take for example $F(u) = u(u - 1/2)$. The solution belongs to the conventional spaces, $H^2(\mathbb{R})$ and $C^{2+\alpha}(\mathbb{R})$. All shifted functions $u_h(x) = u_0(x + h)$, $h \in \mathbb{R}$ are also solutions of this problem. The norm in these spaces does not depend on h, that is all these solutions are uniformly bounded. On the other hand, this family of solutions is not compact. As a consequence, the operator is not proper.

Let h_k be a sequence such that $h_k \to \infty$ as $k \to \infty$. The functions u_{h_k} form a solution in the sense of sequence defined in the previous section. If we choose a locally convergent subsequence of this sequence, we obtain in the limit the trivial solution $u \equiv 0$. Its existence is obvious and is not interesting for applications.

This example explains the origin of the difficulties in the study of elliptic problems in unbounded domains. The situation is different if we consider weighted spaces. Let $C_\mu^{2+\alpha}(\mathbb{R})$ be the space of functions from $C^{2+\alpha}(\mathbb{R})$ for which the norm $\|u\|_{C_\mu^{2+\alpha}(\mathbb{R})} = \|u\mu\|_{C^{2+\alpha}(\mathbb{R})}$ is bounded. Here $\mu(x) = \sqrt{1+x^2}$. In the example above, the family of solutions $u_h(x)$ becomes unbounded in the norm of the space $C_\mu^{2+\alpha}(\mathbb{R})$. Therefore, for each bounded domain D in the function space, $u_h \in D$ only for a finite interval of h. This provides properness in the weighted spaces and the possibility to define topological degree.

The advantages of the weighted spaces should be payed off by the necessity to obtain a priori estimates of solutions in a stronger norm. We will consider two possible situations. In the next section we will discuss problems with solutions invariant with respect to translation in space. In this section, we will briefly consider non-autonomous problems without space invariance. We will restrict ourselves to a particular class of operators, which gives a simple illustration of a general situation.

Consider a nonlinear elliptic operator $A(u) : E(\Omega) \to F(\Omega)$ in an unbounded domain $\Omega \subset \mathbb{R}^n$. Suppose that there are a priori estimates of solutions in the $E_\infty(\Omega)$ norm. We can impose additional conditions in order to provide a priori estimates in the weighted space $E_{\mu,\infty}(\Omega)$. Let us illustrate it with the following example:

$$A(u) = u'' + F(u,x),$$

where F is a sufficiently smooth function of both variables such that

$$F(u,x) \begin{cases} < 0, & u > 1, \quad x \in \mathbb{R}, \\ > 0, & u < -1, \quad x \in \mathbb{R}. \end{cases}$$

From the maximum principle it follows that $|u(x)| \leq 1$, $-\infty < x < \infty$ for any bounded solution $u(x)$ such that $u(\pm\infty) = 0$. Suppose that the function F is linear for x sufficiently large:

$$F(u,x) = -a^2 u, \quad |x| \geq N$$

with some positive constants a and N. Since the solution is exponential for $|x| \geq N$, then

$$|u(x)| \begin{cases} \leq e^{-a(|x|-N)}, & |x| \geq N, \\ \leq 1, & |x| < N. \end{cases}$$

An estimate of the norm $\|u\|_{C_\mu^{2+\alpha}(\mathbb{R})}$ follows from this inequality.

This example characterizes the general situation. If the operator is linear or close to linear for x in an exterior domain, then the behavior of solutions there is exponential. Together with the estimate in the $E_\infty(\Omega)$, it can provide the estimate in the $E_{\mu,\infty}(\Omega)$ norm.

4.3.4. Travelling waves

Definitions. A travelling wave solution of the parabolic equation

$$\frac{\partial u}{\partial t} = \frac{\partial^2 u}{\partial x^2} + F(u)$$

considered for $-\infty < x < \infty$, is a solution of the form $u(x,t) = w(x - ct)$. Here c is a constant called the wave speed. It is unknown together with the function $w(x)$ which satisfies the equation

$$w'' + cw' + F(w) = 0. \qquad (4.37)$$

We look for solutions of this equation with some given limits at infinity,

$$w(\pm\infty) = w_\pm, \quad w_+ \neq w_-, \qquad (4.38)$$

where $F(w_\pm) = 0$. Similar definitions can be given for systems of equations and in the multi-dimensional case.

There is an extensive literature devoted to travelling wave solutions of parabolic systems (see, e.g., [568] and the references therein). Our purpose here is to discuss some properties of problem (4.37), (4.38) in relation with elliptic problems in unbounded domains.

We recall the classification of nonlinearities. If $F'(w_\pm) < 0$, then it is called the bistable case, if $F'(w_\pm) > 0$ then it the unstable case, if one of the derivatives $F'(w_+)$ and $F'(w_-)$ is positive and another one is negative, then it is the monostable case. This classification is related to the stability of the stationary points $w_\pm$ with respect to the equation $dw/dt = F(w)$. It is also possible that some of these derivatives equal zero. For the sake of simplicity, we suppose here that $F'(w_\pm) \neq 0$.

This classification does not include the case where $w_+ = w_-$ and $F'(w_+) < 0$ where a nonconstant solution can exist for $c = 0$.

Fredholm property of linearized operators. Let $(w_0(x), c_0)$ be a solution of problem (4.37), (4.38). Consider the operator linearized about this solution

$$Lu = u'' + cu' + b(x)u,$$

where $b(x) = F'(w_0(x))$, acting from $E = C^{2+\alpha}(\mathbb{R})$ into $E' = C^\alpha(\mathbb{R})$. Its essential spectrum consists of two parabolas on the complex plane:

$$\lambda(\xi) = -\xi^2 + c_0 i\xi + F'(w_\pm).$$

In the bistable case, both of them lie in the left half-plane. The operator satisfies the Fredholm property, and its index equals zero.

In the unstable case, since $F'(w_\pm) > 0$, both of them are partially in the right half-plane. If $c_0 = 0$, then the essential spectrum passes through the origin and the operator does not satisfy the Fredholm property. If $c \neq 0$, then this property is satisfied and the index of the operator is also zero. Though we do not discuss here the existence of solutions, it should be noted that solutions of problem (4.37), (4.38) in the unstable case do not exist. So the considerations about the Fredhom property and index are justified for a function $b(x)$ with positive limits at infinity but they do not have sense for the operator linearized about the wave.

Consider, finally, the monostable case. Let us suppose for certainty that $F'(w_+) > 0$ and $F'(w_-) < 0$. It is known that c_0 is positive in this case. The operator satisfies the Fredholm property and its index equals 1.

It can be easily verified that the operator L has a zero eigenvalue with the eigenfunction $w_0'(x)$. This is related to the invariance of solutions with respect to translation in space.

Persistence of solutions. Suppose that the nonlinearity F depends continuously on the parameter τ and consider the problem

$$w'' + cw' + F_\tau(w) = 0, \quad w(\pm\infty) = w_\pm. \tag{4.39}$$

We suppose that it has a solution $(w_0(x), c_0)$ for $\tau = 0$. Is it possible to affirm that the solution persists under small variation of the parameter? A conventional tool to study this question is the implicit function theorem. However, it cannot be directly applied for the problem under consideration because the linearized operator has a zero eigenvalue and is not invertible. We will show how it can be applied after certain rearrangement of the problem. We will use the solvability conditions formulated in term of formally adjoint operators.

Let us begin with the bistable case. Suppose that the wave $w_0(x)$ exists and that it is monotone with respect to x. In fact, the monotonicity of the solution can be proved. Its existence holds under certain conditions on the function F_τ. It can be for example $F_\tau(w) = w(w - 1/2 - \tau)(1 - w)$ with $w_+ = 0, w_- = 1$.

The principal eigenvalue of the operators L and of the homogeneous, formally adjoint, operator L^*,

$$L^* v \equiv v'' - c_0 v' + b(x)v$$

is zero. It is simple and the corresponding eigenfunctions $v_0(x) = w_0'(x)$ and $v_0^*(x)$ are positive up to a constant factor [568]. Equation

$$Lu \equiv u'' + c_0 u' + b(x)u = f \tag{4.40}$$

is solvable if and only if

$$\int_{-\infty}^{\infty} f(x) v_0^*(x) dx = 0.$$

Consider, next, the linearization of the operator A with respect to both variables, w and c:

$$M(u, c) = u'' + c_0 u' + b(x)u + cw_0'.$$

4. Existence and bifurcations of solutions

The operator M acts from $E \times \mathbb{R}$ into E'. We write the equation

$$M(u, c) = f, \qquad (4.41)$$

as

$$u'' + c_0 u' + b(x)u = f - cw_0'$$

and apply the same solvability condition as above. This equation is solvable if and only if

$$\int_{-\infty}^{\infty} f(x) v_0^*(x) dx = c \int_{-\infty}^{\infty} w_0'(x) v_0^*(x) dx.$$

The integral in the right-hand side is different from zero since both functions under the integral are positive up to a constant factor. Therefore, we can choose a value of c such that the solvability condition is satisfied. Hence, equation (4.41) is solvable for any right-hand side. In fact, it has a one-dimensional family of solutions $u(x) + tv_0(x)$ and, consequently, the operator M is not invertible. Let us consider the subspace

$$E_0 = \{u \in E, \ \int_{-\infty}^{\infty} u(x) v_0^*(x) dx = 0\}.$$

Then the solution of equation (4.41) becomes unique on this subspace, and the operator $M : E_0 \times \mathbb{R} \to E'$ is invertible.

Let us now return to the nonlinear operator $A(u, c)$ which corresponds to the left-hand side of (4.39):

$$A_\tau(u, c) = (u + w_0)'' + c(u + w_0)' + F_\tau(u + w_0).$$

We consider it as acting from $E_0 \times \mathbb{R}$ into E'. The operator equation

$$A_\tau(u, c) = 0 \qquad (4.42)$$

has a solution $u = 0, c = c_0$ for $\tau = 0$. The operator satisfies the conditions of the implicit function theorem. Therefore, there exists a solution (u_τ, c_τ) of equation (4.42) for all τ sufficiently small. If the dependence of the nonlinearity on the parameter is differentiable, then this is also true for the dependence of the solution on the parameter.

Thus, if problem (4.39) has a solution for some value of τ, then it persists for close values of the parameter. The value of the speed also depends on the parameter. The persistence of travelling waves for the scalar parabolic equation can be directly verified by means of the existence results. It remains true for systems of equations and in the multi-dimensional case where the existence may not be proved by other methods. We recall that we use the bistability which determines the Fredholm property and the zero index of the operator, and the simplicity of the zero eigenvalue from which it follows that the inner product $(v_0, v_0^*)_{L^2(\mathbb{R})}$ of the zero eigenfunctions for the direct and adjoint operators is different from zero. The latter ensures the solvability of equation (4.41) under the variation of the parameter. We use the zero index of the operator in order to conclude that the codimension of the image, which equals the dimension of the kernel of the formally adjoint operator, equals 1. Therefore, there exists a unique linearly independent solvability condition.

The solvability conditions are formulated here in terms of orthogonality to solutions of a homogeneous formally adjoint equation. In Chapter 6 they are obtained for Sobolev spaces. They remain the same for Hölder spaces.

Let us now discuss the monostable case. The operator L satisfies the Fredholm property, its index equals 1, the dimension of its kernel is greater than or equal to 1 because w_0' is the eigenfunction corresponding to the zero eigenvalue. It can be verified that the kernel of the operator L^* is empty. Indeed, the solutions of the equation $L^*v = 0$ behave exponentially at $+\infty$,

$$v(x) \sim ke^{\mu x}, \quad \mu = \frac{c_0}{2} \pm \sqrt{\left(\frac{c_0}{2}\right)^2 - F'(w_+)}.$$

Since $c_0 > 0$ and $F'(w_+) > 0$, then μ is positive (or have a positive real part) and there are no bounded solutions. Thus, the dimension of the kernel of the operator L equals 1, the codimension of the image equals 0, and equation (4.40) is solvable for any right-hand side for the fixed value of $c = c_0$. This is the difference with respect to the bistable case where we need to vary c in order to provide solvability.

We note that the subspace E_0 here should be defined in a different way since the eigenfunction corresponding to the zero eigenvalue of the formally adjoint operator does not exist. We can consider the subspace of functions orthogonal in the sense of L^2 to the eigenfunction v_0.

We have discussed here the simplest examples of the persistence of solutions under small variations of the problem. More complex examples can involve reaction-diffusion systems with non-Fredholm operators [137], [138] and reaction-diffusion problems with convection [52].

Topological degree. A topological degree for travelling waves was introduced in [561] (see also [562], [568], [567]). Some particular features of travelling waves are related to the invariance of solutions with respect to translation in space. This signifies that the solutions are not isolated and the usual degree constructions are not directly applicable. Moreover, this results in the existence of a zero eigenvalue of the linearized problem. One of the possibilities to deal with such families of solutions is to introduce a functionalization of the parameter c. This means that instead of an unknown constant c we consider some given functional $c(w)$. The translation of the solution $w(x)$ changes the value of the functional. Hence, if $(w(x), c(w(\cdot)))$ is a solution of problem (4.39), then $(w(x+h), c(w(\cdot +h)))$ does not satisfy it. Thus, we can eliminate the translational invariance.

We will discuss here the application of the Leray-Schauder method to travelling waves from the point of view of a priori estimates of solutions in weighted

4. Existence and bifurcations of solutions

spaces. Consider, as before, problem (4.39) with the nonlinearity $F_\tau(w)$ which satisfies the following conditions: $F_\tau(w_\pm) = 0$, $F'_\tau(w_\pm) < 0$ and

$$F_\tau(w) < 0 \text{ for } w_+ < w < w_0, \quad F_\tau(w) > 0 \text{ for } w_0 < w < w_-$$

with some $w_0 \in (w_+, w_-)$. This means that we consider the bistable case where the essential spectrum of the linearized operator is completely in the left half-plane and all conditions for the degree construction are satisfied. In the monostable case, the essential spectrum is partially in the right half-plane and the conditions of the degree construction are not satisfied. In some cases, we can introduce special weighted spaces with an exponential weight which moves the essential spectrum to the left half-plane.

It follows from the maximum principle that any solution of problem (4.39) satisfies the estimate $0 < w(x) < 1, -\infty < x < \infty$. Therefore, we can estimate the norm $\|w\|_{C^{2+\alpha}(\mathbb{R})}$ for some $\alpha \in (0,1)$. However, in order to use the topological degree, we need to obtain a priori estimates in the weighted space $C_\mu^{2+\alpha}(\mathbb{R})$. It appears that it is a special type of a priori estimates determined by the global behavior of solutions. We explain it below.

First of all, we introduce the representation $w(x) = \psi(x) + u(x)$, where $\psi(x)$ is a C^∞ function such that

$$\psi(x) = \begin{cases} w_+, & x \leq -1, \\ w_-, & x \geq 1. \end{cases}$$

This allows us to work with functions $u(x)$ converging to zero at infinity. Since $F'(w_\pm) < 0$, then a solution $u(x)$ of the equation

$$(u+\psi)'' + c(u+\psi)' + F_\tau(u+\psi) = 0$$

decays exponentially at infinity. This means that for some positive constants N, k, and δ we have the estimate

$$|u(x)| \leq k e^{-|\delta|x}, \quad |x| \geq N.$$

It holds if $u(x)$ is sufficiently small such that $F_\tau(u+\psi)$ can be approximated by linear functions $F'(w_\pm)u$. Therefore the value of N is such that the solution $|u(x)| \leq \epsilon$ for $|x| \geq N$ and for some positive ϵ, that is $w = u + \psi$ is sufficiently close to $w_\pm$. The exponential estimate at infinity allows us to estimate $|u\mu|$ for $|x| \geq N$. The estimate of $|u\mu|$ for $|x| \leq N$ depends on N.

We can see here the principal difference of travelling waves in comparison with non-autonomous problems discussed in Section 4.5.3 where the value of N was fixed by the assumptions on the nonlinearity. Here the value of ϵ is fixed but not N which is not a priori known and can depend on the solution and on the value of the parameter τ. Moreover, it is possible that $N(\tau) \to \infty$ as $\tau \to \tau_0$ for some τ_0. In this case, the norm $\|u\|_{C_\mu^{2+\alpha}(\mathbb{R})}$ becomes unbounded.

Thus, in order to obtain a priori estimates of solutions, we need to show that the value of $N(\tau)$ is bounded. We recall that it is chosen in such a way that $w_\tau(x)$ remains outside some given ϵ-neighborhoods of the points w_+ and w_- for $-N(\tau) < x < N(\tau)$. If $N(\tau)$ tends to infinity, then this means that the solution $w_\tau(x)$ is "attracted" by an intermediate stationary point or by another invariant manifold. Such global behavior is difficult to control and there are few results of this type. However, there are some classes of reaction-diffusion systems for which such estimates can be obtained [561], [568]. In this case, it is possible to use the topological degree and the Leray-Schauder method to prove the existence of travelling waves. Some other methods to prove the existence of waves are known, such as phase space analysis or limiting passage from problems in bounded intervals to the whole axis. In spite of the difference in the approaches, the main difficulty remains to control the global behavior of solutions.

4.3.5. Decay and growth of solutions We proved in Chapters 4 and 5 that solutions of the homogeneous equation $Lu = 0$ with a normally solvable operator $L : E_q \to F_q$ decay exponentially at infinity. Let us now return to this question and study linear nonhomogeneous equations and nonlinear equations. Consider an infinitely differentiable positive function $\mu(x)$ and assume that

$$\frac{D^\beta \mu(x)}{\mu(x)} \to 0, \quad |x| \to \infty \qquad (4.43)$$

for any multi-index $\beta = (\beta_1, \ldots, \beta_n)$ such that $|\beta| = \beta_1 + \cdots + \beta_n > 0$. This condition is satisfied if $\mu(x)$ has polynomial decay or growth at infinity. We introduce the operator $L_\mu = \frac{1}{\mu} L\mu : E_\infty \to F_\infty$ acting in the same spaces as the operator L. Since

$$\frac{1}{\mu} D^k(u\mu) = D^k u + \frac{1}{\mu} \sum_{\alpha+\beta=k, |\beta|>0} a_{\alpha\beta} D^\alpha u D^\beta \mu,$$

and the second term in the right-hand side tends to 0 by virtue of (4.43), then limiting operators for the operators L and L_μ are the same. Therefore, if Condition NS is satisfied for the operator L and, as a consequence, it is normally solvable with a finite-dimensional kernel, then this is also true for the operator L_μ.

Let Conditions NS and NS* be satisfied for the operator L. Then the equation

$$Lu = f \qquad (4.44)$$

is solvable if and only if $(v, f) = 0$ for any solution v of the homogeneous adjoint equation $L^* v = 0$. We recall that such solutions v are exponentially decaying at infinity. Denote $u_\mu = u/\mu$, $f_\mu = f/\mu$. Then $L(\mu u_\mu) = \mu f_\mu$ or

$$L_\mu u_\mu = f_\mu. \qquad (4.45)$$

Assume that $f_\mu \in F_\infty$. Then this equation is solvable if and only if $(w, f_\mu) = 0$ for all solutions w of the equation $L_\mu^* w = 0$. Since $L_\mu^* = \mu L^*(1/\mu)$, then $w = \mu v$,

4. Existence and bifurcations of solutions

where $L^*v = 0$. Hence
$$(w, f_\mu) = (\mu v, f/\mu) = (v, f)$$
and solvability conditions for equations (4.44) and (4.45) coincide. We note that $w = v\mu \in (F_\infty)^*$ since v decays exponentially and μ grows slower than any exponential. The homogeneous adjoint equations $L^*v = 0$ and $L^*_\mu w = 0$ have the same number of linearly independent solutions. We can now formulate the following theorem.

Theorem 4.12. *If the operator $L : E_\infty \to F_\infty$ satisfies the Fredholm property, then it is also true for the operator $L_\mu : E_\infty \to F_\infty$ where the weight function μ satisfies (4.43). If $f, f_\mu \in F_\infty$, then equation (4.45) is solvable in E_∞ if and only if equation (4.44) has a solution $u \in E_\infty$ and $u_\mu = u/\mu \in E_\infty$ is a solution of equation (4.45).*

Proof. It remains to show that $u_\mu \in E_\infty$. Suppose that this is not the case. Since $u_\mu = u/\mu, u \in E_\infty$, then u_μ has at most polynomial growth. More precisely, there exists a function $\nu(x) \geq 1$, $\nu(x) \to \infty$ as $|x| \to \infty$, which satisfies (4.43) and such that the function u_μ/ν belongs to the space E_∞ but does not converge to 0 as $|x| \to \infty$.

On the other hand, we know that equation (4.45) has a solution $\tilde{u} \in E_\infty$. Write $z = u_\mu - \tilde{u}$. Then $L_\mu z = 0$. Let $z_\nu = z/\nu$. Then $z_\nu \in E_\infty$ and z_ν does not converge to 0 since $\tilde{u}/\nu \to 0$ as $|x| \to \infty$. Let us introduce the operator $L_\nu = \frac{1}{\nu} L_\mu \nu$. We have
$$L_\nu z_\nu = \frac{1}{\nu} L_\mu \nu z_\nu = \frac{1}{\nu} L_\mu z = 0.$$
Thus, equation $L_\nu z = 0$ has a solution $z_\nu \in E_\infty$, which does not converge to zero at infinity. This contradicts the Fredholm property of the operator L_ν. The theorem is proved. $\square$

From this theorem we immediately obtain an estimate of the decay rate of solutions at infinity. If $f \in F_\infty$ satisfies solvability conditions and $f(x)(1+|x|^k) \leq M_1$ for all x and some constant M_1, then $u(x)(1+|x|^k) \leq M_2$ for all x and some other constant M_2. Indeed, put $\mu(x) = 1/(1 + |x|^k)$. Then $f_\mu = f/\mu \in F_\infty$. Hence $u_\mu = u/\mu \in E_\infty$ is a solution of equation (4.45). Therefore, $u(x)(1 + |x|^k) = u/\mu \in E_\infty$.

Suppose now that $f \in F_\infty$ but at least one of the solvability conditions is not satisfied. Then equation (4.44) does not have solutions in E_∞. Suppose that it has an unbounded solution u such that $u_\mu = u/\mu \in E_\infty$, where $\mu(x) = 1 + |x|^k$, $|k| > 0$. Then $f_\mu = f/\mu \in F_\infty$. Thus $L_\mu u_\mu = f_\mu$, that is equation (4.45) is solvable in E_∞. On the other hand, f_μ does not satisfy at least one solvability condition. This contradiction shows that equation (4.44) cannot have solutions with polynomial growth, that is solutions for which $u/\mu \in E_\infty$. The same reasoning can be used for a small exponential weight since it preserves the Fredholm property and solvability conditions. It allows us to prove exponential growth of solutions if the solvability conditions are not satisfied. In fact, exponential weight may not be necessarily small. The restriction on the weight is that the operator L_μ should belong to the same connected component of Fredholm operators as the operator L (see Chapter 5, Section 4).

Thus, solutions of equation (4.44) with $f \in F_\infty$ can be either bounded, if the solvability conditions are satisfied, or grow exponentially, if they are not satisfied. Let us illustrate this property with the example of the second-order operator

$$Lu = a(x)u'' + b(x)u' + c(x)u.$$

Assume that the formally adjoint equation has two bounded linearly independent solutions v_1 and v_2. Then solution $u(x)$ of equation (4.44) can be represented in the form

$$u(x) = u_1(x) \int_{-\infty}^{x} v_1(y)f(y)dy + u_2(x) \int_{-\infty}^{x} v_2(y)f(y)dy.$$

Here u_1 and u_2 are exponentially growing solutions of the equation $Lu = 0$. If the solvability conditions

$$\int_{-\infty}^{\infty} v_i(y)f(y)dy = 0, \quad i = 1, 2$$

are satisfied, then the solution is bounded. Otherwise, if at least one of the integrals is different from zero, then $u(x)$ grows exponentially at infinity.

Thus, we have an alternative of Phragmén-Lindelöf type: either the solution is bounded or exponentially growing (see Historical and bibliographical comments). As we discussed above, it is related to the Fredholm alternative. These results do not hold if the Fredholm property is not satisfied. This can be already seen for the simplest example of the equation $u'' = 0$ where the solution $u(x) = x$ is neither bounded nor exponentially growing.

The results presented above allow us to study behavior of solutions of nonlinear equations. Let us consider the operator

$$A(u) = Lu + B(u),$$

where $L : E_\infty \to F_\infty$ is a linear Fredholm operator, $B(u)$ is nonlinear. Suppose that the equation $A(u) = 0$ has a solution $u_0 \in E_\infty$ such that $u_0 \to 0$ as $|x| \to \infty$, and $|B(u_0(x))| \le K|u_0(x)|^2$ for some positive constant K and for $|x|$ sufficiently large. We will estimate the decay rate of the solution at infinity.

Consider the equation $Lu = f$, where $f = -B(u_0) \in F_\infty$. It has a solution $u = u_0$. We note that $|f| \le K|u_0|^2$ for large $|x|$. Let us choose an infinitely differentiable function $\mu(x) \le 1$ such that $\mu(x) \to 0$ as $|x| \to \infty$, $f_\mu = f/\mu \in F_\infty$ but $f_\mu(x)$ does not converge to zero at infinity. Suppose that μ satisfies (4.43). Then Theorem 4.12 is applicable and the function $u_\mu = u_0/\mu \in E_\infty$ is a solution of the equation $L_\mu u = f_\mu$. We have

$$f_\mu = \frac{f}{\mu} \le K\frac{|u_0|^2}{\mu} = K\mu\frac{|u_0|^2}{\mu^2} \le K_1\mu.$$

4. Existence and bifurcations of solutions

Hence, f_μ converges to zero at infinity. This contradicts the assumption above. Hence the function $\mu(x)$, which determines the decay rate of f, cannot satisfy (4.43). Therefore the decay of f should be faster than polynomial.

In fact, we can obtain a stronger result. It is not necessary to assume that $\mu(x)$ satisfies (4.43). We use this condition in order to prove that the operator L_μ satisfies the Fredholm property, has the same index and the same dimension of the kernel. This is true for exponential weights with some restrictions on the exponent (Chapter 5, Section 4).

4.3.6. Non-Fredholm operators. We have discussed in Chapter 10 solvability conditions for some classes of linear elliptic operators without the Fredholm property. Even if such conditions can be established, usually it is not clear how to apply them for nonlinear problems because the nonlinearity does not preserve them. We present here a rare example of nonlinear operators for which we can apply non-Fredholm solvability conditions.

Consider the integro-differential operator

$$A(u) = -\Delta u - au + \int_{\mathbb{R}^n} s(x-y) F(u(y), y) dy$$

acting from $H^2(\mathbb{R}^n)$ into $L^2(\mathbb{R}^n)$. Here a is a positive constant. The operator

$$Lu = -\Delta u - au$$

does not satisfy the Fredholm property. Indeed, the unique limiting operator coincides with the operator L, and the limiting problem $Lu = 0$ has nonzero bounded solutions. Nevertheless, solvability of the equation $Lu = g$ in $L^2(\mathbb{R}^n)$ can be easily obtained due to the particular form of the operator. Since it has constant coefficients, we can apply the Fourier transform and conclude that it is solvable if and only if $\hat{g}/(|\xi|^2 - a) \in L^2(\mathbb{R}^n)$, where $\hat{g}$ is the Fourier transform of the function g.

We suppose that $s \in L^2(\mathbb{R}^n)$, $F(u,x)$ satisfies the Lipschitz condition with respect to u uniformly in x,

$$|F(u_1, x) - F(u_2, x)| \leq \kappa |u_1 - u_2| \quad \forall x \in \mathbb{R}^n$$

and

$$|F(u, x)| \leq K|u| + h(x) \quad \forall x \in \mathbb{R}^n. \tag{4.46}$$

Here κ and K are some positive constants, $h(x) \in L^2(\mathbb{R}^n)$.

Let $v \in L^2(\mathbb{R}^n)$. Consider the linear equation

$$\Delta u + au = \int_{\mathbb{R}^n} s(x-y) F(v(y), y) dy. \tag{4.47}$$

By virtue of (4.46), $F(v(y), y) \in L^2(\mathbb{R}^n)$. Denote by $\hat{s}$ the Fourier transform of the function s. Suppose that for some positive constant M,

$$\left| \frac{\hat{s}(\xi)}{a - |\xi|^2} \right| \leq M, \quad \forall \xi \in \mathbb{R}^3. \tag{4.48}$$

Applying the Fourier transform to (4.47), we conclude that this equation has a solution $u \in L^2(\mathbb{R}^n)$.

Thus, we can define the operator T acting in $L^2(\mathbb{R}^n)$ which puts in correspondence the unique solution u of equation (4.47) to each function v. Fixed points of this operator are solutions of the equation $A(u) = 0$. We will show that under some conditions the existence of fixed points can be proved by the principle of contracting mappings.

Let $v_1, v_2 \in L^2(\mathbb{R}^n)$ and u_1, u_2 be the corresponding solutions of equation (4.47). We have

$$\hat{u}_1(\xi) - \hat{u}_2(\xi) = \frac{\hat{s}(\xi)}{a - |\xi|^2} (\hat{f}_1(\xi) - \hat{f}_2(\xi)),$$

where $\hat{f}_i(\xi)$ is the Fourier transform of $F(v_i(y), y)$. Then

$$\|u_1 - u_2\|_{L^2(\mathbb{R}^n)} \leq M \left(\int_{\mathbb{R}^3} |F_1(v_1(y), y) - F_2(v_2(y), y)|^2 dy \right)^{1/2}$$

$$\leq \kappa M \|v_1 - v_2\|_{L^2(\mathbb{R}^n)}.$$

Thus we have proved the following theorem.

Theorem 4.13. *If condition (4.48) is satisfied and $\kappa M < 1$, then equation $A(u) = 0$ has a unique solution in $L^2(\mathbb{R}^n)$.*

This approach can be applied for other operators, which represent a sum of a linear operator with constant coefficients and a convolution.

4.4 Bifurcations

4.4.1. Local bifurcations. One of the applications of the topological degree concerns local bifurcations of solutions. The degree construction presented in this chapter allows us to use it for general elliptic problems in bounded or unbounded domains. Consider the operator equation

$$A_\tau(u) = 0 \qquad (4.49)$$

depending on the parameter τ, $A_\tau : E \to F$. Suppose that it has the solution $u = 0$ for all values of the parameter, $A_\tau(0) \equiv 0$. The value $\tau = \tau_0$ of the parameter is called the bifurcation point if for any $\epsilon > 0$ there exists such $\tau \in (\tau_0 - \epsilon, \tau_0 + \epsilon)$ that equation (4.49) has a solution $u(\tau) \neq 0$ for this value of the parameter, and $\|u(\tau)\|_E \leq \epsilon$.

The implicit function theorem (Section 4.1.2) gives a necessary condition of bifurcation. The value τ_0 can be a bifurcation point only if the linearized operator $A'_{\tau_0}(0)$ is not invertible.

Theorem 4.14. *Suppose that the operator $A_\tau(u) : E \to F$ satisfies the conditions of the implicit function theorem (Theorem 4.3). If $A'_{\tau_0}(0)$ is a Fredholm operator*

4. Existence and bifurcations of solutions

with the zero index and τ_0 is a bifurcation point, then the operator $A'_{\tau_0}(0)$ has zero eigenvalue.

In the case of elliptic operators in unbounded domains, their Fredholm property follows from Conditions NS and NS*, and their zero index from the ellipticity with a parameter. This theorem does not require the existence of the degree. In order to formulate sufficient condition of bifurcation, let us introduce the notion of the index of a solution. Suppose that the topological degree for the operator $A_\tau(u)$ is defined and that the ball $B_\epsilon = \{\|u\|_E \leq \epsilon\}$ contains the unique solution $u = 0$. Then the degree $\gamma(A_\tau, B_\epsilon)$ is called the index of the solution $u = 0$. We denote it by $\mathrm{ind}(\tau)$. Similarly, the index is defined for nonzero solutions.

Theorem 4.15. *If the index* $\mathrm{ind}(\tau)$ *of the solution* $u = 0$ *of equation* (4.49) *changes at* τ_0, *then it is a bifurcation point.*

Proof. Suppose that τ_0 is not a bifurcation point. Then there exists a positive ϵ such that the ball B_ϵ contains only the solution $u = 0$ of equation (4.49) in some neighborhood of the point τ_0. Therefore the degree $\gamma(A_\tau, B_\epsilon)$ does not change at τ_0. This contradicts the assumption of the theorem that the index $\mathrm{ind}(\tau)$ changes for this value of the parameter. The theorem is proved. □

In the case of elliptic operators the index of a solution can be expressed through the number of positive eigenvalues (or negative, depending on the definition of the operator) of the linearized operator, $\mathrm{ind}(\tau) = (-1)^\nu$ (Section 3.5). If a simple real eigenvalue crosses the origin, then the index changes, and the corresponding value of the parameter is a bifurcation point. Let us consider some examples of local bifurcations.

Second-order equation in a bounded interval. Consider the Dirichlet problem in the interval $0 \leq x \leq 1$:

$$u'' + F_\tau(u) = 0, \quad u(0) = u(1) = 0, \tag{4.50}$$

where $F_\tau(u)$ is a sufficiently smooth function of u such that $F_\tau(0) = 0$ for all $\tau \in [0, 1]$ and

$$|F'_{\tau_1}(u) - F'_{\tau_2}(u)| \leq K|\tau_1 - \tau_2|,$$

where K is some positive constant. The function $u = 0$ is a solution of this problem for all τ. In order to study bifurcations of solutions, we introduce the operator

$$A_\tau(u) = u'' + F_\tau(u)$$

acting from the space E_0 of functions from $C^{2+\alpha}[0, 1]$ with the zero boundary condition into the space $E = C^\alpha[0, 1]$. This operator is continuous with respect to the parameter τ in the operator norm. The spectrum of the operator linearized about the trivial solutions

$$L_\tau v = v'' + F'_\tau(0)v$$

consists of real simple eigenvalues which can be easily found explicitly. Let $\lambda(\tau)$ be an eigenvalue of the operator L_τ such that $\lambda(\tau_0) = 0$ and $\lambda'(\tau_0) \neq 0$. Then it is a bifurcation point.

In the case of the Neumann boundary condition, this assertion remains the same. It should be modified in the case of the periodic boundary condition since the eigenvalues have multiplicity 2. If $\lambda_k = F'_\tau(0) - (2\pi k)^2$ with an integer k is the eigenvalue, which crosses the origin, and the corresponding eigenfunctions are $\sin(2\pi k x)$ and $\cos(2\pi k x)$, then we consider the subspace of functions orthogonal to one of the eigenfunctions,

$$\widehat{E}_0 = \left\{ u \in E, \int_0^1 u(x) \sin(2\pi k x) dx = 0 \right\}.$$

We can apply the construction described above for the operator $A_\tau : \widehat{E}_0 \to E$ and prove bifurcation of nontrivial solutions in the subspace $\widehat{E}_0$.

Second-order equation on the axis. The example considered above illustrates bifurcation of solutions when a simple eigenvalue crosses the origin. This is applicable for general elliptic problems when the degree is defined. In the case of unbounded domains, we should take into account the essential spectrum of the corresponding operator and introduce appropriate function spaces. Consider the operator

$$A_\tau(u) = u'' + F_\tau(x, u)$$

acting from the space $E_0 = C^{2+\alpha}_\mu(\mathbb{R})$ into the space $E = C^\alpha_\mu(\mathbb{R})$, where $\mu = \sqrt{1 + x^2}$ is a weight function. We assume that the function $F_\tau(x, u)$ is sufficiently smooth with respect to the variables x and u and to the parameter τ. Exact conditions are given in Sections 1–3.

Consider, as example, $F_\tau(x, u) = -au + \tau \phi(x) f(u)$, which is convenient for what follows. Here a is a positive constant, $\phi(x) \geq 0$ has a bounded support, $f(0) = 0$ and $f(u) > 0$ for $u > 0$. The essential spectrum of the linearized operator fills the negative half-axis, $\lambda \leq -a$. Hence the degree can be defined. Consider the operator L_τ linearized about the trivial solution:

$$L_\tau v = v'' - av + \tau \phi(x) f'(0) v.$$

It can be shown that the principal eigenvalue of this operator, that is the eigenvalue with the maximal real part, crosses the origin for some positive $\tau = \tau_0$. This result is based on the characterization of the principal eigenvalue of second-order scalar elliptic operators [563]. As in the previous example, $\tau = \tau_0$ is a bifurcation point.

Some additional aspects should be taken into account for bifurcations of travelling waves. Travelling waves are solutions of problem (4.37), (4.38) on the whole axis (Section 4.3.4). They are invariant with respect to translation in space. Therefore, the linearized operator has a zero eigenvalue. In order to study bifurcations of solutions, we need to remove it. This can be done by functionalization of the parameter [561], [568] or by introduction of an appropriate subspace orthogonal to the corresponding eigenfunction.

4. Existence and bifurcations of solutions

Turing structures. Turing structures (diffusion instability, dissipative structures) are spatially inhomogeneous solutions, which appear due to the diffusion terms [522]. Such structures cannot be observed for the single reaction-diffusion equation. Let us consider the following example of two equations starting directly with the eigenvalue problem:

$$u'' + a_{11}u + a_{12}v = \lambda u,$$
$$dv'' + a_{21}u + a_{22}v = \lambda v.$$

If we consider it in a bounded interval with the periodic boundary condition, then we can extend its solution to the whole axis by periodicity and look for the solution in the form $p \exp(i\xi x)$, where p is a two-dimensional vector and $\xi \in \mathbb{R}$. Therefore, we need to find eigenvalues of the matrix

$$A(\xi) = \begin{pmatrix} a_{11} - \xi^2 & , & a_{12} \\ a_{21} & , & a_{22} - d\xi^2 \end{pmatrix}.$$

Suppose that the matrix $A(0)$ has both eigenvalues with a negative real part, that is

$$a_{11} + a_{22} < 0, \quad \det A(0) > 0. \tag{4.51}$$

Then the homogeneous in space solution is stable without diffusion. Can it lose stability due to diffusion? In this case, the matrix $A(\xi)$ should have one zero and one negative eigenvalue for some $\xi \neq 0$. Hence

$$\det A(\xi) \equiv d\xi^4 - (da_{11} + a_{22})\xi^2 + \det A(0) = 0.$$

If $d = 1$, then this equation does not have real solutions by virtue of (4.51). However, if $d \neq 1$, such solutions can exist. Exact conditions are well known and can be easily found.

Nonlocal reaction-diffusion equations. Another interesting example of the emergence of spatial structures is given by the integro-differential equation

$$du'' + u\left(a - \int_{-\infty}^{+\infty} \phi(x-y)u(y)dy\right) = 0. \tag{4.52}$$

Here $\phi(x)$ is a bounded function such that $\int_{-\infty}^{+\infty} \phi(x)dx = 1$, a and d are positive constants. This equation has a homogeneous in space solution $u = a$. Linearizing the corresponding operator about this solution, we obtain the eigenvalue problem

$$dv'' - a\int_{-\infty}^{+\infty} \phi(x-y)v(y)dy = \lambda v.$$

Applying the Fourier transform, we find the explicit expression for the spectrum:

$$\lambda(\xi) = -d\xi^2 - a\tilde{\phi}(\xi),$$

where $\tilde{\phi}(\xi)$ is the Fourier transform of the function ϕ. It can be verified that $\lambda(0) < 0$ and $\lambda(\xi)$ is negative for $|\xi|$ sufficiently large. Depending on the function ϕ, there can exist intermediate values of ξ for which $\lambda(\xi)$ becomes positive [25], [83], [208], [194]. This results in the emergence of spatial structures. They have interesting applications in population dynamics.

Natural convection. One of the classical problems in hydrodynamics is related to convection in a layer of an incompressible liquid heated from below. It is also called the Rayleigh-Benard problem. It is described by the Navier-Stokes equations and the heat equation. In the two-dimensional case:

$$u\frac{\partial u}{\partial x} + v\frac{\partial u}{\partial y} = -\frac{\partial p}{\partial x} + \Delta u, \tag{4.53}$$

$$u\frac{\partial v}{\partial x} + v\frac{\partial v}{\partial y} = -\frac{\partial p}{\partial y} + \Delta v + R\theta, \tag{4.54}$$

$$u\frac{\partial \theta}{\partial x} + v\frac{\partial \theta}{\partial y} = \Delta \theta, \tag{4.55}$$

$$\frac{\partial u}{\partial x} + \frac{\partial v}{\partial y} = 0. \tag{4.56}$$

Here u is the horizontal and v the vertical component of the velocity, θ the dimensionless temperature, p the pressure, R the Rayleigh number. This system of equations is considered in the two-dimensional strip $\{0 \leq y \leq 1, -\infty < x < \infty\}$ with the boundary conditions

$$y = 0 : \theta = 0, \; \frac{\partial u}{\partial y} = 0, \; v = 0; \quad y = 1 : \theta = 1, \; \frac{\partial u}{\partial y} = 0, \; v = 0. \tag{4.57}$$

Other boundary conditions can also be considered. Linearizing problem (4.53)–(4.57) about the solution $\theta(x, y) = y, u = v = 0$, we obtain an eigenvalue problem, which can be explicitly solved. When the principal eigenvalue crosses the origin, another solution with a nonzero velocity field bifurcates.

4.4.2. Branches of solutions. Analysis of the index of solutions can give some information about their number and stability.

Suppose that the index of the solution $u = 0$ equals 1 for $\tau < \tau_0$ and -1 for $\tau > \tau_0$. If we consider all solutions for each value of the parameter, then the sum of their indices should be independent of τ by virtue of the homotopy invariance of the degree. Figure 12 presents some examples of bifurcation diagrams. Solutions with ind = 1 are shown by solid lines while solutions with ind = -1 by dashed lines. In the case of the supercritical bifurcation (Figure 12 a)), the sum of the indices equals 1, in the case of the subcritical bifurcation -1, in the last case it is 0.

Assume that the whole spectrum of the operator $A'_\tau(0)$ lies in the left half-plane for $\tau < \tau_0$ and that the bifurcation occurs due to the principal eigenvalue, that is the eigenvalue with the maximal real part, crossing the origin. Then the

4. Existence and bifurcations of solutions

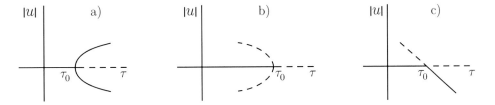

Figure 12: Three types of local bifurcations: a) supercritical, b) subcritical, c) change of stability; $|u|$ denotes a norm of the solution. Solid lines ind $= 1$, dashed lines ind $= -1$.

operator linearized about the solutions with the index 1 has the whole spectrum in the left half-plane. If the index equals -1, then it has one eigenvalue in the right half-plane. In the first case, the stationary solution of the corresponding evolution problem is stable, in the second case unstable.

The sets of solutions of equation (4.49) can form continuous branches. This should be understood in the following sense. Suppose that $A_\tau(u)$ is a Fredholm and proper operator with the zero index, and the conditions of the degree construction are satisfied (Section 3). Let u_n be a sequence of solutions of this equation corresponding to the values τ_n of the parameter, $\|u_n\|_E \leq K$ for some positive constant K. Assume that $\tau_n \to \tau_*$. Then there exists a solution u_* of this equation for $\tau = \tau_*$ and a subsequence $u_{n_k} \to u_*$ as $n_k \to \infty$. Indeed, this follows from the properness of the operator.

Let further $u(\tau)$ be a branch of solutions of equation (4.49) defined for $\tau \leq \tau_*$, the index of the solution $u(\tau_*)$ be defined and different from zero. Then this branch can be continued for larger values of τ. This means that for any $\delta > 0$ sufficiently small there exists $\epsilon > 0$ such that the solution $u(\tau)$ exists for $\tau_* < \tau < \tau_* + \epsilon$ and belongs to the δ-neighborhood of the solution $u(\tau_*)$. Indeed, since the index of the solution $u(\tau_*)$ is defined, then for any $\delta > 0$ sufficiently small, the ball $B_\delta = \{u \in E, \|u - u(\tau_*)\|_E \leq \delta\}$ does not contain other solutions of equation (4.49) for $\tau = \tau_*$. Therefore, there exists a positive ϵ such that there are no solutions at the boundary of the ball for $\tau_* < \tau < \tau_* + \epsilon$. Hence the degree $\gamma(A_\tau, B_\delta)$ is defined for these values of τ and it is different from zero. Consequently, there exists a solution of equation (4.49) inside the ball for each such τ.

Similarly, the branches of solutions can be extended for smaller values of the parameter. If the index of a solution is zero, then it may disappear and the branch may not be extended.

Equation of branching. Properties of Fredholm operators allow us to derive the equation of branching in the vicinity of the bifurcation point. Let $L : E \to F$ be a linear Fredholm operator, E and F Banach spaces. Denote by n the dimension of its kernel E_0, and by $u_1, \ldots, u_n$ its basis. Suppose that the equation $Lu = f$ is solvable if and only if $(f, \psi_i) = 0$, $i = 1, \ldots, m$, where $\psi_i \in F^*$.

We can introduce a system of functionals $\phi_i \in E^*$, $i = 1, \ldots, n$ biorthogonal to the functions u_i: $(u_i, \phi_j) = \delta_{ij}$, $i, j = 1, \ldots, n$ and the projector P on the subspace E_0: $Pu = \sum_{i=1}^n (u, \phi_i) u_i$. This projector determines the representation $E = E_0 \oplus E_1$ of the space E as a direct sum of the kernel E_0 and of the complementary subspace E_1.

Similarly, we can introduce a system of functions $f_1, \ldots, f_m \in F$ biorthogonal to the functionals ψ_i: $(f_i, \psi_j) = \delta_{ij}$, $i, j = 1, \ldots, m$ and the projector $Qf = \sum_{i=1}^m (f, \psi_i) f_i$. It determines the representation of the space F as a direct sum of the image F_1 of the operator L and of the complementary finite-dimensional subspace F_0 spanned on the functions f_i. The restriction $\hat{L} : E_1 \to F_1$ of the operator L is an isomorphism. Due to the Banach theorem it has a bounded inverse.

Let us now return to equation (4.49) with a Fredholm operator A_τ. As above, we assume that $A_\tau(0) = 0$ for all τ, and $\tau = \tau_0$ is a bifurcation point. The operator $L = A'_{\tau_0}(0)$ is not invertible. Put $B(u, \tau) = -A_\tau(u) + Lu$ and $u = v + w$, where $v \in E_0$ and $w \in E_1$. Then we can write equation (4.49) as $Lu = B(u, \tau)$ or

$$\hat{L}w = B(v + w, \tau), \quad QB(v + w, \tau) = 0, \tag{4.58}$$

where Q is the projector introduced above. If $v = 0, \tau = \tau_0$, then $w = 0$. From the implicit function theorem it follows that the first equation in (4.58) has a unique solution $w = w(v, \tau)$ for sufficiently small v and $\tau - \tau_0$. Then from the second equation in (4.58) we obtain the equation

$$QB(v + w(v, \tau), \tau) = 0$$

with respect to v. This is the Liapunov-Schmidt equation which determines behavior of solutions in a small neighborhood of the bifurcation point. We note that the index of the operator L can be different from zero. Moreover, this construction does not use properness or topological degree. Therefore we do not need to introduce weighted spaces when considering elliptic problems in unbounded domains.

Example. Consider the scalar reaction-diffusion equation

$$u'' + F_\tau(u, x) = 0$$

in $\mathbb{R}$. We suppose that the function F_τ is sufficiently smooth with respect to u, x and τ, and $F_\tau(0, x) \equiv 0$. The operator $A_\tau(u)$ corresponds to the left-hand side of this equation and acts from the Hölder space $E = C_0^{2+\alpha}(\mathbb{R})$ of functions with zero limit at infinity into the space $F = C_0^\alpha(\mathbb{R})$, $0 < \alpha < 1$. We assume that the linearized operator

$$L_\tau v = v'' + F'_\tau(0, x) v$$

satisfies Conditions NS and NS* and has its essential spectrum in the left halfplane. Then it is a Fredholm operator with zero index. Let it have a simple zero

4. Existence and bifurcations of solutions

eigenvalue for $\tau = \tau_0$ and v_0 be the corresponding eigenfunction. Then the kernel E_0 of the operator $L = L_{\tau_0}$ is a one-dimensional subspace composed by the functions $u = cv_0$, where c is a real constant. Since the operator L is formally self-adjoint, then v_0 is the eigenfunction of the formally adjoint operator L^* corresponding to the zero eigenvalue. The equation $Lu = f$ is solvable if and only if

$$\int_{-\infty}^{\infty} f(x)v_0(x)dx = 0.$$

The integral here is well defined since v_0 decays exponentially at infinity. The first equation in (4.58) can be written as

$$w'' + F_{\tau_0}(0)w = B(cv_0 + w, \tau),$$

where $B(u, \tau) = -F_\tau(u) + F'_{\tau_0}(0)u$. Its solution $w = w(c, \tau)$ exists for sufficiently small c and $\tau - \tau_0$. From the solvability condition we obtain the equation

$$\int_{-\infty}^{\infty} B(cv_0 + w(c, \tau), \tau)v_0 dx = 0,$$

which gives the relation between c and τ. This is the equation of branching. Taylor's expansion of the function F_τ can be used in order to simplify it and to find an approximate analytical solution.

Supplement. Discrete Operators

Infinite systems of algebraic equations can arise in some applications or as a result of discretization of differential equations in unbounded domains. Though the results obtained for elliptic problems in unbounded domains cannot be directly applied to infinite-dimensional discrete operators, the methods developed in this book can be adapted for them. In particular, we will define limiting problems in order to formulate conditions of normal solvability. We will discuss solvability conditions and some other properties of discrete operators. One of the results concerns the generalization of the Perron-Frobenius theorem about the principal eigenvalue of the matrices with non-negative off-diagonal elements to infinite matrices. We will also see that conditions of normal solvability are related to stability of finite difference approximations of differential equations. The representation below follows the works [28]–[30]. Some related questions are discussed in [443].

1 One-parameter equations

1.1 Limiting operators and normal solvability

Let E be the Banach space of all bounded real sequences

$$E = \left\{ u = \{u_j\}_{j=-\infty}^{\infty}, \ u_j \in \mathbb{R}, \ \sup_{j \in \mathbb{Z}} |u_j| < \infty \right\} \qquad (1.1)$$

with the norm

$$\|u\| = \sup_{j \in \mathbb{Z}} |u_j|,$$

and $L : E \to E$ be the linear difference operator

$$(Lu)_j = a_{-m}^j u_{j-m} + \cdots + a_0^j u_j + \cdots + a_m^j u_{j+m}, \ j \in \mathbb{Z}, \qquad (1.2)$$

where $m \geq 0$ is an integer and $a_{-m}^j, \ldots, a_0^j, \ldots, a_m^j \in \mathbb{R}$ are given coefficients. In some cases, it will be also convenient for us to consider complex coefficients. This operator acts on sequences of numbers depending on one integer parameter j. In this sense, we call such operators and the corresponding equations one-parameter operators and equations. They can arise as a result of discretization of differential equations on the real axis.

Denote by $L^+ : E \to E$ the limiting operator

$$(L^+ u)_j = a^+_{-m} u_{j-m} + \cdots + a^+_0 u_j + \cdots + a^+_m u_{j+m}, \; j \in \mathbb{Z}, \qquad (1.3)$$

where

$$a^+_l = \lim_{j \to \infty} a^j_l, \; l \in \mathbb{Z}, \; -m \leq l \leq m. \qquad (1.4)$$

We are going to define the associated polynomial for the operator L^+. To do this, we are looking for the solution of the equation $L^+ u = 0$ under the form $u_j = \exp(\mu j), \; j \in \mathbb{Z}$ and obtain

$$a^+_{-m} e^{-\mu m} + \cdots + a^+_{-1} e^{-\mu} + a^+_0 + a^+_1 e^{\mu} + \cdots + a^+_m e^{\mu m} = 0.$$

Let $\sigma = e^{\mu}$ and

$$P^+(\sigma) = a^+_m \sigma^{2m} + \cdots + a^+_0 \sigma^m + \cdots + a^+_{-m}. \qquad (1.5)$$

We present without proof the following auxiliary result (see [28]).

Lemma 1.1. *The equation $L^+ u = 0$ has nonzero bounded solutions if and only if the corresponding algebraic polynomial P^+ has a root σ with $|\sigma| = 1$.*

We will find conditions in terms of P^+ for the limiting operator L^+ to be invertible. We begin with an auxiliary result concerning continuous deformations of the polynomial P^+. Without loss of generality, we can assume that $a^+_m = 1$. Consider the polynomial with complex coefficients

$$P(\sigma) = \sigma^n + a_1 \sigma^{n-1} + \cdots + a_{n-1} \sigma + a_n. \qquad (1.6)$$

Lemma 1.2. *Suppose that a polynomial $P(\sigma)$ does not have roots with $|\sigma| = 1$ and it has k roots with $|\sigma| < 1$, $0 \leq k \leq n$. Then there exists a continuous deformation $P_\tau(\sigma)$ $0 \leq \tau \leq 1$, such that*

$$P_0(\sigma) = P(\sigma), \quad P_1(\sigma) = (\sigma^k - a)(\sigma^{n-k} - \lambda),$$

and the polynomial $P_\tau(\sigma)$ does not have roots with $|\sigma| = 1$ for any $0 \leq \tau \leq 1$. Here $\lambda > 1$ and $a < 1$ are real numbers.

Proof. Let us represent the polynomial $P(\sigma)$ in the form

$$P(\sigma) = (\sigma - \sigma_1) \ldots (\sigma - \sigma_n),$$

where the roots $\sigma_1, \ldots, \sigma_k$ are inside the unit circle, and the other roots are outside it. Consider the polynomial

$$P_\tau(\sigma) = (\sigma - \sigma_1(\tau)) \ldots (\sigma - \sigma_n(\tau))$$

1. One-parameter equations

that depends on the parameter τ through its roots. This means that we change the roots and find the coefficients of the polynomial through them. We change the roots in such a way that for $\tau = 0$ they coincide with the roots of the original polynomial, for $\tau = 1$ it has the roots $\sigma_1, \ldots, \sigma_k$ with $(\sigma_i)^k = a$, $i = 1, \ldots, k$ (inside the unit circle) and $n-k$ roots $\sigma_{k+1}, \ldots, \sigma_n$ such that $(\sigma_i)^{n-k} = \lambda$, $i = k+1, \ldots, n$ (outside of the unit circle). This deformation can be done in such a way that there are no roots with $|\sigma| = 1$. The lemma is proved. □

Using the associated polynomials P^+ and P^- of L^+ and L^-, we can study normal solvability of operator L.

Theorem 1.3. *The operator L is normally solvable with a finite-dimensional kernel if and only if the corresponding algebraic polynomials P^+ and P^- do not have roots σ with $|\sigma| = 1$.*

Proof. Suppose that the polynomials P^+, P^- do not have roots σ with $|\sigma| = 1$. Let $\{f^n\}$ be a sequence in the image $\operatorname{Im} L$ of the operator L such that $f^n \to f$ and let $\{u^n\}$ be such that $Lu^n = f^n$.

Suppose in the beginning that $\{u^n\}$ is bounded in E. We construct a convergent subsequence. Since $\|u^n\| = \sup_{j \in \mathbb{Z}} |u_j^n| \leq c$, then for every positive integer N, there exists a subsequence $\{u^{n_k}\}$ of $\{u^n\}$ and $u = \{u_j\} \in E$ such that

$$\sup_{-N \leq j \leq N} |u_j^{n_k} - u_j| \to 0, \quad (1.7)$$

that is $u^{n_k} \to u$ as $k \to \infty$ uniformly on each bounded interval of j. Using a diagonalization process, we extend u_j to all $j \in \mathbb{Z}$. It is clear that $\sup_{j \in \mathbb{Z}} |u_j| \leq c$, that means $u \in E$. Passing to the limit as $k \to \infty$ in the equation $Lu^{n_k} = f^{n_k}$, we get $Lu = f$, so that $f \in \operatorname{Im} L$.

We show that the convergence in (1.7) is uniform with respect to all $j \in \mathbb{Z}$. If, by contradiction, there exists $j_k \to \infty$ such that $|u_{j_k}^{n_k} - u_{j_k}| \geq \varepsilon > 0$, then the sequence $y_j^k = u_{j+j_k}^{n_k} - u_{j+j_k}$ verifies the inequality $|y_0^k| = |u_{j_k}^{n_k} - u_{j_k}| \geq \varepsilon$ and the equation

$$a_{-m}^{j+j_k} y_{j-m}^k + \cdots + a_0^{j+j_k} y_j^k + \cdots + a_m^{j+j_k} y_{j+m}^k = f_{j+j_k}^{n_k} - f_{j+j_k}, \ j \in \mathbb{Z}. \quad (1.8)$$

Since the sequence $\{y^k\}$ is bounded in E, there exists a subsequence $\{y^{k_l}\}$ which converges to some $y^0 \in E$ uniformly with respect to j on bounded intervals. We pass to the limit as $k_l \to \infty$ in (1.8) and obtain

$$a_{-m}^+ y_{j-m}^0 + \cdots + a_0^+ y_j^0 + \cdots + a_m^+ y_{j+m}^0 = 0, \ j \in \mathbb{Z}.$$

Thus, the limiting equation $L^+ u = 0$ has a nonzero bounded solution and Lemma 1.1 leads to a contradiction. Therefore the convergence $u_j^{n_k} - u_j \to 0$ is uniform with respect to all $j \in \mathbb{Z}$. Since $Lu = f$, then $\operatorname{Im} L$ is closed.

We note that in order to prove that $\ker L$ has a finite dimension, it suffices to show that every sequence u^n from $B \cap \ker L$ (where B is the unit ball) has a convergent subsequence. We prove this using the same reasoning with $f^n = 0$.

We analyze now the case where $\{u^n\}$ is unbounded in E. Then we write $u^n = x^n + y^n$, with $\{x^n\} \in \operatorname{Ker} L$ and $\{y^n\}$ in the supplement of $\operatorname{Ker} L$. Then $Ly^n = f^n$. If $\{y^n\}$ is bounded in E, then it follows as above that $\operatorname{Im} L$ is closed. If not, then we repeat the above reasoning for $z^n = y^n/\|y^n\|$ and $g^n = f^n/\|y^n\|$. Passing to the limit on a subsequence n_k (such that $z^{n_k} \to z^0$) in the equality $Lz^{n_k} = g^{n_k}$ and using the convergence $g^{n_k} \to 0$, one obtains that z^0 belongs to the kernel of the operator L and to its supplement. This contradiction finishes the proof of the closeness of the image.

Assume now that $\operatorname{Im} L$ is closed and $\dim \operatorname{Ker} L$ is finite. Suppose, by contradiction, that one of the polynomials, for certainty P^+, has a root on the unit circle. Then there exists a solution $u = \{u_j\}_{j=-\infty}^{\infty}$ of the equation $L^+u = 0$, where $u_j = e^{i\xi j}$, $\xi \in \mathbb{R}$, $j \in \mathbb{Z}$.

Let $\alpha = \{\alpha_j\}_{j=-\infty}^{\infty}$, $\beta^N = \{\beta_j^N\}_{j=-\infty}^{\infty}$, $\gamma^N = \{\gamma_j^N\}_{j=-\infty}^{\infty}$ be a partition of unity ($\alpha_j + \beta_j^N + \gamma_j^N = 1$) given by

$$\alpha_j = \begin{cases} 1, & j \leq 0 \\ 0, & j \geq 1 \end{cases}, \quad \beta_j^N = \begin{cases} 1, & 1 \leq j \leq N \\ 0, & j \leq 0, j \geq N+1 \end{cases}, \quad \gamma_j^N = \begin{cases} 1, & j \geq N+1 \\ 0, & j \leq N \end{cases}.$$

Consider a sequence $\varepsilon_n \to 0$ as $n \to \infty$. For a fixed ε_n, put

$$u_j^n = e^{i(\xi+\varepsilon_n)j}, \quad v_j^n = (1-\alpha_j)\left(u_j^n - u_j\right), \quad f_j^n = Lv_j^n, \quad j \in \mathbb{Z}.$$

It is clear that $u_j^n \to u_j$ as $n \to \infty$ uniformly on every bounded interval of integers j.

It is sufficient to prove that $f^n \to 0$. Indeed, in this case, since the image of the operator is closed and the kernel is finite dimensional, then $v^n \to 0$. But this is in contradiction with

$$\|v^n\| = \sup_{j>0} \left| e^{i(\xi+\varepsilon_n)j} - e^{i\xi j} \right| \geq m > 0,$$

for some m.

In order to show that $f^n \to 0$ as $n \to \infty$, we represent f_j^n in the form

$$\begin{aligned} f_j^n &= (\alpha_j + \beta_j^N + \gamma_j^N)\left(L\left[(\beta^N + \gamma^N)(u^n - u)\right]\right)_j \\ &= \alpha_j \left(L\left[(\beta^N + \gamma^N)(u^n - u)\right]\right)_j + \beta_j^N \left(L\left[(\beta^N + \gamma^N)(u^n - u)\right]\right)_j \\ &\quad + \gamma_j^N \left(L\left[\beta^N(u^n - u)\right]\right)_j + \gamma_j^N \left((L - L^+)\left[\gamma^N(u^n - u)\right]\right)_j \\ &\quad + \gamma_j^N \left(L^+[\gamma^N(u^n - u)]\right)_j. \end{aligned} \quad (1.9)$$

1. One-parameter equations

A simple computation shows that the first three terms in the right-hand side of the last equality tend to zero as $n \to \infty$ uniformly with respect to all integer j.

Next, condition (1.4) and the boundedness of the norms $\|u^n\|$ and $\|u\|$ lead to the convergence

$$\left|\gamma_j^N \left((L - L^+)\left[\gamma^N (u^n - u)\right]\right)_j\right| \leq |\gamma^N (L - L^+)|_0 \cdot \|\gamma^N (u^n - u)\| \to 0,$$

as $N \to \infty$, where $|\cdot|_0$ is the norm of the operator. For a given N, one estimates the last term in the right-hand side of (1.9). Since $u_j = e^{i\xi j}$, $j \in \mathbb{Z}$ is a solution of the equation $L^+ u = 0$, then

$$(L^+(u^n - u))_j = (L^+ u^n)_j = (L^+ u^n)_j - e^{i\varepsilon_n j}(L^+ u)_j$$
$$= e^{i(\xi + \varepsilon_n)j}[a_{-m}^+ e^{-i\xi m}\left(e^{-i\varepsilon_n m} - 1\right) + \cdots + a_{-1}^+ e^{-i\xi}\left(e^{-i\varepsilon_n} - 1\right)$$
$$+ a_1^+ e^{i\xi}\left(e^{i\varepsilon_n} - 1\right) + \cdots + a_m^+ e^{i\xi m}(e^{i\varepsilon_n m} - 1)],$$

so that

$$(L^+(u^n - u))_j = i\varepsilon_n e^{i(\xi + \varepsilon_n)j}[a_{-m}^+ (-m) e^{-i\xi m} e^{ib_{-m}} + \cdots + (-1)a_{-1}^+ e^{-i\xi} e^{ib_{-1}}$$
$$+ a_1^+ e^{i\xi}.e^{ib_1} + \cdots + a_m^+ m e^{i\xi m} e^{ib_m}], \; j \in \mathbb{Z},$$

where b_j, $j = 0, \pm 1, \ldots, \pm m$ are some numbers. Thus, the last term in (2.9) goes to zero as $n \to \infty$ and, therefore, $f^n \to 0$. This completes the proof. $\square$

We are now ready to establish the invertibility of L^+.

Theorem 1.4. *If the operator L^+ is such that the corresponding polynomial $P^+(\sigma)$ does not have roots with $|\sigma| = 1$, then it is invertible.*

Proof. Lemma 1.2 applied for $P^+(\sigma)$ implies the existence of a continuous deformation $P_\tau(\sigma)$, $0 \leq \tau \leq 1$, from the polynomial $P_0 = P^+$ to

$$P_1(\sigma) = \left(\sigma^k - a\right)\left(\sigma^{2m-k} - \lambda\right)$$

such that $P_\tau(\sigma)$ does not admit solutions with $|\sigma| = 1$. Here $\lambda > 1$, $a < 1$ are given. The operator which corresponds to P_1 is L_1^+ defined by the equality

$$(L_1^+ u)_j = u_{j+k} - a u_j - \lambda u_{j+2k-2m} + a\lambda u_{j+k-2m}.$$

Indeed, looking for the solution of the equation $L_1^+ u = 0$ in the form $u_j = e^{\mu j}$, we obtain

$$e^{\mu k} - a - \lambda e^{\mu(2k-2m)} + a\lambda e^{\mu(k-2m)} = 0.$$

We put $\sigma = e^\mu$ and get

$$\left(\sigma^k - a\right)\left(\sigma^{2m-k} - \lambda\right) = 0,$$

so P_1 is the above polynomial. Taking $a = 1/\lambda$, we have

$$\left(L_1^+ u\right)_j = (Mu)_j - (1/\lambda)\, u_j,$$

where

$$(Mu)_j = u_{j+k} - \lambda u_{j+2k-m} + u_{j+k-2m}.$$

The operator M is invertible for large $\lambda \geq 0$. Indeed,

$$M = -\lambda(T - \frac{1}{\lambda} S),$$

where $(Tu)_j = u_{j+2k-m}$, $(Su)_j = u_{j+k} + u_{j+k-2m}$. Since T is invertible, then $T - S/\lambda$ is also invertible for λ large enough.

Hence L_1^+ is also invertible for sufficiently large λ and its index is zero. Since the polynomial P_τ does not have solutions σ with $|\sigma| = 1$, for any $0 \leq \tau \leq 1$, then the corresponding continuous deformation of the operator L_τ^+ does not admit nonzero bounded solutions (see Lemma 1.1). By Theorem 1.3, one obtains that L_τ^+ is normally solvable with a finite-dimensional kernel. From the general theory of Fredholm operators, we know that the index does not change in the process of such deformation. Since the index of L_1^+ is $\kappa\left(L_1^+\right) = 0$, we deduce that $\kappa\left(L^+\right) = 0$. This, together with the fact that the kernel of the operator L^+ is empty, implies that it is invertible. The theorem is proved. □

A similar result can be stated for L^-. As a consequence, we can study the Fredholm property of L with the help of the polynomials P^+ and P^-.

Corollary 1.5. *If the limiting operators L^+ and L^- for an operator L are such that the corresponding polynomials $P^+(\sigma)$ and $P^-(\sigma)$ do not have roots with $|\sigma| = 1$ and have the same number of roots inside the unit circle, then L is a Fredholm operator with the zero index.*

Proof. We construct a homotopy of L in such a way that L^+ and L^- are reduced to the operator in Theorem 1.4. This is a homotopy in the class of the normally solvable operators with finite-dimensional kernels. Since the operators L^+ and L^- coincide, we finally reduce L to an operator with constant coefficients. According to Theorem 1.4, it is invertible. Therefore L is a Fredholm operator and has the index 0, as claimed. □

We note that if the polynomials $P^\pm(\sigma)$ do not have roots with $|\sigma| = 1$, then solutions of the equation $Lu = 0$ decay exponentially at infinity. This can be proved employing the properties of the holomorphic operator-functions similar to the proof in the case of elliptic operators (cf. Section 4, Chapter 5).

1. One-parameter equations

1.2 Solvability conditions

In this section, we establish solvability conditions for the equation

$$Lu = f. \qquad (1.10)$$

Let $\alpha(L) = \dim \operatorname{Ker} L$ and $\beta(L) = \operatorname{codim} \operatorname{Im} L$, (u, v) the inner product in l^2,

$$(u, v) = \sum_{j=-\infty}^{\infty} u_j v_j.$$

We define the formally adjoint L^* of the operator L by the equality

$$(Lu, v) = (u, L^* v).$$

Let $L^{\pm}$ and $L_{\pm}^*$ be the limiting operators associated with L and L^*, respectively. We suppose that the following condition is satisfied.

(H) The polynomials P^+, P^- corresponding to L^+ and L^- do not have roots with $|\sigma| = 1$ and have the same number of roots with $|\sigma| < 1$. Similarly for the polynomials P_+^* and P_-^* corresponding to L_+^* and L_-^*.

Corollary 1.5 implies that L and L^* are Fredholm operators with the index zero.

Lemma 1.6. *If Condition H is satisfied, then $\beta(L) \geq \alpha(L^*)$.*

Proof. By the definition of Fredholm operators, equation (1.10) is solvable if and only if

$$\varphi_k(f) = 0, \ k = 1, \ldots, \beta(L) \qquad (1.11)$$

for some linearly independent functionals $\varphi_k \in E^*$, $k = 1, \ldots, \beta(L)$. On the other hand, consider the functionals ψ_l given by

$$\psi_l(f) = \sum_{j=-\infty}^{\infty} f_j v_j^l, \ l = 1, \ldots, \alpha(L^*), \qquad (1.12)$$

where v^l, $l = 1, \ldots, \alpha(L^*)$ are linearly independent solutions of the homogeneous equation $L^* v = 0$. Since v_j^l are exponentially decreasing with respect to j, then the functionals ψ_l are well defined.

In order to prove that $\beta(L) \geq \alpha(L^*)$, suppose that it is not true. Then among the functionals ψ_l there exists at least one functional (say ψ_1) which is linearly independent with respect to all φ_k, $k = 1, \ldots, \beta(L)$. This means that $(\exists) f \in E$ such that (1.11) holds, but

$$\psi_1(f) = \sum_{j=-\infty}^{\infty} f_j v_j^1 \neq 0. \qquad (1.13)$$

From (1.11) it follows that equation (1.10) is solvable. We multiply it by v^1 and find $(Lu, v^1) = (f, v^1)$. By (1.13) observe that the right-hand side is different from zero. But since v^1 is a solution of the equation $L^*v = 0$, we deduce that $(Lu, v^1) = (u, L^*v^1) = 0$. This contradiction proves the lemma. □

Since L is formally adjoint to L^*, then similarly to the lemma we obtain $\beta(L^*) \geq \alpha(L)$. Therefore, if $\kappa(L) = \alpha(L) - \beta(L)$ is the index of the operator L, then
$$\kappa(L) + \kappa(L^*) \leq 0. \tag{1.14}$$
Since in our case $\kappa(L) = \kappa(L^*) = 0$, then it follows that
$$\beta(L) = \alpha(L^*), \quad \beta(L^*) = \alpha(L). \tag{1.15}$$

Theorem 1.7. *Equation* (1.10) *is solvable if and only if*
$$\sum_{j=-\infty}^{\infty} f_j v_j^l = 0, \; l = 1, \ldots, \alpha(L^*), \tag{1.16}$$
where $v^l = \{v_j^l\}_{j=-\infty}^{\infty}$, $l = 1, \ldots, \alpha(L^*)$ *are linearly independent solutions of the equation* $L^*v = 0$.

Proof. Equation (1.10) is solvable if and only if (1.11) holds for some functionals $\varphi_k \in E^*$, $k = 1, \ldots, \beta(L)$. Consider the subspaces Φ and Ψ of E^* generated by the functionals φ_k, $k = 1, \ldots, \beta(L)$ and by ψ_l from (1.12), $l = 1, \ldots, \alpha(L^*)$, respectively. By (1.15) we deduce that their dimensions coincide. We show that actually $\Phi = \Psi$. We first verify that $\Psi \subseteq \Phi$. Indeed, if it is not the case, then there exists $\psi \in \Psi$, $\psi \notin \Phi$. Then $\exists f \in E$ such that (1.11) holds, but at least one $\psi_l(f) \neq 0$, so we get the same contradiction as in the proof of Lemma 1.6. Therefore, $\Psi \subseteq \Phi$ and since they have the same dimensions, we get that $\Psi = \Phi$. The theorem is proved. □

1.3 Spectrum of difference and differential operators

Consider the difference operator
$$(Lu)_j = a_j(u_{j+1} - 2u_j + u_{j-1}) + b_j(u_{j+1} - u_j) + c_j u_j,$$
where a_j, b_j, c_j are real numbers. It can be considered as a discretization of the second-order differential equation on the real axis:
$$Mu = a(x)u'' + b(x)u' + c(x)u.$$
We will discuss how the essential spectrum of the difference and of the differential operators are related to each other. Let us assume that the sequences a_j, b_j, c_j

2. First-order systems

converge to a, b, c, respectively, as $i \to \infty$, and consider the infinite system of equations
$$a(u_{j+1} - 2u_j + u_{j-1}) + b(u_{j+1} - u_j) + cu_j = \lambda u_j.$$
We substitute $u_j = e^{i\xi j}$ and obtain
$$\lambda(\xi) = (2a + b)\cos\xi + ib\sin\xi - 2a - b + c.$$
Here ξ is a real parameter, $\lambda(\xi)$ is the essential spectrum. If it crosses the origin, the operator L does not satisfy the Fredholm property.

Let $a = \eta^2\alpha$, $b = \eta\beta$,
$$\lambda(\xi, \eta) = (2\eta\alpha + \beta)\eta\cos\xi + i\eta\beta\sin\xi - 2\eta^2\alpha - \eta\beta + c.$$
Here η is a large parameter. This scaling corresponds to a finite difference approximation of the first and second derivatives. If we consider λ as a function of ξ for a fixed η, then, as before, we obtain the essential spectrum of the operator L. We will now consider λ as a function of η and will show that it converges to the essential spectrum of the operator M as $\eta \to \infty$. Put $\lambda = \mu + i\nu$. Equating real and imaginary parts in the last equality, we can express μ through ν and λ and exclude ξ:
$$\nu = b\eta\sqrt{-\frac{(c-\mu)^2}{(2a\eta^2 + b\eta)^2} + \frac{2(c-\mu)}{2a\eta^2 + b\eta}}.$$
In the limit of large η, we obtain
$$\nu = b\sqrt{\frac{c-\mu}{a}}.$$
Therefore, $\mu = c - a\nu^2/b^2$,
$$\lambda = \mu + i\nu = -a\frac{\nu^2}{b^2} + i\nu + c.$$
We finally put $\eta = \nu/b$ and obtain
$$\lambda = -a\eta^2 + ib\eta + c.$$
If the coefficients of the operator M converge to a, b, and c at infinity, then the last formula gives the essential spectrum of the operator M.

2 First-order systems

Consider the linear algebraic system of equations
$$U(j) - U(j-1) = A(j)U(j), \qquad (2.1)$$

where $A(j)$ are $n \times n$ matrices, $U(j)$ are n-vectors, $j \in \mathbb{Z}$. We call such a system, for which only two consecutive values of the parameter j are present, first-order systems. Denote by $\Phi(j)$ the fundamental matrix of this system, that is the matrix whose columns are linearly independent solutions of (2.1). Suppose that there are n linearly independent solutions. Let

$$\Psi(j) = (\Phi^{-1}(j))^T,$$

where the superscript T denotes the transposed matrix. Therefore, $\Psi(j)$ is the fundamental matrix of the system

$$V(j) - V(j-1) = -A(j)^T V(j-1), \tag{2.2}$$

which is adjoint to system (2.1). We note that V in the right-hand side of (2.2) is taken at $j-1$, while U in (2.1) is taken at j.

Consider next the nonhomogeneous equation

$$W(j) - W(j-1) = A(j)W(j) + f(j-1). \tag{2.3}$$

Its solution can be given by the formula

$$W(j) = \Phi(j) \sum_{i=0}^{j-1} \Psi(i)^T f(i). \tag{2.4}$$

We will also use another form of the solution:

$$W(j) = -\Phi(j) \sum_{i=j}^{\infty} \Psi(i)^T f(i). \tag{2.5}$$

2.1 Solvability conditions

Assume that the elements $\varphi_{hk}(j)$ ($h, k = 1, \ldots, n$) of the fundamental matrix $\Phi(j)$ ($j \in \mathbb{Z}$) of the homogeneous system behave exponentially at infinity: $\varphi_{hk}(j) \sim a_{hk}^{\pm} e^{\lambda_k^{\pm} \cdot j}$ as $j \to \pm\infty$, where $\lambda_k^{\pm} \neq 0$, $\lambda_k^{\pm}$ are different for different k and $a_{hk}^{\pm}$ are such that the limit matrix of $\Phi(j)$ is invertible for all j. Therefore,

$$\Phi(j) = (\varphi_{hk}(j))_{h,k=\overline{1,n}} \sim \begin{pmatrix} a_{11}^{\pm} e^{\lambda_1^{\pm} j} & a_{12}^{\pm} e^{\lambda_2^{\pm} j} & \cdots & a_{1n}^{\pm} e^{\lambda_n^{\pm} j} \\ a_{21}^{\pm} e^{\lambda_1^{\pm} j} & a_{22}^{\pm} e^{\lambda_2^{\pm} j} & \cdots & a_{2n}^{\pm} e^{\lambda_n^{\pm} j} \\ \cdots & \cdots & \cdots & \cdots \\ a_{n1}^{\pm} e^{\lambda_1^{\pm} j} & a_{n2}^{\pm} e^{\lambda_2^{\pm} j} & \cdots & a_{nn}^{\pm} e^{\lambda_n^{\pm} j} \end{pmatrix} \tag{2.6}$$

as $j \to \pm\infty$. Then

$$\Psi(j) = (\Phi^{-1}(j))^T \sim \begin{pmatrix} b_{11}^{\pm} e^{-\lambda_1^{\pm} j} & b_{12}^{\pm} e^{-\lambda_2^{\pm} j} & \cdots & b_{1n}^{\pm} e^{-\lambda_n^{\pm} j} \\ b_{21}^{\pm} e^{-\lambda_1^{\pm} j} & b_{22}^{\pm} e^{-\lambda_2^{\pm} j} & \cdots & b_{2n}^{\pm} e^{-\lambda_n^{\pm} j} \\ \cdots & \cdots & \cdots & \cdots \\ b_{n1}^{\pm} e^{-\lambda_1^{\pm} j} & b_{n2}^{\pm} e^{-\lambda_2^{\pm} j} & \cdots & b_{nn}^{\pm} e^{-\lambda_n^{\pm} j} \end{pmatrix}, \tag{2.7}$$

2. First-order systems

as $j \to \pm\infty$. If $f_m(j)$ $(m = 1, \ldots, n)$ are the elements of $f(j)$, it follows that $\Psi(i)^T f(i)$ behaves like

$$\begin{pmatrix} s_1^\pm(i) e^{-\lambda_1^\pm i} \\ s_2^\pm(i) e^{-\lambda_2^\pm i} \\ \ldots \\ s_n^\pm(i) e^{-\lambda_n^\pm i} \end{pmatrix}, \text{ where } s_p^\pm(i) = b_{1p}^\pm f_1(i) + \cdots + b_{np}^\pm f_n(i), \ p = 1, \ldots, n.$$

Let k, h, l, q be integers, $0 \le k, h, l, q \le n$, such that $k + h + l + q = n$ and:

(i) $\lambda_p^+ < 0$, $\lambda_p^- > 0$, $(\forall)\, p = 1, \ldots, k$;
(ii) $\lambda_p^+ < 0$, $\lambda_p^- < 0$, $(\forall)\, p = k+1, \ldots, k+h$;
(iii) $\lambda_p^+ > 0$, $\lambda_p^- > 0$, $(\forall)\, p = k+h+1, \ldots, k+h+l$;
(iv) $\lambda_p^+ > 0$, $\lambda_p^- < 0$, $(\forall)\, p = k+h+l+1, \ldots, k+h+l+q(=n)$.

If this is not the case, we can rearrange the order of the numbers λ_p in such a way that (i)–(iv) hold. If one or several of the numbers k, h, l, q is zero, then we omit the corresponding line.

Denote by $A_1(j), \ldots, A_n(j)$ the columns of $\Phi(j)$ and by $B_1(j), \ldots, B_n(j)$ the columns of $\Psi(j)$. Then $A_1(j), \ldots, A_k(j)$ are bounded at $+\infty$ and $-\infty$, $A_{k+1}(j), \ldots, A_{k+h}(j)$ are bounded at $+\infty$ and grow at $-\infty$, $A_{k+h+1}(j), \ldots, A_{k+h+l}(j)$ grow at $+\infty$ and decay at $-\infty$, while $A_{k+h+l+1}(j), \ldots, A_n(j)$ grow at both $+\infty$ and $-\infty$. Since $\ker L$ is the subspace generated by the bounded (at both $+\infty$ and $-\infty$) columns of $\Phi(j)$, we find that $\dim \ker L = k$.

As a consequence of the behavior of $A_1(j), \ldots, A_n(j)$, we conclude that $B_1(j), \ldots, B_k(j)$ are exponentially growing at $+\infty$ and $-\infty$, $B_{k+1}(j), \ldots, B_{k+h}(j)$ are unbounded at $+\infty$ and bounded at $-\infty$ (decaying to 0), $B_{k+h+1}(j), \ldots, B_{k+h+l}(j)$ are bounded at $+\infty$ and unbounded at $-\infty$, and $B_{k+h+l+1}(j), \ldots, B_n(j)$ are bounded at both $+\infty$ and $-\infty$.

We put

$$W(j) = \Phi(j) \left[\sum_{i=0}^{j-1} \Psi_{1,k}(i)^T f(i) + \sum_{i=-\infty}^{j-1} \Psi_{k+1,k+h}(i)^T f(i) \right.$$
$$\left. - \sum_{i=j}^{\infty} \Psi_{k+h+1,k+h+l}(i)^T f(i) + \sum_{i=-\infty}^{j-1} \Psi_{k+h+l+1,n}(i)^T f(i) \right], \quad (2.8)$$

where $\Psi_{\alpha,\beta}(i)^T$ is the $n \times n$ matrix which has the lines $\alpha, \alpha+1, \ldots, \beta$ $(\alpha \le \beta)$ as the matrix $\Psi(i)^T$ and all the other lines zero. If one of the numbers k, l, h, q is zero, then the corresponding $\Psi_{\alpha,\beta}(i)^T$ is considered zero.

We verify that $W(j)$ is a solution of (2.3). Indeed, denoting by $S(j)$ the square bracket in (2.8), we can write

$$W(j) - W(j-1) = (\Phi(j) - \Phi(j-1))S(j) + \Phi(j-1)(S(j) - S(j-1))$$
$$= A(j)\Phi(j)S(j) + \Phi(j-1)\Psi(j-1)^T f(j-1)$$
$$= A(j)W(j) + f(j-1).$$

Remark that for every $p = 1, \ldots, n$ and $j \to \pm\infty$,

$$B_p(j)^T f(j) \sim s_p^\pm(j) e^{-\lambda_p^\pm j} = \left(\sum_{m=1}^n b_{mp}^\pm f_m(j)\right) e^{-\lambda_p^\pm j}. \tag{2.9}$$

From (2.8), we can easily see that for each $m = 1, \ldots, n$, the elements $w_m(j)$ of the vector

$$W(j) = \begin{pmatrix} w_1(j) \\ w_2(j) \\ \ldots \\ w_n(j) \end{pmatrix}$$

are given by

$$w_m(j) = \sum_{p=1}^{k} \varphi_{mp}(j) \sum_{i=0}^{j-1} B_p(i)^T f(i) + \sum_{p=k+1}^{k+h} \varphi_{mp}(j) \sum_{i=-\infty}^{j-1} B_p(i)^T f(i)$$
$$- \sum_{p=k+h+1}^{k+h+l} \varphi_{mp}(j) \sum_{i=j}^{\infty} B_p(i)^T f(i) + \sum_{p=k+h+l+1}^{n} \varphi_{mp}(j) \sum_{i=-\infty}^{j-1} B_p(i)^T f(i). \tag{2.10}$$

If k or h or l or q is zero, then the corresponding sum is zero.

By hypothesis (i) and (2.9), observe that for the first k terms of $w_m(j)$, we have the estimates

$$\begin{cases} \sum_{i=0}^{j-1} |B_p(i)^T f(i)| \leq M e^{-\lambda_p^+ j}, & \forall j \geq 0 \\ \sum_{i=0}^{j-1} |B_p(i)^T f(i)| \leq M e^{-\lambda_p^- j}, & \forall j \leq 0 \end{cases}, \tag{2.11}$$

$p = 1, \ldots, k$. Here $|\cdot|$ means the matrix norm. Therefore the sum

$$\sum_{p=1}^{k} (\varphi_{mp}(j) \sum_{i=0}^{j-1} B_p(i)^T f(i))$$

from (2.10) is bounded for any bounded f.

2. First-order systems

We now study the second sum from (2.10). For any p from $k+1$ to $k+h$, condition (ii) holds. So $\varphi_{mp}(j) \sim a_{mp}^{\pm} e^{\lambda_p^{\pm} j}$ are bounded at $+\infty$ and grow at $-\infty$ and $s_p^{\pm}(i) e^{-\lambda_p^{\pm} i}$ are exponentially growing at $+\infty$ and exponentially decaying at $-\infty$. It follows that the sum $\sum_{i=-\infty}^{j-1} B_p(i)^T f(i)$ is well defined and, using estimates similar to (2.11), we conclude that the sum

$$\sum_{p=k+1}^{k+h} (\varphi_{mp}(j) \sum_{i=-\infty}^{j-1} B_p(i)^T f(i))$$

is bounded for all bounded f.

For $p = k+h+1, \ldots, k+h+l$, by hypothesis (iii), remark that $a_{mp}^{\pm} e^{\lambda_p^{\pm} j}$ are unbounded at $+\infty$ and bounded at $-\infty$, while $\sum_{i=j}^{\infty} B_p(i)^T f(i)$ is well defined. Estimates similar to (2.11) hold again and therefore the sum

$$\sum_{p=k+h+1}^{k+h+l} (\varphi_{mp}(j) \sum_{i=j}^{\infty} B_p(i)^T f(i))$$

in (2.10) is bounded for all bounded f.

If $p = k+h+l+1, \ldots, k+h+l+q (=n)$, by (iv) we get that the sum $\sum_{i=-\infty}^{j-1} B_p(i)^T f(i)$ is well defined and estimates of (2.11) type hold. Hence,

$$\sum_{p=k+h+l+1}^{n} (\varphi_{mp}(j) \sum_{i=-\infty}^{j-1} B_p(i)^T f(i))$$

is bounded if and only if

$$\sum_{i=-\infty}^{\infty} B_p(i)^T f(i) = 0, \quad p = k+h+l+1, \ldots, k+h+l+q(=n). \tag{2.12}$$

These equalities provide solvability conditions for (2.3).

Consequently, if the solvability conditions (2.12) hold, then equation (2.3) is solvable. The solution is given by (2.8). The codimension of the operator is q and the index is $k - q$. All the reasoning remains valid also when one or more of the numbers k, h, l, q is zero. Therefore we have proved the following theorem.

Theorem 2.1. *Suppose that the fundamental matrix* $\Phi(j) = (\varphi_{hk}(j))_{h,k=1,\ldots,n}$ *of system* (2.1) *is invertible and behaves exponentially at* $\pm\infty$, $\varphi_{hk}(j) \sim a_{hk}^{\pm} e^{\lambda_k^{\pm} j}$ *as* $j \to \pm\infty$, *where* $a_{hk}^{\pm} \neq 0$, $\lambda_k^{\pm} \neq 0$ *and* $\lambda_k^{\pm}$ *are different for different* k. *Then the* operator L corresponding to system (2.1) is Fredholm. If there are k values of λ_p with $\lambda_p^+ < 0$, $\lambda_p^- > 0$ and q values of λ_p with $\lambda_p^+ > 0$, $\lambda_p^- < 0$ ($0 \leq k, q \leq n$), then the index of L is $k - q$ and the solvability conditions for (2.3) are

$$\sum_{i=-\infty}^{\infty} B_p(i)^T f(i) = 0,$$

for p corresponding to the q values for which $\lambda_p^+ > 0$, $\lambda_p^- < 0$.

2.2 Higher-order equations

In Section 1.1 we proved normal solvability of higher-order difference operators. We can reduce them to first-order systems in order to use for them the solvability conditions obtained in the previous section. They are applicable not only in the case of zero index considered in Section 1.2. Consider the operator

$$(Lu)_j = a_0^j u_j + a_1^j u_{j-1} + \cdots + a_{2m}^j u_{j-2m}, \; j \in \mathbb{Z},$$

where $m \in \mathbb{N}^*$ and the coefficients $a_k^j \in \mathbb{R}$ ($0 \leq k \leq 2m$) are given.

We show that the equation $Lu = 0$ can be transformed into a first-order difference system of the form (2.1). Indeed, writing

$$u_j - u_{j-1} = v_j^1, \; v_j^1 - v_{j-1}^1 = v_j^2, \ldots, v_j^{2m-2} - v_{j-1}^{2m-2} = v_j^{2m-1},$$

one easily observes that

$$\begin{cases} u_j - u_{j-1} = v_j^1 \\ (v_j^1 - v_{j-1}^1 =) v_j^2 = u_j - 2u_{j-1} + u_{j-2} \\ \quad \cdots \cdots \cdots \cdots \\ (v_j^{2m-2} - v_{j-1}^{2m-2} =) \\ v_j^{2m-1} = u_j - C_{2m-1}^1 u_{j-1} + C_{2m-1}^2 u_{j-2} - \cdots - C_{2m-1}^{2m-1} u_{j-2m+1} \\ v_j^{2m-1} - v_{j-1}^{2m-1} = u_j - C_{2m}^1 u_{j-1} + C_{2m}^2 u_{j-2} - \cdots + C_{2m}^{2m} u_{j-2m}. \end{cases}$$

We want to write the right-hand side of the last formula as a function of u_j, $v_j^1, \ldots, v_j^{2m-1}$. To do this, remark that the above equalities imply that $u_{j-1} = u_j - v_j^1$, $u_{j-2} = v_j^2 - u_j + 2(u_j - v_j^1)$ and so on. Analogously, from the penultimate equation, it follows that u_{j-2m+1} can be written as

$$u_{j-2m+1} = -v_j^{2m-1} + \alpha_0^j u_j + \alpha_1^j v_j^1 + \cdots + \alpha_{2m-2}^j v_j^{2m-2},$$

for some $\alpha_k^j \in \mathbb{R}$ ($0 \leq k \leq 2m-2$). This, together with the equation $Lu = 0$, leads to a first-order difference system of the form

$$\begin{cases} u_j - u_{j-1} = v_j^1 \\ v_j^1 - v_{j-1}^1 = v_j^2 \\ \cdots\cdots\cdots \\ v_j^{2m-2} - v_{j-1}^{2m-2} = v_j^{2m-1} \\ v_j^{2m-1} - v_{j-1}^{2m-1} = A_0^j u_j + A_1^j v_j^1 + \cdots + A_{2m-1}^j v_j^{2m-1}, \end{cases}$$

for some coefficients A_k^j, $0 \leq k \leq 2m-1$, $j \in \mathbb{Z}$. Writing

$$U(j) = \begin{pmatrix} u_j \\ v_j^1 \\ \cdots \\ v_j^{2m-1} \end{pmatrix}, \quad A(j) = \begin{pmatrix} 0 & 1 & 0 & \cdots & 0 \\ 0 & 0 & 1 & \cdots & 0 \\ \cdots & \cdots & \cdots & \cdots & \cdots \\ 0 & 0 & 0 & \cdots & 1 \\ A_0^j & A_1^j & A_2^j & \cdots & A_{2m-1}^j \end{pmatrix},$$

we conclude that the equation $Lu = 0$ can be written in the form (1.1). Therefore we can apply the solvability conditions obtained above.

3 Principal eigenvalue

Finite matrices with non-negative off-diagonal elements possess some special spectral properties given by the Perron-Frobenius theorem. Their eigenvalue with the maximal real part is real and the corresponding eigenvector is positive. We discuss here similar properties for infinite matrices. We will use the solvability conditions obtained above.

Consider the Banach space E of infinite sequences $u = (\ldots, u_{-1}, u_0, u_1, \ldots)$ with the norm

$$\|u\| = \sup_j |u_j|$$

and the operator L acting in E,

$$(Lu)_j = a_{-m}^j u_{j-m} + \cdots + a_0^j u_j + \cdots + a_m^j u_{j+m}, \quad j = 0, \pm 1, \pm 2, \ldots,$$

where m is a positive integer and $a_k^j \in \mathbb{R}$, $-m \leq k \leq m$ are given coefficients. We assume that there exist the limits

$$a_k^\pm = \lim_{j \to \pm\infty} a_k^j, \quad k = 0, \pm 1, \ldots, \pm m. \tag{3.1}$$

Consider the limiting operators $L^\pm$,

$$(L^\pm u)_j = a_{-m}^\pm u_{j-m} + \cdots + a_0^\pm u_j + \cdots + a_m^\pm u_{j+m}, \quad j = 0, \pm 1, \pm 2, \ldots.$$

Let
$$a^{\pm}_{-m} \neq 0, \quad a^{\pm}_{m} \neq 0, \qquad (3.2)$$
and suppose that the equations
$$L^{\pm}u - \lambda u = 0$$
do not have nonzero bounded solutions for any real $\lambda \geq 0$. We will call it Condition NS(λ). If it is satisfied, then from the results of Section 1 it follows that L is a Fredholm operator with the zero index.

Consider the polynomials
$$P^{\pm}_{\lambda}(\sigma) = a^{\pm}_m \sigma^{2m} + \cdots + a^{\pm}_1 \sigma^{m+1} + (a^{\pm}_0 - \lambda)\sigma^m + a^{\pm}_{-1}\sigma^{m-1} + \cdots + a^{\pm}_{-m}.$$

From Lemma 1.1 it follows that Condition NS(λ) is satisfied if and only if the polynomials $P^{\pm}_{\lambda}(\sigma)$ do not have roots with $|\sigma| = 1$. As a consequence we can obtain the following result.

Lemma 3.1. *If Condition NS(λ) is satisfied, then*
$$a^{\pm}_{-m} + \cdots + a^{\pm}_m < 0,$$
that is $L^{\pm} q < 0$, where q is a sequence with all elements equal to 1.

Proof. Suppose that the assertion of the corollary does not hold. Then $P^{\pm}_0(1) \geq 0$. On the other hand, for λ sufficiently large $P^{\pm}_{\lambda}(1) < 0$. Therefore for some λ, $P^{\pm}_{\lambda}(1) = 0$. We obtain a contradiction with Lemma 1.1. □

We recall that the formally adjoint operator L^* is defined by the equality
$$(Lu, v) = (u, L^*v).$$

If we consider L as an infinite matrix, then L^* is the adjoint matrix. Let $\alpha(L^*)$ be the dimension of $\ker L^*$ and $f = \{f_j\}_{j=-\infty}^{\infty} \in E$. In Section 1.2 we proved that the equation $Lu = f$ is solvable if and only if
$$\sum_{j=-\infty}^{\infty} f_j v^l_j = 0, \quad l = 1, \ldots, \alpha(L^*),$$
where v^l are linearly independent solutions of the equation $L^*v = 0$.

In what follows we say that u is positive (non-negative) if all elements of this sequence are positive (non-negative). From now on we suppose that
$$a^j_k > 0, \quad k = \pm 1, \pm 2, \cdots, \pm m, \quad j = 0, \pm 1, \pm 2, \ldots \qquad (3.3)$$
and that there exists a positive solution w of the equation
$$Lu = 0. \qquad (3.4)$$

3. Principal eigenvalue

This means that L has a zero eigenvalue and the corresponding eigenvector is positive. We will show that the zero eigenvalue is simple and all other eigenvalues lie in the left half-plane. Moreover, the homogeneous adjoint equation has a positive solution, which is unique up to a constant factor. It is a generalization of the Perron-Frobenius theorem for infinite matrices. The method of the proof follows the method developed for elliptic problems in unbounded domains. Similarly to elliptic problems it is assumed that the essential spectrum lies to the left of the eigenvalue with a positive eigenvector. We note that the operator L can be considered as an infinite-dimensional $(2m+1)$-diagonal matrix with positive elements in all nonzero diagonals except for the main diagonal where the signs of the elements are not prescribed.

3.1 Auxiliary results

Suppose conditions (3.1)–(3.3) are satisfied. We begin with the positiveness of the solution of the equation $Lu = f$ for $f \leq 0$. We will use the notation

$$U_-(N) = (u_{N-m}, \ldots, u_{N-1}), \quad U_+(N) = (u_{N+1}, \ldots, u_{N+m}).$$

Lemma 3.2. *Let* $Lu = f$, *where* $f \leq 0$, $u \geq 0$, $u \not\equiv 0$. *Then* $u > 0$.

Proof. Suppose that $u_j = 0$ for some j. Since $u \not\equiv 0$, there exists i such that $u_i = 0$, and either $u_{i+1} \neq 0$ or $u_{i-1} \neq 0$. The equation $(Lu)_i = f_i$ gives a contradiction in signs. The lemma is proved. □

Lemma 3.3. *If the initial condition* u^0 *of the problem*

$$\frac{du}{dt} = Lu, \quad u(0) = u^0 \tag{3.5}$$

is non-negative, then the solution $u(t)$ *is also non-negative for all* $t \in (0, \infty)$.

Proof. Consider the auxiliary problem

$$\frac{du_i}{dt} = (Lu)_i, \quad -N \leq i \leq N, \ t \geq 0,$$
$$U_-(-N) = 0, \ U_+(N) = 0, \ t \geq 0,$$
$$u(0) = u^0,$$

where the unknown function is $u = (u_{-N}, u_{-N+1}, \ldots u_0, \ldots, u_{N-1}, u_N)$.

Since $u^0 \geq 0$ and Lu has non-negative off-diagonal coefficients, it follows that the solution $u^N = \left(u_{-N}^N, u_{-N+1}^N, \ldots u_0^N, \ldots, u_{N-1}^N, u_N^N\right)$ of the above problem is non-negative.

If we compare the solution u^N at the interval $[-N, N]$ and the solution u^{N+1} at the interval $[-N-1, N+1]$, we find $u^{N+1} \geq u^N$. Indeed, the difference $u^{N+1} - u^N$ verifies a problem similar with the above one, but with a non-negative initial condition and with zero boundary conditions. The solution of this problem is non-negative, i.e., $u^{N+1} \geq u^N$. So the sequence is monotonically increasing with respect to N. The sequence is also bounded with respect to N: $\|u^N(t)\| \leq M$, for all N and $t \in [0,T]$, where T is any positive number, $M > 0$ depends on u^0 and on the coefficients a_k^i of L, which are bounded. Being bounded and monotone, u^N is convergent as $N \to \infty$ in $C([0,T];E)$ to some u. Then u verifies problem (3.5) and $u \geq 0$, as claimed. □

Corollary 3.4. (Comparison theorem). *Let $u^1(t)$ and $u^2(t)$ be solutions of the equation*
$$\frac{du}{dt} = Lu$$
with the initial conditions $u^1(0)$ and $u^2(0)$, respectively. If $u^1(0) \leq u^2(0)$, then $u^1(t) \leq u^2(t)$ for $t \geq 0$.

Lemma 3.5. *If the initial condition u^0 of the problem*
$$\frac{du}{dt} = L^+u, \quad u(0) = u^0 \tag{3.6}$$
is constant (independent of j), then the solution $u(t)$ is also constant. For any bounded initial condition the solution of problem (3.6) converges to the trivial solution $u = 0$.

The proof of this lemma follows from Lemma 3.1 and Corollary 3.4.

Lemma 3.6. *If u is a solution of the problem*
$$Lu = f, \quad j \geq N, \ U_-(N) \geq 0, \tag{3.7}$$
where $f \leq 0$, $u_j \to 0$ as $j \to \infty$, and N is sufficiently large, then $u_j \geq 0$ for $j \geq N$.

Proof. By virtue of Lemma 3.1 there exists a constant $\epsilon > 0$ such that $L^+q < -\epsilon$. Let us take N large enough such that
$$|((L - L^+)q)_j| \leq \frac{\epsilon}{2}, \quad j \geq N. \tag{3.8}$$

Suppose that $u_j < 0$ for some $j > N$. Due to the assumption that $u_j \to 0$ as $j \to \infty$, we can choose $\tau > 0$ such that $v_j = u_j + \tau q_j \geq 0$ for all $j \geq N$, and there exists $i > N$ such that $v_i = 0$. Since $V_-(N) > 0$ and $v_j > 0$ for all j sufficiently large, there exists $k > N$ such that $v_k = 0$ and either $v_{k+1} \neq 0$ or $v_{k-1} \neq 0$ (that is $v_{k+1} > 0$ or $v_{k-1} > 0$).

We have
$$Lv = Lu + \tau L^+q + \tau(L - L^+)q = f + \tau L^+q + \tau(L - L^+)q. \tag{3.9}$$

In view of (3.8), $L^+q < -\epsilon$ and $f \leq 0$, the right-hand side of this equality is less than or equal to 0 for $j \geq N$. We obtain a contradiction in signs in the equation corresponding to k. The lemma is proved. □

3. Principal eigenvalue

Remark 3.7. The assertion of the lemma remains true if we replace (3.7) by

$$Lu \leq \alpha u, \ j \geq N, \ U_-(N) \geq 0, \tag{3.10}$$

for some positive α. Indeed, one obtains $Lv \leq \alpha u + \tau L^+ q + \tau (L - L^+) q$ instead of (3.9) where $(Lv)_k > 0$ and $\alpha u_k + \tau L^+ q_k + \tau (L - L^+) q_k < \alpha u_k - \epsilon \tau / 2 = -\tau (\alpha + \epsilon/2) < 0$ because $v_k = 0$.

3.2 Location of the spectrum

The main result of this section is given by the following theorem.

Theorem 3.8. *Let Condition NS(λ) be satisfied and equation (3.4) have a positive bounded solution w. Then:*

1. *The equation*

$$Lu = \lambda u \tag{3.11}$$

   does not have nonzero bounded solutions for Re $\lambda \geq 0$, $\lambda \neq 0$.
2. *Each solution of equation (3.4) has the form $u = kw$, where k is a constant.*
3. *The equation*

$$L^* u = 0 \tag{3.12}$$

   has a positive solution unique up to a constant factor.

Proof. We first consider the case where $\lambda = \alpha + i\beta$, $\alpha \geq 0$, $\beta \neq 0$. Suppose by contradiction that there exists a bounded nonzero solution $u = u^1 + iu^2$ of this equation. Then $Lu^1 = \alpha u^1 - \beta u^2$ and $Lu^2 = \beta u^1 + \alpha u^2$. Consider the equation

$$\frac{dv}{dt} = Lv - \alpha v, \ v(0) = u^1. \tag{3.13}$$

Its solution is

$$v(t) = u^1 \cos \beta t - u^2 \sin \beta t. \tag{3.14}$$

For the sequence $u = \{u_j\} = \{u_j^1 + iu_j^2\}$, we write $\hat{u} = \{|u_j|\}$. Let us take the value of N as in Lemma 3.6 and choose $\tau > 0$ such that

$$\hat{u}_j \leq \tau w_j, \ |j| \leq N, \tag{3.15}$$

where at least for one j_0 with $|j_0| \leq N$, we have the equality

$$\hat{u}_{j_0} = \tau w_{j_0}. \tag{3.16}$$

For $j \geq N$ consider the problem

$$\begin{cases} \dfrac{dy}{dt} = Ly - \alpha y, \\ y_{N-k}(t) = \hat{u}_{N-k}, \ k = 1, \ldots, m, \ y_\infty(t) = 0, \end{cases} \tag{3.17}$$

$$y(0) = \hat{u}, \tag{3.18}$$

and the corresponding stationary problem

$$L\bar{y} - \alpha\bar{y} = 0, \ \bar{y}_{N-k} = \hat{u}_{N-k}, \ k = 1,\ldots,m, \ \bar{y}_\infty = 0. \tag{3.19}$$

The operator corresponding to problem (3.19) satisfies the Fredholm property and has the zero index. Indeed, its normal solvability can be proved similar to Theorem 1.3. Condition NS(λ) implies that its index is zero. Moreover, the corresponding homogeneous problem has only the zero solution. It follows from Lemma 3.6 applied to u and $-u$. Therefore problem (3.19) is uniquely solvable.

We show that the solution $y(t)$ of problem (3.17), (3.18) converges to $\bar{y}$ as $t \to \infty$. For this we consider the solution $y^*(t)$ of problem (3.17) with the initial condition $y^*(0) = \rho q$, where ρ is such that

$$\rho q_j \geq \hat{u}_j, \ j \geq N.$$

By Lemma 3.1, we have $L^\pm q < 0$. Since L^+ is close to L for $j \geq N$, with N large enough, it follows that $(Lq)_j < 0, \ j \geq N$. Then $y^*(t)$ monotonically decreases in t for each $j \geq N$ fixed. From the positiveness and the decreasing monotonicity of y^*, we deduce that $y^*(t)$ converges as $t \to \infty$ to some $x = \lim_{t\to\infty} y^*(t) \geq 0$. It satisfies the equation $Lx - \alpha x = 0$. Taking the limit also in the boundary conditions, one obtains that $x_{N+k} = \hat{u}_{N+k}$, for $k = 1,\ldots,m$ and $x_\infty = 0$, so x is a solution of problem (3.19). By the uniqueness, we get $x = \bar{y}$, i.e., there exists the limit $\lim_{t\to\infty} y^*(t) = \bar{y}$.

On the other hand, let y_* be the solution of (3.17) with the initial condition $y_*(0) = 0$. It can be shown that y_* increases in time and it has an upper bound. As above, we can deduce that y_* converges to $\bar{y}$. Therefore

$$\lim_{t\to\infty} y_*(t) = \lim_{t\to\infty} y^*(t) = \bar{y}.$$

By virtue of the comparison theorem applicable in this case (because $0 \leq \hat{u}_j \leq \rho q_j$, $j \geq N$), we have

$$y_*(t) \leq y(t) \leq y^*(t), \ j \geq N.$$

Hence

$$\lim_{t\to\infty} y_j(t) = \bar{y}_j, \ j \geq N.$$

One can easily verify that

$$v_j(t) \leq \hat{u}_j \ \text{for all} \ j \in \mathbb{Z}. \tag{3.20}$$

Then it follows from the comparison theorem that

$$v_j(t) \leq y_j(t), \ j \geq N, \ t \geq 0.$$

From this we have

$$v_j(t) = v_j(t + 2\pi n/\beta) \leq y_j(t + 2\pi n/\beta).$$

3. Principal eigenvalue

Passing to the limit as $n \to \infty$, we obtain

$$v_j(t) \leq \bar{y}_j, \quad j \geq N, \ t \geq 0.$$

Observe that $L(\tau w - \bar{y}) \leq \alpha(\tau w - \bar{y})$, $j \geq N$ and $\tau w_N - \bar{y}_N \geq 0$. We can apply Remark 3.7 to $\tau w - \bar{y}$. Therefore

$$\bar{y}_j \leq \tau w_j, \quad j \geq N.$$

Hence
$$v_j(t) \leq \tau w_j \qquad (3.21)$$

for $j \geq N, t \geq 0$. A similar estimate can be obtained for $j \leq -N$. Together with (3.15), these prove (3.21) for all $j \in \mathbb{Z}$.

The sequence $z(t) = \tau w - v(t)$ is a solution of the equation

$$\frac{dz}{dt} = Lz - \alpha z + \alpha \tau w.$$

Since $z(t) \geq 0$ (via (3.21) for all $j \in \mathbb{Z}$), z is not identically zero, and is periodic in t, it follows that $z_j(t) > 0$ for all j and $t \geq 0$. Indeed, suppose that for some $t = t_1$ and $j = j_1$, $z_{j_1}(t_1) = 0$. Consider first the case where $\alpha > 0$. Since $(dz_{j_1}/dt)(t_1) \leq 0$ and $w_{j_1} > 0$ we obtain a contradiction in signs in the equation for z_{j_1}. If $\alpha = 0$, then the equation becomes

$$\frac{dz}{dt} = Lz. \qquad (3.22)$$

Assuming that $z(t)$ is not strictly positive, we easily obtain that it is identically zero for all j. We have $(dz_{j_1}/dt)(t_1) \leq 0$ and $(Lz)_{j_1}(t_1) \geq 0$. Then $(Lz)_{j_1}(t_1) = 0$, so all $z_j(t_1) = 0$. Since z_{j_1} verifies $dz_{j_1}/dt = (Lz)_{j_1}$, $z_{j_1}(t_1) = 0$, by the uniqueness we find $z_{j_1}(t) = 0$, $t \geq t_1$. Combining this with $z_j(t_1) = 0$, $(\forall) j \in \mathbb{Z}$, we get $z_j(t) = 0$, $(\forall) j \in \mathbb{Z}$, $(\forall) t \in (0, \infty)$.

Thus, in both cases $z_j(t)$ is positive for all j and t. We take $t \geq 0$ such that

$$e^{-i\beta t} = \frac{u_{j_0}}{|u_{j_0}|},$$

with j_0 from (3.16), i.e., $\cos \beta t = u_{j_0}^1/|u_{j_0}|$ and $\sin \beta t = -u_{j_0}^2/|u_{j_0}|$. Then, $v_{j_0}(t) = u_{j_0}^1 \cos \beta t - u_{j_0}^2 \sin \beta t = |u_{j_0}|$, hence with the aid of (3.16) we obtain the contradiction

$$z_{j_0}(t) = \tau w_{j_0} - |u_{j_0}| = 0.$$

The first assertion of the theorem is proved for nonreal λ.

Assume now that $\lambda \geq 0$ is real and that u is a nonzero bounded solution of (3.11). We suppose that at least one of the elements of the sequence $\{u_j\}$ is negative. Otherwise we could change the sign of u. We consider the sequence

$v = u + \tau w$, where $\tau > 0$ is chosen such that $v \geq 0$ for $|j| \leq N$, but $v_{j_0} = 0$ for some j_0, $|j_0| \leq N$. We have
$$Lv = \lambda v - \lambda \tau w, \tag{3.23}$$
and therefore $v_j \geq 0$ for all j by virtue of Lemma 3.6. Indeed, for $|j| \leq N$, the inequality holds because of the way we have chosen τ. For $j \geq N$, one applies Lemma 3.6 for (3.23) written in the form $(L - \lambda I)v = -\lambda \tau w$, $j \geq N$, with $v_N \geq 0$. If $j \leq -N$, the reasoning is similar.

If $\lambda > 0$, then the equation for v_{j_0} leads to a contradiction in signs. Thus equation (3.11) cannot have different from zero solutions for real positive λ.

2. If $\lambda = 0$, then we define $v = u + \tau w$ as above. Here u is the solution of (3.11) with $\lambda = 0$, i.e., $Lu = 0$. Using the above reasoning for $\lambda \geq 0$, we have $v_j \geq 0$, $(\forall)\, j \in \mathbb{Z}$, but it is not strictly positive (at least $v_{j_0} = 0$). In addition, v satisfies the equation $Lv = 0$. It follows from Lemma 3.2 that $v \equiv 0$. This implies $u_j = -\tau w_j$, $(\forall)\, j \in \mathbb{Z}$.

3. The limiting operators $L^{\pm}$ are operators with constant coefficients. The corresponding matrices are $(2m+1)$-diagonal matrices with constant elements along each diagonal. The matrices associated to the limiting operators $L^*_{\pm}$ of L^* are the transposed matrices, which are composed by the same diagonals reflected symmetrically with respect to the main diagonal. Therefore the polynomials $(P^*_\lambda)^{\pm}(\sigma)$ for the operator L^* will be the same as for the operator L. As indicated in the beginning of Section 3, the operator L^* satisfies the Fredholm property and it has the zero index.

We note first of all that equation (3.12) has a nonzero bounded solution v. Indeed, if such solution does not exist, then by virtue of the solvability conditions, the equation
$$Lu = f \tag{3.24}$$
is solvable for any f. This implies $\operatorname{Im} L = E$ and hence $\operatorname{codim}(\operatorname{Im} L) = 0$. Since the index of L is zero, it follows that $\dim(\ker L) = 0$. But by part two of the theorem, we get $\dim(\ker L) = 1$. This contradiction shows that a nonzero bounded solution v of equation (3.12) exists. Moreover, it is exponentially decreasing at infinity (see the end of Section 1.1).

We recall next that equation (3.24) is solvable if and only if
$$(f, v) = 0. \tag{3.25}$$

If $v \geq 0$, then from Lemma 3.2 for equation $L^* v = 0$, it follows that v is strictly positive, as claimed.

If we assume that a non-negative solution of equation (3.12) does not exist, then it has an alternating sign. Then we can find a bounded sequence $f < 0$ such that (3.25) is satisfied.

Let u be the corresponding solution of (3.24). There exists a τ (not necessarily positive), such that $\tilde{u} = u + \tau w \geq 0$ for $|j| \leq N$, but not strictly positive. Since

$L\tilde{u} = f$ and $f < 0$, $\tilde{u}_N \geq 0$, and $\tilde{u}_j \to 0$ as $j \to \infty$, by virtue of Lemma 3.6, one finds $\tilde{u} \geq 0$ for all j. But for those j where $\tilde{u}$ vanish, this leads to a contradiction in signs in the equation. Therefore $\tilde{u} > 0$. The theorem is proved. □

4 Stability of finite difference schemes

Finite difference approximation of differential equations in $\mathbb{R}^2$ leads to infinite-dimensional difference operators acting on sequences u_{ij} that depend on two indices i and j. In this sense, we call the corresponding equations multi-parameter equations. Consider the problem

$$\sum_{k=-m}^{m}\sum_{l=-n}^{n} a_{i+k,j+l} u_{i+k,j+l} = f_{ij}, \quad -\infty < i,j < \infty, \tag{4.1}$$

where m and n are some given integers. Let us introduce the notation

$$U = \{u_{ij}, \ -\infty < i,j < \infty\}, \quad F = \{f_{ij}, \ -\infty < i,j < \infty\},$$

$$(LU)_{ij} = \sum_{k=-m}^{m}\sum_{l=-n}^{n} a_{i+k,j+l} u_{i+k,j+l}.$$

Then equation (4.1) can be written as $LU = F$. We consider the function space

$$E = \{U = (u_{ij}), \ \sup_{ij} |u_{ij}| < \infty\}$$

with the norm $\|U\| = \sup_{ij} |u_{ij}|$. We will suppose that $\sup_{ij} |a_{ij}| < \infty$. It can be easily seen that if this condition is satisfied, then the operator $L : E \to E$ is bounded.

Definition 4.1. Let (i_n, j_n) be a sequence such that $|i_n| + |j_n| \to \infty$ as $n \to \infty$. Write $a_{ij}^n = a_{i+i_n, j+j_n}$. Suppose that a_{ij}^n converges locally, that is on every bounded set of the indices i, j to some $\hat{a}_{ij}$. Then the operator

$$(\hat{L}U)_{ij} = \sum_{k=-m}^{m}\sum_{l=-n}^{n} \hat{a}_{i+k,j+l} u_{i+k,j+l}$$

is called the limiting operator.

Condition NS. Any limiting problem $\hat{L}U = 0$ does not have nonzero solutions in E.

Theorem 4.2. *The operator L is normally solvable with a finite-dimensional kernel if and only if Condition NS is satisfied.*

The proof is similar to the proof in the case of one-parameter operators.

Condition NS can be formulated explicitly if limiting operators have constant (independent of i, j) coefficients. We restrict ourselves to the example of the difference operator obtained as a discretization of the elliptic equation $\Delta u = f$, $x \in \mathbb{R}^2$:

$$u_{i-1,j} + u_{i,j-1} - 4u_{i,j} + u_{i+1,j} + u_{i,j+1} = f_{ij}, \quad -\infty < i, j < \infty. \tag{4.2}$$

Put

$$v(\xi) = \sum_{i,j=-\infty}^{\infty} e^{i(i\xi_1 + j\xi_2)} u_{ij}, \quad g(\xi) = \sum_{i,j=-\infty}^{\infty} e^{i(i\xi_1 + j\xi_2)} f_{ij},$$

where $\xi_1, \xi_2 \in [0, 2\pi)$ and i is the imaginary unit. We multiply each equation in (4.2) by $\exp(i(i\xi_1 + j\xi_2))$ and take the sum with respect to i, j:

$$(e^{i\xi_1} + e^{i\xi_2} - 4 + e^{-i\xi_1} + e^{-i\xi_2}) v(\xi) = g(\xi).$$

We will call the expression

$$P_0(\xi_1, \xi_2) = \left(e^{i\xi_1/2} - e^{-i\xi_1/2}\right)^2 + \left(e^{i\xi_2/2} - e^{-i\xi_2/2}\right)^2$$
$$= -4\left(\sin^2(\xi_1/2) + \sin^2(\xi_2/2)\right)$$

the symbol of the operator

$$(L_0 U)_{ij} = u_{i-1,j} + u_{i,j-1} - 4u_{i,j} + u_{i+1,j} + u_{i,j+1}.$$

For $|\xi| = |\xi_1| + |\xi_2|$ sufficiently small,

$$P_0(\xi_1, \xi_2) \approx -(\xi_1^2 + \xi_2^2),$$

that is the symbol of the difference operator is approximated by the symbol of the corresponding differential operator.

The operator L_0 coincides with its unique limiting operator. Since the limiting problem $L_0 U = 0$ has nonzero solutions, which correspond to solutions of the equation $P_0(\xi_1, \xi_2) = 0$, then the operator L_0 is not normally solvable with a finite-dimensional kernel. Consequently, it does not satisfy the Fredholm property. The operator $L_0 - \sigma$, where σ is a positive constant is invertible.

Consider next the operator

$$(L_1 U)_{ij} = -u_{i,j} + u_{i,j-1} + a(u_{i-1,j} - 2u_{i,j} + u_{i+1,j}) - \sigma u_{i,j},$$

where a and σ are some positive constants. We have

$$P_1(\xi_1, \xi_2) = e^{i\xi_2} - 1 + a\left(e^{i\xi_1/2} - e^{-i\xi_1/2}\right)^2 - \sigma$$
$$= -4a \sin^2(\xi_1/2) + \cos(\xi_2) + i \sin(\xi_2) - 1 - \sigma,$$
$$\operatorname{Re} P_1(\xi_1, \xi_2) < 0 \ \forall \, \xi_1, \xi_2 \in \mathbb{R}.$$

4. Stability of finite difference schemes

If $\sigma = 0$, then the symbol P_1 has zeros and the operator does not satisfy the Fredholm property. We note that the operator L_1 corresponds to the finite different approximation of the parabolic equation

$$\frac{\partial u}{\partial t} = \frac{\partial^2 u}{\partial x^2} - \delta u$$

implicit with respect to time. Here $a = h_t/(h_x)^2$, $\sigma = \delta\, h_t$, where h_t is the time step and h_x is the space step.

For the operator

$$(L_2 U)_{ij} = -u_{i,j+1} + u_{i,j} + a(u_{i-1,j} - 2u_{i,j} + u_{i+1,j}) - \sigma u_{i,j}$$

we obtain

$$P_2(\xi_1, \xi_2) = -e^{-i\xi_2} + 1 + a\left(e^{i\xi_1/2} - e^{-i\xi_1/2}\right)^2 - \sigma$$
$$= -4a\sin^2(\xi_1/2) - \cos(\xi_2) + i\sin(\xi_2) + 1 - \sigma.$$

If $a < 1/2$, then for all $\sigma > 0$ sufficiently small, $P(\xi_1, \xi_2) \neq 0$. The operator L_2 arises as an explicit finite difference approximation of the same parabolic equation. The implicit scheme is unconditionally stable while the explicit scheme is stable if $a < 1/2$. Therefore, Condition NS is related to stability of the finite difference approximation.

The last example is related to discretization of the equation

$$\frac{\partial u}{\partial t} = \frac{\partial^2 u}{\partial x^2} + c\frac{\partial u}{\partial x} - \delta u.$$

The symbol of the operator

$$(L_3 U)_{ij} = -u_{i,j} + u_{i,j-1} + a(u_{i-1,j} - 2u_{i,j} + u_{i+1,j}) + c(u_{i+1,j} - u_{i,j}) - \sigma u_{i,j}$$

is

$$P_3(\xi_1, \xi_2) = -4a\sin^2(\xi_1/2) + c(\cos(\xi_1) - i\sin(\xi_1) - 1) + \cos(\xi_2) + i\sin(\xi_2) - 1 - \sigma.$$

If $c > 0$, then $P_3(\xi_1, \xi_2) \neq 0$. This corresponds to stability of the upwind discretization scheme. If $c < 0$, this is not the case.

Thus, in the examples above stability of finite difference schemes occurs when the corresponding difference operator is normally solvable with a finite-dimensional kernel.

Historical and Bibliographical Comments

The theory of elliptic equations was developed over more than two centuries. Various methods were suggested in the framework of this theory or came from other areas of mathematics. The theory of potential, the method of Green's functions, applications of holomorphic functions, and variational methods were used or developed in relation with elliptic equations already in the XIXth century. An important development started at the end of the XIXth beginning of the XXth centuries. Several methods were suggested to prove existence of solutions (Schwarz's method, method of successive approximations, some others); it was the beginning of the spectral theory, of the method of Fredholm integral equations, singular integral equations and boundary problems for analytical functions, and the method of a priori estimates, which received further development from the 1930s. Then the Leray-Schauder method and other topological methods, development of functional analysis and of the theory of function spaces formed our actual understanding of elliptic boundary value problems. Methods of numerical analysis were strongly developed in relation with computer simulations. We can also mention various asymptotic methods, the maximum principle, and some others. Many types of equations have been introduced and studied. Among them Laplace and Poisson equations, biharmonic, Navier-Stokes, Monge-Ampere, minimal surface, reaction-diffusion, Cauchy-Riemann, various degenerate or mixed equations, and some others. Combined with different boundary conditions, Dirichlet, Neumann, Robin, mixed, nonlinear boundary conditions, this creates a big variety of elliptic problems. Existence, uniqueness or nonuniqueness of solutions, and regularity are among traditional questions about solutions of elliptic problems. Another range of questions concerns solvability conditions, Fredholm property, and index. Furthermore, spectral properties and bifurcations of solutions; decay and growth, positiveness and various properties specific for particular applications.

All these methods, types of problems and the questions to study form the structure of the theory of elliptic boundary value problems. Historical and bibliographical comments presented below can help to follow its development (see also the reviews by Brezis and Browder [80], Nirenberg [385], the monographs by Miranda [352] and Sologub [501][2], the historical essay by Grattan-Guinness [211]).

1 Historical Notes

1.1 Beginning of the theory

Studies of gravimetrical problems prepared the way for appearance of the theory of potential and of the theory of elliptic partial differential equations. In 1686 Newton solved the problem of attraction between various bodies: a homogeneous ball and a ball composed of spherical layers, an ellipsoid and a material point at its axis [379]. In 1742 Maclaurin continued to study attraction of material points by ellipsoids [331]. Lagrange [287] (1773) and Laplace [296] (1782) investigated gravimetrical problems in a more general formulation and introduced the notion of potential. Euler [157], [159] (1736, 1765) and Bernoulli [62] (1748) were close to the notion of potential in their works on the motion of material points under the action of some forces even before it was introduced by Lagrange.

The second-order partial differential equation, later called the Laplace equation,

$$\frac{\partial^2 S}{\partial x^2} + \frac{\partial^2 S}{\partial y^2} + \frac{\partial^2 S}{\partial z^2} = 0 \qquad (1.1)$$

was first written by Euler in 1756 [158] when he studied potential motion of an incompressible fluid[3]; S is the velocity potential, which is related to the components of the velocity:

$$u = \frac{\partial S}{\partial x}, \quad v = \frac{\partial S}{\partial y}, \quad w = \frac{\partial S}{\partial z}.$$

He looked for solutions of this equation in the form of polynomials, $S = (Ax + by + Cz)^n$. Later, similar problems in fluid mechanics were studied by Lagrange (1788).

Laplace derived this equation in 1782 when he studied the properties of potentials in the problems of gravimetry [296]. First he wrote it in spherical coordinates and later in Descartes coordinates [297] (1787). He was mistaken in the derivation and considered this equation not only in the case where the point was outside the attracting body but also inside it. This was corrected later by Poisson. In [427] (1813) he derived the equation

$$\frac{\partial^2 V}{\partial x^2} + \frac{\partial^2 V}{\partial y^2} + \frac{\partial^2 V}{\partial z^2} = -4\pi\rho, \qquad (1.2)$$

---
[2]Sologub referred to [42], [93], [502] as to the only other general works devoted to the development of the theory of potential and elliptic equations in the XVIII–XIXth centuries.
[3]Euler used the notation dd/dx^2 for the second derivative [501].

1. Historical Notes

where ρ is the mass density distribution. Later he gave other derivations of this equation [428], [429]. Existing theory at that time could not treat a singularity under the integral which was necessary to justify the derivation. This equation was studied by Green, Ostrogradskii, Gauss. The first rigorous investigation was done by Gauss in 1840 [193]. He derived this equation in the case of continuous density, studied the properties of the potential and obtained the relation between the volume and surface integrals.

In 1828 Green published his "Essay on the application of mathematical analysis to the theory of electricity and magnetism" [212], [171] where he introduced his formulas and function. Green formulated the first boundary value problem for the Laplace equation and suggested a method of its solution. For a long time his work was unknown to the scientific community. In 1845 Thomson fortuitously obtained two copies of Green's work and sent it to Crelle for publication in his journal where it appeared in three parts from 1850 to 1854. Among the first who used Green's method were B. Riemann and Helmholtz [31].

Thomson investigated properties of the potentials and solutions of elliptic equations. In [516] (1847) he formulated theorems on the existence and uniqueness of solutions of the equation $\nabla \cdot (\alpha^2 \nabla V) = 4\pi\zeta$ in the whole space and of the Laplace equation $\Delta V = 0$ in a bounded domain with the boundary condition $\frac{\partial V}{\partial n} = F$. The proofs, which may not exactly correspond to nowadays standards, are based on the variational method. Gauss already used a similar method before [193] for the Laplace equation with a given value of the function at the boundary.

Further investigations of the first boundary value problem were carried out by Dirichlet. He proved the following assertion [301]. *For any bounded domain D in $\mathbb{R}^3$ there exists a unique function $u(x,y,z)$ continuous together with its first partial derivatives, which satisfies the equation $\Delta u = 0$ inside the domain and takes some given values at the boundary of the domain.* He considered the integral

$$U = \int_D \left(\left(\frac{\partial u}{\partial x}\right)^2 + \left(\frac{\partial u}{\partial y}\right)^2 + \left(\frac{\partial u}{\partial z}\right)^2 \right) dxdydz.$$

Assuming that there exists at least one function which minimizes the integral, Dirichlet proved that it is unique and that it is a solution of the boundary value problem. The Dirichlet principle of minimization was used by Riemann when he developed the theory of analytical functions [454].

A variational principle based on minimization of some functionals and used by Gauss, Thomson, Dirichlet, Riemann was postulated and not proved. In 1870 Weierstrass showed that it may not be true [578], that is that the minimizing function may not exist. His counterexample shocked the mathematical world of that time. Only in 1900 could Hilbert justify some of the existence results based on this approach [234].

As before, when Green formulated the first boundary problem for the Laplace equation studying some problems in electrostatics, problems of electrodynamics led Kirchhoff in 1845–1848 to introduce boundary value problems of the second type [261]. Later F. Neumann formulated them in the form as it is used now and gave a solution based on Green's formula [377]. Similar to Green, he was more interested in physical aspects and some mathematical details were ignored.

The third boundary value problem, together with the first one, was formulated by Fourier in his works on heat conduction [180], [181]. He gave their solution in the form of trigonometrical series[4].

Thus, at the first stage of the development of the theory of elliptic equations, boundary value problems for the Laplace and Poisson equations were formulated and some approaches to their solution were suggested. They were not yet rigorously justified from the mathematical point of view.

1.2 Existence of solutions of boundary value problems

In the second part of the XIXth century, existence of solutions of boundary value problems for the Laplace equation was investigated by Schwarz [476, 481] and Neumann [376] who gave the first rigorous proofs, Poincaré [420], Harnack [227] and other authors. In 1888, J. Riemann summarized and generalized preceding works by Schwarz and Harnack [453] (see also Paraf [400]). We briefly discuss below some of the methods developed at the time.

At the same time, the theory of potential experienced further development. In 1861 C. Neumann introduced logarithmic potential in order to study the first boundary value problem for the Laplace equation in plane domains [375], [376]. Laplace and Lame had used it before as a Newton potential of an infinite cylinder. Double layer potential was introduced in 1853 by Helmholtz in relation with some problems in electrodynamics [229]. It was later investigated by Lipschitz [320] and C. Neumann [376]. C. Neumann used the double layer potential in his method of arithmetic means which he developed to solve first and second boundary value problems. This method is based on successive approximations.

Let us describe in more detail two methods which had an important influence on further development of the theory of elliptic equations.

Schwarz's method. Schwarz applied conformal mappings in order to study existence of solutions of elliptic boundary value problems. However, he had to begin with a rigorous proof of Riemann's theorem. The proof given by his teacher was not complete because of the counterexample of Weierstrass to the variational principle. First of all, he proved that a polygon and then an arbitrary simply connected convex domain can be mapped to a circle [476] (1869), [477] (1870). As a consequence of these results, boundary value problems for the Laplace equation in simply connected convex domains were reduced to similar problems in the circle. Next, he gave an explicit solution for the first boundary problem in the circle in

---

[4]The method of representation of solutions in trigonometrical series was first used by Bernoulli for the problem of cord oscillation.

1. Historical Notes

the form of the Poisson integral [481]:

$$u(r,\phi) = \frac{1}{2\pi}\int_0^{2\pi} f(\psi)\frac{(1-r^2)d\psi}{1-2r\cos(\psi-\phi)+r^2}, \quad 0 \le r \le 1, \quad u(1,\phi) = f(\phi),$$

where $f(\phi)$ is a continuous periodic function given at the boundary of the circle. He used Fourier series. Similar results were obtained before by C. Neumann with the method of Green's function.

The next step of his construction deals with existence of solutions in domains which are not convex [478]. Let D_1 and D_2 be two convex domains with a nonempty intersection and with the boundaries S_1 and S_2. Schwarz's method allows the proof of the existence of solutions in the domain $D = D_1 \cup D_2$. Denote by Γ_1 the part of the boundary S_1 inside D_2 and by Γ_2 the part of the boundary S_2 inside D_1. The method is based on successive approximation where at each step the problems in domains D_1 and D_2 are solved by the method indicated above. Denote these solutions by $u_i^{(1)}$ and $u_i^{(2)}$. In order to solve the problem at the next step we complete the boundary conditions at the parts S_1 and S_2. The values of $u_i^{(2)}$ at S_1 and $u_i^{(1)}$ at S_2 are taken. Schwarz proved convergence of these approximations. A similar approach can be used for a union of several convex domains.

Successive approximations (Picard). In 1890 Picard published his study of nonlinear elliptic problems where he developed the method of successive approximations [411]. He considered the Dirichlet problem for the equation

$$A(x,y)\frac{\partial^2 u}{\partial x^2} + 2B(x,y)\frac{\partial^2 u}{\partial x \partial y} + C(x,y)\frac{\partial^2 u}{\partial y^2} = F\left(u, \frac{\partial u}{\partial x}, \frac{\partial u}{\partial y}, x, y\right)$$

assuming that $B^2 - AC \ne 0$. If we denote the linear operator in the left-hand side by L, then the method of successive approximations can be written as

$$Lu_n = F\left(u_{n-1}, \frac{\partial u_{n-1}}{\partial x}, \frac{\partial u_{n-1}}{\partial y}, x, y\right).$$

At each step, this linear equation completed by the boundary conditions can be solved. The main question is, of course, about convergence of the sequence of functions u_n. At this point, probably for the first time in the theory of elliptic equations, Picard came by necessity to use estimates of solutions. He obtained estimates of solutions together with their second derivatives in the case of small domains or for general domains (including unbounded) under more restrictive conditions on F. To pass from small to big domains, he used the method of Schwarz. Thus he proved the existence of solutions of a nonlinear elliptic boundary value problem.

The method of successive approximations was later used and generalized by Picard [412, 413], Le Roy [302], Lindeberg [317] (Neumann boundary condition),

Giraud [201, 202, 203] (in $\mathbb{R}^m$), and other authors. We note that Bernstein also used this method in order to prove analyticity of solutions [63] (see below).

The estimates obtained by Picard were not yet a priori estimates, which would appear later in the works by Bernstein. At each step of successive approximations he used existence of a solution in a more or less explicit form through Green's function. An interesting point to be emphasized is that to estimate second derivatives of solutions of the linear equation $\Delta u = f$, Picard assumed that f had continuous first derivatives [411]. After this assumption he remarked, with a reference to Harnack [227], who in his turn referred to Hölder, that this assumption was excessive and it would be sufficient to require that f satisfied the Hölder condition. On the other hand, it is not sufficient to assume that f is continuous. The estimate of the Hölder norm of the second derivative was obtained only in 1934 in the work by Schauder [470]. This "small" improvement of the estimate appeared to be crucial.

Other methods. Poincaré developed *méthode de balayage* (sweeping method) [420] based on construction of equivalent potentials. Let D be a bounded domain with the boundary S and with a point mass m at some point $P \in D$. Is it possible to construct a simple layer potential with some density ρ given on S in such a way that the value of the potential outside D equals m/r? Poincaré suggested a method of construction of such potential and used it to solve the Dirichlet problem for the Laplace equation. We will not discuss here this construction but will only mention that existence of such potential is equivalent to existence of Green's function [501]. If ρ is the desired density of the simple layer potential, then

$$G = \frac{1}{r} - \int_S \frac{\rho}{r}\, d\sigma$$

is Green's function of the Dirichlet problem. Hence it ensures the existence of a solution. Let us also mention that Poincaré generalized Neumann's method of arithmetic means for nonconvex domains.

Studying some problems of electrostatics, Robin came to the integral equation

$$\rho = \frac{1}{2\pi} \int_S \frac{\rho \cos \phi}{r^2}\, d\sigma,$$

where ϕ is the angle between the inner normal vector to the surface S at a given point P and a straight line connecting P with a point of the element $d\sigma$. He gave a solution of this equation based on the method of successive approximations [455], [456] (1886–1887). It is possible that Robin's method stimulated Fredholm in his method of integral equations [501].

We can also indicate the method by Kirchhoff for the first boundary value problem for convex domains [262], the works by Liapunov [313] (1898–1902) who justified the methods by Green, C. Neumann, Robin and the works by Steklov [505]–[508] (1897–1902) who obtained further generalizations of the existence results.

1. Historical Notes

1.3 Other elliptic equations

Helmholtz equation. The equation

$$\Delta u + k^2 u = 0$$

with various boundary conditions was intensively studied in relation with heat conduction problems. Ostrogradskii [393] (1828–1829) used the Fourier method and looked for a solution of the equation

$$\frac{\partial v}{\partial t} = a\Delta v$$

in the form of series,

$$v(x,y,z,t) = \sum_{i=1}^{\infty} A_i e^{-ak_i^2 t} u_i(x,y,z).$$

He assumed the existence of an infinite sequence of eigenvalues k_i and eigenfunctions u_i. Though Ostrogradskii had not yet introduced these notions precisely, he proved the orthogonality of eigenfunctions corresponding to different eigenvalues [501].

Helmholtz studied this equation in relation with propagation of sound [228] (1860). He introduced oscillating potentials, proved an analogue of Green's formula and represented the solution as a sum of oscillation and Newton potentials. These works were continued by Mathieu [335] in 1872. Weber studied the Helmholtz equation in the two-dimensional case [577] (1869). He proved existence of solutions of boundary value problems and existence of an infinite sequence of eigenvalues and eigenfunctions. His results were not completely justified because he used variational methods assuming the existence of functions minimizing the Dirichlet integral. Further important contributions to investigation of eigenvalues and eigenfunctions were made by Schwarz [480] (1885) and Poincaré in [420], [425] (1890, 1895).

Equation $\Delta u + p(x,y)u = 0$. This equation was first studied by Schwarz in 1872 for a particular form of the function p [479]. In 1885 he studied it for an arbitrary positive function p and with the boundary condition $u = 1$ at the boundary S of the domain T [480]. He considered a sequence of functions $u_0 = 1, u_1, u_2, \ldots$ which satisfy the equations

$$\Delta u_i + p u_{i-1} = 0, \quad u|_S = 0, \quad i = 1, 2, \ldots.$$

Then he proved that under certain conditions the series $u = u_0 + u_1 + \cdots$ converges and it gives a solution of the problem. For this purpose, he considered the integrals

$$W_n = \int_T p u_n \, dx dy, \quad n = 1, 2, \ldots$$

and the ratios
$$c_1 = \frac{W_1}{W_0}, \quad c_2 = \frac{W_2}{W_1}, \ldots, \quad c_n = \frac{W_n}{W_{n-1}}, \ldots.$$
He proved that this sequence grows and tends to some c. If $c < 1$, then the series u converges and represents a solution continuous in T together with its first derivatives. For an arbitrary c, the series
$$u = u_0 + tu_1 + t^2 u_2 + \cdots \tag{1.3}$$
converges for $|t| < 1/c$, and u is a solution of the equation
$$\Delta u + tp(x,y)u = 0$$
satisfying the boundary condition $u|_S = 1$. If $|t| = 1/c$, then the series diverges. In this case there exists an eigenfunction of the problem
$$\Delta u + k^2 p(x,y)u = 0, \quad u|_S = 0,$$
where $k^2 = 1/c$. Schwarz obtained also a variational representation of the eigenvalue and proved that its dependence on the domain T is continuous.

This equation was also studied by Picard and Poincaré. Picard showed that a wide class of elliptic equations can be reduced to this form by a change of variables. Poincaré used the method developed by Schwarz to prove existence of all eigenfunctions.

The equation $\Delta u - k^2 u = 0$ was first studied by Mathieu in relation with the Helmholtz equation [336], [337]. He used the variational principle to prove existence of solutions of the first boundary value problem.

Classification of the equations. Classification of second-order equations
$$A\frac{\partial^2 u}{\partial x^2} + 2B\frac{\partial^2 u}{\partial x \partial y} + C\frac{\partial^2 u}{\partial y^2} + F = 0$$
was introduced by Dubois-Reymond in 1889 [136]. Depending on the relation between the coefficients,
$$B^2 - AC > 0, \quad B^2 - AC = 0, \quad B^2 - AC < 0,$$
this equation is hyperbolic, parabolic or elliptic. It can be reduced to the corresponding canonical form by a change of variables.

Biharmonic equations. The biharmonic equation $\Delta^2 u = 0$ was first introduced and studied by Mathieu [334], [335], [337] (1869–1885) to solve some problems of elasticity. He proved existence and uniqueness of solutions of the corresponding boundary value problems. In the 1890s these works were continued for the biharmonic and n-harmonic equations by Almansi [14], [15], Boggio [73], [74], Gutzmer [224], Lauricella [298], Levi-Civita [307], [308], Venske [534].

1. Historical Notes

1.4 Analyticity

During the Mathematical Congress in Paris in 1900, Hilbert formulated the following problem.

Let z be a function of x and y bounded and continuous together with its derivatives up to the third order. If z is a solution of the equation

$$F\left(x, y, z, \frac{\partial z}{\partial x}, \frac{\partial z}{\partial y}, \frac{\partial^2 z}{\partial x^2}, \frac{\partial^2 z}{\partial x \partial y}, \frac{\partial^2 z}{\partial y^2}\right) = 0, \qquad (1.4)$$

where the function F is analytic and satisfies the inequality

$$4 F'_{\frac{\partial^2 z}{\partial x^2}} F'_{\frac{\partial^2 z}{\partial y^2}} - \left(F'_{\frac{\partial^2 z}{\partial x \partial y}}\right)^2 > 0,$$

then z is also analytic.

This problem was solved by Bernstein in 1904 [63] (a short communication was published in Note des Comptes Rendus, 1903). By that time, Picard [411] had proved analyticity of the solution of the linear equation

$$\frac{\partial^2 z}{\partial x^2} + \frac{\partial^2 z}{\partial y^2} + a\frac{\partial z}{\partial x} + b\frac{\partial z}{\partial y} + cz = 0,$$

and for some other equations (1895), Lutkemeyer (Göttingen, Dissertation 1902) and Holmgren (Mathemat. Annalen 1903), studied independently of each other the case where

$$F = \frac{\partial^2 z}{\partial x^2} + \frac{\partial^2 z}{\partial y^2} - f\left(x, y, z, \frac{\partial z}{\partial x}, \frac{\partial z}{\partial y}\right). \qquad (1.5)$$

Bernstein's works were important not only because of the proof of analyticity of the solution but, even more, because they influenced further development of the theory of elliptic problems and operator equations. Let us first briefly discuss the proof of analyticity. At each step of successive approximations, we need to solve a linear equation. Following Picard [413], Bernstein started with the equation

$$\Delta v = F(x, y)$$

in a circle of radius R with zero boundary conditions. He also used this example to explain the difference of his approach. The right-hand side and the solution are represented in the form of trigonometrical series:

$$F(x, y) = A_0(\rho) + \sum_n A_n(\rho)\cos(n\theta) + \sum_n B_n(\rho)\sin(n\theta),$$

$$v(x, y) = C_0(\rho) + \sum_n C_n(\rho)\cos(n\theta) + \sum_n D_n(\rho)\sin(n\theta).$$

Here (ρ, θ) are polar coordinates,

$$A_n(\rho) = \rho^n \sum_{p=0}^{\infty} \alpha_{np} \rho^{2p}, \quad B_n(\rho) = \rho^n \sum_{p=0}^{\infty} \beta_{np} \rho^{2p}.$$

Then

$$C_n(\rho) = \frac{\rho^n}{R^{n-2}} \sum_{p=0}^{\infty} \left(\frac{\rho^{2p+2}}{R^{2p+2}} - 1 \right) \frac{\alpha_{pn}}{(2p+2)(2p+2n+2)},$$

$$D_n(\rho) = \frac{\rho^n}{R^{n-2}} \sum_{p=0}^{\infty} \left(\frac{\rho^{2p+2}}{R^{2p+2}} - 1 \right) \frac{\beta_{pn}}{(2p+2)(2p+2n+2)}.$$

These representations allow one to estimate the solution and its first derivative. In the semi-linear problem (1.5), this proves convergence of the series for the solution and its derivatives and proves its analyticity. However, this is not sufficient for the nonlinear problem (1.4) because the series for the second derivatives of solutions may not converge. This is why Bernstein introduced what he called normal series instead of the usual Taylor expansion above:

$$A_n(\rho) = \rho^n \sum_{p=0}^{\infty} \sum_{q=0}^{\infty} \alpha_{pq} \rho^{2p} (R^2 - \rho^2)^q, \quad B_n(\rho) = \rho^n \sum_{p=0}^{\infty} \sum_{q=0}^{\infty} \beta_{pq} \rho^{2p} (R^2 - \rho^2)^q.$$

Then he obtained

$$C_n(\rho) = \rho^n \sum_{p=0}^{\infty} \sum_{q=0}^{\infty} c_{pq} \rho^{2p} (R^2 - \rho^2)^q, \quad D_n(\rho) = \rho^n \sum_{p=0}^{\infty} \sum_{q=0}^{\infty} d_{pq} \rho^{2p} (R^2 - \rho^2)^q.$$

The difference with the previous expressions is that $R^{2p+2} - \rho^{2p+2} \sim R - \rho$ as $\rho \to R$, while $(R^2 - \rho^2)^q \sim (R - \rho)^q$. This more rapid convergence to 0 near the boundary allowed him to obtain more precise estimates and also estimates of the second derivatives. This is important because they enter the right-hand sides in the method of successive approximations for the nonlinear equation. Thus, the proof is based on investigation of convergence of trigonometrical series with a special form of expansion for the coefficients.

This approach can be simplified if the circle is replaced by an annulus [67]. Much later, in 1959, Bernstein proved a priori estimates of solutions in Sobolev spaces [68] and remarked that they followed from his early works.

In the next work [64], Bernstein came back to these questions with a slightly modified approach. He developed and formulated more precisely some ideas which appeared already in the previous work. Namely, about a priori estimates of solutions and about introduction of a parameter in the equation. He proved the following theorem.

Theorem. *Given an analytic equation*

$$F(r,s,t,p,q,z,x,y,\alpha) = 0,$$

where $F'_r F'_z \leq 0$[5], the Dirichlet problem is solvable for any α between α_0 and α_1 if it is solvable for $\alpha = \alpha_0$ and if, assuming a priori existence of solution, it is possible to obtain a priori estimate of the modulus of z and of its first two derivatives by the boundary values.

In order to understand this result, we can think about the solution in the form of series with respect to the powers of $\alpha - \alpha_0$. A priori estimates provide convergence of the series with the radius of convergence independent of α_0. Hence, by a finite number of steps we can move from α_0 to α_1.

Bernstein formulated these results not only for analytic but also for regular and even irregular solutions. So the question was not already about analyticity but about a general method to study existence of solutions. He expected that it would open a wide field of research. This was confirmed in the 1930s when Leray and Schauder defined topological degree and applied it for elliptic equations. The Leray-Schauder method, which employs Bernstein's idea about deformation of a given problem to some model problem with a priori estimates of solutions, is now one of the most powerful and widely used methods to study operator equations.

1.5 Eigenvalues

Linear elliptic equations were intensively studied in the last years of the XIXth century in works by Poincaré, Neuman, Lyapunov, Steklov and other authors. The basis of the spectral theory was developed. Its beginning can be related to the work by Weber (1869) who studied the Dirichlet problem for the equation $\Delta u = \lambda u$. He stated the existence of a sequence of numbers λ_i for which this problem has a nonzero solution [577] (see also Mathieu [336]). Rigorous proof of the existence of one eigenvalue was first given by Schwarz (Fenn. Acta, XV, 1885). Picard proved existence of a second eigenvalue (C.R. CXVII, 1893) and Poincaré of infinity of eigenvalues (Rendiconti Palermo, VIII, 1894). The proof is based on the method of successive approximations. These works were continued by Poincaré, Steklov, Le Roy, and Zaremba. Expansion in the series with respect to eigenfunctions was obtained (Steklov [506, 507], Zaremba [588, 589], see [508] and the references therein). The solvability condition for the Laplace equation with the Neumann boundary condition was formulated by Zaremba and Steklov in the works cited above. All these questions had important development several years later due to the work by Fredholm on integral equations.

Let us briefly described the works by Poincaré where he proved the existence of all eigenvalues of the Laplace operator [423], [424]. Following Schwarz [480], he

---

[5]Technical condition which, according to Bernstein, can be removed though the proof becomes more complex.

considered the equation
$$\Delta u + tu + f = 0,$$
which depends on a complex parameter t. He studied the first boundary value problem in a 3D domain with an analytical boundary. He looked for the solution in the form of series (1.3) and proved that it is a meromorphic function. Its poles $k_1^2, k_2^2, \ldots$ are the eigenvalues, the residues $U_1, U_2, \ldots$ correspond to the eigenfunctions.

1.6 Fredholm theory

Fredholm in his work [183] (1900) suggested a new method to study integral equations and applied it to the Dirichlet problem for the Laplace equation. In the first part of the work he considered the equation

$$\phi(x) + \lambda \int_0^1 f(x, s)\phi(s)ds = \psi(x), \tag{1.6}$$

called later the Fredholm equation. Here λ is a real parameter, f and ψ are continuous functions of their arguments. The expression

$$D(\lambda) = 1 + \lambda \int_0^1 f(x_1, x_1)dx_1 + \frac{\lambda^2}{2}\int_0^1\int_0^1 f\begin{pmatrix} x_1, & x_2 \\ x_1, & x_2 \end{pmatrix} dx_1 dx_2 + \cdots$$

$$= \sum_{n=0}^{\infty} \frac{\lambda^n}{n!} \int_0^1 \cdots \int_0^1 f\begin{pmatrix} x_1, & x_2, & \ldots, & x_n \\ x_1, & x_2, & \ldots, & x_n \end{pmatrix} dx_1 dx_2 \ldots dx_n,$$

where

$$f\begin{pmatrix} x_1, & x_2, & \ldots, & x_n \\ y_1, & y_2, & \ldots, & y_n \end{pmatrix} = \begin{vmatrix} f(x_1, y_1) & f(x_1, y_2) & \cdots & f(x_1, y_n) \\ f(x_2, y_1) & f(x_2, y_2) & \cdots & f(x_2, y_n) \\ & \cdots & \cdots & \\ f(x_n, y_1) & f(x_n, y_2) & \cdots & f(x_n, y_n) \end{vmatrix}$$

is called the determinant of the equation. From the estimate

$$\left| f\begin{pmatrix} x_1, & x_2, & \ldots, & x_n \\ y_1, & y_2, & \ldots, & y_n \end{pmatrix} \right| < \sqrt{n^n} \sup |f(x,y)|$$

follows the convergence of the series in the definition of $D(\lambda)$ for all λ. The series

$$D_1(\xi, \eta) = f(\xi, \eta) + \lambda \int_0^1 f\begin{pmatrix} \xi, & x_1 \\ \eta, & x_1 \end{pmatrix} dx_1$$

$$+ \frac{\lambda^2}{2} \int_0^1 \int_0^1 f\begin{pmatrix} \xi, & x_1, & x_2 \\ \eta, & x_1, & x_2 \end{pmatrix} dx_1 dx_2 + \cdots$$

$$= \sum_{n=0}^{\infty} \frac{\lambda^n}{n!} \int_0^1 \cdots \int_0^1 f\begin{pmatrix} \xi, & x_1, & x_2, & \ldots, & x_n \\ \eta, & x_1, & x_2, & \ldots, & x_n \end{pmatrix} dx_1 dx_2 \ldots dx_n$$

1. Historical Notes

also converges for all λ. Fredholm proved the following relation:
$$D_1(\xi, \eta) = f(\xi, \eta) D(\lambda) - \lambda \int_0^1 f(\xi, \tau) D_1(\tau, \eta) d\tau.$$
From this formula it follows that the function
$$\Phi(x) = \psi(x) D(\lambda) - \lambda \int_0^1 D_1(x, t) \psi(t) dt \tag{1.7}$$
satisfies the equation
$$\Phi(x) + \lambda \int_0^1 f(x, s) \Phi(s) ds = \psi(s) D(\lambda). \tag{1.8}$$
Therefore, if $D(\lambda) \neq 0$, then equation (1.6) has a unique continuous solution
$$\phi(x) = \frac{\Phi(x)}{D(\lambda)} = \psi(x) - \lambda \int_0^1 \frac{D_1(x, t) \psi(t)}{D(\lambda)} dt.$$
Next, assuming that λ_0 is a zero of $D(\lambda)$ of multiplicity ν, that is $D(\lambda) = (\lambda - \lambda_0)^\nu D_0(\lambda)$, Fredholm obtained $D_1(\xi, \eta) = (\lambda - \lambda_0)^{\nu_1} D_1(\xi, \eta)$, where $\nu \geq \nu_1 + 1$. Hence, from (1.7) it follows that $\Phi(x) = (\lambda - \lambda_0)^{\nu_1} \Phi_1(x)$, and from (1.8) that $\Phi_1(x)$ is a solution of the homogeneous equation
$$\phi(x) + \lambda_0 \int_0^1 f(x, s) \phi(s) ds = 0.$$
Thus, either equation (1.6) has a unique solution or the homogeneous equation has a nonzero solution.

In the second part of the work [183], Fredholm applied this result to study the first boundary value problem for the Laplace equation in plane domains. Let L be a closed curve given parametrically by the functions $\xi = \xi(s), \eta = \eta(s)$, where s is the length of the arc. It is assumed that these functions are sufficiently smooth and the length of L equals 1. Fredholm formulated the problem to find a double layer potential w such that it satisfies the relation
$$v - v' = \lambda(v + v') + 2\psi$$
on L. Here v and v' are the limiting values of the potential from inside and outside of the curve, respectively, and ψ is a continuous function given on L. The value $\lambda = -1$ corresponds to the internal problem, $\lambda = 1$ to the external one. Denote by $\phi(s)/\pi$ the density of the potential. Then
$$w(x, y) = \frac{1}{\pi} \int_0^1 \phi(s) d\left(\arctan \frac{\eta - y}{\xi - x}\right),$$
$$v = \phi(s_0) + \frac{1}{\pi} \int_0^1 \phi(s) \frac{\partial}{\partial s} \left(\arctan \frac{\eta(s) - \eta(s_0)}{\xi(s) - \xi(s_0)}\right),$$
$$v' = \phi(s_0) - \frac{1}{\pi} \int_0^1 \phi(s) \frac{\partial}{\partial s} \left(\arctan \frac{\eta(s) - \eta(s_0)}{\xi(s) - \xi(s_0)}\right).$$

Writing
$$f(s_0, s) = \frac{1}{\pi} \frac{\partial}{\partial s}\left(\arctan \frac{\eta(s) - \eta(s_0)}{\xi(s) - \xi(s_0)}\right),$$
Fredholm obtained the integral equation
$$\phi(s_0) - \lambda \int_0^1 \phi(s) f(s_0, s) ds = \psi(s_0).$$

Hence the results on the solvability of equation (1.6) become applicable to the first boundary problem for the Laplace equation. This work had an important influence on further development of the theory of elliptic equations.

In 1903, Fredholm published his work [184] devoted to the integral equation
$$\phi(\xi, \eta) + \lambda \int_a^b \int_a^b f(\xi, \eta, x, y) \phi(x, y) dx dy = \psi(\xi, \eta), \tag{1.9}$$
where λ is a parameter, ψ and f are bounded functions. Determinant D of this equation is defined as
$$D = 1 + \sum_{k=1}^{\infty} d_k \lambda^k,$$
where
$$d_k = \frac{1}{k!} \int_a^b \cdots \int_a^b F_k(x_1, \ldots, x_k, y_1, \ldots, y_k)\, dx_1 \ldots dx_k\, dy_1 \ldots dy_k,$$
$$F_k = \begin{vmatrix} f(x_1,y_1,x_1,y_1) & f(x_1,y_1,x_2,y_2) & f(x_1,y_1,x_3,y_3) & \cdots & f(x_1,y_1,x_k,y_k) \\ f(x_2,y_2,x_1,y_1) & f(x_2,y_2,x_2,y_2) & f(x_2,y_2,x_3,y_3) & \cdots & f(x_2,y_2,x_k,y_k) \\ \cdots & \cdots & \cdots & \cdots & \cdots \\ f(x_k,y_k,x_1,y_1) & f(x_k,y_k,x_2,y_2) & f(x_k,y_k,x_3,y_3) & \cdots & f(x_k,y_k,x_k,y_k) \end{vmatrix}.$$

Another function of λ,
$$M_n = \sum_{k=1}^{\infty} \delta_k^n \lambda^{k+n},$$
is called the minor of order n. The coefficients δ_k^n are explicitly given by formulas similar to the coefficients d_k. He proved the following theorem.

Theorem. *Nonzero solutions of the homogeneous equation*
$$\phi(\xi, \eta) + \lambda \int_a^b \int_a^b f(\xi, \eta, x, y) \phi(x, y) dx dy = 0$$
exist if and only if λ is a root of D. If n is the order of the first minor which does not vanish for this value of λ, then there are n linearly independent solutions

1. Historical Notes

Φ_m, $m = 1,\ldots,n$ of this equation. In this case, nonhomogeneous equation (1.9) has a solution if and only if the right-hand side ψ satisfies the equalities[6]

$$\int_a^b \int_a^b \psi(x,y)\Psi_m(x,y)dxdy = 0, \quad m = 1,\ldots,n.$$

Hilbert generalized this result for the case where the function f had a logarithmic singularity[7]. This generalization, used by Mason in the paper [333], appeared in 1904 in order to apply these results for elliptic equations. He studied the equation

$$\Delta u + \lambda A(x,y)u = f(x,y) \qquad (1.10)$$

in a bounded domain Ω with the boundary condition $u = \sigma(s)$ at the boundary S. The functions A and f are supposed to be bounded, the boundary S is composed by a finite number of arcs of analytic curves. It was known that in this case there exists Green's function G of the Laplace operator with the zero boundary condition. Applying Green's formula, he obtained the equation

$$u(\xi,\eta) + \lambda \int_a^b \int_a^b \frac{A(x,y)}{2\pi} G(x,y,\xi,\eta)u(x,y)dxdy = F(\xi,\eta), \qquad (1.11)$$

where

$$F(\xi,\eta) = -\frac{1}{2\pi}\int_S \sigma \frac{\partial G}{\partial n}ds - \frac{1}{2\pi}\int_\Omega f(x,y)G(x,y,\xi,\eta)dxdy.$$

Domain Ω contains in the square $[a,b] \times [a,b]$ and Green's function is extended by zero. Fredholm's theorem can now be applied to equation (1.11). It gives the following result.

Theorem. *There exists a unique solution of equation (1.10) for any λ different from the roots of the function D. If λ is a root of this function and n is the order of the first nonzero minor, then there are n linearly independent solutions $\Phi_k(x,y)$ of the homogeneous problem ($f - 0, \sigma - 0$). The nonhomogeneous problem is solvable if and only if the following conditions are satisfied[8]:*

$$\int_\Omega f(x,y)\Phi_k(x,y)dxdx - \int_S \sigma\frac{\partial \Phi_k}{\partial n}ds = 0, \quad k = 1,2,\ldots,n.$$

Mason proved existence of an infinite number of λ_i for which the homogeneous problem has a nonzero solution. Other boundary conditions were also considered. Picard used reduction to Fredholm equations in [414]–[416] (see Section 2.2.1 for other references).

---

[6] Explicit expressions for the functions Φ_m, Ψ_m and for the solution are given. We do not present them here because of their complexity.

[7] Mason [333] referred to Hilbert's lectures in Göttingen in 1901–1902 and to the dissertations of Kellog (1902) and Andrae (1903) who generalized Hilbert's proof for the case of n variables.

[8] Since the problem is self-adjoint, solvability conditions are formulated in terms of solutions of the homogeneous equation which coincides with the homogeneous adjoint equation

2 Linear equations

2.1 A priori estimates

In 1929 Hopf published the work [238] devoted to the elliptic system of the first order
$$au'_x + bu'_y - v'_y = f, \quad bu'_x + cu'_y + v'_x = g,$$
where $ac - b^2 > 0$ and the coefficients a, b, and c satisfy Hölder's condition. Using Green's formula, he obtained an interior estimate of the first derivatives of the solution through the Hölder norms of the right-hand sides and the maxima of $|u|$ and $|v|$. In his notation,
$$|u'_x|, |u'_y| \leq \beta_1 R^{-1}([u] + [v]) + \beta_2([f] + [g]) + \beta_3 R^\alpha (H_\alpha[f] + H_\alpha[g]),$$
where $R \leq d$, and d is the distance to the boundary. In his next work, [239], he proved that certain derivatives of the solution of a linear second-order equation are Hölder continuous. More precisely, it is assumed that the coefficients and the right-hand side of the equation (notation is changed)
$$Lu = \sum a_{ik}(x_1, \ldots, x_n)\frac{\partial^2 u}{\partial x_i \partial x_k} + \sum b_j(x_1, \ldots, x_n)\frac{\partial u}{\partial x_j} + c(x_1, \ldots, x_n)u = f \tag{2.1}$$
are m-Hölder continuous, that is have continuous derivatives up to order m and the mth derivatives satisfy the Hölder condition. Then the solution is $(m+2)$-Hölder continuous. The estimates of the Hölder norm of the solution were not given.

The same year, in [467], among other results Schauder presented the estimate
$$\|\omega\|_{2,\alpha} \leq M(\|\rho\|_\alpha + \|\phi\|_{2,\alpha})$$
for the equation
$$A(x,y)\frac{\partial^2 \omega}{\partial x^2} + B(x,y)\frac{\partial^2 \omega}{\partial x \partial y} + C(x,y)\frac{\partial^2 \omega}{\partial y^2} = \rho(x,y)$$
with the boundary condition $\omega = \phi$. The coefficients satisfy the Hölder condition with the exponent α, and $4AC - B^2 > 0$. He did not give the proof and referred to the work by L. Lichtenstein [314]. In this work, the estimate was not explicitly formulated but followed from the method of proof of the existence of a solution which used the results by Levi [306]. In the next work [468], Schauder proved a similar estimate for the same equation in $\mathbb{R}^n$.

In 1934 Schauder published his work [470] devoted specifically to a priori estimates for the second-order linear elliptic equation (2.1) in a bounded domain G with the boundary R of the class $C^{2+\alpha}$ and with the Dirichlet boundary condition, $u = \phi$ at R. The coefficients b_j and c are α-Hölder continuous as well as the function f, the coefficients a_{ik} satisfy Hölder condition with the exponent $\alpha + \epsilon$,

2. Linear equations

$\det(a_{ik}) = 1$. He proved that if a solution u is twice α-Hölder continuous then the following estimate holds:

$$\|u\|_{2,\alpha}^G \leq K\left(\|f\|_\alpha^G + \|\phi\|_{\alpha,2}^R + \max_G |u|\right).$$

The method of proof is based on the estimates of the potentials in the $(2+\alpha)$-Hölder norm. Commented analysis of original Schauder's proof (with English translation) is presented by Barrar [49].

These works by Hopf and Schauder, based on the prevailing theory of elliptic boundary value problems at that time, determined to some extent its further development, emphasizing the role of a priori estimates. In subsequent works, Morrey obtained estimates of Hölder norms for second-order equations and systems [355, 356, 357] (see also Miranda [352], [353], Yudovich [584]), interior estimates for general elliptic systems were proved by Douglis and Nirenberg [134], interior estimates for second-order systems by Nash [370] and Morrey [358], interior and boundary estimates for higher-order equations by Browder [85] and by Agmon, Douglis, Nirenberg [7]; estimates in maximum norms, Miranda [354], Agmon [6]; second-order equations in unbounded domains, Oskolkov [395]. This cycle of works was concluded by Agmon, Douglis, Nirenberg [8] who studied general elliptic systems in bounded or unbounded domains.

In parallel to these works, estimates in Sobolev spaces were investigated. Second-order or higher-order equations and systems with the Dirichlet or other types of boundary conditions, interior and boundary estimates in L^p-norms with $p=2$ or $p \neq 2$ were studied by Ladyzhenskaya [283], Guseva [220], Nirenberg [380, 381], Browder [85, 86], Koselev [267]–[270], Hörmander [241], Slobodetskii [495], Agmon, Douglis, Nirenberg [7, 8] Peetre [402], Yudovich [584]–[586], Volevich [542]. Estimates for general elliptic systems were obtained in [8] and [542].

Further investigations of a priori estimates were devoted to various generalizations: non smooth and unbounded domains and coefficients, degenerate equations, more general notions of elliptic problems.

Weighted spaces. Consider the equation $u'' = f$ on the real line. Suppose that $|f(x)| \sim |x|^p$ as $|x| \to \infty$ for $p \neq -1, -2$. Then it follows from the equation that $|u'(x)| \sim |x|^{p+1}$ and $|u(x)| \sim |x|^{p+2}$. The values $p = -1, -2$ are singular because integration of $1/x$ gives $\ln x$ and not a power of x. Such behavior of solutions suggests that we introduce weighted spaces with polynomial weights different for u and for f. Similar to Chapter 4, a priori estimates of solutions in these spaces allow us to prove that the operator is normally solvable with a finite-dimensional kernel. It is important to note that the operator satisfies the Fredholm property in the properly chosen weighted spaces while this is not true in the spaces without weight. The dimension of the kernel and the codimension of the image of the operator $L_0 = u''$ can be found explicitly. If the operator contains lower-order terms, $Lu = u'' + Bu$, where $Bu = a(x)u' + b(x)u$, with the coefficients $a(x)$ and

$b(x)$ decaying at infinity, and the decay is sufficiently fast, then the operator B is compact. If L_0 satisfies the Fredholm property, then it is also valid for the operator L, and it has the same index.

Consider now the class of operators $A = A_\infty + A_0$ in $\mathbb{R}^n$, where A_∞ is a homogeneous elliptic operator with constant coefficients and A_0 contains lower-order terms with the coefficients converging to zero at infinity. Then this operator does not satisfy the Fredholm property in the usual Sobolev or Hölder spaces. Indeed, the limiting problem $A_\infty u = 0$ in $\mathbb{R}^n$ has a nonzero solution $u =$ const. However, it can be satisfied in some special weighted spaces. In [386] the following a priori estimate is obtained:

$$\sum_{|\alpha|\le m} \left\| |x|^{|\alpha|+\rho} \frac{\partial^\alpha u}{\partial x^\alpha} \right\|_p \le C_0 \left\| |x|^{m+\rho} A_\infty u \right\|_p.$$

Here m is the order of the operator, $\|\cdot\|_p$ denotes the $L^p(\mathbb{R}^n)$ norm, $u \in H^{m,p}(\mathbb{R}^n)$, $A_\infty u \in L^p(\mathbb{R}^n)$. This estimate holds if and only if $-n/p < \rho < r - m + n/p$ and $\rho + m - n/p'$ is not a nonnegative integer, where r is the smallest nonnegative integer greater than $(m - n/p')$, p' is determined by the equality $1/p + 1/p' = 1$.

This estimate allows the authors to obtain the estimate

$$\sum_{|\alpha|\le m} \left\| \sigma^{|\alpha|+\rho} \frac{\partial^\alpha u}{\partial x^\alpha} \right\|_p \le C \left(\left\| \sigma^{m+\rho} A u \right\|_p + \|\sigma^\rho u\|_p \right)$$

for the operator A and to prove that its kernel has a finite dimension. Here $\sigma(x) = (1 + |x|^2)^{1/2}$ and the coefficients of the operator A_0 have certain decay rate at infinity.

The Fredholm property of this class of operators is studied in [323], [324]. It is assumed that if $\rho > -n/p$ then $\rho + m - n/p' \notin \mathbb{N}$ (cf. above), if $\rho \le -n/p$ then $-(\rho + n/p) \notin \mathbb{N}$. The kernel and cokernel of the operator consist of polynomials and their dimensions equal to $d(-\rho-n/p)-d(-\rho-m-n/p)$ and $d(\rho+m-n/p')-d(\rho-n/p')$, respectively. Here $d(k)$ is the dimension of the space of polynomials of degree less than or equal to k. It is supposed to be 0 if $k < 0$. Similar problems for elliptic operators in Hölder spaces are investigated in [57], [75], [255]. Exterior problems for the Laplace operator in weighted Sobolev spaces are considered in [20], [21] (see also [524]), and for more general operators in Hölder spaces in [76]. The dimension of the kernel and the Fredholm property are studied in [575], [576] in the case where A_∞ is a first-order operator and the coefficients of the operator A_0 vanish outside of some ball. Invertibility of the operator $(-1)^m \Delta^m u + V(x)u$ in weighted spaces was studied by Kondratiev [265].

2.2 Normal solvability and Fredholm property

2.2.1. Reduction to integral equations. After publication of Fredholm's papers, the theory of integral equations and its applications to elliptic problems had an important development. Several monographs appeared: Heywood and Fréchet [230],

2. Linear equations

Hilbert [236], Lalescu [288] (all three published in 1912), Volterra et al. [572] (1913), Goursat [209] (1917), Muskhelisvili [364] (1922). Later these methods were presented in numerous papers and monographs (see, e.g., [528], [406], [78], [145]). Let us recall a method of reduction of elliptic boundary value problems to Fredholm equations (cf. Section 1.6). Consider the Dirichlet problem

$$\Delta u = 0, \quad u|_S = f(s) \tag{2.2}$$

or the Neumann problem for the Laplace equation

$$\Delta u = 0, \quad \frac{\partial u}{\partial n}|_{\partial \Omega} = g(x). \tag{2.3}$$

Here $\Omega \subset \mathbb{R}^2$ is either a bounded simply connected domain with a sufficiently smooth boundary[9] S, or the exterior domain. Consider the simple layer potential $V(x)$ and the double layer potential $W(x)$ defined by the equalities

$$V(x) = \int_S \frac{\rho(s)}{r} ds, \quad W(x) = \int_S \sigma(s) \frac{\partial \ln r}{\partial n} ds,$$

where ρ and σ are the densities of the potentials, $r = |x - s|$, n is the outer normal vector, ds the element of the arc's length. Then

$$W_i(x) = -\pi \sigma(x) - \int_S \sigma(s) \frac{\cos(r,n)}{r} ds, \quad W_e(x) = \pi \sigma(x) - \int_S \sigma(s) \frac{\cos(r,n)}{r} ds,$$

$$\frac{\partial V_i}{\partial n} = \pi \rho(x) + \int_S \rho(s) \frac{\cos(r,n)}{r} ds, \quad \frac{\partial V_e}{\partial n} = -\pi \rho(x) + \int_S \rho(s) \frac{\cos(r,n)}{r} ds,$$

where the subscript i signifies the limit of the corresponding function or of the derivative from inside of the domain, e from outside. Looking for the solution of the Dirichlet problem in the form of the double layer potential and taking into account the boundary condition, we obtain the integral equations

$$\sigma(x) \pm \frac{1}{\pi} \int_S \sigma(s) \frac{\cos(r,n)}{r} ds = \mp \frac{1}{\pi} f(x), \quad x \in S \tag{2.4}$$

with respect to the density of the potential. The upper sign corresponds to the interior problem, the lower sign to the exterior one. Similarly, for the Neumann problem

$$\rho(x) \mp \frac{1}{\pi} \int_S \rho(s) \frac{\cos(r,n)}{r} ds = \mp \frac{1}{\pi} f(x), \quad x \in S. \tag{2.5}$$

---

[9] In fact, it is supposed to be a Liapunov contour defined by the following conditions: 1. there exists a tangent and a normal vector at each point of the boundary, 2. the angle θ between the normal vectors at two points can be estimated by Ar^α, where r, the distance between the points A, is a positive constant, $0 < \alpha \leq 1$, 3. The boundary can be represented as a univocal function in a δ-neighborhood with the same δ for each point.

Since $|\cos(r,n)/r| < c/r^{1-\alpha}$, where α is the constant in the definition of Liapunov's contour, then the kernel has a weak singularity, and the theory of Fredholm equations is applicable. The interior Dirichlet and exterior Neumann problems are adjoint to each other, the same as the exterior Dirichlet problem and interior Neumann problem. This allows one to obtain solvability conditions. This method has numerous generalizations including multiple connected domains, multi-dimensional problems and so on.

2.2.2 Singular integral equations. Singular integral equations were introduced by Poincaré when he studied some problems in fluid mechanics (rising tides) [426] and by Hilbert in his works on boundary problems for analytical function [235], [236]. Normal solvability and index of one-dimensional singular integral equations were obtained by Nöther [389] in 1921[10]. Normal solvability for systems of equations was proved by Giraud [204] in 1939 and several years later by Muskheleshvili and Vekua [365] who also obtained a formula for the index (see also [366], [532]).

Elliptic boundary value problems can be reduced to singular integral equations. If we look for the solution of the Dirichlet problem in the form of the double layer potential, as it is done in the previous section, then we obtain an integral equation of Fredholm type. If we look for the solution of this problem in the form of the simple layer potential, then we obtain the singular integral equation

$$\int_S \rho(t) \frac{\cos \alpha(x,t)}{r} ds = f(x), \qquad (2.6)$$

where x and t are points on the contour S, r the distance between them, and $\alpha(x,t)$ the angle between the vector $\overline{x}t$ and the positive tangent at t. This equation can also be written in the complex form

$$\int_S \frac{\rho(t)}{t-x} dt - i \int_S \rho(t) \frac{\cos(r,n)}{r} ds = f(x)$$

with the singular part given by the Cauchy integral and its regular part similar to the integral term in equation (2.5). Equation (2.6) was obtained by Bertrand [69] in 1923. Its complete study was possible only later [366] due to development of the theory of singular integral equations.

Reduction of some second-order elliptic equations and systems of equations to singular integral equations is discussed in the monograph by I.N. Vekua [528] (see also [527], [529], [530]). Normal solvability and a formula for the index of general elliptic boundary value problems in bounded plane domains were obtained in the 1950s by A.I. Volpert [544]–[552]. Agranovich and Dynin proved normal solvability, homotopy invariance of index and its reduction to the index of singular

---

[10] Normally solvable operators with a finite-dimensional kernel and a finite codimension of the image were called Nötherian operators, while Fredholm operators were used for the particular case of zero index. Nowadays the term Fredholm operators is more often used independently of the value of the index.

2. Linear equations

integral equations for multi-dimensional problems [9], [10], [12]. Dzhuraev studied multiple connected domains [146]. Reduction of elliptic boundary value problems to integral equations are also discussed in more recent papers and monographs [419], [583], [344].

Boundary problems. Elliptic boundary value problems can be reduced to boundary problems for analytical functions directly (see, e.g., Section 8.2 in Chapter 8) or through singular integral equations. The method of reduction of singular integral equations to boundary problems for analytical functions was suggested by Carleman [105]. It was later developed and generalized by Gakhov, Muskhelishvili, and many other authors (see [190], [366], [532] and the references therein). Consider the singular integral equation

$$a(t)u(t) + \frac{b(t)}{\pi i} \int_S \frac{u(\zeta)}{\zeta - t} \, d\zeta = f(t), \tag{2.7}$$

where S is a closed simple contour, the coefficients $a(t), b(t)$ satisfy the Lipschitz condition. Denote by D^+ the domain inside the contour, by D^- outside it and consider the Cauchy integral

$$\Phi(z) = \frac{1}{2\pi i} \int_S \frac{u(\zeta)}{\zeta - z} \, d\zeta.$$

It determines two holomorphic functions Φ^+ and Φ^-, respectively in the inner and outer domains. Taking into account their limiting values at the contour, we reduce equation (2.7) to the equation

$$\Phi^+(t) - \frac{a(t) - b(t)}{a(t) + b(t)} \Phi^-(t) = \frac{f(t)}{a(t) + b(t)}, \quad t \in S. \tag{2.8}$$

Thus, the singular integral equation (2.7) is reduced to the Hilbert problem[11]: find two holomorphic functions Φ^+ and Φ^- defined in the domains D^+ and D^- and satisfying some linear relation at the contour S.

The Riemann-Hilbert problem is to find a holomorphic function $\Phi(z) = u + iv$ in a bounded or unbounded domain D continuous up to the boundary S and satisfying the boundary condition $au - bv = c$ on S[12]. Here a, b, and c are real-valued continuous functions, the contour S is closed and simple. It can be reduced to the Hilbert problem and it is clearly related to the Dirichlet problem ($b = 0, c = f$). Solvability conditions and index of these boundary problems and of their various generalizations are known. Solvability conditions are formulated in terms of adjoint problems. The index is expressed as a rotation of some vector field at the contour S (cf. Chapter 8).

Some other examples which show the connection between elliptic problems, singular integral equations and boundary problems for analytical functions are

---

[11] It is sometimes also called the Riemann problem. It was introduced by Hilbert in 1905 (see [236]). He referred to it as the Riemann problem.
[12] It is a particular case of the problem formulated by Riemann in his dissertation [454].

presented in Section 9 of Chapter 8. More detailed presentation of this topic can be found elsewhere. We will return to multi-dimensional singular integral equations below when we discuss the index of elliptic problems.

2.2.3 Elliptic operators. A priori estimates of solutions of elliptic boundary value problems determine their normal solvability. Consider an operator L acting from a Banach space E into another Banach space F and suppose that the following estimate holds:

$$\|u\|_E \leq K \left(\|Lu\|_F + \|u\|_{E_0}\right) \tag{2.9}$$

for any $u \in E$ and with a constant K independent of u. Here E_0 is a wider space such that E is compactly embedded in it. To be more specific, let L be a second-order elliptic operator, $E = C^{2+\alpha}(\bar{\Omega})$ and $F = C^{\alpha}(\bar{\Omega})$, $E_0 = C^2(\bar{\Omega})$, Ω be a bounded domain with a $C^{2+\alpha}$ boundary. In this case (2.9) is equivalent to the Schauder estimate by virtue of the estimate

$$\|u\|_{E_0} \leq \epsilon \|u\|_E + C_\epsilon \|u\|_{C^0(\bar{\Omega})},$$

where ϵ is positive and arbitrarily small, the constant C_ϵ depends on ϵ.

If u_n is a bounded sequence in E, then it has a subsequence u_{n_k}, which converges to some u_0 in the C^2-norm. It can be easily verified that $u_0 \in E$. Indeed, it is sufficient to pass to the limit in the estimate $|D^2 u_{n_k}(x_1) - D^2 u_{n_k}(x_2)|/|x_1 - x_2|^\alpha \leq M$ as k tends to infinity.

We can now verify that the image of the operator is closed. If $Lu_n = f_n$ and $f_n \to f_0$ in F for some f_0, then $Lu_0 = f_0$. By virtue of (2.9), $u_{n_k} \to u_0$ in E. This simple proof remains valid for general elliptic problems (see Chapter 4 for more detail). In order to show that the kernel is finite dimensional, it is sufficient to note that it follows from compactness of the unit ball in the subspace of functions for which $Lu = 0$. This can be proved in a similar way. A priori estimates for adjoint or formally adjoint operators imply the finite codimension of the image.

In addition to the works devoted to a priori estimates cited above in Section 2.1, let us indicate the papers by Vishik [536], [537] where he introduced strongly elliptic operators and proved their Fredholm property. Related questions were studied later by Browder [84]. Schechter obtained solvability conditions for normal systems (Section 1, Chapter 6) in terms of formally adjoint problems [473]. Agranovich and Dynin proved normal solvability in Sobolev spaces and showed that the index of elliptic problems equals the index of some singular integral operator [12].

2.2.4 Limiting problems and unbounded domains. In the case of unbounded domains, the usual a priori estimates (Section 2.1) are not sufficient for normal solvability. We need to obtain some stronger estimates (Section 2.5 of Chapter 1 and Chapter 4) or, equivalently, to impose Condition NS in addition to the usual estimates.

Elliptic operators in the sense of Petrovskii in the whole $\mathbb{R}^n$ were studied by Mukhamadiev [360]–[362]. The author introduced the notion of limiting op-

2. Linear equations

erators and proved that the Fredholm property and solvability conditions were satisfied if and only if all limiting operators are invertible. Solvability conditions were formulated in terms of orthogonality to solutions of the homogeneous adjoint problem. These results were generalized for operators with bounded measurable coefficients except for the coefficients of the principal terms which were supposed to be Hölder continuous. Mukhamadiev introduced a priori estimates where the norm in the right-hand side was taken with respect to some bounded domain (Section 3.5 of Chapter 1 and Chapter 4). Such estimates imply normal solvability in unbounded domains (see also [523]). They follow from Condition NS. This condition allows also a direct proof of normal solvability, without proving first a priori estimates.

Limiting operators and their inter-relation with solvability conditions and with the Fredholm property were first studied by Favard [160], Levitan [309], [310] (see also Shubin [491]) for differential operators on the real axis, and later for some classes of elliptic operators in R^n (Mukhamadiev [361], [362], Barillon, Volpert [48]), in cylindrical domains (Collet, Volpert [114], [567]), or in some specially constructed domains (Bagirov and Feigin [45], [46]). Some of these results were obtained for the scalar case, some others for the vector case, under the assumption that the coefficients of the operator stabilize at infinity or without this assumption. Limiting operators were also used for some classes of pseudo-differential operators [153], [295], [440]–[442], [475], [490], discrete operators [28]–[30], [443], and some integro-differential equations [26], [27].

Limiting domains and operators for general elliptic problems and the corresponding function spaces are introduced in [564], [566], [570]. The presentation of a priori estimates, normal solvability and the Fredholm property in Chapters 3–5 follows these papers. The method to obtain a priori estimates of solutions is based on an isomorphism of pseudo-differential operators obtained as a modification of elliptic differential operators.

2.2.5 Other studies of the Fredholm property. Elliptic boundary value problems in domains with non-smooth boundary were studied by Eskin [154], Kondratiev [264], Feigin [168], Grisvard [214], [215], Moussaoui [359], Mazya and Plamenevskii [342]. These first works were followed by numerous papers (see, e.g., [462], [372], [141], [444], [213]) and several monographs [216], [217], [124], [373], [341].

Investigation of degenerate elliptic problems begins in the 1920s with the works by Tricomi [521], Holgrem [237]; see the literature review in the monograph by Smirnov [497]. Let us mention the cycle of works by Vishik (see [538] and [540] among them), more recent works by Levendorskii [304], [305], and by other authors [225], [247], [113], [259], [125], [260], [535].

Problems with the tangential derivative in the boundary conditions were studied by Egorov and Kondratiev [147], [148], Hörmander [242], Paneah [397], [398] (see also more recent papers [407], [582], [580] and monographs [431], [399]). Fredholm property of various nonlocal problems was investigated in [222], [494], [272], [581], [221], [219].

2.3 Index

2.3.1 Problems in the plane. Elliptic boundary value problems in the plane can be studied by reduction to singular integral equations in one space dimension. This method was developed by I.N. Vekua for certain classes of elliptic problems [528], [530], [531]. It allowed him to prove normal solvability of boundary value problems and to find their index. Further development of these works was due to A.I. Volpert [544]–[546]. He used fundamental matrices of elliptic systems of equations constructed by Ya.B. Lopatinskii [325], [326]. In [547] normal solvability was proved and the index was computed for general first-order systems and in [552] for general higher-order systems in the plane (see also [550]–[553]). The Dirichlet problem for elliptic systems was studied in [546]. It was shown that the index of this problem can be equal to an arbitrary even number and a formula for the index was given. It was proved that the index is a homotopy invariant and the formula for the index was obtained in terms of this invariant [547]. A class of canonical matrices was suggested for which 1) the elliptic system of first-order equations can be reduced to the Cauchy-Riemann system by a continuous deformation of the coefficients in such a way that the number of linearly independent solutions of the homogeneous direct and adjoint problems and the index do not change, 2) any elliptic first-order system with smooth coefficients can be reduced to the canonical system by a linear nonsingular transformation [547]. The index of a two-dimensional elliptic problem is discussed in a more recent work by Rowley [460].

2.3.2 Multi-dimensional problems. The formula for the index of boundary value problems for systems of harmonic functions in a three-dimensional domain was obtained by A.I. Volpert in 1960 [553] (see also [555]). It was also related to the index of multi-dimensional singular integral equations and to elliptic systems on a sphere. The method consists of several steps. First of all, the index of first-order elliptic systems on the two-dimensional sphere is found by reduction to a boundary value problem in a plane domain. This allows a computation of the index for some special class of elliptic boundary value problems in three-dimensional domains. This last result is used to find the index of two-dimensional singular integral equations [554] and of multi-dimensional equations [556]. Finally, the index of boundary value problems for harmonic functions is found by reduction to singular integral equations. Let us present these results in more detail since they are important for understanding of the index theories.

First-order elliptic systems on a sphere. Consider the first-order system

$$A^1(\xi)\frac{\partial u}{\partial \xi^1} + A^2(\xi)\frac{\partial u}{\partial \xi^2} + A^0(\xi)u = f(\xi), \qquad (2.10)$$

where (ξ^1, ξ^2) are local coordinates on the surface S (see [555] for the exact definitions) in $\mathbb{R}^3$ homeomorphic to a two-dimensional sphere, $A^j(\xi), j = 0, 1, 2$ are continuous complex square matrices of the order p, u and f are vectors

2. Linear equations

of the order p. It is assumed that the ellipticity condition is satisfied, that is $\det(A^1(\xi)\alpha_1 + A^2(\xi)\alpha_2) \neq 0$ for any nonzero vectors (α_1, α_2) and $\xi \in S$.

In order to find the index of this system, it is reduced to a Dirichlet problem in a plane region. Let us briefly describe this construction. The surface S is represented as a union of two simply connected parts S_1 and S_2. They are projected to the unit disc $D = \{x_1^2 + x_2^2 \leq 1\}$. Then we obtain two elliptic systems in D,

$$B^1(x)\frac{\partial v}{\partial x^1} + B^2(x)\frac{\partial v}{\partial x^2} + B^0(x)v = g(x), \quad x \in D, \tag{2.11}$$

$$C^1(x)\frac{\partial w}{\partial x^1} + C^2(x)\frac{\partial w}{\partial x^2} + C^0(x)w = h(x), \quad x \in D, \tag{2.12}$$

where the matrices B^j and C^j, $j = 0, 1, 2$ and the vectors g, h can be expressed through the matrices A^j and the vector f and the corresponding mappings. Solutions v and w of these two systems have the same values at the boundary ∂D. Therefore, we obtain the Dirichlet problem for the system of $2p$ equations:

$$T^1(x)\frac{\partial z}{\partial x^1} + T^2(x)\frac{\partial z}{\partial x^2} + T^0(x)z = q(x), \quad x \in D, \tag{2.13}$$

$$b\, z = 0, \quad x \in \partial D. \tag{2.14}$$

Here

$$T^j = \begin{pmatrix} B^j & , & 0 \\ 0 & , & C^j \end{pmatrix}, \quad z = \begin{pmatrix} v \\ w \end{pmatrix}, \quad q = \begin{pmatrix} g \\ h \end{pmatrix}, \quad b = (E_p, -E_p),$$

E_p is the unit matrix of the order p. The index of problem (2.13), (2.14) can be found (see the previous section and Section 8 of Chapter 8). In order to give an explicit formula the for index, let us consider the following ordinary differential systems of equations, which correspond to the principal terms of (2.11), (2.12):

$$iB^1(x)w + B^2(x)\frac{dw}{dt} = 0, \quad -iC^1(x)w + C^2(x)\frac{dw}{dt} = 0. \tag{2.15}$$

Let $w_1(x, t)$ be a continuous stable fundamental matrix of the first system in (2.15). This means that the columns of this matrix form a linearly independent system of solutions and they tend to zero as $t \to +\infty$. Moreover, $w_1(x, 0)$ is continuous for $x \in D$. Denote by $w_2(x, t)$ continuous stable fundamental matrix of the second system in (2.15). Put

$$\chi = \frac{1}{2\pi}\left(\arg\det\left(\overline{w_2'(x,0)}w_1(x,0)\right)\right)_\gamma,$$

where γ is the contour $x_1^2 + x_2^2 = 1$ with a positive orientation. Prime denotes transposed matrix and bar complex conjugate. Then the value of the index κ of system (2.10) is given by the formula $\kappa = -2\chi + p$. Let us mention two important properties of an index: two homotopic systems have the same index; for any given integer, there is a system with this value of the index.

Boundary value problems in $\mathbb{R}^3$. Consider the elliptic boundary value problem

$$\Delta u = 0, \quad x \in \Omega, \tag{2.16}$$

$$B\left(x, \frac{\partial}{\partial x}\right) u \equiv \sum_{j=1}^{3} B_j(x) \frac{\partial u}{\partial x} = f(x), \quad x \in S \tag{2.17}$$

in a domain $\Omega \subset \mathbb{R}^3$ with a sufficiently smooth boundary S homeomorphic to a sphere. $B_j(x), j = 1, 2, 3$ are sufficiently smooth complex square matrices of the order p. The Lopatinskii condition

$$\det B(x, \nu(x) + i\tau) \neq 0, \quad x \in S \tag{2.18}$$

is supposed to be satisfied. Here $\nu(x)$ is the normal vector at the point $x \in S$, τ is any tangent vector.

Let us begin with the particular case where p is even and $B(x, \nu(x)) = 0$ for all $x \in S$, that is the normal component of the derivative in the boundary condition is zero. Under this assumption, we can introduce in a natural way a first-order system of equations on S. In order to define it, we cover the surface S with a finite number of subsets S' of the surface S homeomorphic to a plane domain D' with a homeomorphism $\phi : D' \to S'$. For each such subset we obtain the system

$$A^1(\xi) \frac{\partial \hat{u}}{\partial \xi^1} + A^2(\xi) \frac{\partial \hat{u}}{\partial \xi^2} = \hat{f}(\xi), \quad \xi \in D', \tag{2.19}$$

where $\hat{u}(\xi) = u(\phi(\xi))$, $\hat{f}(\xi) = f(\phi(\xi))$,

$$\sum_{k=1}^{3} B_k(x) \frac{\partial u}{\partial x_k} = A^1(\xi) \frac{\partial \hat{u}}{\partial \xi^1} + A^2(\xi) \frac{\partial \hat{u}}{\partial \xi^2}.$$

By virtue of the Lopatinskii condition for the boundary value problem, the ellipticity condition for the first-order system is satisfied. Therefore its index is well defined and it can be found by the formula given above, $\kappa = -2\chi + p$.

There is a one-to-one correspondence between solutions of problem (2.16), (2.17) and of system (2.19). Indeed, if $u(x)$ is a solution of problem (2.16), (2.17), then $\hat{u}(\xi) = u(\phi(\xi))$ is a solution of system (2.19). Conversely, if $\hat{u}(\xi)$ is a solution of system (2.19), then the function $U(x) = \hat{u}(\phi^{-1}(x))$ satisfies boundary conditions (2.17). Then solution of problem (2.16), (2.17) can be found as a solution of the following Dirichlet problem: $\Delta u = 0, u|_S = U$. Hence, we can determine the index of problem (2.16), (2.17).

Singular integral equations. Consider the system of singular integral equations

$$a(x)\mu(x) + \int_S b(x, y - x)\mu(y) d_y S + T\mu = f(x), \tag{2.20}$$

2. Linear equations

where S is a surface in $\mathbb{R}^3$ homeomorphic to a sphere, the functions a, b and the surface S satisfy conventional regularity conditions, T is a regular integral operator. The symbol $\Phi(\tau)$ of this system is a square matrix of order p, defined and continuous on the set P of all unit tangent vectors τ to the surface S. If

$$\det \Phi(\tau) \neq 0, \quad \tau \in P, \tag{2.21}$$

then system (2.20) is normally solvable, the dimensions of the kernel of the homogeneous and homogeneous adjoint systems are finite [346], [347]. Hence its index is well defined.

Boundary value problem (2.16), (2.17) can be reduced to a singular integral equation with the simple layer potential,

$$u(x) = \frac{1}{2\pi}\int_S \frac{1}{r}\mu(y)d_y S,$$

where $r = |x-y|$. Substituting it in the boundary condition, we obtain the equation

$$B(x,\nu(x))\mu(x) + \frac{1}{2\pi}\int_S \frac{1}{r^3} B(x, y-x)\mu(y)d_y S = f(x). \tag{2.22}$$

The symbol of this system is $\Phi(\tau) = B(x,\nu(x) + i\tau)$. Therefore conditions (2.18) and (2.21) are equivalent. Hence, using the results on the singular integral equations, we conclude that problem (2.16), (2.17) has a finite-dimensional kernel and a finite number of solvability conditions. Its index equals the index of system (2.22).

The index $\kappa(\Phi)$ of the singular integral equation (2.20) is a homotopy invariant. This means that if two symbols Φ_1 and Φ_2 can be reduced to each other by a continuous deformation in such a way that condition (2.21) is satisfied, then $\kappa(\Phi_1) = \kappa(\Phi_2)$. The symbol of the product of two operators equals the product of the symbols. Since the index of the product of two operators equals the sum of their indices, then $\kappa(\Phi_1\Phi_2) = \kappa(\Phi_1) + \kappa(\Phi_2)$.

Let us construct a function $l(A)$ defined on invertible matrices A of order p. Consider first an invertible complex matrix $A(\tau)$ of order 2. Let $(a_1(\tau) + ia_2(\tau), a_3(\tau) + ia_4(\tau))$ be one of its rows. Put

$$a(\tau) = (a_1(\tau), a_2(\tau), a_3(\tau), a_4(\tau)), \quad a_0(\tau) = \frac{a(\tau)}{|a(\tau)|},$$

where $|a|$ denotes the length of the vector a. Thus, a_0 maps the space of unit vectors tangent to S to the unit three-dimensional sphere. We define $l(A)$ as the degree of this mapping. We note first of all that this function does not depend on the choice of the row of the function A. Indeed, if $\tilde{a}_0$ is the mapping corresponding to the second row, it is sufficient to consider the linear homotopy $\sigma a_0 + (1-\sigma)\tilde{a}_0$. Since the matrix A is invertible, then this mapping does not vanish for any $\sigma, 0 \leq \sigma \leq 1$. Hence the value of the degree does not depend on σ.

We define next this function for $p > 2$. It can be reduced by a continuous deformation which preserves invertibility to the matrix $\begin{pmatrix} E & 0 \\ 0 & A_0(\tau) \end{pmatrix}$, where E is the unit matrix of the order $p - 2$, $A_0(\tau)$ is a matrix of the second order. By definition, $l(A) = l(A_0)$. This function possesses the following three properties: 1. if A and B are homotopic, then $l(A) = l(B)$, 2. $L(AB) = l(A) + l(B)$, 3. l can take any integer values.

We can now express the function $\kappa(\Phi)$ through $l(\Phi)$. If $l(\Phi) = 0$, that is Φ is homotopic to a constant matrix, then $\kappa(\Phi) = 0$. This and the group properties of these functions indicated above allow one to affirm that $\kappa(\Phi) = \gamma l(\Phi)$ where γ is an integer number. Therefore, in order to determine the index, we need to specify the value of γ. Let $p = 2$. Consider a first-order system on S with sufficiently smooth coefficients and such that $\chi = 0$. It can be directly verified that for the corresponding mapping Φ, $l(\Phi) = 2$. On the other hand, consider the singular integral equation (2.22) and the corresponding boundary value problem (2.16), (2.17) in the particular case where $B(x, \nu(x)) = 0$. It was shown above that $\kappa(\Phi) = 2$. Hence $\gamma = 1$.

Thus, we obtain the index of the singular integral equation (2.20), $\kappa = l(\Phi)$, and of the boundary value problem (2.16), (2.17), $\kappa = l(B)$.

Further works. Index of multi-dimensional singular integral equations was studied by Michlin [348], Boyarski [79], Seeley [482], Calderon [101]. Normal solvability and index of elliptic boundary value problems and of singular integral operators were studied by Agranovich and Dynin. In [142], [12] it was shown that the index of elliptic boundary value problems could be reduced to the index of singular integral operators, the homotopic invariance of the index was discussed in [9], [142].

Important development of the index theory was due to the works by Atiyah and Singer [36], Atiyah and Bott [37] (see also [38], [39], [40]). They were followed by many other publications (see [143], [144], [396] and the references therein).

2.3.3 Other methods. The method of computation of the index using regularizors of elliptic operators was developed by Fedosov. If $A : H_1 \to H_2$ is a Fredholm operator and $R : H_2 \to H_1$ is its regularizor such that $I_1 - RA$ and $I_2 - AR$ are nuclear operators, then the abstract formula $\text{ind } A = \text{tr } (I_1 - RA) + \text{tr } (I_2 - AR)$ can provide an analytic expression for the index [161], [162], [163], [164]. This method was also used to find the index for random elliptic operators [167] and for operators on a wedge [166]. Rabier generalized this approach for elliptic operators in $\mathbb{R}^n$ under weaker assumptions on the coefficients [434].

This approach is applicable for some problems in unbounded domains. For the boundary value problem

$$-\Delta u + u = g \ (y \geq 0), \quad \cos\alpha(x)\frac{\partial u}{\partial y} + \sin\alpha(x)\frac{\partial u}{\partial x} = h \ (y = 0)$$

in the half-plane, where $\cos\alpha(x) = 1$ for $|x|$ sufficiently large, the index equals $-\frac{1}{\pi}\alpha(x)|_{-\infty}^{\infty}$ [162] (cf. Section 6 of Chapter 8). The construction of a regularizor

2. Linear equations

and application of this approach can be less clear if the coefficients of the operator at infinity are not constant. In the example above, this is the case if $\cos\alpha(x) = 1$ for x sufficiently large and $\cos\alpha(x) = -1$ for $-x$ sufficiently large. This case is essentially different. The index cannot be computed directly by the method of expanding domains presented in Chapter 8. We need to use reduction to the Cauchy-Riemann system.

The index theories are discussed in the recent monographs [165], [451], [579]. The index of some classes of elliptic operators in weighted spaces is found in [323], [324], [57], [75]. Computation of the index of elliptic problems in unbounded cylinders is carried out in [114] (see Chapter 9).

2.4 Elliptic problems with a parameter

After the first papers by Agmon [5] and by Agranovich, Vishik [13] ellipticity with a parameter for the problem

$$A(x, \lambda, D)u = f, \quad x \in \Omega, \tag{2.23}$$
$$B(x, \lambda, D)u = g, \quad x \in \partial\Omega, \tag{2.24}$$

or for some of its special cases in bounded domains Ω, was studied by Geymonat, Grisvard [196], Roitberg [457], Agranovich [11], and Denk, Volevich [129]. The most general problem considered in unbounded domains Ω, which was studied before [571], is the following problem:

$$Au := \sum_{|\alpha| \leq 2m} a_\alpha(x) D^\alpha u - \lambda^{2m} u = f \quad \text{in } \Omega, \tag{2.25}$$

$$Bu := \sum_{|\alpha| \leq m_j} b_{j\alpha}(x) D^\alpha u = g \quad \text{on } \partial\Omega, \ j = 1, \ldots, r, \tag{2.26}$$

where m and m_j are some integers, and the sector S is of the form:

$$S = \{\lambda : |\arg \lambda| \leq \theta\}. \tag{2.27}$$

These results embrace parameter-elliptic problems, which are obtained from parabolic systems in the sense of Petrovskii, but some important classes of parameter elliptic problems are excluded, for example, first-order systems ($m = 1/2$), in which case the sector S is

$$S = \left\{\lambda : |\arg \lambda| \leq \theta, \ |\pi - \arg \lambda| \leq \theta, \ \theta < \frac{\pi}{2}\right\}.$$

There are three known methods to prove existence of solutions of problem (2.23), (2.24). The first method uses formally adjoint problems, the second one is based on the theory of sectorial operators, and the third method relies on the direct construction of the inverse operator.

The first method stems from the paper [5]. In the case of scalar equation ($N = 1$) it is supposed that the boundary conditions are normal. This means that the boundary $\partial\Omega$ is non-characteristic with respect to the boundary operator B_j, $m_j < 2m$, and the orders of the boundary operators are distinct. The operator A with domain

$$D(A) = \{u \in W^{2m,p}(\Omega),\ B_j = 0,\ j = 1,\ldots,r,\ 1 < p < \infty\}$$

is considered and a priori estimates with a parameter are obtained. It follows from these estimates that the operator A has a zero kernel and a closed range. As it was proved by Schechter [471], then there exists a formally adjoint elliptic problem with normal boundary conditions. By the above arguments it is obtained that the formally adjoint operator has a zero kernel and hence the range of A coincides with $L^p(\Omega)$.

A priori estimates and existence theorems in the case of unbounded domains are proved for these problems by Higuchi [233] and by Freeman, Schechter [186].

For systems of equations ($N \geq 1$) a priori estimates and existence results for problem (2.25), (2.26) were obtained by Amann [16] under the assumption that $u \in W^{2m+s,p}(\Omega)$, where s is a real nonnegative number. Instead of the normality mentioned above in the scalar case in the proof of the existence he supposed that the operator A was uniformly strongly elliptic and the boundary operators had the form

$$B_j u = \frac{\partial^{k+j-1} u}{\partial \beta_j^{k+j-1}},\ j = 1,\ldots,m,\ k = 0,\ldots,m$$

with the lower-order boundary operators, where

$$(\beta_j(x), \nu(x)) \geq c > 0,\ \forall x \in \Gamma,\ j = 1,\ldots,m,$$

and $\nu(x)$ was the normal vector. It is shown in Chapter 7 that the existence results hold true without these restrictions though this class of operators may be interesting in other considerations.

There is a large number of works devoted to sectorial operators in relation with elliptic problems (2.25), (2.26) (see [329], [128], [121] and the references therein). In the paper by Denk, Hieber and Prüs [128] problem (2.25), (2.26) is considered for equations with operator coefficients. Using methods based on the theory of sectorial operators, they have obtained existence results for the case where $\Omega = \mathbb{R}^n$, $\Omega = \mathbb{R}^n_+$, and $\Omega \subset \mathbb{R}^n$ is a domain with a compact boundary. It is supposed that the sector S is of the type (2.27) and that there are some restrictions on the function spaces (for example, in the case of $W^{l,p}$ the domain of the operator belongs to $W^{2m,p}$; $l > 2m$ is excluded).

The third of the methods mentioned above is the method introduced by Agranovich and Vishik [13] for bounded domains Ω. This method does not require restrictions which are needed for other methods. In Sobolev spaces they introduced norms depending on the parameter, obtained a priori estimates with constants independent of the parameter and proved existence of solutions of problem (2.25), (2.26) (the parameter λ entered the operator in a more general way) by a direct construction of the inverse operator. A generalization of these results for mixed-order systems in bounded domains has been obtained by Agranovich [11].

The presentation of Chapter 7 follows the work [571]. The Agranovich-Vishik method is developed for unbounded domains. We note that these results are obtained for general mixed-order parameter-elliptic problems in uniformly regular unbounded domains Ω without any additional restrictions. Compared with the theory of sectorial operators (for the case (2.25), (2.26)) the boundary $\partial\Omega$ is not supposed to be compact and the operators are proved to be sectorial in the spaces $L_q^p (1 < p < \infty, 1 \leq q \leq \infty)$, which are more flexible than the spaces $L^p = L_p^p$. We note that a priori estimates do not take place in the space $W^{2m,p}$ for $p = 1$ and $p = \infty$ (see, e.g., [487] for counterexamples even for the Laplace operator). However they are obtained for the spaces $W_q^{2m,p} (1 < p < \infty, 1 \leq q \leq \infty)$. The behavior of the functions $u \in W_q^{2m,p}$ at infinity is determined by the value of q. For example, the behavior at infinity of the functions from $L_1^p (1 < p < \infty)$ is similar to that in L^1.

3 Decay and growth of solutions

We discussed decay and growth of solutions of elliptic boundary value problems in Chapters 4, 5 and 11 [566]. We can briefly summarize them as follows. If the Fredholm property is satisfied, then solutions of linear homogeneous equations decay exponentially at infinity. Decay rate of solutions of nonhomogeneous equations is determined by the right-hand side.

Consider the equation $Lu = f$, where the operator L acts from a space E_∞ into F_∞. The Fredholm alternative affirms that either f satisfies solvability conditions, and then this equation has a solution, or it does not satisfy the solvability conditions, in which case there is no solution in E_∞. Taking into account the results of Section 4.3.5 of Chapter 11, we can reformulate this alternative: either the solution is bounded or it grows exponentially at infinity. This shows the relation between the Fredholm property and Phragmén and Lindelöf type results. The latter are intensively studied for elliptic boundary value problems.

Phragmén and Lindelöf wrote in their work [410] published in 1908:

On connaît le rôle que joue dans l'Analyse le principe suivant: Soient dans le plan de la variable complexe x un domaine connexe, T, et une fonction monogène[13], $f(x)$, régulière à l'intérieur de ce domaine. Supposons que le module $|f(x)|$ est uniforme dans le domaine T et vérifie pour tout point ξ de son contour cette condition:

---

[13] holomorphic

(A) *Quelque petit qu'on se donne le nombre positive ϵ l'inégalité*

$$|f(x)| < C + \epsilon,$$

où C désigne une constante, est vérifié dès que x, restant à l'intérieur de T, est suffisamment rapproché du point ξ.

Dans ces conditions on aura, pour tout point pris dans l'intérieur de T,

(I) $$|f(x)| \leq C,$$

l'égalité étant d'ailleurs exclue si la fonction ne se réduit pas à une constante.

After that they generalized this assertion:

Principe général. *Admettons que le module de la fonction monogène $f(x)$, qui est supposée rélulière à l'intérieur du domaine T, soit uniforme dans ce domains et vérifie la condition (A) sur son contour, en exceptant les points d'un certain ensemble E.*

Admettons d'autre part qu'il existe une fonction monogène, $\omega(x)$, rélulière et différente de zéro dans T et jouissant en outre les propriétés suivantes:

(a) *A l'intérieur de T le module $|\omega(x)|$ est uniforme et vérifie la condition*

$$|\omega(x)| \leq 1.$$

(b) *En désingant par σ, ϵ des nombres positifs aussi petits qu'on voudra et par ξ un point quelconclue de l'ensemble E, on aura*

$$|\omega^\sigma f(x)| < C + \epsilon,$$

dès que x restant à l'intérieur de T, sera suffisamment rapproché du point ξ.

Dans ces conditions, la conclusion (I) rests valable pour tout point x intérieur au domaine T.

As example of applications of this principle given in the paper, let us present the following theorem:

Soient un domaine connexe, T, faisant partie d'une bande de largeur π/α, et une fonction monogène, $f(x)$, réguliée á l'intérieur de ce domaine, dont le module est uniforme dans T et vérifie la condition (A) sur son contour (à distance finie). On suppose d'ailleurs que l'expression $\exp(-\epsilon e^{ar})f(x)$ quelque petit que soit le nombre positif ϵ, tend uniformément vers zéro dans le domaine en question lorsque r croît indéfiniment. Cela étant, on aura $|f(x)| \leq C$ pour tout point intérieur à T.

This theorem indicates the connection between growth rate of a holomorphic function with the facts that it is bounded and that it has its maximum at the boundary. Results by Phragmén and Lindelöf [410] attracted much attention in the

3. Decay and growth of solutions

theory of elliptic partial differential equations beginning from the 1950s. Gilbarg [199] studied the inequality $Lu \leq 0$ for the operator $L(u) = au_{xx} + 2bu_{xy} + cu_{yy} + du_x + eu_y$ in an unbounded domain D. He proved that if u is nonnegative at the boundary and $\liminf_{r\to\infty} m(r)/r = 0$, where $m(r) = \min_{|z|=r} u(z)$, then $u(z) \geq 0$ in D. Related problems were studied by Huber [245], Serrin [483], Hopf [240], Friedman [187], Lax [299]. In the 1960s, there was a number of works devoted to various estimates of growth rate of solutions of linear second-order elliptic equations in unbounded domains (Cegis [108], Arson and Evgrafov [34], Arson and Iglickii [35], Fife [173]), first-order elliptic systems were considered in [33]. Maximum principle in connection with the growth rate of solutions was studied by Landis [290], Solomencev [500], Oddson [390]. Second-order systems with periodic coefficients were studied by Landis and Panasenko [294], the degenerate equation was studied by Mamedov and Guseynov [332].

Theorems of Phragmén-Lindelöf type for higher-order linear elliptic equations were proved by Landis [291]–[293] and later by Doncev [132], [133]. They studied uniformly elliptic operators $P(x,D) = \sum_{|\alpha|\leq m} a_\alpha(x) D^\alpha$ in cylinders. A typical result says that the solution is either bounded or exponentially growing. Nadirashvili [367], [368] considered the same operator in $\mathbb{R}^n$. Oleinik and Iosifyan [391] generalized the Saint-Venant principle for the two-dimensional biharmonic equation and used it to prove a Phragmén-Lindelöf theorem for energy. Cai and Lin [100] studied the biharmonic operator in an infinite strip and proved that the energy either decays or grows exponentially.

Liouville and Phragmén-Lindelöf type theorems for general elliptic problems in the Douglis-Nirenberg sense were studied by Oleinik and Radkevich [392]. The typical result is that if a solution $u(x)$ of the homogeneous problem in $\mathbb{R}^n$ or in an unbounded cylinder admits some exponential estimate $|u(x)| \leq \exp(\delta|x|)$, then it is identically zero. This result is obviously related to the uniqueness of solutions of elliptic boundary value problems. It is interesting to discuss how it is related to the Fredholm property. Even the simplest equation $\Delta u = 0$ has a nonzero solution, which is not exponentially growing. This is because the corresponding operator does not satisfy the Fredholm property. The operators considered in [392] are operators with parameters. In the example above this gives $\Delta u - \mu^2 u = 0$ with certain conditions on μ which provide the unique solvability.

A semi-linear second-order equation was considered by Herzog [232]. Many works devoted to Phragmén-Lindelöf theorems for semi-linear and quasi-linear second-order elliptic equations appear beginning from the 1980s (Mikljukov [350], Gubachev [218], Granlund [210], Novruzov [388], Aviles [41] and the others; see also Jin and Lancaster [251] and the references therein). Phragmén-Lindelöf type theorems for the minimal surface equation

$$\mathrm{div}\,(\nabla u/\sqrt{1+|\nabla u|^2}) = 0$$

were studied by Nitsche [387], Miklyukov [351], Hsieh et al. [244] and other authors. Maximum principle and positive solutions for equations with the p-Laplacian

$\Delta_p u = \operatorname{div}(|\nabla u|^{p-2}\nabla u)$ were studied by Kurta [281], Liskevich et al. [321] (see also the references therein). Maximum principle and Phragmén-Lindelöf theorems for the fully nonlinear second-order equation were studied by Capuzzo and Vitolo [103], [104]. Energy estimates for higher-order quasi-linear elliptic equations in unbounded domains were obtained by Shishkov [488], [489] and Gadzhiev [189]; semilinear equations of the fourth order were considered by Celebi et al. [106].

To summarize this literature, it estimates growth and decay of solutions at infinity and relates these estimates to the maximum and comparison principles, boundedness, positiveness and uniqueness of solutions of linear and nonlinear equations. The methods of energy estimates and of sub- and super-solutions are often used and are not necessarily related to Phragmén-Lindelöf and Liouville theorems or the Saint-Venant principle.

Decay rate of positive solutions of semi-linear and quasi-linear second-order equations in $\mathbb{R}^n$ is studied by Johnson et al. [252], [253], Kawano et al. [257], Ren and Wei [452], Flucher and Muller [179], Bae et al. [44], Deng et al. [130]; in unbounded cylinders by Payne et al. [401]; for a system of two equations by Hulshof and van derVorst [246]. Egorov and Kondratiev study linear second-order elliptic operators in various unbounded domains with nonlinear boundary conditions [149], [150]. It is shown that if growth of solutions at infinity is not too fast, then they converge to zero. Growth rate and uniqueness of solutions of a semi-linear second-order equation in $\mathbb{R}^n$ are studied by Diaz and Letelier [131] and Dai [122]; McKenna and Reichel [343] prove multiplicity of radial solutions of biharmonic equations with different growth rates.

4 Topological degree

In his book published in 1955 [352], Miranda wrote that the beginning of the modern theory of nonlinear elliptic equations could be related to the International Mathematical Congress in Paris in 1900 where Hilbert suggested, among other problems, that any solution of an analytical elliptic equation was analytical (the 19th problem). This problem had initiated numerous works. Several years after that Bernstein had proved this theorem in the case of two space variables and continued the investigation of the Dirichlet problem for nonlinear equations. His basic idea was that the existence of solutions followed from appropriate a priori estimates. Only in the 1930s did the real meaning of these results and their applicability become clear due to works by Schauder, Leray, Caccioppoli where they developed new approaches on the basis of functional analysis in abstract spaces and of new a priori estimates. These works resulted in the development of the topological degree theory. "The class of Fredholm and proper mappings was the first class of nonlinear mappings in infinite-dimensional spaces for which it appeared to be possible to generalize the degree theory by Brouwer-Hopf in finite-dimensional spaces" [276].

4. Topological degree

In the general setting, topological degree $\gamma(A, \Omega)$ is an integer which depends on the operator $A : E \to F$ and on the domain $\Omega \subset E$. It should satisfy certain properties, namely, homotopy invariance, additivity, normalization. There exist different degree constructions adapted to some particular function spaces and operators. Various degree constructions and reviews are given by Nagumo [369], Zeidler [590], Krasnoselskii, Zabreiko [276], Mawhin [340], Skrypnik [493] and by other authors.

Finite-dimensional spaces. The degree theory for finite-dimensional mappings comes back to Kronecker, Poincaré, Brouwer, Hopf. Let Ω be a bounded domain in $\mathbb{R}^n$ and Φ a mapping acting from the closure $\bar{\Omega}$ of the domain Ω into $\mathbb{R}^n$. The degree $\gamma(\Phi, \Omega)$ can be first defined for smooth mappings Φ. Its definition is based on the notion of regular points. A point y is regular if its inverse image $\Phi^{-1}(y)$ contains a finite number of points $x_1, \ldots, x_n$ in Ω (it can be empty) and the Jacobian matrices $\Phi'(x_i)$ are non-degenerate. A point is singular if it is not regular.

Assuming that $\Phi(x) \neq 0$ at the boundary $\partial \Omega$ and $y = 0$ is a regular point, we put by definition,

$$\gamma(\Phi, \Omega) = \sum_{i=1}^{n} \text{sign det } \Phi'(x_i) \qquad (4.1)$$

for all $x_i \in \Omega$ such that $\Phi(x_i) = 0$. According to Sard's lemma, the Lebesgue measure of singular points is zero in the image space. Hence, if $y = 0$ is singular, then it can be approximated by regular points. This allows definition of the degree in the case where y is not regular. Finally, if the mapping is not smooth but continuous, then it can be approximated by a smooth mapping, and the degree can also be defined. This is the so-called Brouwer degree.

Consider the following example of its application. Suppose that the function $F_\tau(u) : \mathbb{R}^n \to \mathbb{R}^n$ is continuous with respect to u and τ, $F_\tau(0) = 0$ for all $\tau \in [0, 1]$, and for some $R > 0$,

$$F_\tau(u) \neq 0 \text{ if } |u| = R, \ \tau \in [0, 1]. \qquad (4.2)$$

Suppose, next, that for $\tau < \tau_0$ the equation $F_\tau(u) = 0$ has a unique solution $u = 0$ in the ball $B_R : \{|u| \leq R\}$, and all eigenvalues of the matrix $F'_\tau(0)$ have negative real parts; for $\tau = \tau_0$ a simple real eigenvalue crosses zero and becomes positive. Then for $\tau < \tau_0$,

$$\gamma(F_\tau, B_R) = \text{sign det } F'_\tau(0) = (-1)^n.$$

The value sign det $F'_\tau(0)$ is called the index of the stationary point $u = 0$. For $\tau > \tau_0$ the index is different, sign det $F'_\tau(0) = (-1)^{n-1}$. On the other hand, the degree $\gamma(F_\tau, B_R)$ remains the same because of the condition (4.2). From (4.1) we conclude that there are other solutions of the equation $F(u) = 0$ for $\tau > \tau_0$. The value τ_0 of the parameter τ is called the bifurcation point.

The index of stationary points is related to their stability with respect to ordinary differential systems of equations $du/dt = F(u)$.

Another application of the degree concerns the existence of solutions. If, for example, all vectors $F_\tau(u)$ at the boundary of the ball B_R are directed inside the ball (or outside it), then the degree $\gamma(F_\tau, B_R)$ is different from zero. From the principle of nonzero rotation it follows that there exists a solution of the equation $F_\tau(u)$ inside B_R. Here we do not assume that $F_\tau(0) = 0$.

Leray-Schauder degree. Consider a real Banach space E and an operator $A = I + B$, where I is the identity operator and $B : E \to E$ is a continuous compact operator. Let $\Omega \subset E$ be a bounded domain.

The operator B can be approximated by a finite-dimensional operator $\mathcal{B}$. We consider a finite-dimensional subspace E_0 which contains some internal points of the domain Ω, $\Omega_0 = \Omega \cap E_0$, $\mathcal{B}_0$ is the restriction of the operator $\mathcal{B}$ to E_0. We suppose that $\mathcal{B}_0 \Omega_0 \subset E_0$.

We assume next that $A(u) \neq 0$ for $u \in \partial\Omega$. Then for appropriately chosen approximating operators, $\mathcal{A}_0(u) = u + \mathcal{B}_0(u) \neq 0$ for $u \in \partial\Omega_0$. Therefore we can define the degree $\gamma(\mathcal{A}_0, \Omega_0)$. By definition, the value of the degree $\gamma(A, \Omega)$ equals the degree of the approximating finite-dimensional mappings. It can be verified that the definition is correct, that is the value of the degree does not depend on the choice of approximating operators.

This is the Leray-Schauder degree [303]. One of its applications is related to elliptic problems in bounded domains. Consider, as example, the problem

$$\Delta u + F(u) = 0, \quad u|_{\partial D} = 0$$

in a bounded domain $D \in \mathbb{R}^n$. Here the function $F(u)$ and the boundary of the domain are sufficiently smooth. We can introduce the operator $A = I + \Delta^{-1} F$ acting in the Hölder space $C^\alpha(\bar{\Omega})$ for some $0 < \alpha < 1$. Here Δ^{-1} corresponds to the resolution of the Poisson equation with the Dirichlet boundary condition. This is a compact operator in $C^\alpha(\bar{\Omega})$. Therefore, the operator A represents compact perturbation of the identity operator. Its relation to the elliptic boundary value problem is obvious. The Leray-Schauder degree can be used to study existence and bifurcations of solutions.

This approach cannot be used if the domain D is unbounded since the operator Δ^{-1} is not compact.

Generalized monotone operators. Consider an operator A acting from a real separable reflexive Banach space E into its dual E^*. Then we can define monotone and pseudo-monotone mappings and some other classes of mappings related to the notion of monotonicity. The definitions and the references can be found, e.g., in [493]. Following [88], [492], [493] we introduce the class of operators which satisfy the condition:

4. Topological degree

Condition 4.1. Let Ω be a domain in E, $A : \overline{\Omega} \to E^*$ and for each sequence $u_n \in \overline{\Omega}$ from the weak convergence $u_n \rightharpoonup u_0$ and from the inequality

$$\overline{\lim}_{n \to \infty} \langle Au_n, u_n - u_0 \rangle \leq 0$$

it follows that the convergence u_n to u_0 is strong.

Let v_i, $i = 1, 2, \ldots$ be a complete system of the space E and suppose that $v_1, \ldots, v_n$ are linearly independent for any n. Denote by E_n the linear hull of these elements and by A_n the finite-dimensional approximation of the operator A:

$$A_n u = \sum_{i=1}^{n} \langle Au, v_i \rangle v_i, \quad u \in \overline{\Omega}_n,$$

where $\Omega_n = \Omega \cap E_n$.

We consider a class of operators satisfying Condition 4.1. Suppose that $A(u) \neq 0$ for $u \in \partial\Omega$. Then it can be shown that, for n sufficiently large, $A_n u \neq 0$ if $u \in \partial\Omega_n$. Hence we can define the degree $\gamma(A_n, \Omega_n)$ for the finite-dimensional mapping A_n. It is proved that it does not depend on n for n sufficiently large. The degree $\gamma(A, \Omega)$ is defined as $\gamma(A_n, \Omega_n)$ for large n. The degree can also be defined in the case of nonseparable spaces [492].

This construction is applied in [493] in order to define the degree for general scalar nonlinear elliptic problems

$$F(x, u, \ldots, D^{2m} u) = f(x), \quad x \in \Omega, \tag{4.3}$$

$$G_j(x, u, \ldots, D^{m_j} u) = g_j(x), \quad j = 1, \ldots, m, \ x \in \partial\Omega. \tag{4.4}$$

Here $\Omega \subset \mathbb{R}^n$ is a bounded domain with the boundary of the class C^l, F and G_j are some given functions having continuous derivatives with respect to all arguments up to orders $l - 2m + 1$ and $l - 2m_j + 1$, respectively, l is an integer number such that $l \geq l_0 + n_0$, $l_0 = \max(2m, m_1 + 1, \ldots, m_m + 1)$, $n_0 = [n/2] + 1$. The boundary value problem (4.3), (4.4) is considered in the space $H^l(\Omega)$ with $f \in H^{l-2m}(\Omega)$, $g_j \in H^{l-m_j-1/2}(\partial\Omega)$. The operator corresponding to this problem is introduced as acting from $H^l(\Omega)$ into $(H^l(\Omega))^*$. It is proved that it satisfies Condition 4.1 ([493], Theorem 2.4, p. 76), which allows one to define the degree.

Fredholm and proper operators. Fredholm and proper mappings was the first class of nonlinear mapping in infinite-dimensional spaces for which it was possible to generalize the Brouwer-Hopf theory developed for finite-dimensional spaces ([276], p. 285). This was done by Caccioppoli (see the bibliography in [352]) who defined the degree modulus 2. The important development of this theory was due to the work by Smale [496] who generalized Sard's lemma for Fredholm operators and defined the degree as the number of solutions of the operator equation $f(x) = y$ modulus 2. For almost all y these solutions are regular and their number is finite.

Based on the results by Smale, Elworthy and Tromba [151], [152] defined the oriented degree for Fredholm and proper operators of the zero index as

$$\deg f = \sum_{x \in f^{-1}(y)} \operatorname{sgn} T_x f, \qquad (4.5)$$

where $\operatorname{sgn} T_x f = \pm 1$ depending on whether the orientation on the manifold is preserved. This degree is homotopy invariant modulus 2.

Various notions of orientation were used for the degree construction for Fredholm and proper mappings in [77], [54], [55], [174]–[178]. A simple description of the orientation can be given following [54]. Let E and F be real vector spaces, $L : E \to F$ be a Fredholm operator. A linear operator $A : E \to F$ is called a corrector if its range is finite dimensional and $L + A$ is an isomorphism. For two different correctors A and B, it is possible to define the operator $K = I - (L+B)^{-1}(L+A)$ with a finite-dimensional range. If $E_0 \subset E$ is a finite-dimensional subspace which contains the range of K, then the determinant $\det(I - K)|_{E_0}$ is well defined. The two correctors are called equivalent if the determinant is positive. Orientation of Fredholm operators with the zero index is defined as one of two equivalence classes. This approach comes back to the work [339] where correctors are defined with the help of projectors onto the kernel of the operator L. In [405] (and subsequent works [174]–[178]) correctors are compact operators. Therefore, $I - K$ is a compact perturbation of the identity operator, and the Leray-Schauder degree is used to define the orientation. When the orientation is introduced, the degree can be defined similar to (4.5).

Another approach to define the orientation is suggested in [170], [123], [248]. Assuming that the operator $L + \lambda I$ satisfies the Fredholm property for all real $\lambda \geq 0$ and that it has only a finite number ν of positive eigenvalues (together with their multiplicities), we can define the orientation as $(-1)^\nu$. This construction is well adapted for elliptic boundary value problems because it is naturally related to the spectrum of the linearized operator. Similar to other degree constructions, this one requires a precise specification of operators and function spaces [567], [565].

Degree can also be defined in the case of operators with a positive index. In this case it is not an integer number but a cobordism class [151], [152], [77], [382]. It has limited applications to elliptic problems [448], [595], [382], [384].

Properness of elliptic boundary value problems is studied in [595] for bounded domains and in [438], [564]–[566] for unbounded domains. Various approaches to the construction of the topological degree and its application for elliptic problems in unbounded domains were used in [123], [509], [510], [518] in the one-dimensional case. In [561], [562] the degree was constructed in the one-dimensional case and for elliptic problems in cylinders using the method of [493]; in [567] the degree for Fredholm and proper operators is constructed for elliptic problems in cylinders; [438] uses the degree construction of [177], [178] for elliptic problems in $\mathbb{R}^n$. Finally, topological degree for general elliptic problems in unbounded domains is constructed in [565]. The presentation of Chapter 11 follows the ideas of this work.

Approximation-proper mappings. The class of A-proper mappings was introduced by Petryshyn [408]. Browder and Petryshyn constructed a multi-valued degree for such mappings [89], [90], [409].

Let X and Y be separable real Banach spaces, $X_n \subset X$ and $Y_n \subset Y$ their finite-dimensional subspaces, $\dim X_n = \dim Y_n$, $P_n : X_n \to X$ and $Q_n : Y_n \to Y$ be continuous, generally nonlinear mappings.

Suppose that $G \subset X$ is an open set. A mapping $T : \bar{G} \to Y$ is called A-proper if for any $y \in Y$ and any sequence x_{n_j} such that

$$x_{n_j} \in X_{n_j}, \quad P_{n_j} x_{n_j} \in \bar{G}, \quad \|Q_{n_j} T P_{n_j} x_{n_j} - Q_{n_j} y\|_Y \to 0 \text{ as } n_j \to \infty$$

there exists $x \in X$ and a subsequence $x_{n_{j_k}}$ such that $Tx = y$ and $P_{n_{j_k}} x_{n_{j_k}} \to x$.

We can define the degree γ_n for the finite-dimensional mapping $T_n = Q_n T P_n :$ $\bar{G}_n \to Y_n$, where $G_n = P_n^{-1}(G)$. A convergent subsequence of the sequence γ_n is taken, by definition, as the degree of mapping T. Since the limiting value γ may not be unique, the degree is multi valued. It possesses the property of nonzero rotation, that is if its value is different from zero, then there exists a solution of the corresponding operator equation, and homotopy invariance. This construction is applicable to elliptic boundary value problems ([493], p. 67). It is established that A-proper operators are proper and satisfy the Fredholm property ([409], pp. 26, 27).

5 Existence and bifurcation of solutions

Bifurcations of solutions. Bifurcation theory developed under the influence of numerous applications. It brings together the theory of Fredholm operators, spectral theory, methods of small parameter, variational methods, topological degree. Branching of solutions was investigated already by Newton who studied the question of determination of all solutions of the equation $f(x,y) = 0$ in the vicinity of a point (x_0, y_0) where $f'_y(x_0, y_0) = 0$. He looked for the solution in the form of power series [378]. Later Lagrange invented the method of small parameter [286]. In the beginning of the XXth century Lyapunov [330] and Schmidt [474] developed the theory of branching of solutions of functional equations (see also Poincaré [421], [422]). They showed that bifurcations of solutions of nonlinear integral equations could be reduced to the analogous problems for implicit functions. Further development of this theory is presented in the monographs by Lichtenstein [316], Vainberg and Trenogin [525], and in the collection of papers edited by Keller and Antman [258].

Application of the topological degree to study bifurcations of solutions begins in the 1950s with works by Cronin [118]–[120] and Krasnoselskii [273]. It was used by Velte [533] and Yudovich [587] for some problems in hydrodynamics and later by many authors for various functional equations and elliptic problems (see [258], [276], [340] and references therein). Topological degree for continuous branches of solutions was first applied probably by Krasnoselskii [274], some time later by Rabinowitz [445], [446] and other authors.

Development of the methods of the bifurcation theory initiated, beginning from the 1970s, numerous works on bifurcations of solutions of elliptic problems[14]. The works by Crandall and Rabinowitz [117], Rabinowitz [445], Ambrosetti and Prodi [19], Brezis and Turner [82], P.-L. Lions [319], Brezis and Nirenberg [81] devoted to local bifurcations, global branches and multiplicity of solutions were followed by many others (see recent papers [56], [95], [109], [300], [447], [592] and monographs [18], [135] and the references therein). Symmetry of solutions and symmetry breaking were studied by Gidas [198], Smoller and Wasserman [498],[499], Vanderbauwhede [526], Shih [486], Budd [92] (see also [94] [169], [485] and the references therein). Bifurcation of solutions of elliptic equations in $\mathbb{R}^n$ were investigated by Toland [519], Stuart [512], Rumbos and Edelson [461], Brown and Stavrakakis [91], Cingolani and Gamez [112], Deng and Y. Li [127], Stavrakakis [503], H. Jeanjean et al [250], Polacik [430], del Pino [126]; in particular, bifircations from the essential spectrum by Stuart [511], Benci and Fortunato [53], Cao [102], Rother [458], L. Jeanjean [249], Badiale [43]. Problems in exterior domains or in unbounded cylinders were considered by Furusho [188], Lachand-Robert [282], Sun [513]. Bifurcations of travelling waves, which are solutions of elliptic problems in unbounded domains, have some specific features because these are families of solutions invariant with respect to translation (see [568] and the references therein).

Let us also mention that bifurcations of solutions are studied in the case of nonlinear boundary conditions, for systems of equations, for various problems arising in hydrodynamics, combustion, elasticity and other applications, for degenerate equations, and in some other cases.

Existence of solutions. Investigation of nonlinear elliptic equations begins with works by Picard who developed the method of successive approximations (see Section 1, Historical Notes). Bernstein used this method to prove analyticity of solutions and invented, in fact, the method of a priori estimates of solutions and continuation with respect to a parameter. This approach was fully developed later in the Leray-Schauder method.

Topological methods to prove existence of solutions were first used by Birkhoff and Kellog [70]. The main idea of their method was formulated by Schauder as a fixed point theorem [466]. It was generalized by Tikhonov for linear topological spaces [517]. Fixed point theorems were used by Nemytskii, and then by many other authors, to prove existence of solutions of nonlinear integral equations (see [273] and the references therein).

The works by Birkhoff and Kellog [70], Levy [311], Schauder [463]-[467], Leray and Schauder [303], Caccioppoli [96]-[98] published in the 1920-30s introduced topological methods in analysis and were applied to integral equations and to partial differential equations. After these works, analysis of the existence of solutions is reduced, in the appropriate functional setting where some fix point theorems or the topological degree can be used, to obtain a priori estimates of

---

[14]There are hundreds of papers devoted to bifurcations of solutions of elliptic equations. This short literature review is necessarily incomplete

solutions. Topological degree for compact perturbations of the identity operator was constructed in [303] and was applied to elliptic problems. A detailed review of the existence results obtained by the 1950s is given in the monograph by Miranda [352]. Further development of this field is presented in the monographs by Ladyzhenskaya and Uraltseva [285], Gilbarg and Trudinger [200], Skrypnik [493], [492], Temam [514], [515], Krylov [278], Koshelev [271], Begehr and Wen [51], Chen and Wu [110], Caffarelli and Cabre [99], Gasinski and Papageorgiou [192], Ambrosetti and Malchiodi [18], Apreutesei [24], Drabek and Milota [135].

It should be noted that during the last several decades the number of papers devoted to nonlinear elliptic problems has grown exponentially, doubling every ten years (see the concluding remarks below). The order of magnitude for the next decade can be thousands of papers every year. A possible way to face this situation is to introduce a unified system of key words used in all papers.

6 Concluding remarks

The theory of elliptic partial differential equations developed over about two and a half centuries. After some initial period when the foundation of this theory was established, it had an important development from the last quarter of the XIXth century till the middle of the XXth century: new methods of analysis were suggested and the theory of linear equations was basically created. The next stage of this development, which continues nowadays, is characterized by the intensive investigation of various nonlinear problems. The main methods to study nonlinear problems were proposed already in the 1930s. However, a boisterous development of this field began in the 1960s. Taking into account the exponential growth of the number of papers (see the tables below), we can expect that it will continue for at least several decades more.

Mathscinet, the database of the American Mathematical Society provides an interesting tool to analyze the evolution of this field. The following tables show the number of publications in all fields of mathematics and in elliptic equations (as of February 2010).

years	all mathematics	ell. eq. (*)	35J prim/second	% of all math.
1960–69	161169	1514	37 (**)	–
1970–79	316336	3419	3390	1.08
1980–89	474302	5910	7572	1.60
1990–99	600225	7483	11419	1.90
2000–09	776871	9220	17385	2.24

(*) "elliptic equation" or "elliptic system" or "elliptic operator" or "elliptic problem"
(**) such a low number is probably because this classification was only recently introduced

The number of mathematical papers published in 1960–1969 and in 2000–2009 increased almost five times. At the same time, the world population increased only twice, from 3 bln in 1960 to 6 bln in 2000. From 1970–1979 to 2000–2009 the number of all mathematical papers increased about 2.5 times, the number of papers on elliptic equations (35J) about 5.5 times. Thus, the number of mathematical papers grows approximately twice as fast as the world population, the number of papers on elliptic equations twice as fast as of all mathematical papers.

One of the most rapidly expanding topics in mathematics is nonlinear elliptic equations.

years	35J primary	35J6* primary (*)	35J 'and' nonlinear
1960–69	35	3	2
1970–79	2286	456	356
1980–89	4087	1506	1026
1990–99	5551	2889	1653
2000–09	9493	5802	3089

(*) 35J60 – Nonlinear elliptic equations,
35J61 – Semilinear elliptic equations,
35J62 – Quasilinear elliptic equations,
35J65 – Nonlinear boundary value problems for linear elliptic equations,
35J66 – Nonlinear boundary value problems for nonlinear elliptic equations,
35J67 – Boundary values of solutions to elliptic equations.

The AMS subject classification 35J6* contains almost exclusively nonlinear equations. The classification number 35J67 (primary), for which this may not be the case, concerns only about one hundred papers for all years. It does not have much influence on the statistics. Though 35J6* is not the only classification number for nonlinear equations, its evolution shows the general tendency. The number of papers from 1970–1979 to 2000–2009 increased about 13 times. During the last three decades the number of papers has approximately doubled every ten years. A system of unified key words can help to structure and to manage this avalanche of works.

The number of publications or citations in a given field are not the only criteria of its importance. If the theory of Fredholm operators has been almost completely developed, properties of linear and nonlinear non-Fredholm operators are not yet sufficiently well studied and will require further investigations. It seems to be one of the most important open questions in analysis.

Acknowledgement

As I have noted in the preface, most of the results of this book were obtained from joint works with Aizik Volpert. His role in it cannot be overestimated.

Some parts of this book are based on joint works with my friends and colleagues to whom I express my profound gratitude. We worked together with J.F. Collet on the index of elliptic operators in cylindrical domains, with S. Kryzhevich on non-Fredholm operators on the axis, with B. Kazmierczak and V. Vougalter on non-Fredholm operators in $\mathbb{R}^n$, with N. Apreutesei on discrete operators.

I was honored to meet M. Agranovich, L. Nirenberg, M. Vishik, L. Volevich whose works on the theory of elliptic problems were our primary source of inspiration.

Finally, I am grateful to my employer, Centre National de la Recherche Scientifique (CNRS) which provides a unique opportunity for scientific research.

Bibliography

[1] R.A. Adams. Sobolev spaces. Academic Press, New York, 1975.

[2] L.Ya. Adrianova. Introduction to linear systems of differential equations. Translations of Mathematical Monographs V. 146. American Mathematical Society. Providence, Rhode Island. 1995.

[3] S. Agmon. Multiple layer potentials and Dirichlet problem for higher-order equations in the plane. Comm. Pure Appl. Math., 10 (1957), 179–239.

[4] S. Agmon. The coerciveness problem for integro-differential forms. J. Analyse Math., 6 (1958), 183–223.

[5] S. Agmon. On the eigenfunctions and on the eigenvalues of general elliptic boundary value problems. Comm. Pure Appl. Math., 1959, 12, 623–727.

[6] S. Agmon. Maximum theorems for solutions of higher-order elliptic equations. Bull. Amer. Math. Soc. 66 (1960) 77–80.

[7] S. Agmon, A. Douglis, L. Nirenberg. Estimates near the boundary for solutions of elliptic partial differential equations satisfying general boundary conditions. Comm. Pure Appl. Math., 12 (1959), 623–727.

[8] S. Agmon, A. Douglis, L. Nirenberg. Estimates near the boundary for solutions of elliptic partial differential equations satisfying general boundary conditions II. Comm. Pure Appl. Math., 17 (1964), 35–92.

[9] M.S. Agranovich. On the index of elliptic operators. Soviet Math. Dokl., 3 (1962), 194–197.

[10] M.S. Agranovich. General boundary-value problems for integro-differential elliptic systems. (Russian) Dokl. Akad. Nauk SSSR, 155 (1964), 495–498.

[11] M.S. Agranovich. Elliptic boundary problems. Encyclopaedia Math. Sci., vol. 79, Partial Differential Equations, IX, Springer, Berlin, 1997, pp. 1–144.

[12] M.S. Agranovich, A.D. Dynin. General boundary value problems for elliptic systems in higher-dimensional regions. Dokl. Akad. Nauk SSSR, 146 (1962), 511–514.

[13] M.S. Agranovich, M.I. Vishik. Elliptic problems with a parameter and parabolic problems of general type. Uspekhi Mat. Nauk 1964, 19 (3(117)), 53–161. English translation: Russian Math. Surveys 1964, 19 (3), 53–157.

[14] E. Almansi. Sull'integrazione dell'equazione differenziale $\Delta^2\Delta^2 = 0$. Ann. Mat. Pura ed Appl., Ser. 3, t. 2 (1899), 1–51.

[15] E. Almansi. Sull'integrazione dell'equazione differenziale $\Delta^{2n} = 0$. Atti Accad. Sci. Torino, 31 (1895–1896), 527–534.

[16] H. Amann. Existence and regularity for semilinear parabolic evolution equations. Ann. Sc. Norm. Sup. Pisa, Serie IV, XI (1984), 593–676.

[17] H. Ammari, H. Kang, H. Lee. Layer potential techniques in spectral analysis. AMS, Providence, 2009.

[18] A. Ambrosetti, A. Malchiodi. Nonlinear analysis and semilinear elliptic problems. Cambridge Studies in Advanced Mathematics, 104. Cambridge University Press, Cambridge, 2007.

[19] A. Ambrosetti, G. Prodi. On the inversion of some differentiable mappings with singularities between Banach spaces. Ann. Mat. Pura Appl., 93 (1972), No. 4, 231–246.

[20] C. Amrouche, V. Girault, J. Giroire. Dirichlet and Neumann exterior problems for the n-dimensional Laplace operator. An approach in weighted Sobolev spaces. J. Math, Pures Appl., 76 (1997), 55–81.

[21] C. Amrouche, F. Bonzom. Mixed exterior Laplace's problem. J. Math. Anal. Appl., 338 (2008), 124–140.

[22] A.R.A. Anderson, M.A.J. Chaplain, K.A. Rejniak. Single-cell-based models in biology and medicine. Birkhäuser, Basel, 2007.

[23] A.R.A. Anderson, K.A. Rejniak, P. Gerlee and V. Quaranta Modelling of Cancer Growth, Evolution and Invasion: Bridging Scales and Models, Math. Model. Nat. Phenom., 2 (2007), No. 3, 1–29.

[24] N. Apreutesei. Nonlinear second-order evolution equations of monotone type and applications. Pushpa Publishing House, 2007.

[25] N. Apreutesei, N. Bessonov, V. Volpert, V. Vougalter. Spatial structures and generalized travelling waves for an integro-differential equation. DCDS B, 13 (2010), No. 3, 537–557.

[26] N. Apreutesei, A. Ducrot, V. Volpert. Competition of species with intraspecific competition. Math. Model. Nat. Phenom., 3 (2008), No. 4, 1–27.

[27] N. Apreutesei, A. Ducrot, V. Volpert. Travelling waves for integro-differential equations in population dynamics. DCDS B, 11 (2009), No. 3, 541–561.

[28] N.C. Apreutesei, V.A. Volpert. Some properties of infinite-dimensional discrete operators. Topol. Meth. Nonlin. Anal., 24(1) (2004), 159–182.

[29] N.C. Apreutesei, V.A. Volpert. Solvability conditions for some difference operators. Adv. Difference Eq., 1 (2005), 1–13.

[30] N. Apreutesei, V. Volpert. On the eigenvalue of infinite matrices with nonnegative off-diagonal elements, Proceedings of the Conference on Differential and Difference Equations and Applications (R. Agarwal and K. Perera – editors), 81–90, 2006

[31] T. Archibald. Connectivity and smoke-ring: Green's second identity in its first fifty years. Mathematics Magazine, 62 (1989), pp. 219–232.

[32] N. Aronszajn, A.N. Milgram. Differential operators on Riemannian manifolds. Rend. Circ. Mat. Palermo, 2 (1952), 1–61.

[33] I.S. Arson. A Phragmén-Lindelöf theorem for a linear elliptic systems whose coefficients depend on a variable. (Russian) Mat. Sb. (N.S.), 61(103) (1963), 362–376

[34] I.S. Arson, M.A. Evgrafov. On the growth of functions, harmonic in a cylinder and bounded on its surface together with the normal derivative. Dokl. Akad. Nauk SSSR, 142 (1962), 762–765 (Russian).

[35] I.S. Arson, M.A. Iglickii. On the decrease of harmonic functions in a cylinder. Dokl. Akad. Nauk SSSR, 152 (1963), 775–778 (Russian).

[36] M.F. Atiyah, I.M. Singer. The index problem for manifolds with boundary. Bull. Amer. Math. Soc., 69 (1963), 422–433.

[37] M.F. Atiyah, R. Bott. The index of elliptic operators on compact manifolds. Bombay Colloquium on Differential Analysis, Oxford Univ. Press, 1964, 175–186.

[38] M.F. Atiyah, I.M. Singer. The index of elliptic operators. I. Ann. of Math., (2) 87 (1968), 484–530.

[39] M.F. Atiyah, G.B. Segal. The index of elliptic operators. II. Ann. of Math., (2) 87 (1968), 531–545.

[40] M.F. Atiyah, I.M. Singer. The index of elliptic operators. III. Ann. of Math., (2) 87 (1968), 546–604.

[41] P. Aviles. Phragmén-Lindelöf and nonexistence theorems for nonlinear elliptic equations. Manuscripta Math., 43 (1983), no. 2-3, 107–129.

[42] M. Bacharach. Abriss der Geschichte der Potentialtheorie. Göttingen, 1883.

[43] M. Badiale. A note on bifurcation from the essential spectrum. Adv. Nonlinear Stud., 3 (2003), No. 2, 261–272.

[44] S. Bae, T. Chang, D.H. Pahk. Infinite multiplicity of positive entire solutions for a semilinear elliptic equation. J. Differential Equations, 181 (2002), no. 2, 367–387.

[45] L.A. Bagirov. Elliptic equations in an unbounded domain. Mat. Sb. (N.S.), 86 (128) (1971), 121–139.

[46] L.A. Bagirov, V.I. Feigin. Boundary value problems for elliptic equations in domains with an unbounded boundary. Dokl. Akad. Nauk SSSR, 211 (1973), No. 1. Translation in Soviet Math. Dokl., 14 (1973), No. 4, 940–944.

[47] S. Banach. Sur les opérations dans les ensemble abstraits et leur applications aux équations intégrales. Fund. Math., 3 (1922), 133–181.

[48] C. Barillon, V. Volpert. Topological degree for a class of elliptic operators in R^n. Top. Methods of Nonlinear Analysis, 14 (1999), No. 2, 275–293.

[49] R.B. Barrar. On Schauder's paper on linear elliptic differential equations. J. Math. Anal. Appl., 3 (1961), 171–195.

[50] J. Barros-Neto. Inhomogeneous boundary value problems in a half-space. Ann. Sc. Norm. Sup. Pisa, 19 (1965), 331–365.

[51] H.G.W. Begehr, G.C. Wen. Nonlinear elliptic boundary value problems and their applications. Pitman Monographs and Surveys in Pure and Applied Mathematics, 80. Longman, Edinburgh, 1996.

[52] M. Belk, B. Kazmierczak, V. Volpert. Existence of reaction-diffusion-convection waves in unbounded cylinders. Int. J. Math. and Math. Sciences, 2005, No. 2, 169–194.

[53] V. Benci, D. Fortunato. Does bifurcation from the essential spectrum occur? Comm. Partial Differential Equations, 6 (1981), No. 3, 249–272.

[54] P. Benevieri, M. Furi. A simple notion of orientability for Fredholm maps of index zero between Banach manifolds and degree theory. Annales des Sciences Mathématiques de Québec, 22 (1998), 131–148.

[55] P. Benevieri, M. Furi. On the concept of orientability for Fredholm maps between real Banach manifolds. Topol. Methods Nonlinear Anal., 16 (2000), no. 2, 279–306.

[56] E. Benincasa, A. Canino. A bifurcation result of Bohme-Marino type for quasilinear elliptic equations. Topol. Methods Nonlinear Anal., 31 (2008), No. 1, 1–17.

[57] N. Benkirane. Propriété d'indice en théorie Höldérienne pour des opérateurs elliptiques dans R^n. CRAS, 307, serie I (1988), 577–580.

[58] H. Berestycki, L.A. Caffarelli, L. Nirenberg. Symmetry for elliptic equations in a half-space. Boundary value problems for partial differential equations and applications, 27–42, RMA Res. Notes Appl. Math., 29, Masson, Paris, 1993.

[59] H. Berestycki, L.A. Caffarelli, L. Nirenberg. Inequalities for second-order elliptic equations with applications to unbounded domains. I. A celebration of John F. Nash, Jr. Duke Math. J., 81 (1996), No. 2, 467–494.

[60] H. Berestycki, L.A. Caffarelli, L. Nirenberg. Further qualitative properties for elliptic equations in unbounded domains. Dedicated to Ennio De Giorgi. Ann. Scuola Norm. Sup. Pisa Cl. Sci., (4) 25 (1997), No. 1-2, 69–94.

[61] H. Berestycki, M. Esteban. Existence and bifurcation of solutions for an elliptic degenerate problem. J. Differential Equations, 134 (1997), No. 1, 1–25.

[62] D. Bernoulli. Remarques sur le principe de la conservation des forces vives. Hist. et Mém. Acad. Sci. et Bell. Lettr., Berlin, 1748 (1750), 356–364.

[63] S.N. Bernstein. Sur la nature analytique des solutions de certaines équations aux dérivées partielles du second ordre. Math. Ann.,59 (1904), 20–76.

[64] S.N. Bernstein. Sur la généralisation du problème de Dirichlet (Deuxième Partie). Math. Ann., 69 (1910), 82–136.

[65] S.N. Bernstein. Sur les surfaces définies au moyen de leur courbure moyenne ou totale. Annales de l'Ecole Normale, 27 (1910), 233–256.

[66] S.N. Bernstein. Sur les équations du calcul des variations. Annales de l'Ecole Normale, 29 (1912), 431–485.

[67] S.N. Bernstein. Démonstration du theorème de M. Hilbert sur la nature analytique des solutions des équations du type elliptique sans l'emploi des séries normales. Math. Z.

[68] S.N. Bernstein. Some a priori estimates in Dirichlet's generalized problem. (Russian) Dokl. Akad. Nauk SSSR, 124 (1959), 735–738.

[69] G. Bertrand. Le problème de Dirichlet et le potentiel de simple couche. Bull. des Sciences Math., 2-eme série, XLVII (1923), 282–307.

[70] G.D. Birkhoff, O.D. Kellog. Invariant points in function space. Trans. Amer. Math. Soc., 23 (1922), 96–115.

[71] A.V. Bitsadze. Boundary problems for systems of linear differential equations of elliptic type. Communication of Georgian Academy of Sciences, V (1944), No. 8, 761–770 (Russian).

[72] A.V. Bitsadze. Boundary value problems for elliptic equations of second order. Moscow, Nauka, 1966. English translation: North-Holland, Amsterdam, 1968.

[73] T. Boggio. Integrazione dell'equazione $\Delta^2\Delta^2 = 0$ in una corona circolare e in uno strato sferico. Atti Ist. Veneto Sci. Let. ed Arti, t. 59 (ser. 8, t. 2), par. 2 (1899–1900), 497–508.

[74] T. Boggio. Un teorema di reciprocità sulle funzioni di Green d'ordine qualunque. Atti Accad. sci. Torino, 35 (1899–1900), 498–509.

[75] P. Bolley, T.L. Pham. Propriété d'indice en théorie Höldérienne pour des opérateurs différentiels elliptiques dans R^n. J. Math. Pures Appl., 72 (1993), 105–119.

[76] P. Bolley, T.L. Pham. Propriété d'indice en théorie Höldérienne pour le problème extérieur de Dirichlet. Comm. Partial Differential Equations, 26 (2001), no. 1-2, 315–334.

[77] J. Borisovich, V. Zvyagin, J. Sapronov. Nonlinear Fredholm maps and the Leray-Schauder theory. Uspekhi Mat. Nauk, 32 (1977), 3–54; English translation in Russian Math. Surveys, 32 (1977), 1–54.

[78] B.V. Boyarski. On the Dirichlet problem for a system of elliptic equations in space. Bull. Acad. Polon. Sci. Sér. Sci. Math. Astr. Phys., 8 (1960) 19–23 (Russian).

[79] B.V. Boyarski. On the index problem for systems of singular integral equations. Bull. Acad. Polon. Scienc. Ser. Math. Astr. Phys., 11 (1963), No. 10.

[80] H. Brezis, F. Browder. Partial differential equations in the 20th century. Advances in Mathematics, 135 (1998), 76–144.

[81] H. Brezis, L. Nirenberg. Positive solutions of nonlinear elliptic equations involving critical Sobolev exponents. Comm. Pure Appl. Math., 36 (1983), No. 4, 437–477.

[82] H. Brezis, R.E.L. Turner. On a class of superlinear elliptic problems. Comm. Partial Differential Equations, 2 (1977), no. 6, 601–614.

[83] N.F. Britton. Spatial structures and periodic travelling waves in an integro-differential reaction-diffusion population model. SIAM J. Appl. Math., 6 (1990), 1663–1688.

[84] F.E. Browder. The Dirichlet problem for linear elliptic equations of arbitrary even order with variable coefficients. Proc. Nat. Acad. Sci. U.S.A., 38 (1952), 230–235.

[85] F.E. Browder. On the regularity properties of solutions of elliptic differential equations. Comm. Pure Appl. Math., 9 (1956), 351–361.

[86] F.E. Browder. Estimates and existence theorems for elliptic boundary value problems. Proc. Nat. Acad. Sci. USA, 45 (1959), 355–372.

[87] F.E. Browder. On the spectral theory of elliptic differential operators. Math. Annalen, 142 (1961), 22–130.

[88] F.E. Browder. Nonlinear elliptic boundary value problems and the generalized topological degree. Bull. Amer. Math. Soc., 76 (1970), 999–1005.

[89] F.E. Browder, W.V. Petryshhyn. The topological degree and Galerkin approximations for noncompact operators in Banach spaces. Bull. Amer. Math. Soc., 74 (1968), 641–646.

[90] F.E. Browder, W.V. Petryshhyn. Approximation methods and the generalized topological degree for nonlinear mappings in Banach spaces. J. Funct. Analysis, 3 (1969), 217–245.

[91] K.J. Brown, N. Stavrakakis. Global bifurcation results for a semilinear elliptic equation on all of R^N. Duke Math. J., 85 (1996), No. 1, 77–94.

[92] C. Budd. Symmetry breaking and semilinear elliptic equations. Continuation techniques and bifurcation problems. J. Comput. Appl. Math., 26 (1989), No. 1–2, 79–96.

[93] H. Burkhardt, F. Meyer. Potentialtheorie. Encykl. Math. Wiss., Bd. II, T. 1, H. 4, Leipzig, 1900, 464–503.

[94] J. Busca, M. Esteban, A. Quaas. Nonlinear eigenvalues and bifurcation problems for Pucci's operators. (English summary) Ann. Inst. H. Poincaré Anal. Non Linéaire, 22 (2005), No. 2, 187–206.

[95] X. Cabré, A. Capella. Regularity of radial minimizers and extremal solutions of semilinear elliptic equations. J. Funct. Anal., 238 (2006), No. 2, 709–733.

[96] R. Caccioppoli. Sugli elementi uniti delle transformazioni funzionali: un teorema di esistenza e di unicità e alcune sue appplicazioni. Rend. Sem. Mat. Padova, 3 (1932), 1–15.

[97] R. Caccioppoli. Un principo d'inversione per le corrispondenze funzionali e sue applicazioni alle equazioni a derivate parziali. Rend. Acc. Lincei, 16 (1932), 390–395, 484–489.

[98] R. Caccioppoli. Sulle corrispondenze funzionali inverse diramate: teoria generale e applicazioni ad alcune equazioni non lineari e al problema di Plateau. Rend. Acc. Lincei, 24 (1936), 258–263, 416–421.

[99] L.A. Caffarelli, X. Cabré. Fully nonlinear elliptic equations. AMS Colloquium Publications, Vol. 43, AMS, Providence.

[100] C.X. Cai, C.H. Lin. A Phragmén-Lindelöf alternative theorem for the biharmonic equation. Acta Sci. Natur. Univ. Sunyatseni, 35 (1996), no. 2, 15-20.

[101] A.P. Calderon. The analytic calculation of the index of elliptic equations. Proc. Nat. Acad. Sci.U.S.A., 57 (1967), 1193–1194.

[102] D.M. Cao. Positive solution and bifurcation from the essential spectrum of a semilinear elliptic equation on R^n. Nonlinear Anal., 15 (1990), No. 11, 1045–1052.

[103] I.D. Capuzzo, A. Vitolo. On the maximum principle for viscosity solutions of fully nonlinear elliptic equations in general domains. Matematiche (Catania) 62 (2007), no. 2, 69–91.

[104] I.D. Capuzzo, A. Vitolo. A qualitative Phragmén-Lindelöf theorem for fully nonlinear elliptic equations. J. Differential Equations 243 (2007), no. 2, 578–592.

[105] T. Carleman. Sur la résolution de certaines équations intégrales. Arkiv för Mathematik, Astronomi och Physik, 16 (1922), No. 26, 1–19.

[106] A.O. Celebi, V.K. Kalantarov, F. Tahamtani. Phragmén-Lindelöf type theorems for some semilinear elliptic and parabolic equations. Demonstratio Math. 31 (1998), no. 1, 43–54.

[107] S. Chandrasekhar. Hydrodynamics and hydromagnetic stability. Clarendon Press, New York, 1961.

[108] I.A. Cegis. A Phragmén-Lindelöf type theorem for functions harmonic in a rectangular cylinder. Soviet Math. Dokl., 2 (1961) 113–117.

[109] M. Chaves, J. Garcia Azorero. On bifurcation and uniqueness results for some semilinear elliptic equations involving a singular potential. J. Eur. Math. Soc., 8 (2006), No. 2, 229–242.

[110] Y.-Z. Chen, L.-C. Wu. Second-order elliptic equations and elliptic systems. Translation of Math. Monographs, Vol. 174, AMS, Providence, 1998.

[111] M. Chipot. l goes to infinity. Birkhäuser, Basel, 2002.

[112] S. Cingolani, J.L. Gamez. Positive solutions of a semilinear elliptic equation on R^N with indefinite nonlinearity. Adv. Differential Equations, 1 (1996), No. 5, 773–791.

[113] A. Cioffi. Zero-index problems for a class of degenerate elliptic operators. Ricerche Mat. 32 (1983), no. 2, 237–261.

[114] J.F. Collet, V. Volpert. Computation of the index of linear elliptic operators in unbounded cylinders. J. Funct. Analysis, 164 (1999), 34–59.

[115] W.A. Coppel. Dichotomies in stability theory. Lecture Notes in Mathematics, No. 629, Springer-Verlag, Berlin, 1978.

[116] C. Corduneanu, N. Gheorghiu, V. Barbu. Almost periodic functions. Second English Edition, Chelsea, New York, 1989.

[117] M.G. Crandall, P.H. Rabinowitz. Bifurcation from simple eigenvalues. J. Functional Analysis, 8 (1971), 321–340.

[118] J. Cronin. The existence of multiple solutions of elliptic differential equations. Trans. Amer. Math. Soc., 68 (1950), No. 1, 105–131.

[119] J. Cronin. Branch points of solutions in Banach space. Trans. Amer. Math. Soc., 69 (1950), No. 2, 208–231.

[120] J. Cronin. Analytic functional mappings. Annals of Math., 58 (1953), No. 1, 175–181.

[121] G. Da Prato, P.C. Kunstmann, I. Lasiecka, A. Lunardi, R. Schnaubelt, L. Weis. Functional analysis methods for evolution equations. Springer, Berlin, 2004.

[122] Q. Dai. Entire positive solutions for inhomogeneous semilinear elliptic systems. Glasg. Math. J., 47 (2005), no. 1, 97–114.

[123] E.N. Dancer. Boundary value problems for ordinary differential equations on infinite intervals. Proc. London Math. Soc., 30 (1975), No. 3, 76–94.

[124] M. Dauge. Elliptic boundary value problems on corner domains. Smoothness and asymptotics of solutions. Lecture Notes in Mathematics, 1341. Springer-Verlag, Berlin, 1988.

[125] S.P. Degtyarev. On the optimal regularity of solutions of the first boundary value problem for a class of degenerate elliptic equations. Ukr. Mat. Visn. 3 (2006), no. 4, 443–466, 584; translation in Ukr. Math. Bull. 3 (2006), no. 4, 423–446.

[126] M. del Pino. Supercritical elliptic problems from a perturbation viewpoint. Discrete Contin. Dyn. Syst., 21 (2008), No. 1, 69–89.

[127] Y. Deng, Y. Li. Existence of multiple positive solutions for a semilinear elliptic equation. Adv. Differential Equations, 2 (1997), No. 3, 361–382.

[128] R. Denk, M. Hieber, J. Prüss. R-boundedness, Fourier multipliers and problems of elliptic and parabolic type. Memoirs AMS, Vol. 166 (2003), No. 788, 1–113.

[129] R. Denk, L. Volevich. Elliptic boundary value problems with large parameter for mixed-order systems. AMS Translations, 206 (2002), Eds. M. Agranovich, M. Shubin, pp. 29–64.

[130] Y. Deng, Y. Guo, Y. Li. Existence and decay properties of positive solutions for an inhomogeneous semilinear elliptic equation. Proc. Roy. Soc. Edinburgh Sect. A 138 (2008), no. 2, 301–322.

[131] G. Diaz, R. Letelier. Uniqueness for viscosity solutions of quasilinear elliptic equations in R^N without conditions at infinity. Differential Integral Equations, 5 (1992), no. 5, 999–1016.

[132] T. Doncev. The behavior of the solution of a higher-order elliptic equation in infinite domains. Vestnik Moskov. Univ. Ser. I Mat. Meh., 26 (1971), no. 5, 12–15.

[133] T. Doncev. The behavior of the solutions of higher-order elliptic equations. Godisnik Vis. Tehn. Ucebn. Zaved. Mat., 9 (1973), no. 3, 71–77.

[134] A. Douglis, L. Nirenberg. Interior estimates for elliptic systems of partial differential equations. Comm. Pure Appl. Math., 8 (1955), 503–538.

[135] P. Drabek, J. Milota. Methods of nonlinear analysis. Applications to differential equations. Birkhäuser Advanced Texts. Birkhäuser Verlag, Basel, 2007.

[136] P. Dubois-Reymond. Über lineare partielle Differentialgleichungen zweiter Ordnung. J. reine und angew. Math., 104 (1889), 241–301.

[137] A. Ducrot, M. Marion, V. Volpert. Systèmes de réaction-diffusion sans propriété de Fredholm. CRAS, 340 (2005), 659–664.

[138] A. Ducrot, M. Marion, V. Volpert. Reaction-diffusion-convection problems with non Fredholm operators. Int. J. Pure and Applied Mathematics, 27 (2006), No. 2, 179–204.

[139] A. Ducrot, M. Marion, V. Volpert. Reaction-diffusion problems with non-Fredholm operators. Adv. Diff. Eq., 13 (2008), No. 11–12, 1151–1192.

[140] A. Ducrot, M. Marion, V. Volpert. Reaction-diffusion waves (with the Lewis number different from 1). Publibook, Paris, 2008.

[141] R. Duduchava, B. Silbermann. Boundary value problems in domains with peaks. Mem. Differential Equations Math. Phys., 21 (2000), 1–122.

[142] A.S. Dynin. Multidimensional elliptic boundary value problems with a single unknown function. Dokl. Akad. Nauk SSSR, 141 (1961) 285–287. Translated in Soviet Math. Dokl., 2 (1961), 1431–1433.

[143] A.S. Dynin. The index of an elliptic operator on a compact manifold. Uspekhi Mat. Nauk, 21 (1966), No. 5 (131), 233–248.

[144] A.S. Dynin. On the index of families of pseudodifferential operators on manifolds with a boundary. Dokl. Akad. Nauk SSSR, 186 (1969) 506–508. English translation: Soviet Math. Dokl., 10 (1969) 614–617.

[145] A. Dzhuraev. General boundary-value problems for elliptic equations with non-analytic coefficients. (Russian) Sibirsk.Mat. Zh. 4 (1963), 539–561.

[146] A. Dzhuraev. Investigation of a boundary problem of the plane for an elliptic equation. Sibirsk. Mat. Zh., 6 (1965), 484–498 (Russian).

[147] Yu.V. Egorov, V.A. Kondratiev. On an oblique derivative problem. Dokl. Akad. Nauk SSSR, 170 (1966) 770–772.

[148] Yu.V. Egorov, V.A. Kondratiev. The oblique derivative problem. Mat. sb. (N.S.), 78 (120) (1969), 148–176.

[149] Yu.V. Egorov, V.A. Kondratiev. On the behavior of solutions of a nonlinear boundary value problem for a second-order elliptic equation in an unbounded domain. Tr. Mosk. Mat. Obs., 62 (2001), 136–161; translation in Trans. Moscow Math. Soc. 2001, 125–147.

[150] Yu.V. Egorov, V.A. Kondratiev. On the asymptotic behavior of solutions to a semilinear elliptic boundary problem. (English summary) International Conference on Differential and Functional Differential Equations (Moscow, 1999). Funct. Differ. Equat., 8 (2001), no. 1-2, 163–181.

[151] K.D. Elworthy, A.J. Tromba. Degree theory on Banach manifolds. In: Nonlinear Functional Analysis, F.E. Browder, Ed., Proc. Symp. Pure Math., Vol. 18, Part I, 1970, 86–94.

[152] K.D. Elworthy, A.J. Tromba. Differential structures and Fredholm maps on Banach manifolds. In: Global Analysis, S.S. Chern, S. Smale, Eds., Proc. Symp. Pure Math., 15 (1970), 45–94.

[153] A.K. Erkip, E. Schrohe. Normal solvability of elliptic boundary value problems on asymptotically flat manifolds. J. Funct. Analysis, 109 (1992), 22–51.

[154] G.I. Eskin. General boundary value problems for equations of principal type in a plane domains with angular points. Uspekhi Mat. Nauk, 18 (1963), No. 3, 241–242.

[155] G. Eskin. Boundary value problems for elliptic pseudodifferential equations. Translations of Math. Monographs, Vol. 52, AMS, Providence, 1981.

[156] M.J. Esteban, P.L. Lions. Existence and nonexistence results for semilinear elliptic problems on unbounded domains. Proc. Roy. Soc. Edinburgh, 93 (1983), 1–14.

[157] L. Euler. Mechanica sive motus scientia analytice exposita, t. 2. Petropoli, 1736.

[158] L. Euler. Principia motus fluidorum. Novi comment. Acad. scient. Petropolitanae, t. 6, 1756–1757 (1761), 271–311.

[159] L. Euler. Theoria motus corporum solidorum seu rigidorum. Rostok, 1765.

[160] J. Favard. Sur les équations différentielles linéaires à coefficients presque-périodiques. Acta Math., 51 (1928), 31–81.

[161] B.V. Fedosov. Analytic formulae for the index of elliptic operators. Trudy Moskov. Mat. Obsch., 30 (1974), 159–241.

[162] B.V. Fedosov. An analytic formula for the index of an elliptic boundary value problem. Math. USSR Sbornik, 22 (1974), No. 1, 61–89.

[163] B.V. Fedosov. Analytic formula for the index of an elliptic boundary value problem. III. (Russian) Mat. Sb. (N.S.), 101 (143) (1976), No. 3, 380–401, 456.

[164] B.V. Fedosov. The index of elliptic families on a manifold with boundary. Dokl. Akad. Nauk SSSR, 248 (1979), No. 5, 1066–1069.

[165] B.V. Fedosov. Index theorems. Partial differential equations, VIII, 155–251, Encyclopaedia Math. Sci., 65, Springer, Berlin, 1996.

[166] B.V. Fedosov, B.W. Schulze, N.N.Tarkhanov. On the index of elliptic operators on a wedge. J. Funct. Anal., 157 (1998), No. 1, 164–209.

[167] B.V. Fedosov, M.A. Shubin. The index of random elliptic operators and of families of them. Dokl. Akad. Nauk SSSR, 236 (1977), No. 4, 812–815.

[168] V.I. Feigin. Elliptic equations in domains with multidimensional singularities on the boundary, Uspekhi Mat. Nauk, 27 (1972), no. 2, 183–184.

[169] P. Feng, Z. Zhou. Multiplicity and symmetry breaking for positive radial solutions of semilinear elliptic equations modelling MEMS on annular domains. Electron. J. Differential Equations, (2005), No. 146, 14 pp.

[170] C. Fenske. Analytische Theorie des Abbildungrades für Abbildungen in Banachräumen. Math. Nachr., 48 (1971), 279–290.

[171] N.M. Ferrers, Ed. Mathematical papers of George Green. London, 1871; Chelsea Publ. Company, New York, 1970.

[172] R.A. Fisher. The wave of advance of advantageous genes. Ann. Eugenics, 7 (1937), 355–369.

[173] P. Fife. Growth and decay properties of solutions of second-order elliptic equations. Ann. Scu. Norm. Sup. Pisa, 20 (1966), 675–701.

[174] P.M. Fitzpatrick. The parity as an invariant for detecting bifurcation of the zeroes of one parameter families of nonlinear Fredholm maps. In: Topological Methods for Ordinary Differential Equations, M. Furi, P. Zecca, Eds., Lecture Notes in Math., Vol. 1537, 1993, 1–31.

[175] P.M. Fitzpatrick, J. Pejsachowicz. Parity and generalized multiplicity. Trans. Amer. Math. Soc., 326 (1991), 281–305.

[176] P.M. Fitzpatrick, J. Pejsachowicz. Orientation and the Leray-Schauder degree for fully nonlinear elliptic boundary value problems. Mem. Amer. Math. Soc., 483 (1993).

[177] P.M. Fitzpatrick, J. Pejsachowicz, P.J. Rabier. The degree of proper C^2 Fredholm mappings. J. Reine Angew., 427 (1992), 1–33.

[178] P.M. Fitzpatrick, J. Pejsachowicz, P.J. Rabier. Orientability of Fredholm families and topological degree for orientable nonlinear Fredholm mappings. J. Funct. analysis, 124 (1994), 1–39.

[179] Flucher, M., Muller, S. Radial symmetry and decay rate of variational ground states in the zero mass case. SIAM J. Math. Anal. 29 (1998), no. 3, 712–719.

[180] J. Fourier. Théorie du mouvement de la chaleur dans les corps solides. Mémoire de l'Académie Royale des Sciences, t. 5 (1821–1822), 153–246.

[181] J. Fourier. Théorie analytique de la chaleur. Mémoire de l'Académie Royale des Sciences, t. 8 (1825), 581–622.

[182] D.A. Frank-Kamenetskii. Diffusion and heat transfer in chemical kinetics. Third edition. Nauka, Moscow, 1987.

[183] I. Fredholm. Sur une nouvelle méthode pour la résolution du problème de Dirichlet. Öfversigt af Kongl. Vetenskaps–Akad. Förhandlingar, Arg.57, No. 1, Stockholm, 39–46.

[184] I. Fredholm. Sur une classe d'équations fonctionnelles. Acta Math., 27 (1903), 365–390.

[185] R.S. Freeman. On the spectrum and resolvent of homogeneous elliptic differential operators with constant coefficients. Bull. Amer. Math. Soc., 72 (1966), 538–541.

[186] R.S. Freeman, M. Schechter. On the existence, uniqueness and regularity of solutions to general elliptic boundary-value problems. J. Diff. Equat., 15 (1974), 213–246.

[187] A. Friedman, On two theorems of Phragmén-Lindelöf for linear elliptic and parabolic differential equations of the second order, Pacific J. Math., 7 (1957), 1563–1575.

[188] Y. Furusho. Existence and bifurcation of positive solutions of asymptotically linear elliptic equations in exterior domains. Nonlinear Anal., 8 (1984), No. 6, 583–593.

[189] T.S. Gadzhiev. On the behavior of solutions of a mixed boundary value problem for higher-order quasilinear elliptic equations of divergence type. (Russian) Differentsialnye Uravneniya, 26 (1990), no. 4, 703–706, 735.

[190] F.D. Gakhov. Boundary value problems. Fizmatlit, Moscow, 1958 (Russian).

[191] F.R. Gantmakher. The theory of matrices. Chelsea Publishing, Providence, RI, 1998.

[192] L. Gasiński, N.S. Papageorgiou. Nonsmooth critical point theory and nonlinear boundary value problems. Chapman and Hall/CRC, Boca Raton, 2005.

[193] C.-F. Gauss. Théorèmes généraux sur les forces attractives et répulsives qui agissent en raison inverse du carré des distances. Journal de mathématiques pures et appliquées 1-ère série, tome 7 (1842), 273–324.

[194] S. Genieys, V. Volpert, P. Auger, Pattern and waves for a model in population dynamics with nonlocal consumption of resources, Math. Model. Nat. Phenom., 1 (2006), no. 1, 65–82.

[195] G.Z. Gershuni, E.M. Zhuhovitskii. Convective instability of incompressible fluid. Nauka, Moscow, 1972.

[196] G. Geymonat, P. Grisvard. Alcuni risultati di teori spettrale per i problemi ai limiti lineari ellittici. Rend. Sem. Mat. Univ. Padova, 38 (1967), 121–173.

[197] J. Giacomoni. Global bifurcation results for semilinear elliptic problems in R^N. Comm. PDE, 23 (1998), no. 11-12, 1875–1927.

[198] B. Gidas. Symmetry properties and isolated singularities of positive solutions of nonlinear elliptic equations. Nonlinear partial differential equations in engineering and applied science (Proc. Conf., Univ. Rhode Island, Kingston, R.I., 1979), pp. 255–273, Lecture Notes in Pure and Appl. Math., 54, Dekker, New York, 1980.

[199] D. Gilbarg. The Phragmén-Lindelöf theorem for elliptic partial differential equations, J. Rat. Mech. Anal., 1 (1952), 411–417.

[200] D. Gilbarg, N.S. Trudinger. Elliptic partial differential equations. Second edition. Springer, Berlin, 1983.

[201] G. Giraud. Sur le problème de Dirichlet généralisé. Equations non linéaires à m variables. Annales Scientifiques de l'ENS, 3-ème série, tome 43 (1926), 1–128.

[202] G. Giraud. Sur le problème de Dirichlet généralisé (deuxième mémoire). Annales Scientifiques de l'ENS, 3-ème série, tome 46 (1929), 131–245.

[203] G. Giraud. Sur certains problèmes non linéaires de Neumann et sur certains problèmes non linéaires mixtes (suite). Annales Scientifiques de l'ENS, 3-ème série, tome 49 (1932), 245–309.

[204] G. Giraud. Sur une classe d'équations intégrales où figurent des valeurs principales d'intégrales simples. Ann. Scientifique de l'ENS, 3-ème série, 56 (1939), 119–172.

[205] S.K. Godunov. Equations of mathematical physics. Moscow, Nauka, 1971 (Russian).

[206] I. Gohberg, S. Goldberg, M.A. Kaashoek, Classes of linear operators, Vol.1, Operator theory: Advances and applications, Birkhäuser, 1990.

[207] I. Gohberg, M. Krein. The basic propositions on defect numbers and indices of linear operators. Amer. Math. Soc. Transl., 13 (1960), No. 2, 185–264.

[208] S.A. Gourley. Travelling front solutions of a nonlocal Fisher equation. J. Math. Biol., 41 (2000), 272–284.

[209] E. Goursat. Cours d'analyse mathématique. Paris, Gauthier-Villars, 1917. Volumes 1,2,3.

[210] Granlund, S. A Phragmén-Lindelöf principle for subsolutions of quasilinear equations. Manuscripta Math., 36 (1981/82), no. 3, 355–365.

[211] I. Grattan-Guinness. Why did George Green Write His Essay of 1828 on Electricity and Magnetism? The American Mathematical Monthly, 102, No. 5. (1995), 387–396.

[212] G. Green. An essay on the application of mathematical analysis to the theory of electricity and magnetism. Nottingham, 1828; Journal für die reine und

angewandte Mathematik (Crelle's Journal) 39 (1850) 73–79; 44 (1852) 356–374; 47 (1854) 161–221.

[213] J.A. Griepentrog, L. Recke. Linear elliptic boundary value problems with non-smooth data: normal solvability on Sobolev-Campanato spaces. Math. Nachr., 225 (2001), 39–74.

[214] P. Grisvard. Alternative de Fredholm relative au problème de Dirichlet dans un polygone ou un polyèdre. Boll.Un. Mat. Ital., (4) 5 (1972), 132–164.

[215] P. Grisvard. Alternative de Fredholm relative au problème de Dirichlet dans un polyèdre. Ann. Scuola Norm. Sup. Pisa Cl. Sci. (4) 2 (1975), no. 3, 359–388.

[216] P. Grisvard. Elliptic Problems in Nonsmooth Domains, Monographs and Studies in Mathematics 24, Pitman, Boston, 1985.

[217] P. Grisvard. Singularities in boundary value problems. Recherches en Mathématique Appliquée, 22. Masson, Paris; Springer-Verlag, Berlin, 1992.

[218] A.P. Gubachev. A Phragmén-Lindelöf theorem for quasilinear elliptic equations in a cylindrical domain. Mat. Zametki, 30 (1981), no. 2, 197–201, 314 (Russian).

[219] P.L. Gurevich, A.L. Skubachevski. On the Fredholm and unique solvability of nonlocal elliptic problems in multidimensional domains. Tr. Mosk. Mat. Obs., 68 (2007), 288–373; translation in Trans. Moscow Math. Soc., 2007, 261–336.

[220] O.V. Guseva. On boundary problems for strongly elliptic systems. Dokl. Akad. Nauk SSSR (N.S.), 102 (1955), 1069–1072 (Russian).

[221] A.K. Gushchin. A condition for the complete continuity of operators arising in nonlocal problems for elliptic equations. Dokl. Akad. Nauk, 373 (2000), no. 2, 161–163 (Russian).

[222] A.K. Gushchin, V.P. Mikhailov. On the solvability of nonlocal problems for a second-order elliptic equation. Mat. Sb., 185 (1994), no. 1, 121-160; translation in Russian Acad. Sci. Sb. Math., 81 (1995), no. 1, 101–136.

[223] B. Gustaffson, A. Vasiliev. Conformal and potential analysis in Hele-Shaw cells. Birkhäuser, Basel, 2006.

[224] A. Gutzmer. Remarques sur certaines équations aux différences partielles d'ordre supérieur. J. math. pures et appl., sér. 4, t. 6 (1890), 405–422.

[225] B. Hanouzet. Espaces de Sobolev avec poids application au problème de Dirichlet dans un demi espace. Rend. Sem. Mat. Univ. Padova, 46 (1971), 227–272.

[226] G.H. Hardy, J.E. Littlewood, G. Polya. Inequalities. Cambridge. 1934.

[227] A. Harnack Die Grundlagen der Theorie des logarithmischen Potentiales und der eindeutigen Potentialfunktion in der Ebene. Leipzig, 1887.

[228] H. Helmholtz. Theorie der Luftschwingungen in Röhren mit offenen Enden. J. reine und angew. Math., Bd. 57 (1860), 1–72.

[229] H. Helmholtz. Über einige Gesetze der Vertheilung elektrischer Ströme in körperlichen Leitern ... H. Helmholtz. Wiss. Abhandlungen, Bd. 1, Leipzig, 1882, 475–519.

[230] H.B. Heywood, M. Fréchet. L'équation de Fredholm et ses applications à la physique mathématique. Paris, A. Hermann, 1912.

[231] D. Henry. Geometric theory of semilinear parabolic equations, Lecture Notes Math. 840, Springer-Verlag, Berlin, 1981.

[232] J.O. Herzog. Phragmén-Lindelöf theorems for second-order quasilinear elliptic partial differential equations. Proc. Amer. Math. Soc., 16 (1964), 721–728.

[233] Y. Higuchi. A priori estimates and existence theorem on elliptic boundary value problems for unbounded domains. Osaka J. Math., 5 (1968), 103–135.

[234] D. Hilbert. Über das Dirichletsche Prinzip. Jahresber. Dtsch. Math. Ver., 1900.

[235] D. Hilbert. Über eine Anwendung der Integralgleichungen auf ein Problem der Funktionentheorie. Verhandl. des III Internat. Mathematiker Kongresses, Heidelberg, 1904.

[236] D. Hilbert. Grundzüge einer allgemeinen Theorie der linearen Integralgleichungen. Leipzig, B.G. Teubner, 1912.

[237] E. Holmgren. Sur un problème aux limites pour l'équation $y^m z_{xx} + z_{yy} = 0$. Arkiv Mat.,Astr., Fysik, 19B (1926), No. 4.

[238] E. Hopf. Zum analytischen Charakter der Lösungen regulärer zweidimensionaler Variationsprobleme. (German) Math. Z. 30 (1929), no. 1, 404–413.

[239] E. Hopf. Über den funktionalen, insbesondere den analytischen Charakter der Lösungen elliptischer Differentialgleichungen zweiter Ordnung. (German) Math. Z. 34 (1932), no. 1, 194–233.

[240] E. Hopf. Remarks on the preceding paper by D. Gilbarg. J. Rational Mech. Anal. 1, (1952), 419–424.

[241] L. Hörmander. On regularity of the solutions of boundary problems. Acta Math., 99 (1958), 225–264.

[242] L. Hörmander. Pseudo-differential operators and non-elliptic boundary problems. Ann. of Math., (2) 83 (1966), 129–209.

[243] L. Hörmander. Linear partial differential operators. Springer, Heidelberg 1969.

[244] C.-C. Hsieh, J.-F. Hwang, F.-T. Liang. Phragmén-Lindelöf theorem for minimal surface equations in higher dimensions. Pacific J. Math., 207 (2002), no. 1, 183–198.

[245] A. Huber. A theorem of Phragmén-Lindelöf type. Proc. Amer. Math. Soc., 4 (1953). 852–857.

[246] J. Hulshof, R.C.A.M. van der Vorst. Asymptotic behaviour of ground states. (English summary) Proc. Amer. Math. Soc. 124 (1996), no. 8, 2423–2431.

[247] R. Infantino, M. Troisi. Degenerate elliptic operators in domains of R^n. Boll. Un. Mat. Ital., B (5) 17 (1980), no. 1, 186–203.

[248] C.A. Isnard. Orientation and degree in infinite dimensions. Notices Amer. Math. Soc., 19 (1972), A-514.

[249] L. Jeanjean. Local conditions insuring bifurcation from the continuous spectrum. Math. Z., 232 (1999), No. 4, 651–664.

[250] H. Jeanjean, M. Lucia, C.A. Stuart. Branches of solutions to semilinear elliptic equations on R^N. Math. Z., 230 (1999), 79–105.

[251] Z. Jin, K. Lancaster. Theorems of Phragmén-Lindelöf type for quasilinear elliptic equations. J. Reine Angew. Math., 514 (1999), 165–197.

[252] R.A. Johnson, X.B. Pan, Y. Yi. Singular ground states of semilinear elliptic equations via invariant manifold theory. Nonlinear Anal., 20 (1993), No. 11, 1279–1302.

[253] R.A. Johnson, X.B. Pan, Y. Yi. Positive solutions of super-critical elliptic equations and asymptotics. Comm. Partial Differential Equations, 18 (1993), no. 5-6, 977–1019.

[254] L.V. Kantorovich, G.P. Akilov. Functional analysis. Second edition. Nauka, Moscow, 1977.

[255] L. Karp. On Liouville type theorems for second-order elliptic differential equations. Ann. Scuola Norm. Sup. Pisa Cl. Sci. (4) 22 (1995), no. 2, 275–298.

[256] T. Kato. Perturbation theory for linear operators. Springer, Heidelberg, 1966.

[257] Kawano, N., Yanagida, E., Yotsutani, S. Structure theorems for positive radial solutions to $\mathrm{div}(|Du|^{m-2}Du) + K(|x|)u^q = 0$ in R^n. J. Math. Soc. Japan, 45 (1993), no. 4, 719–742.

[258] J.B. Keller, S. Antman. Bifurcation theory and nonlinear eigenvalue problems. W.A. Benjamin, Inc., New York, 1969.

[259] Sh.B. Khalilov. Solvability of the Dirichlet problem for multidimensional elliptic systems. (Russian) Differentsialnye Uravneniya 26 (1990), no. 9, 1621–1626, 1655; translation in Differential Equations 26 (1990), no. 9, 1207–1212.

[260] Sh.B. Khalilov. The Dirichlet problem for a Petrovskii-elliptic system of second-order equations. Differ. Uravn., 42 (2006), no. 3, 416–422; translation in Differ. Equ., 42 (2006), no. 3, 444–451.

[261] G. Kirchhoff. Über die Anwendbarkeit der Formeln für die Intensitäten der galvanischen Ströme ... Ann. Physik und Chemie, Bd. 75 (1848), 189–205.

[262] G. Kirchhoff. Beweis der Existenz des Potentials, das an der Grenze des betrachteten Raumes gegebene Werthe hat ... Acta Math., Bd. 14, (1890–1891), 179–183.

[263] A.N. Kolmogorov, I.G. Petrovsky, N.S. Piskunov. Étude de l'équation de la diffusion avec croissance de la quantité de matière et son application à un

problème biologique. B. Univ. d'Etat à Moscou, Sér. Intern. A, 1 (1937), 1–26.

[264] V.A. Kondratiev. Boundary value problems for elliptic equations in domains with conical points. Trudy Mosk. Mat. Obshch., 16 (1967), 209–292.

[265] V.A. Kondratiev. Invertibility of Schrödinger operators in weighted spaces. Russian J. Math. Phys., 1 (1993), no. 4, 465–482.

[266] D.J. Korteweg. Sur la forme que prennent les équations du mouvement des fluides si l'on tient compte des forces capillaires causées par des variations de densité considérables mais connues et sur la théorie de la capillarité dans l'hypothèse d'une variation continue de la densité, Archives Néerlandaises des Sciences Exactes et Naturelles, 6 (1901), 1–24.

[267] A.I. Koshelev. On the boundedness in L_p of the derivatives of solutions of elliptic differential equations. Matem. Sbornik, 38 (80) (1956), No. 3, 359–372 (Russian).

[268] A.I. Koshelev. On the boundedness in L_p of the derivatives of solutions of elliptic differential equations and systems. Doklady AN SSSR, 110 (1956), No. 3, 323–325 (Russian).

[269] A.I. Koshelev. On differentiability of solutions of elliptic differential equations. (Russian) Dokl. Akad. Nauk SSSR (N.S.), 112 (1957), 806–809.

[270] A.I. Koshelev. On the boundedness in L_p of derivatives of solutions of elliptic equations and elliptic systems. (Russian) Dokl. Akad. Nauk SSSR (N.S.), 116 (1957) 542–544

[271] A.I. Koshelev. Regularity of solutions of elliptic equations and systems. Nauka, Moscow, 1986 (Russian).

[272] O.A. Kovaleva, A.L. Skubachevskii. Solvability of nonlocal elliptic problems in weighted spaces. Mat. Zametki 67 (2000), no. 6, 882–898 (Russian); translation in Math. Notes, 67 (2000), no. 5–6, 743–757.

[273] M.A. Krasnoselskii. Topological methods in the theory of nonlinear integral equations. GITTL, Moscow, 1956 (Russian).

[274] M.A. Krasnoselskii. Positive solutions of operator equations. GIFML, Moscow, 1962 (Russian).

[275] M.A. Krasnoselskii, G.M. Vaineko, P.P. Zabreiko, Ya.B. Rutitskii, V.Ya. Stetsenko. Approximate solution of operator equations. Nauka, Moscow, 1969 (Russian).

[276] M.A. Krasnoselskii, P.P. Zabreiko. Geometrical methods of nonlinear analysis. Moscow, Nauka, 1975 (Russian).

[277] S.G. Krein. Linear differential equations in Banach spaces. Amer. Math. Soc., Providence, 1971.

[278] N.V. Krylov. Nonlinear elliptic and parabolic equations of the second order. Nauka, Moscow, 1985.

[279] S.G. Kryzhevich, V.A. Volpert. Different types of solvability conditions for differential operators, Electron. J. Diff. Eqns., (2006), No. 100, 1–24.

[280] P. Kuchment. Floquet theory for partial differential equations. Birkhäuser Verlag, Basel, 1993.

[281] V.V. Kurta. Behavior of solutions of second-order quasilinear elliptic equations in unbounded domains (Russian). Ukranian. Mat. Zh. 44 (1992), no. 2, 279–283; translation in Ukrainian Math. J. 44 (1992), no. 2, 245–248.

[282] T. Lachand-Robert. Bifurcation et rupture de symétrie pour un problème elliptique sur-linéaire dans un cylindre. C. R. Acad. Sci. Paris Sér. I Math., 314 (1992), No. 13, 1009–1014.

[283] O. Ladyzhenskaya. On the closure of the elliptic operator. Doklady Akad. Nauk SSSR (N.S.), 79 (1951), 723–725 (Russian).

[284] O.A. Ladyzhenskaya. Mathematical questions of dynamics of viscous incompressible fluid. Moscow, Fizmatgiz, 1961 (Russian).

[285] O.A. Ladyzhenskya, N.N. Uraltseva. Linear and quasilinear problems of elliptic type. Moscow, Nauka, 1973 (Second edition).

[286] J.L. Lagrange. Solution de différents problèmes de calcul integral. Miscellaneu Taurinensia, t. 3 (1762–1765), t. 4 (1760).

[287] J.L. Lagrange. Sur l'attraction des sphéroides elliptiques. Nouveaux mém. Acad. Sci. et Bell. Lettr., Berlin 1773. Oeuvres de Lagrange, t. 3, Paris, 1869, 619–649.

[288] T. Lalescu. Introduction à la théorie des équations intégrales. Paris, A. Hermann, 1912.

[289] L.D. Landau, E.M. Lifschitz. Hydrodynamics. Nauka, Moscow, 1986.

[290] E.M. Landis. Some questions in the qualitative theory of second-order elliptic equations (case of several independent variables). Uspekhi Mat. Nauk, 18 (1963) No. 1 (109) 3–62 (Russian).

[291] E.M. Landis. Theorems of Phragmén-Lindelöf type for solutions of elliptic equations of high order. Soviet Math. Dokl., 11 (1970), 851-855.; translated from Dokl. Akad. Nauk SSSR, 193 (1970), 32–35.

[292] E.M. Landis. Second-order equations of elliptic and parabolic type. Nauka, Moscow, 1971 (Russian). English translation: AMS, Providence, 1997.

[293] E.M. Landis. The behavior of the solutions of higher-order elliptic equations in unbounded domains. Trudy Moskov. Mat. Ob., 31 (1974), 35–58.

[294] E.M. Landis, G.P. Panasenko. A variant of a theorem of Phragmén-Lindelöf type for elliptic equations with coefficients that are periodic in all variables but one. (Russian) Trudy Sem. Petrovsk., No. 5 (1979), 105–136.

[295] B. Lange, V. Rabinovich. *Pseudodifferential operators in $\mathbb{R}^n$ and limit operators,* Mat. Sb. (N.S.), 129 (171) (1986), No. 2, 175–185.

[296] P.S. Laplace. Théorie des attractions des sphéroides et de la figure des planètes. Hist. et Mém. Acad. Sci. Paris, 1782 (1785), 113–196. Oeuvres, t. 10, Paris, 1894, 341–419.

[297] P.S. Laplace. Mémoire sur la théorie de l'anneau de Saturne. Hist. et Mém. Acad. Sci. Paris, 1787 (1789), 249–257. Oeuvres, t. 11, Paris, 1895, 275–292.

[298] G. Lauricella. Integrazione dell'equazione $\Delta^2(\Delta^2 u) = 0$ in un campodi forma circolare. Atti Accad. Sci. Torino, 31 (1895–1896), 610–618.

[299] P. Lax. A Phragmén-Lindelöf theorem in harmonic analysis and its application to some questions in the theory of elliptic equations. Comm. Pure Appl. Math., 10 (1957), 361–389.

[300] V.K. Le. Some existence and bifurcation results for quasilinear elliptic equations with slowly growing principal operators. (English summary) Houston J. Math. 32 (2006), No. 3, 921–943.

[301] P.G. Lejeune-Dirichlet. Vorlesungen über die im umgekehrten Verhältnis des Quadrats der Entfernung wirkenden Kräfte. Leipzig, 1876.

[302] E. Leroy. Sur l'intégration des équations de la chaleur. Annales Scientifiques de l'E.N.S., 3-ème série, tome 14 (1897), 379–465.

[303] J. Leray, J. Schauder. Topologie et équations fonctionnelles. Annales Scientifiques de l'Ecole Normale Supérieure Sér. 3, 51 (1934), 45–78.

[304] S. Levendorskii. On the types of degenerate elliptic operators. Mat. Sb., 180 (1989), no. 4, 513–528, 559 (Russian); translation in Math.USSR-Sb. 66 (1990), no. 2, 523–540.

[305] S. Levendorskii. Degenerate elliptic equations. Mathematics and its Applications, 258. Kluwer Academic Publishers Group, Dordrecht, 1993.

[306] E.E. Levi. Sulle equazioni lineari totalmente ellittiche alle derivate parziali. Rendiconti del Circolo Matematico di Palermo, Bd. XXIV, 1907, 275–317.

[307] T. Levi Civita. Sull'integrazione dell'equazione $\Delta_2 \Delta_2 u - 0$. Atti Accad. Sci. Torino, 33 (1897–1898), 932–956.

[308] T. Levi-Civita. Sopra una transformazione in sé stressa della equazione $\Delta_2 \Delta_2 u = 0$. Atti Ist. Veneto sci. Let. ed Arti, ser. 7, t. 9 (1897–1898), 1399–1410.

[309] B.M. Levitan. Almost periodic functions. Gostekhizdat, Moscow, 1953 (Russian).

[310] B.M. Levitan, V.V. Zhikov. Almost periodic functions and differential equations. Cambridge Univ. Press, 1982.

[311] P. Levy. Sur les fonctions de lignes implicites. Bull. Soc. Math. de France, 48 (1920), 13–27.

[312] L.S. Li, S. Yan. Eigenvalue problem for quasilinear elliptic equations on R^N. Comm. PDE, 14 (1989), 1291–1314.

[313] A.M. Liapunov. Works on the theory of potential. GITTL, Moscow, 1949 (Russian).

[314] L. Lichtenstein. Randwertaufgaben der Theorie der linearen partiellen Differentialgleichungen zweiter Ordnung vom elliptischen Typus. Die erste Randwertaufgabe. Allgemein ebene Gebiete. Journal für die reine und angewandte Mathematik, 142 (1913), 1–40,

[315] L. Lichtenstein. Neuere Entwicklung der Theorie partieller Differentialgleichungen zweiter Ordnung vom elliptischen Typus. Encykl. Math. Wiss., Bd. II, 3. Heft, 8 (1924), 1277–1334.

[316] L. Lichtenstein. Vorlesungen über einige Klassen nichtlinearer Integralgleichungen und Integro-Differentialgleichungen nebst Anwendungen. Berlin, 1931.

[317] J.W. Lindeberg. Sur l'intégration de l'équation $\Delta u = fu$. Annales Scientifiques de l'ENS, 3-ème série, tome 18 (1901), 127–142.

[318] J.L. Lions, E. Magenes. Problèmes aux limites non homogènes et applications. Volume 1. Dunod, Paris, 1968.

[319] P.-L. Lions. On the existence of positive solutions of semilinear elliptic equations. SIAM Rev., 24 (1982), No. 4, 441–467.

[320] R. Lipschitz. Beiträge zur Theorie der Vertheilung der statischen und der dynamischen Electricität in leitenden Körpern. J. reine und angew. Math., Bd. 58 (1861), 1–53.

[321] V. Liskevich, S. Lyakhova, V. Moroz. Positive solutions to nonlinear p-Laplace equations with Hardy potential in exterior domains. J. Differential Equations, 232 (2007), no. 1, 212–252.

[322] P.I. Lizorkin, (L_p, L_q)-multipliers of Fourier integral. Dokl. Akad. Nauk SSSR, 152 (1963), 808–811 (Russian); English translation: Soviet Math. Dokl., 4 (1963), 1420–1424.

[323] R.B. Lockhart. Fredholm property of a class of elliptic operators on noncompact manifolds. Duke Math. J., 48 (1981), 289–312.

[324] R.B. Lockhart, R.C. McOwen. On elliptic systems in R^n. Acta Mathematica, 150 (1983), 125–135.

[325] Ya.B. Lopatinskii. Fundamental system of solutions of elliptic systems of linear differential equations. Ukrain. Mat. Zhurnal, III (1951), No. 1, 3–38.

[326] Ya.B. Lopatinskii. Fundamental solutions of systems of differential equations of elliptic type. Ukrain. Mat. Zhurnal, III (1951), No. 3, 290–316.

[327] Ya.B. Lopatinskii. On a method of reducing boundary problems for a system of differential equations of elliptic type to regular equations. Ukrain. Mat. Zurnal, V (1953), 123–151.

[328] Ya.B. Lopatinskii. The theory of general boundary problems. Naukova Dumka, Kiev, 1984.

[329] A. Lunardi. Analytic semigroups and optimal regularity in parabolic problems. Birkhäuser, Basel, 1995.

[330] A.M. Lyapunov. Sur les figures d'équlibre peu différentes des ellipsoides d'une masse liquide homogène donnée d'un movement de rotation. Zap. Akad. Nauk (Comptes rendus de l'Académie des Sciences), Saint Petersbourg, 1906.

[331] C. Maclaurin. A treatise of fluxion. V. 1,2. Edinburgh, 1742.

[332] I.T. Mamedov, S.T. Guseynov. Behavior in unbounded domains of solution of degenerate elliptic equations of the second order in divergence form. Trans. Acad. Sci. Azerb. Ser. Phys.-Tech. Math. Sci., 19 (1999), no. 5, Math. Mech., 86–97 (2000).

[333] M. Mason. Sur les solutions satisfaisant à des conditions aux limites données de l'équation différentielle $\Delta u + \lambda A(x,y)u = f(x,y)$. Journal de mathématiques pures et appliquées 5e série, 10 (1904), 445–489.

[334] E. Mathieu. Mémoire sur l'équation aux différences partielles du quatrième ordre $\Delta\Delta u = 0$ et sur l'équilibre d'élasticité d'un corps solide. Journal de mathématiques pures et appliquées, 2e série, tome 14 (1869), 378–421.

[335] E. Mathieu. Mémoire sur l'équation aux différences partielles de la physique mathématique. Journal de mathématiques pures et appliquées, 2e série, tome 17 (1872), 249–323.

[336] E. Mathieu. Etude des solutions simples des équations aux différences partielles de la physique mathématique. Journal de mathématiques pures et appliquées, 3e série, 5 (1879), 5–20.

[337] E. Mathieu. Théorie du potentiel et ses applications à l'électrostatiques et magnetisme. Part 1, 2. Paris, 1985, 1986.

[338] J. Mawhin. Equivalence theorems for nonlinear operator equations and coincidence degree theory for some mappings in locally convex topological vector spaces. J. Diff. Equations, 12 (1972), 610–636.

[339] J. Mawhin. Topological degree and boundary value problems for nonlinear differential equations. In: Topological methods for ordinary differential equations, M. Furi, P.Zecca. Eds., Lecture Notes in Math., 1527 (1993), 74–142.

[340] J. Mawhin. Topological degree methods in nonlinear boundary value problems. Conference Board of Math. Sciences, AMS, No. 40.

[341] V.G. Mazya, V. Kozlov, J. Rotmann. Point Boundary Singularities in Elliptic Theory, AMS, Providence, 1997.

[342] V.G. Mazya, B.A. Plamenevskii. Elliptic boundary value problems on manifolds with singularities, Problems of Mathematical Analysis, Vol. 6, Univ. of Leningrad, Leningrad, 1977, 85–142.

[343] P.J. McKenna, W. Reichel. Radial solutions of singular nonlinear biharmonic equations and applications to conformal geometry. Electron. J. Differential Equations, 2003, No. 37, 13 pp.

[344] W. McLean. Strongly elliptic systems and boundary integral equations. Cambridge University Press, Cambridge, 2000.

[345] H. Meinhardt. Models of biological pattern formation. Academic Press, London, 1982.

[346] S.G. Michlin. Singular integral equations. Uspekhi Mat. Nauk, III, 3 (25) (1948), 29–112.

[347] S.G. Michlin. Multi-dimensional singular integrals and integral equations. Fizmatgiz, Moscow, 1962.

[348] S.G. Michlin. On the index of a system of singular integral equations. Dokl. Akad. Nauk SSSR, 152 (1963), No. 3, 555–558.

[349] S.G. Michlin. Sur les multiplicateurs des intégrales de Fourier. Dokl. Akad. Nauk SSSR, 109 (1956), 701–703.

[350] V.M. Mikljukov. Asymptotic properties of subsolutions of quasilinear equations of elliptic type and mappings with bounded distortion. Mat. Sb. (N.S.), 111 (153) (1980), no. 1, 42–66, 159 (Russian).

[351] V.M. Miklyukov. Some peculiarities of the behavior of solutions of minimal surface type equations in unbounded domains. Mat. Sb. (N.S.), 116 (158) (1981), no. 1, 72-86 (Russian).

[352] C. Miranda. Equazioni alle derivate parziale di tipo elliptico. Springer-Verlag, Berlin, 1955.

[353] C. Miranda. Sul problema misto per le equazioni lineari ellittiche. Ann. Mat. Pura Appl., Ser., 4, 39 (1955), 279–303.

[354] C. Miranda. Teorema del massimo modulo e teorema di esistenza e di unicità per il problema di Dirichlet relativo alle equazioni ellittiche in due variabili. Ann. Mat. Pura Appl., (4) 46 (1958), 265–311.

[355] C.B. Morrey. On the solutions of quasi-linear elliptic partial differential equations. Trans. Amer. Math. Soc., 43 (1938), no. 1, 126–166.

[356] C.B. Morrey. Second-order elliptic systems of differential equations. Proc. Nat. Acad. Sci.U.S.A., 39 (1953), 201–206.

[357] C.B. Morrey. Second-order elliptic systems of differential equations. Contribution to the theory of partial differential equations. Ann. of Math. Studies, No. 33, Princeton University Press, 1954, 101–159.

[358] C.B. Morrey. Second-order elliptic equations in several variables and Hölder continuity. Math. Z., 72 (1959/1960), 146–164.

[359] M. Moussaoui. Régularité de la solution d'un problème à dérivée oblique. C. R. Acad. Sci. Paris Sér. A, 279 (1974), 869–872.

[360] E.M. Mukhamadiev. On the invertibility of elliptic partial differential operators. Soviet Math. Dokl., 13 (1972), 1122–1126.

[361] E.M. Mukhamadiev. Normal solvability and the Noethericity of elliptic operators in spaces of functions on R^n. Part I. Zap. Nauch. Sem. LOMI, 110

(1981), 120–140; English translation: J. Soviet Math., 25 (1984), No. 1, 884–901.

[362] E.M. Mukhamadiev. Normal solvability and Noethericity of elliptic operators in spaces of functions on R^n. II. Zap. Nauchn. Sem. LOMI, 138 (1984), 108–126.

[363] J.D. Murray. Mathematical biology. Springer, Berlin, 1989.

[364] N.I. Muskhelishvili. Applications des intégrales analogues à celles de Cauchy à problèmes de la physique mathématique. Tiflis, Imprimerie de l'état, 1922.

[365] N.I. Muskhelishvili, N.P. Vekua. Riemann problem for several unknown functions and its application to systems of singular integral equations. Trudy of Tbilisi Math. Institute, 12 (1943), 1–46.

[366] N.I. Muskhelishvili. Singular integral equations. OGIZ, Moscow, 1946 (Russian).

[367] N.S. Nadirashvili. A Phragmeń-Lindelöf theorem for a class of higher-order elliptic equations. Uspekhi Mat. Nauk 33 (1978), no. 6 (204), 210. English translation: Russian Math. Surveys 33 (1978), no. 6, 251–252.

[368] N.S. Nadirashvili. Estimation of the solutions of elliptic equations with analytic coefficients which are bounded on some set. Vestnik Moskov. Univ. Ser. I Mat. Mekh. No. 2 (1979), 42–46, 102 (Russian).

[369] M. Nagumo. Degree of mapping in linear local convex topological spaces. Amer. J. of Math., 73 (1951), No. 3, 497–511.

[370] J. Nash. Continuity of solutions of parabolic and elliptic equations. Amer. J. Math., 80 (1958) 931–954.

[371] S.A. Nazarov, K. Pileckas. On the Fredholm property of the Stokes operator in a layer-like domain. J. Analysis and Appl., 20 (2001), No. 1, 155–182.

[372] S.A. Nazarov, B.A. Plamenevskii. Elliptic problems with radiation conditions on the edges of the boundary. Mat. Sb. 183 (1992), no. 10, 13–44; translation in Russian Acad. Sci. Sb. Math. 77 (1994), no. 1, 149–176.

[373] S.A. Nazarov, B.A. Plamenevskii. Elliptic problems in domains with piecewise smooth boundaries. De Gruyter Exposition in Mathematics 13, Berlin New York 1994.

[374] V.V. Nemytskii. Solution of equations of elliptic type in small domains. Mat. Sbornik, 1 (43) (1936), No. 4, 501.

[375] C. Neumann. Über die Integration der partiellen Differentialgleichung $\partial^2\Phi/\partial x^2 + \partial^2\Phi/\partial y^2 = 0$. J. reine und angew. Math., Bd. 59 (1861), 335–366.

[376] C. Neumann. Untersuchungen in dem Gebiete des logarithmischen und Newtonschen Potentiales. Leipzig, 1877.

[377] F. Neumann. Vorlesungen über die Theorie des Potentials und der Kugelfunktionen. Leipzig, 1887.

[378] I. Newton. Mathematical works. ONTI. Moscow, 1937 (Russian).

[379] I. Newton. Mathematical principals of natural philosophy. Translation from Latin by A.N. Krylov. In: A.N. Krylov. Collection of works, Volume VII, Moscow, 1936 (Russian).

[380] L. Nirenberg. Remarks on strongly elliptic partial differential equations. Comm. Pure Appl. Math., 8 (1955), 649–675.

[381] L. Nirenberg. Estimates and existence of solutions of elliptic equations. Comm. Pure Appl. Math., 9 (1956), 509–530.

[382] L. Nirenberg. An application of generalized degree to a class of nonlinear problems. Colloq. Anal. Fonct. Liège Centre Belge de Rech. Math., 1971, 57–73.

[383] L. Nirenberg. An application of generalized degree to a class of nonlinear elliptic equations. J. Anal. Math., 37 (1980), 248–275.

[384] L. Nirenberg. Variational and topological methods in nonlinear problems. Bulletin of the AMS, 4 (1981), No. 3, 267–302.

[385] L. Nirenberg. Partial differential equations in the first half of the century. Development of mathematics 1900-1950 (Luxembourg, 1992), 479–515, Birkhäuser, Basel, 1994.

[386] L. Nirenberg, H.F. Walker. The null spaces of elliptic partial differential operators in R^n. J. Math. Analysis and Appl., 42 (1973), 271–301.

[387] J.C.C. Nitsche. The isoperimetric inequality for multiply-connected minimal surfaces. Bull. Amer. Math. Soc., 71 (1965), 195–270.

[388] Novruzov, A.A. Theorems of Phragmén-Lindelöf type for solutions of second-order linear and quasilinear elliptic equations with discontinuous coefficients. Dokl. Akad. Nauk SSSR 266 (1982), no. 3, 549–552 (Russian).

[389] F. Nöther. Über eine Klasse singulärer Integralgleichungen. Math. Ann., 82 (1921), 42–63.

[390] J. Oddson. Phragmén-Lindelöf and comparison theorems for elliptic equations with mixed boundary conditions. Arch. Rat. Mech. Anal., 26 (1967), 316-334.

[391] O.A. Oleinik, G.A. Iosifyan. The Saint-Venant principle in two-dimensional theory of elasticity and boundary problems for a biharmonic equation in unbounded domains. Sib. Mat. Zh., Vol. 19 (1978), No. 5, 1154–1165.

[392] O.A. Oleinik, E.V. Radkevich. Analyticity and theorems of Liouville and Phragmén-Lindelöf type for elliptic systems of differential equations. Mat. Sb., 9 (55) (1974), No. 1, 130–145.

[393] M.V. Ostrogradskii. Collections of works. Vol. 1. AN USSR, Kiev, 1959 (Russian).

[394] H.G. Othmer, K. Painter, D. Umulis, C. Xue. The intersection of theory and application in elucidating pattern formation in developmental biology. Math. Model. Nat. Phenom., 4 (2009), No. 4, 3–82.

[395] A.P. Oskolkov. On the solution of a boundary value problem for linear elliptic equations in an unbounded domain. (Russian. English summary) Vestnik Leningrad. Univ., 16 (1961) no. 7, 38–50.

[396] R. Palais. Seminar on the Atiyah-Singer index theorem. Ann. of Math. Studies, No. 57, Princeton Univ. Press, Princeton, 1965.

[397] B.P. Paneah. On the theory of the solvability of the oblique derivative problem. Mat. Sb. (N.S.), 114 (156) (1981), no. 2, 226–268, 335.

[398] B.P. Paneah. Some boundary value problems for elliptic equations, and related Lie algebras. Mat. Sb. (N.S.), 126 (168) (1985), no. 2, 215–246 (Russian).

[399] B.P. Paneah. The oblique derivative problem. The Poincaré-problem. Mathematical Topics, 17. Wiley-VCH Verlag Berlin GmbH, Berlin, 2000.

[400] A. Paraf. Sur le problème de Dirichlet et son extension au cas de l'équation linéaire générale du second ordre. Annales de la faculté des sciences de Toulouse, 1-ère série, 6 (1892), No. 2, 1–24, No. 3, 25–75.

[401] L.E. Payne, P.W. Schaefer, J.C. Song. Bounds and decay results for some second-order semilinear elliptic problems. Math. Methods Appl. Sci., 21 (1998), no. 17, 1571–1591.

[402] J. Peetre. Théorèmes de régularité pour quelques classes d'opérateurs différentiels. Medd. Lunds Univ. Mat. Sem., 16 (1959), 1–122.

[403] J. Peetre. Another approach to elliptic boundary problems. Comm. on Pure and Appl. Math., 14 (1961), 711–731.

[404] J. Pejsachowicz, P. Rabier. Degree theory for C^1 Fredholm mapping of index 0. Journal d'Analyse Mathématique, 76 (1998), 289–319.

[405] J. Pejsachowitz, A. Vignoli. On the topological coincidence degree for perturbations of Fredholm operators. Boll. Unione Mat. Ital., 17-B (1980), 1457–1466.

[406] B. Pettineo. Sul problema di derivata obliqua per le equazioni lineari a derivate parziali del secondo ordine di tipo ellittico in due variabili. Atti Accad. Sci. Lett. Arti Palermo. Parte I, (4) 16, (1955/1956), 5–26.

[407] M. Pettineo. On the nonregular oblique derivative problem with respect to elliptic linear partial differential equations. Rend. Mat. Appl., (7) 11 (1991), no. 2, 341–349.

[408] W.V. Petryshyn. On the approximate solvability of nonlinear equations. Math. Ann., 177 (1968), 156–164.

[409] W.V. Petryshyn. Generalized topological degree and semilinear equations. Cambridge Univ. Press, 1995.

[410] E. Phragmén, E. Lindelöf. Sur une extension d'un principe classique de l'analyse et sur quelque propriétés des fonctions monogènes dans le voisinage d'un point singulier. Acta Math., 31 (1908), 381–406.

[411] E. Picard. Mémoire sur la théorie des équations aux dérivées partielles et la méthode des approximations successives. Journal de mathématiques pures et appliquées, 4e série, tome 6 (1890), 145–210.

[412] E. Picard. Sur l'équation $\Delta u = e^u$. Journal de mathématiques pures et appliquées, 5-ème série, tome 4 (1898), 313–316.

[413] E. Picard. Sur la généralisation du problème de Dirichlet. Acta Mathematica, 1902.

[414] E. Picard. Sur la solution du problème généralisé de Dirichlet relatif à une équation linéaire du type elliptique au moyen de l'équation de Fredholm. Annales Scientifiques de l'ENS, tome 23 (1906), 509–516.

[415] E. Picard. Sur la détermination des intégrales des équations linéaires aux dérivées partielles par les valeurs des dérivées normales sur un contour. Annales Scientifiques de l'ENS, tome 24 (1907), 335–340.

[416] E. Picard. Sur une équation aux dérivées partielles du second ordre relative à une surface fermée, correspondant à une équation calorifique. Annales Scientifiques de l'ENS, tome 26 (1909), 9–17.

[417] V.A. Pliss. Uniformly bounded solutions of linear differential systems. Diff. equations, 13 (1977) no. 5, 883–891.

[418] V.A. Pliss. Integral sets of periodic differential systems. Moscow. Nauka, 1977 (Russian).

[419] Yu.K. Podlipenko. Boundary value problems for the Helmholtz equation in certain domains with infinite boundaries. Akad. Nauk Ukrain. SSR Inst. Mat. Preprint 1990, no. 47, 59 pp (Russian).

[420] H. Poincaré. Sur les équations aux dérivées partielles de la physique mathématique. Amer. J. Math., 12 (1890), no. 3, 211–294.

[421] H. Poincaré. Sur l'équilibre d'une masse fluide animée d'un mouvement de rotation. Comptes rendus de l'Académie des Sciences t. 100 (1885), 346–348.

[422] H. Poincaré. Les méthodes nouvelle de la mécanique céleste. 1. Gauthier-Villars, Paris, 1892.

[423] H. Poincaré. Sur les équations de la physique mathématique. Rend. Circolo mat. Palermo, t. 8 (1894), 57–155. Oeuvres, t. 9, 123–196.

[424] H. Poincaré. Sur l'équation des vibrations d'une membrane. Compte Rend. Acad. Sciences, t. 118 ((1894), 447–451. Oeuvres, t. 9, 119–122.

[425] H. Poincaré. Théorie analytique de la propogation de la chaleur. Paris, 1895.

[426] H. Poincaré. Leçons de mécanique céleste, tome III. Paris, 1910.

[427] S.D. Poisson. Remarques sur une équation qui se présente dans la théorie des attractions des sphéroides. Nouveau Bull. Sci. Soc. Philom., t. 3, 1813, 388–392.

[428] S.D. Poisson. Mémoire sur la théorie du magnétisme en mouvement. Mém. Acad. Sci. Inst. France, t. 6, 1823 (1827), 441–570.

[429] S.D. Poisson. Mémoire sur l'attraction des sphéroides. Paris, 1826, 329–379.
[430] P. Polacik. Morse indices and bifurcations of positive solutions of $\Delta u + f(u) = 0$ on R^N. Indiana Univ. Math. J., 50 (2001), No. 3, 1407–1432.
[431] P.R. Popivanov, D.K. Palagachev. The degenerate oblique derivative problem for elliptic and parabolic equations. Mathematical Research, 93. Akademie Verlag, Berlin, 1997.
[432] F. Quinn, A. Sard. Hausdorff conullity of critical images of Fredholm maps. Amer. J. Math., 94 (1972), 1101–1110.
[433] P.J. Rabier. Asymptotic behavior of the solutions of linear and quasilinear elliptic equations on R^n. Trans. AMS, 356 (2003), no. 5, 1889–1907.
[434] P.J. Rabier. On the Fedosov-Hörmander formula for differential operators. Integr. Equ. Oper. Theory, 62 (2008), 555–574.
[435] P.J. Rabier, C.A. Stuart. C^1-Fredholm maps and bifurcation for quasilinear elliptic equations on RN, Recent Trends in Nonlinear Analysis, Progress in Nonlinear Differential Equations and Their Applications, vol. 40, Birkhäuser, Basel, 2000, 249–264.
[436] P.J. Rabier, C.A. Stuart. Exponential decay of the solutions of quasilinear second-order equations and Pohozaev identities, J. Diff. Equat., 165 (2000), 199–234.
[437] P.J. Rabier, C.A. Stuart. Fredholm properties of Schrödinger operators in $L^p(R^N)$. J. Diff. Integral Equat., 13 (2000), 1429–1444.
[438] P.J. Rabier, C.A. Stuart. Fredholm and properness properties of quasilinear elliptic operators on R^N. Math. Nachr., 231 (2001), 129–168.
[439] P.J. Rabier, C.A. Stuart. Global bifurcation for quasilinear elliptic equations on R^N. Math. Z., 237 (2001), 85–124.
[440] V. Rabinovich. Pseudodifferential operators in unbounded domains with conical structure at infinity. Mat. Sb. (N.S.), 80 (122) (1969), 77–96.
[441] V.S. Rabinovich. The Fredholm property of general boundary value problems on noncompact manifolds, and limit operators. Dokl. Akad. Nauk, 325 (1992), No. 2, 237–241. English translation: Russian Acad. Sci. Dokl. Math., 46 (1993), No. 1, 53–58.
[442] V. Rabinovich, S. Roch. Fredholmness of convolution type operators. Operator theoretical methods and applications to mathematical physics, 423–455, Operator Theory: Advances and Applications, 147. Birkhäuser Verlag, Basel, 2004.
[443] V. Rabinovich, S. Roch, B. Silbermann. Limit operators and their applications in operator theory. Operator Theory: Advances and Applications, 150. Birkhäuser Verlag, Basel, 2004.
[444] V. Rabinovich, B.W. Schulze, N. Tarkhanov. Boundary value problems in domains with non Lipschitz boundary. Proceedings of the Third World

Congress of Nonlinear Analysis, Part 3 (Catania, 2000). Nonlinear Anal. 47 (2001), no. 3, 1881–1891.

[445] P.H. Rabinowitz. Some global results for nonlinear eigenvalue problems. J. Functional Anal., 7 (1971), 487–513.

[446] P.H. Rabinowitz. Some aspects of nonlinear eigenvalue problems. Rocky Mountain J. Math., 3 (1973), 161–202.

[447] P.H. Rabinowitz, J. Su, Z.-Q. Wang. Multiple solutions of superlinear elliptic equations. Atti Accad. Naz. Lincei Cl. Sci. Fis. Mat. Natur. Rend. Lincei (9) Mat. Appl., 18 (2007), No. 1, 97–108.

[448] N.M. Ratiner. Application of the degree theory to a problem with oblique derivative. Izvestiya VUZ, (2001), No. 4 (467), 43–52 (Russian).

[449] M. Reed, B. Simon. Methods of modern physics, Vol. IV: Analysis of operators. Academic Press, New York, 1978.

[450] D. O'Regan, R. Precup. Theorems of Leray-Schauder type and applications. Gordon and Breach Science Publishers, Amsterdam, 2001.

[451] S. Rempel, B.W. Schulze. Index theory of elliptic boundary problems. Akademie-Verlag, Berlin, 1982.

[452] Ren, X., Wei, J. On a semilinear elliptic equation in R^2 when the exponent approaches infinity. J. Math. Anal. Appl., 189 (1995), no. 1, 179–193.

[453] J. Riemann. Sur le problème de Dirichlet. Annales Scientifiques de l'ENS, Tome 5 (1888), 327–410.

[454] B. Riemann. Grundlagen für eine allgemeine Theorie der Funktionen einer veränderlichen complexen Grösse. Werke, Leipzig, 1876.

[455] G. Robin. Distribution de l'électricité à la surface des conducteurs fermées et ouvertes. Ann. Ecole Normale, sér. 3, t. 3 (1886).

[456] G. Robin. Distribution de l'électricité sur une surface des conducteurs fermée convexe. Compte Rendu Acad. Sciences, t. 104 (1887), 1834–1836.

[457] Ya. Roitberg. Elliptic boundary value problems in the spaces of distributions. Kluwer Academic Publishing, 1996.

[458] W. Rother. Bifurcation for some semilinear elliptic equations when the linearization has no eigenvalues. (English summary) Comment. Math. Univ. Carolin., 34 (1993), No. 1, 125–138.

[459] B. Rowley. Matrix polynomials and the index problem for elliptic systems. Trans. of AMS, 349 (1997), No. 8, 3105–3148.

[460] B. Rowley. An index formula for elliptic systems in the plane. Trans. of AMS, 349 (1997), No. 8, 3149–3179.

[461] A.J. Rumbos, A.L. Edelson. Bifurcation properties of semilinear elliptic equations in R^n. Differential Integral Equations, 7 (1994), No. 2, 399–410.

[462] T.O. Shaposhnikova. Applications of multipliers to the problem of coercivity inWlpof the Neumann problem. Translated in J. Soviet Math. 64 (1993), no.

6, 1381–1388. Nonlinear equations and variational inequalities. Linear operators and spectral theory, 237–248, 253, Probl. Mat. Anal., 11, Leningrad. Univ., Leningrad, 1990.

[463] J. Schauder. Zur Theorie stetiger Abbildungen in Funktionalräumen. Math. Zeitschr., 26 (1927), No. 1, 47–65.

[464] J. Schauder. Bemerkungen zu meiner Arbeit "Zur Theorie stetiger Abbildungen in Funktionalräumen". Math. Zeitschr., 26 (1927), 417–431.

[465] J. Schauder. Invarianz des Gebietes in Funktionalräumen. Studia Math., 1 (1929), 123–139.

[466] J. Schauder. Der Fixpunktsatz in Funktionalräumen. Studia Math., 2 (1930), 170–179.

[467] J. Schauder. Über den Zusammenhang zwischen der Eindeutigkeit und Lösbarkeit partieller Differentialgleichungen zweiter Ordnung vom elliptischen Typus. Math. Z., 36 (1932), 661–721.

[468] J. Schauder. Über das Dirichletsche Problem im Grossen für nicht lineare elliptische Differentialgleichungen. Mathematische Zeitschrift, 37 (1933).

[469] J. Schauder. Über lineare elliptische Differentialgleichungen zweiter Ordnung. Math. Z., 38 (1934), 257–282.

[470] J. Schauder. Numerische Abschätzungen in elliptischen linearen Differentialgleichungen. Studia Math., 5 (1934), 34–42.

[471] M. Schechter. General boundary value problems for elliptic partial differential equations. Comm. Pure Appl. Math., 12 (1959), 457–482.

[472] M. Schechter. Mixed boundary problems for general elliptic systems. Comm. Pure Appl. Math., 13 (1960), No. 2, 183–201.

[473] M. Schechter. Various types of boundary conditions for elliptic equations. Comm. Pure Appl. Math. 13 1960, 407–425.

[474] E. Schmidt. Zur Theorie linearer und nicht linearer Integralgleichungen. Teil 3. Über die Auflösungen der nicht linearen Integralgleichungen und die Verzweigung ihrer Lösungen. Math. Ann., 65 (1908).

[475] E. Schrohe. Fréchet algebra techniques for boundary value problems on noncompact manifolds: Fredholm criteria and functional calculus via spectral invariance. Math. Nachr., 199 (1999), 145–189.

[476] H.A. Schwarz. Über einige Abbildungsaufgaben. Gesammelte mathematische Abhandlungen, Bd. 2, Springer, Berlin, 1890, 65–83.

[477] H.A. Schwarz. Zur Theorie der Abbildung. Gesammelte mathematische Abhandlungen, Bd. 2, Springer, Berlin, 1890, 108–132.

[478] H.A. Schwarz. Über einen Grenzübergang durch alternierendes Verfahren. Gesammelte mathematische Abhandlungen, Bd. 2, Springer, Berlin, 1890, 133–143.

[479] H.A. Schwarz. Beitrag zur Untersuchung der zweiten Variation des Flächeninhalts von Minimalflächenstücken im Allgemeinen und von Theilen der Schraubenfläche im Besonderen. Gesammelte mathematische Abhandlungen, Bd. 1, Springer, Berlin, 1890, 151–167.

[480] H.A. Schwarz. Über ein die Flächen kleinsten Flächeninhalts betreffendes Problem der Variationsrechnung. Gesammelte mathematische Abhandlungen, Bd. 1, Springer, Berlin, 1890, 223–269.

[481] H.A. Schwarz. Zur Integration der partiellen Differentialgleichung $\frac{\partial^2 u}{\partial x^2} + \frac{\partial^2 u}{\partial y^2} = 0$. Journal reine angew. Math., 74 (1872), 218–253.

[482] R.T. Seleey. Integro-differential operators on vector bundles. Trans. Amer. Math. Soc., 117 (1965), No. 5, 167–204.

[483] J. Serrin. On the Phragmén-Lindelöf principle for elliptic differential equations. J. Rat. Mech. Anal. 3 (1954), 395–413.

[484] Z.Ya. Shapiro. On general boundary value problems for equations of elliptic type. Izvestia AN SSSR, 17 (1953), No. 6, 539–562 (Russian).

[485] J. Shi. Semilinear Neumann boundary value problems on a rectangle. Trans. Amer. Math. Soc., 354 (2002), No. 8, 3117–3154.

[486] Y.-W. Shih. Symmetry breaking and existence of many positive nonsymmetric solutions for semi-linear elliptic equations on finite cylindrical domains. Nonlinear Anal. 31 (1998), No. 3–4, 465–474.

[487] N. Shimakura, Partial differential operators of elliptic type, Transl. of Math. Monographs, Vol. 99, AMS, Providence, 1991.

[488] A.E. Shishkov. Behavior of the solutions of the Dirichlet problem for high-order quasilinear divergence form elliptic equations in unbounded domains. Sibirsk. Mat. Zh., 28 (1987), no. 6, 134–146, 220 (Russian).

[489] A.E. Shishkov. The Phragmén-Lindelöf principle for higher-order quasilinear elliptic equations in divergence form. Uspekhi Mat. Nauk, 43 (1988), no. 4(262), 231–232 (Russian); translation in Russian Math. Surveys, 43 (1988), no. 4, 237–238.

[490] M.A. Shubin. The Favard-Muhamadiev theory, and pseudodifferential operators. Dokl. Akad. Nauk, 225 (1975), No. 6, 1278–1280.

[491] M.A. Shubin. Almost periodic functions and partial differential operators. Russian Math. Surveys, 33 (1978), No. 2, 1–52.

[492] I.V. Skrypnik. Nonlinear elliptic boundary value problems. Teubner-Verlag, Leipzig, 1986.

[493] I.V. Skrypnik. Methods for analysis of nonlinear elliptic boundary value problems. Translation of Math. Monographs, Vol. 139, AMS, Providence, 1994.

[494] A.L. Skubachevskii. Nonlocal elliptic problems and multidimensional diffusion processes. Russian J. Math. Phys., 3 (1995), no. 3, 327–360.

[495] L.N. Slobodetskii. Estimates in L_p of solutions of elliptic systems. Doklady AN SSSR, 123 (1958), 616–619.

[496] S. Smale. An infinite-dimensional version of Sard's theorem. Amer. J. Math., 87 (1965), 861–866.

[497] M.M. Smirnov. Degenerate elliptic and hyperbolic equations. Nauka, Moscow, 1966 (Russian).

[498] J.A. Smoller, A.G. Wasserman. Symmetry breaking and nondegenerate solutions of semilinear elliptic equations. Nonlinear functional analysis and its applications, Part 2 (Berkeley, Calif., 1983), 397–400, Proc. Sympos. Pure Math., 45, Part 2, Amer. Math. Soc., Providence, RI, 1986.

[499] J.A. Smoller, A.G. Wasserman. Bifurcation from symmetry. Nonlinear diffusion equations and their equilibrium states, II (Berkeley, CA, 1986), 273–287, Math. Sci. Res. Inst. Publ., 13, Springer, New York, 1988.

[500] E.D. Solomencev. A Phragmén-Lindelöf type theorem for harmonic functions in space. (Russian) Dokl. Akad. Nauk SSSR, 155 (1964), 765–766.

[501] V.S. Sologub. The development of the theory of elliptic equations in the eighteenth and nineteenth centuries. Naukova Dumka, Kiev, 1975 (Russian).

[502] A. Sommerfeld. Randwertaufgaben in der Theorie der partiellen Differentialgleichungen. Encykl. Math. Wiss., Bd. II, T. 1, H. 4, 5. Leipzig, 1900, 1904, 504–570.

[503] N.M. Stavrakakis. Global bifurcation results for semilinear elliptic equations onRN: the Fredholm case. J. Differential Equations, 142 (1998), No. 1, 97–122.

[504] E. Stein, G. Weiss. Introduction to Fourier analysis on Euclidean spaces. Princeton Univ. Press, Princeton, 1971.

[505] W. Stekloff. Sur le problème de la distribution de l'électricité et le problème de Neumann. Compte Rendu Académie Sciences, t. 125 (1897).

[506] W. Stekloff. Sur les problèmes fondamentaux de la physique mathématique. Annales Scientifique de l'ENS, 3e série, 19 (1902), 191–259.

[507] W. Stekloff. Sur les problèmes fondamentaux de la physique mathématique (suite et fin). Annales Scientifique de l'ENS, 3e série, 19 (1902), 455–490.

[508] W. Stekloff. Théorie générale des fonctions fondamentales. Annales de la faculté des sciences de Toulouse, 2e série, 6 (1904), No. 4, 351–475.

[509] C.A. Stuart. Some bifurcation theory for k-set constructions. Proc. London Math. Soc., 27 (1973), 531–550.

[510] C.A. Stuart. Global properties of components of solutions of nonlinear second-order ordinary differential equations on the half-line. Ann. Scuola N. Pisa, ii (1975), 265–286.

[511] C.A. Stuart. Bifurcation from the continuous spectrum in the L^2-theory of elliptic equations on R^n. Recent methods in nonlinear analysis and applications (Naples, 1980), 231–300, Liguori, Naples, 1981.

[512] C.A. Stuart. Bifurcation in $L^p(\mathbb{R}^N)$ for a semilinear elliptic equation. Proc. London Math. Soc., (3) 57 (1988), No. 3, 511–541.

[513] S.M. Sun. Bifurcation theory for semi-linear elliptic equations in a two- or three-dimensional cylindrical domain. J. Math. Anal. Appl., 187 (1994), No. 3, 887–918.

[514] R. Temam. Navier-Stokes equations. Theory and numerical analysis. North-Holland, Amsterdam, 1977.

[515] R. Temam. Navier-Stokes equations and nonlinear functional analysis. Regional Conference Series in Applied Mathematics. SIAM, Philadelphia, 1983.

[516] W. Thomson. Note sur une équation aux différences partielles qui se présente dans plusieurs questions de physique mathématique. Journal de mathématiques pures et appliquées, 1re série, tome 12 (1847), 493–496.

[517] A.N. Tikhonov. Ein Fixpunktsatz. Math. Ann., 111 (1935).

[518] J.F. Toland. Global bifurcation theory via Galerkin's method. Nonlinear Anal. TMA, 1 (1977), 305–317.

[519] J.F. Toland. Positive solutions of nonlinear elliptic equations-existence and nonexistence of solutions with radial symmetry in Lp(RN). Trans. Amer. Math. Soc., 282 (1984), No. 1, 335–354.

[520] H. Triebel. Interpolation theory, function spaces, differential operators. Berlin, Deutscher Verlag, 1978.

[521] F. Tricomi. Ancora sull'equazione $yz_{xx} + z_{yy} = 0$. Rend. Acc. Lincei, 1927, Ser. VI, 6.

[522] A.M. Turing. Chemical basis of morphogenesis. Phil. Trans. R. Soc. London B, 237 (1952), 37–72.

[523] V.M. Tyurin. On the Fredholm property of linear operators of elliptic type on R^n. Differ. Uravn. 40 (2004), no. 2, 251–256, 287; translation in Differ. Equat., 40 (2004), no. 2, 265–270.

[524] B.R. Vainberg. Asymptotical methods in equations of mathematical physics. Moscow University, Moscow 1982.

[525] M.M. Vainberg, V.A. Trenogin. Theory of branching of solutions of nonlinear equations. Nauka, Moscow, 1969 (Russian).

[526] A. Vanderbauwhede. Symmetry-breaking at positive solutions of elliptic equations. Bifurcation: analysis, algorithms, applications (Dortmund, 1986), 349–353, Internat. Schriftenreihe Numer. Math., 79, Birkhäuser, Basel, 1987.

[527] I.N. Vekua. Boundary value problems of the theory of linear elliptic equations with two independent variables. Mitt. Georg. Abt. Akad.Wiss. USSR [Soobshchenia Gruzinskogo Filiala Akad. Nauk SSSR] 1, (1940). 497–500 (Russian).

[528] I.N. Vekua. New methods of solution of elliptic equations. OGIZ, Moscow, 1946 (Russian).

[529] I.N. Vekua. On a representation of the solutions of differential equations of elliptic type. Soobshch. Akad. Nauk Gruzin. SSR. 11, (1950), 137–141 (Russian).

[530] I.N. Vekua. Systems of differential equations of the first order of elliptic type and boundary value problems with applications to the theory of shells. Matem. Sbornik, 31 (73) (1952), No. 2, 217–314 (Russian).

[531] I.N. Vekua. Generalized analytical functions. Fizmatgiz, Moscow, 1959 (Russian).

[532] N.P. Vekua. Systems of singular integral equations. Moscow, GITTL, 1950 (Russian).

[533] W. Velte. Stabilität und Verzweigung stationärer Lösungen der Navier-Stokesschen Gleichungen beim Taylorproblem. Arch. Rat. Mech. Anal., 22 (1966), 1–14.

[534] O. Venske. Zur Integration der Gleichung $\Delta\Delta u = 0$ für ebene Bereiche. Nachr. Ges. Wiss. Göttingen, 1891, 27–34.

[535] O.A. Vikhreva. Generalized and Fredholm solvability of a mixed boundary value problem for a degenerate elliptic equation. Vestn. Samar. Gos. Univ. Estestvennonauchn. Ser. 2007, no. 6, 194–203 (Russian).

[536] M.I. Vishik. On strongly elliptic systems of differential equations. Doklady Akad. Nauk SSSR (N.S.), 74, (1950), 881–884.

[537] M.I. Vishik. On strongly elliptic systems of differential equations. Mat. Sb., 29 (1951), 615–676.

[538] M.I. Vishik. On the first boundary values problem for elliptic equations degenerate at the boundary. Dokl. Akad. Nauk, 93 (1953), No. 1, 9–12.

[539] M.I. Vishik, G.I. Eskin. Normally solvable problems for elliptic systems of convolution equations. Mat. Sb., 74 (116) (1967), 326–356; English translation: Math. USSR Sb., 3 (1967), 303–330.

[540] M.I. Vishik, V.V. Grushin. Lectures on degenerate elliptic problems. Seventh Mathematical Summer School (Kaciveli, 1969), pp. 3–143. Izdanie Mat. Akad. Nauk Ukrain. SSR, Kiev, 1970.

[541] L.R. Volevich. A problem of linear programming arising in differential equations. Uspekhi Mat. Nauk, 18 (1963), No. 3, 155–162.

[542] L.R. Volevich. Solvability of boundary problems for general elliptic systems. Mat. Sbor., 68 (1965), 373–416; English translation: Amer. Math. Soc. Transl., 67 (1968), Ser. 2, 182–225.

[543] L.R. Volevich, B.P. Paneyah. Certain spaces of generalized functions and embedding theorems. Russian Math. Surveys, 20 (1965), No. 1, 1–73.

[544] A.I. Volpert. Dirichlet problem for second-order elliptic system on plane. Doklady Akad. Nauk SSSR, 79, (1951), No. 2, 185–187.

[545] A.I. Volpert. Investigation on boundary value problems for elliptic system of differential equations on plane. Doklady Akad. Nauk SSSR, 114 (1957), No. 3, 462–464.

[546] A.I. Volpert. On computation of index of Dirichlet problem. Doklady Akad. Nauk SSSR, 10 (1958), 1042–1044.

[547] A.I. Volpert. Normal solvability of boundary value problems for elliptic systems of differential equations on the plane. Theor. and Appl. Math., 1 (1958), 28–57 (Russian).

[548] A.I. Volpert. On the first boundary value problem for elliptic system of differential equations. Doklady Akad. Nauk SSSR, 127 (1959), No. 3, 487–489.

[549] A.I. Volpert. Boundary value problems for higher-order elliptic systems on plane. Doklady Akad. Nauk SSSR, 127 (1959), No. 4, 739–741.

[550] A.I. Volpert. On the index and normal solvability of boundary value problems for elliptic systems of differential equations on plane. Uspekhi Mat. Nauk, 15 (1960), No. 3, 189–191.

[551] A.I. Volpert. On the reduction of boundary value problems for elliptical systems of equations of higher order to problems for systems of first order. Doklady Akad. Nauk USSR, 9 (1960), 1162–1166.

[552] A.I. Volpert. On the index and normal solvability of boundary value problems for elliptic systems of differential equations on plane. Trudy Mosc. Mat. Obshch., 10 (1961), 41–87 (Russian).

[553] A.I. Volpert. On the index of boundary value problems for the system of harmonic functions with three independent variables. Doklady Akad. Nauk SSSR, 133 (1960), No. 1, 13–15.

[554] A.I. Volpert. On the index of the system of two-dimensional singular integral equations. Doklady Akad. Nauk SSSR, 142 (1962), No. 4, 776–777.

[555] A.I. Volpert. Elliptic systems on a sphere and two-dimensional singular integral equations. Matem. Sb., 57 (1962), 195–214.

[556] A.I. Volpert. On the index of the system of multidimensional singular integral equations. Doklady Akad. Nauk SSSR, 152 (1963), No. 6, 1292–1293.

[557] V. Volpert. The spectrum of elliptic operators in unbounded cylinders. Dokl. Akad. Nauk Ukrain. SSR Ser A, 1981 (9), 9–12.

[558] A.I. Volpert, S.I. Hudjaev. Analysis in classes of discontinuous functions and equations of mathematical physics. Martinus Nijhoff Publishers, Dordrecht, 1985.

[559] V. Volpert, B. Kazmierczak, M. Massot, Z. Peradzynski. Solvability conditions for elliptic problems with non Fredholm operators. Appl. Math. (Warsaw), 29 (2002), no. 2, 219–238.

[560] V. Volpert, S. Petrovskii. Reaction-diffusion waves in biology. Physics of Life Reviews, 10.1016/j.plrev.2009.10.002

[561] A.I. Volpert, V.A. Volpert. Applications of the rotation theory of vector fields to the study of wave solutions of parabolic equations. Trans. Moscow Math. Soc., 52 (1990), 59–108.

[562] A.I. Volpert, V.A. Volpert. The construction of the Leray-Schauder degree for elliptic operators in unbounded domains. Annales de l'IHP. Analyse non linéaire, 11 (1994), No. 3, 245–273.

[563] V. Volpert, A. Volpert. Spectrum of elliptic operators and stability of travelling waves. Asymptotic Analysis, 23 (2000), 111–134.

[564] A. Volpert, V. Volpert. Normal solvability and properness of elliptic problems. Partial differential equations, 193–237, Amer. Math. Soc. Transl. Ser. 2, 206, Amer. Math. Soc., Providence, RI, 2002.

[565] V. Volpert, A. Volpert. Properness and topological degree for general elliptic operators. Abstract and Applied Analysis, (2003), No. 3, 129–181.

[566] A. Volpert, V. Volpert. Fredholm property of elliptic operators in unbounded domains. Trans. Moscow Math. Soc. 67 (2006), 127–197.

[567] V. Volpert, A. Volpert, J.F. Collet. Topological degree for elliptic operators in unbounded cylinders. Adv. Diff. Eq., 4 (1999), No. 6, 777–812.

[568] A.I. Volpert, V.A. Volpert, V.A. Volpert. Travelling wave solutions of parabolic systems. Translation of Mathematical Monographs, Vol. 140, Amer. Math. Society: Providence, 1994.

[569] A. Volpert, V. Volpert. Formally adjoint problems and solvability conditions for elliptic operators. Russian Journal of Mathematical Physics, 11 (2004), No. 4, 474–497.

[570] A. Volpert, V. Volpert. Normal solvability of general linear elliptic problems. Abstract and Applied Analysis, (2005), no. 7, 733–756.

[571] A. Volpert, V. Volpert. Elliptic problems with a parameter in unbounded domains. Adv. Diff. Eq., 12 (2007), No.5, 573–600.

[572] V. Volterra, M. Tomasetti, F.S. Zarlatti. Leçons sur les équations intégrales et les équations intégro-différentielles. Leçons professées à la Faculté des sciences de Rome en 1910. Paris, Gauthier-Villars, 1913.

[573] V. Vougalter, V. Volpert. Solvability conditions for some non-Fredholm operators. Proceedings of the Edinburgh Mathematical Society, FirstView Article doi: 10.1017/S001309150900023.

[574] V. Vougalter, V. Volpert. On the solvability conditions for some non-Fredholm operators. Int. J. Pure and Applied Math., 60 (2010) No. 2, 169–191.

[575] H.F. Walker. On the null-space of first-order elliptic partial differential operators in R^n. Proc. of AMS, 30 (1971), No. 2, 278–286.

[576] H.F. Walker. A Fredholm theory for a class of first-order elliptic partial differential operators in R^n. Trans. Amer. Math. Soc., 165 (1972), 75–86.

[577] H. Weber. Über die Integration der partiellen Differentialgleichung: $\frac{\partial^2 u}{\partial x^2} + \frac{\partial^2 u}{\partial y^2} + k^2 u = 0$. Math. Ann., 1 (1869), 1–36.

[578] K. Weierstrass. Über das sogenannte Dirichletsche Princip. Math. Werke, Bd. 2, Berlin, 1895.

[579] J.T. Wloka, B. Rowley, B. Lawruk. Boundary value problems for elliptic systems. Cambridge Univ. Press, 1995.

[580] S.Ya. Yakubov. Noncoercive boundary value problems for elliptic partial differential and differential operator equations. Results Math., 28 (1995), no. 1-2, 153–168.

[581] S. Yakubov. A nonlocal boundary value problem for elliptic differential-operator equations and applications. Integral Equations Operator Theory, 35 (1999), no. 4, 485–506.

[582] A. Yanushauskas. On the non-Fredholm oblique derivative problem. Geometry, analysis and mechanics, 227–241, World Sci. Publ., River Edge, NJ, 1994.

[583] A.I. Yanushauskas. The Dirichlet problem for a Petrovskii-elliptic system of second-order equations. Sibirsk. Mat. Zh. 40 (1999), no. 1, 226–234; translation in Siberian Math. J. 40 (1999), no. 1, 195–203.

[584] V.I. Yudovich. Some estimates connected with integral operators and with solutions of elliptic equations. Dokl. Akad. Nauk SSSR, 138 (1961), 805–808 (Russian).

[585] V.I. Yudovich. Some bounds for solutions of elliptic equations. Mat. Sb. (N.S.), 59 (101) (1962), suppl. 229–244 (Russian).

[586] V.I. Yudovich. On a bound for the solution of an elliptic equation. Uspekhi Mat. Nauk, 20 (1965) no. 2 (122), 213–219 (Russian).

[587] V.I. Yudovich. On bifurcation of rotating fluid. Dokl. Akad. Nauk SSSR, 169 (1966), No. 2, 306–309 (Russian).

[588] S. Zaremba. Sur l'équation aux dérivées partielles $\Delta u + \xi u + f = 0$ et sur les fonctions harmoniques. Annales scientifiques de l'E.N.S., 3e série, 16 (1899), 427–464.

[589] S. Zaremba. Sur l'intégration de l'équation $\Delta u + \xi u = 0$. Journal de mathématiques pures et appliquées, 5e série, 8 (1902), 59–118.

[590] E. Zeidler. Theory of nonlinear operators. Proc. Summer School, Akademie-Verlag, Berlin, 1974, 259–311.

[591] Ya.B. Zeldovich, G.I. Barenblatt, V. Librovich, G.M. Makhviladze. The mathematical theory of combustion and explosion. Plenum, New York, 1985.

[592] Y. Zhao, Y. Wang, J. Shi. Exact multiplicity of solutions and S-shaped bifurcation curve for a class of semilinear elliptic equations. J. Math. Anal. Appl., 331 (2007), No. 1, 263–278.

[593] V.V. Zhikov. *G*-convergence of elliptic operators. Matematicheskie Zametki, 33 (1983), No. 3, 345–356 (English translation).

[594] V.V. Zhikov, S.M. Kozlov, O.A. Oleinik, Ha Ten Ngoan. Averaging and *G*-convergence of differential operators. Usp. Mat. Nauk, 34 (1979), No. 5, 65–133.

[595] V.G. Zvyagin, N.M. Ratiner. Oriented degree of Fredholm maps of non-negative index and its application to global bifurcation of solutions. Lect. Notes Math., 1520 (1992), 111–137.

[596] V.G. Zvyagin, V.T. Dmitrienko. Properness of nonlinear elliptic differential operators in Hölder spaces, Global Analysis – Studies and Applications, V, Lecture Notes in Mathematics, vol. 1520, Springer, Berlin, 1992, 261–284.

Notation of Function Spaces

$W^{l,p}(\mathbb{R}^n)$, 10
$L^p(\mathbb{R}^n)$, 10
$H^{l,p}(\mathbb{R}^n)$, 11
$B^{l,p}(\mathbb{R}^n)$, 11
$D(\mathbb{R}^n)$, 11, 48
$D'(\mathbb{R}^n)$, 47
$W^{l,p}(\Omega)$, 11
$H^{l,p}(\Omega)$, 11
$B^{l,p}(\Omega)$, 11
$B^{l-1/p,p}(\Gamma)$, 11
$C^{k+\alpha}(\bar{\Omega}) \left(= C^{(k+\alpha)}(\bar{\Omega})\right)$, 11

$W_q^{s,p}$, 47, 93
$W_\infty^{s-1/p,p}(\partial\Omega)$, 116, 146
$W_\infty^{s,p}(\Omega)$, 116, 146
$W_q^{-s,p'}$, 93
$\dot{W}^{-s,p'}(\Omega)$, 121
L_q^p, 93

$E_q, E_q(\Omega)$, 47, 50, 54, 229
$E_\infty, E_\infty(\Omega)$, 47, 55, 59, 61, 74, 229
$E_p(\Gamma)$, 66
$E_\infty(\Gamma)$, 70
$E_{loc}(\Gamma)$, 66
E^*, 78

$E^*(\mathbb{R}^n)$, 78
$(E^*)_\infty$, 78
$(E^*)_p$, 78
E^*_{loc}, 78
$(E_\infty)^*_\omega$, 80
$(E_\infty)^*_0$, 80
E_D, 85
$(E^{**})_\infty$, 88
$((E^*)_1)^*$, 88
$[E(\Omega)]^* (= (E(\Omega))^*)$, 89
$\hat{E}_0^*(\Omega)$, 89
$E_0^*(\Omega)$, 91
$((E(\Omega))^*)_\infty$, 92
$((E(\Omega))_1)^*$, 92
E_∞, F_∞, 117, 147
E^*, F^*, 120
$E_\infty(\Omega), F_\infty(\Omega)$, 130
$(E^*(\Omega))_\infty, (F^*(\Omega))_\infty$, 130
$(F^*_{-1}(\Omega))_\infty$, 130
F^*_{-1}, 121

$C_0^\infty(\Omega)$, 89
$\hat{C}_0^\infty(\Omega)$, 89
$\tilde{C}_0^\infty(\Omega)$, 91
$M(E)$, 48

Index

additivity, 469
analytic semigroups, 284
analyticity of solutions, 561
a priori estimates, 105, 116, 568
    Schauder, 20
    for adjoint operators, 183
    in Hölder spaces, 22
    in Sobolev spaces, 23
    in unbounded domains, 42
    with condition NS 154, 231
    with condition NS*, 183
    with condition NS(seq), 297
    with condition NS*(seq), 298

bifurcations
    local, 46, 518
    subcritical, 523
    supercritical, 523

conditions
    Lopatinskii, 15, 253, 347
    Condition D, 24, 117, 137, 228, 253, 293
    Condition NS, 37, 136, 154, 172, 230, 247, 297
    Condition NS*, 38, 183, 230, 247, 297
    Condition NS(seq), 41, 297
    Condition NS*(seq), 41, 298
    Condition R, 142
convergence
    local weak, 62
    of domains, 138, 294
    of solutions, 319

equations
    biharmonic, 560
    diffusion-convection, 443
    fourth-order, 221
    fully nonlinear, 497
    Helmholtz, 559
    higher order, 540
    Navier-Stokes, 498
    nonlocal reaction-diffusion, 8, 521
    of branching, 523
    one-parameter, 527
    quasi-linear, 495
    Schrödinger, 443
    semi-linear, 493
    singular integral, 572
    Stokes, 18
exponential
    decay, 203, 231
    dichotomy, 404

finite difference schemes, 549
Fredholm
    property, 25, 183, 571
        in Hölder spaces, 217
    theory, 564

homotopy invariance, 469, 479

index, 25, 28, 39, 291
    in bounded domains, 350
    in unbounded domains, 352
    of Cauchy-Riemann system, 329, 345
    of first-order systems on a sphere, 576
    of first-order systems on the axis, 365
    of Laplace operator with oblique derivative, 333, 345

of Poincaré problem, 356
of second-order equations on the axis, 366
of second-order equations in cylinder, 388
of singular integral equations, 578
stabilization, 41, 302
interpolation inequality, 255

limiting
   domain, 138, 294
      of the first type, 32, 296
      of the second type, 34, 296
   operator, 153, 295
   problem, 35, 295
lower estimates, 388

method
   Galerkin, 490
   Leray-Schauder, 45, 492
   Newton's, 491
   Projection, 490
   Schwarz's, 556
   successive approximations, 557
   sweeping, 558
multipliers, 48
normalization, 469

operator
   adjoint, 120
   discrete, 527
   elliptic, 13
   formally adjoint, 228
   Fredholm, 25, 400
   in holder spaces, 170
   in the Douglis-Nirenberg sense, 14
   Laplace, 16, 221, 331
      with oblique derivative, 29, 355
   local, 100
   non-Fredholm, 42, 401, 517
      weakly, 408
      strongly, 442
   nonlinear Fredholm, 447
   normally solvable, 25, 163
   non-proper, 454
   proper, 448

properly elliptic, 13, 15
semi-Fredholm, 26
uniformly elliptic, 254, 285
uniformly elliptic with a parameter, 254, 285
with a nonzero index, 499
with a parameter, 252, 285
with a parameter at infinity, 285
orientation, 470

potential
   double layer, 5, 21
   logarithmic, 556
   Newton, 4, 21
   simple layer, 4, 558
principal eigenvalue, 541
problem
   elliptic in the ADN sense, 15
   elliptic in the sense of Petrovskii, 16
   elliptic in unbounded domains, 30, 501
   elliptic with a parameter, 252, 581
   formally adjoint, 38, 228
   Hilbert, 355
   Poincaré, 356
   regular elliptic, 228
   uniformly elliptic, 16
properness, 448
   in Hölder spaces, 449
   in Sobolev spaces, 455

solution
   in the sense of sequence, 403
   smoothness, 246
solvability conditions
   for non-Fredholm operators, 408
   for discrete operators, 536
   for second-order operators, 368
   with formally adjoint operator, 230
spaces
   Besov, 11
   Bessel potentials, 11
   dual, 78, 89, 176
   Hölder, 12, 170
   Sobolev 10, 146
   Sobolev-Slobodetskii, 11
   weighted, 180, 215, 397, 507, 569

Index

system
    Cauchy-Riemann, 18, 324
    canonical, 347
    first order, 360, 428

theorem
    fixed point, 489
    implicit function, 489
    Michlin's, 109
    on Fredholm property, 195, 198, 200, 202, 212, 218, 237
    on normal solvability, 163, 193, 529
    on solvability conditions, 195, 198, 200, 209, 230, 241, 242, 245
    on stabilization of the index, 313, 317
    on unique solvability, 282
topological degree, 43, 468
    for approximation-proper mappings, 591
    for elliptic operators, 484
    for Fredholm and proper operators, 474, 589
    for travelling waves, 512
    for monotone operators, 588
    in finite dimensional spaces, 587
    Leray-Schauder, 44, 588
    non-existence, 487
    uniqueness, 480
travelling waves, 509
Turing structures, 521

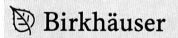

Monographs in Mathematics

The foundations of this outstanding book series were laid in 1944. Until the end of the 1970s, a total of 77 volumes appeared, including works of such distinguished mathematicians as Carathéodory, Nevanlinna and Shafarevich, to name a few. The series came to its name and present appearance in the 1980s. In keeping its wellestablished tradition, only monographs of excellent quality are published in this collection. Comprehensive, in-depth treatments of areas of current interest are presented to a readership ranging from graduate students to professional mathematicians. Concrete examples and applications both within and beyond the immediate domain of mathematics illustrate the import and consequences of the theory under discussion.

Managing Editors
Amann, H. , Universität Zürich, Switzerland
Bourguignon, J.-P., IHES, France
Grove, K. , University of Maryland, USA
Lions, P.-L., Paris-Dauphine, France

■ **Vol. 96: Arendt, W., Batty, C.J.K., Hieber M. and Neubrander, F.**, Vector-valued Laplace Transforms and Cauchy Problems , 2nd edition
2011. 552 pages. Hardcover.
ISBN 978-3-0348-0086-0

This monograph gives a systematic account of the theory of vector-valued Laplace transforms, ranging from representation theory to Tauberian theorems. In parallel, the theory of linear Cauchy problems and semigroups of operators is developed completely in the spirit of Laplace transforms. Existence and uniqueness, regularity, approximation and above all asymptotic behaviour of solutions are studied. Diverse applications to partial differential equations are given. The book contains an introduction to the Bochner integral and several appendices on background material. It is addressed to students and researchers interested in evolution equations, Laplace and Fourier transforms, and functional analysis.

This authoritative work is likely to become a standard reference on both the Laplace transform and its applications to the abstract Cauchy problem. ... The book is an excellent textbook as well. Proofs are always transparent and complete. All this makes the text very accessible and self-contained.

J. van Neerven, *Nieuw Archief voor Wiskunde*, No. 3, 2003

■ **Vol. 100: Triebel, H.**, Theory of Function Spaces III
2006. 438 pages. Hardcover.
ISBN 978-3-7643-7581-2

This book deals with the recent theory of function spaces as it stands now. Special attention is paid to some developments in the last 10–15 years which are closely related to the nowadays numerous applications of the theory of function spaces to some neighbouring areas such as numerics, signal processing and fractal analysis. In particular, typical building blocks as (non-smooth) atoms, quarks, wavelet bases and wavelet frames are discussed in detail and applied afterwards to some outstanding problems of the recent theory of function spaces such as a local smoothness theory, fractal measures, fractal analysis, spaces on Lipschitz domains and on quasi-metric spaces.

The book is essentially self-contained, although it might also be considered as the continuation of the two previous books of the author with the same title which appeared as volumes 78 and 84 in this book series. It is directed to mathematicians working in analysis, numerics and fractal geometry, and to (theoretical) physicists interested in related subjects such as signal processing.